8-7-89

D1242141

THE CHEMICAL ANALYSIS OF WATER

General Principles and Techniques

Second Edition

The Chemical Analysis
of Water

General Principles and Techniques

Second Edition

By

D.T.E. Hunt and A.L. Wilson

Water Research Centre
Medmenham Laboratory
Buckinghamshire

ROYAL
SOCIETY OF
CHEMISTRY

British Library Cataloguing in Publication Data

Wilson, A.L.
The chemical analysis of water: general principles and techniques. — 2nd ed.
1. Water — Analysis
I. Title   II. Hunt, D.T.E.   III. Royal Society of Chemistry
546'.22   QD142
ISBN 0-85186-797-9

Reprinted 1988
First Edition 1974

Published by The Royal Society of Chemistry,
Burlington House, London W1V 0BN

Printed at The Alden Press, Oxford.

Foreword to the Second Edition

The first edition of this book, published in 1974, was prepared because the growing importance attached to the analysis of many types of water suggested that the time was ripe to attempt a critical appraisal of generally important principles in the measurement of water quality. In the years since that time, interest in water quality has continued to develop, and a number of countries, including the United Kingdom, have developed new or extended existing legal requirements on the qualities of various types of water. These requirements commonly call for the measurement of many different substances, a large proportion of which must be measured and controlled at very small concentrations. There have been considerable developments in the analytical techniques and procedures applied to meet these needs, and semi- and fully-automatic procedures have been increasingly employed to meet the challenge of greater numbers of samples. Coupled with these changes has also been a much greater emphasis than before on the importance of errors in analytical results, and on the need to estimate and control such errors. In view of the above developments during the last decade, we are grateful to the Royal Society of Chemistry for their invitation to revise and up-date the first edition.

This second edition, like the first, attempts an integrated discussion of all aspects of the measurement of water quality as a guide to those concerned with defining, establishing, controlling, and reviewing programmes of water sampling and analysis. We have also tried to provide a reasonably complete picture of the current state of the field for anyone wishing to become acquainted with it or any of its component aspects. The emphasis throughout is on fresh, natural waters and drinking waters because these represent our main interests; we hope, however, that much of the book will be found useful for other types of waters.

In view of the above-mentioned emphasis on the errors of analytical results and the various problems that we have met in estimating and controlling the sizes of those errors, we have greatly expanded all parts of the text of the first edition that bore on this subject. Our aim has been to provide a comprehensive discussion of all types of error so that the book may be used by analysts as an aid to ensuring the appropriate accuracy of analytical results. In addition, in response to the many developments in analytical techniques and procedures, we have up-dated and expanded many parts of the first edition that dealt with these topics, and new sections have been added where necessary. One of us (ALW) has been largely responsible for Sections 1 to 10 and 12, and the other (DTEH) for Sections 11 and 13, but we have discussed each other's drafts in detail so that the final text represents our joint views.

We should like to express our thanks to: Mr. O.D. Hydes (Department of the Environment) and Dr. K.C. Thompson (Yorkshire Water Authority) for permission to cite their paper on the variation, between instruments, of interference effects in flame atomic-absorption spectrometry, published in the Water Research Centre's Analytical Newsletter; again to Mr. O.D. Hydes for helpful information concerning Section 3.2.4.$(f)(v)$; to Mr. L.R. Pittwell (Department of the Environment) for information on the publication of the Standing Committee of Analysts; to

v

Mr. D.S. Hiorns (Station Design Department, Central Electricity Generating Board) for supplying a copy of CEGB Standard 500123, Water Analysis Equipment; to Mr. M.J. Gardner and Mrs I. Wilson (both of the Water Research Centre) for reading, and providing helpful comments on, the sections on plasma emission spectrometry and chromatography, respectively; to Mr. J.C. Ellis (of the Water Research Centre) for many valuable discussions on statistical points; and to the four manufacturers of atomic-absorption equipment who provided performance data for the Tables in Section 11.3.5. Of course, any deficiencies and errors remain our sole responsibility. We should also like to thank Pergamon Press for permission to reproduce Tables 5.1 and 5.2, and the Controller of Her Majesty's Stationery Office for permission to reproduce Figure 9.1. We are grateful to Dr. S.C. Warren, Director of WRc Environment, for permission to use library facilities, and to Mr. K.V. Bowles and Mrs. M. Blake, the librarians, for help in that use. We are also grateful to Amersham International for the use of the library facilities of that company. Needless to say, thanks are also due to all our colleagues, past and present, discussions with whom have always been of great value. Finally, we wish to thank our wives, Mair and Ivana, for their help and encouragement while we were preparing the text.

DTEH

ALW

A.L. Wilson: An Appreciation

On 17th March 1985, while the proofs of this book were being prepared, Tony Wilson died suddenly at his home in Flackwell Heath, Buckinghamshire. This untimely loss of an outstanding scientist and personality brought great sadness to his many friends and professional colleagues, both in the United Kingdom and overseas.

Born in Brighton in 1929, Anthony Leslie Wilson was educated at Vardean Grammar School and Kings College, London, where he took an honours degree in chemistry. He worked, for eighteen years, at the Atomic Energy Research Establishment, Salwick and the Central Electricity Research Laboratories, Leatherhead, before joining the Water Research Association at Mcdmenham – later to become a constituent laboratory of the Water Research Centre – in 1968. He remained with the Centre until his retirement in 1980, when he held the position of Manager of the Analysis and Instrumentation Division.

His considerable reputation as an analytical chemist was the product of a prodigious capacity for work and the painstaking application of his considerable intellect, not only to the development of a wide range of methods, but also to the fundamental principles of analysis and analytical quality control. His work on the latter was especially pioneering and its importance has become very widely recognised.

His approach to the specification and assessment of analytical performance and to the control of analytical errors forms the basis of the standard practices of both the electricity generating and water industries in the U.K. The Department of the Environment, in its Harmonised Monitoring Scheme, and the World Health Organisation, in its Global Environmental Monitoring Scheme, both adopted his procedures for ensuring comparability of results from groups of laboratories and the Department of the Environment's Standing Committee of Analysts has incorporated his ideas on performance characterisation in its published methods.

Tony Wilson authored many papers on analysis and analytical quality control and, of course, was singlehandedly responsible for the first edition of this work, published by the then Society for Analytical Chemistry in 1974. In 1975 he was awarded the Louis Gordon Memorial Prize for the best paper of the year in the journal *Talanta* (one of a series in which he drew together in a coherent manner the important factors to be considered in characterising the performance of analytical methods).

It was, however, not only in the field of chemical analysis that Tony gained distinction. By taking advantage of early retirement he was able to pursue more fully a keen interest in archaeology. An amateur in the field of Cretan prehistory only in the true sense of working for the pleasure and intellectual reward, he wrote several papers on Minoan civilisation and was, at the time of his death, working on a very different book from this – a catalogue of all the locations in Crete where Neolithic and Minoan finds had been reported, including reviews of the evidence and analyses of its interpretation.

The reader of the first ten sections of this book will, I am sure, come to feel the presence of a remarkable intellect, able to comprehend fundamental principles and details of application alike, yet convey both with concise clarity.

Tony Wilson's keenness of mind, strength of purpose and vitality will be missed by all who knew him and it is not merely a duty but a personal pleasure to record here my indebtedness to him for all his kindnesses and patient encouragement during the writing of this book. He will, of course, be most missed by his widow Ivana, herself an accomplished analytical chemist, whose help in seeing the book through the final stages of publication is also gratefully acknowledged.

DTEH
1st September 1985

CONTENTS

Contenu

1 Introduction

1.1 THE IMPORTANCE OF WATER ANALYSIS

Water* has always been a vital material for Man's existence. Its uses for drinking, cooking, agriculture, transport, industry, and recreation show immediately the extent to which it is an integral part of our life; even romantic properties have been ascribed to this unique substance. No matter what the purpose for which water is required, it has long been recognised that its suitability for that purpose can be affected by other substances dissolved and/or suspended in the water. For example, the chemical analysis of water intended for human consumption has formed the subject for analytical texts for over a hundred years. Such examples need not be laboured for many will readily occur to all readers. The main point to be stressed here is that the last few decades have seen a large and steady increase in the importance attaching to many aspects of water quality for all of the uses to which it is put. The number of impurities of concern has been continually increasing (and there appears to be no sign that this process is ending), and smaller and smaller concentrations of many substances are becoming of interest. These trends were initially concentrated in the more developed countries of the world, but the rapid and continuing advancement of many other countries has meant that concern with water quality is now a topic of importance in virtually all countries, especially with respect to drinking water.

Many countries now have national legislation stipulating the maximum acceptable concentrations (or some equivalent concept) for specified impurities in drinking water and, often, other types of water. Since the first edition of this book, the United Kingdom has become one of these countries through its membership of the European Community and that body's various Directives dealing with the quality of waters, including drinking water and also surface water used to prepare drinking water (see Section 2.9 for references to water-quality standards). There has also been growing concern with the need to protect the environment from the various forms of pollution that may be increased by growing populations, industrialisation, modern agricultural methods, *etc.* That this problem is now of global concern is exemplified by the United Nations Environment Programme which has established a Global Environmental Monitoring System (GEMS) whose function is exactly what its name suggests, *i.e.*, to monitor the quality of the environment on a global basis so that any important systematic changes may be detected. The World Health Organisation has been given the task of arranging the

---

* The word *water* is commonly used in two different ways. In the purely chemical sense, the word represents the chemical compound H_2O. In the second and everyday usage, the word connotes a wide range of different aqueous solutions, *e.g.*, river-water, rainwater, drinking water, boiler-water. This latter usage is rather arbitrary as it is not clear when one should stop using the word 'water' and begin to use the term 'aqueous solution'. This everyday usage, however, is useful, and, in practice, seldom leads to ambiguity; it is, therefore, the everyday sense of 'water' that is used throughout this book. Where ambiguity could arise, the text will contain adequate clarification.

necessary programme for fresh surface waters,[1] and many countries throughout the world are participating.

Such developments in the field of water quality have, of course, had important effects on water analysis. Thus, the number of laboratories making routine analyses of water has shown a steady increase as has also the number of analysts concerned with water. In addition, a much wider range of analytical techniques and more sophisticated methods and instrumentation than hitherto are being employed to meet analytical requirements of growing volume and difficulty. This has called for considerable changes to the text of the first edition. Further, the number of analytical methods described for a given impurity is increasing, particularly as more laboratories become concerned with the development and application of new and/or improved analytical methods. An associated development has been the world-wide proliferation of collections of 'Standard Methods' for the analysis of water. Such collections have been issued by many local, national and international organisations, and are now available in almost bewildering profusion (see Section 9.5).

Another particularly important development within the last decade or so has been the growing recognition that analytical results are subject to errors whose magnitude may be so large, especially in trace analysis, that the results become essentially worthless if not positively misleading. The need for analysts *and* those who use analytical results to recognise this, and for analysts to estimate and, when necessary, control the magnitude of such errors has been increasingly stressed in recent years by many authors and organisations concerned with water quality. Indeed, a new branch of the field with a rapidly growing literature has appeared, *i.e.*, analytical quality control (or analytical quality assurance or cognate terms). This tendency has, of course, been markedly stimulated by the developments mentioned above, *i.e.*, the growing need for water quality to meet legally defined standards, and the various regional, national, and international organisations requiring comparison of results from different laboratories. In view of these developments, the topics of errors and their estimation have been treated in much greater detail than in the first edition.

Of course, the analytical process is not the only source of inaccurate results; the various factors summarised under the term 'sampling' may also lead to grossly unrepresentative results if appropriate procedures are not employed. Again, in the last decade there has been increasing recognition of the many difficult problems in designing and applying satisfactory sampling procedures for waters of various types, and publications dealing with such problems have appeared with greater frequency than hitherto.

In view of all these developments, it was thought that the time was ripe to revise and up-date the first edition of this work. Before describing the organisation of this second edition, a brief account of its aims is given in Section 1.2.

1.2 AIMS OF THIS EDITION

The aims of this edition remain essentially the same as those of the first, and the following is very largely what was written in the first edition.

A monograph of the present size cannot attempt to give the detailed written procedures for all the analytical methods for all the substances that are of interest in waters of various types. In any event, as mentioned above, there are already, in the authors' view, more than enough such comprehensive collections of methods, and we have no wish to add to the number. Detailed analytical procedures are, therefore, not given, but references are quoted throughout the text both to the many available collections of methods and also to other useful papers and publications. However, although there is no shortage of published analytical methods, there have been few critical and detailed appraisals of all the analytical principles important for water. In view of the developments mentioned in Section 1.1, these principles seem to us of particular importance because laboratories are increasingly concerned with questions such as: 'when and where should samples be collected?', 'which analytical method or technique is appropriate for a particular purpose?', and 'how accurate are the analytical results?' It is to questions of this nature that this book is primarily addressed. Although there is little originality in the treatment of the various topics, the book remains, to the best of the authors' knowledge, one of the very few attempts at a detailed integration of all aspects involved in the measurement of water quality. The treatment is particularly intended for those concerned with defining, establishing, controlling, and reviewing programmes of water analysis. However, because of the detailed treatment of all aspects of the measurement of water quality, the book should also provide the means by which anyone may achieve a good appreciation of the current 'state-of-the-art' of the whole field or any of its components. In this context, we should mention here a reference handbook recently issued by the World Health Organization (Regional Office for Europe).[2] This book is in three volumes, the first of which deals with many of the topics covered here, while the other two provide detailed descriptions of a very wide range of physical, chemical, radiological, biological, bacteriological, and virological measurement procedures of relevance to water. This is a valuable reference work, but its first volume provides a much less detailed treatment of most of the subjects discussed in this book.

It will be seen that quite large portions of the text deal with subjects that may be thought to fall outside the scope of 'analysis'. This has been done quite deliberately because, in the authors' opinion, it is unwise to view analysis solely as those operations leading to estimates of the concentrations of a number of impurities in particular samples. These operations are, of course, vital, but it is just as important to ensure that such work is fruitfully applied to real problems, particularly when, as is often the case, there is a tendency to require more and more analyses without concomitant increase in the resources allocated to obtain those analyses. For this reason, analysis should deal with all aspects of the need for, use of, and provision of analytical results.

Given the very wide range of applications in which information on water quality is required, we have made no attempt to consider every conceivable analytical situation that could arise, and then to consider how the principles described in the text should be implemented. The aim has been to state the principles and to give practical examples in such a way that analysts can apply them in a common-sense manner to their particular tasks.

This book is intended to deal solely with the analysis of those aqueous solutions to which the term 'water' is commonly applied. Thus, potable waters and naturally occurring fresh waters are included, but trade-effluents, sewage, process-solutions, *etc.*, are generally excluded. An attempt has been made throughout the text to formulate the principles described in such a way that they will have maximal relevance to other types of sample. In addition, techniques and procedures relevant to waters will sometimes also be relevant to other types of aqueous samples, but this is not necessarily so. Analysts should, therefore, be particularly cautious when considering the suitability of a technique or procedure – satisfactorily applied to one type of sample – for a different type. Throughout the text emphasis has been placed on the analysis of natural fresh waters and potable waters. This partially reflects the experience of the authors, but also mirrors the growing interest in, and importance of, such waters. No attempt has been made to provide a comprehensive discussion of the various problems that beset many analyses of sea and estuarine waters, but we have tried to make clear the nature of these problems and to include references to recent publications dealing with suitable procedures. By this approach, it is hoped that the text should be useful to all those concerned with any type of water.

The wide range of waters to be analysed and the varied reasons for the analysis lead to very large numbers of determinands* of potential interest. In keeping with our general approach, no attempt has been made explicitly to consider all substances that analysts may have to determine. The references to other publications should enable analysts to locate methods for other substances they have to measure, and in Section 9.5 we have summarised the methods recommended for a number of commonly important determinands by two of the more well established collections of 'Standard Methods' in the United Kingdom and the United States. When examples and practical problems are discussed, they are usually in terms of impurities of most common interest for one or more types of water. In this way it is hoped to illustrate – in a practical way and to the greatest possible number of analysts – the principles discussed in the text.

Finally, many of the topics discussed will be illustrated by examples drawn from analysis in the concentration range 0.001 to 10 mg l^{-1}. To some extent this again reflects the experience of the authors, but it also stems from the greater difficulties often met in measuring these rather than higher concentrations. The determination of such low concentrations is, therefore, a fruitful field for practical examples of problems that may be encountered. However, as before, the principles are intended to be applicable to any concentration. It should not be thought, therefore, that this book deals solely with trace analysis.

1.3 ARRANGEMENT OF THE TEXT

In establishing a programme of sampling and analysis, it is useful to consider the total process as composed of a number of sequential stages as shown in Figure 1.1 which is based on similar diagrams appearing in references 2 and 5. Each of these stages is treated in detail in subsequent sections, and a summary of the main contents of each section is given below. However, it is important to note that the separate consideration of these various aspects is rather artificial because, in

---

* The term 'determinand' is used throughout to signify 'that which is to be determined'.[3,4]

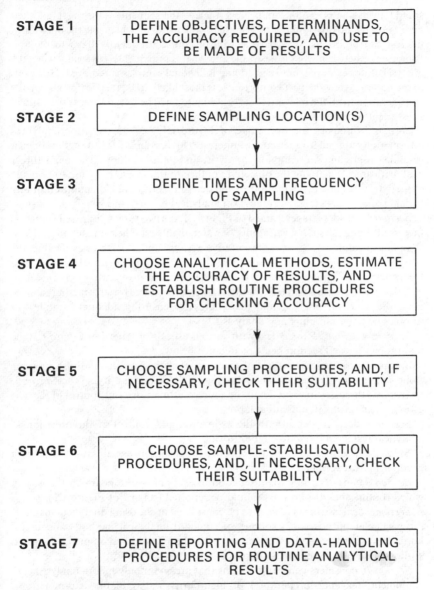

Figure 1.1 *Sequential consideration of important factors in establishing programmes of sampling and analysis*

practice, all aspects tend to be rather intimately inter-connected. Indeed, a decision made at a later stage, *e.g.*, the choice of an analytical method, may require a change in a decision made in a preceding stage, *e.g.*, the frequency of sampling. This 'feedback' from one stage to one or more of the earlier stages is very important and should always be borne in mind. Nevertheless, the sequential approach adopted here is useful in providing a reasonable framework in which all the factors of interest can be discussed. Analysts will, of course, frequently have to make a compromise between what is desirable and what is practicable, especially when the analysis of, for example, unexpected and individual samples is required. However, in the authors' view the best compromise is most likely to be reached by giving due consideration to all the relevant factors in Figure 1.1. The broad topics dealt with in subsequent sections of the book are as follows.

Section 2 discusses the first stage of Figure 1.1, *i.e.*, deciding the analytical information required for the water of interest. This is essential for the definition of a suitable programme of sampling, analysis, and data-handling. The main topics dealt with are: (1) defining the objectives of the programme and the use to be made of analytical results, (2) definition of the determinands, and (3) definition of any special requirements (*e.g.*, accuracy) on analytical results.

Section 3 considers stages 2 and 3 of Figure 1.1, the two topics discussed being: (1) factors affecting the spatial variability of water quality and their impact on choice of sampling locations to give representative results, and (2) factors affecting the temporal variability of water quality and their influence on the times and frequencies of sampling necessary to provide representative samples. This section also deals with stages 5 and 6 of Figure 1.1, *i.e.*, procedures for abstracting samples from the water of interest, and procedures for ensuring adequate stability of samples between sampling and analysis.

Sections 4 to 7 and 9 to 12 deal with various aspects of stage 4 of Figure 1.1, the main topics in each section being as detailed below.

Section 4 describes the nature of errors in analytical results and also presents some simple statistical ideas and techniques of value in quantifying these errors.

Section 5 discusses in great detail the sources, estimation, and control of the bias of analytical results in individual laboratories.

Section 6 deals in depth with the estimation and control of the precision of analytical results in individual laboratories.

Section 7 describes a detailed procedure for ensuring that all laboratories of a group achieve a specified accuracy.

Section 8 considers how the discussion in Sections 4 to 6 bears on the way in which routine analytical results should be reported (stage 7 of Figure 1.1).

Section 9 deals with: (1) factors to be considered in choosing analytical methods for particular purposes, (2) sources of standard analytical methods and useful books and journals, and (3) a summary of the methods recommended in two authoritative collections of standard methods.

Section 10 considers general problems that are commonly met in analysis, *e.g.*, equipment, reagents, preparation of high-purity water, blank determinations, contamination.

Section 11 provides: (1) a description and comparison of the principles, performances, and relative advantages and disadvantages of analytical techniques of general importance in modern water analysis, and (2) a description of a number of interesting and/or important applications of other techniques to water analysis.

Section 12 discusses: (1) the advantages and disadvantages of automatic and on-line analysis, (2) the types of equipment that are currently available for such analyses, and (3) some generally important points that must be considered in applications of these types of analysis.

Section 13 gives brief consideration to the very important aspects of the use of computers and associated equipment in modern analysis, and of the procedures to be used for data-handling purposes.

1.4 REFERENCES

[1] R. Helmer, *Nat. Resour.* 1981, **17**, 7.
[2] M. J. Suess, 'Examination of Water for Pollution Control', Volumes 1 to 3, Pergamon, Oxford, 1982.
[3] A. L. Wilson, *Talanta*, 1965, **12**, 701.
[4] International Organization for Standardization, ISO 6107/2-1981, ISO, Geneva, 1981 (BS 6068:Part 1:Section 1.2:1982).
[5] Standing Committee of Analysts, 'General Principles of Sampling and Accuracy of Results', HMSO, London, 1980.

2 Information Requirements of Measurement Programmes

In this section we stress the importance of detailed planning of measurement programmes. In particular, unambiguous definition of the information a programme is required to give seems to us essential, and we describe a number of aspects that should be considered in defining these information requirements. Although analysts are not, in general, responsible for this definition of needs, we think it highly desirable for them to be intimately involved with it whenever possible and well aware, therefore, of the various aspects considered below.

2.1 NEED FOR DETAILED PLANNING

Most laboratories are on occasion required suddenly and unexpectedly to analyse samples, and it will then often be impossible to apply some or all of the detailed considerations described below. As measurement programmes become larger and/or lengthier, however, their detailed design clearly becomes more important in terms of their efficiency and economy of effort. It is suggested, therefore, that the approach noted below should be considered for any regular programme of water-quality measurement.

In planning a measurement programme, seven questions generally need to be considered:

(1) what are the objectives of the programme?
(2) which determinands are of interest?
(3) where and when are samples to be taken?
(4) which analytical methods are to be used?
(5) how are samples to be taken and subsequently handled?
(6) how are results to be reported?
(7) what is to be done with the reported results?

These questions provide a framework for defining measurement programmes. Mancy[1] and Mancy and Allen[2] have described similar frameworks, and they made the important point that the results from a programme should be regularly reviewed to decide if changes (e.g., of determinands or sampling frequency) in the programme are necessary. A similar point has been made increasingly by many other authors; see, for example, references 3 to 7. The importance of such reviews cannot be overemphasised – without them much wasted time and effort are likely and, indeed, occur. The ability to make these reviews is greatly aided by ready accessibility of past analytical results and the ability quickly and economically to process those results. These needs are not generally met by the more traditional systems of data-handling such as log books, filing cabinets, manual calculation, etc., but modern computing systems are very suitable, and have been increasingly adopted by many organisations. These aspects are taken up in more detail in

Section 13. For the present, it is simply suggested that these reviews should be an integral part of any measurement programme.

Some measurement programmes are so large and complex (*e.g.*, study on international, national, or regional scales of the quality of surface waters) that very detailed planning is essential.[2-7] This seems a suitable application for systems-analysis techniques, and, indeed some ten years ago a few publications dealt with such applications.[3-8] As far as we know, however, these techniques have not found any general or widespread use. For those who may be interested to read more of this approach, reference 8 gives a very detailed account of the procedure.

2.2 MANAGERIAL AND ANALYTICAL ASPECTS

The seven questions listed in Section 2.1 will be considered in detail in subsequent sections. The aim of the present section is to stress that questions (1), (2), and (7) are, in essence, not analytical questions at all. They represent managerial rather than analytical decisions, and they have generally to be made by someone (or some group) requiring information on water quality in order to achieve one or more objectives. In the following, this person (or group) is termed 'the user of analytical results' or simply 'the user'. Basically it is the user's responsibility to define the information he requires and it is the analyst's task to provide that information; see also references 9 and 10 on this and related points. It should also be noted that the needs of the user are also intimately concerned in answering questions (3) to (6) in Section 2.1; this will be amplified in the following sub-sections. These simple ideas lead to clear views on the relative roles of the user and the analyst in planning measurement programmes, and these views are stated in the next two paragraphs. In practice, attempts to apply these ideas may lead to several problems which are considered subsequently.

A very important role of the analyst is to satisfy himself that the user's requirements are stated clearly and unambiguously; otherwise, all subsequent planning will be on an unsound basis. Once the analyst is satisfied with this aspect of the user's statement of requirements, he must then consider whether or not the analytical results needed to provide the required information can be obtained. Three answers to this question are possible: (1) some or all of the information cannot be obtained by currently available analytical techniques; (2) the information can be obtained but only with more or better resources, *e.g.*, staff, space, equipment; (3) the information can be obtained with existing resources. The first two answers will require detailed discussions between the user and the analyst in order to reach a compromise between what is ideally required and what is practically achievable. It is important that such discussions be approached objectively by the analyst on the basis that the user's interests are of prime concern. Equally, the user must be prepared to justify requests for analytical effort on the basis of the value of the information for his purposes. It is the authors' view that the planning of quality-measurement programmes will be aided if these different responsibilities of user and analyst are borne in mind.

Of course, it is not suggested that the user and analyst carry out these functions without joint discussions. Indeed, a continuous dialogue between the two is highly desirable during the planning stage of any programme and also during the reviews of the programme and its efficiency. The point to be emphasised is that the user should approach this planning unbiased by any suspicions he may have of the information capable of being supplied by the analyst. The user must concentrate initially on defining his information needs, and rely on the analyst to tell him of the practicability of obtaining this information. Moody[11] has made a similar point in discussing the organisation of a data-collection programme for water-resources planning, development, and management. When, as is common, the user is in fact a group or committee, the continuous dialogue between user and analyst can (and, in the authors' view, should) be readily achieved by ensuring that analysis is adequately represented on the group from its inception. Currie[10] and Taylor[12] have recently stressed this point. Although there seems to be a tendency for such an arrangement to be becoming more common, we feel that this movement still has a long way to go.

Several points on the suggestions in the two preceding paragraphs may cause difficulties:

(1) It is sometimes argued that, in planning a programme, there is no point in attempting to define the information ideally required because there is never sufficient analytical effort. Such an argument is dangerous because, if the user does not define the information he needs to achieve his objectives, he will certainly never get it. On the other hand, if the analyst knows the information ideally required he may well be able to work gradually towards this goal even if it cannot be achieved immediately. Of course, a responsible approach by the user is essential in that he should, in principle, balance his requests for analytical data (and the consequent costs) against the benefits the information is expected to provide. At present it is true that, though the costs of supplying information can be quite accurately assessed, the value of the benefits appears to be virtually impossible to quantify for many types of programme (see Section 3.4.3). In the authors' view, this again is not a valid argument against the approach suggested here; rather it represents the identification of an important problem calling for further investigation.

(2) When the two functions of user and analyst are fulfilled by one and the same person, it will usually be more, rather then less, important for that person to bear the two functions clearly in mind. Otherwise, efficient executions of both roles may be impeded by the natural tendency to continue doing what has been done before.

(3) It is not uncommon that a user, when requested by an analyst to define his information needs more clearly, asks for the advice of the analyst, simultaneously creating the impression that the user has no clear ideas himself. This seems to us a highly undesirable situation in which to plan a measurement programme because the implication is that the user does not have a sufficiently clear idea of his own real objectives and the information needed to achieve them. We feel that the analyst has a responsibility in such situations to suggest more detailed consideration by the user of these two aspects rather than to take the possibly easier path of providing the advice requested on points of detail. Otherwise, all subsequent planning by the user, perhaps when the analyst will have no opportunity to advise, is likely to be on an equally unsound basis, We realise, of course, that this suggestion is easier made than implemented.

A final point to be emphasised is that just as analytical expertise should be involved in the planning of programmes so too should expertise in other fields whenever appropriate. The complexity of processes affecting the quality of many water systems[2] will usually require experts in topics such as estuarine processes, limnology, river pollution, or others as appropriate to a particular programme (see also Sections 3.1, 3.2, and 3.3.1). For analysts wishing to pursue such aspects further, we have quoted references to a number of relevant publications in Section 2.9. Statistical expertise will often be invaluable in helping to decide how best to formulate objectives in quantitative terms (see Section 2.3) and in evaluating the consequences of such quantification in terms of factors such as the required sampling frequency and accuracy of analytical results. Similarly, experts in computing systems may well be required when considering computerised data-handling systems.

2.3 DEFINING THE OBJECTIVES OF MEASUREMENT PRO-GRAMMES

The following paragraphs essentially reproduce the text of the first edition; since they were written, the various suggestions have been adopted by a Working Group of the World Health Organization[6] and by the Standing Committee of Analysts* of the United Kingdom's Department of the Environment.[5]

It is an obvious point that the objectives of a programme should be clearly and precisely formulated by the user. If this is not done, inappropriate analytical data may well be provided and/or the user is likely to call for needlessly large or unduly small numbers of results. Furthermore, if the objectives are not precisely expressed, it will be difficult or impossible to decide the extent to which they are achieved. In the past, this aspect has often been given only superficial attention, but there appears to be a tendency in recent years for this undesirable situation to improve. Nevertheless, the importance of the topic is such that the suggestions made in the first edition are repeated here.

First, the objectives of measurement programmes should be put in writing. This was recommended by Kittrell[14] and Thomann[15] (see also more recently references 5 to 7) and their reasons may be summarised as follows: (1) careful consideration of what the objectives should be is encouraged, (2) the chance of misunderstanding is reduced, (3) the pursuit of interesting but inessential paths is discouraged, and (4) assessment of the value of the measurement programme is facilitated, and this is particularly relevant to reviews of the need for changes in the programme (see Section 2.1). However, the writing of objectives is not in itself a sufficient defence against imprecise ideas, and an additional device is next considered.

It is suggested that analytical information should be requested on a regular basis only when the user knows beforehand that the results will be used – in a precisely known fashion – to answer one or more defined questions on quality. Requests for analyses based on the thought that the results may ultimately prove to be useful should be avoided, particularly when – as is increasingly the case – analytical and sampling effort is limited. There is an almost infinite number of analyses that might

* For the organisation and functions of this Committee, see Dick.[13]

be useful in most situations, but it is completely impracticable to attempt any such comprehensive coverage. Selection from all the possible objectives and determinands is essential, and the principle suggested at the beginning of this paragraph seems a useful way of deciding what information is justifiably requested. Many other authors have made similar points on the need to concentrate effort on the essentials; see, for example, references 1, 2, 5–7, and 16.

As a further means of optimising measurement programmes, the user should formulate his information needs as quantitatively as possible. As an extreme and perhaps rather artificial example – though such real situations have occurred – of a badly defined requirement, consider the statement 'to obtain information on the quality of rivers'. Such a statement is almost completely useless as a basis for the design of a measurement programme for the following reasons:

(1) the determinands of interest are not specified so that the analyses required are not known;
(2) the particular rivers and the locations on those rivers are not defined so that inappropriate sampling positions may well be chosen;
(3) the required accuracy of results is neither stated nor implied so that inappropriate methods of sampling and analysis may well be used;
(4) no indication is given of the time-scale or sampling frequency so that too few or too many samples may be collected and analysed;
(5) no indication is given of how quality is to be expressed quantitatively (*e.g.*, as an average or median or maximum), and the tolerable uncertainty on any such statistics is not stated so that again inappropriate sampling and analytical techniques may well be used;
(6) as a result of all the above lacks, there is no indication of the amount of data that will need to be processed and the nature of the required data-treatment so that appropriate data-handling techniques cannot be defined.

Users of analytical results must seek to avoid uncertainties such as those in the above example by careful and quantitative definition of every aspect of their requirements. Thus, an objective such as the above would be better expressed by a statement of the following form: 'to estimate each year the annual average concentration of ammonia ($NH_3 + NH_4^+$) at all river sites used for the production of potable water, the annual averages to be in error (95% confidence level) by no more than 0.1 mg l^{-1} of nitrogen or 20% of the ammonia concentration whichever is the greater'. Appropriate statements of this type for the other determinands of interest then provide a set of quantitative targets essential in optimising the choice of sampling, analytical, and data-handling techniques. Plant-control requirements often call for a rather different type of statement, but the aim should also be to formulate a precise and quantitative expression of the information needs. Of course, in formulating these needs, it is essential to review all relevant past data.

Definition of objective and information requirements as suggested above will certainly involve appreciable time and effort if the user has not clearly identified the exact purposes for which he needs, and proposes to use, information. If, however, this clear identification has already been made, as obviously it should have been, then the quantitative detail proposed above should be readily derived. In any event, for any substantial measurement programme, any time spent on the planning stage will usually be an extremely small fraction of the total effort that will subsequently be devoted to the programme and the evaluation of its results. All subsequent activities rest on this definition of information requirements, and it is vital that it be accorded the importance that it warrants.

2.4 DEFINING THE DETERMINANDS OF INTEREST

The particular determinands appropriate to a programme are, of course, critically dependent on the type of water and the objective. This aspect is beyond the scope of this book, but some relevant references are assembled in Section 2.9. There are, however, two important general aspects that should always be considered in defining the determinands for any programme, and these are discussed in Sections 2.4.1 and 2.4.2. For both of these aspects, detailed discussions between the user and the analyst will often be especially important.

2.4.1 Need for Unambiguous Definition of Determinands

The first and obvious point, but one well worth stating, is that each determinand must be unambiguously defined; otherwise, there is no guarantee that the analytical method chosen will be appropriate. This aspect is sometimes not given the attention that it warrants. Its importance arises from the facts that: (1) many substances in water may exist in various chemical and physical forms, and (2) most analytical systems* respond to varying degrees to different forms of the same substance. Thus, in general, before an appropriate analytical method can be chosen, it is necessary to know which forms of a particular substance the user requires to be determined. Detailed discussions between the user and the analyst are usually of great value when considering these matters. An example will help to indicate the points generally to be considered.

Suppose an analyst were asked to determine iron in water. Such a request can be interpreted in several ways which would entail the choice of different analytical methods whose use could well lead to different results for samples. This problem for iron is well known in the analysis of both high-purity boiler feed-water[17] and also natural waters.[18] The problem arises because iron compounds may exist in dissolved, colloidal, and particulate forms; these compounds may also contain iron in different oxidation states and complexed/chelated by various substances. The particulate and colloidal forms may cover a large size range, and may also be composed of a number of different compounds of iron. Finally, compounds of iron and/or its ions may be sorbed on other particulate materials. Which of this rather large number of substances are to be determined should clearly not be decided by the choice of analytical procedure favoured by an analyst merely because of, for example, its convenience or speed. Rather, those forms relevant to the particular objectives should be stated unambiguously and a method adopted, whenever possible, that ensures the accurate measurement of all those forms. Of course, situations will arise where suitable analytical methods are not available, and some modification of objectives may then be necessary after joint discussion between the

---

* See Section 4 for definition of terms such as analytical system, method, *etc.*

user and the analyst. A recent example of a lack of suitable, clear definition for our example of iron arises in a Directive of the European Community concerning the quality of water for human consumption.[19] Here, the determinand is defined merely as 'Iron' with no further information even though the Directive requires Member States to ensure that the iron content does not exceed a specified concentration. This Directive shows many other such examples, but it is by no means the only example that could be quoted of the prevalence of loose definitions.

This problem of vague definition of determinands can be particularly important when a programme requires analyses by a number of independent laboratories, perhaps on an international basis, and the results from the various laboratories are required to be comparable. For those determinands that may exist in different forms, this desired comparability is virtually certain not to be achieved if the determinands are ambiguously defined. This problem is common because many determinands frequently of interest in waters may exist in a number of different forms, *e.g.*, phosphorus, nitrogen, carbon, silicon, metals. Each substance will tend to have its own range of different forms and each should, therefore, be considered individually in the light of analytical knowledge and the information required by the user. There are, however, certain classifications of such general interest that a few words about them are useful.

A distinction between dissolved and undissolved forms is often called for, but whatever theoretical considerations may suggest, in practice an arbitrary decision is required to define the criterion to be used in making this distinction. The experimental approach almost invariably adopted is to filter the sample and to estimate the concentrations of the determinand in the filtrate and retained by the filter. Of course, filters may allow particles to pass through them, and it is, therefore, essential for user and analyst to agree on the type of filter (*e.g.*, with respect to its effective pore-size) to be used. This question of the choice of the filter is discussed in Section 11.2.1. When filters are used, the terms 'filterable' and 'non-filterable' are preferable to 'dissolved' and 'non-dissolved', respectively, in emphasising the operational nature of the analytical differentiation of the two forms (see p. 90 of ref. 20).

Another common requirement is for the determination of a particular chemical form of a substance when a number of different forms are present. Although some analytical systems may be completely selective for a particular form, most are not. For the latter systems, it is useful to name the determinand in such a way that, again, the operational nature of the differentiation is clear. For example, the determination of silicate and orthophosphate is often required in natural and industrial waters, and is usually made by absorptiometric or spectrophotometric procedures (involving the formation of heteropoly compounds with molybdate) that respond to several other forms of silicon and phosphorus. Terms such as 'molybdate-reactive' silicon and phosphorus are desirable[21] to emphasise that the analytical results reflect all those forms that react with molybdate. Many publications provide examples of schemes for this operational differentiation of various forms of a substance; see, for example, the schemes for phosphorus in refs. 20 and 22, but see also Section 11.5.

In the past, certain determinands were often defined as a class of compounds, *e.g.*, phenolic substances, pesticides, polycyclic aromatic hydrocarbons. The idea behind this seems to have been that most, if not all, members of the class possessed undesirable properties; for example, on chlorination of water containing phenolic compounds, undesirable tastes and odours are produced by many chlorinated phenols. Such definitions, however, may lead to large problems, especially when the particular compounds present and their relative concentrations may vary from sample to sample. Not only is it usually impracticable to measure all the large number of such compounds that might be present, but even if such methods were available, the interpretation of the analytical results would almost certainly be ambiguous because the undesirable properties of the various compounds will, in general, vary quantitatively from one compound to another. As an artificially simplified example (but one realistically reflecting the problem), suppose a water contains two different phenolic compounds, A and B, and that A gives five times more intense tastes than the same mass concentration of B. Suppose also that an analytical procedure is available which responds identically to A and B. Then for a sample containing A at three times the concentration of B, the analytical response would be equivalent to 4B, and the taste equivalent to 16B, *i.e.*, a ratio, taste/ analytical response, of 4. If, however, a sample contained A at one-third the concentration of B, the analytical response would be 1.3B and the taste 2.7B, *i.e.*, a ratio of 2. Thus, the analytical response in such situations does not generally provide a valid measure of the undesirable property of interest in the water. This problem may well be even greater if more than two compounds are present, and if the analytical system responds differently to the different compounds. For determinands of this type, two approaches are desirable: (1) to make a direct assessment of the undesirable property of interest, *e.g.*, in the above example by tasting or smelling the water, or (2) by defining particular compounds of interest and using analytical systems that measure those compounds individually.

A further requirement that is commonly stated is the determination of the total concentration of a substance. We would suggest that users should be cautious in requesting such measurements because they may lead to difficult analytical problems that may not be justified by the value of the information obtained. For example, long and cumbersome procedures for the pretreatment of samples may be required to convert all forms of the determinand to forms that are analytically measurable. Iron and phosphorus are two common examples of substances for which such pretreatment problems arise. Furthermore, pretreatment procedures often worsen the precision of analytical results, and this will frequently decrease the value of the results, to the user. Thus, the user should always be adequately confident that he really does need to know the total concentration of *all* forms of a particular substance before requesting such analyses. Yet again, detailed discussions between the user and the analyst on such points are invaluable.

2.4.2 Non-specific Determinands

Some determinands of common interest correspond not to particular chemical species such as chloride, orthophosphate, sodium, *etc.*, but rather to a group of varied species with a common property of interest. Allen and Mancy[23] used the term 'non-specific water-quality parameters' for such determinands, examples of which are: biochemical and chemical oxygen demands; electrical conductivity; taste;

odour; filterability; *etc.* Many of these determinands are either operational (*e.g.*, the propensity of a water to block a filter, to cause corrosion, or to consume dissolved oxygen) or aesthetic (*e.g.*, taste, odour, appearance). For such determinands, two points are worth bearing in mind.

First, these determinands have no independent existence but, rather, are defined only by the analytical procedure used to measure them. It is important, therefore, to ensure that uses of these determinands valid in some applications are also valid for the user's own particular interests. The determination of B.O.D., for example, has often been applied in circumstances where other determinands would be more appropriate.

Second, there is often a tendency to attempt to replace non-specific determinands by specific ones that can be measured more conveniently or more rapidly or more accurately. Careful consideration should be given to such replacements because the specific determinand may not be capable of giving all the information supplied by its non-specific counterpart. For example, the propensity of a water to block a filter may be affected by the concentration of aluminium in the water, so that measurement of that concentration may provide a simple and useful means of estimating filter life. However, other factors may also be involved, and their combined effects may only be revealed by a filterability test.

The determination of non-specific determinands has its own problems. In particular, as the quantitative value for a particular sample is governed by the experimental conditions used in the measurement, it is essential to define and control these conditions rigidly if reliable and comparable results are to be obtained. This problem of comparability may be especially marked when results from a number of laboratories are involved. Such situations are implied when, as is common, non-specific determinands are specified in mandatory water-quality standards. Nonetheless, because such determinands may give the best indication of the aspect of quality of real interest, they should be specified by the user whenever considered necessary. Once again, detailed discussions between the analyst and user should consider if the analytical problems involved in the measurement of non-specific determinands are such that alternative determinands should replace those of primary interest.

2.5 LOCATION, TIME, AND FREQUENCY OF SAMPLING

2.5.1 Sampling Locations

The various considerations involved in selecting the exact positions from which samples are to be collected are discussed in detail in Section 3.2. It suffices to note here that the general locations at which the quality of the water are of interest should always be included in statements of objectives and information requirements. Sometimes such statements will also define the exact locations for sampling, *e.g.*, when the quality of the influent to, or effluent from, a water-treatment plant is of interest. On other occasions the objectives may be such that the exact locations of interest remain to be decided, *e.g.*, when the effect of an effluent discharge on the quality of the receiving water is of interest. Whenever possible, though, the statement of information needs should contain as detailed a specification as possible of the locations at which quality needs to be measured.

In all instances, it is, of course, important to ensure that the exact positions from which samples are collected are adequately representative of the water quality; this is discussed in detail in Section 3.2.

2.5.2 Time and Frequency of Sampling

These two related topics are discussed in detail in Section 3.3. Within the present context of defining information requirements, we again simply note that if the user has any important needs for the times when quality is of interest or for the frequency with which quality should be measured, these should be clearly defined. For example, in process-control applications, the need to control changes in quality may be so important that a minimal sampling frequency should be defined. In other situations, the frequency in itself will be of no interest provided that it is sufficient to ensure an adequately accurate representation of quality during some defined period (see Section 2.8). The time of sampling rather than the frequency can also be of particular concern when the quality of the water of interest shows more or less regular variations, *e.g.*, diurnal variations in the concentration of dissolved oxygen in rivers and variations of quality on start-up and shut-down of industrial plant. Again, any such particular times should be clearly defined when stating information needs.

It should be noted that although the ideal approach might often be to obtain a continuous record of the concentrations of the substances of interest, this is, at present, generally impracticable. Suitable analytical instrumentation for all determinands is either not available or else is too costly or insufficiently reliable or accurate. However, when suitable instruments can be obtained, their use can be advantageous. This and other aspects of on-line analysis are discussed in Sections 12.1 and 12.3.

2.6 REQUIREMENTS CONCERNING ANALYTICAL RESULTS

To ensure that appropriate methods of sampling and analysis are selected, the user should specify any essential requirements he may have concerning analytical results. Experience shows that it is quite inadequate for the user to rely on uniform interpretation of his broad needs by analysts. Three aspects are of general importance, and are considered in Sections 2.6.1. to 2.6.3.

2.6.1 Range of Concentrations to be Determined

The concentration* range of interest can markedly affect the selection of an analytical method. For example, as is well known, procedures suitable for the determination of concentrations in the mg l^{-1} range are often quite unsuitable for concentrations of μg l^{-1}. The most important point here is for the user to specify the smallest concentration, C_L, in which he is interested because this will govern the

---

* When the determinand is not expressed in concentration units, *e.g.*, electrical conductivity, the range of values of interest should be understood.

precision required of the analytical system. In choosing this lower limit, the user must bear in mind that demands for extremely low limits will often require greater analytical sophistication and effort. He should not, therefore, generally specify lower limits than his particular objectives justify.

Thus, suppose an impurity is thought to have no appreciable adverse effects on water quality if present at concentrations less than, say, 0.1 mg 1^{-1}. The user would clearly require a lower limit of less than 0.1 mg^{-1}, but to specify a value of, say, 0.0001 mg 1^{-1} would usually be unnecessary. As a general suggestion for such situations, a lower limit of approximately 10% of the smallest concentration of importance would seem to be reasonable, *i.e.*, in the above example the lower limit would be 0.01 mg 1^{-1}. Of course, other considerations should be used if they are more appropriate. Existing standards for water quality (see Section 2.9) are, therefore, of value in choosing the lower limits for measurement. When such standards do not exist, the user should normally still specify a lower limit, but bearing in mind the point made above that as limits are decreased, greater analytical complexity and cost usually result. This definition of lower limits should, in the authors' view, seldom present difficulty to the user if he has already defined how he will use the analytical results (see Section 2.3). It is of interest to note that a Directive of the European Community on the sampling and analysis of surface waters[24] specifies the limits of detection required of analytical methods.

The upper concentration limit of interest is preferably also defined, but this aspect is usually relatively unimportant in selecting appropriate methods of analysis.

2.6.2 Accuracy Required of Analytical Results

(a) The Need to Define the Tolerable Error of Analytical Results
The subject of the errors of analytical results is discussed in Section 4 and those unfamiliar with this topic are advised to read that section before continuing with the present section.

So far as the user of analytical results is concerned, the important point is that an analytical result must be regarded as only an estimate of the true concentration. This estimate generally differs from the true value as a result of random and systematic errors arising in the sampling and analytical systems. The resultant error of the analytical result affects the ability of the user to derive the information he requires. As a simple example, suppose the analyses were made to check whether or not a particular standard for water quality were met, *e.g.*, that the lead content of a drinking water should not exceed 50 μg 1^{-1}. If an analytical result of 45 μg 1^{-1} was obtained and the random error was $\pm$ 10 μg 1^{-1}, the user would clearly be unable to decide whether or not the standard was met. The effect would be the same if the result was subject to a systematic error of unknown or uncertain magnitude. The maximum tolerable sizes of these two types of error should, therefore, be specified by the user in the light of his particular uses of the results. Other authors have stressed this need to specify the tolerable uncertainties of results; see, for example, refs. 5 to 7 and 9–12. If the user has properly defined his objectives, it will always be possible to determine the effects of errors on his ability to achieve these objectives. Some simple examples of this are given in Section 4.6, but statistical advice will often be invaluable and should be sought whenever possible. Again, the user should note that greater analytical complexity and cost will generally result as demands for the accuracy required of results become more stringent. Accordingly, joint discussions between the user and the analyst will often be beneficial to both. Having decided the appropriate values for these maximum tolerable errors, the user should state them clearly [see (*b*) below] so that they will form clear, quantitative targets to assist the analyst in his choice of suitable sampling and analytical procedures.

It is interesting to note that the European Community has issued a Directive[24] specifying the precisions and accuracies required for the measurement of many determinands in surface waters. Similar types of error specification have been included in the United Nations Environment Programme/World Health Organization scheme for the international monitoring of natural waters,[25,26] and in the United Kingdom's Harmonised Monitoring Scheme for River Waters.[27,28] Similarly, German proposals[29] on the monitoring of drinking water have recommended maximum tolerable standard deviations for a number of determinands.

(b) Specification of Errors for General Monitoring of Waters

As mentioned in (*a*) above, the maximum tolerable values for random and systematic errors depend on the particular application, and each case should be considered individually. For many applications, however, involving general, long-term monitoring of water quality for determinands whose concentrations may cover a relatively wide range (*e.g.*, environmental monitoring), an approach previously proposed by the authors is thought to be of common value. This approach also illustrates a number of points generally involved in specifying the tolerable magnitudes of errors, and it is, therefore, described below. It has been applied in the United Kingdom's Scheme for the Harmonised Monitoring of River Waters[27,28] and in other types of programmes in this country.

(i) Need to specify both random and systematic errors—Separate values for the maximum tolerable random and systematic errors of individual analytical results should be defined because the effects of these two types or error on decision making differ (see Section 4.6). The magnitude of random error can be defined only with respect to a chosen probability level, and for the general applications considered here the 0.05 probability level (equivalent to 95% confidence limits) is considered reasonable. Of course, other levels may be chosen if they are thought to be more appropriate. For example, the 0.01 level (equivalent to 99% confidence limits) might be more suitable in the control of crucial aspects of water quality. It is also assumed here that the random errors arising in the analytical process follow the normal distribution. Experience suggests that this is usually a reasonable assumption in the present context, but, again, it can be modified if knowledge shows it not to be valid. Given this basis, it follows that the maximum size of the random error can be equated to $2\sigma_t$, where σ_t is the standard deviation of the analytical results.*

It is next suggested that, in the absence of any particular needs to the contrary, it is generally useful to specify equal numerical magnitudes for the maximum tolerable random and systematic errors. This division of the total error between its random and systematic components is to some extent arbitrary, but experience

---

* The numerical factor should strictly be 1.96; the difference from the quoted value of 2 is negligible in contexts such as the present where one is defining target values for analytical errors.

shows that it is often useful. Of course, other values for the ratio, random error/ systematic error, may be specified if desired. Caution is needed, however, if values appreciably greater than unity are considered for this ratio because this may well imply the need for prohibitively large numbers of replicate measurements in any experimental tests to estimate the actual magnitude of the systematic error. Currie has stressed the importance of this ratio[10] and the underlying statistical ideas are described in Section 4.6.2.

(ii) Quantitative expression of tolerable error—The expression of the tolerable error is sometimes made by statements such as 'the error should not exceed a stated proportion (*e.g.*, 10%) of the results'. For determinands whose concentrations are not essentially constant, such statements often overlook the fact that analytical errors, when expressed as a proportion of the result, increase markedly as the lower concentration limit of an analytical system is approached. For example, at the limit of detection, the random error (95% confidence limits) is approximately 60% of the result [see Section 8.3.1(*c*)]. In addition, as concentrations decrease towards the lower limit of interest it is often the case that the tolerable percentage error increases; indeed, this may well always be so for the type of measurement programme being considered here. Users should, therefore, consider whether their requirements are better expressed in a different way. The type of statement suggested here is of the form: 'the error should not exceed c mg l^{-1} or $p\%$ of the concentration, whichever is the greater'. Given the suggestions in (*i*) above, this method of specifying tolerable error would apply to both random and systematic error, both being specified by the same values of c and p. Note that as the total error is the sum of the random and systematic errors, these suggestions are equivalent to a maximum tolerable total error of $2c$ mg l^{-1} or $2p\%$ of the concentration, whichever is the greater.

(iii) Choice of numerical values for c and p—As before, the numerical values to be specified for c and p may be chosen for each programme, but again experience suggests some commonly useful values that represent a reasonable compromise between what can be routinely achieved analytically and the desire to specify very small errors. A value of $p = 10$ for a number of determinands has been found suitable in the Harmonised Monitoring Scheme for River Waters,[27,28] but this value may be difficult to achieve (and to show to have been achieved) for many determinands present in only trace concentrations and/or requiring relatively complicated analytical procedures for their determination. Given that $p = 10$ and that the random error (95% confidence limits) is equal to $2\sigma_t$ [see (*i*) above], it follows that the maximum tolerable standard deviation, σ_t, is $0.5p = 5\%$. The value for c is usefully equated to half the smallest concentration of interest, C_L, (see Section 2.6.1). Though again arbitrary, this has the effect of ensuring that the maximum tolerable total error of a result at a concentration, C_L, is equal to C_L. Thus, $c = 0.5C_L$, and it follows that the maximum tolerable standard deviation is $0.5c = 0.25C_L$.

(iv) Resulting numerical values for the maximum tolerable errors—From the suggestions in Section 2.6.1 and (*i*) to (*iii*) above, the maximum tolerable values for the standard deviation and systematic error of analytical results can be deduced for

any concentration of the determinand. It is emphasised that these numerical values are best seen as firm, quantitative targets at which the analyst can aim, but they need to be applied with commonsense. For example, if an analyst achieved a standard deviation of 11 mg l^{-1} when the target value was 10 mg l^{-1}, it would be rare for this to represent unacceptable precision. Examples of the application of the above calculations are given in Table 2.1 and illustrated in Figure 2.1 for determinands with different numerical values for their maximum allowable concentrations.

(v) Other aspects of this scheme for error specification—In the above-mentioned scheme, the numerical values of c, p, and the desired probability level provide all the criteria needed by an analyst in judging the actual or potential suitability of analytical systems with respect to accuracy. In some circumstances it may also be useful to consider another criterion based on the following idea. The user will generally require a high probability that the determinand will be detected in samples in which the concentrations are greater than or equal to C_L. This will be achieved provided that the limit of detection of an analytical system is not greater than C_L, and this condition represents another criterion for judging the performance of analytical systems. The statistical ideas involved in defining the limit of detection are discussed fully in Section 8.3. It should be noted, however, that if the required value of the limit of detection, L_D, is equated to C_L, a slight numerical inconsistency may arise between the two requirements: (1) $L_D = C_L$, and (2) $\sigma_t = 0.25C_L$. For example, under many circumstances $L_D = 3.29\sigma_t$ [for 95% confidence of detection – see Section 8.3.1(c)], so that σ_t should not exceed $0.3L_D = 0.3C_L$. On the other hand, the scheme described above specifies that σ_t should not exceed $0.25C_L$. This inconsistency could be eliminated, for example, by using a slightly different confidence level to define the limit of detection, but this complication does not seem worthwhile. If such target values are interpreted with commonsense, no difficulty should be caused. In any event, we would stress that, in our view, the primary targets should be those directly concerned with the required precision and bias of analytical results. The equation of C_L and L_D provides a check on the reasonableness of those targets with respect to the power of detection.

TABLE 2.1

Calculation of Maximum Tolerable Errors in Analytical Results

Maximum allowable concentration*	Smallest concentration* of interest, C_L	Maximum tolerable error*		
		Standard deviation, σ_t†	Systematic error†	Total error†
100	10.0	2.5 or 0.05C	5.0 or 0.1C	10.0 or 0.2C
50	5.0	1.25 or 0.05C	2.5 or 0.1C	5.0 or 0.2C
10	1.0	0.25 or 0.05C	0.5 or 0.1C	1.0 or 0.2C
5.0	0.5	0.125 or 0.05C	0.25 or 0.1C	0.5 or 0.2C
1.0	0.1	0.025 or 0.05C	0.05 or 0.1C	0·1 or 0.2C

* All expressed in the same concentration units: *e.g.*, if the maximum allowable concentration were 100 μg l^{-1}, the smallest concentration of interest and the maximum tolerable errors are also expressed in units of μg l^{-1}.

† In the quoted numerical values, C denotes the concentration of determinand in the sample. Of the two values tabulated for each error, that one applies which has the larger value for a given determinand concentration; see Figure 2.1.

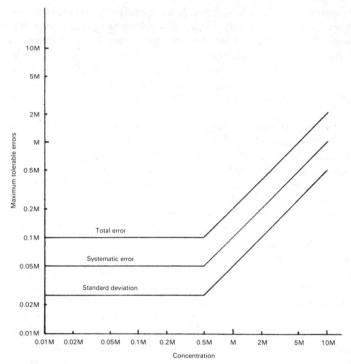

Figure 2.1 *Dependence of maximum tolerable errors on the concentration of the determinand*
(The approach is described in Section 2.6 of the text; M is the maximum tolerable concentration of a substance, $c = 0.5C_L$, and $p = 10\%$)

A few additional points on the scheme are also worth noting.

(1) The concept of 'a smallest concentration of interest' may not be directly relevant for certain determinands, *e.g.*, pH value. When this is so, the expressions for the tolerable errors will usually need to specify only fixed proportional or absolute errors.

(2) As mentioned above, the values for c and p should be chosen appropriately for each measurement programme. As their numerical values are reduced, however, the analytical effort and cost needed to achieve them will usually increase and, at some point, become impracticably large. Care is required, therefore, to specify errors whose magnitudes are both tolerable and achievable. The values for c and p suggested above are equivalent to a maximum total error (95% confidence limits) of 20% at the higher concentration of a determinand. This may seem excessively large to many readers, but the author's experience, as mentioned above, is that this level of accuracy may be difficult to achieve – and perhaps even more difficult to show to have been achieved – for many determinands. In this connection it is interesting that McFarren *et al.*,[30] in considering the results from any interlaboratory studies

of different methods for analysing water, concluded that for only a relatively small fraction of the methods studied were total errors of 25% or less achieved by all or most of the laboratories. See also refs. 31 to 34 for experiences in the United Kingdom in the analysis of river waters and drinking water.

(3) In specifying a particular method of analysis in order effectively to define a non-specific determinand (see Section 2.4.2), care is also necessary to ensure that the method allows achievement of the required accuracy and limit of detection.

(4) Circumstances may be conceived where individual analytical results are to be obtained over a period of time with the *sole* intention of using them to form an estimate of the mean (or some other statistic) concentration of a determinand during the period. Given such a situation the specification of the maximum tolerable errors of individual analytical results is no longer directly relevant because it is only the error of the finally derived mean that is of importance. The total error of this statistic will be governed by two sources of variation: that due to the inherent variability of the water of interest during the period of concern (see Section 3.3), and that due to the errors involved in measuring the quality of the water on each of the occasions selected during the period. Situations can certainly arise where the former is much greater than the latter so that, at the sampling frequency achievable, the measurement errors are relatively unimportant. The question then arises – can the maximum tolerable errors specified for the individual measurements be substantially relaxed relative to the general magnitudes of errors suggested above? Strictly, the answer to this seems to be 'yes', but we would still not generally suggest such relaxation for the following reasons. First, given a constant systematic error (for a given concentration) on an individual result from the measurement process, this error is unaffected by averaging a number of results; the systematic error of the mean of n results is the same as the systematic error of an individual result whatever the value of n. Thus, whenever the systematic error of a statistic derived from a number of analytical results is of interest (and it seems to us that there is usually, if not always, interest in this error), it may be dangerous substantially to relax any target values for the systematic error of individual measurements. Second, if this target for systematic error is not substantially relaxed, it will usually be impracticable to relax the target value for random error (the standard deviation) substantially because this could require prohibitively large effort in any experimental attempts to estimate the magnitude of the systematic error (see Section 4.6.2). Finally, there is the point that it would perhaps be unlikely that users would guarantee not to make use of individual analytical results.

2.6.3 Time of Analysis

In some measurement programmes, particularly those where analytical results are required to allow short-term control of water quality, it is important that each result be obtained within a given time of sampling. Whenever this is so, the user should specify the maximum tolerable period between the beginning of a sampling occasion and the time when he has the result for that sample. This time may play an important role in the choice of analytical and sampling procedures, and is one of the factors that may indicate the desirability of on-line analysis (see Sections 12.1 and 12.3).

2.7 EXPRESSION OF ANALYTICAL RESULTS

Factors to be considered in relation to the reporting of analytical results are generally well known and only three aspects are mentioned here.

2.7.1 Units in Which Results are Expressed

Different units can be used for reporting results for determinand concentrations and other types of measurements. For example, nitrate may be quoted as mg NO_3 1^{-1} or mg N 1^{-1}. Although uniformity in such matters is desirable from many points of view, the use to be made of results may also indicate a clear preference for a particular unit. For example, when considering mass-balances for ammonia, nitrate, and nitrite, calculations are facilitated if all are reported as mg N 1^{-1}. It seems to the authors best for the appropriate units to be decided by the user for each application and for all results to be accompanied by an unambiguous statement of the units in which they are expressed.

2.7.2 Rounding-off Results

It is common practice for the analyst to round-off his results with the intention that the number of figures in a reported result indicates its accuracy. Detailed consideration of this practice is reserved for Section 8.1; here we simply state our views that:

(1) rounding should not be used as a means of indicating the accuracy of results (see Section 2.7.3);
(2) the degree of rounding should basically be decided by the *user of the results* from consideration of the uses to which the results will be put, but discussion with the analyst will, as always, be worthwhile.

2.7.3 Reporting the Accuracy of Analytical Results

Rather than rounding results, a much more direct and informative method for indicating their accuracy is the simple and obvious one of attaching to each result a statement of its error. From the user's point of view, if this is done, he has all the information available. There are a number of problems involved in obtaining an estimate of this error, and these are discussed in detail in Sections 5 and 6. Detailed consideration of the reporting of analytical results will be found in Sections 8.2 and 8.3.2.

2.8 USE TO BE MADE OF ANALYTICAL RESULTS

As mentioned above (Section 2.3), it is strongly recommended that the user decide exactly what use he will make of analytical results before any are obtained. This is, of course, not to say that he should not review this use once some results have been obtained. Detailed planning of the use of results will always help to ensure optimal design and efficiency of any measurement programme, and will often reveal important aspects that need more detailed consideration.

To a large extent, the use of results will already have been fixed in considering the objectives and information requirements of a programme. For example, in process-control applications, each analytical result will usually be compared with one or more target values and action taken on the basis of the difference between the result and the target(s). The use to be made of results in such programmes follows directly from the objective. Some programmes, however, are intended to characterise, rather than directly to control, water quality. For such applications there is an almost invariable need to calculate (from the individual analytical results) summarising statistics to provide quantitative measures of quality. The user should, therefore, decide which statistic(s) (*e.g.*, mean, median, 95 percentile) he requires as well as the period for which they are to refer. It is also essential that the user decide the uncertainty he can tolerate in such statistics, and that this uncertainty be stated very clearly as part of the overall definition of information needs. Without such specified tolerable uncertainties, a rational, quantitative approach to the estimation of the required sampling frequency is impossible. This need has been stressed by many authors and is considered in detail in Section 3.3.

In many measurement programmes, very large numbers of individual analytical results are produced. Computer techniques for storing and processing these results are usually of great value, if not essential, in allowing the required information to be

TABLE 2.2

Books on Water Quality and Related Topics

Subject	Ref.
Natural Waters:	
General	4, 35–42
Groundwaters	4, 43–46
Lakes and reservoirs	4, 47–49
Streams and rivers	4, 50–53
Estuaries	4, 54–58, 88
Seas and oceans	59–61
Pollution	41, 42, 46, 50, 62–67
Biological aspects	55, 64, 68
Drinking water and health	38, 66, 69*, 70*, 71–75
Production of drinking water	38, 42, 66, 76–79, 89
Sewage and sewage treatment	42, 78
Industrial plant	37, 79, 80
Environmental chemistry, geochemistry, chemistry of aquatic processes, *etc.*	36, 40, 45, 81–87
Water quality standards for natural and drinking waters†	19, 24, 53, 67, 69*, 70*, 75, 90–98
Review on sources of information on water chemistry	99

* These two publications are being revised to be replaced by a single publication; see ref. 100.
† Implementation in the United Kingdom of the Directives of the European Community is discussed in the Biennial Reports of the Standing Technical Advisory Committee on Water Quality; see most recently ref. 101.

extracted with maximum speed and efficiency, and the use of such systems has grown markedly in the last decade. These techniques have other advantages, and their use is briefly considered in Section 13. The analysis of information requirements suggested above is clearly essential in defining the computer programs that will be required for a particular measurement programme.

2.9 FURTHER READING ON WATER QUALITY

Many aspects of water quality are or may be relevant to the design of measurement programmes, *e.g.*, processes affecting quality, means of controlling quality, and mandatory standards for water quality. Knowledge of such matters is usually essential in defining the positions and times at which samples are to be collected (see Sections 3.1 to 3.4) and the determinands of interest for any measurement programme. As noted above, such aspects are primarily the responsibility of the user of results, but for analysts wishing to pursue these and related topics we have assembled a number of useful references in Table 2.2. There is an extremely large number of relevant publications, which continue to appear at an increasing rate, and we have been unable to attempt a comprehensive review of them. Our aim has rather been to provide analysts with useful introductions to each topic. Accordingly, the non-appearance of a particular publication in Table 2.2 by no means implies that it has been thought not worthy of mention.

2.10 REFERENCES

[1] K. H. Mancy in 'Proceedings of a Seminar on the Design of Environmental Information Systems', Katowice, Poland, 11–20 January, 1973; ed. R. A. Deininger, Ann Arbor Science Publishers, Ann Arbor, Michigan, 1974, pp. 173–197.
[2] K. H. Mancy and H. E. Allen, in 'Examination of Water for Pollution Control', Volume 1, ed. M. J. Suess, Pergamon, Oxford, 1982, pp. 1–22.
[3] G. B. Murdock, in 'Proceedings of a National Symposium on Data and Instrumentation for Water Quality Management', ed. J. E. Kerrigan, University of Wisconsin, Madison, 1970, pp. 57–75.
[4] IHD-WHO Working Group, 'Water Quality Surveys' (Studies and reports in hydrology 23), UNESCO/WHO, Paris, 1978.
[5] Standing Committee of Analysts, 'General Principles of Sampling and Accuracy of Results, 1980', H.M.S.O., London, 1980.
[6] A. L. Wilson, in ref. 2, pp. 23–77.
[7] American Chemical Society, *Anal. Chem.*, 1980, **52**, 2242.
[8] Federal Water Quality Administration, 'Design of Water Quality Surveillance Systems', US Government Printing Office, Washington, 1970.
[9] P. J. Elving and H. Kienitz, in 'Treatise on Analytical Chemistry', Part 1, Volume 1, ed. I. M. Kolthoff and P. J. Elving, Second Edition, Wiley, Chichester, 1978, pp. 53–94.
[10] L. A. Currie, in ref, 9, pp. 95–242.
[11] D. W. Moody, in ref. 3, pp. 325–335.
[12] J. K. Taylor, *Anal. Chem.*, 1981, **53**, 1588A.
[13] T. A. Dick, *Water Pollut. Control*, 1982, **81**, 185.
[14] F. W. Kittrell, 'A Practical Guide to Water Quality Studies of Streams', US Government Printing Office, Washington, 1969.
[15] R. V. Thomann, in ref. 3, pp. 76–86.
[16] A. Key, 'River Water Quality', Water Resources Board, Reading, 1967.
[17] A. L. Wilson, *Analyst*, 1964, **89**, 402.
[18] W. K. Dougan and A. L. Wilson, Technical Report TR 83, Water Research Centre, Medmenham, Bucks, 1972.
[19] Council of the European Communities, *Off. J. Eur. Communities*, 1980, **23**, No. L-229, 11.
[20] American Public Health Association *et al.*, 'Standard Methods for the Examination of Water and Wastewater', Fifteenth Edition, American Public Health Association, Washington, 1981.

21 J. D. H. Strickland and T. R. Parsons, 'A Practical Handbook of Sea-water Analysis', Fisheries Research Board of Canada, Ottawa, 1968.
2 Standing Committee of Analysts, 'Phosphorus in Water, Effluents and Sewages, 1980', H.M.S.O., London, 1981.
23 H. E. Allen and K. H. Mancy, in 'Water and Water Pollution Handbook', Volume 3, ed. L. L. Ciaccio, Dekker, London, 1972, pp. 971–1020.
24 Council of the European Communities, *Off. J. Eur. Communities*, 1979, **22**, No. L-271, 44.
25 R. Helmer, *Nat. Resour.*, 1981, **17**, 7.
26 S. Barabas, in 'Advances in the Identification and Analysis of Organic Pollutants in Water', Volume 1, ed. L. H. Keith, Ann Arbor Science Publishers, Ann Arbor, Michigan, 1981, pp. 481–494.
27 E. A. Simpson, *J. Inst, Water Eng, Scient.*, 1978, **32**, 45,
28 A. L. Wilson, *Analyst*, 1979, **104**, 273.
29 I. Gans and M. Sonneborn, in 'Die Trinkwasser-Verordnung', ed. K. Aurand *et al.*, Erich Schmidt Verlag, Berlin, 1977, pp. 181–195.
30 E. F. McFarren, R. J. Lishka and J. H. Parker, *Anal. Chem.*, 1970, **42**, 358.
31 L. Ranson and A. L. Wilson, Technical Report TR 28, Water Research Centre, Medmenham, Bucks, 1976.
32 Analytical Quality Control (Harmonised Monitoring) Committee, *Analyst*, 1979, **104**, 290.
33 Analytical Quality Control (Harmonised Monitoring) Committee, *Analyst*, 1982, **107**, 680.
34 Analytical Quality Control (Harmonised Monitoring) Committee, *Analyst*, 1982, **107**, 1407; 1983, **108**, 1365; and Technical Reports TR190, 1983, and TR196, 1984, Water Research Centre, Medmenham, Bucks; The Severn Estuary Chemists' Sub-Committee, *Analyst*, 1984, **109**, 3.
35 T. R. Camp, 'Water and its Impurities', Chapman and Hall, London, 1963.
36 J. D. Hem, 'Study and Interpretation of the Chemical Characteristics of Natural Water', Second Edition, US Government Printing Office, Washington, 1970.
37 C. E. Hamilton, ed., 'Manual on Water', American Society for Testing and Materials, Philadelphia, 1978.
38 W. S. Holden, ed., 'Water Treatment and Examination', Churchill, London, 1970.
39 M. J. Hammer and K. A. MacKichan, 'Hydrology and Quality of Water Resources', Wiley, Chichester, 1981.
40 S. D. Faust and O. M. Aly, 'Chemistry of Natural Waters', Ann Arbor Science Publishers, Ann Arbor, Michigan, 1981.
41 D. Hammerton, ed. 'Organic Micropollutants in Water', Institute of Biology, London, 1982.
42 N. D. Bedding, A. E. McIntyre, R. Perry, and J. N. Lester, *Sci. Total Environ.*, 1982, **25**, 143 and 1983, **26**, 255.
43 R. A. Freeze and J. A. Cherry, 'Groundwater', Prentice-Hall, Englewood Cliffs, N.J., 1979.
44 R. E. Jackson, ed., 'Aquifer Contamination and Protection' (Studies and reports in hydrology 30), U.N.E.S.C.O., Paris, 1980.
45 G. Matthess, 'The Properties of Groundwater', Wiley, Chichester, 1982.
46 W. van Duijvenbooden, P. Glasbergen, and H. van Lelyveld, ed., 'Quality of Groundwater', Elsevier, Oxford, 1981.
47 G. E. Hutchinson, 'A Treatise on Limnology', Volumes 1 and 2, Wiley, London, 1957 and 1967.
48 J. M. Symons, ed., 'Water Quality Behaviour in Reservoirs', US Government Printing Office, Washington, 1969.
49 Organization for Economic Co-operation and Development, 'Eutrophication of Waters, Monitoring, Assessment and Control', O.E.C.D., Paris, 1982.
50 M.J. Stiff, ed., 'River Pollution Control', Ellis Horwood, Chichester, 1980.
51 A. M. Gower, ed., 'Water Quality in Catchment Ecosystems', Wiley, Chichester, 1980.
52 J. Lewin, ed., 'British Rivers', Allen and Unwin, London, 1981.
53 National Water Council, 'River Water Quality, the Next Stage. Review of Discharge Consent Conditions', National Water Council, London, 1978.
54 Natural Environment Research Council, 'Bibliography of Estuarine Research' (Publication Series C, No. 17), N.E.R.C., London, 1976.
55 E. J. Perkins, 'The Biology of Estuaries and Coastal Waters', Academic, London, 1974.
56 J. D. Burton and P. S. Liss, ed., 'Estuarine Chemistry, Academic, London, 1976.
57 C. B. Officer, 'Physical Oceanography of Estuaries and Associated Coastal Waters', Wiley, London, 1976.
58 E. Olausson and I. Cato, eds., 'Chemistry and Biogeochemistry of Estuaries', Wiley, London, 1980.
59 J. P. Riley and R. Chester, 'Introduction to Marine Chemistry', Academic, London, 1971.
60 J. P. Riley and G. Skirrow, ed., 'Chemical Oceanography', Second Edition, Volumes 1–6, Academic, London, 1975.
61 E. K. Duursma and R. Dawson, ed., 'Marine Organic Chemistry: Evolution, Composition, Interactions and Chemistry of Organic Matter in Seawater', Elsevier, Oxford, 1981.
62 S. J. Eisenreich, ed., 'Atmospheric Pollutants in Natural Waters', Ann Arbor Science Publishers, Ann Arbor, Michigan, 1981.

63 V. Novotny and G. Chesters, 'Handbook of Nonpoint Pollution', Van Nostrand-Reinhold, London, 1981.
64 C. F. Mason, 'Biology of Freshwater Pollution', Longman, London, 1981.
65 E. A. Laws, 'Aquatic Pollution: An Introductory Text', Wiley, Chichester, 1981.
66 R. L. Jolley, W. A. Brungs, and R. B. Cummings, ed., 'Water Chlorination: Environmental Impact and Health Effects', Volumes 1–3, Ann Arbor Science Publishers, Ann Arbor, Michigan, 1980.
67 S. P. Johnson, 'The Pollution Control Policy of the European Communities', Graham and Trotman, London, 1979.
68 Uhlmann, D., 'Hydrobiology', Wiley, London, 1980.
69 World Health Organization, 'European Standards for Drinking Water', Second Edition, W.H.O., Geneva, 1970.
70 World Health Organization, 'International Standards for Drinking Water', Third Edition, W.H.O., Geneva, 1971.
71 World Health Organization, 'Surveillance of Drinking Water Quality', W.H.O., Geneva, 1976.
72 R. G. Ainsworth, Technical Report TR 167, Water Research Centre, Medmenham, Bucks, 1981.
73 H. van Lelyveld and B. C. J. Zoetman, ed., 'Water Supply and Health', Elsevier, Oxford, 1981.
74 J. D. McKinney, ed., 'Environmental Health Chemistry', Ann Arbor Science Publishers, Ann Arbor, Michigan, 1981.
75 US National Academy of Sciences, 'Drinking Water and Health', National Academy of Sciences, Washington, 1977.
76 W. O. Skeat, ed., 'Manual of British Water Engineering Practice', Volumes 1–3, Fourth Edition, The Institution of Water Engineers, London, 1969.
77 C. R. Cox, 'Operation and Control of Waste Processes', Second Edition, World Health Organization, Geneva, 1969.
78 E. W. Steel, and T. J. McGhee, 'Water Supply and Sewage', Fifth edition, McGraw-Hill, London, 1980.
79 P. Hamar, J. Jackson, and E. F. Thurston, ed., 'Industrial Water Treatment Practice', Butterworths, London, 1961.
80 H. E. Hömig, 'Physikochemische Grundlagen der Speisewasserchemie', Vulkan Verlag, Essen, 1959.
81 'Environmental Chemistry' (Specialist Periodical Reports), Volume 1, ed, G. Eglinton, 1975 and Volume 2, ed. H. J. M. Bowen, 1972, The Royal Society of Chemistry, London.
82 H. J. M. Bowen, 'Environmental Chemistry of the Elements', Academic, London, 1979.
83 P. Beneš and V. Majer, 'Trace Chemistry of Aqueous Solutions', Elsevier, Oxford, 1980.
84 A. Lerman, 'Geochemical Processes: Water and Sediment Environments', Wiley, London, 1979.
84 W. Stumm and J. J. Morgan, 'Aquatic Chemistry', Second Edition, Wiley, Chichester, 1981.
86 J. I. Drever, 'The Geochemistry of Natural Waters', Prentice-Hall, Englewood Cliffs, N.J., 1982.
87 O. Hutzinger, ed., 'The Handbook of Environmental Chemistry,' Volumes 1–3, Springer, Heidelberg, 1982.
88 E. K. Roberts, ed., 'Estuaries and Coastal Waters of the British Isles: a bibliography of recent scientific papers. No. 3', Marine Biological Association of the United Kingdom, Plymouth, 1979.
89 W. M. Lewis, ed., 'Developments in Water Treatment', Applied Science Publishers, London, 1980.
90 Council of the European Communities, *Off. J. Eur. Communities*, 1975, **18**, No. L-194, 26.
91 Council of the European Communities. *Off. J. Eur. Communities*, 1976, **19**, L-31, 1.
92 Council of the European Communities, *Off. J. Eur. Communities*, 1976, **19**, L-129, 30.
93 Council of the European Communities, *Off. J. Eur. Communities*, 1978, **21**, L-222, 1.
94 Council of the European Communities, *Off. J. Eur. Communities*, 1979, **22**, L-281, 47.
95 Council of the European Communities, *Off. J. Eur. Communities*, 1980, **23**, L-20, 43.
96 J. Smeets and R. Amavis, *Water Air Soil Pollut.*, 1981, **15**, 483.
97 R. E. Train, 'Quality Criteria for Water', Castle House Publications, London, 1979.
98 J. A. Winter, in 'Chemistry in Water Reuse', Volume 1, ed. W. J. Cooper, Ann Arbor Science Publishers, Ann Arbor, Michigan, 1981, pp. 187–205.
99 J. J. Delfino, *Environ. Sci. Technol.*, 1977, **11**, 669.
100 W. M. Lewis, in ref. 73, pp. 285–292.
101 Standing Technical Advisory Committee on Water Quality, 'Third Biennial Report' (DOE/NWC Standing Technical Committee reports No. 33), H.M.S.O., London, 1983.

3 Sampling

3.1 BASIC PROBLEMS AND AIMS OF SAMPLING

When all the decisions discussed in Section 2 have been made, the basic strategy, as it were, of the measurement programme has been defined. The tactics involved in implementing the programme require detailed consideration of the procedures to be used in sampling, analysis, and data-handling. As already mentioned in Sections 1 and 2, the various factors relevant to the design of programmes of sampling and analysis are usually closely inter-related so that, for example, difficulties arising from these tactical matters may necessitate changes in the strategy. A simple example of this is that insufficient analytical resources may be available to make all the measurements implied by a given strategy. This inter-relationship of factors should always be borne in mind, but for simplicity we deal only with sampling in the present section.

One of the basic problems of sampling arises because it is not generally feasible to analyse all the waters whose quality is of interest; indeed, in the vast majority of applications this is impossible. Thus, one must generally rely on the analysis of selected portions (*i.e.*, samples) of the water of interest, and the quality of this water must then be inferred from that of the samples. An additional point of importance is that the samples almost always represent only an extremely small proportion of the water-body under consideration. Of course, if the quality were essentially constant in time and space, this inference would present no problems. Such constancy is, however, rarely if ever observed; in most circumstances virtually all waters show both spatial and temporal variations in quality. Accordingly, the quality at one position in a whole-body may inadequately represent the qualities at other positions; similarly, the quality at one moment in time may not allow an adequately accurate inference of the quality at other times. Although, as will be seen below, other problems may arise in sampling, this problem of the degree of spatial and temporal representativeness of samples is basic to all applications. Clearly, the selection of the times and positions of sampling generally requires great care if they are to provide the information (of appropriate accuracy) required by the user of analytical results. It follows immediately, as already stressed in Section 2, that clear, quantitative definition of the information required is essential to allow a rational approach to the establishment of optimal sampling programmes; the importance of this point cannot be over-emphasised.

Given that the water of interest shows spatial and temporal variability, the accuracy with which its quality can be inferred can clearly be improved by increasing the number of sampling positions throughout the water-body and/or the number of occasions on which samples are obtained. In general, however, such increases lead to greater costs of the measurement programme. There is, therefore, a general need to attempt to define the minimal number of sampling positions and occasions that will provide the desired information.

With this background, the basic aims of sampling may be summarised as follows:

(1) to select the positions (see Section 3.2) and times (see Section 3.3) of sampling such that the required information on quality can be inferred with adequate accuracy at minimal cost;

(2) to ensure that the concentrations of determinands in samples are identical* to those in the water of interest at the positions and times of sampling (see Section 3.5);

(3) to ensure that the concentrations of determinands in samples do not change in the period between sampling and analysis (see Section 3.6). A number of effects may cause difficulties – sometimes very large – in achieving these aims; such problems are discussed in the noted sections. A particular problem that may well arise when the desired sampling positions and times have been decided is that the total number of samples in a given period is greater than the available resources can handle. Possible approaches to reducing the number of samples and to the overall design of sampling programmes are considered in Section 3.4.

It should also be noted that the above aims implicitly assume that samples are removed from the water-body and then subsequently analysed elsewhere, *e.g.*, in a laboratory. Of course, some of the problems can be eliminated by the use of *in-situ* measurement techniques in which the analytical sensor is immersed in the water-body at the desired position; collection of samples is thereby avoided and problems arising in relation to aims (2) and (3) will be minimised and often eliminated. Similarly, the use of continuous-measurement techniques (in which a continuous stream of sample is abstracted from the water-body and analysed continuously) may also reduce or eliminate problems arising from temporal variability and in relation to aim (3). The potential value of such techniques should, therefore, always be borne in mind, and appropriate use made of them. At present, however, several reasons (see Section 12) prevent the general use of this approach, and this position seems likely to continue for some time to come. Accordingly, the vast bulk of water-quality measurements are based on the removal of samples from the water-body and their subsequent analysis; it is this situation which is primarily addressed here.

Two further points should be stressed at this stage. First, for a particular measurement programme the appropriate sampling and analytical techniques are very closely inter-related. For example, the number of samples to be analysed in a given period may have a profound effect on the choice of analytical methods, and the volume of sample specified by an analytical method may markedly affect the choice of sampling procedure. It is essential, therefore, that both sampling and analytical aspects and their inter-relation are properly considered in the planning and subsequent stages of measurement programmes. This should present no problem when the analyst is responsible for the selection and use of sampling and

---

* In principle, it would suffice to ensure that the concentrations in samples were a constant and known proportion of those in the water of interest; analytical results could then be appropriately corrected. In practice, however, such an approach will generally lead to greater difficulties, and it is not further considered.

analytical procedures, but it is a point to be very carefully watched when this is not so [see Section 3.5.1(*b*)(*iii*)].

Second, for the reasons noted in the preceding paragraph and below, in planning sampling programmes it is important – whether or not the analyst has responsibility for their design and execution – that all relevant disciplines and expertise are properly represented. Other authors[1-3] also stress this point (see also the guidelines for data-acquisition and data-quality evaluation in environmental chemistry recommended by a committee of the American Chemical Society[4]).

Although some of the problems of sampling are discussed in many publications referred to below, very few present a systematic treatment of all the relevant aspects. Of those known to us, two of the most recent and comprehensive are those published by the World Health Organization (Regional Office for Europe)[5] and the United Kingdom Standing Committee of Analysts.[6] Lest it should be thought that these two provide a completely independent approach to the present book, we feel we must point out that the former was prepared by one of us in collaboration with a Working Group, and it shows a number of resemblances to the present section, which is, however, considerably more detailed. The second of the above publications is to a large extent based on the former, though it deals with a larger range of types of water. An excellent, comprehensive (and independent) treatment of all aspects of sampling has recently been provided by Berg,[3] and reference will frequently be made to this in subsequent sections. Finally, we would mention a joint publication by the United Educational, Scientific and Cultural Organization and World Health Organization;[7] this also provides an independent, valuable and integrated discussion of many aspects of programmes concerned with rivers, lakes, estuaries, and groundwaters, but its treatment of many of the topics considered below tends to be less detailed than that of refs. 3, 5, and 6, and the present section.

3.2 SAMPLING LOCATION AND POINT

By sampling location and point we mean, respectively, the general position within a water-body and the exact position at a sampling location at which samples are obtained. The choice of sampling positions is complicated by the great variety of processes that can affect the spatial heterogeneity of water-bodies. This problem is made still more complex because these processes may vary markedly in relative importance from one water-body to another, from one determinand to another, and from time to time. Therefore, local and detailed knowledge of the character-istics of a water-body is indispensable for the optimal selection of sampling positions. Nonetheless, there are several important, useful principles and points that are applicable to all water-bodies, and these are considered in Section 3.2.1. In addition, to aid the consideration of particular water-bodies, we have grouped instances of spatial heterogeneity into two classes that are dealt with in Sections 3.2.2 and 3.2.3 together with the various types of processes leading to such heterogeneity. Finally, in Section 3.2.4 we have summarised a number of more detailed points and references relevant to particular types of waters, *e.g.*, rivers, estuaries, groundwaters, and drinking waters.

3.2.1 General Principles

(a) Sampling Locations – Importance of Precise Formulation of Objectives of Measurement Programmes

As noted in Section 2.5.1, the objectives of a programme will sometimes immediately define the sampling locations. For example, when the concern is to measure the efficiency of a chemical plant for purifying water, sampling locations will be required before and after the plant. Similarly, when the effect of an effluent discharge on the water-quality of a receiving river is of interest, samples will be required from locations upstream and downstream of the discharge. Many similar types of application commonly occur. For larger scale water-bodies (*e.g.*, a river basin, a large estuary, a large, urban drinking-water distribution system), however, the objectives may be defined in terms that provide essentially no indication of sampling locations. For example, objectives such as 'to measure river quality within a river basin' or 'to measure the quality of water in a distribution system' give no indication of which of the virtually infinite number of possible sampling locations are of interest. As stressed in Section 2.5.1, such broadly expressed objectives are completely inadequate as a basis for the detailed planning of efficient programmes of sampling and analysis, and should always be sharpened so that they do indicate the position of sampling locations. A commonly useful device for helping in this respect is to consider the intended use(s) of the water since this will aid in indicating those positions in a water-body where quality is of key importance. For example, if one is concerned with the quality of water drunk by people in a town, sampling locations will, in general, need to be at the taps from which those people obtain their drinking water. This is a simple and self-evident example, but it is mentioned because the principle seems not always to be followed. Furthermore, when formulating objectives in this more precise manner, it may well be possible to group the sampling locations into several classes of different, relative importances. This may also help the design of efficient sampling programmes because it seems that there is seldom if ever sufficient effort available to make all the analyses at *all* the sampling locations that may ideally be required.

To summarise, therefore, we suggest that sampling locations should always be explicitly defined by the statement of the objectives of a programme. Given that this has been done, it remains to decide the sampling point(s) from which to withdraw samples at each location; this aspect is considered in (*b*) below.

(b) Selecting the Sampling Point(s) at Each Location

Many effects may lead to local, non-homogencous spatial distribution of determinands in the water at the chosen sampling locations. Thus, the exact positions of sampling points at each location should be carefully considered if adequately representative samples are to be obtained. The complexities that are possible in most water-bodies, variations in the information requirements from one application to another, and the need to attempt to minimise the cost of sampling and analysis preclude exact recommendations on sampling points appropriate for all applications. All authors agree on this point, and all emphasise that sampling points are best selected on the basis of local knowledgc [see (*i*) below] of the water-body of interest and the information required on quality. Nevertheless, several principles and points appear to be of very broad applicability in making such decisions, and are summarised below.

(i) Preliminary survey of, and tests on, the water-body of interest—No matter what theoretical or practical considerations suggest for the positions of sampling points, most authors concur that there is no adequate substitute for:

(1) a preliminary examination of the water-body of interest to establish both the factors likely to lead to spatial heterogeneity of determinands and also the sampling points least affected by such heterogeneity, and following this

(2) preliminary experimental tests of the adequacy of the proposed sampling points.

To aid this approach, Sections 3.2.2 and 3.2.3 discuss various types and causes of heterogeneity; Section 3.2.4 summarises a number of generally important aspects of the tests noted in (2), and also considers some more detailed aspects concerning the sampling of particular types of waters.

(ii) Avoiding boundaries of water-bodies—The concentrations of determinands at the boundaries of water-bodies (*e.g.*, the walls of pipes, the banks, surfaces, and bottoms of natural waters) are often not representative of the vast bulk of the water. Thus, sampling from such boundaries should usually be avoided except when these regions are of direct interest.

(iii) Convenience and accessibility—Sampling points seem sometimes to be chosen more on the basis of their convenience and accessibility than on their suitability as sources of representative samples. The two former factors are, of course, important, but it is always worth seeking an alternative point if the one that is convenient is likely to provide inadequately representative samples and thereby information of uncertain value. Berg[3] also stresses such cautions.

(iv) Need for information on the flow-rate of water—Some measurement programmes require information on the mass-flows of determinands in flowing streams rather than, or as well as, their concentrations. Whenever this is so, the choice of sampling point(s) has the additional constraint that the flow-rate of the water at those points must be known with adequate accuracy. See also Section 3.4.2.

(v) Installation of sampling equipment—In some programmes, equipment required for sampling (and possibly also analysis) may need to be permanently installed at or near the sampling point. This may involve substantial expenditure not only for the equipment itself but also for the appropriate housing of the equipment. If the initial choice of sampling point proves unsatisfactory, the cost and effort involved in moving the existing equipment or installing new may well make it difficult to rectify the situation. It is clearly desirable, therefore, that all those concerned with the initial installation and use of sampling equipment should carefully consider the choice of sampling point before any installation is made.

(vi) Sample-compositing—If adequately small heterogeneity cannot be assumed

at a required sampling location, it will generally be necessary to sample from a number of points at that location, the number depending on the degree of heterogeneity and the accuracy required. Provided that the spatially averaged quality at that location is of interest, some economy of analytical effort is possible by mixing portions of the individual samples from the sampling points on each sampling occasion, and analysing the composite sample so formed. Various precautions need to be observed if this approach is adopted, and are considered in Section 3.3.2(d).

3.2.2 Heterogeneity Arising from Incompletely Mixed Waters

Clearly, if two or more waters are incompletely mixed at a sampling location, determinands may be heterogeneously distributed there. It is useful to distinguish two extreme situations though there is no clear-cut boundary between them in practice.

(a) Static Situations

In these situations, the water-body of interest is continually composed, over a given period, of two or more incompletely mixed waters. Such situations are exemplified by the thermal stratification of lakes and reservoirs, the stratification of salt and fresh water in estuaries, and the different composition of water at different points within the steam/water drum of boilers. It is impossible in these situations to characterise the quality of the entire water-body by using only one sampling point, and the number of points required depends on the size of the water-body, the degree of heterogeneity, and the exact information required on quality.

Note also that the degree of heterogeneity may vary with time. For example, the thermal stratification of lakes may disappear completely during the winter, and one sampling point might then suffice. Thus, if this type of heterogeneity is known or suspected to be present, it is desirable to check its magnitude under different conditions so that the most appropriate number of sampling points is used. It is also worth remembering that the degree of heterogeneity may depend on the direction considered in the water-body. For example, lakes can be stratified vertically but be essentially homogeneous laterally, and *vice versa*.

The presence of this type of heterogeneity can usually be relatively simply detected by measuring the concentration of one (or at most a few) determinands known to be indicative, *e.g.*, dissolved oxygen in thermally stratified lakes and chloride or salinity in estuaries (see Section 3.2.4).

(b) Dynamic Situations

The term 'dynamic' is intended here to denote that one or more waters are in process of mixing with another water. Examples of common importance are the dispersion of an effluent or a tributary stream in a river. Analogous situations occur in industrial plant when one liquid stream flows or is injected into another, *e.g.*, in the fluoridation of potable water, in pH-control of process-waters. In such situations, and when the effect of one stream on the average quality of the other is of interest, it is usually desirable to choose sampling points at positions where the determinands of interest are homogeneously distributed over the cross-section of the water of interest. If such cross-sections can be found, one sampling point suffices for representative sampling. On the other hand, if only one point is sampled at a non-homogeneous cross-section, not only will the results be unrepresentative of the fully mixed water, but also temporal variations in the results may merely reflect variations in the degree of mixing rather than true variations in average quality. As in (a) above, the degree of mixing may vary with time because of variations in, for example, flow-rates, temperatures, and densities of the component streams. It is worth stressing that the mixing of two liquid streams can, depending on circumstances, be very slow so that lateral heterogeneity may persist for long distances downstream of the initial point of mixing of, for example, two rivers.[8] The phenomenon of mixing is rather complex in that it is affected by many factors some of which are difficult to quantify, particularly for natural water-bodies such as rivers, estuaries, *etc.* It is desirable, therefore, that tests of the degree of heterogeneity at a chosen cross-section should be made under various conditions.

This type of heterogeneity can often be relatively simply detected by measuring the concentration of a determinand present in appreciable amount in only one of the waters undergoing mixing. It should, however, be noted that a special problem may arise (depending on the information required) when a determinand in one water is dispersing into another water and, at the same time, is also reacting chemically. This situation has been discussed by Ruthven[9] for the case of biochemical oxygen demand in rivers, and by Jaffe *et al.*[10] for chlorinated organic pesticides. An alternative approach to using a naturally present determinand to investigate the degree of mixing is to inject a material which acts as a convenient tracer; this is considered in Section 3.2.4(a)(v).

If the results of such tests show that the determinands of interest are heterogeneously distributed to an important degree at the chosen cross-section, it is best to seek and test another cross-section where mixing is more complete. If no other location is feasible, sufficient sampling points at the tested location must be used to ensure adequately representative results. A minimum of three points (at approximately 1/4, 1/2, and 3/4 channel-width) is often likely to be reasonable for waters flowing in pipes and relatively simple channels, but more may be needed, particularly in less simple water-bodies such as estuaries and large rivers. When more than one sampling point is used at a given cross-section of a channel, the variation of flow-rate over the cross-section of the channel needs to be considered in deciding how to calculate either the average quality from the results obtained from the individual sampling points or the proportions in which the individual samples are to be mixed when a composite sample is to be prepared.

The rate at which mixing occurs in channels depends on the direction considered. For example, when an effluent is dispersing in a river, vertical mixing occurs most rapidly, longitudinal most slowly, and lateral at an intermediate rate.[11,12] Depending on the objectives of the programme, the distance downstream of the position where cross-sectional mixing is complete may sometimes be chosen with advantage.[13] For example, if interest centred on short-term variations of quality caused by variations in an effluent, this distance should be zero as this will give the smallest longitudinal dispersion of the effluent, and hence the variations in quality will be greatest. If, on the other hand, the measurements were intended to give a long-term average for the river quality, it could be preferable to make the sampling location as far downstream as possible as this would reduce the temporal variability of results and thereby reduce the required sampling frequency (see Section 3.3). Of course, the validity of this latter approach is governed by the stability of the determinand and the presence of other sources of it downstream of the effluent. The example, however, illustrates that it is useful to remember the dependence of mixing-rate on direction in the water-body.

It is sometimes possible to improve the representativeness of samples by promoting the mixing of waters. There will usually be little that can be done in natural water-bodies though sampling after features such as weirs may be useful for some determinands. There are, however, sometimes possibilities for installing, or making use of existing, devices in pipes.

3.2.3 Heterogeneity Caused by Non-homogeneous Distribution of Determinands

Even when neither of the situations considered in Section 3.2.2 occurs, an otherwise homogeneous body of water may show heterogeneous distribution of certain determinands as a result of two main effects. First, undissolved materials will tend to be heterogeneously distributed if their densities differ from that of water. Second, chemical, physical, and biological reactions may occur to different extents in different parts of the water-body, and thereby lead to heterogeneity of the determinands involved in, and produced by, these reactions. These two effects are considered in (*a*) and (*b*), respectively, below. It should be noted that when this class of heterogeneity is present, each determinand needs, in general, to be considered individually.

(a) Undissolved Materials

There are many practically important examples of this type of heterogeneity. The size of any effects depends on the difference in density of the undissolved materials and the water, the size of the particles or bubbles of the materials, and various hydrodynamic factors such as the degree of turbulence in the water; see, for example, refs. 3 and 14. Thus, undissolved inorganic materials in rivers and other natural water-bodies tend to increase in concentration with increasing depth because the particles tend to settle. On the other hand, certain biological detritus may tend to rise towards the surface of the water because its density is less than that of water; oils also commonly demonstrate this effect markedly. Similar effects occur in industrial plant. Separation in directions other than the vertical may also occur, *e.g.*, when the water flows round a bend [see Section 3.5.2(*c*)].

The extent of this type of heterogeneity may be so large in extreme situations (*e.g.*, oil-films on the surface of water) that it is not meaningful to speak of a sample representative of the entire water-body. In such situations, special sampling techniques may be required, *e.g.*, for surface oil-films, special surface-sampling techniques [see Section 3.5.4(*a*)(*iii*)]. More usually the heterogeneity is not so extreme, and an attempt to achieve representative sampling, though difficult, is necessary. Two approaches are possible.

(1) In flowing water-bodies there may be a region where the determinand is essentially homogeneously distributed. For example, undissolved materials are likely to be more or less uniformly distributed over the cross-section of a

pipe in which water is flowing vertically upwards. Equally, certain plant components, *e.g.*, pumps, may induce sufficient turbulence to ensure a transient homogeneity, and weirs in rivers may produce a similar effect. It is often worth trying to find such regions in a water-body, checking experimentally the degree of heterogeneity, and if this is adequately small, using just one sampling point.

(2) More than one sampling point is used, their positions being chosen so that the resulting samples provide an adequate assessment of the variation of determinand concentration with position; from this the average concentration (or other required information) can be derived. This approach is sometimes used, for example, in measuring the concentration of suspended solids in rivers,[15,16] and in obtaining samples of saturated steam.[17] Clearly, the number and position of sampling points depends on the degree of heterogeneity. It is, therefore, still worth considering the hydrodynamics of the situation so that a sampling location can be selected where the degree of heterogeneity is likely to be minimal because this will decrease the number of sampling points required. For example, the sampling of undissolved solids in horizontal pipes or immediately before or after bends should be avoided whenever possible since such locations may well accentuate heterogeneity. Having decided the sampling location exactly, the extent of heterogeneity should be checked experimentally whenever possible so that the appropriate number of sampling points can be decided. In calculating the average concentration at a location in a flowing stream, the degree of variation of the flow-velocity over the cross-section must be borne in mind, and appropriate allowance made for this when necessary. An alternative approach to the use of several fixed sampling points at a given location is to sample continuously while the inlet of the sample collection device is moved at constant speed over the cross-section of interest. This latter approach is sometimes used for sampling suspended solids in rivers.[15,16]

(b) Substances Affected by Chemical, Physical, and Biological Reactions

Many types of reaction may affect the concentrations of many determinands. If these reactions proceed to different extents at different positions in a water-body, spatial heterogeneity at a given sampling location, as well as between locations, may well result. For example, the growth of algae in the upper layers of water-bodies may lead to the latter's depletion in nutrient substances such as phosphate, nitrogen species, and silicate, while other determinands, *e.g.*, dissolved oxygen, alkalinity, pH, may increase as a result of algal respiration. These and the many other possible reactions are often confined to or concentrated in the boundaries of water-bodies, and it is usually advisable to avoid sampling in such regions unless there is direct interest in such reactions. Some reactions, however, may proceed throughout the entire body of water *e.g.*, the reaction between carbonaceous materials and oxygen dissolved in a river water.

The extent to which reactions occur is, in general, governed by a number of factors, *e.g.*, temperature, light intensity, presence of catalysts, *etc.* Thus, it is possible that the degree of heterogeneity will vary from season to season and time to time as such factors change.

3.2.4 Particular Types of Water-bodies

The preceding sections show that the selection of sampling locations and points requires, in general, detailed knowledge of the water-body of interest and of the chemical, physical, and biological processes that may affect its quality. As a result, this selection is another important example of the need for a collaborative approach by the user, analyst, and any other experts (e.g., limnologists, biologists, hydrologists, geologists) with knowledge of the relevant aspects. Of course, the manager of the water-body should also be represented if he is not also the user of results. Unless this type of approach is followed, it is likely, particularly for the larger and more complex water-bodies, that adequately representative sampling will not be achieved.

For the types of measurement programmes with which this book is primarily concerned, it will rarely if ever be the case that the analyst will himself possess all the desirable knowledge of the water-body of interest and of all the relevant processes. Accordingly, the detailed consideration and selection of sampling locations and points is often likely to fall mainly on experts other than the analyst. Nevertheless, it seems to us of considerable value for the analyst to be well aware of at least the more important of the various considerations involved. We have, therefore, attempted in (b) to (f) below to summarise what we believe to be generally applicable and important aspects for several types of waters of common interest. The references cited will allow such points to be followed further should this prove necessary. In this connection, texts dealing with the properties and behaviours of, and processes occurring in, various types of water-body will also be useful (see Section 2.9).

As emphasised in Sections 3.2.1 to 3.2.3, experimental tests to assess the degree of heterogeneity are generally essential or highly desirable in selecting sampling locations and points. There are several generally applicable considerations for such tests, and these are noted in (a) to avoid repetition in (b) to (f).

(a) Tests to Assess the Degree of Heterogeneity
In essence, the nature of the tests is simply to collect samples at a number of relevant positions, and to analyse the individual samples so that the results obtained show the degree of heterogeneity. Several precautions are necessary, however, if such tests are to achieve their objective, and these are summarised below together with some other points that may aid the execution of the tests.

(i) Choice of determinands to be measured—In principle, because of the variety of processes that may affect spatial heterogeneity, it is desirable that all the determinands to be measured routinely are also measured in these tests. Reduction of the analytical effort involved, however, will facilitate the tests. Accordingly, careful preliminary consideration of the various possible types of heterogeneity (Sections 3.2.2 and 3.2.3) and their relevance to each of the determinands of interest is useful in that this may allow one determinand to be used as an 'indicator' for a number of others. For example, if heterogeneity arose only through incomplete mixing of two or more waters, it may often be possible to use just one determinand as an 'indicator'; alternatively, a special 'tracer' substance may be added to one of the waters to facilitate the tests [see (v) and (vi) below].

(ii) Times at which the tests are made—As noted in Sections 3.2.2(*a*) and (*b*) and 3.2.3(*b*) the degree of heterogeneity may vary with time. If it is possible to decide for a given water-body when heterogeneity will be most marked, it may suffice to make the tests at only that time. Usually, however, it seems difficult to determine such a time with confidence, and it is, therefore, desirable to make the tests on at least two occasions when the factors causing heterogeneity are likely to have substantially different effects. For example, two such occasions of common relevance for many water-bodies are summer and winter because of the marked seasonal variation of climatic factors.

(iii) Statistical aspects of the tests—If the analytical procedures to be used for the tests were such that repeated analyses on the same sample gave results varying by, say, $\pm 20\%$, there would clearly be some difficulty in attempting to distinguish differences in quality of, say 5% between different sampling points. Consequently, it is generally essential to ensure that the tests are designed statistically so that their objectives can be achieved. Readers unfamiliar with statistical ideas and their application may consult Section 4, but note that detailed consideration of the various experimental designs relevant to such tests is beyond the scope of this book. References to such statistical approaches are quoted in the following sections. In general, they involve the statistical technique known as 'Analysis of Variance' (ANOVA) which is described in virtually every book on statistical techniques for scientists (see Section 4.8). Whenever possible, readers are advised to seek statistical advice in planning, and interpreting the results from, tests of spatial heterogeneity.

It should be noted that the proper design of the tests requires an estimate of the precision* of analytical results for each determinand to be measured (see Sections 4.3 and 6). In general, therefore, it is best not to attempt such tests until relevant estimates are available; otherwise, wasted effort may well result. See also (*v*) and (*vi*) below.

(iv) Allowance for temporal variability of water quality—It is generally desirable that, on each test occasion, the samples from the various positions are all collected at the same time; this eliminates complications in the interpretation of the results if the water-quality varies during the period of sampling. Such an ideal, however, will often be impracticable. If, therefore, it is thought that temporal variations in water-quality may be large enough to be important, some allowance for this additional source of variation must be made. When this is necessary, it will usually best be achieved by appropriate modification of the statistical design of the experiment so that the effects of temporal and spatial variability can be separated.

(v) Use of 'tracer' substances—When heterogeneity arises through incomplete mixing of two or more waters, the tests can often be simplified by mixing a carefully chosen 'tracer' with one of the waters, and measuring only this tracer in the collected samples.[18] Dyes[19] or radioactive isotopes[20,21] are commonly used for such purposes, though other inorganic or organic substances naturally present in, or specially added to, one of the waters may also be used.[21–25] The use of dyes and

---

* It is assumed that, for these tests, any possible bias of analytical results will be essentially the same for every sampling point, but this is not necessarily so (see Sections 4.4 and 5).

isotopes is often advantageous in that their measurement is simple, precise, and rapid, and does not require specialised analytical expertise and facilities. Indeed, when dyes are used, the mixing may sometimes be observed visually and photographed. It is generally desirable, of course, that such tracers are present in only one of the waters undergoing mixing. It is also essential to ensure that specially added tracers introduce no important hazard either to those using them or to those using or coming into contact with the waters to which the tracers are added.

(vi) Use of continuously analysing instruments for measurement—An alternative approach to the collection of samples from a number of positions and their subsequent analysis is to move the inlet of a sampling line [see Section 3.5.1(*a*)] through the water of interest, the outlet of the line being connected to one or more instruments continuously measuring the determinands of interest.[26—29] Provided that temporal variations in quality are negligible in the period required to traverse the region of interest, this approach can be of value in rapidly providing a detailed view of spatial heterogeneity. Indeed, even when appreciable temporal variations occur, the approach may still be useful in that a rapid 'reconnaissance' may well serve as a very useful basis for more efficient tests of the type described above.

(b) Rivers and Streams

Sampling locations should, whenever possible, be at cross-sections where vertical and lateral mixing of any effluents or tributaries are complete, and the processes of mixing are, therefore, important,[3,5—7] and are discussed in many publications. Three recent papers[30—32] deal particularly with sampling aspects, while refs. 11, 12, 33, and 34 contain more theoretical discussions of mixing processes.

Procedures for testing the degree of heterogeneity at a river cross-section and for interpreting the results obtained using analysis of variance techniques are described in, for example, refs. 8, 10, 32, 35, and 36.

To avoid non-representative samples caused by surface films and/or the entrainment of bottom deposits, it has often been recommended [5,6,37] that samples should, whenever possible, be collected not closer than 30 cm to the surface or the bottom; see also refs. 3 and 7 for rather different suggestions. It may sometimes be useful to obtain an average sample from a water-column in a river cross-section; for this, the Lund-tube sampler has been recommended[6,38] (see also ref. 39).

When the quality of river water abstracted for a particular purpose (*e.g.*, the production of drinking water) is of interest, the sampling point should, in general, be at or near the point of abstraction. Note, however, that changes of quality may occur between the actual point of abstraction and the inlet to the treatment plant.[40] If, however, one wishes to control the amount or time of abstraction on the basis of the water quality, an additional sampling location upstream of the abstraction point will usually be needed, the distance upstream being dependent on the travel-time of the river, the speed with which the relevant analyses can be made, and the upstream locations of sources of the determinands [see Section 12.3.3(*a*)].

When the broad aim is to assess the quality of a complete river or river basin, the number of potentially relevant sampling locations is usually extremely large. It is, therefore, usually necessary to assign different priorities to the various locations in order to arrive at a practicable sampling programme.[3,5,6] Such considerations are very closely connected with the question of sampling frequency, and a number of approaches have been described in many publications dealing with the overall design of sampling programmes for river systems (see Section 3.4). Many authors have stressed the value of, and need for, identification of locations where quality problems are or may be most acute. For example, Hines *et al.*[41] discuss in detail the chemical and hydrological processes that may lead to such problems; see also ref. 42 for a detailed comparison of the merits of fixed-location monitoring and intensive, short-term surveys at selected locations for the routine assessment of rivers. Chamberlain *et al.*[43] give a rather more theoretical and statistically oriented approach with similar aims to those of Hines *et al.* In recent years, several highly sophisticated, statistical/mathematical approaches have also appeared; see, for example, refs. 44 to 47. Sharp[48] has proposed a quite different approach to assigning relative priorities to different sampling locations. It is also worth noting that the procedure known as 'catchment quality control', though intended for a different purpose, includes the identification of most important effluents entering a river system from the viewpoint of water quality.[49] Finally, the analysis of river sediments has been suggested[50] as a convenient means of reconnaissance of river systems to decide the locations where water quality is of particular interest with respect to pollutants. A Working Group of the World Health Organization has recently issued a useful, introductory report[51] on the significance of sediments in rivers.

(c) Lakes and Reservoirs

The exact positions of sampling points depends heavily on the precise aims of the measurement programme. For example, if the aim were to characterise the water drawn off from a reservoir, samples should be collected at one or more positions representative of the quality at the draw-off point(s). Often, however, there is need to study the various physical, chemical, and biological processes that may markedly affect quality and its spatial variations. When this is so, a number of sources of heterogeneity within the water-body need careful consideration in selecting sampling points. These sources include thermal stratification, which leads to variations of quality with depth, and the effects of influent streams, lake morphology, and wind, which can all lead to both lateral and vertical heterogeneity. Texts on limnology (*e.g.*, refs. 52 and 53, but see also 54) should be consulted for details of such processes. The main points relevant to the choice of positions for sampling are noted in many publications, *e.g.*, refs. 3, 5—7, 52, 53, 55, and 56, Note also that shallow and/or relatively isolated embayments of lakes and reservoirs may show marked differences in quality from the main body of water.[6,57]

Detailed accounts of tests to assess the degree of heterogeneity in lakes and reservoirs are presented in refs. 57 to 60. In water-bodies of sufficient depth in temperate climates, thermal stratification is often the most important source of vertical heterogeneity from spring to autumn. Measurement of dissolved oxygen and temperature is a convenient means of following the development of such stratification, and has the advantage that both measurements can be made automatically and continuously *in situ*.

Various physical effects may lead to slow mixing of streams entering lakes, and reservoirs, and care should, in general, be taken not to sample in incompletely mixed regions. In many instances, general run-off from the surrounding land may also be an important source of various pollutants,[61] and the possible contribution from atmospheric contaminants should also be borne in mind.[62]

The number of algae in the surface layers of the water-body may have a marked effect on the concentrations of nutrient and other substances, and it should be remembered that the lateral distribution of algae can, in turn, be much affected by the direction and velocity of the wind.

It is worth noting here that, in a national survey of the water quality of lakes of the U.S.A., helicopters were used for sampling in order to allow rapid coverage of the large geographic areas involved.[63] Care is needed, however, that turbulence in the water caused by a helicopter's blades does not invalidate such sampling[5,6] [see Section 3.5.2(*b*)].

(d) Groundwaters

Normally, sampling locations will correspond to the points of abstraction when routine monitoring of groundwater abstracted for a particular purpose is required. However, for: (1) long-term monitoring of the overall quality of an aquifer, and (2) investigation of the contamination of groundwaters, a number of sampling locations is generally necessary. Other existing boreholes may be appropriate for (1), but it will usually be necessary to arrange for special sampling wells to be drilled for (2). Given the cost of such wells, everything possible should be done to ensure that they are located in the most appropriate positions. To this end, all authors agree (see, for example, refs. 3, 6, 7, and 64) that the maximum possible use must be made of all existing knowledge of both the hydrogeology of the area concerned and also the potential sources and natures of contaminants. Berg[3] suggests that sampling wells be constructed one at a time so that the results obtained from one can be used to help decide the location of the next. Given this need for hydrogeological expertise, we simply note here that groundwater quality may vary laterally and vertically, that the degree of the variation may change with time and with the rate at which water is abstracted, and that these variations may themselves vary in absolute and relative magnitudes from one determinand to another. Texts on groundwater (see Section 2.9) and the proceedings[65-67] of several recent conferences contain much information on the various effects that may occur; see also ref. 68.

Many publications contain general discussions of the choice of sampling locations and points; see, for example, refs. 6, 7, 55, and 56. More detailed considerations will be found in refs. 68 to 73, and ref. 74 describes a sophisticated statistical approach to the selection of sampling locations for an aquifer. A useful paper on the spatial patterns of contaminants in groundwaters makes a number of valuable points concerning sampling locations,[75] and the increasingly important topic of the movement of organic compounds in aquifers has been recently reviewed.[76] Further, more detailed descriptions of procedures (involving the statistical evaluation of results) to establish the degree of spatial variability of quality and the number of sampling locations required for long-term monitoring purposes are given in refs. 77 to 79, and see the comment on the last in ref. 80.

(e) Estuaries and Coastal Waters

The complexity that may be shown by these water-bodies with respect to spatial heterogeneity (lateral and vertical – and both are time-dependent) again makes it essential that the choice of sampling locations and points be made by those with knowledge of the relevant, basic processes involved.[6,7] In addition to the books cited in Section 2.9, ref. 7 gives a good introductory discussion of problems and approaches, and stresses the importance of preliminary tests in assessing appropriate sampling locations. Detailed accounts of surveys in particular estuaries have been reported, for example, by Hammerton *et al.*[81] and Sharp *et al.*[82] Ref. 83 provides an example of a preliminary test to determine suitable sampling locations in an estuary.

(f) Drinking Water Distribution Systems

(i) Introduction—Several publications, *e.g.*, refs. 6, 55, and 84—89, give some broad suggestions on sampling locations and points in distribution systems. It will, of course, be useful to consult these and other relevant publications, but those known to us tend to deal with only relatively straightforward points which may be summarised as follows:

(1) sample at the positions where each source of water enters a distribution system;
(2) sample at positions where contamination may enter a system;
(3) sampling may be necessary at positions where different waters mix within a distribution system;
(4) local, expert knowledge of the characteristics of a system is important in selecting sampling locations.

Ref. 3 covers similar ground, but in a rather more detailed and helpful manner, different types of distribution system also being considered. None of the references cited, however, attempts any detailed discussion or rationalisation of the important and difficult problem mentioned below.

The growing emphasis on the need to control the chemical quality of drinking water in order to protect the health of those consuming it is leading to increasing interest in, and need for, sampling the water emerging from consumers' taps.* Given the vast numbers of such taps in normal distribution systems, and that the amount of effort available for sampling and analysis is generally limited, only a very small proportion of the total number of taps can usually be sampled within a reasonable period. Therefore, a very important and often difficult problem arises immediately, namely, which taps should be sampled? It is surprising that this question is addressed so seldom in the published literature, and we know of only one published paper[91] which considers the general problem and suggests principles and procedures for dealing with it. We understand that the problem will be discussed in some detail in the forthcoming revision[92] of the World Health Organization's Drinking Water Standards, and that the treatment there will be based on a paper prepared by one of us (ALW) in 1977, and subsequently modified in the light of

---

* For example, the EEC Directive[90] relating to the quality of water intended for human consumption stipulates that all water intended for human consumption shall be monitored at the point where it is made available to the user. For practical reasons, sampling at the stop-tap where the consumer's service-pipe is connected to the water supply is not usually feasible, and sampling from the consumer's tap is, therefore, necessary; see, for example, Short.[91]

comments from a number of colleagues. For the present, we thought it would be useful to outline below a rationale that we believe may be helpful in tackling this problem of selecting taps from which to sample. The discussion is based on the 1977 paper mentioned above, and the treatment concentrates on principles for reasons made clear in the text. The paper by Short[91] lays more emphasis on the practical approach to a particular problem, and should also be consulted.

(ii) Classification of determinands of interest—It is useful initially to classify the determinands of interest into four types:

Type 1—The treated water(s) passing into the distribution system* normally contain unimportant concentrations of the determinand, *and* the distribution system contributes negligible amounts to the water. Organochlorine insecticides, for example, will often be of this type.

Type 2—The treated water(s) passing into the distribution system normally contain appreciable concentrations of the determinand which are negligibly affected in passing through the system. Chloride, for example, is usually of this type.

Type 3—The concentration of the determinand within the distribution system is governed largely by its concentration in the treated water(s), but the determinand is also subject to processes within the distribution system that affect its concentration. Common examples of this type are chlorine (affected by reaction with organic and inorganic materials as well as biological species) and aluminium (affected by deposition of undissolved forms with possible re-entrainment of such forms).

Type 4—The treated water(s) passing into the distribution system normally contain unimportant concentrations of the determinand, but the distribution system contributes substantial concentrations to the water. Lead, for example, is often of this type.

The type to which a given determinand is assigned may vary from one distribution system to another depending on the quality of the raw water, the nature of the treatment of the raw water, and the nature of the distribution system. For example, when these systems contain inappreciable sources of lead, this determinand will often be of Type 1 (assuming, as is common, that the treated water contains a negligible concentration); otherwise, lead will usually be of Type 4. Thus, again, preliminary surveys of the nature of the distribution system and measurements of water quality at a number of locations may well be necessary to allow firm classification of all the determinands of interest.

(iii) Choice of sampling locations—From the definitions in *(ii)*, it follows that the concentrations of determinands of Type 1 at consumers' taps can be reasonably estimated by sampling the treated water(s) going into supply. Thus, for these determinands the choice of sampling location is simplified because prime reliance may be placed on sampling the treated water(s). Of course, complete reliance on this approach may be unwise, and occasional samples from consumers' taps are also desirable as a continuing check that the determinands assigned to this type are correctly classified.

The preceding paragraph applies also to determinands of Type 2 if there is only

---

* For simplicity, the term 'distribution system' here includes service-pipes and domestic plumbing through which the water passes in reaching the consumers' taps.

one source of treated water for the distribution system. In this situation, the difference between the two types is relevant to sampling frequency rather than sampling location. If, however, there is more than one treated water supply, additional sampling locations within the distribution system will, in general, be required (depending on the differences in the qualities of the waters) in order to be able to assess the degree of mixing of the waters at appropriate parts of the system; see, for example, refs. 3 and 91. Relevant sampling locations can be decided only from detailed knowledge of a particular system, but there seems, in principle, to be no great difficulty involved.

In contrast, determinands of Types 3 and 4 are liable to changes in concentration between the treated water supply inlets and consumers' taps as a result of many potentially important and interdependent processes whose effects depend on the determinand. Examples of such processes are the corrosion/erosion of pipes and pipe-linings, the leaching of materials from pipe-linings, the deposition of materials (scale, sediment) in pipes and tanks, the chemical and physical re-entrainment of deposited materials, the chemical reactions between substances in the water, and biological processes. These processes are also often affected by other factors, and the important example of lead well indicates the complexity that may arise. Thus, the concentration of lead in the water emerging from a tap may be affected by:

(1)   the length and diameter of lead service and domestic pipes, and possibly also by the metallurgical properties of the particular batch of metal concerned;
(2)   the quality of the water;
(3)   the degree of vibration to which lead pipes are subject;
(4)   the hydrodynamic conditions such as rate of flow through the pipes;
(5)   the temporal pattern of water usage at a particular tap and its associated part of the distribution system;
(6)   the temperature of the water and the pipes.

Of course, not all determinands are subject to all such effects, but there is a general inability to estimate the concentrations of determinands of Types 3 and 4 at consumers' taps from samples taken at locations such as those for determinands of Types 1 and 2. In general, therefore, prime reliance should be placed on samples from consumers' taps for determinands of Types 3 and 4 when the concentrations at such taps are to be estimated. The main difference between Types 3 and 4 is that for the former it is also generally important to sample at the treated water supply points relatively frequently whereas relatively infrequent sampling there is adequate for Type 4 determinands. It is possible, in principle, for the processes affecting the concentrations of determinands of Types 3 and 4 to occur in restricted parts of a distribution system so that samples collected from locations after those parts could be used to infer the quality at a number of consumers' taps. Provided that such parts, and the corresponding sampling locations, can be identified and remain constant, this approach would simplify the choice of sampling location. Nevertheless, we feel that such an approach is subject to several uncertainties that may affect the accuracy with which the quality at consumers' taps is inferred, and it should, therefore, be used only if one has adequate evidence of its reliability.

If the concentration of each Type 3 and 4 determinand in the water from consumers' taps were the same at all taps, there would, of course, be no problem; in principle, a sample collected from any tap would then be fully representative of all taps. In practice, such a simple situation seems never to be found, and – given the above considerations – could hardly be expected. Thus, the very important question arises – from which of the very large number of taps should samples be collected?

(iv) Selection of taps to be sampled—This selection involves consideration of: (1) the number of taps that should be sampled in a given period, and (2) the positions of the taps within the distribution system. If only a limited amount of effort is available for sampling and analysis, the former question cannot be separated from the further question of whether or not each tap should be sampled more than once in the period of interest (to check temporal variability* – see Section 3.3). Two extreme strategies for selecting the locations of taps to be sampled can usefully be distinguished:

(1) the tap-locations are selected on a completely random basis;
(2) the tap-locations are selected systematically from knowledge of the spatial variations of factors affecting the concentrations of the relevant determinands.

Which of these two strategies, if not a combination of both, is the more effective depends on both the nature and magnitude of the spatial variations of quality *and also the information required.* This raises important statistical questions and we recommend that statistical advice be obtained whenever possible. Ellis and Lacey,[93] however, give a simple, introductory but very useful account of some of the important differences between random and systematic sampling, and their paper is well worth consulting; see also Ellis.[94]

Random sampling is usually appropriate either when the magnitude of any systematic spatial variations are small in relation to random variations or when nothing is known of the nature of spatial variations. Random sampling may, however, be far from optimal[93] (in the sense of minimising the number of taps to be sampled) if there are important, systematic spatial variations within the distribution system. For example, in a distribution system in which 1% of all service and domestic pipes were of lead, random sampling would, on average, lead to 99% of samples being collected from taps at which lead concentrations are likely to be very small (assuming that lead is a Type 4 determinand). Such a programme would provide an unbiased value (with confidence limits) for the lead concentration expected at an 'average tap', but this concept in the present example seems of little value if not positively misleading. In this example (depending on the exact information required[94]), it might well be more fruitful to follow strategy (2) more closely, devoting relatively more effort to sampling from taps at which undesirably high concentrations of lead might be present; see also (*v*) below.

Systematic sampling schemes imply detailed knowledge of the spatial variability of quality within a distribution system. Thus, preliminary surveys of such systems are of general value in indicating the extent to which systematic variations occur and need to be considered. Complete reliance on systematic sampling is, however, probably unwise – certainly in the initial stages of a sampling programme – because it provides little or no confirmation of the correctness with which all relevant systematic effects have been identified.

---

* If only the average quality of the water drawn by consumers from their taps is of interest, a semi-automatic device for providing a composite sample of this water may be useful; see Section 3.5.4(*b*).

We know of no way in which more positive and helpful suggestions can be given because each combination – information required/determinand/distribution system – appears at present to require individual consideration. An example of a sampling programme to obtain a particular type of information is briefly described in (*v*) below.

(v) An example of a sampling programme—An interesting example of the design of a sampling programme is provided by investigations in the United Kingdom to identify areas in which the proportion of houses – with greater than desirable concentrations of lead in the drinking water – is sufficiently great to warrant remedial action. For this purpose, an expert group of the Department of the Environment/National Water Council's Standing Technical Advisory Committee on Water Quality designed and recommended[95] a sampling programme which has been used by Water Authorities throughout the country. The details of this programme and related aspects have not been published, but the basis is a 'sequential sampling' design which is similar to that described on pp. 383—387 of ref. 96. The approach is generally applicable to any water supply area and determinand, and consists, in essence, of the stages described below.

(1)   A number (previously chosen – see below) of houses is selected at random from all those in the area of interest, and a list of the selected houses is prepared.

(2)   A sample of drinking water is obtained from each of the first *n* houses on the list from (1), and analysed for the determinands of interest. The value of *n* can be chosen to suit local conditions; a value corresponding to approximately 10% of all the selected houses may often be suitable.

(3)   For a given determinand, the number of houses from (2) giving a concentration greater than a previously decided maximum acceptable value is compared with two statistically derived limits (see ref. 96 for details of the calculation of these limits). If the observed number of houses is greater than one of these limits, the area is concluded to have an unacceptably high proportion of houses with unacceptably high concentrations of the determinand, and no further houses are sampled. If the observed number is less than the second limit, the area is concluded to have a tolerable proportion of such houses, and no further houses need be sampled. Finally, if the observed number falls between the two limits, the next *n* houses from those remaining on the list from (1) are sampled, and the above comparisons made again for all the houses sampled so far. This process is repeated until either it has been concluded that the area is acceptable or unacceptable or all the originally selected houses have been sampled without reaching such a conclusion. In the latter event, a decision must be made as to whether to expand the programme to include still more houses or to attempt some other approach to reach a definite conclusion.

This sequential scheme has the merit, compared with sampling programmes based on a fixed number of houses, that it allows, on average, more rapid (and with a smaller number of sampled houses) identification of areas which have either a much smaller proportion or a much larger proportion of houses (with undesirable water quality) than some just tolerable proportion. When, however, the true proportion of houses with undesirable water quality falls approximately midway between the two statistically derived limits, it is likely that the observed proportion will continually also fall between these two limits as the number of houses sampled increases. For this reason, a maximum number of houses to be sampled generally needs to be set at the beginning of the programme.

Note that the scheme as outlined above employs a random selection of the houses to be sampled. This approach ensures that unbiased estimates of the proportion of houses with unacceptable water quality are obtained. Sometimes, however, it may be possible to divide an area of interest into two or more sub-areas, some of which are very likely to have a tolerable proportion of houses with undesirable water quality while the others are much more likely to have an unacceptably high proportion of such houses. For example, such areas might correspond to those with and without lead service and domestic plumbing pipes. When such sub-areas can be identified, it will usually be economic of sampling and analytical effort to examine each sub-area separately by the above programme, *i.e.*, to introduce an element of systematic sampling into the overall plan.

3.3 TIME AND FREQUENCY OF SAMPLING

3.3.1 The Nature of the Problem

It is seldom if ever that the quality of a water at *only* one instant of time is of interest; normally, information is required for a time-period during which quality may vary. The basic problem arises, therefore, of deciding the times at which to collect samples so that they will adequately represent the quality during the period of interest (see Section 3.1). The nature and size of any temporal variability are clearly of key importance in this connection.

The best technical solution to this problem would often be to use an automatic, on-line instrument providing continuous analysis of the water of interest. This approach can be of great value in that, in principle, a continuous record of quality is obtained and the problems of selecting particular times for sampling does not arise.* Such instruments are, of course, used in a number of applications for both natural and industrial waters (see Section 12.3). As noted in Section 3.1, however, this approach is not generally applicable, and at present the bulk of water analyses are, and are likely to continue to be, made on discrete samples. Thus, consideration of the time and frequency of sample collection is necessary.

We should note here, in passing, that for those determinands for which suitable on-line instruments are available, it may sometimes be economically preferable to use them even when their high sampling frequency is not technically necessary. Such instruments may provide both more and also cheaper analytical results than when samples are collected manually and returned to a laboratory for analysis.

3.3.2 General Considerations

As noted previously, it is important to attempt to ensure that all relevant disciplines

---

* In general, continuously analysing instruments do not provide a continuous record of the true concentrations in the water of interest because of the finite response time of an instrument and its associated sampling system (see Section 12.3). In practice, however, this effect is negligible in many applications.

and experience are involved in the selection of times and frequencies of sampling. Temporal variability arises through the many processes affecting quality, and knowledge of those processes is essential in planning a sampling programme. A particularly important aspect is that of statistical techniques for these have a very important role in this topic as will be seen below, and this is especially true as the size and/or duration of a measurement programme increase. As noted in Section 2.2, the definition of a sampling programme is not generally an analytical question, though it has, of course, very important analytical implications. Consequently, our aim has simply been to indicate what seem to us the major factors requiring consideration with appropriate references to the literature so that readers may pursue further any topic that is of interest. No attempt has been made to delve deeply into the various statistical aspects that arise. Indeed, many of the papers appearing in recent years involve much more complex statistical and mathematical techniques than in earlier studies. Detailed, critical discussion of these more advanced approaches is beyond our competence, and we think it probable that most analysts will require professional statistical advice if the application of such techniques were to be considered.

In (a) to (g) below we deal with the following points of general importance in choosing the times of sampling whatever the exact purposes of the measurement programme:

(a) definition of the information required;
(b) random and systematic variations of water quality;
(c) establishing and reviewing times of sampling;
(d) need to consider periods of abnormal variability;
(e) duration of sampling occasions and the use of composite samples;
(f) simultaneous consideration of sampling positions and times;
(g) general types of sampling programme.

We then consider (Sections 3.3.3 and 3.3.4) two commonly occurring types of sampling programme from among the various types that may arise.

(a) Importance of Quantitative Definition of the Required Information

We make no apology for repeating the essential need to define the information required on quality precisely, including a statement of the tolerable uncertainty of this information (see Section 2). Without such a definition, a rational approach to the optimal design of sampling programmes is impossible.

(b) Random and Systematic Variations of Water Quality

(i) Introduction—It is useful to consider a simple model in which the temporal variability of water quality is attributed to systematic and/or random variations. In the present context, random variations may be described by saying that the concentration at a given instant of a determinand subject to such variations cannot be exactly predicted.* Such prediction is possible, in principle, for determinands

---

* Readers unfamiliar with statistical ideas may consult Section 4.3 for a simple account of the broad concepts or, of course, statistical texts (see Section 4.8). The paper by Ellis and Lacey[93] is a very readable but useful introduction to the topics involved.

subject to only systematic variations such as gradual trends and cyclic variations. In the simplest case, random variations can be described by the normal distribution, but other distributions may arise, *e.g.*, the log normal.[3,37,93,97—99] Ref. 3, 37, 98, and 99 describe simple procedures that help to decide the type of distribution best representing a particular set of data. Natural waters provide many examples of systematic variability. For example, increasing use of a chemical in the catchment area of a river may lead to an increasing trend in the concentration of that chemical in river waters; the use of nitrates in artificial fertilisers is one well known instance. Cyclic variations are exemplified by the daily and annual cycles that many determinands show in surface waters as a result of daily and annual variations in temperature and sunlight, *e.g.*, the diurnal variation of dissolved oxygen in rivers and the commonly occurring annual cycle of silicon concentrations in rivers associated with algal activity. Many other examples could be quoted, but the above suffice for illustration. Note, however, that cycles of other periods are possible, *e.g.*, twice daily and monthly cycles may occur in estuaries through tidal effects, weekly cycles may arise through weekly cycles of industrial activity, *etc.* Industrial waters may also be subject to trends and cycles of quality though generalisation is difficult because of the great dependence on the particular plant and its methods of operation and control.

It is also very important to bear in mind that the nature and magnitude of temporal variability may well be quite different for different determinands in a given water. As a result, the optimal times and frequencies of sampling may well vary from one determinand to another.

This simple model of temporal variability allows several generally useful points to be made, and these are dealt with in (*ii*) to (*vi*) below. We have centred the discussion on the estimation of the mean concentration during a given period so that the various points may be seen in a concrete context which is of common interest. The general approach, however, is applicable to the estimation of statistics (*e.g.*, percentiles) other than the mean.

(ii) Random and systematic sampling—A number of sampling strategies are possible, but for present purposes it suffices to distinguish two extremes – random and systematic sampling. In random sampling, the times of sampling are selected at random, *i.e.*, each small time-interval of the whole period has an equal chance of being sampled.* In systematic sampling, a particular pattern of sampling times is chosen and used, *e.g.*, samples are collected daily or weekly with an essentially constant interval between successive samples. Random sampling has the advantage that it always provides unbiased estimates of statistics such as the mean concentration during the period of interest, and has been often recommended. It has, however, the severe disadvantage of being extremely inconvenient, if not impossible, in most applications. Systematic sampling does not suffer from that disadvantage, and *provided that* the times of sampling are suitably chosen (see below) it may also allow the required information to be obtained with a smaller number of samples than required for random sampling.[93] In general, therefore,

---

* This is true when concentrations are of interest. For other applications, *e.g.*, when the mass-flow of a determinand is of interest, see refs, 37 and 93.

systematic sampling seems preferable to us; the various precautions to be observed are considered below.

   (iii) Random variations are the dominant source of variability—When random variations are the only or major source of variability, the times of sampling are not important, and systematic sampling may be used (see Figures 3.1 a and b). The total number of samples required during the period of interest is governed by the tolerable uncertainty of the statistic of interest as discussed in Section 3.3.4(*a*). It is stressed that data from the water-body of concern should be examined before assuming the dominance of random errors [see (*c*) below].

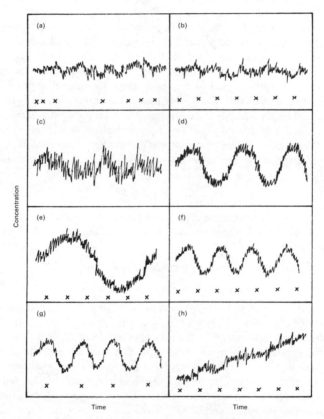

Figure 3.1 *Effect of the nature of temporal variability on sampling times (× denotes sampling times). (a) Random variations only – random sampling; (b) random variations only – systematic sampling; (c) random variations greater than cyclic; (d) cyclic variations greater than random; (e) sampling when only one cycle is of interest; (f) sampling for average over many cycles; (g) sampling when only one part of a cycle is of interest; (h) sampling when random variations and trend are present*

(iv) Cyclic variations occur—When random variations are much greater than cyclic fluctuations (Figure 3.1 c), the points made in *(iii)* apply. When the cyclic variations are of similar size to, or greater than, the random (Figure 3.1 d), the times of sampling are important, and should be chosen so that representative sampling of the cycle(s) is achieved. For example, if the quality throughout the whole of the cycle is of interest, the times of sample collection should be chosen so that different parts of the cycle are sampled (Figures 3.1 e and f). It has been suggested[37] that approximately six equally spaced points on a cycle will usually suffice. On the other hand, if the greatest (or smallest) concentration is of interest, samples should be taken at only the corresponding point on a cycle (Figure 3.1 g). When the period of interest will include a number of repeated cycles and information on quality throughout the cycles is required, it is particularly important to arrange the sampling times so that the same part of the cycle is not sampled on each occasion; this would clearly lead to a biased result (Figure 3.1 g). Thus, the sampling period should not be an integral multiple of the period of the cyclic variation of quality.

When cyclic variations are present, therefore, the timing of sample-collection must be suitably chosen; provided that this is done, systematic sampling will generally be preferable for the reasons noted in *(ii)* above. The total number of samples required in the period of interest is then determined approximately by similar statistical considerations to those mentioned in *(iii)*. That only the random variations need be considered can be illustrated by a simple model. Suppose that an impurity showed a sinusoidal variation of concentration, that its mean concentration was independent of time, and that it was also affected by random variations. The concentration, C_t, at any time, t, would then be given by an equation of the form

$$C_t = \mu + \alpha \sin(\beta t - \gamma) + \varepsilon_t \qquad (3.1)$$

where μ, α, β and γ are constants, and ε_t is the random variation at time t. The observed mean, $\bar{C}$, for n equally spaced samples over the period, T, of one cycle is then given by

$$\bar{C} = \Sigma(\mu + \alpha \sin(\beta t_i - \gamma) + \varepsilon_{ti})/n \qquad (3.2)$$

so that

$$\bar{C} = \mu + [\Sigma \alpha \sin(\beta t_i - \gamma)]/n + (\Sigma \varepsilon_{ti}/n) \qquad (3.3)$$

Provided that n is not too small, the term involving the sine function on the right-hand side of equation (3.3) will be approximately zero,* so that the variability of $\bar{C}$ is dominantly determined by the random variations, $\Sigma \varepsilon_{ti}/n$. Ellis and Lacey[93] discuss, with numerical examples, this feature of systematic sampling.

Cyclic variations are sometimes readily apparent on visual examination of the data. In general, however, to obtain the best estimates of the natures and magnitudes of cyclic effects when the data are also subject to random variations, it is necessary to use statistical techniques. Note also that, if the interval between successive samples is T_s, cyclic variations of period less than $2T_s$ cannot be detected in the data. Two main techniques have been applied. First, one may examine the data quite generally with the aim of determining the periods and amplitudes of any statistically significant cyclic variations; this approach is usually referred to as

---

* For example, when $n = 5$ and $t_i = (\gamma/\beta) + (i-1)T/4$, the term is exactly zero as may also be seen intuitively from Figure 3.1 e.

spectral analysis. The second approach is to determine from the data the parameters that best describe a particular cycle, *e.g.*, an annual cycle; this approach is commonly called harmonic analysis. The statistical details of both techniques are beyond the scope of this book, but are described in a number of publications dealing with water quality; see, for example, refs. 3, and 99—107. Both these techniques call for a computer for the calculations and a body of past data covering a time period at least three or four times (preferably more) the length of the longest period cycle of interest. Spectral analysis is recommended in ref. 3 as the basic approach to determining sampling frequency.

(v) Trends as well as cyclic and random variations of quality may occur—When a gradual linear trend in quality may also occur during the period of interest, it is again usually preferable to use systematic sampling (Figure 3.1 h). Again, the random variations are then of dominant importance in deciding the number of samples required during the period; this may be shown by the same approach as in *(iv)* but using the model $C_t = \mu + \delta t + \varepsilon_t$, where δ is another constant. The number of samples required may, therefore, be calculated as in Section 3.3.4(*a*). Of course, the interval between successive samples should, in general, be chosen so that the same part of any cyclic variation is not always sampled.

A gradual trend may sometimes be apparent from simple visual inspection of the data. In general, however, to obtain the best estimate of the size of the trend, particularly when other types of variation are of appreciable size, statistical techniques are required, *e.g.*, regression analysis. The use of linear regression analysis for estimating the slope of a linear calibration curve has been briefly described in Section 5.4.2(*e*), and equation 5.35 may be used for calculating an estimate of δ in the above model—provided, of course, that it is a linear trend that is present. See also refs. 3 and 378, the latter of which deals in detail with estimation of a linear trend in rivers when results are obtained monthly. We would recommend, however, that statistical advice be sought in estimating the size of a trend when both random and cyclic variations are also present.

(vi) Statistical independence of successive samples—In *(iii)* to *(v)* above, we noted that the total number of samples required during the period of interest can be calculated, at least approximately, by statistical techniques. Much of the discussion in the literature on these techniques as applied to the design of sampling programmes has been on the basis that successive samples are statistically independent, *i.e.*, that there is no correlation between the random variations of two successive samples no matter how small the interval between them. For example, in terms of the model in *(iv)* above, the size of ε_{ti} would be independent of $\varepsilon_{t(i-1)}$. Under this condition, the appropriate statistical techniques are essentially simple as illustrated in Section 3.3.4(*a*). If, however, this condition is not satisfied, *i.e.*, if ε_{ti} is correlated with $\varepsilon_{t(i-1)}$, then more complex statistical procedures are required to obtain reasonable estimates of the number of samples necessary. The correlation of ε_{ti} and $\varepsilon_{t(i-1)}$ is often called auto-correlation (sometimes serial correlation), and the simplest model for such a situation is of the form:

$$C_{ti} = \rho C_{t(i-1)} + \varepsilon_{ti} \qquad (3.4)$$

where ρ is a constant for a given sampling interval, and the other symbols have the meanings noted above. In general, the numerical value of ρ varies between 1 and 0 as the sampling interval increases.

The potential importance of auto-correlation in the estimation of sampling frequency was mentioned in the first edition of this book, but at that time no quantitative studies of the practical importance of the effect were known. Within the last few years, however, a number of authors have considered this problem in detail in terms of sampling programmes for rivers;* see, for example, refs. 44, 45, and 108—118. It is beyond our competence to discuss this topic in detail, and the original papers should be consulted by those wishing to pursue the subject. In the present context, however, two points should be noted.

First, two quite distinct statistical approaches have been used; one is found in refs. 44, 45, and 108—113, and the other in refs. 114—118. Neither group of papers refers to the other. This seems to us particularly unfortunate since the underlying assumptions, theoretical approaches, and basic equations derived by the two groups appear to be quite different, and yet critical evaluation of the validity and applicability of the two approaches is likely to be beyond the statistical ability of many chemists. For example, the auto-correlation considered by the first group of papers is that remaining when the effect of systematic variations such as annual cycles has been eliminated. In contrast, the second group apparently ignores any such systematic effects.

Second, both groups of papers conclude that the degree of auto-correlation for various determinands in rivers can be such that the simple statistical procedures such as those in Section 3.3.4(*a*) for estimating sampling frequency are not, in general, valid. For example, with daily sampling, values of ρ between 0.75 and 0.9 are reported in the first group of papers for a number of determinands in several rivers.[108,109] Such a degree of auto-correlation is concluded to lead to substantial increase of the confidence interval (derived from the results from *n* samples) for the long-term population mean concentration of a determinand in a river. Lettenmaier also concludes[45,108] that, in estimating this mean, relatively little value accrues from sampling more frequently than once a week to a fortnight. The second group of papers[114,116,118] concludes that far fewer samples are needed to estimate (with a given precision) the mean concentration in a particular year in the presence of auto-correlation than in its absence. The latter studies also quote practical results for the number of samples needed to estimate annual mean concentrations of ammonia and nitrate in the River Rhine. It is not clear, however, to what extent the marked decrease in numbers (when auto-correlation is allowed for) is due to the presence of annual cyclic variation [see (*iv*) above] rather than auto-correlation in the sense used here.

Given the conclusions of the groups and also the differences between them, it seems to us that there is urgent need for a detailed review of all relevant statistical aspects of auto-correlation in the context of sampling programmes. What is required is clear advice, in readily understandable terms, on the assumptions, applicabilities and procedures of the various possible approaches. Until such advice

---

* We know of no studies for water-treatment plant or drinking water; the topic is briefly considered for sewage-treatment plant in refs. 119 and 120. It has been suggested[109] that river samples may be regarded as approximately independent for monthly or greater sampling intervals; see also refs. 8, 110, and 111.

is available, we can only suggest that the best statistical advice possible be sought in the planning of each sampling programme.

(c) Establishing and Reviewing Sampling Times and Frequencies

From the discussion in (*b*) above, it is clear that information on the nature and magnitude of temporal variability is required to define optimal sampling programmes. When insufficient data on variability is available at the planning stage, it will usually be very desirable to make, whenever possible, at least some short-term preliminary tests to provide a tentative assessment of variability that may be refined as the programme proceeds. In any event, unless one is very confident of being able to define the optimal programme at the planning stage, it will often be advantageous to *begin* the routine programme with as great a sampling frequency as possible; the information obtained will be invaluable in subsequently deciding if the times and frequencies of sampling require change. If, at the start of a routine programme, the frequency is too small adequately to reveal the nature and size of temporal variability, it is unlikely that an optimal sampling programme will ever be achieved.

Once the times and frequencies of sampling have been finally decided upon and implemented in a routine programme, there seems often to be a tendency to maintain them unchanged. All authors agree, however, that it is essential to review past data to decide whether or not changes in the programme are indicated. Such reviews should include aspects such as:

(1) whether or not the required information is being obtained or needs to be changed;
(2) the nature and magnitude of temporal variability (see also Section 3.4.2).

This review of the mass of data commonly generated by routine programmes is greatly facilitated when the data are routinely stored in, and evaluable by, computer systems (see Section 13). Indeed, without such systems, certain interpretive techniques such as investigating the extent of auto-correlation, harmonic and spectral analysis are essentially impossible.

Another facet of reviewing sampling programmes concerns possible changes in time of the nature of temporal variability. For example, in planning a programme initially, the data used may have shown that a potentially important source of variation (*e.g.*, diurnal variation) was of negligible size. On this basis, it would be reasonable to design the programme in such a way that the time of sampling within a day was unimportant. However, diurnal variation might become more important in the future, but would not then be revealed if samples had always been collected at the same time of day. Accordingly, unless one feels very confident that sources of variability will remain constant, it is prudent to undertake special tests occasionally to check the sizes of those cyclic effects that the routine programme cannot detect. In the present example, such special tests could involve the collection of a number of samples at approximately equally spaced intervals during one day. In this context, it is useful to note that, if systematic sampling is used routinely with an interval between successive samples of T_s, cyclic variations of period smaller than $2T_s$ cannot be detected in the routine data.

(d) Periods of Abnormal Variability

Most water-bodies are subject at times to conditions in which quality is markedly affected and may be very atypical of commonly prevailing conditions, *e.g.*, flood conditions in rivers, algal blooms in lakes, commissioning and start-up operations of industrial plant, drinking-water quality when a domestic tap is first used in the morning after an overnight-period of stagnation (lead and some other metals are particularly subject to this last effect[212]). The possibility of such periods should be considered when planning sampling programmes and, even more importantly, when considering the information required on quality. In particular, it is generally necessary to decide if, and exactly what, information on the abnormal quality is required. This point is stressed since uncertainty or ambiguity on these aspects is likely to lead to great difficulties in the planning and execution of sampling programmes. If information on the abnormal quality is required, it will usually be better to regard this as the subject of a special programme for which sampling frequency will usually need to be substantially greater than that appropriate to more usual quality conditions.

(e) Duration of Each Sampling Occasion and Sample-compositing

(i) Maximum tolerable duration of a sampling occasion—The time taken to collect a sample of the required volume means that variations during this time period cannot be detected. This causes no problems in many applications, but it should be borne in mind whenever the rate of change of quality is either large or of interest. For example, when drinking water passes through lead pipes and/or tanks before reaching domestic taps, it is common for the lead concentration, on first opening the tap in the morning, to be relatively high but then to decrease rapidly.[212] One consequence is that the duration of a sampling occasion may determine whether or not the concentration of lead in the sample is greater or smaller than the maximum acceptable concentration specified in water quality standards [see also Section 3.5.3(*d*)]. In such instances, decisions on the maximum tolerable duration for the collection of a sample depend crucially on precisely what information is required on quality.

(ii) Composite samples—When only the average quality and not its variability during a period is of interest, the commonly used technique of preparing a composite sample from a series of discrete samples may be of value. This approach has the advantages that the effects of temporal variability are reduced and only one analysis is required. Such compositing is subject to several restrictions which may be summarised as follows.

(1)   The concentrations of determinands in the samples must not change appreciably between the collection of the first sample and the analysis of the composite sample. This can be a very important restriction for many determinands; see Section 3.6.

(2)   The concentrations of determinands in the composite sample must be the same as the appropriate average of the concentrations in the individual samples, *i.e.*, composite samples should not be used for determinands whose concentrations may be affected by compositing through any process other than dilution.[37]

(3) The proportions in which the individual samples are mixed must be appropriate to the information required. For example, if each individual sample were collected over an equal time period and the average concentration were of interest, equal volumes of all the samples should be used. If, on the other hand, the mass-flows rather than concentrations of determinands were required, the volumes of samples should be proportional to the flow-rate of the water on each sampling occasion. Such points are considered in detail by many authors; see, for example, refs. 3, 37, and 121. An interesting approach to obtaining composite samples of drinking water drawn by consumers from domestic taps is described in Section 3.5.4(*b*).

(4) The imprecision of the result obtained for the average quality will be worse than that derived from the analysis of the individual samples whenever this imprecision is governed by random analytical errors rather than the temporal variability of the water of interest. Thus, if σ_s is the standard deviation characterising the temporal variability of the water, and σ_t is the standard deviation of individual analytical measurements, the standard deviation, σ_m, of the mean result from the separate analyses of n samples is given by

$$\sigma_m = [\sigma_s^2 + \sigma_t^2)/n]^{1/2},$$

while the standard deviation, σ_c, of the result from one analysis of the composite sample is given by

$$\sigma_c = [(\sigma_s^2/n) + \sigma_t^2]^{1/2}.$$

Thus, so long as σ_t^2 is small compared with σ_s^2/n, compositing has negligible effect on the precision of the result for the average concentration.

(iii) Alternative approaches to sample-compositing—An alternative approach to obtaining what is, in effect, a composite sample is simply to increase the duration of each sampling occasion. This also provides the benefit of reducing the number of analyses required and, if the sampling is continuous, may well provide a more representative sample. The approach is subject to essentially the same restrictions as sample-compositing, the problem of sample instability requiring especially careful consideration. The ideal is, in general, to make the duration of sampling equal to the period of interest, but this will be impracticable in many applications. Sonneborn [122] has recently reviewed the role of this approach in many commonly important types of water quality measurement, and has considered the problems to which it may be liable. He concluded that it could be of great value. Theoretical treatment of the effects of the duration of each sampling occasion in relation to the period of interest will be found in a group of relatively complex statistical studies [114—118] that have been applied to river-water analysis.

Practical approaches to extending the duration of a sampling occasion have commonly been based on the use of capillary inlets for the water or capillary outlets for the air (in the sample-collection container) in order to restrict the flow-rate of water into the container. Such devices, some of which are designed to be left in the water of interest at the sampling point, have been described by Sonneborn [122] and Schmitz. [123] Note that the flow-rate through, and detailed design of, such sample-collection devices need careful consideration since they may, through a number of effects, lead to changes in the concentrations of determinands (see Sections 3.5.2 and 3.5.3). To avoid possible problems in achieving representative sampling for undissolved materials at very small sample flow-rates [see Section 3.5.2(*a*)], an alternative approach has sometimes been adopted. In this, water is abstracted from the water of interest at a satisfactorily large flow-rate, and a device (a 'sample splitter') is placed at the end of the sampling line to divert only a small proportion of the sample stream into the sample-collection container. Again, great care is needed in the design and operation of sample splitters[124a,b] to ensure that they do not affect the concentrations of determinands (see Sections 3.5.2. and 3.5.3).

Another approach to extending the duration of a sampling occasion is usefully noted here, though in the terminology used in this book [see Section 3.5.1(*a*)] the approach involves part of what we consider to be the analytical procedure. In essence, a stream of sample is passed through a unit that removes the determinand completely from the sample stream, and the determinand is subsequently recovered from the unit (*e.g.*, an adsorbing column) before completing the analytical procedure. This type of technique has the additional advantage of effectively concentrating the determinand so that the measurement of small concentrations is facilitated. Further, the determinand may well be more stable in the concentrated form than in the original sample so that problems of sample instability are eased;[125–128] see also Section 3.6. This approach is, of course, subject to the same general problems of sampling as those mentioned above, and special care is generally required to avoid problems of contamination from the equipment, reagents, *etc.*, used for concentrating the determinand (see Sections 3.5.2, 10.8, and 11.2). For example, the continuous 'bleed' of trace amounts of organic substances from the adsorbant, XAD-2 (commonly used and recommended for trace organic compounds in waters[3,128]), has been reported to restrict the detection limits achievable.[129] Other possible difficulties include: (1) simultaneous concentration of substances interfering in subsequent stages of the analytical procedure, (2) degradation of organic substances in the concentrated form, *e.g.*, by bacteria,[130] and (3) inability to concentrate all forms of the determinand completely, this being a particular problem for trace metals.[127]

In principle, any procedure for sample concentration (see Section 11.2.2) capable of being applied on-line to a sample stream may be used for this approach. There is a massive literature on concentration techniques, and it is not necessary to attempt to deal with this here, particularly since a large proportion of the studies deal primarily with means of increasing analytical sensitivity rather than of extending the duration of sampling occasions. Accordingly, we mention only a few references, concerned explicitly with the sampling context, to illustrate the types of application that have been reported. Sonneborn[122] and Josefsson[131] have given short reviews of the topic, and described a number of practical devices; the latter author deals purely with trace organic compounds. Meranger *et al.*[127] critically review the subject for trace metals.

Continuous evaporation,[122] continuous centrifugation,[132–134] and filtration[135] have all been used. By far the most commonly used procedure, however, involves a column packed with an adsorbing material, many types of which have been employed for a variety of determinands. One of the most well known and widely used procedures has been the adsorption of a wide range of organic compounds on activated carbon.[136] This procedure and the problems that may be experienced have been reviewed by many authors; see, for example, refs. 137—140. The growing interest in organic compounds in waters has, however, led to much investigation and application of other adsorbing materials. Refs. 139, and 141—143 provide useful reviews of such materials and their uses. Other adsorbents have been used for inorganic determinands, *e.g.*, cation-exchange resins for metals in boiler feed-water,[144] rain water,[380] and natural waters,[381] and activated alumina for radio-isotopes in sea water and fresh waters.[145]

Josefsson[131] concluded that continuous solvent extraction may often have advantages over adsorbing columns for organic compounds; both he and Yohe *et al.*[146] have described suitable equipment for this purpose.

For trace metals, Meranger *et al.*[127] concluded that three techniques had greatest potential – coprecipitation followed by flotation, ion exchange and adsorption, and electrodeposition – but that no procedure yet designed met all the ideal requirements for even a limited number of metals. A recent application of the first of these techniques is described in ref. 147, and examples of the last may be found in refs. 304 and 305. The use of enzymes to concentrate mercury has also been reported.[303]

An interesting variant of this general approach has been reported[148] (for trace elements in rivers) in which a dialysis cell is immersed in the river at the sampling point, the internal compartment of the cell also containing an efficient adsorbing material for the trace elements of interest. Under these conditions, the rate of diffusion of dissolved trace elements through the membrane of the cell is expected to be directly proportional to their concentration in the river water. Analysis of the adsorbing material thereby provides a measure of the average concentration in the water during the period of sampling. A similar approach, but without the adsorbing material, has been reported[149] for dissolved oxygen in waters. Another type of *in situ* technique for extending the duration of sampling when polynuclear aromatic hydrocarbons are of interest consists in the immersion of sheets of commercially available oil-adsorbent cloth in the water for one to four days.[150] The adsorbed hydrocarbons are subsequently extracted from the sheets and then measured.

(f) Simultaneous Consideration of Sampling Points and Times
Identification of the sampling points and times required for a programme may well lead to the situation where there are insufficient resources to allow the required sampling frequencies at all the sampling points. This problem is discussed in Section 3.4.2.

(g) Types of Sampling Programme
Sampling programmes may be classified in a number of ways depending on their broad objectives and the particular information they are intended to produce. Examples of different classifications may be found in refs. 37, 41, 55, 118, and 151—153. For simplicity, we feel that the classification suggested in the first edition of this book remains useful. In this, only two broad types of programme are distinguished: (1) quality control and (2) quality characterisation, both terms being defined below. This classification has also been adopted in refs. 5, 6, and 55. Of course, like many classifications, this one serves primarily to aid consideration of

the principles involved. In practice, actual programmes for waters may contain elements of both types, but the above classification should help to focus attention on the main purpose(s) of a given programme, and thereby aid its planning.

The term '*quality control programme*' is used here to signify that there is need to control the instantaneous concentrations of one or more determinands, and that, in principle, the need for corrective action to improve water quality is shown by individual analytical results. Programmes of this type usually involve the control of quality at one particular location, *e.g.*, as in normal process control, though water-bodies other than water-treatment plant may be involved. Sampling programmes for quality control purposes are discussed in Section 3.3.3.

The term '*quality characterisation programme*' here denotes the situation where a number of analytical results are required to characterise the quality of a water over a period. Such information may be required for many purposes, *e.g.*, in research investigations, for long-term assessment of water quality as in environmental monitoring, and also in indicating the adequacy of water quality over a longer time-scale than that involved in quality-control programmes. Sampling programmes for quality characterisation purposes are considered in Section 3.3.4.

3.3.3 Quality-control Programmes

(a) Types of Control Programme
Following Kateman and Pijpers,[118] it is useful to distinguish two types of control: (1) threshold control and (2) process control.

In *threshold control*, the primary aim is to ensure that the concentrations of one or more determinands do not exceed specified limits (thresholds); provided that this aim is achieved, the determinand concentrations are unimportant. Thus, any water quality standard specifying that a determinand must not exceed a specified limit implies the need for threshold control. Situations do occur where a determinand concentration should not be less than a specified limit, its concentration above that limit being unimportant; dissolved oxygen in river water sometimes offers an example of this situation. It is also possible, in principle, to have a double-threshold requirement, *i.e.*, where a determinand must neither exceed nor fall below specified limits. The principles involved in all threshold control situations are the same.

In *process control*, the aim is generally to control the concentrations of one or more determinands within narrow limits of specified values, these values ideally being achieved continuously with negligible variations. An example is the control of the fluoride concentration in the fluoridation of drinking water.

In practice, as noted above, actual sampling programmes may fulfil more than one function. For example, when threshold control is required, it may also be necessary to characterise the water quality even when threshold limits are not exceeded. Thus, one sampling programme may be composed of elements of both threshold control and also quality characterisation. It is, nevertheless, useful to maintain this distinction between types of programme since it aids their optimisation.

There appears to be a growing tendency to express water quality standards in probabilistic terms; see ref. 154 for a good discussion of this. Thus, rather than specifying that a stated concentration of a determinand shall not be exceeded, such

standards specify the average concentration o
that shall not be exceeded over a given period
imply longer term control based on a number o
period, and the sampling programmes for such
of the quality characterisation type.

The above definitions of threshold control ar
types of programme represent rather different
the objective is to minimise the proportion o
concentration transgresses a specified limit. I
rather to minimise the variability of the conce
specified value. It is useful, therefore, to consid
selection of sampling times and frequencies for these two types of programme.

(b) Threshold Control

(i) Primary factors governing sampling times and frequencies—Many authors have stressed that the only general way of achieving certain detection (so that corrective action may be taken) of all deviations beyond a threshold is to measure the concentrations of relevant determinands continuously. When this approach cannot be applied, there will be some risk that the desired degree of control will not be achieved. The times and frequencies of collecting samples for analysis should, therefore, be selected so that the risks of important deviations beyond thresholds are adequately small. For simplicity, we speak hereafter of the control of only one determinand since the considerations apply equally to all determinands needing to be controlled; note, however, that the optimal times and frequency of sampling may well vary from one determinand to another.

On the above basis, two primary factors govern the times and frequency of sample collection:

(1) the economic and/or other penalties incurred when the concentration of a determinand exceeds its threshold limit, and the dependence of these penalties on the magnitudes and durations of such deviations;
(2) the probabilities of occurrence of deviations of various magnitudes and durations.

Definition of factor (1) is the responsibility of the user of results, though it is apparently very difficult to quantify the penalties associated with deviations beyond the threshold for many water quality situations (see Section 3.4.3). Nevertheless, as noted in ref. 155, the optimal sampling policy may depend crucially on these penalties, and the best attempt possible should be made to quantify them. In any event, a control policy should be formulated explicitly by the user of results; otherwise, there can be no rational basis for planning the control programme. Of course, the control policy may have marked effects on the routine operation of the water-body of interest; see, for example, Section 4.6.2(*d*)(*ii*).

The extent to which factor (2) can be taken into account depends on the extent to which the past concentrations (and their temporal variability) of a determinand can be validly used as indications of their future values. If, and this would be an exceptional situation, the user decides that past data cannot be used to indicate the

appear to be no alternative but to require the interval between
...es to be rather shorter than the shortest duration considered
...a deviation beyond the threshold limit. Usually, however, existing
...e used as a reasonable indication of the temporal variability of water
...n the near future; long-term extrapolation is not necessary if data are
...rly reviewed as suggested in Section 3.3.2(c). Of course, confidence in the
...dity of this approach increases with increasing knowledge of the sources of the
determinands of interest and of the factors governing their temporal variability.
Such aspects should be carefully considered in planning control programmes. For
example, in the control of the quality of water produced by a water-treatment plant
whose raw water is abstracted from a river, it is important to know of upstream
effluents that may be important sources of a determinand to be controlled. We
assume now that existing relevant data on the concentrations of a determinand may
be used for assessing the probabilities of deviations beyond a threshold as discussed
in (ii) below. In the absence of such data, an alternative approach is to construct a
mathematical model of the water-body of interest so that these probabilities may be
assessed from the model; such an approach is described for rivers in ref. 155.

(ii) Assessing the probability of deviations beyond a threshold—Given a set of n
results, R_i $(i=1$ to $n)$, for a determinand concentration, they should be examined and
analysed to determine the nature and magnitude of temporal variability. The
possibility of gradual trends and cyclic variations should be considered in addition
to random variations [see Section 3.3.2(b)] because the probability of a threshold
being exceeded will generally depend on the sizes of both systematic and random
variations. For example, diurnal variations of dissolved oxygen in a river may
require sampling to be done at a particular time of the day. Similarly, dissolved
oxygen is often likely to fall below its threshold limit in rivers during the warmer
months of the year so that a greater frequency of sampling may well be desirable in
those months. The possible importance of abnormal conditions [see Section
3.3.2(d)] should also be considered at this stage. Having decided if there are any
such periods in which sampling should be concentrated, it is then useful to examine
the data for those periods to determine the mean, $\bar{R}$, and standard deviation, s, of
the results. Assuming that the aim is to maintain the concentration below a
threshold, λ, and that water quality has been reasonably satisfactory in the past,
three main types of situation may be revealed by past results as illustrated in Figure
3.2 a to c.

Case 1: $\bar{R} < \lambda$, $s/(\lambda - \bar{R}) \ll 1$ *(Figure 3.2a)*—In this situation, there is an extremely
small probability that λ will be exceeded, and relatively infrequent sampling will
suffice. Such a situation could correspond to the control of a rather uncommon
impurity, *e.g.*, selenium, in drinking water. So long as the water-treatment plant, the
source of the raw water, and the sources of any selenium in the raw water did not
change appreciably, a small sampling frequency would suffice. Of course, if any of
these factors changed, or if the measured concentrations of selenium began to
increase, more frequent sampling would be wise. In passing, it is worth noting that
this example illustrates the value of starting a sampling programme with a relatively
large sampling frequency which is then progressively changed as more information
is obtained from the analytical results [see section 3.3.2(c)].

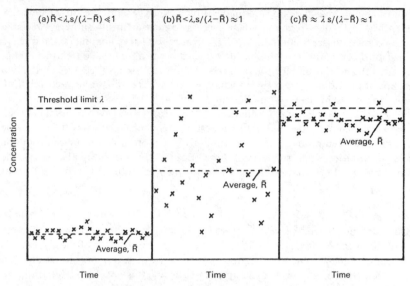

Figure 3.2 *Assessing probability of a threshold being exceeded* (× *denotes analytical results;* s = *standard deviation of analytical results*)

The concept of adjusting the sampling frequency to match the probability of threshold transgressions is discussed in more detail by Kateman and Pijpers[118] and Heidtke and Armstrong;[155] see also (*iii*) below.

Case 2: $\bar{R} < \lambda$, $s/(\lambda - \bar{R}) \approx 1$ *(Figure 3.2b) and*

Case 3: $\bar{R} \approx \lambda$, $s/(\lambda - \bar{R}) \approx 1$ *(Figure 3.2c)*—In both these cases, there is an appreciable probability that the threshold, λ, will be exceeded from time to time. As a first step in considering such situations, it is useful to consider the size of analytical errors if this has not already been done before any measurements were obtained. The observed variability of past results, and hence the apparent probability of threshold transgressions could be caused by analytical errors alone, particularly for Case 3, rather than by the variability of the true concentrations of the determinand. The effects of analytical errors and their implications for threshold control are considered in Section 4.6. If this consideration of errors dictates some change in the analytical procedure or in the control of the water-body of interest, the past data are clearly not directly relevant to the assessment of a future sampling programme. It may, therefore, be better to operate for a period under the new conditions with as high a sampling frequency as possible in order to obtain more-relevant data for estimating the required sampling frequency [see (*iii*) below].

(*iii*) *Estimating the optimal sampling frequency*—For Case 1 in (*ii*) above, the conclusions reached in that section may suffice for many purposes. Procedures for quantitative estimation of the optimal sampling frequency for this case are, however, described by Kateman and Pijpers.[118] For Cases 2 and 3, we would recommend that statistical advice be obtained whenever possible because estimation of the optimal sampling frequency involves, in general, the rather complex statistical techniques required when successive measurements show appreciable auto-correlation [see Section 3.3.2(*b*)(*vi*)]. Procedures for estimating this frequency when auto-correlation is negligible have been described for sewage-treatment plant in refs. 119 and 120, and for rivers in refs. 43 and 155; see also ref. 156. Of the two groups of procedures (see p. 54) in which allowance is made for auto-correlation, only one deals in detail with threshold control.[117,118] According to these two references, as the value of ρ in equation (3.4) increases for a given interval between successive samples, the required sampling frequency decreases because the measured concentration at a given time gives a more accurate estimate of the concentration at some subsequent time. Both these references also stress the economy of effort that may result from their allowance for this effect of auto-correlation.

(*c*) *Process Control*
Kateman and Pijpers[118] give a very useful discussion of the fundamental factors involved in achieving a given efficiency of process control. One of these factors is sampling frequency, and the equations presented by these authors allow quantitative calculation of the effects of sampling frequency. It is possible to determine an optimal frequency if the penalties incurred by deviations from the desired concentration can be quantified. The reference cited should be consulted for details, but one of the points made is well worth repeating here.

   The effect of the procedures used for sampling and analysis on the overall efficiency of the control of the process is governed by five factors: (1) the time delay between sampling and producing the analytical result, (2) the sampling frequency, (3) the volume of sample collected, (4) the time constant of the analytical system, and (5) the precision of the analytical results. These factors should not be considered in isolation because compensating effects are possible between good and poor values of different factors. For example, the same overall efficiency of control may be achieved with a range of sampling frequencies provided that the precision of analytical results is correspondingly improved as the sampling frequency is decreased.

3.3.4 Quality Characterisation Programmes
In these applications, the aim is to estimate one or more parameters characterising the concentrations of one or more determinands during a defined period. For example, the mean concentration is often of interest, and is estimated as the mean of the results from a number of samples. Other parameters may, however, be of concern; for example, increasing interest has been shown in recent years in the estimation of percentiles.* Whatever the parameter of interest, if the uncertainty tolerable in its estimated value has been defined [see Sections 2.8 and 3.3.2(*a*)] *and* an estimate of the nature and magnitude of temporal variability is available, statistical techniques then allow objective estimation of the minimal sampling frequency. This topic has been discussed by many authors of whom Ellis and Lacey[93] give a particularly thorough and readable account. In the following, for

---

* In general, the P percentile is the concentration that is exceeded for $(100 - P) \%$ of the total period of interest.

purposes of illustration, we deal mainly with the estimation of mean concentrations [(a) and (b) below]; estimation of percentiles is briefly considered in (c).

(a) Systematic Variations are Negligibly Small
In this situation, changes in quality are assumed to arise solely through random variations. Given the parameter to be estimated (*e.g.*, the mean) and its tolerable imprecision at a given confidence level, the number of samples, n, to be collected and analysed in the period of interest is chosen so that the confidence interval for the true mean does not exceed the tolerable imprecision. For example, if the random variations of the measured concentrations follow a normal distribution with standard deviation, σ, the 95% confidence limits for the true mean are given by [see Section 4.3.3(b)]

$$\bar{R} \pm 1.96\sigma/\sqrt{n} \tag{3.5}$$

where $\bar{R}$ is the mean of the n results. Thus, if the tolerable imprecision of the mean were ε, n is given by

$$n = (1.96\sigma)^2/\varepsilon^2 \tag{3.6}$$

As an example, suppose that ε were required to be not greater than 5% of the mean, and that σ were equivalent to 20% of the mean; n would then be approximately 64.

Several points should be noted for this approach. First, the value of σ is governed by the random variations of the true concentrations in the water-body *and* by the random errors arising from sampling and analysis of samples. Second, it should not be thought that this type of approach allows exact prediction of a suitable sampling frequency. The inevitable random error of the estimate, s, that must be used in place of σ [see Section 4.3.5(a)] will sometimes cause the calculated value of n to be greater than necessary and sometimes smaller. The approach does, however, provide an objective means of estimating a reasonable value for n which will tend to be correct on average over many applications; this is much to be preferred to inexplicit and subjective approaches. The procedure also quantifies the relationship between ε and n; this is of help if the originally chosen value for ε proves to require an impracticably large value for n. Finally, equation (3.5) is correct only if the random variations of the measured concentrations are independent, *i.e.*, when there is negligible auto-correlation between successive measurements [see Section 3.3.2(b)(vi)]. Given the calculated value of n, the samples are, in general, collected at approximately equally spaced times during the period of interest, but see also Section 3.3.2(b).

Allowance can be made in these calculations for distributions other than the normal; Montgomery and Hart[37] discuss in detail the case when the log normal distribution is appropriate. When, however, the mean is to be estimated, it will usually suffice to assume the normal distribution because the means of n results follow a distribution which approximates to the normal more and more closely as n increases whatever the distribution of individual results.[93,157] The approximation is sufficiently close for most practical purposes for values of n as small as 5. This represents a very attractive feature of the mean as a parameter for estimation.

Similar calculations can be made to estimate the sampling frequency when percentiles of a distribution are of interest; see, for example, reference 37. However, as Ellis and Lacey stress,[93] the validity of the values then obtained for n is markedly dependent on the exact shape of the low and high tails of the distribution, and the assumption of the normal distribution may be quite inappropriate when small or large percentiles are of interest. These authors recommend a quite different approach to the estimation of sampling frequency for percentiles, and this is summarised in (c) below.

One other approach that may be helpful in assessing the effect of sampling frequency is worth brief mention here. When the period of interest is T, and past measurements of quality are available for a period of at least several times T, the effect of reductions of sampling frequency can be examined empirically (assuming that the causes of temporal variation are constant) by considering sub-sets of the results. For example, suppose that the annual mean concentration of a determinand were required, and that past records for, say, 5 years were available at a sampling frequency of weekly. For each of these years, the data can be grouped into sub-sets corresponding to the results that would have been obtained by fortnightly, monthly, quarterly, *etc.*, sampling. The means of these sub-sets in relation to the mean of all results for a year provide some indication of the increase in the uncertainty of mean results obtained at a smaller sampling frequency. This approach has been used by many authors – see, for example, refs. 35, 37, 60, and 61 – and can be useful even though it lacks any firm statistical basis; note, though, that it does make some allowance for the effect of auto-correlation.

(b) Systematic Variations are not Negligible
Provided that any systematic variations (trends and cycles) are appropriately and representatively sampled, the approach described in (a) may be used to estimate sampling frequency [see Section 3.3.2(b)]. The value of s can be estimated from the variations observed over periods short compared to the periods of any systematic effects. In addition to any limitations noted in (a), we stress that appropriate and representative sampling of cyclic variations is crucial; see Section 3.3.2(b)(iv).

(c) Estimation of Percentiles
There appears to be a tendency for water quality standards to be defined increasingly in terms of a specified percentile that should not be exceeded, *e.g.*, that the 95 percentile concentration for an impurity in a water should not exceed a specified concentration. Ellis and Lacey[93] provide a valuable discussion with numerical examples of various approaches to the estimation of percentiles and their relative advantages and disadvantages. They stress that the confidence limits calculated for percentiles are markedly dependent on the exact nature of the frequency distribution of the measured values, and are also very wide for small or large percentiles even when the correct frequency distribution is known [see (a) above]. They conclude that such confidence limits '. . . are of limited value except in emphasising the statistical hazards in this area.'

The same authors go on to make the important point that, when the estimation of a percentile is required in checking the compliance of a water with a water quality standard, the objective of the measurement programme can be reformulated in such a way that a simple yet rigorous approach to the achievement of the objective is made possible and, at the same time, the estimation of the required sampling frequency is facilitated. Suppose that a standard for water quality requires that the *true* 95 percentile does not exceed a concentration, λ.* Then, rather than test whether the 95 percentile estimated from a measurement programme is significantly greater than λ, it is equivalent to test whether the proportion of measured values greater than λ is significantly greater than 5%. This proportion always follows a precisely defined distribution (the Binomial distribution) whose properties allow the appropriate number of samples in a period to be estimated. The authors' paper[93] should be consulted for the exact methods of calculation and other details which may also be found in many statistical texts; see, for example, ref. 96.

3.4 OVERALL DESIGN OF SAMPLING PROGRAMMES

When the detailed considerations of the locations/points and times of sampling have been completed, they must, of course, be integrated to produce the overall sampling programme. References to publications dealing with this total design of programmes for various types of water are collected in Section 3.4.1.

One may often meet the problem that the resources available for sampling and analysis are insufficient to allow all the desired samples to be collected and analysed. Several approaches are possible for reducing the size of this difficulty, and are considered in Section 3.4.2. Finally, in Section 3.4.3 we touch on a related and important aspect, namely, the consideration of the cost-benefit relations of sampling and analysis programmes.

3.4.1 References to the Design of Complete Sampling Programmes

The references are gathered in Table 3.1 according to the type(s) of water to which they relate, but, of course, procedures described for one type of water may well be of value for other types. The number of publications dealing *in detail* with particular types of water varies greatly with the type. Thus, we know of few such publications for drinking water, lakes, and water-treatment plant. On the other hand, sampling programmes for rivers and river basins and, to a smaller extent, groundwaters have been considered by many authors. Ward *et al.*[158] have reviewed the literature on rivers for the period up to 1975–6, and we have quoted only later publications for

TABLE 3.1
References to the Design of Complete Sampling Programmes

Type of water	Refs.
Rivers and river basins	3, 5–7, 41–46, 158–164
Lakes and reservoirs	3, 5–7, 53, 56, 60, 160
Groundwaters	3, 6, 7, 64, 68–74, 160, 242
Estuaries	5–7, 81, 82, 160
Drinking water (distribution systems)	3, 84–86†, 91, and see Section 3.2.4(*f*)

† Refs. 84 and 85 are in process of revision; see ref. 92. The International Organization for Standardization is also preparing detailed recommendations on sampling programmes for these and other types of water.

* Some standards are expressed in terms of the percentiles of the measured values rather than of those of the population of interest; see Ellis and Lacey[93] for the difficulties caused by this approach.

this water type. Of these later publications, many describe relatively complex statistical procedures which Lettenmaier[159] has reviewed up to 1979.

3.4.2 Possible Approaches to Reducing the Number of Samples

When the desired number of samples is greater than available resources can deal with, either the number of sampling points or the number of samples at some or all of these points must be reduced. Since the originally chosen points and times of sampling were selected to ensure that the required information was obtained, reduction of the sampling programme will require modification of the detailed objectives of the programme. In this position, consultation between the user of results, those planning the sampling programme, and those responsible for the routine sampling and analysis is essential.

In decreasing the total number of samples generated by a programme, virtually all authors agree that it is usually preferable to reduce the number of sampling locations rather than the sampling frequency. Otherwise, substantial reductions in sampling frequency may easily lead to such markedly increased uncertainty that the information from all sampling locations becomes essentially worthless. In addition, however, to any compromise between the number of sampling locations and the number of samples at each, there are two main approaches to decreasing the total of effort required for sampling and analysis.* The first involves the use of composite samples as discussed in Section 3.3.2(*e*). The second is based on the prediction of the concentration of a determinand from other experimental measurements. The latter approach clearly requires the statistical analysis of past data in order to determine the most suitable prediction procedures and their accuracy; three main variants are possible.

First, the concentrations of a determinand at two or more sampling locations may be sufficiently well correlated that measurements on samples from one location may be used to predict concentrations at others. Ref. 165 gives a relatively simple example for a river, and a much more complex approach involving multi-variate statistical procedures is reported in ref. 47. This approach is recommended in ref. 3.

Second, the concentration of a determinand may be correlated sufficiently well with flow conditions (or other physical factors affecting water quality) that knowledge of the latter provides an adequately accurate estimate of the former. Many authors have studied the relationship between concentration and river discharge; see, for example, refs. 107 and 166—170. It should, however, be noted that a number of authors have stated that concentrations predicted in this way may have relatively large errors; see, for example, refs. 107, 168, and 169. In addition, the relationship between concentration and discharge may itself vary with time depending on previous variations of discharge. Clearly, for this approach to be feasible, a reasonably accurate means for measuring discharge is required at the sampling location. As a variant of this approach, knowledge of the discharge at a sampling location may be used to decide the most important times for sampling, and this may allow some relaxation of the routine frequency of sampling. For example, it is commonly found that the concentrations of undissolved, solid

* As previously noted, we are not considering on-line analysis because, at present, this is not a general solution to the problem. Of course, such analysis may well be invaluable when it is practicable.

materials tend to be much higher during periods of high discharge. Thus, if the annual mean concentration were of interest, the total number of samples could possibly be reduced by concentrating most effort on periods of high discharge.[171] This aspect is closely related to the topic of estimating mass-flows rather than concentrations of determinands; this has been studied again by many authors – see, for example, refs. 170—175.

Finally, the concentrations of one or more determinands may be used to predict the concentration of another determinand.[3] This approach can be especially helpful when the measured determinands are determined by on-line analytical techniques providing essentially continuous records. For example, Boes[176] has shown that the total hardness of a river water could be predicted quite accurately from the measured values of electrical conductivity, pH, and chloride. The well known use of electrical conductivity to estimate 'total dissolved solids' is another example.[177]

3.4.3 Cost–Benefit Relationships for Measurement Programmes

Several papers have considered the cost-effectiveness aspects of measurement programmes for river water, and ref. 178 provides a typical example of the type of approach that has been adopted and which could also be applied to other types of water. In essence, such studies attempt to quantify the dependence of the cost of the measurement programme on the uncertainty of the information derived from the measurements. For example, when the aim is to estimate the annual mean concentration of a determinand, the cost of a programme is related to the confidence interval for the mean; as this interval must be made smaller and smaller, so the number of samples must be increased [see Section 3.3.4(a)] and the cost of the programme increases. Such information may well be helpful in considering how cost varies with effectiveness, but this is essentially irrelevant to any decisions on the much more fundamental aspect – the economic value of the information provided by a programme. There might well be little or no point in operating a highly refined and cost-effective programme at great cost if the information derived allowed only a minor saving in other expenditure. Clearly, the basic criterion should be the justifiable cost of a programme in terms of the benefits derived from its results, *i.e.*, it is a cost–benefit, rather than a cost-effectiveness, analysis which is crucial in this context.[5,6,32,382] The problem then immediately arises, however, that it appears to be very difficult, if not impossible, in many applications to quantify the economic value of information on water quality. For example, Ludwig and Storrs[179] state that it is very difficult or impossible to produce a rational, quantitative assessment of the benefits deriving from regional water quality management. We are aware of very few papers[32,180,382] that even attempt to address this problem in any detail, and we have no general solution to offer. We thought it appropriate to mention the point here, and to suggest that it forms a very important subject for future investigation. Without some reasonable basis for judging the relative values of different programmes, the rational allocation of scarce resources (money, manpower, laboratory space, *etc.*) among competing programmes seems to us severely hampered if not prevented. In this present position, one may well find in practice that the resources available for a number of programmes are fixed. In such circumstances, the cost-effectiveness analyses mentioned above may be useful in helping to ensure the best possible use of the available resources.

3.5 PROCEDURES FOR OBTAINING SAMPLES OF WATERS

3.5.1 Introduction

(a) Basic Considerations and Definitions
We shall not discuss the various problems involved in the precise definition of terms that must be used when sampling procedures are considered. A variety of terms, however, is used in different senses by different authors, and to avoid confusion we have summarised below the meanings of the terms as used in this book. There are International and British Standards[180a] that give definitions of many relevant terms, and we have tried to follow these recommendations whenever possible. For a number of reasons, however, we have sometimes felt obliged to use additional terms or to adopt slightly different connotations of terms included in those Standards.

(i) Sample—This term denotes a portion of the water of interest when it has been transferred from a sampling point in a water-body to another position (the *'sample-delivery point'*) where either analysis is started or one or more *'samples for analysis'* [see (2) below] are prepared.* We think it useful to maintain as clear a distinction as possible between sampling and analytical processes, and for this reason we regard a sample as having received no treatment of any kind apart from the physical processes involved in the transference of the water of interest. The term *'sample stream'* denotes a continuous stream of sample. Note that 'sample' refers only to a portion of the *water* of interest.

(ii) Sample for analysis This term denotes all or part of a sample after it has been prepared for analysis. This preparation is referred to as *'sample preparation'*. When such preparation is necessary, we regard it as an integral part of the total sampling process. Sample preparation is, however, not always required; for example, analysis may be made of a sample or sample stream directly it becomes available at the sample delivery point. Note that 'sample for analysis' refers only to a portion of the *water* of interest, and that we include only three types of procedure under 'sample preparation'.

(1)   The appropriate identification of samples, *e.g.*, by placing suitable identifying labels on *'sample-collection containers'* [see (*iv*) below] or on *'sample containers'* [see (*v*) below].
(2)   The transfer of all or part of a sample to one or more 'sample containers'.
(3)   The physical and/or chemical treatment of samples to minimise any changes in the concentrations of one or more determinands in the period between the instants when the samples first become available at the sample delivery point and when their analysis begins. Such treatments are referred to as *'sample stabilisation'*, and are considered in detail in Section 3.6. The term 'sample preservation' is also very commonly used in the same sense.

---

* Situations arise where a sample first becomes available at one position but is then taken to another position for analysis or preparation of samples for analysis. For simplicity, we ignore such situations since they do not affect any of the subsequent discussion.

Sample stabilisation procedures may have important consequences for analytical procedures. For example, samples are often acidified to stabilise the total concentrations of each of a number of trace metals, but this acidification also tends to dissolve undissolved forms of metals in the original samples; analytical procedures for measuring the total concentrations of metals may thereby be simplified. Similarly, analytical procedures carried out at a sample-delivery point directly a sample or sample stream becomes available can be an important means of overcoming problems caused by the instability of determinands [see Sections 3.3.2. (*e*) (*iii*) and 3.6.3(*c*)]. Note, however, that a consequence of our definition of sample preparation is that procedures for the continuous concentration of one or more determinands in a sample stream are regarded as part of the analytical and not the sampling process.

In the following we shall often use the term 'sample' in place of 'sample for analysis' when the distinction between the two is unimportant.

(iii) Sampling device   This term denotes the one or more pieces of equipment used to provide a sample or sample stream at the sample-delivery point. Sampling devices vary widely in complexity; they range from a simple bottle held by a person to a specially drilled sampling well into an aquifer together with any components required to bring a portion of the groundwater to the surface. The need for sampling devices is often obviated when *in situ* measurement techniques can be used, and this is an attractive feature of such techniques (see Section 12.1).

(iv) Sample-collection device   This term denotes that part of a sampling device by means of which a portion of the water of interest is brought to a sample-delivery point. There are two types of sample-collection devices.

(1)   *Sample-collection container* refers to a container that is immersed at a sampling point in a water-body, and is then partially or completely filled with that water and brought back to the sample-delivery point. Note that sample-collection containers may also function as sample containers.
(2)   *Sampling line* refers to what is, in essence, a tube or pipe (plus associated components such as valves, taps, cooling coils as necessary) with its inlet at a sampling point and its outlet at the sample-delivery point.

Since both types fulfil the same function, they are subject to many identical considerations notwithstanding their different natures as noted also in ref. 187.

When considering the effects of sampling devices on the representativeness of samples, the sample-collection device is often of crucial importance. It is very necessary to remember, however, that sampling devices generally include other components which may also affect the accuracy of samples.

(v) Sample container   This term denotes a container in which a sample for analysis is stored after sampling is completed and before analysis begins. Sample containers may have important effects on the stability of many determinands, and are considered in detail in Section 3.6.2.

(vi) Sampling method   This term denotes a written specification of a sampling

device, the details of its installation and use, and the procedures for sample preparation.

(vii) Sampling system   This term denotes an assembly of all the components necessary to provide samples for analysis for all determinands of interest from a given sampling point in a water-body. Such components include a sampling method and a sampling device, and also often include sampling personnel and sample containers. Special facilities at the sample-delivery point may also be required, *e.g.*, to allow sample-stabilisation procedures specified by the sampling method, to provide protection from contamination and environmental effects, *etc.*

(b) General Observations on Sampling Systems

(i) Importance of recognising the problems caused by sampling systems   Incorrect choice, assembly, and operation of sampling systems can lead to markedly non-representative samples. In general, therefore, great care should be devoted to these aspects if the measured determinand concentrations in samples are to be reliable estimates of the concentrations in a water-body. Although a number of sampling requirements may, in fact, present no important problems, we have deliberately emphasised below the difficulties that may arise. The main reasons for this emphasis lie in the broad purpose of this book and the tendency for sampling systems sometimes not to be given the critical attention they merit. A striking example of the importance of sampling systems is the observation that past results for a number of trace metals in sea-water are one or more magnitudes greater than those obtained more recently when more rigorous precautions were taken in sampling[181,182] – but see also ref. 183.

In *(ii)* to *(vii)* below we note a number of general points concerning sampling systems. General problems associated with the nature and design of sampling devices are considered in Section 3.5.2, and problems arising in their operation are discussed in Section 3.5.3. Brief details of particular sampling devices for a number of common sampling requirements are given in Section 3.5.4. Finally, sample preparation is discussed in Section 3.6 which provides detailed treatment of the very important aspect, sample stabilisation.

(ii) Safety   The collection of samples may sometimes expose sampling personnel to risk, *e.g.*, of drowning. Such risks should be carefully identified and appropriate precautions taken. Refs. 5 and 6 give some helpful suggestions on this aspect of sampling.

(iii) Involvement of analysts with sampling systems   It is – fortunately in our opinion – increasingly common for the analyst to be intimately involved with one aspect of sampling systems, namely, sample stabilisation. His involvement, however, with other aspects such as the choice and operation of sampling devices is often rather small since he commonly has no responsibility for such matters. This seems to us an undesirable situation for two reasons. First, the analyst will often have expert knowledge and experience of factors that may be crucial with respect to sampling devices and their use, *e.g.*, contamination of samples, reactions between

determinands and other materials (see below). Second, as many authors have emphasised, there is little point in the analyst expending considerable effort on the pursuit of analytical accuracy if inadequate attention is paid to the accuracy of sampling. The analyst will usually be best qualified to judge the relative importances of the errors from sampling and analysis. It seems to us, therefore, essential that the analyst should strive to be directly involved in all aspects of the selection and operation of sampling systems.

(iv) Knowledge of the errors caused by sampling systems The scientific literature is full of accounts of investigations of the errors (in analytical results) caused by analytical systems. In addition, large and growing numbers of analytical methods contain statements on the magnitudes of both systematic and also random errors that may be expected when the procedures specified by the methods are followed. To judge from the literature of which we are aware, there is a relative scarcity of such investigations that deal with sampling systems for natural and drinking waters; it is also relatively rare for a published sampling method to attempt any quantification of errors that may arise in its use. Fortunately, in recent years more attention seems to have been paid to redressing this imbalance between sampling and analysis, but such efforts have been predominantly devoted to sample stabilisation. The quantitative experimental investigation of errors associated with sampling devices and their operation appears to have received much less attention. We would, therefore, strongly urge the need for, and value of, such studies and their publication.

In connection with the last point, a few additional observations are worth making here; they may at first sight seem rather abstruse, but we believe that they have very important, practical consequences. In essence, it seems to us that the experimental estimation of errors is, in general, much more difficult for sampling systems than for analytical systems. This difference stems from the fact that the estimation of errors caused by any system basically involves supplying the system with a known and/or constant input, and then observing the output of the system. The variability of the output gives an estimate of the precision (see Section 4.3), and comparison of the value of the output with that of the input (when the latter is known) estimates the bias (see Section 4.4). It is usually much easier to provide such a controlled input for analytical systems because, in principle, one needs only a portion of water; the input to a sampling system, however, is a water-body, *e.g.*, a river, an aquifer, *etc.* of which neither the water-quality nor its spatial and temporal constancy are generally known. Of course, just as standard solutions may be used to simulate samples, so too may water-bodies be simulated by using appropriate test-rigs or other equipment, but again it is usually much more difficult to achieve adequately close simulation of the relevant features of water-bodies than of samples. Finally, even when a water-body can be well simulated, there is a further problem that the output of a sampling system is a sample, and an analytical system is required, therefore, to allow comparison of the input and output. Inadequate performance of the analytical system may prevent valid conclusions on the errors caused by the sampling system.

An immediate consequence of these observations is that it is extremely difficult and usually impracticable to estimate the sizes of all the errors that may be caused by sampling systems during routine programmes of water quality measurement (see also Section 3.5.5). The implications of this are considered in (*v*) below.

(v) Selection and operation of sampling systems From the points made in (*iv*), we conclude that the choice, assembly, and operation of sampling systems generally require individual consideration of each requirement. The relative lack of knowledge and many potential sources of error call for strenuous efforts at the planning stage to choose the most appropriate sampling system for a given purpose. Lee and Jones[184] have recently made a similar point; they write: 'Investigations should never assume that commonly used sampling and analytical methods are applicable in all situations.' In this context, the general sources of error identified in Sections 3.5.2, 3.5.3, and 3.6 may be used as a 'check-list' in the critical consideration of potentially applicable sampling procedures. For this, it is important to bear in mind that the suitability of a given sampling system may depend markedly on the determinand(s) of interest.

In considering potentially suitable sampling systems, we would also suggest that the aim should generally be to select systems that are capable of providing negligibly biased samples [see Sections 4.6.1(*c*) and 5.5], *i.e.*, the concentrations of all determinands are the same in the samples (at the start of analysis) as in the water-body at the times and points of sampling. Achievement of this aim will also generally minimise any random variations in the representativeness of samples.

Even when a potentially suitable sampling system can be identified, several factors may prevent adequately representative samples from being obtained. For example, the sampling device may be inaccurately constructed or be made from inappropriate materials, the method of operation of the sampling device may be inadequately controlled, excessive contamination may occur at the sample-delivery point, *etc.* To minimise such problems, it is essential that the sampling method be as completely and unambiguously specified as possible, and that this method be followed faithfully in all subsequent work. Other authors have stressed the importance of these points; see, for example, refs. 2–4, 17, and 185.

As noted above, it is likely that analysts will frequently meet requirements where it is not possible to decide the sizes of one or more potential errors. We would make two suggestions for such situations. First, the general aim should be to err on the side of safety. That is to say, if there is a choice to be made between, say, two types of sampling device, one of which is much easier to use than the other, we suggest selection of the device for which errors are thought to be smaller, whether or not this is the more convenient to use. Second, a careful note should be made of any potential errors whose sizes are not known, and strenuous efforts made to estimate experimentally their sizes, preferably before a sampling system is applied routinely. It may well be impossible for the analyst to make these tests himself, but we suggest that he strive to ensure that they are then made by others.

Note also that the sizes of errors caused by sampling systems are not necessarily constant in time. Many factors that may affect errors vary themselves with time, *e.g.*, sampling personnel, contamination at the sample-delivery point, the amounts of materials and biological species deposited and sorbed within sample-collection devices, deterioration (*e.g.*, corrosion) of components of sampling devices, *etc.* Two points follow from this. First, a regular programme of inspection, maintenance and, when necessary, replacement of components of sampling devices should be arranged. Second, this programme should be accompanied by a programme of regular testing to estimate the sizes of as many errors as possible with the aim of confirming the adequacy of the former programme. Both programmes should be regarded as an integral part of a routine sampling programme as other authors have also recommended; see, for example, refs. 3, 4, and 186 and Section 3.5.5.

(vi) Sampling personnel as components of sampling systems The personnel involved in the collection and preparation of samples are a very important component of sampling systems. Accordingly, everything possible should be done to ensure that they faithfully follow the procedures specified by a sampling method, and that they receive appropriate training and instruction as an aid to the achievement of this aim;[3] see also Section 10.3. Automatic sampling procedures not requiring human intervention have the advantage[127] of eliminating one component of sampling systems that may cause difficulties, but this benefit may be accompanied by other dangers; each situation should be considered individually.

(vii) Special sampling procedures for particular determinands Certain determinands, *e.g.*, dissolved gases and highly volatile organic compounds, often require special sampling procedures that are normally specified and described in the relevant analytical methods. We have made no attempt, therefore, to discuss such procedures in detail.

3.5.2 Problems Associated with the Design of Sampling Devices

(a) Types of Problem
Satisfactory sampling devices should meet three criteria which, though obvious, are worth emphasis.

(1) A sampling device should not affect the concentrations of determinands at the sampling point in a water-body. This is discussed in *(b)* below.
(2) A sample-collection device should be designed and constructed in such a way that the concentrations of determinands entering it are the same as those in the water-body at the time and point of sampling. This is discussed in *(c)* below.
(3) A sample-collection device should not affect the concentrations of determinands in the water once it has entered the device; see *(d)* below.

As noted above, there are two types of sample-collection device, *i.e.*, sampling lines and sample-collection containers. Both types are subject to many essentially identical problems, and are often considered under the general title of sample-collection devices in *(c)* and *(d)* below. Note also that much of the discussion of sample containers in Section 3.6.2 may also be relevant to such devices.

(b) Effect of Sampling Devices on the Water of Interest
Any component of a sampling device that is not normally present in the water-body may affect the concentrations of determinands in the water of interest through two main effects: (1) by disturbance of physical, chemical and biological processes and equilibria, and (2) by its contaminating effect on the water.

A simple example of (1) arises when one wishes to obtain a sample from a particular part of a water-body, and the sampling device causes mixing with water of different quality from other parts of the body. For this reason, caution has been urged[5,6] in using helicopters as bases from which to collect samples from the surface layers of stratified water-bodies; the down-draught from the blades of the helicopter may cause mixing of the surface and lower layers. A further example is provided by van Duijvenbooden[208] who notes that the drilling of a well may disturb the quality of groundwater around the well. Depending on the hydrogeological conditions, several weeks to more than a year may be required before the original water-quality is restored, and in this period sampling is thought by the author to be useless.

In considering the effect (2) – contamination – it is important to bear in mind that *all* components of a sampling device are potential contaminants, and that devices suitable for certain needs may not be suitable for others with more stringent requirements for accuracy. Just as for the control of contamination in analysis (see Section 10.8), so too for the control of contamination from sampling devices there is no substitute for critical selection of all components of a device and for experimental tests of the degree of contamination caused by a device. We realise that this will often not be easy, but this does not change the importance of the points. The contamination of the water *within a sample-collection device* is discussed in (*d*) below; here, we simply quote some examples of important contamination effects caused by sampling devices on the water of interest.

When a boat or ship must be used as a base from which to collect samples, the boat is part of the sampling device, and may cause several types of contamination. Such effects and measures to overcome them are described in many papers, particularly those concerned with sea-water analysis; see, for example, refs. 126, 188, and 189.

In many situations, a sample-collection container must be used at depth in a water-body, and is lowered into the water on a cable. Many authors have stressed that the cable may cause contamination, and that its material should be chosen carefully, *e.g.*, bare stainless steel should not be used when determining trace metals in sea water[126] – though see also ref. 183.

As a final example, we mention the possible contaminating effects of drilling fluids and materials of construction of sampling wells when samples of groundwater are required.[3,6] The use of PVC for the well-casing is recommended or adopted by many authors, but this may cause contamination when trace organic compounds and possibly certain trace metals are of interest.[3,72,184] Lee and Jones[184] recommend vigorous pumping of water from sampling wells before collecting samples in order to minimise such problems.

The above examples show the range and complexity of problems that may occur, and also emphasise the importance of experimental tests whenever possible to check the extent of contamination. The need for such tests has been increasingly voiced by many authors in recent years; see, for example, refs. 3, 5, 6, 126, 187, and 190–192. See also Section 3.5.1(*b*)(*v*).

(c) Effect of the Inlet to the Sample-collection Device

There is usually no problem in satisfying condition (2) in (*a*) above except when the determinands consist of undissolved materials with densities different from that of the water. When such materials are present, the rate of withdrawal of the water through a sampling line may affect their concentration in the water entering the line. This effect arises because the undissolved materials tend (as a result of their different inertia) not to follow the streamlines of the water as it enters the inlet to the sampling line. (Although the velocity of withdrawal may be considered more appropriately dealt with in Section 3.5.3, the effect is considered here as it also has important implications for the design and installation of the sampling line.) Badzioch[193] has discussed this effect in detail for solids in gas streams, and its implications for natural waters[3] and water in power-stations[194] have been considered. The effect is illustrated diagrammatically in Figure 3.3 for the case of a flowing stream of water containing particles whose density is greater than that of the water. It can be seen that when the velocity in the sampling line is greater than that of the water of interest, the concentration of particles in the water entering the line will tend to be falsely small whereas the reverse is true when the velocity in the line is smaller than that of the water being sampled. Ideally, the velocity of the sample stream should equal the velocity of the water being sampled as the latter flows past the inlet to the sampling line; this condition is referred to as '*iso-kinetic sampling*'. The errors arising from deviations from this ideal velocity depend on several factors including the size and shape of the particles and their density in relation to that of the water, the velocities of the water of interest and the sample stream, and the size and shape of the inlet to the sampling line. Although failure to achieve iso-kinetic sampling may cause negligible error in some situations,[3] it seems generally desirable to aim for at least approximately iso-kinetic conditions when using a sampling line in order to minimise the possibility of this source of error. The sampling of suspended solids and iso-kinetic aspects are considered in detail in ref. 3. The need for iso-kinetic sampling of waters and wastewaters has also been

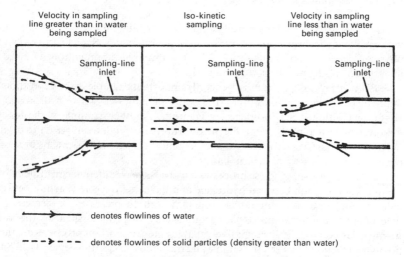

Figure 3.3 *Iso-kinetic sampling*

stressed by the United States Geological Survey,[15] Shelly,[195] and Reed,[196] and a similar recommendation to ours above has also been included in refs. 5, 6, and 55.

For the same reason, when undissolved materials are of interest in a flowing stream of water, the inlet of the sampling line should face into the water flow.[3,5,6] Thus, the practice of simply joining the sampling line to the wall of a pipe to be sampled is generally undesirable [see also Sections 3.2.1(b)(ii) and 3.2.3]; rather, the sampling line should project into the pipe with its inlet(s) facing upstream. Ref. 3 suggests that the inlet should be within $\pm 20°$ of facing directly upstream.

The implications of this effect when static water-bodies are sampled using a sampling line appears not to have been investigated though it would seem that errors could again arise. Experimental investigation of the effect of rate of sample-abstraction may well be helpful in these situations, but caution in interpreting the results is necessary because the velocity of the sample-stream may also cause other effects considered in (d) below. It appears that the effect may also occur when discrete samples are collected in sample-collection containers; see ref. 15.

A quite different type of effect arises when sample-collection containers are used to collect samples from depth in water-bodies. Some containers are so designed that, as they are lowered through the water-body, the water passes through them until they reach the desired depth for sampling [see Section 3.5.4(a)(i)] at which point they are sealed. Thus, much of the water entering the container is not that of which a sample is required, and the possibility arises that materials in the water that is not of interest may remain in the container [see (d) below] and contaminate the sample. Again, we know of no quantitative studies of this effect. Fletcher and Polson,[197] however, have also noted its potential importance, and cited it as one reason for preferring sample-collection containers that are *opened* and closed only at the desired sampling point.

(d) Effects within the Sample-collection Device

(i) Types of effect   Many processes may occur within sample-collection devices with resultant effects on the concentrations of determinands. Such processes can be physical, chemical, and biological, and a few examples are mentioned below as illustrations of the problems that may arise.

(1) Physical processes. The most generally important example is the deposition of undissolved, solid materials onto the walls of a device when hydrodynamic conditions within it are insufficiently vigorous. Another example is the loss of dissolved gases that may occur if the pressure decreases while a sample is in a device. Sorption of surface-active substances is another possibility. Such processes generally lead to losses of determinands.

(2) Chemical processes. These processes may be between different constituents of a sample, e.g., reaction between hydrazine and dissolved oxygen, reaction between chlorine and organic compounds, and may lead to increases or decreases of determinand concentrations. Unless an unusually long time is spent by the water within a device, however, it seems likely that the more important processes are those involving reactions between the sample and the sample-collection device itself. Such processes may lead to increased determinand concentrations when determinands

are leached from the device (*e.g.*, copper from copper pipes, organic compounds from plastic materials). Decreased concentrations may result from direct reaction of the determinand with materials of construction of the device, *e.g.*, dissolved oxygen can react with mild steel or copper.

(3) Biological processes. Especially when natural waters are sampled, various biological species may settle on the surfaces of the sample-collection device, and then react with constituents of the sample. For example, large proportions of the ammonia in a sample may be oxidised by bacterial films with simultaneous release of other nitrogen species into the sample. The general metabolism of biota may also release various substances into the sample, *e.g.*, dissolved oxygen and organic substances. The sudden detachment of such biological films may also lead to unacceptable samples.

Effects such as these can be important for both sample-collection containers and sampling lines, but the permanent installation and relative inaccessibility of sampling lines may well aggravate these difficulties. Whichever of the two types of device that is used, certain general precautions are useful, and are noted in (*ii*) to (*v*) below; see also Section 3.6.2 on sample containers.

(ii) Contact time with water　It is generally useful to minimise this contact time. When sampling lines are used, therefore, their lengths should be as short as possible, and a high linear velocity through them will usually be advantageous – subject to any requirements on iso-kinetic sampling. Recommended minimum velocities of sample streams range from 0.3 to 0.6 m s^{-1} for several dissolved determinands.[17] Deposition of suspended materials in sampling lines carrying natural waters can be a particular problem, and a number of suggestions have been made for the velocities necessary to prevent this source of error. The values quoted vary somewhat from author to author – as might be expected since the required velocity depends on the properties of the particles, and the diameter and geometry of the sampling lines. Relevant theoretical considerations are discussed in refs. 3, 14, and 195, and velocities in the range 0.6 to 3.0 m s^{-1} have been suggested.[3,17,121,195,196,198,199] Deposition of impurities in sampling lines used for high-pressure steam can also be important, and special devices are needed to overcome this problem.[17]

Unusually long sampling lines (as may be required, for example, when all the outlets of a number of sampling lines are brought to the same sample-delivery point) are best regarded with suspicion until confirmation of their suitability has been obtained. Experimental tests to assess the effects of variations of the velocity of a sample stream around the velocity tentatively chosen are useful in revealing potential sources of error, but the effect noted in (*c*) above must be considered in interpreting the results for undissolved materials.

It is also worth noting that longitudinal mixing of the sample-stream in a sampling line will tend to smooth out temporal variations of quality at the sampling point in the water-body. This effect will commonly be unimportant, but its magnitude will increase as the length of the sampling line increases and as its design encourages mixing.

(iii) Surface/volume ratio　It might be expected that this ratio would sometimes be an important factor, the effects due to interaction between the sample and the

sample-collection device tending to become less important as the ratio decreased. Thus, the diameter of sampling lines or the volume and shape of sample-collection containers may sometimes have a role to play. We know, however, of no general studies of this aspect of sampling lines; for sample-collection containers, see Section 3.6.2(*c*).

(iv) Materials of construction These materials should be chosen so that contamination of the sample is not expected to be a problem. Plastic materials are often suitable, particularly when trace amounts of metals are of interest, but they may need extensive preliminary cleaning [see Section 3.6.2(*b*)] and/or use to ensure that negligibly small concentrations of determinands are leached from them. Though Teflon is often regarded as a particularly good material from this point of view, the presence of metallic inclusions in the walls of Teflon equipment has been reported.[197,200,201] Such potential sources of contamination are very likely to vary from one application to another. Plastic materials are usually unsuitable when organic compounds are of interest in samples since many authors have reported the leaching of various organic materials from a range of plastics in contact with water; see, for example, ref. 3 and 202–204. Under these circumstances, metal[190] or glass devices seem preferable for organic determinands though the fragility of the latter will often preclude its use for sampling lines; Teflon lines are recommended in ref. 3. Stainless steel is often suitable for many applications, and has also commonly been recommended for sampling waters at high temperature and pressure. Note though that Pebler[205] has reported that a number of dissolved gases may react with stainless steel unless it is passivated; he recommends monel as an alternative to stainless steel.

Clearly, each sampling situation should be considered individually since the choice of material(s) for sample-collection devices is governed by the determinands of interest and their concentrations, the tolerable inaccuracy of results, the nature of the water of interest, and the environment in which the device will be used. It is always desirable to check that the selected device (whatever its nature) does not cause appreciable contamination of samples (see also Section 10.8). This suggestion has been increasingly made by many authors in recent years; see, for example, refs. 3, 5, 6, 126, 187, and 190–192, and Section 3.5.1(*b*)(*v*).

Finally, it is worth noting a rather unusual type of contamination in sampling water for the determination of small concentrations of dissolved oxygen. Everitt *et al.*[206] showed that many plastic materials, commonly used for flexible tubing in sample-collection devices, exhibited appreciable permeability to oxygen from the air with resultant contamination of sample streams. Their work was primarily concerned with power-station waters, but the same effect has been reported[207] in sampling groundwater.

(v) Cleanness of internal surfaces Sample-collection devices should be kept clean, particularly with regard to undissolved solid materials and biological films on surfaces. Opaque devices are generally desirable for waters in which algal growths may occur; the exclusion of light from the sample helps to prevent the multiplication of algae on surfaces. The ability to maintain clean surfaces is also aided by simple design. Thus, smooth surfaces and, in sampling lines, the absence of bends and stagnant areas are all likely to be of value.[3]

3.5.3 Problems Associated with the Manner of Operation of Sampling Devices

(a) Eliminating the Effect of a Sampling Device on Water Quality

Whenever the presence of a sampling device in a water-body can affect the water quality at a sampling point, it will generally be necessary to attempt to abstract sufficient water from that position at such a rate that this effect of the device is at least temporarily eliminated. Once this has been achieved, a valid sample or sample stream may be obtained. The main use of this approach of which we are aware arises in the sampling of groundwater.

When sampling groundwater from wells that are not pumped continually, most authors recommend that several well-volumes at least should be pumped before beginning the sampling; see, for example, Nacht.[383] The aim is to remove stagnant water which is unrepresentative of that in the aquifer. Lee and Jones[184] suggest that detailed tests are needed to determine the volume of water that must be pumped before sampling, and that these tests should be made twice a year for the first 3 years of operation of a well. Keely,[209] however, suggests that, in some situations, it may be useful to collect a series of samples at approximately logarithmically spaced time-intervals beginning immediately pumping is started. The patterns of temporal variation observed for different determinands may then provide useful information on the sources of contaminants in the water. Clearly, the best procedure needs, in general, to be established for each application.

(b) Use of Sample-collection Devices

(i) Sampling lines   Water containing undissolved solids may produce deposits in the water-body, in parts of the sampling device, and in the sampling lines if the velocity of the sample stream through the last is too small. When such deposits occur, the flow disturbance caused by suddenly allowing a sample stream to flow through a sampling line may re-entrain some of the deposited material so that its concentration in the sample is falsely high (see also the next paragraph). The cross-contamination of successive samples for suspended solids has been reported through such effects.[210] Thus, when undissolved materials are of interest, the flow conditions during sampling should, in general, be standardised, and an adequate volume of sample run to waste before any is collected.[3] Opening taps and valves in the sampling line once sampling has begun is particularly to be avoided. The flushing of sampling lines before *and* after each sampling occasion is recommended in ref. 3 as a means of keeping the lines clean. Generally desirable also is a regular programme of inspection and maintenance of sampling lines to the maximum extent possible. In some applications, a permanent flow of sample is advantageous in helping to ensure equilibrium between the sample stream and the sampling line. On the other hand, when the water being sampled is liable to cause biological films within the line, this undesirable effect may well be aggravated if the sample stream flows continually. Indeed, this effect has led to the suggestion that a sampling line should be removed from wastewaters when samples are not required.[211] Once again, given such competing requirements, each situation needs to be considered individually.

There is at least one important application – the sampling of drinking water from consumers' taps – where special consideration is required. Many workers have shown that when water is first run from a tap in the early morning (after an overnight stagnation period), the concentrations of several metals derived from plumbing materials (*e.g.*, lead, cadmium, copper, zinc) may be unusually high for a short period and then rapidly decrease as water continues to flow from the tap; see, for example, ref. 212. Depending on the objectives of the measurements, it may be necessary to sample this 'first-draw' water immediately the tap is opened. The size, duration, and reproducibility from occasion to occasion of this transient effect may depend on a number of factors, and this has important implications for the precise sampling procedures that should be adopted, *i.e.*, rate of opening the tap, flow-rate of water from the tap, *etc.* The appropriate procedure for this purpose is crucially dependent on the exact objectives of the measurements, and, once more, each application requires individual consideration; see also (*c*) below.

(ii) Sample-collection containers   Containers are also subject to the accumulation of undissolved materials and biological species on their internal surfaces, and the suggestion in (*i*) of a regular programme of inspection and maintenance is equally apposite for these containers.

(c) Volume of Sample Collected
For many purposes, the volume of sample is unimportant provided that it is sufficient to allow the required analyses to be made. There are, however, some situations where the volume may affect the representativeness of samples for analysis.

(1)   When contact of the sample with air is to be avoided, sample-collection containers and sample containers should be completely filled. This need is usual when dissolved gases and other determinands affected by them are of interest; it arises also for volatile substances whose vapours may pass into the air-space of an incompletely filled container, *e.g.*, certain organic compounds.[384] For such determinands, special containers of both types are often necessary, and are described in the appropriate sampling and analytical methods.

(2)   When undissolved materials, which settle when a sample is allowed to stand, are of interest, it is generally desirable not to fill either type of container completely.[3] This will allow better agitation of samples and re-mixing of undissolved materials before withdrawing any portions of a sample for sample-preparation or analysis. Errors arising in such withdrawals are discussed in ref. 3. Of course, (1) and (2) are incompatible, so that sampling procedures are further complicated.

(3)   When small concentrations of materials present as relatively large undissolved particles are of interest, it may be necessary to stipulate a minimum volume of sample to be collected *and* analysed in order to control errors arising from statistical variations in the number of particles in a given volume of sample. For example, suppose that the water of interest contained, on average, ten particles per litre. Samples of 100 ml would not then all contain one particle; some would contain none, some one, some two, *etc.* The relative variability of the number of particles decreases with increasing sample volume. If desired,

calculation of the importance of this effect can be made on the basis that the number of particles in a given volume will follow the Poisson distribution; see, for example, ref. 96.

(4) Whenever the quality of the water of interest changes rapidly in relation to the time during which a sample is collected, this time – and hence the sample volume – may affect the concentration of a determinand in a sample. In many applications this effect is unimportant, but there is at least one important instance, already mentioned in (*b*) above, where the effect may require detailed consideration, *i.e.*, when 'first-draw' samples of drinking water from consumers' taps are required. The relatively high concentrations of some metals that may be found on first opening a tap may decrease so rapidly that the volume of sample collected has a very marked effect on its metal concentrations. As before, when such effects are appreciable, the appropriate sample volume can be decided only from the exact purpose(s) of the measurements.

(d) Rinsing of Containers with Samples

Many publications recommend that each sample-collection container and sample container be rinsed two or three times with sample before finally being filled. This is sometimes a reasonable practice, but it may lead to errors when undissolved materials – and perhaps also readily adsorbed substances – are of interest. Some or all of such materials in the portions of sample used for rinsing containers may be left in them, and thus cause contamination of the final samples. Rinsing containers may also cause problems when it is desired to collect the sample directly into a stabilising reagent [see Section 3.6.3(*d*)]. In general, therefore, unless there is knowledge to the contrary, it seems to us better to adopt the approach of ensuring that both types of container are adequately clean and free of water so that samples may be collected directly into them without the need for rinsing. For virtually all purposes, water remaining in containers after cleaning in a laboratory can be adequately removed simply by inverting them and, preferably, shaking them vigorously a few times. Note that ref. 3 proscribes rinsing containers with the sample when trace organic compounds are of interest. See also Section 3.6.2(*b*) on the cleaning of sample containers.

(e) External Contamination at the Sample-delivery Point

Contamination of samples from the environment (including sampling personnel) around a sample-delivery point can be important when trace impurities are to be determined. This problem has been stressed[3,213] when samples of drinking water are collected from consumers' taps, and the provision of special 'clean-areas' has been reported by many authors for handling samples of sea-water on board ships; see, for example, refs. 126 and 189. Sample-delivery points should always be carefully inspected for possible sources of contamination so that appropriate precautions can be taken. Regular tests of the importance of contamination are also useful, and are discussed in Section 3.5.5. See also Section 10.8 on contamination in general.

3.5.4 Commonly Used Sampling Systems

Almost all the many compilations of analytical methods for one or more types of water (see Section 9.5) include one or more sections in which particular sampling systems and their use are described. Similar treatments are provided in many other publications dealing with various aspects of water quality, for example, refs. 3, 5–7, 52, 53, and 68. In addition, many organisations have issued documents concerned solely with sampling systems for waters; see, for example, refs. 15, 17, 55, 72, 126, 187, 195, and 213a. In view of all these sources, we have made no attempt here to describe and discuss all the many sampling systems that have been and are used. Our aim is rather to note briefly the main types of system in common use, and to provide references to recent and/or interesting applications so that readers may pursue particular points if they wish. To this end, we have divided sampling systems into two types: manual and automatic. For the first, a person must visit a sample-delivery point on each occasion a sample is required; this is considered in (*a*). Automatic systems collect a number of samples without requiring the presence of a person on every sampling occasion, and these are discussed in (*b*). Of course, all the general points made in Sections 3.5.1 to 3.5.3 apply to both types.

(a) Manual Sampling Systems

(i) Systems involving sample-collection containers For many purposes, specially designed and installed sampling devices are not required when natural waters are of interest. It often suffices simply to immerse a bottle in the water of interest, and this technique may be applicable also for some purposes in water-treatment plant. Of course, this procedure is often extended in the sense that a relatively large container is thus filled with water, and then used to fill a series of sample containers. When this type of system is used, it may often be advantageous to use some form of frame or line (non-contaminating materials) to hold the sample-collection container so that it can be immersed away from the immediate position of the person collecting the sample. By this means, contaminating and other undesirable effects of the site on which he is standing (or the boat, *etc.*, from which sampling is undertaken) can be minimised.

When it is necessary to sample from a particular depth in waters where this simple technique cannot be used, special sample-collection containers (sometimes called 'depth-samplers') are available that can be lowered into the water on a cable, and which collect a sealed sample at the required depth. Many such devices are described in various publications; see, for example, refs. 3, 6, 7, 52, 53, 68, 72, 126, 136, 190, 191, and 214. When trace organic compounds are to be measured, glass containers mounted on special frames have been used.[215] Riley[190] stresses the importance of using containers with very positive sealing mechanisms to prevent mixing of water in the container with water from smaller depths as the container is brought back to the surface.

When special depth-samplers of this type are used, it is usually necessary to transfer samples from them to one or more sample containers; this transfer may be a source of problems [Section 3.5.3(*c*)]. Further, if such samplers are sealed only when they reach the desired depth, other problems may arise [Section 3.5.2(*c*)]. Various approaches have been described for overcoming such difficulties, usually by employing the one container for both sample collection and sample storage before analysis, and by arranging that it is opened only at the required depth and then subsequently sealed again; see, for example, refs. 197 and 215.

Essentially the same principles of construction and use are often applied to sample collection containers for sampling from particular depths in boreholes and sampling wells; the containers are then sometimes referred to as 'bailers'. For this application, however, the limited diameters of many sampling wells usually require specially constructed sample-collection containers; see, for example, refs. 3, 64, 68, 72, 207, and 216. In sampling groundwater, it is also important to bear in mind, as many authors have stressed, that the water in an aquifer is not, in general, in equilibrium with the temperature and pressure at the well-head. Since the concentrations of dissolved gases such as oxygen, carbon dioxide, and hydrogen sulphide in water – and the speciation of other substances (*e.g.*, calcium, trace metals) affected by such gases and related factors – depend on temperature and pressure, extreme care is commonly necessary to ensure that potential problems caused by sample instability are overcome. For example, prevention of contact of samples with the atmosphere may well be required for a number of waters and determinands. Examples of procedures and equipment used to deal with such problems are provided in refs. 3, 68, 72, 217–219, 243, and 383. Other means of obtaining samples of groundwater are considered in (*ii*) below, but it is useful to note here that a quite different approach may be of value, *i.e.*, the extraction of the interstitial water from core samples obtained by drilling; see, for example, refs. 3 and 64. The extraction by centrifugation of the relatively small volume of water usually contained in appropriate lengths of such cores is described in refs. 220 and 221. Of course, this procedure is not suitable for routine, long-term measurement programmes, and the small volumes of samples from such cores may cause analytical difficulties.

(ii) Systems using sampling lines Sampling lines are commonly required in water-treatment and other plant, and are also employed in a number of applications for natural waters. Descriptions of a variety of lines are given in many publications; see, for example, refs. 3, 5, 6, 17, 195, 213, and 222.

In some systems, the flow of sample through the line is achieved by the natural pressure differential between the water-body and the outlet of the sampling line. In others, the sample must be pumped, sucked (by vacuum), or pressurised (by a gas) through the line; many different arrangements have been used, and it may often be merely a matter of convenience which governs the particular procedure adopted. The following points, however, should be noted. When dissolved gases and volatile organic compounds and possibly other determinands whose chemical forms and concentrations may be affected by dissolved gases are of interest, it is generally desirable to ensure that the sample is slightly pressurised to prevent gases coming out of solution. Thus, suction-type pumps and suction by vacuum should generally be avoided for such purposes. Losses of volatile organic substances when samples are taken with use of peristaltic pumps have been studied by Ho.[384] Riley[190] discusses various pumping procedures, and points out that care is required to ensure that the pump does not contaminate the water nor damage particulate material and biota when these are of interest. The pump must also be capable of giving the required rate of flow of sample.

Sampling lines for natural waters such as rivers should be carefully selected and installed to minimise blockage of their inlets by coarse debris in the water. The practice of facing the inlet downstream to reduce such problems should, in general, be avoided if undissolved materials are to be measured [see Section 3.5.2(c)]. It seems to be better to rely on relatively fine screening systems placed around the inlet *and frequently cleaned.*[3] The need for pumps to provide a sample stream in such systems can cause difficulties affecting their installation when river levels fluctuate markedly. Care is also necessary to protect sampling lines against damage by boats and vandalism. These and related problems are discussed by many authors – see, for example, refs. 17 and 223.

Several types of sampling line have been used for groundwater. These are reviewed in refs. 3, 64, 68, and 72, and only one approach is mentioned here. If it is required to sample at a particular depth in a sampling well with minimal effect from the water at other depths, a pair of inflatable 'bellows' (or 'packers') may be lowered down the well to the desired depth and then inflated. This procedure isolates the water of interest which may then be sampled by means of a sampling line.[3,64,224] Of course, the materials used for the bellows and their support must be chosen to avoid contamination. See also refs. 184, 225, and 226.

The flow-rate (and related factors) in, and materials of construction of, sampling lines may be very important factors affecting the representativeness of samples (see Sections 3.5.2 and 3.5.3).

The design of sampling devices that may operate at chosen depths while being towed by a ship is described in refs. 29 and 227.

(iii) Sampling the surface layers of water   When there is direct interest in investigating the thin surface film in which a number of substances may concentrate, special sampling devices are required. These techniques have been recently reviewed by Baier.[228]

(b) Automatic Sampling Systems
These systems consist, in general, of a sampling device (incorporating a sampling line) and a unit (installed at the sample-delivery point) which: (1) automatically controls the timing of the collection of a series of samples and (2) houses the appropriate number of sample containers. Such systems have been reviewed by a number of authors; see, for example, refs. 3, 15, 121, 195, 196, 198, 199, 229, and 230. The control unit usually provides the ability to vary several factors such as the number of samples in a given time period, the length of that period, and the time period over which each sample is collected. Some units also allow sample collection to be based on the flow-rate of the water of interest rather than time. One common example of the application of such systems is the collection of 12 or 24 samples in a period of 24 h, portions of the individual samples subsequently being used to prepare one composite sample [see Section 3.3.2(e)] for analysis. Automatic systems can be invaluable for many purposes, *e.g.*, studying variations in quality, obtaining samples at inaccessible sites, preparing composite samples.

Since these systems generally involve a sampling line providing a sample stream as required, they are liable to the various problems associated with sampling lines (see Sections 3.5.2 and 3.5.3). It appears that many of the early systems may well have provided poor samples (sometimes grossly unrepresentative) through failure to achieve: (1) a reasonable approximation to iso-kinetic sampling, (2) inadequate flow-velocity to prevent deposition of undissolved materials in the sampling line and control unit, and (3) prevention of the development of biological films within the sampling line and control unit. Detailed investigation of such effects in recent years has led to improvements in the design and operation of automatic sampling systems, but very close attention is still required to the various sources of errors identified in the sections noted above. Schofield[121] notes, for example, that many of the commercially available systems are still not good for suspended solids. Detailed consideration of, and recommendations on, such aspects have recently been given in ref. 3.

A further problem to which insufficient attention appears to have been given in initial applications is that of changes in the concentrations of determinands in samples in the period between sampling and analysis (see Section 3.6). For many determinands, this problem may be minimised by adding stabilising reagents to the empty sample containers, but for other determinands it appears that the only practicable approach is to ensure that the sample containers are stored in a refrigerated/heated compartment maintained at approximately 4 °C. Even this precaution may not ensure satisfactory sample stability, and the problem requires very careful consideration. In any event, it should also be noted that not all commercially available units have this facility, and that even when it is provided, there remains a need, in general, to check the stability of samples (see Section 3.6).

Automatic systems of this type have the valuable feature that they can relatively easily be arranged to begin collecting samples only when certain unpredictable events occur, *e.g.*, when the flow of water in a channel exceeds a given value. Many publications describe such applications in which a signal from an instrument responding to the event of interest, *e.g.*, a flow recorder, is used to activate the automatic sampling system; see, for example, refs. 231 to 235. Such an approach can also be used to advantage when some determinands in a water are measured continuously. Provided that good correlation can be established between other determinands and those measured continuously, valuable economy of analytical effort may be achieved by using the signals from the instruments to initiate sample collection for those determinands that must be measured manually in the laboratory. Such automatically triggered sampling systems are also highly desirable for the collection of samples of rainwater,[187] the system being opened to the atmosphere only when rain is falling; contamination of the samples from dust during dry periods is thereby minimised. See, for example, refs. 235–238 and 385.

Of course, *all* the materials of construction coming into contact with samples should be such that no important contamination occurs.[3] Most if not all of the commercially available units seem not to have been designed from this point of view for sampling for trace organic compounds. A system designed for this purpose and constructed essentially of glass and Teflon has been recently described.[239] Finally, ref. 240 gives a detailed account of a system designed to prevent the loss of volatile organic compounds, *e.g.*, halomethanes, from samples.

An automatic sampling device has been reported[241] for the filtration of sea-water when the particulate fraction of trace metals is of interest.

An interesting example of what amounts to an automatic means of obtaining a composite sample [see Section 3.3.2(*e*)(*ii*)] of drinking water drawn by consumers from domestic taps has been described by Haring.[241a] The device consists of a small plastic unit which is attached to a tap. The unit provides two outflows, one for the use of the consumer and the other (a small proportion of the total flow) is diverted to a sample-collection container. By this means, the composite sample provides, in principle, a method of assessing the average concentrations of determinands in the water used by consumers. This may be of particular value for determinands (such as metals from plumbing materials) subject to large temporal variations in concentration. The unit also contains a simple switch which shuts off the sample stream when desired. Provided, therefore, that the consumer is prepared to operate this switch, the composite sample may be made to represent only the water drawn for consumption. Brief mention of the use of this device in the U.K. by the Water Research Centre is made in ref. 212 where the performance of the device is said to be technically satisfactory though inconvenient. We know of no detailed study of the importance of the various factors – considered in Sections 3.5.2 and 3.5.3 – that may affect the representatives of samples.

A final and important point is that protection must be provided for automatic sampling units when they are placed in environments that may have adverse effects on their operation, *e.g.*, rain, snow, dust, *etc.*; protection against vandalism may also be necessary.

3.5.5 Routine Control and Checking of Errors Caused by Sampling Systems

(a) Introduction
In Section 3.5.1(*b*)(*v*) we noted the need for regular programmes of: (1) inspection and maintenance of sampling devices, and (2) tests to estimate the magnitudes of errors caused by sampling systems; both programmes were also suggested to be integral parts of any routine sampling programme. In Section 3.5.1(*b*)(*iv*) we also observed that there are, in general, some basic difficulties in estimating the errors arising in sampling systems. In the present section, therefore, we wish to consider this estimation of errors in a little more detail.

We consider in (*b*) some general precautions to be observed in planning a routine programme of tests, and discuss in (*c*) several types of test that may be useful. We stress, however, that the execution of such a programme in no way lessens the importance of the inspection and maintenance of sampling devices. These latter aspects – given a sound initial choice and assembly of a sampling system – are the basic means of control of errors, and the programme of tests should be seen as primarily a means of attempting to confirm the adequacy of that control.

(b) Planning a Routine Programme of Tests
A number of factors must be considered in planning any programme of tests, *e.g.*, the accuracy required of routine analytical results, the frequency of sampling, the man-power available, *etc.* Here, we wish to stress only three technical points that seem to us of particular importance in the present context.

First, we have already suggested that the general aim should be to use sampling systems that lead to no systematic differences between the concentrations of determinands in the water samples and the samples; see Section 3.5.1(*b*)(*v*). When this aim is achieved, it is likely that any random differences will be negligible.

Accordingly, the emphasis in the tests should be on the detection and estimation of bias.

Second, the sources and sizes of bias depend on the determinand and its chemical and physical forms present in the water of interest, and on the sampling system. Thus, for a given system, tests appropriate for estimating bias for one determinand may be inappropriate for another determinand. Ideally, critical consideration of the sources of bias for each determinand should have been made before routine application of a sampling system [Section 3.5.1(*b*)(*v*)], and this provides an excellent basis for identifying the most useful tests for each determinand. We make this point because effort will seldom if ever be available to make all the tests ideally desirable, and priorities will almost certainly have to be assigned to particular determinands and tests. For example, it may well be preferable to concentrate effort on only those determinands most subject to problems, *e.g.*, undissolved materials.

Third, analyses of samples are generally required in estimating the errors caused by sampling systems. Unless the errors caused by the analytical system are known, it is impossible to estimate the sampling errors. It will usually be the analytical precision that is crucial in this respect, but the possible effects of analytical bias should be considered when planning the tests (see Section 4.6.1). The effect of analytical precision on the ability to detect and estimate bias is discussed in Section 4.6.2. That section emphasises two points:

(1) statistical techniques are essential in interpreting the results from tests on sampling systems;
(2) there may well be little point in making any tests if the analytical precision (or practicable degree of replication of the analyses) is not sufficiently good to allow detection of bias of the magnitude of interest.

To decide if precision is adequate, therefore, it is necessary to decide the size of bias that any test is required to detect. This decision should be governed by the maximum tolerable bias in analytical results (see Section 2.6.2). These points seem to us indisputable, but their consequences in terms of the required analytical precision or degree of replication may often be alarming. Nevertheless, we strongly recommend that these considerations be made when planning tests since there seems little point in devoting effort to tests only capable of detecting a bias of, say, 50 % if a bias of 5 % were considered intolerable. This aspect of testing reinforces the vital role played by the inspection and maintenance programme.

(c) Types of Test that can be Used
We consider here only tests concerned with those parts of sampling systems involved in producing 'samples for analysis'; tests of sample stability are discussed in some detail in Section 3.6.4. Our discussion is also restricted to tests that we feel may be practicable as a regular and integral part of routine sampling programmes, and that are capable of providing, in principle, unambiguously interpretable results. For these reasons we have excluded the following three broad types of test though they may well be of value in estimating errors caused by sampling systems:

(1) tests* in which one sampling system is compared with one or more other systems of the same or different type;

* Tests of this type are recommended for preliminary investigations of sampling systems in ref. 186, but see also Section 5.3.

(2) tests involving the use of special 'test-rigs';
(3) tests* relying on the spatial and/or temporal constancy of the water quality at a selected sampling point.

These restrictions limit the applicable tests to two types considered in (*i*) and (*ii*) below; some concluding remarks are made in (*iii*).

(i) Tests of sample-collection devices When a solution for which the concentrations of one or more determinands are known with negligible error can be taken to, or near, a sampling point, and then used in place of the water of interest as the input to a sample-collection device, analysis of the resulting sample(s) acts as a direct test of bias. Of course, the manner in which the solution flows into the device should closely simulate that obtaining in routine use of the device. This approach may often be applicable for sample-collection containers; whether or not it is useful for sampling lines depends on several factors such as the accessibility of the inlet to a line and the ability to achieve the normal velocity of the sample stream in the line. It may well be impracticable adequately to simulate normal conditions when undissolved materials in flowing waters are of interest, and each situation requires individual consideration. The test has the advantage that the samples obtained of the test solution are subject to many – for some sampling systems, all – sources of error arising in sampling. The need to transport what may well be large volumes of the test solution is, however, a disadvantage.

When contamination is thought to be the dominant source of error, the test solution should preferably contain negligible concentrations of determinands. Such a solution will, however, be of no value when losses of determinands in the sample collection device are the main source of bias. In this event, since losses are likely to depend on the concentration, it will usually be desirable to use two test solutions with determinand concentrations of approximately $0.1C_G$ and C_G, where C_G is the greatest concentration of interest in the water of interest. The nature of the test solution also needs careful thought. Ideally, its general composition and the chemical and physical forms of the substances in it should correspond to those normally present in the water of interest since these factors may affect any bias arising in sampling. There are likely to be, however, important difficulties in measuring the concentrations of a number of determinands in a large volume of this water with negligible error, and in ensuring that these concentrations remain stable during the period between the analysis and use of the solution. Standard solutions are not subject to the former problem, but may not adequately reproduce the forms of determinands and other substances.

Since this type of test is potentially subject to a number of sources of error, there is the possibility of its results not revealing errors arising in different stages of the procedure and counter balancing each other. For example, losses of determinands occurring in a sample-collection device may be balanced by gains through contamination at the sample-preparation stage. Exact balancing is perhaps unlikely, but it is as well to bear this possibility in mind.

Clearly, this type of test is not always practicable and – even when it is – often

* See footnote to p. 89.

faces a number of difficulties. In spite of this, we feel that it is always worth considering. The test is sometimes referred to as a 'field-blank', and has been recommended by a number of authors for a variety of water types and snow; see, for example, refs. 3, 4, 126, 376, 377, and 383.

(ii) Tests of sample preparation The approach in (*i*) may be applied solely to the sample-preparation stage of sampling [see Section 3.5.1(*a*)(*ii*)], and all the remarks in (*i*) apply equally. When, however, the test solution contains negligibly small concentrations of determinands, the analytical results for the samples of this solution provide an unambiguous estimate of the size of contamination during sample preparation and storage before analysis, assuming that contamination during preparation is not lost through subsequent sample instability. Carefully purified water is the usual test solution, but note the difficulties that may arise in ensuring that determinand concentrations in this water are, and remain, negligibly small; these problems are discussed in Sections 5.4.4(*b*) and 10.7.

A second type of test that has been recommended[186] is to prepare two 'samples for analysis' from one sample. If the difference between the analytical results is greater than would be expected from analytical errors alone, the test indicates the presence of random errors in the sample-preparation stage. Although this does not, therefore, provide a direct test of bias, detection of random errors in the sampling process is likely to be a reasonable indication of the presence of systematic errors also.

(iii) Concluding remarks From the above brief discussion, it is clear that it will be rather exceptional for routine tests to be able to provide unambiguous and adequately precise estimates of all sources of error that may arise in sampling systems. This stresses yet again the importance of devoting critical attention to the initial selection, assembly, operation, and maintenance of all components of sampling systems.

3.6 PREPARATION, TRANSPORT, STORAGE, AND STABILITY OF SAMPLES

3.6.1 Introduction

(a) Sample Preparation, Transport, and Storage
The three types of procedure we include under the heading 'sample preparation' are given in Section 3.5.1(*a*)(*ii*). Of these procedures, sample identification is discussed in detail in refs. 3, 5, and 6, and needs no discussion here. Various aspects of the other two procedures are considered in (*b*) below and Sections 3.6.2 to 3.6.5. Here, we would stress only the importance of ensuring that adequate facilities are provided at the sample-delivery point to allow proper execution of the procedures involved in sample preparation; protection from contamination and the environment are commonly desirable, and services such as compressed gases, electricity, and suction lines may also be required.

Factors to be considered in organising the transport of samples to a laboratory and the storage of samples at the laboratory until analysis can begin are again considered in refs. 3, 5, and 6, and, with the exception of those concerning sample stability, no discussion is attempted here. Of particular importance is the need to ensure that all conditions necessary for the stabilisation of samples are achieved during the transport of samples, and that this transport is made as speedily as possible.

(b) Sample Stability
In this book, samples are said to be stable with respect to a given determinand if the concentration of that determinand does not change in the period between sample collection and analysis.* In practice, there is usually some delay between sampling and analysis, and most determinands may be liable to instability. Thus, it is generally essential to: (1) take appropriate steps to minimise instability and (2) make the analyses before any important changes in concentration occur. It is suggested here that the general aim should be, whenever possible, to attempt to ensure that any bias in analytical results caused by instability is negligible in relation to the maximum tolerable bias of results [see Sections 3.5.1(*b*)(*v*) and 5.5].

In the past, this problem of sample instability has not always received the attention it warrants, but there has been growing appreciation of its importance, and increasing numbers of publications are appearing on investigations of instability and means of decreasing its magnitude. The tendency of earlier work was perhaps to seek universally applicable procedures for ensuring adequate stability, and some success has been achieved along these lines. As will be seen below, however, the degree of instability of a given determinand may depend markedly on the nature of the sample and many other factors that are completely controllable only with great difficulty if at all. It is, therefore, not surprising that the results of different investigators on the degree of instability of a particular determinand – and on the efficiencies of different procedures for reducing such instability – often show important differences.† As a result, there is a growing tendency in recent years for authors to recommend the approach that was also suggested in the first edition of this book, namely, that it is always desirable for each laboratory to ensure that adequate precautions are adopted for its particular samples, experimental tests of instability (see Section 3.6.4) being generally necessary for this purpose; see, for example, refs. 5, 6, 191, 192, and 244–246. Notwithstanding such precautions and tests, it is also always prudent to begin analysis as soon after sampling as possible.

Provided that sample containers are transported and stored before analysis in such a way that their contamination (see Section 10.8) is avoided, the main factors affecting stability are:

(1) the nature of the sample;
(2) the sample-container;
(3) the procedures adopted to minimise sample instability.

The many aspects of (1) are noted throughout Sections 3.6.2 and 3.6.3, but one

* Note that it is generally impossible to prove experimentally that a sample is stable in this sense (see Section 3.6.4). In practice, however, it is sufficient to show that any instability does not exceed a specified value. Note also that, with this meaning of stability, contamination of samples is a source of instability.
† The procedures used for interpreting and reporting the results of such investigations are another source of such differences: see Section 3.6.4.

general point should be stressed here. It appears to be essentially universal experience that, no matter what the determinand, problems caused by instability become relatively more severe as the concentrations of interest decrease. Although we know no way of quantitatively correlating the degree of instability with concentration, the general point is well worth bearing in mind. For example, a laboratory may have successfully overcome any problems of instability for samples containing, say, 100 μg l^{-1} of a determinand; the laboratory should certainly not expect that the same procedures will necessarily prove suitable for samples containing 1 to 10 μg l^{-1}.

Discussions of sample instability and of stabilisation procedures are included in most texts dealing with the analysis of water (see Section 9.5), and individual analytical methods are increasingly including detailed specification of procedures recommended for minimising instability and its effects. Nevertheless, we think it useful to consider here a number of commonly important aspects in order to illustrate the complexity of the topic and the problems that must be faced; such information is of value also when considering the experimental design of tests of instability. Accordingly, we deal with sample containers and sample-stabilisation procedures in Sections 3.6.2 and 3.6.3, respectively. To indicate the type of procedures that are involved, we summarise in Section 3.6.3(e) some of the recommendations on determinands of common interest in two recent authoritative collections of methods for analysing waters. Finally, in Section 3.6.4 we make some brief observations on the design of, and interpretation of results from, experimental tests of instability. This last section, which was not included in the first edition, has been added because of the growing need to estimate and control the errors of analytical results. We would also add that the reporting of results from tests of instability is by no means always unambiguous, and this certainly does not help to bring order to this complex topic.

3.6.2 Sample Containers

(a) Commonly Used Materials
Special materials (*e.g.*, stainless steel, monel[205]) may be required for particular purposes, and such instances should be specified in the sampling and/or analytical method. There are also a relatively small number of reported uses of polycarbonate and PVC plastic bottles. At present, however, the great majority of normally used sample containers appears to be made of one of three materials: glass, polyethylene, and polypropylene, though Teflon is favoured by some authors for certain determinands; see, for example, refs. 247 and 248. It should be noted that each of the above names of the three main materials is somewhat ambiguous. For example, both borosilicate and soda glass are recommended and used; similarly, high- and low-density polyethylene bottles are available, and the term 'polyethylene' may be applied to bottles composed of a mixture of polyethylene and polypropylene. As will be seen below, the exact nature of the container material may lead to different degrees of instability of a given determinand. It is, therefore, highly desirable that authors specify as clearly as possible the exact nature of the sample containers they either recommend or reject.

For a number of determinands and purposes, any of the three main materials appear to be equally satisfactory so far as sample stability is concerned. Glass bottles have the advantages that it is easier to see the condition of their internal surfaces, and that they can, in general, be cleaned more vigorously than can plastic materials. On the other hand, the smaller liability of plastic containers to breakage is often of great practical value in the field and during transport. This same property also allows samples to be frozen in plastic bottles [as a means of achieving stability – see Section 3.6.3(*b*)] – a process not without risk in glass. This illustrates the important general point that the container material cannot be chosen independently of the sample-stabilisation procedure. The cost and availability of suitable containers are, of course, also relevant considerations, the first of which would appear to rule out the general use of Teflon. Plastic materials have a much greater permeability to water vapour [and other gases and vapours too – see (*b*) below] than glass. Provided, however, that caps and stoppers of bottles are properly sealing, the loss of water from samples in polyethylene or polypropylene containers may be ignored for all normal periods between sampling and analysis. Miles and Cook[192] have noted that polycarbonate bottles should not be used for long-term storage of samples because of this material's high permeability to water vapour. A number of publications provide information on the properties of various materials used for sample containers; see, for example, refs. 200, 249, and 250.

The two most important effects on sample stability that may be caused by, or associated with, sample containers are: (1) contamination of the sample by materials from the container or passing through the container, and (2) losses of determinands from the sample through sorption by the container, reaction with the container, and passage through the container. These two aspects are considered in (*b*) and (*c*), respectively.

As a broad indication of the sample containers commonly used in water analyses, we have summarised in Table 3.2 a number of recommendations made in two authoritative collections of analytical methods for waters.

(b) Contamination of Samples
The possibility of sample contamination by substances leached from the container should always be considered. There is an extremely large literature on this point, and it suffices here to mention two books,[200,251] a recent review[252] and a collection of papers;[253] these provide a host of further references. Sometimes, one type of bottle is obviously more suitable than another, *e.g.*, polyethylene (or polypropylene) should be used when traces of sodium or silicon are to be measured, glass is commonly essential for trace concentrations of organic compounds; note also that in the latter instance, amber-glass is often recommended[3] in order to reduce photochemical and biological effects. In addition, most authors now appear to favour polyethylene containers when trace metals are of interest. Contamination is not necessarily confined to major constituents of the container material. Thus, iron, manganese, zinc, boron and lead may, for example, be leached from glass, and contamination by a number of metals from plastic containers has also been reported and studied. In considering such contamination, it is important not to overlook the caps or stoppers of bottles; these are sometimes made from or contain materials other than that of the bottle, and may be troublesome sources of contamination, *e.g.*, zinc from rubber inserts in screw-caps and organic compounds from cork.[254] It is generally

safer to ensure that caps and stoppers are made solely of the container material. As a result of such contamination, many investigations of procedures for cleaning bottles have been reported; these are discussed below, but one other important general point on contamination is mentioned first.

The amounts and availabilities of potential contaminants in container materials may clearly depend on factors such as the raw materials and processes used in their manufacture. Bottles from different manufacturers may, therefore, show different effects – indeed, this may also occur for bottles from the same manufacturer. Solid particles impressed into the walls of bottles during manufacture may be a particularly variable source of contamination from one bottle to another.[197,200,201] Thus, the fact that one worker finds particular bottles to be satisfactory for a given purpose by no means necessarily implies that all bottles of similar type will be suitable for that purpose. It is prudent, therefore, to be very careful in selecting and putting into use the bottles to be used as sample containers. In trace analysis (say, rather arbitrarily, for concentrations less than 1 mg l^{-1}), it is suggested that a sound procedure is to assume that sample containers may cause contamination until it has been proved otherwise. Such caution seems doubly necessary because the degree of contamination of samples may depend on other factors that vary from one laboratory to another, *e.g.*, temperature, nature of samples, the ratio volume of sample/area of container in contact with the sample, contact time between sample and container, *etc.* In addition, the many non-quantitative and/or ambiguous reports in the literature may well be misleading if their conclusions are accepted unquestioningly – especially when the needs for accuracy of a laboratory are more stringent than those of the publishing laboratories (see Section 3.6.4).

Given the potential importance of contamination in trace analysis, virtually all authors stress the need for thorough cleaning of new containers before they are put into use, and many papers contain detailed descriptions of recommended cleaning procedures, particularly for the polyethylene bottles so widely used for inorganic trace analysis. For a given determinand, there are often considerable differences between these procedures. For example, for cleaning polyethylene bottles when trace metals are to be determined, the recommended cleaning solutions include: 10% nitric acid,[255] 20% nitric acid,[3] 50% nitric acid,[136] and the same strengths of hydrochloric acid (refs. 6, 136, 191, 234, 244, and 256). The recommended cleaning times vary from more than 12 days[234,256] to 2 days,[191,255] one paper dealing with sea-water analysis also recommending a subsequent contact of 1 week with sea-water before using the bottles; some papers do not specify a time. Cleaning at room temperature seems most often to be used, but one worker cleans bottles at 70 °C.[256] It has also been reported that the cleaning of bottles can be greatly speeded by placing them, filled with acid, in an ultrasonic bath.[257] Many other references could be cited, but the above suffice to show the wide variations that exist in published procedures. Few papers provide quantitative data showing the need for the cleaning procedures described, but see, for example, refs. 200 and 249 for such studies. Whatever the reasons for these variations, they emphasise the value of preliminary tests on cleaned bottles to determine whether or not contamination of samples is likely to be satisfactorily small (see also below).

A few general points on the solutions used for cleaning may also be made here. Clearly, these solutions should not contain large amounts of determinands.

Chromic acid is often suitable for glass bottles to be used for trace organic compounds, and as noted above diluted mineral acids are generally used for polyethylene bottles to be used for trace metals. In this latter application, it has been reported that dilute hydrochloric and nitric acids may show different relative efficiencies for leaching different elements from the bottles.[249] Care is also essential to ensure that the cleaning solution is washed from the containers so that any residual amount does not affect subsequent samples. The purity of the water(s) used for this rinsing also needs to be carefully controlled. Finally, although the above considerations have been in terms of cleaning new bottles, the cleaning of used sample containers must also be approached with equal care. Of course, once bottles have been adequately cleaned, they must be stored before use under conditions such that they do not become contaminated (see Section 10.8); many authors recommend that bottles be stored in plastic bags when not in use.

It is worth noting here that Mart[256,258] has reported that leaching of lead from polyethylene bottles was reduced to unmeasurable values when samples were frozen in the bottles.

One other type of contamination may arise with plastic bottles, *i.e.*, the diffusion of gases and vapours through the walls of the bottles.[200,259] This effect has been noted for oxygen[206,207] and mercury,[260] the interesting point being made for the latter that the size of the effect was increased when the sample contained a stabilising reagent that converts elemental mercury to the dissolved mercuric form. Contamination by this route from gases and vapours in the air around sample containers should, therefore, be borne in mind. The use of high- rather than low-density polyethylene decreases the rate of diffusion. When such effects may occur, however, it would seem generally safer to use glass containers (though see ref. 261) and/or to control the environment in which the bottles are stored. In this connection, ref. 262 discusses the contamination by volatile organic substances of samples stored in containers specially designed for sampling for trace organic compounds. Of course, contamination from the air may also enter samples through ill fitting caps and stoppers. A different type of atmospheric contamination arises when dust settles on bottles, particularly their shoulders and necks. Protection against this effect is generally useful, and may be simply achieved by placing the bottles in plastic bags.

The care needed to ensure that the containers do not cause contamination will make it generally desirable to reserve specially cleaned and tested bottles solely for this type of analysis and possibly for each determinand. Many authors have noted the value of this; see, for example, Thompson.[263]

As noted above, tests of the degree of contamination caused by sample containers are generally desirable.[5,6,17,187] In principle, such tests are easily made by placing high-purity water (see Section 10.7) in each of the bottles, storing them as samples are to be stored, and analysing the contents of each bottle at the beginning and end of an appropriate period. Since the magnitude of the contamination may be affected by the stabilisation procedure (see Section 3.6.3), the procedure to be used for samples should also be applied to the high-purity water in these special tests. In practice, such tests face a potentially important problem. The solutions in each of the bottles must be analysed at two different times, and it follows that the analytical system used for this purpose must be calibrated at essentially zero concentration of the determinand on each occasion. It will be impossible, in general, to draw unambiguous conclusions from the results for the stored solutions unless [see Sections 5.4.4(*b*) and 10.7.2]:

(1) either it is known that the determinand content of the water used to prepare the calibrating solutions was identical on both occasions, or

(2) this determinand content is accurately determined on both occasions so that an appropriate correction can be made.

Tests such as these (see also Sections 10.7 and 10.8) are very usefully included in collaborative inter-laboratory studies intended to validate new analytical methods. The involvement of a number of laboratories provides an excellent opportunity for validation or rejection of procedures intended to overcome variable effects such as contamination. The results obtained should allow detailed and appropriate procedures for the cleaning of sample containers to be included in the final recommended analytical method with some reasonable certainty that most laboratories using the method will not meet great difficulties.

(c) Losses of Determinands

Losses of determinands caused by the nature of the sample container may arise in a number of ways.

First, as noted in (*b*) above, gases and vapours may diffuse through the walls or caps of plastic containers. This can, in general, be overcome by using glass containers, but the caps and stoppers of bottles must also provide effective seals to prevent losses by this route. It is also generally important to ensure that the design of bottle and its cap or stopper is such that the bottle can be completely filled when this is necessary; see, for example, ref. 261 and Section 3.5.3(*c*).

Second, direct reactions between the determinand and the container, *e.g.*, the reaction of fluoride with glass, may lead to losses. In addition, indirect effects may occur. For example, the dissolution of soda glass by a sample may affect the pH value of a weakly buffered sample. An interesting example of this type of effect has been reported by Heiden and Aikens[204] who were studying the stability of solutions containing mercury (II). They showed that the losses of mercury were quite different in polyethylene bottles of different types, and the effect appears probably due to differences in the natures and amounts of organic substances leached from the containers, these differences leading to different rates of loss of the stabilising reagent, potassium dichromate, added to the solutions; see also Section 3.6.3(*d*).

Third, and of greatest general importance, many determinands may be sorbed by the container from samples. This effect is often especially important when small concentrations of metallic impurities are to be determined, and there is an extremely large literature on this topic. Fortunately, there appears to be essentially universal agreement that the effect for metals can be eliminated by the acidification of samples, and this is considered in Section 3.6.3(*d*); see also Section 3.6.3(*b*) for the use of deep-freezing for the same purpose. Accordingly, no discussion of the many aspects of this topic is attempted here. A recent collection of papers[253] contains a number of relevant studies and many references to earlier papers; see also ref. 245. Although we certainly support the value of acidification when metals are to be measured, this procedure is sometimes not suitable (*e.g.*, when the speciation of the metals is of interest), and may also cause problems, *e.g.*, in inter-laboratory analytical studies [see Section 7.2.10(*b*)]. We have, therefore, noted what seem to us the most important points to bear in mind for unacidified samples.

(1) The proportion of the metal concentration lost in a given time depends on the metal, the chemical and physical forms of the metal present, and the concentrations of these forms.

(2) The pH and other components of the sample may affect the losses through effects on the speciation/complexation of the metal and through competition for sorption sites on the container surface. Mart[258] notes that losses from samples such as sea-water (containing relatively large concentrations of cations competing for sorption sites) are less than from fresh waters. Indeed, one paper[264] reports that cadmium and lead in samples were stabilised by pretreatment of containers with a solution containing 5 mg l^{-1} of aluminium for 6 h. Thus, losses from pure, standard solutions should not be relied on as indications of the losses from samples.

(3) Losses may depend on the nature of the container material.

(4) The rates of sorption of metals may be such that important losses occur within a few hours[256] though periods of up to one week have also been noted[256] as tolerable for lead and cadmium in samples of filtered sea-water, but see also ref. 244.

(5) Losses to glass surfaces may be reduced by forming a hydrophobic silicone-coating on them; see Batley and Gardner[191] and references therein.

(6) The factor, volume of solution/area of bottle in contact with the solution, seems to have received rather little study. Cheam and Agemian[265,266] concluded that losses of arsenic and selenium were reduced when this ratio was increased by using larger bottles. Massee *et al.*[245] have also checked the effect of this factor for a number of metals, and report a very important effect for cadmium.

(7) The temperature of the solution may affect losses.[267]

(8) Many workers note differences between their results and those from other studies. As prediction of the magnitude of losses appears to be difficult if not impossible, many authors recommend experimental tests for every application (see Section 3.6.4).

Losses by sorption have also been reported for orthophosphate.[268,269]

Although many authors have noted losses of organic compounds by sorption on the surfaces of glass bottles, published quantitative studies appear to be much less common than for metals. Refs. 270 to 274 are examples showing that appreciable losses may occur, particularly for chlorinated insecticides, polychlorinated biphenyls, and polynuclear aromatic hydrocarbons. This source of inaccuracy is widely recognized for chlorinated insecticides, and analytical methods for such substances usually include procedures intended to ensure that sorbed determinands are recovered from sample containers; see, for example, refs. 136 and 275. Many authors have noted a similar tendency for sorbable determinands also to be sorbed by undissolved solid materials in samples, and care is necessary to ensure the unbiased determination of such forms of determinands. It is of interest that Carlsberg and Martinsen[270,271] found that the amounts of chlorinated insecticides and polynuclear aromatic hydrocarbons sorbed by glass bottles were very much smaller in water containing high concentrations of humic substances. They attributed this effect to the formation of association compounds between the determinands and humic substances; on this point, see also refs. 276 and 277. Sullivan *et al.*[272] also found that the degree of sorption of a compound could depend on the concentrations of others. These observations provide yet another example of how the degree of instability depends on the nature of the sample, and indeed it seems likely that many of the general points made above on the sorption of metals are also applicable to organic compounds. Our knowledge of such effects for trace organic compounds seems sure to increase with the growing emphasis being attached to their *quantitative* determination in waters.

3.6.3 Sample Stabilisation

(a) Introduction

Many chemical, biological and physical processes may occur in samples between sampling and analysis, and such processes are a common source of instability of many determinands. The rates of the various reactions are sometimes sufficient to cause important changes in concentration within a few hours and often within a day. Numerous examples could be quoted, but the following suffice to indicate the types of reaction that may occur.

For chemical reactions, we can mention that between chlorine and humic materials in samples of drinking water with resultant loss of chlorine and formation of halomethanes. Depolymerisation of condensed inorganic phosphates and polymeric silicic acid may occur, and cause increases in orthophosphate and monomeric silicic acid. Precipitation processes may also occur, *e.g.*, of calcium carbonate with resultant changes in related determinands such as pH, alkalinity, hardness, *etc.* Oxidation and reduction processes are also possible. An example of the former is the reaction between oxygen and Fe^{II} to form Fe^{III} which may then precipitate, while the latter may be illustrated by the reduction of Hg^{II} to elemental mercury which may then be lost by volatility. Such reactions are generally affected by the concentrations of the reacting species, by other components of the sample that affect those concentrations, and by the temperature of the sample. The other components of the sample may also affect the reactions through their effect on the ionic strength of the sample, and through catalytic and inhibitive effects. Clearly, the importance of such reactions and effects varies from one type of sample to another, and also from one measurement programme to another depending on the accuracy required of analytical results. Thus, tolerably small instability of a determinand in one type of sample and application by no means implies the same for other sample types and applications, and *vice versa*.

Similar points also apply to instability caused by biological activity which is often particularly important in samples of natural waters. Bacteria and algae are very commonly present in such waters and can consume, partially or completely, a number of substances required for their growth, *e.g.*, nitrogen, phosphorus, silicon, and organic compounds are often especially liable to instability though many other substances may be affected. Losses of determinands, thus, result, but metabolic products of the processes may enter the sample and lead to increases in other determinands, *e.g.*, consumption of ammonia by certain bacteria may lead to increases in nitrate and nitrite. Although these processes occur within the sample itself, many biological species tend to attach to, and grow on, the internal surfaces of sample containers. Under certain circumstances, therefore, it may be worth considering the sterilisation of bottles by heat. This is a standard procedure for the bacteriological examination of waters, but appears to be very rarely used in routine chemical analysis. The numbers and types of organisms present in samples may vary markedly from one sample type to another, and from time to time in a given type of sample. The extent of biological activity also depends, in general, on other factors such as temperature, degree of illumination, and the presence of toxic substances in samples. A recent review on organic contaminants in water contains much information on relevant degradation and decomposition processes.[278]

An example of a physical process that can be very important for some determinands is the loss by volatility of determinands.

Given the many potential sources of instability, a prudent approach is to assume that one's own samples will be unstable, and to select an appropriate procedure for sample stabilisation to minimise this source of error – if a particular procedure is not already specified in the analytical and sampling methods that have been selected. Many such procedures have been developed and applied, and the three main types are discussed separately in (*b*) to (*d*) below though, in practice, they may be combined. Of course, the use of special stabilisation procedures may be less convenient and more demanding of effort [see (*e*)(*i*) below] in sample collection and preparation, but the elimination of potentially important errors and the greater analytical flexibility allowed by less unstable samples will generally be worthwhile.

As an indication of the stabilisation procedures that may be required for a number of determinands of common interest, we have summarised in Section 3.6.4(*e*) the recommendations in two recent authoritative collections of analytical methods for waters.

(b) Special Storage Conditions for Samples

(i) Refrigeration   Biological activity is sometimes prevented or reduced by storing samples in the dark and at low temperatures in refrigerators (4 °C is commonly recommended). This procedure may also help to decrease some of the other sources of instability, *e.g.*, the vapour pressure of volatile compounds and rates of many chemical reactions are decreased. The approach is widely recommended and used for natural waters [see (*e*) below], and one source[6] suggests that it is often worth using it as a minimal precaution when the stabilities of determinands in samples are unknown. In general, it seems best to regard the procedure as a means of reducing rather than eliminating instability; see, for example, refs. 279 and 280. This effect is nevertheless of value because it may then well allow samples to be analysed before any instability has caused too great changes in determinand concentrations. Of course, if this stabilisation procedure is used, it should, in general, be applied immediately after sample collection.

(ii) Deep-freezing   An extension of the above approach is to freeze the samples in their sample containers immediately after collection, and to store the frozen samples in deep-freeze units (−20 to −10 °C are commonly recommended) until

required. Like the previous procedure, this approach has the useful feature that no reagents and minimal manipulation of samples are involved so that the possibility of contamination is minimised. Mart has also reported[266,258] that leaching of trace metals from polyethylene bottles and sorption of trace metals by such bottles are essentially eliminated in frozen samples. It also appears that, once samples are frozen, many of the processes causing instability are prevented so that some determinands are stabilised for many months. There is, of course, the need for means to freeze samples rapidly directly after collection and deep-freeze storage units are required, but these present no great problem in many situations.

The changes in the concentrations of determinands in the liquid phase that occur during the freezing and thawing of samples may well affect the speciation of many substances, and a number of authors have noted that the procedure should not be used when speciation is either of interest or can affect the concentrations of determinands; see, for example, refs. 3, 191, and 281–284. Thus, the procedure is not suitable for the stabilisation of the suspended solids content of samples.[3,6,284,285] It has also been recommended that deep-freezing should not generally be used for waters with high concentrations of humic materials.[282] The freezing process may cause rupture of the cells of biota with release of various substances into the sample, and this has been noted[6] as another possible disadvantage of the procedure.

Several workers have reported more variable results when frozen and then thawed samples were analysed, and have warned against the use of deep-freezing for silicon[136,286] and phosphate;[287–290] a similar effect was noted for aluminium,[291] but it disappeared when samples were set aside for 2 days after thawing and before analysis. A detailed investigation[292] of the effect of freezing for nitrate, phosphate, and silicate in coastal and estuarine waters concluded that the procedure satisfactorily stabilised nitrate, was reasonable for phosphate, and satisfactorily preserved silicate in the saline samples though losses of 5 to 25 % were observed for silicate in the low-salinity estuarine waters. All determinands showed a tendency to give less precise results after freezing and thawing, but this tendency was smaller when samples were frozen more quickly. Losses of mercury have also been reported[191] when samples were frozen. Some authors have found freezing to preserve the biochemical and chemical oxygen demands of samples,[264] but others have not confirmed this.[6,283,284] Notwithstanding such problems, the procedure is well worth consideration for some determinands, and has been recommended in a number of collections of analytical methods; see, for example, (e) below.

The apparent simplicity of the procedure should not be allowed to conceal several factors that may be of importance, *e.g.*, the rate of cooling, the volume of the sample, the shape and size of the sample container, the rate of thawing, *etc.* There appear to be few quantitative studies of such effects, but two recent papers[284,292] indicate both the potential complexity and the need for further studies. The need for very thorough shaking of the thawed samples before taking any portions for analysis has also been stressed;[293] otherwise, the solution at the top of the sample container may be essentially pure water.

(c) Immediate Treatment of Sample not Involving Addition of Stabilising Reagents

(i) Filtration Algal and bacterial activity in samples can often be adequately

reduced by filtering the sample during or immediately after collection. Use of filters with pore sizes of approximately 0.5 μm is usually recommended for this purpose, and they give efficient removal of most if not all important biota. Both membrane and glass-fibre filters are used, and the technique is widely recommended and used for determinands such as nitrogen, phosphorus, and silicon species which are especially liable to change through biological processes. Although this procedure has been used for many years as a means of sample stabilisation for these substances, it should not be forgotten that bacteria may consume or degrade other substances; an interesting account of the value of filtration in stabilising a number of trace organic compounds has appeared.[279]

As an additional precaution, it is commonly recommended that filtered samples be stored in the dark at approximately 4 °C.

A variety of filtration systems is available commercially to allow the filtration to be made in the field at the sample-delivery point; see Section 11.2.1. This procedure is clearly inapplicable to determinands that are partially or completely removed by filtration. It is also important to ensure that impurities leached from the filter and filtration system do not cause appreciable contamination of the sample; extensive pre-washing of the filters is sometimes needed to eliminate such contamination.[6,214,294-296] Note also that sorption of ammonia by glass frits and glass-fibre filter papers has been reported.[297] Filtration and the problems it may cause are further discussed in Section 11.2.1. It is always a wise precaution to check the magnitude of any contamination by filtration before applying this technique routinely.

(ii) Beginning analysis directly after sample collection An alternative approach to sample stabilisation is to start the analysis of samples directly after sample collection so that the determinand is converted to a form in which it is adequately stable. Relatively simple manipulations, which can be made at the sample-delivery point, may suffice for this purpose, and the analysis may be completed later in the laboratory. One example of this approach arises in the Winkler procedure for determining dissolved oxygen; addition of the manganese and alkali reagents leads to the formation of the oxidised manganese precipitate which is stable for relatively long periods.[136,298] As other examples, phosphate,[290,299] ammonia,[300] iron(II),[301] and aluminium[302] may be quoted.

A variant of this approach is to treat the sample by a preliminary concentration procedure at the sample delivery point; determinands are often more stable in the concentrating medium, *e.g.*, a column of adsorbing material, than in the sample. This approach has been discussed in Section 3.3.2(*e*)(*iii*).

(d) Addition of Stabilising Reagents to Samples

(i) General precautions Adequate stabilisation of samples may often be achieved by the addition of a chemical reagent to them so that the processes causing instability are prevented or reduced. Many different reagents have been and are recommended and used, the final concentration of a given reagent in the sample often showing substantial variations from one author to another. Again, as for other aspects of sample stabilisation, the conclusions reached by different workers

on the relative efficiencies of reagents do not always agree. Before considering some of the more commonly used reagents, it is useful to note some general precautions and limitations in their use.

(1) Stabilising reagents cannot be used when they affect the speciation of determinands *and* this speciation is of interest. For example, acidification of samples cannot be used to stabilise metals whose speciation is to be studied because the acid may cause dissolution of solid species, desorb metals adsorbed on solid particles,[306] change the complexation of metals, *etc.*

(2) The determinand content of a reagent must be sufficiently small that the accuracy of analytical results is not substantially affected. In general, high-purity reagents are necessary in trace analysis.

(3) Stabilising reagents may affect the performance of analytical systems. For example, the addition of acid may prevent the formation of the desired coloured compound in spectrophotometric procedures. In other instances less extreme effects may occur; for example, the reagent may affect the slope of the calibration curve of the analytical system so that this curve must be established in the presence of the reagent. The combination stabilising reagent/analytical procedure must, therefore, be carefully chosen, and calibration made appropriately. Sometimes, such effects may be overcome by adding another reagent at the start of analysis. For example, acid may be neutralised by adding an alkali, but the addition of another reagent may aggravate problems of contamination.[291]

(4) Appropriate correction must be made for the dilution of the sample caused by the addition of a reagent in solution. The addition of small volumes of relatively concentrated reagents is, therefore, useful.

(5) In general, it is preferable[3,5,6,136] to add the reagent to the empty sample container before sample is added, but in some instances it appears to be valid to add the reagent some time after the sample enters the container. For example, in using nitric acid to stabilise metals, it has been reported that the acid may be added several days after sampling,[307] though see also ref. 308. Similarly, it has been found that addition of EDTA 40 days after sampling rapidly and completely desorbed lead sorbed on borosilicate-glass and polyethylene containers.[309] Nonetheless, it seems to us prudent to aim to add the reagent to the empty sample container until evidence that this is not necessary for the samples of interest is obtained.

(6) Relatively large amounts of stabilising reagents are often required, and great care is necessary when one or more of their constituents are also to be measured in other samples. The use of reagents such as mercuric chloride and chloroform (see below) may well cause contamination problems in laboratories having also to determine very small concentrations of these two substances.

(ii) Stabilising reagents Some reagents are suitable for a number of determinands. For example, biocidal agents such as mercuric chloride, chloroform, and toluene have often been recommended and used for determinands affected by biological processes, *e.g.*, nitrogen and phosphorus species. Care is required, however, because on killing the biological species, determinands such as orthophos-

phate may be released.[310,311] It is worth noting the use of methylene chloride as a bactericide when trace organic compounds are of interest,[374] and formaldehyde has been used for the same purpose for dissolved alkanes in water.[357]

The use of mineral acids to stabilise metals is another example of a reagent suitable for many determinands, and is probably the most universally agreed stabilisation procedure of all; appropriate acidification can stabilise samples for many months. Nitric, hydrochloric, and sulphuric acids have all been and are used, usually in amounts that lead to pH values between 1 and 3.5 in the stabilised samples. It seems that the required pH value depends on the metal of interest, but values of 1 to 1.5 are suitable for many commonly measured metals. Although acidification has greatest value for trace metals, it is also recommended for magnesium and/or calcium which are often present at much greater concentrations; see, for example, refs. 136, 243, 312, 313, and 313*a*. It appears that rather greater acidity (0.1 to 0.3 M) is required in the sample to stabilise silver; see also refs. 136, 314*a*, and 386 on a completely different reagent – cyanogen iodide – for silver. Mercury presents special problems considered below.

The great growth of interest in mercury in waters during the last 10 to 15 years has also seen the development of a very large literature on the instability of samples and on stabilising reagents to overcome the severe problems often reported; see, for example, the review in ref. 191. Once again, there are some important disagreements among the results of different workers. Some authors have concluded that acidification alone is sufficient to ensure adequate stability (see, for example, ref. 136). There appears, however, to be a more prevalent view that, in addition to acid, a reagent should also be added to ensure that mercury is present as Hg^{II}; otherwise, there is danger that elemental mercury may be formed and then lost by volatility.[315] Strong oxidants and strong complexing agents for Hg^{II} have been studied for this purpose. The importance of biological effects has not normally been noted, but this aspect has been stressed in one study of the problem.[316,317] Although we have not tried to count the numbers of authors favouring the various alternatives reported in the innumerable papers, our impression is that the most widely favoured combination is nitric acid/potassium dichromate (see, for example, refs. 318 to 321), but sulphuric acid/potassium dichromate,[322,323] nitric acid/ammonium thiocyanate,[324] sulphuric acid/sodium chloride,[325] nitric acid/sodium chloride,[326] bromate/bromide,[327] potassium iodide,[328] auric gold,[319] and potassium permanganate[316] have all been reported in recent years. The use of polyethylene bottles may cause problems[204,329] [see Sections 3.6.2(*b*) and (*c*)], and though these may be overcome[329] or may not even occur, it seems preferable to use glass bottles; this practice has been recommended by a number of authors. Finally, if samples contain organomercurial compounds (*e.g.*, methyl mercury, diphenyl mercury), the acid/dichromate reagents have been shown to decompose such compounds,[323,330] and were recommended only when the total mercury concentration was of interest. Goulden and Anthony[323] recommended stabilisation with 1 % sulphuric acid when organomercurial compounds are to be determined.

Yet another general-purpose reagent is sodium thiosulphate which is recommended[375] to prevent changes in many different organic compounds when samples contain free or combined active chlorine; haloforms and polynuclear aromatic hydrocarbons are two examples of compounds for which this procedure is used.

Some reagents are intended to stabilise particular determinands, *e.g.*, cyanide, sulphide, or phenolic compounds. Details of such reagents are normally given in the analytical methods for such determinands; see also (*e*) below.

Finally, a rather unusual method of stabilisation for orthophosphate is worth mentioning since this determinand is commonly found to be rapidly lost from samples containing small concentrations (less than, say, 20 μg l^{-1} of P). Heron[331] concluded that this effect was caused by a bacterial film on the internal surfaces of polyethylene sample containers, and he overcame the problem by treating the insides of the bottles with iodine before sampling; presumably the bactericidal effect of the iodine is responsible. The iodine absorbed by the polyethylene retains its efficiency for relatively long periods so that the same bottles may be used repeatedly without further treatment. This procedure has been recommended by other authors,[214,269,332] and has also been found useful in the authors' laboratory.

(e) Recommended Sample Stabilisation Procedures
Virtually every compilation of analytical methods for waters (see Section 9.5) now includes information on recommended stabilisation procedures, and such recommendations are also commonly given now in individually published analytical methods.* Thus, analysts should have no difficulty in locating such information. Nonetheless, we thought it would be useful to complete this discussion of sample stabilisation by summarising in (*i*) below (Table 3.2) the procedures given in two authoritative compilations. The summary indicates the types of stabilisation procedure of common interest. In addition, we have given in (*ii*) below references to a number of recent investigations of sample instability and stabilisation for a range of determinands. These references and those quoted therein and in Sections 3.6.3(*a*) to (*d*) will help readers wishing to pursue more deeply the stabilisation of particular determinands.

(i) Procedures recommended in refs. 6 and 136 Both these publications contain tables summarising the recommended stabilisation procedures. It should be noted that the former (prepared by the U.K. Standing Committee of Analysts) was issued before many of the individual analytical methods were published. There are occasional slight differences between the recommendations in ref. 6 and those in the analytical methods. When this is so, we have quoted below the latter recommendations,† but have not cited references to all the individual methods consulted.

We would also emphasise that both publications stress the problems involved in recommending stabilisation procedures suitable for a wide range of applications and sample types. Both also emphasise the importance of a critical attitude in selecting and validating these procedures for each application. For example, ref. 136 states:

'Regardless of the sample nature, complete stability for every constituent never can be achieved. At best, preservation techniques only retard chemical and biological changes that inevitably continue after sample collection......

* The International Organization for Standardization is also preparing detailed recommendations on sample stabilisation; see ref. 55.
† The possibility of such differences is noted in ref. 6, and readers are advised always to follow the recommendations in individual analytical methods whenever they have been published.

TABLE 3.2
Recommended Procedures for Sample Stabilisation

Determinand[a]	Sample container[b]		Stabilisation procedure[d]	
	Ref. 6[c]	Ref. 136	Ref. 6[c]	Ref. 136
Alkalinity	P (or G_B)	P or G_B	Fill bottle completely. If storage unavoidable, refrigerate	Refrigerate
Antimony	P	P (or G_B)	Add 3 ml 6M HCl l^{-1} of sample	See 'Metals' below
Arsenic	P or G	P (or G_B)	Add 2 ml 6M HCl l^{-1} of sample	See 'Metals' below
Biochemical oxygen demand*	G	P or G	Fill bottle completely. If storage unavoidable, refrigerate immediately, and test effect of the delay. Do not freeze	Analyse within 2 h or refrigerate and analyse within 6 h
Boron	P (or G_S)	P or G_S	Refrigeration may be needed	No special conditions needed
Bromide	P or G	P or G	No special conditions needed	No special conditions needed
Calcium	P	P or G	Preferable, but not essential, to add 2 ml 5M HCl l^{-1} of sample	See 'Metals' below
Carbon dioxide, free*	G_B	P or G	Analyse immediately or fill bottle completely and store at a temperature less than that of initial sample	As ref. 6
Carbon, total organic	G	G	Fill bottle almost completely and analyse within 2 h. Refrigerate if delay unavoidable	Analyse immediately or refrigerate
Chemical oxygen demand	G	P or G	Refrigerate or add H_2SO_4 to pH 1 to 2. Effective time of stabilisation may vary from few hours to several days, and must be tested	Add H_2SO_4 to pH < 2
Chloride	P or G	P or G	No special conditions needed	No special conditions needed.
Chlorine*	G	P or G	Analyse immediately. (Drinking waters may be stored up to 3 h in amber bottles, filled completely and refrigerated without freezing.) Avoid strong light and agitation	Analyse immediately. Avoid strong light

Determinand	Container	Container	Preservation and storage	Preservation and storage
Colour	G_B	P or G	Store in a cool dark place	Refrigerate
Conductivity, electrical	P	P or G	Fill bottle completely and refrigerate	Refrigerate
Cyanide	P or G_B	P or G	Add NaOH to pH >12 and refrigerate	As ref. 6
Fluoride	P	P	No special conditions needed	No special conditions needed
Halogenated methanes	G or special G containers	Special G containers	Add $Na_2S_2O_3$ and fill bottle completely. (If kept in dark at ambient temperature, stable for at least 24 h in absence of external contamination)	Add $Na_2S_2O_3$ immediately after collection and refrigerate
Hardness, total	P or G	P or G	Fill bottle completely. No special conditions needed	See 'Metals' below
Magnesium	P	P or G	Preferable but not essential to add 2 ml $5M$ HCl l^{-1} of sample	See 'Metals' below
Mercury, total: Non-saline samples	G	G (or P)	Add $K_2Cr_2O_7$ and HNO_3 to pH 1	Add $K_2Cr_2O_7/HNO_3$ and refrigerate or add HNO_3 to pH <2 and refrigerate
Saline samples	—	G	Add 20 ml $4.5M$ H_2SO_4 l^{-1} of sample	—
Metals[e]: Total	P	P (or G_B)	Add 20 ml $5M$ HCl or 2 to 10 ml HNO_3 l^{-1} of sample	Add HNO_3 to pH <2 and refrigerate
Filterable	P	P (or G_B)	Filter immediately and acidify filtrate as for 'total metals'	Filter immediately, acidify filtrate as for 'total metals' and refrigerate
Nitrogen: Ammoniacal*	P or G	P or G	Refrigerate or add HCl to pH 2. Analyse as soon as possible	Add H_2SO_4 to pH <2 and refrigerate
Nitrate* Nitrite* when NO_2^- concn. appreciable	G	P or G	Analyse as soon as possible (no reliable stabilisation method is known) and test to decide maximum tolerable delay	Add H_2SO_4 to pH <2 and refrigerate. Freeze at $-20\ ^\circ C$ or add $HgCl_2$ and refrigerate (No distinction is made between different nitrite concentrations)

TABLE 3.2 (cont.)

Determinand[a]	Sample container[b]		Stabilisation procedure[d]	
	Ref. 6[c]	Ref. 136	Ref. 6[c]	Ref. 136
Nitrogen (cont.) Nitrate* Nitrite* } when NO_2^- concn. inappreciable	P or G	—	Refrigerate or add phenyl mercuric acetate. Analyse as soon as possible	—
Organic*	P or G	P or G	Add H_2SO_4 to pH 2 and refrigerate	As ref. 6
Organochlorine insecticides	G	G	Refrigerate. (Solvent extracts can be stored in a spark-proof refrigerator for months)	Refrigerate
Organophosphorus pesticides	G	G	Add extracting solvent to sample immediately after sampling, shake, and store in a spark-proof refrigerator	Refrigerate
Oxygen, dissolved*	Special G container	Special G container	Fill bottle completely. Analyse immediately or 'fix' sample when Winkler method used	As ref. 6
pH value*	P or G_B (G_B for slightly buffered waters)	P or G	Fill bottle completely. Analyse immediately. If delays unavoidable, refrigerate	Analyse immediately
Phenolic compounds*	G	P or G	Add Na_2AsO_3 and HCl to pH <2	Add H_3PO_4 to pH 4 or H_2SO_4 to pH <2; refrigerate
Phosphorus, 'molybdate-reactive'*[f]	(P or) G_B	G	No generally satisfactory method of long-term stabilisation. Refrigeration (but not freezing) reasonably effective over a few days. Filter immediately when biological activity present	Filter immediately. Refrigerate or freeze at $-10\,°C$.
Polynuclear aromatic hydrocarbons[g]	Special G container	Special G container	Refrigerate. Add $Na_2S_2O_3$ when chlorine present in the sample	As ref. 6

Determinand	Container	Conditions	Container	Conditions
Potassium	P	No special conditions needed	P or G_B	See 'Metals' above
Selenium	—	—	P (or G_B)	See 'Metals' above
Sodium	P	No special conditions needed	P	No special conditions needed
Silica, 'molybdate-reactive'	P	Refrigerate. Filter immediately if biological activity suspected	P	Refrigerate. (Do not freeze)
Solids: Total	G_B	Refrigerate	P or G_B	Refrigerate
Filterable	G_B	No special conditions needed	P or G_B	Sample stabilisation impractical; analyse as soon as possible
Non-filterable*	G_B	No generally suitable method; analyse as soon as possible	P or G_B	Sample stabilisation impractical; analyse as soon as possible
Sulphate	P or G	Refrigerate	P or G	Refrigeration may be needed
Sulphide	P or G_B	Add zinc acetate and NaOH immediately	P or G	Add zinc acetate immediately
Surfactants: Anionic, cationic, and non-ionic	—	Fill bottle completely. Analyse within a few hours or add formaldehyde and refrigerate	—	—
Turbidity*	G_B	No generally suitable method; analyse as soon as possible	P or G	Analyse same day. If longer storage unavoidable, store in dark for up to 24 h

Notes:

a Those determinands particularly liable to instability are indicated by an asterisk (*).

b P = polyethylene for ref. 6 and plastic (polyethylene or equivalent) for ref. 136. G, G_B, G_S = glass, borosilicate glass, and soda glass, respectively.

c See the text for the sources of information used in supplementing ref. 6.

d We have not quoted all details and provisos given in the references which must be consulted before using any of their recommended procedures. The term 'refrigeration' signifies storage at approximately 4 °C in the dark.

e Ref. 6 states that the term 'Metals' includes Al, Ba, Be, Ca, Cd, Co, Cr, Cu, Fe, Mg, Mn, Mo, Ni, Pb, Sr, Sn, U, V, and Zn, and that the quoted stabilisation procedures are thought to be generally useful for this group of metals. It is also stated that the stabilisation procedure must be chosen to be compatible with the analytical method, and that rather different acidification procedures are given in certain individual analytical methods where these are required by those methods. Ref. 136 includes the following under the term 'Metals': Ag, Al, As, Au, Ba, Be, Bi, Ca, Cd, Co, Cr, Cs, Cu, Fe, Ir, K, Li, Mg, Mn, Mo, Na, Ni, Os, Pb, Pt, Rh, Ru, Sb, Se, Si, Sn, Sr, Th, Ti, Tl, V, and Zn. In a few instances, different stabilisation procedures are given in individual analytical methods.

f It is common to apply several analytical procedures to samples to allow the total concentration of phosphorus to be divided among a number of different determinands, e.g., filterable orthophosphate, non-filterable organic phosphorus, etc. Details of such fractionation schemes are discussed in refs. 136 and 335. The need for, and possibilities of, sample stabilisation depend on the particular scheme adopted, and these two references should be consulted.

g Neither ref. 6 nor subsequent publications of the Standing Committee of Analysts available at the time of writing quote details for polynuclear aromatic hydrocarbons – nor are any details given in ref. 136. In view of the interest in these compounds, we have included in the table brief details of the procedure recommended in ref. 3. This last reference gives recommended procedures for a large number of different types of organic compounds.

Clearly it is impossible to prescribe absolute rules for preventing all possible changes. . . . . . to a large degree the dependability of water analyses rests on the experience and good judgment of the analyst.'

Ref. 6 contains similar cautions:

'At present, there is insufficient knowledge to allow recommendation of sample-handling procedures that will unquestionably eliminate these changes of concentration for all determinands in all types of sample. . . . . . To the maximum extent possible, tests of the efficiency of the selected procedures should be made on samples whenever there is doubt on the stability of one or more determinands.'

Given our aim in presenting this information, we have not included all the various provisos and additional data contained in these publications; they or equivalent publications must be consulted by analysts wishing to select stabilisation procedures for their own use. Both references indicate, though in different ways, determinands that are particularly liable to instability, and we have included this in Table 3.2. In general, this choice of relatively unstable determinands seems to us excellent, but readers should bear in mind the various general points made in preceding sections. For example, the instability of nitrate is noted by both publications, and at small concentrations (say, less than 1 mg 1^{-1} of N) this caution is certainly well founded. At appreciably greater concentrations, however, in waters free of biota (*e.g.*, drinking water), the instability of nitrate may well be small; see, for example, ref. 333. Equally, it should not be thought that the recommended procedures necessarily eliminate problems of instability. For example, acidification of samples to reduce the instability of ammonia is included by both publications, but has been reported to be ineffective in at least one paper.[334]

One final caution is necessary. Ref. 136 and some other publications (*e.g.*, refs. 5 and 379) also include, for each determinand, a recommended maximum time between sampling and analysis. Such publications usually – though not always – make clear that these times are intended only as approximate guides, and it seems to us very important that this proviso is always borne in mind; this point is stressed also in refs. 3 and 6.

Note also that the stabilisation of many different types of organic compound is treated in detail in refs. 3 and 375.

An important point emerges immediately from Table 3.2, *i.e.*, when many determinands are to be measured, a relatively large number of sample containers is required for each sampling occasion and point because of the various stabilisation procedures involved. This is undoubtedly inconvenient and effort demanding, and also requires careful organisation and control of the procedures for receiving, cleaning, and preparing sample containers. There appears, however, to be no alternative at present when the degree of instability and the speed with which analyses can be made are such that sample stabilisation is important. It should also be noted that the instability of some determinands is such that essentially immediate analysis is recommended at the sample-delivery point.

(ii) Recent studies of sample stability and stabilisation   In our view, a substantial proportion of published studies provide results that cannot be interpreted

unambiguously through failure apparently to take account of one or more points made in Section 3.6.4. We have included such studies in Table 3.3 because, of course, they may still contain much useful information, but readers should bear in mind the points made in Section 3.6.4.

3.6.4 Design and Interpretation of Tests of Sample Stability

(a) Need for Tests
We have already stressed in previous sections that the large number of factors that may cause sample-instability makes it very desirable for each laboratory to check experimentally that its proposed sample-stabilisation procedures are adequate, *i.e.*, that the initial concentrations of determinands are maintained with negligible error for the longest period likely between sampling and analysis. A wide range of different approaches to the design, execution, interpretation, and publication of such tests has been described in the literature. For a number of reasons, many of these published studies seem to us inadequate in providing the quantitative and unambiguous conclusions so badly needed in this field. The aim of the present section, therefore, is to note a number of points we believe to be of general importance in this connection. We would stress that we are concerned here only with the above purpose; other purposes, *e.g.*, the study and development of new stabilisation-procedures, may well require approaches different from that suggested below.

(b) General Nature of the Experimental Design
The literature shows wide variability in the types of instability observed. The concentrations of some determinands increase while others decrease, and, indeed, the same determinand has been reported to show increase and decrease in the one sample. Further, a wide variety of shapes are reported for the plots of concentration against time. On this basis, it seems to us unwise to plan the tests, in general, with the expectation that a particular type of change will occur, *e.g.*, that any change in concentration, C, will be linearly related to the time, t, between sampling and analysis. Given this view, the aim of the tests should not be seen as the estimation of the parameters of a postulated functional relationship between C and t, *e.g.*, the estimation of α and β in $C = \alpha - \beta t$. In addition, given the general importance of eliminating sources of bias in analytical results [see Sections 3.5.1(*b*)(*v*) and 5.5], we assume that care has been taken to select a sample-stabilisation procedure likely to be effective. In this context, it seems to us that the tests should simply aim to determine the magnitude of the difference, $C_t - C_o$, for one or more times after sampling: C_t is the true concentration at time t after sampling, and C_o is the true concentration immediately after sampling. The tests, therefore, should consist essentially in the analysis of each selected sample immediately after collection and again at one or more subsequent times.

Given the need to analyse directly after sampling, special arrangements may well need to be made for such analyses. In general, it is impossible to make these first analyses without any time delay at all; information in the analytical method or elsewhere, and, if necessary, preliminary tests may help to indicate the maximum tolerable delay before the first analyses. It is, however, clearly desirable to do

TABLE 3.3

Recent Studies of Sample Stability and Stabilisation

Determinand	Ref.*	Determinand	Ref.*
Alkalinity	261	Organic compounds:	
Arsenic	245, 265, 308, 337r	halogenated methanes	353, 356, 359
Asbestos	338	nitrosamines	360
Beryllium	308, 339, 340	organochlorine insecticides	270, 271, 274, 356, 361
Biochemical oxygen demand	283e, 284e, 341e, 352	organophosphorus pesticides	362
Bismuth	342	phenolic compounds	363e, 364
Boron	343	phthalates	270, 271, 272s
Calcium	243, 312	polynuclear aromatic hydrocarbons	273
Carbon:		surfactants	341e, 365
particulate	281	Oxygen, dissolved	341e, 367, 368
total organic	284e, 341e, 344e, 345	Nitrogen:	
Chemical oxygen demand	264, 283e, 284e, 341e	ammonia	280, 311, 334, 341e, 366
Chloride	341e		
Colour	341e	nitrate	280, 292s, 311, 333, 341e, 366
Cyanide	280		
Fluoride	280	nitrite	280, 341e
Germanium	346	organic	280, 341e
Magnesium	243, 312	particulate	281
Mercury	125sr, 191r, 191sr, 204, 260, 321–330, 325s, 349, 350, 351s	Phosphorus:	
		dissolved inorganic phosphates	269, 280, 287, 289, 290, 292s, 299, 311, 341e
Metals, trace (Al, Cd, Cu, Fe, Pb, Zn, *etc.*)	125sr, 126sr, 189s, 190sr, 191r, 191sr, 244s, 245, 247, 255, 256s, 258, 291, 291s, 307–309, 340, 347s, 348s	dissolved organic	280
		condensed inorganic phosphates	369e
		total	341e
		pH value	261, 341e
		Selenium	245, 266, 308, 337r, 370, 371
Organic compounds:		Silicate	292s
acetone	353	Silver	245, 314, 314ae, 372
acetonitrile	353	Sulphate	341e
alkanes	353, 354, 357	Sulphide	280, 341e
amino-acids	355	Suspended solids	283e, 341e
aromatic hydrocarbons	270, 271, 279, 353, 356	Tin	373
	258		

everything possible to make this delay as small as possible. The choice of subsequent times at which to analyse the sample is, for present purposes, best made on the basis of what is expected to be routine practice. If, for example, it is expected that samples will usually be analysed within one day of sampling but that delays of up to 3 days may sometimes occur, then tests at one and 3 days after sampling will provide relevant information. Of course, the use of additional times will provide more information, but at the expense of the greater effort involved – and this may well be substantial [see (*d*) below]. In general, it seems to us better to ensure that adequate testing is made at key operational times rather than to decrease the quality of testing at each of a larger number of different times.

As will be seen in (*d*) below, the degree to which sample instability can be detected and estimated is governed by the precision of the analytical results. Accordingly, it is useful to ensure that as good a precision as possible is achieved. In many circumstances, laboratories will have no choice but to use their routine, analytical system, but if another system providing more precise results were available, it may well be advantageous to use it for these tests. A completely different approach is to add radioactive isotopes of the determinands of interest to the freshly collected sample, and to measure the radioactivity of portions of the sample at the desired times after sampling. This procedure has been used by many workers in studying sample stability (see, for example, ref. 336), and for some determinands and applications may offer the advantages of simplicity and small measurement errors. The technique requires, of course, appropriate counting equipment and laboratory facilities (see Section 11.4.5), is not applicable to all determinands at the concentrations of interest, and also involves the assumption that the added radioisotopes equilibrate very rapidly with the isotopes naturally present in samples; this assumption is by no means necessarily valid. Accordingly, the subsequent considerations of this section are based on the use of the routine analytical system, but the two alternatives should be borne in mind.

(c) Choice of Samples for Testing

For a laboratory dealing with only one, relatively constant source of water, we would suggest that, for a given determinand, two samples – corresponding as far as possible to the upper and lower limits of the range of determinand concentrations of interest – should be tested initially. If satisfactorily small instability is observed, it is probably reasonable to accept the sample stabilisation procedure as sound. In this event, further tests could be limited to regular though infrequent occasions.

The choice of samples is much more difficult for laboratories dealing with many types of sample whose qualities may vary markedly from time to time, and we know no way of deciding with certainty the types of sample and the smallest numbers of each that may be considered to form a reasonable test of a sample-stabilisation procedure. We noted in Sections 3.6.2 and 3.6.3 the many ways in which one or more characteristics of the samples may affect the stability of determinands, and it seems to us impossible to predict with certainty those samples most liable to instability. The decision would be helped if all relevant factors were always positively correlated, *e.g.*, if samples with greatest biotic activity always contained the smallest concentrations of determinands, but this will often not be so. A further complication is that the properties of a given type of sample relevant to instability may vary markedly with time. For example, the instability of silicon in river water may be much smaller in winter than at times in spring and summer when algal blooms occur. Given these complications, we can only suggest that each laboratory critically considers its own needs, the potential sources of instability, and the sample stabilisation procedures to be used, and then makes the best decision possible on the types of sample to be given priority for testing. The results from the first series of tests may help to decide what other tests should be arranged.

Our experience is that routine analytical laboratories often find it difficult to release the large amount of effort generally required for such tests in one concentrated session. Consequently, routine analysis of samples may well proceed without tests of any kind for some determinands. In this position, it is vital that the results of other workers and the efficiencies and limitations of recommended stabilisation procedures be reported – in analytical methods and other publications – as quantitatively and unambiguously as possible because this will be invaluable in deciding the priorities to be assigned to testing the various determinands of interest; see also Section 3.6.4(*f*).

(d) Statistical Aspects

(i) General considerations   It is important that the statistical aspects of these tests be considered in planning them and in interpreting the results obtained. In this section we summarise the considerations involved, but we suggest that Sections 4.3, 4.6, and 6.1 to 6.3 be read before proceeding further.

From the suggestions in (*b*), the aim of the tests – for a particular sample and time after sampling – may be stated as the determination of δ, where $\delta = C_t - C_o$, and C_t and C_o are the true concentrations of the determinand in the sample at times t and zero after sampling. In practice, of course, C_t and C_o are not known, and are estimated as the means of n analytical results for the sample on each analytical occasion; we denote these means by $\bar{R}_t$ and $\bar{R}_o$. The tests, therefore, produce an estimate, d, of δ, where

$$d = \bar{R}_t - \bar{R}_o \qquad (3.7)$$

Both $\bar{R}_t$ and $\bar{R}_o$ are subject to random errors, and neither will, in general, equal the corresponding true concentration (see Section 4.3). It follows from equation (3.7) that d will also be subject to these random errors, and will not generally equal δ. Caution is necessary, therefore, in interpreting the observed value of d in terms of sample instability. For example, a non-zero value of d is very likely to be obtained even when $\delta = 0$. Clearly, in assessing the significance of d, its value in relation to the size of its random error must be considered; this is particularly true when sample instability is small since the random error of d may then well be large compared to d.

Denoting the standard deviations of $\bar{R}_t$ and $\bar{R}_o$ by σ_{mt} and σ_{mo}, respectively, the standard deviation, σ_d, of d is given by [see Section 4.3.2(*b*)]

$$\sigma_d = (\sigma^2_{mt} + \sigma^2_{mo})^{\frac{1}{2}} \qquad (3.8)$$

If we assume that: (1) the analytical system to be used is well controlled (see Section 6.3.1) and (2) C_t differs only slightly from C_o, we may write $\sigma_{mt} = \sigma_{mo} = \sigma_m$. From equation (3.8), therefore,

$$\sigma_d = \sqrt{2}.\sigma_m \qquad (3.9)$$

The statistical ideas involved in assessing the significance of d are discussed in detail in Section 4.6.2. One important consequence of the random error of d is that it is impossible to prove experimentally that $\delta = 0$, *i.e.*, that a sample is stable. Even when $\delta = 0$, the best that can be achieved is to show (at a given probability level) that δ is less that some stated value, δ_c. Such a conclusion will, however, be perfectly acceptable so long as the value of δ_c is acceptably small in relation to the maximum tolerable bias of analytical results [see Sections 2.6.2 and 3.5.1(*b*)(*v*)]. Thus, the detailed design of the tests involves deciding an appropriate value for δ_c for a given sample, and then ensuring that σ_d is sufficiently small that a statistically significant value of d will be obtained (at a given significance level) if $\delta > \delta_c$. For a given analytical system, the only way of ensuring that σ_d is adequately small is to make the value of n sufficiently large. A method for calculating an appropriate value for n is given in Section 4.6.2, but this procedure involves several assumptions whose validity in the present context is considered in (*ii*) to (*iv*) below.

(ii) Frequency distribution of analytical results   The procedure in Section 4.6.2 assumes that $\bar{R}_t$ and $\bar{R}_o$ are normally distributed. We believe that this assumption will usually be reasonable even when $n = 1$, but it should certainly be an adequate approximation when $n > 1$ [see Sections 4.3.1(*a*) and 6.2.1(*e*)(*x*)].

(iii) The value of σ_d when $n = 1$ When $n = 1$, $\sigma_m = \sigma_t$, the total standard deviation of analytical results [see Section 6.2.1(*d*)]. The preliminary tests of precision (Section 6.3) will have provided an estimate, s_t, of σ_t. Thus, from equation (3.9), an estimate of σ_d is given by $s_d = \sqrt{2}.s_t$. These preliminary tests will also show if it is reasonable to expect σ_t to be constant from one analytical occasion to another. For the reasons stated in Section 6.3.1, we would suggest that tests of sample stability should not be attempted until one is confident that σ_t is essentially constant.

Although the procedure in Section 4.6.2 requires knowledge of σ_d to calculate an exact value for n, an approximate value can be obtained by using s_d in place of σ_d. The more advanced statistical texts noted in Section 4.8 describe more accurate procedures for estimating n. It is possible that $n = 1$ will suffice for some tests, but usually $n > 1$ will be necessary.

(iv) The value of σ_d when $n > 1$ The procedure in Section 4.6.2 uses the equality $\sigma_m = \sigma_t/\sqrt{n}$, and this equality is valid provided that each of the n analytical results for a sample on a given analytical occasion is independent of the others [see Section 4.3.2(*c*)]. In general, however, this independence of results does not hold in the present context when the same estimates of calibration parameters are used to calculate all n analytical results on a given occasion (see Section 6.2.1). In terms of the model of random errors described in Section 6.2.1(*d*), σ_m is given by

$$\sigma_m = [(\sigma_w^2/n) + \sigma_b^2]^{\frac{1}{2}} \qquad (3.10)$$

and
$$\sigma_t = [\sigma_w^2 + \sigma_b^2]^{\frac{1}{2}} \qquad (3.11)$$

where σ_w and σ_b are the within- and between-batch standard deviations of analytical results, respectively. Equations (3.10) and (3.11) show that $\sigma_m = \sigma_t/\sqrt{n}$ only when $\sigma_b = 0$. In practice, it will be reasonable to assume that $\sigma_m = \sigma_t/\sqrt{n}$ when σ_b^2 is negligible in comparison with σ_w^2/n. Situations may arise where σ_b is zero, e.g., when standard addition procedures are used [see Sections 6.2.1(b) and 6.2.2]. More usually, however, $\sigma_b \neq 0$, and some means of overcoming the complication represented by equation (3.10) is desirable.

The approach suggested is to consider the sources of random error contributing to σ_b [see Section 6.2.1(c)], and either to make these random errors negligible for a given analytical occasion or to replicate the measurements involved n times on each occasion. As described in Section 6.2.1(c), the practical implications of this approach depend on the nature of the calibration function and the calibration procedure normally used. Each determinand, therefore, needs individual consideration, but we discuss below one commonly occurring situation in order to illustrate the suggested approach.

Suppose that: (1) the analytical response is proportional to determinand concentration, (2) in each batch of analyses a blank determination and a number of standard solutions of different concentrations are analysed, (3) an estimate, k, of the slope of the calibration curve is derived from the analytical responses for the standards, and (4) the analytical result, R, for an analysed portion of a sample is calculated from

$$R = (A_T - A_B)/k \qquad (3.12)$$

where A_T and A_B are the analytical responses for the sample and blank, respectively. Under these conditions, the total standard deviation, σ_t, of individual analytical results is given by [see Section 6.2.1(c)]

$$\sigma_t^2 \approx \sigma_T^2 + \sigma_B^2 + C_o^2 \sigma_k^2/k^2 \qquad (3.13)$$

where the precise meanings of the symbols are as defined in Section 6.2.1(c), but, broadly, σ_T and σ_B represent the standard deviations characterising the random errors of A_T/k and A_B/k, and σ_k is the standard deviation characterising the random error of k. If an independent blank determination and a set of calibration standards were analysed for each portion of sample analysed, and the results were each calculated from equation (3.12), σ_b would then be zero and σ_m then equals $\sigma_t/\sqrt{n}$ [where σ_t is given by equation (3.13)] as desired, *i.e.*, an appropriate value for n may be calculated by the procedure of Section 4.6.2.

The need in the above approach to analyse n sets of standards on each analytical occasion is effort-demanding and may be avoided by ensuring that the number and concentrations of the standards in one set are such that σ_k is sufficiently small to make $C_o^2\sigma_k^2/k^2$ negligible in comparison with $(\sigma_T^2 + \sigma_B^2)/n$. When six or more standards are included in a set, σ_k will usually be sufficiently small that one set of standards in a batch of analyses will suffice; calculation of σ_k may be made if desired by the procedures described in Section 5.4.2(e). Provided that $C_o^2\sigma_k^2/k^2$ may be ignored, $\sigma_t = (\sigma_T^2 + \sigma_B^2)^{\frac{1}{2}}$ and $\sigma_m = [(\sigma_T^2 + \sigma_B^2)/n]^{\frac{1}{2}} = \sigma_t/\sqrt{n}$. For this approach, the n analytical results are calculated from equation (3.12) using the n pairs of sample and blank responses. If σ_B is much smaller than σ_T, it is, in general, possible to analyse fewer than n blanks on each analytical occasion without substantially affecting σ_m.

An alternative approach to ensuring that $\sigma_m = \sigma_t/\sqrt{n}$ is to analyse on each analytical occasion n triads – sample/blank/calibration standard – the n analytical results being calculated from equation (6.10) of Section 6.2.1(*c*)(*ii*). With this procedure, $\sigma_b = 0$, σ_t is given by equation (6.11), and $\sigma_m = \sigma_t/\sqrt{n}$. In general, the concentration of determinand in the calibration standard should be approximately the same as that in the sample. Provided that this equality is closely achieved, the experimental design can be further simplified since only one blank determination need be made in each batch of analyses [see equation (6.11)].

It should be stressed that the required value of n may be large so that tests of stability often call for appreciable effort. For example, with the approach described in the preceding paragraph, n is given by

$$n \approx (7.2\sigma_T/\delta_c)^2 \qquad (3.14)$$

Thus, if δ_c were required to be no larger than $2\sigma_T$, n must be approximately 13. Equation (3.14) shows the value of achieving as small a value as possible for σ_T since this will markedly reduce the value of n for a given value of δ_c.

(e) Sources of Systematic Error in d
If $\bar{R}_t$ and $\bar{R}_o$ are biased with respect to each other (see Section 4.4), the value of d will not be determined solely by changes in the determinand concentration of the sample. Sources of bias in analytical results are considered in detail in Section 5, and may be summarised as follows in the present context.

(i) Calibration and blank correction Since the values for $\bar{R}_t$ and $\bar{R}_o$ are obtained from analyses made on two different occasions, different estimates of calibration parameters will, in general, be used on each occasion. It is very desirable, therefore, that the procedures used to estimate the calibration parameters give unbiased values because this will avoid the possibility of different biases in these estimates on the two occasions. Relevant factors are discussed in Sections 5.4.3 and 5.4.4. Note that, in trace analysis, special care may well be needed to avoid bias caused by the presence of the determinand in the water used to prepare standards or blanks; see Sections 5.4.4(*b*) and 10.7.

(ii) Forms of the determinand It is by no means impossible that, as the sample ages, the chemical and/or physical forms of the determinand change. This could lead to $\bar{R}_t$ being biased with respect to $\bar{R}_o$ (see Section 5.4.5). Control of this source of bias is primarily achieved by appropriate choice of analytical method, but this is not a simple topic and we have no simple advice to offer. Nonetheless, the possibility of this source of error should be borne in mind.

(iii) Interference Given the changes that may occur in the chemical and physical forms of a number of substances as the sample ages, it is possible that the size of interference effects will vary from one test occasion to another; see Section 5.4.6. Again, we have no simple advice, but the problem clearly becomes less potentially important as the liability of the analytical system to interference becomes smaller.

(f) Reporting Results of Stability Tests
Given the above points, we think that little quantitative value attaches to reports

that do not state the experimental design used for stability tests and/or that quote the results in an ambiguous manner. For example, statements such as 'the sample was found to be stable' or 'no important changes of concentrations were observed' are of little help to analysts in establishing their own procedures – nor do such statements contribute greatly to attempts to achieve order in this difficult topic of sample stability. As a minimum in published accounts, we suggest that the experimental design should be unambiguously described and that the results $d+l$ be quoted for each sample and value of t tested, where $d+l$ and $d-l$ are the confidence limits (stated probability level) for δ [see Section 4.3.5(a)].

3.7 REFERENCES

[1] L. A. Currie, *Anal. Lett.*, 1980, **13**, 1.
[2] J. K. Taylor, *Anal. Chem.*, 1981, **53**, 1588A.
[3] E. L. Berg, ed., 'Handbook for Sampling and Sample Preservation of Water and Wastewater', (Report PB 83-124503), U.S. National Technical Information Service, Springfield, Virginia, 1982.
[4] American Chemical Society, *Anal. Chem.*, 1980, **52**, 2242.
[5] A. L. Wilson, in M. J. Suess, ed. 'Examination of Water for Pollution Control', Volume 1, Pergamon, Oxford, 1982, pp. 23–77.
[6] Standing Committee of Analysts, 'General Principles of Sampling and Accuracy of Results. 1980', H.M.S.O., London, 1980, pp. 15–48.
[7] IHD-WHO Working Group on the Quality of Water, 'Water Quality Surveys. A Guide for the Collection and Interpretation of Water-Quality Data', UNESCO/WHO, Paris/Geneva, 1978.
[8] P. H. Whitfield, *Water Resourc. Bull.*, 1983, **19**, 115.
[9] D. M. Ruthven, *Water Res.*, 1971, **5**, 343.
[10] P. R. Jaffe, F. L. Parker, and D. J. Wilson, *J. Environ. Engng Div., Am. Soc. Civ. Engrs.*, 1982, **108**, 639.
[11] N. Yotsukura and W.W. Sayre, *Water Resourc. Res.*, 1976, **12**, 695.
[12] H. B. Fischer, 'Mixing in Inland and Coastal Waters', Academic, London, 1979.
[13] R. V. Thomann, *Water Resourc. Res.*, 1973, **9**, 355.
[14] C. R. O'Melia, *Environ. Sci. Technol.*, 1980, **14**, 1052.
[15] Office of Water Data Coordination, 'National Handbook of Recommended Methods for Water-Data Acquisition', U.S. Geological Survey, Reston, Virginia, 1977, Chapter 3.
[16] H. H. Stevens, G. A. Lutz, and D. W. Hubbell, *J. Hydraul. Div., Am. Soc. Civ. Engrs.*, 1980, **106**, 611.
[17] American Society for Testing and Materials, '1981 Annual Book of ASTM Standards: Part 31. Water', ASTM, Philadelphia, 1981. (Note that this book is issued annually.)
[18] S. N. Davis, G. M. Thompson, H. W. Bentley, and G. Stiles, *Ground Water*, 1980, **18**, 14.
[19] J. C. André and J. Molinari, *J. Hydrol.*, 1976, **30**, 257.
[20] K. E. White, in 'First European Conference on Mixing and Centrifugal Separation, Cambridge, 1974', British Hydromechanics Research Association, Cranfield, 1974, pp. 57–76.
[21] H. Moser, *Vom Wasser*, 1981, **56**, 1.
[22] E. Fogelqvist, B. Josefsson, and C. Roos, *Environ. Sci. Technol.*, 1982, **16**, 479.
[23] L. Brown, *et al.*, *Water Res.*, 1982, **16**, 1409.
[24] J. E. Dorsch and T. F. Bidleman, *Estuarine Coastal and Shelf Sci.*, 1982, **15**, 701.
[25] G. V. Evans, *Int. J. Appl. Radiat. Isotop.*, 1983, **34**, 451.
[26] M. D. Ball, in J. E. Kerrigan, ed. 'Proc. National Symposium on Data and Instrumentation for Water Quality Mangement', University of Wisconsin, Madison, 1970, pp. 283–292.
[27] M. D. Palmer and J. B. Izatt, *Water Res.*, 1970, **4**, 773.
[28] A. Eaton, V. Grant, O. Bricker, and D. Wells, *Estuaries*, 1981, **4**, 379.
[29] P. J. Seiser, *et al.*, *Environ. Sci. Technol.*, 1983, **17**, 47.
[30] W. B. Neely, *Environ. Sci. Technol.*, 1982, **16**, 518A.
[31] T. G. Sanders, D. D. Adrian, and J. M. Joyce, *J. Water Pollut. Contr. Fed.*, 1977, **49**, 2467.
[32] D. D. Adrian, *et al.*, 'Cost Effective Stream and Effluent Monitoring', (Report PB 81-20221 0), U.S. National Technical Information Service, Springfield, Virginia, 1981.
[33] B. M. Chapman, *J. Hydrol.*, 1979, **40**, 139 and 153.
[34] H. T. Shen, *J. Environ. Engng Div., Am. Soc. Civ. Engrs.*, 1978, **104**, 445.
[35] T. G. Sanders, *Water SA*, 1982, **8**, 169.
[36] F. J. J. Brinkmann, B. C. J. Zoetman, and G. J. Piet, *Dtsch. Gewässerkundl. Mitt.*, 1973, **17**, 159.
[37] H. A. C. Montgomery and I. C. Hart, *Water Pollut. Contr. (London)*, 1974, **73**, 3.
[38] J. W. G. Lund and J. F. Talling, *Bot. Rev.*, 1957, **23**, 489.
[39] R. G. Gilbert and J. B. Miller, *Water Resourc. Res.*, 1976, **12**, 812.
[40] J. J. Kubus and D. A. Egloff, *J. Water Pollut. Contr. Fed.*, 1982, **54**, 1592.

41 W. G. Hines, D. A. Rickert, and S. W. McKenzie, *J. Water Pollut. Contr. Fed.*, 1977, **49**, 2031.
42 G. van Belle and J. P. Hughes, *J. Water Pollut. Contr. Fed.*, 1983, **55**, 400.
43 S. G. Chamberlain, C. V. Beckers, G. P. Grimsrud, and R. D. Shull, *Water Resourc. Bull.*, 1974, **10**, 199.
44 D. P. Lettenmaier and S. J. Burges, *J. Environ. Engng Div., Am. Soc. Civ. Engrs.*, 1977, **103**, 785.
45 D. P. Lettenmaier, *Water Resourc. Bull.*, 1978, **14**, 884.
46 A. M. Liebetrau, *Water Resourc. Res.*, 1979, **15**, 1717.
47 M. Lachance, B. Bobée, and D. Goum, *Water Resourc. Res.*, 1979, **15**, 1451.
48 W. E. Sharp, *Water Resourc. Res.*, 1971, **7**, 1641.
49 M. L. Richardson and J. M. Bowron, *Notes on Water Research No. 32*, Water Research Centre, Medmenham, Bucks, 1983.
50 S. R. Aston and I. Thornton, *Water Res.*, 1975, **9**, 189.
51 WHO Working Group, 'Micropollutants in River Sediments', (EURO Reports and Studies No. 61), World Health Organization, Regional Office for Europe, Copenhagen, 1982.
52 G. E. Hutchinson, 'A Treatise on Limnology', Volume 1, Wiley, London, 1957.
53 Organization for Economic Co-operation and Development, 'Eutrophication of Waters, Monitoring, Assessment and Control', O.E.C.D., Paris, 1982.
54 D. E. Ford and H. Stefan, *Water Resourc. Bull.*, 1980, **16**, 243.
55 International Organization for Standardization, ISO 5667/1-1980, I.S.O., Geneva, 1980. (This is identical with BS 6068: Part 6: Section 6.1: 1981).
56 R. D. Pomeroy and G. T. Orlob, 'Problems of Setting Standards and of Surveillance for Water Quality Control', California State Water Quality Control Board, Sacramento, 1967.
57 S. A. Wells and J. A. Gordon, *Water Resourc. Bull.*, 1982, **18**, 661.
58 M. D. Palmer, 'Required Density of Water Quality Sampling Stations at Nanticoke, Lake Erie', Ontario Water Resources Commission, Toronto, 1968.
59 W. J. Maier and W. R. Swain, *Water Res.*, 1978, **12**, 523.
60 K. W. Thornton, R. H. Kennedy, A. D. Magoun, and G. E. Saul, *Water Resourc. Bull.*, 1982, **18**, 471.
61 W. C. Sonzogni, *et al.*, *Environ. Sci. Technol.*, 1980, **14**, 148.
62 S. J. Eisenreich, ed., 'Atmospheric Pollutants in Natural Waters', Ann Arbor Science Publishers, Ann Arbor, Michigan, 1981.
63 Anon., *Environ. Sci. Technol.*, 1973, **7**, 198.
64 W. B. Wilkinson and K. Edworthy, in W. van Duijvenbooden, P. Glasbergen, and H. van Lelyveld, ed., 'Quality of Groundwater', Elsevier, Oxford, 1981, pp. 629–642.
65 J. A. Cole, ed., 'Water Pollution in Europe', Water Information Centre, Port Washington, New York, 1974.
66 Bureau de Recherches Géologiques et Minières, 'Protection des Eaux Souterraines Captées pour l'Alimentation Humaine', Orléans-la-Source, France, 1977.
67 W. van Duijvenbooden, P. Glasbergen, and H. van Lelyveld, ed., 'Quality of Groundwater', Elsevier, Oxford, 1981.
68 R. E. Jackson, ed., 'Aquifer Contamination and Protection', (Studies and Reports in Hydrology No. 30), U.N.E.S.C.O., Paris, 1980.
69 D. K. Todd, R. M. Tinlin, K. D. Schmidt, and L. G. Everett, *J. Am. Water Wks Assoc.*, 1976, **68**, 586.
70 K. H. Zwirnmann and S. Kaden, *Wasserwirtschaft Wassertechnik*, 1982, **32**, 117.
71 D. Reinhold and G. Muller, *Wasserwirtschaft Wassertechnik*, 1982, **32**, 114.
72 M. R. Scalf, *et al.*, 'Manual on Ground Water Sampling Procedures', U.S. Environmental Protection Agency, Washington, 1981.
73 M. Sophocleous, *J. Hydrol.*, 1983, **61**, 371.
74 Y. Bachmat and M. Ben-Zvi, in W. van Duijvenbooden, P. Glasbergen, and H. van Lelyveld, 'Quality of Groundwater', Elsevier, Oxford, 1981, pp. 607–619.
75 K. E. Childs, S. B. Upchurch, and B. Ellis, *Ground Water*, 1974, **12**, 369.
76 P. V. Roberts, M. Reinhard, and A. J. Valocchi, *J. Amer. Water Wks Assoc.*, 1982, **74**, 408.
77 G. L. Feder, *Ann. NY Acad. Sci.*, 1972, **199**, 118.
78 J. M. Rible, P. A. Nash, P. F. Pratt, and L. J. Lund, *Soil Sci. Soc. Am. J.*, 1976, **40**, 566.
79 H. I. Nightingale and W. C. Bianchi, *Water Resourc. Bull.*, 1979, **15**, 1394.
80 L. F. Konikow, *Water Resourc. Bull.*, 1981, **17**, 145.
81 D. Hammerton, A. J. Newton, and T. M. Leatherland, *Water Pollut. Contr. (London)*, 1981, **80**, 189.
82 J. H. Sharp, C. H. Culberson, and T. M. Church, *Limnol. Oceanogr.*, 1982, **27**, 1015.
83 H. Bergmann and B. Hoffmann, *Dtsch. Gewässerkundl. Mitt.*, 1976, **20**, 43.
84 World Health Organization, 'International Standards for Drinking Water', Third Edition, W.H.O., Geneva, 1971.
85 World Health Organization, 'European Standards for Drinking Water', Second Edition, W.H.O., Geneva, 1970.
86 World Health Organization, 'Surveillance of Drinking Water Quality', W.H.O., Geneva, 1976.
87 W. O. Skeat, ed., 'Manual of British Water Engineering Practice', Volume III, Fourth Edition, Institution of Water Engineers, London, 1969.

[88] W. S. Holden, ed., 'Water Treatment and Examination', Churchill, London, 1970.
[89] A. R. Castorina, *J. New Engl. Water Wks Assoc.*, 1978, **92**, 329.
[90] Council of the European Communities, *Off. J. Eur. Communities*, 1980, **23**, No. L-229, 11.
[91] C. S. Short, *Water Services*, 1980, **84**, 529.
[92] W. M. Lewis, in H. van Lelyveld and B. C. J. Zoetman, ed., 'Water Supply and Health', Elsevier, Oxford, 1981, pp. 285–292.
[93] J. C. Ellis and R. F. Lacey, *Water Pollut. Contr. (London)*, 1980, **79**, 452.
[94] J. C. Ellis, *Water (N.W.C.)*, 1980, No. 35, 9.
[95] Standing Technical Advisory Committee on Water Quality, 'Third Biennial Report', H.M.S.O., London, 1983.
[96] O. L. Davies and P. L. Goldsmith, ed., 'Statistical Methods in Research and Production', Fourth Revised Edition, Oliver and Boyd, Edinburgh, 1972.
[97] R. B. Dean, in W. J. Cooper, ed., 'Chemistry in Water Reuse', Volume 1, Ann Arbor Science Publishers, Ann Arbor, Michigan, 1981, pp. 245–258.
[98] P. H. Kemp, 'Statistics in Water Sampling', (C.S.I.R. Research Report 338), National Institute for Water Research, Pretoria, South Africa, 1978.
[99] R. A. Deininger and K. H. Mancy, in M. J. Suess, ed., 'Examination of Water for Pollution Control', Volume 1, Pergamon, Oxford, 1982, pp. 211–264.
[100] T. A. Wastler and C. A. Walter, *J. Sanit. Engng Div., Am. Soc. Civ. Engrs.*, 1968, **94**, 1175.
[101] C. G. Gunnerson, *J. Sanit. Engng Div., Am Soc. Civ. Engrs.*, 1966, **92**, 103.
[102] R. V. Thomann, *J. Sanit. Engng Div., Am. Soc. Civ. Engrs.*, 1967, **93**, 1.
[103] R. V. Thomann, *J. Sanit. Engng Div., Am. Soc. Civ. Engrs.*, 1970, **96**, 819.
[104] V. Kothandaraman, *J. Sanit. Engng Div., Am. Soc. Civ. Engrs.*, 1971, **97**, 19.
[105] F. C. Fuller and C. P. Tsolos, *Biometrics*, 1971, **27**, 1017.
[106] J. S. Shastry, L. T. Fan, and L. E. Erickson, *Water, Air, Soil Pollut.*, 1972, **1**, 233.
[107] A. M. C. Edwards and J. B. Thornes, *Water Resourc. Res.*, 1973, **9**, 1286.
[108] D. P. Lettenmaier, *Water Resourc. Res.*,1976, **12**, 1037.
[109] J. C. Loftis and R. C. Ward, *Water Resourc. Bull.*, 1980, **16**, 501.
[110] J. C. Loftis and R. C. Ward, 'Regulatory Water Quality Monitoring Networks – Statistical and Economic Considerations', (Report EPA 600/4-79-055), U.S. Environmental Protection Agency, Las Vegas, Nevada, 1979.
[111] T. G. Sanders and D. D. Adrian, *Water Resourc. Res.*, 1978, **14**, 569.
[112] T. Schilperoort, S. Groot, B. G. M. van de Wetering, and F. Dijkman, H_2O, 1982, **15**, 82.
[113] B. Bobée and D. Cluis, *J. Hydrol.*, 1979, **44**, 17.
[114] P. J. W. M. Müskens and W. G. J. Hensgens, *Water Res.*, 1977, **11**, 509.
[115] P. J. W. M. Müskens and G. Kateman, *Anal. Chim. Acta*, 1978, **103**, 1.
[116] G. Kateman and P. J. W. M. Müskens, *Anal. Chim. Acta*, 1978, **103**, 11.
[117] P. J. W. M. Müskens, *Anal. Chim. Acta*, 1978, **103**, 445.
[118] G. Kateman and F. W. Pijpers, 'Quality Control in Analytical Chemistry', Wiley, Chichester, 1981.
[119] B. J. Adams and R. S. Gemmell, *J. Water Pollut. Contr. Fed.*, 1973, **45**, 2088.
[120] P. M. Berthouex and W. G. Hunter, *J. Water Pollut. Contr. Fed.*, 1975, **47**, 2143.
[121] T. Schofield, *Water Pollut. Contr. (London)*, 1980, **79**, 468.
[122] M. Sonneborn, *Gewässerschutz Wasser Abwasser*, 1982, **57**, 23.
[123] W. Schmitz, *Verh. Internat. Verein. Limnol.*, 1975, **19**, 2020.
[124] C. E. Hamilton, ed., 'Manual on Water', American Society for Testing and Materials, Philadelphia, 1978.
[124a] J. A. Tetlow, *Effl. Water. Treat. J.*, 1968, **8**, 79.
[125] R. C. Chang and J. S. Fritz, *Talanta*, 1978, **25**, 659.
[126] P. Erickson, 'Marine Trace Metal Sampling and Storage', (Report No. 1, NRCC No. 16472), National Research Council of Canada, Halifax, 1978.
[127] J. C. Meranger, K. S. Subramanian, and C. H. Langford, *Rev. Anal. Chem.*, 1980, **5**, 29.
[128] N. Andersen, *et al.*, Intergovernmental Oceanographic Commission Technical Series 22, U.N.E.S.C.O., Paris, 1982.
[129] R. Care, J. D. Morrison, and J. F. Smith, *Water Res.*, 1982, **16**, 663.
[130] J. C. Loper and M. W. Tabor, in M. D. Waters, ed., 'Short Term Bioassays in the Analysis of Complex Environmental Mixtures', Volume 3, Plenum, New York.
[131] B. Josefsson, A. Bjørseth, and G. Angeletti, ed., 'Analysis of Organic Micropollutants in Water', Reidel, London, 1982, pp. 7–15.
[132] R. F. Packham, D. Roseman, and H. G. Midgley, *Clay Miner. Bull.*, 1961, **4**, 239.
[133] W. T. Lammers, in L. L. Ciaccio, ed., 'Water and Water Pollution Handbook', Volume 2, Dekker, London, 1972, pp. 593–638.
[134] T. S. Bates, S. E. Hamilton, and J. D. Cline, *Estuarine Coastal and Shelf Sci.*, 1983, **16**, 107.
[135] K. Hofmann, in 'Speisewassertagung 1966', Vereinigung der Grosskessel-besitzer, Essen, 1966, p. 2.
[136] American Public Health Association, *et al.*, 'Standard Methods for the Examination of Water and Wastewater', Fifteenth Edition, A.P.H.A., Washington, 1981.

[137] J. B. Andelman and S. C. Caruso, in L. L. Ciaccio, ed., 'Water and Water Pollution Handbook', Volume 2, Dekker, London, 1972, pp. 483–591.

[138] S. D. Faust and I. H. Suffet, in L. L. Ciaccio, ed., 'Water and Water Pollution Handbook', Volume 3, Dekker, London, 1972, pp. 1249–1313.

[139] I. H. Suffet, L. Brenner, J. T. Coyle, and P. R. Cairo, *Environ. Sci. Technol.*, 1978, **12**, 1315.

[140] R. Otson, P. D. Bothwell, and D. T. Williams, *J. Am. Water Wks Assoc.*, 1979, **71**, 42.

[141] M. Dressler, *J. Chromatogr.*, 1979, **165**, 167.

[142] B. A. Glatz, C. D. Chriswell, and G. A. Junk, in D. Schuetzle, ed., 'Monitoring Toxic Substances', (American Chemical Society Symposium No. 94), American Chemical Society, Washington, 1979, pp. 91–100.

[143] R. L. Jolley, *Environ. Sci. Technol.*, 1981, **15**, 874.

[144] E. C. Potter and J. F. Moresby, in 'Ion-exchange and its Applications', Society for Chemical Industry, London, 1955, p. 93.

[145] W. B. Silker, R. W. Perkins, and H. G. Rieck, *Ocean Engng*, 1971, **2**, 49.

[146] T. L. Yohe, I. H. Suffet, and R. J. Grochowski, in C. E. van Hall, ed., 'Measurement of Organic Pollutants in Water and Wastewater', (ASTM Spec. Tech. Publ. 686), A.S.T.M., Philadelphia, 1979, pp. 47–67.

[147] A. Mizuike, M. Hiraide, and K. Mizuno, *Anal. Chim. Acta*, 1983, **148**, 305.

[148] P. Beneš, *Water Res.*, 1980, **14**, 511.

[149] K. V. Slack, *J. Water Pollut. Contr. Fed.*, 1971, **43**, 433.

[150] J. J. Black, T. F. Hart, and P. J. Black, *Environ. Sci. Technol.*, 1982, **16**, 247.

[151] R. C. Ward, *Water Resourc. Bull.*, 1979, **15**, 369.

[152] R. C. Ward and L. R. Freeman, *Water Resourc. Bull.*, 1973, **9**, 1234.

[153] D. G. Walker, *Effl. Water Treat. J.*, 1982, **22**, 184.

[154] R. C. Ward and J. C. Loftis, *J. Water Pollut. Contr. Fed.*, 1983, **55**, 408.

[155] T. M. Heidtke and J. M. Armstrong, *J. Water Pollut. Contr. Fed.*, 1979, **51**, 2916.

[156] J. C. Loftis and R. C. Ward, *Water Resourc. Bull.*, 1981, **17**, 1071.

[157] R. C. Ward, J. C. Loftis, K. S. Nielsen, and R. D. Anderson, *J. Water Pollut. Contr. Fed.*, 1979, **51**, 2292.

[158] R. C. Ward, K. S. Nielsen, and M. Bundgaard-Nielsen, 'Design of Monitoring Systems for Water Quality Management', (Contribution No. 3 from the Water Quality Institute), Danish Academy of Technical Sciences, Copenhagen, 1976.

[159] D. P. Lettenmaier, *Water Resourc. Res.*, 1979, **15**, 1692.

[160] L. G. Everett and K. D. Schmidt, ed., 'Establishment of Water Quality Monitoring Programs', American Water Resources Association, Minneapolis, 1979.

[161] R. W. Koch, T. G. Sanders, and H. J. Morel-Seytoux, *Water Resourc. Bull.*, 1982, **18**, 815.

[162] G. C. Dandy and S. F. Moore, *J. Environ. Engng Div., Am. Soc. Civ. Engrs.*, 1979, **105**, 695.

[163] World Health Organization, 'The Optimization of Water Quality Monitoring Networks', W.H.O. (Regional Office for Europe), Copenhagen, 1977.

[164] J. S. Fenlon and D. D. Young, *Water Pollut. Contr. (London)*, 1982, **81**, 343.

[165] M. A. Tirabassi, *Water Resourc. Bull.*, 1971, **7**, 1221.

[166] D. J. O'Connor, *Water Resourc. Res.*, 1976, **12**, 279.

[167] J. Ceasar, *et al.*, *Environ. Sci. Technol.*, 1976, **10**, 697.

[168] M. C. Feller and J. P. Kimmins, *Water Resourc. Res.*, 1979, **15**, 247.

[169] A. C. Osborne, *J. Hydrol.*, 1981, **52**, 59.

[170] F. A. Johnson and J. W. East, *J. Hydrol.*, 1982, **57**, 93.

[171] R. J. Stevens and R. V. Smith, *Water Res.*, 1978, **12**, 823.

[172] H. Weber, D. Cluis, and B. Bobée, *J. Hydrol.*, 1975, **40**, 175.

[173] R. V. Smith and D. A. Stewart, *Water Res.*, 1977, **11**, 631.

[174] P. H. Whitfield and H. Schreier, *Limnol. Oceanogr.*, 1981, **26**, 1179.

[175] D. C. Grobler, C. A. Bruwer, P. J. Kemp, and G. C. Hall, *Water SA*, 1982, **8**, 121.

[176] R. J. Boes, in J. E. Kerrigan, ed., 'Proc. National Symposium on Data and Instrumentation in Water Quality Management', University of Wisconsin, Madison, 1970, pp. 395–405.

[177] E. Brown, M. W. Skougstad, and M. J. Fishman, 'Methods for Collection and Analysis of Water Samples for Dissolved Minerals and Gases', (Techniques of Water Resources Investigations of the United States Geological Survey – Book 5, Chapter A1), U.S. Government Printing Office, Washington, 1970.

[178] R. C. Ward and D. H. Vanderholm, *Water Resourc. Res.*, 1973, **9**, 536.

[179] H. F. Ludwig and P. N. Storrs, *J. Water Pollut. Contr. Fed.*, 1973, **45**, 2065.

[180] R. J. MacAniff and C. E. Willis, 'The Value of Data Acquisition from Water Quality Monitoring', (Report PB 282 594), U.S. National Technical Information Service, Springfield, Virginia, 1976.

[180a] International Organization for Standardization, ISO 6107/2-1981, I.S.O., Geneva, 1981. (This is identical with BS 6068: Part 1: Section 1.2: 1982.)

[181] D. Taylor, *Anal. Proc.*, 1982, **19**, 561.

[182] L. Mart, *et al.*, *Sci. Total Environ.*, 1982, **26**, 1.
[183] J. M. Bewers and H. L. Windom, *Mar. Chem.*, 1982, **11**, 71.
[184] G. F. Lee and R. A. Jones, *J. Water Pollut. Contr. Fed.*, 1983, **55**, 92.
[185] B. Kratochvil and J. K. Taylor, *Anal. Chem.*, 1981, **53**, 924A.
[186] R. L. Booth, ed., 'Handbook for Analytical Quality Control in Water and Wastewater Laboratories', (Report EPA-600/4-79-019), U.S. Environmental Protection Agency, Cincinnati, 1979.
[187] International Organization for Standardization, ISO 5667/2-1982, I.S.O., Geneva, 1982. (This is identical with BS 6068: Section 6.2: 1983.)
[188] L. Mart, *Fresenius' Z. Anal. Chem.*, 1979, **299**, 97.
[189] K. W. Bruland, R. P. Franks, G. A. Knauer, and J. H. Martin, *Anal. Chim. Acta*, 1979, **105**, 233.
[190] J. P. Riley, in J. P. Riley and C. Skirrow, ed., 'Chemical Oceanography', Second Edition, Volume 3, Academic, London, 1975, pp. 193–514.
[191] G. E. Batley and D. Gardner, *Water Res.*, 1977, **11**, 745.
[192] D. L. Miles and J. M. Cook, in W. van Duijvenbooden, P. Glasbergen, and H. van Lelyveld, ed., 'Quality of Groundwater', Elsevier, Oxford, 1981, pp. 725–731.
[193] S. Badzioch, *Br. J. Appl. Phys.*, 1959, **10**, 26.
[194] A. L. Wilson, *Chem. Ind.*, 1969, 1253.
[195] P. E. Shelley, 'Sampling of Water and Wastewater', (Report EPA-600/4-77-039), U.S. National Technical Information Service, Springfield, Virginia, 1977.
[196] G. D. Reed, *J. Water Pollut. Contr. Fed.*, 1981, **53**, 1481.
[197] W. K. Fletcher and D. Polson, *Limnol. Oceanogr.*, 1982, **27**, 188.
[198] W. J. Keffer, in 'Water Pollution Assessment: Automatic Sampling and Measurement', (A.S.T.M. Spec. Tech. Publ. STP 582), A.S.T.M., Philadelphia, 1975, pp. 2–18.
[199] P. E. Shelley and G. A. Kirkpatrick, in 'Water Pollution Assessment: Automatic Sampling and Measurement', (A.S.T.M. Spec. Tech. Publ. STP 582), A.S.T.M., Philadelphia, 1975, pp. 19–36.
[200] M. Zief and J. W. Mitchell, 'Contamination Control in Trace Element Analysis', Wiley-Interscience, Chichester, 1976.
[201] R. W. Dabeka, A. Mykytink, S. S. Berman, and D. S. Russell, *Anal. Chem.*, 1976, **48**, 1203.
[202] G. A. Junk, H. J. Svec, R. D. Vick, and M. J. Avery, *Environ. Sci. Technol.*, 1974, **8**, 1100.
[203] S. Mori, *Anal. Chim. Acta*, 1979, **108**, 325.
[204] R. W. Heiden and D. A. Aikens, *Anal. Chem.*, 1977, **49**, 668.
[205] A. Pebler, *Anal. Chem.*, 1981, **53**, 361.
[206] G. E. Everitt, E. C. Potter, and R. G. Thompson, *J. Appl. Chem.*, 1965, **15**, 398.
[207] R. W. Gillham, *Ground Water Monitoring Rev.*, 1982, **2**, 36.
[208] W. van Duijvenbooden, in W. van Duijvenboden, P. Glasbergen, and H. van Lelyveld, ed., 'Quality of Groundwater', Elsevier, Oxford, 1981, pp. 617–628.
[209] J. F. Keely, *Ground Water Monitoring Rev.*, 1982, **2**, 29.
[210] R. B. Thomas and R. E. Eads, *Water Resourc. Res.*, 1983, **19**, 436.
[211] R. M. Sykes, A. J. Rubin, S. A. Rath, and M. C. Chang, *J. Water Pollut. Contr. Fed.*, 1978, **50**, 2046.
[212] Water Research Centre, in 'Commission of the European Communities' Second Environmental Research Programme 1976–1980. Reports on Research Sponsored under the Second Phase 1979–1980', Commission of the European Communities, Brussels, 1981, pp. 640–645.
[213] B. P. Hoyt, G. J. Kirmeyer, and J. E. Courchene, *J. Am. Water Wks Assoc.*, 1979, **71**, 720.
[213a] British Standards Institution, BS 1328:1969, B.S.I., London, 1969.
[214] H. L. Golterman, R. S. Clymo, and M. A. M. Ohnstad, 'Methods for Physical and Chemical Analysis of Fresh Waters', Second Edition, Blackwell, Oxford, 1978.
[215] B. H. Gump, *et al.*, *Anal. Chem.*, 1975, **47**, 1223.
[216] R. F. Spalding, M. E. Exner, and J. R. Gormly, *Water Resourc. Res.*, 1976, **12**, 1319.
[217] G. E. Grisak, R. E. Jackson, and J. F. Pickens, in L. G. Everett and K. D. Schmidt, ed., 'Establishment of Water Quality Monitoring Programs', American Water Resources Association, Minneapolis, 1979, pp. 210–232.
[218] W. W. Wood, in 'Techniques of Water Resources Investigations. Book 1', U.S. Geological Survey, Washington, 1981, Chapter D-2.
[219] R. M. Schuller, J. P. Gibb, and R. A. Griffin, *Ground Water Monitoring Rev.*, 1981, **1**, 42.
[220] W. M. Edmunds and A. H. Bath, *Environ. Sci. Technol.*, 1976, **10**, 467.
[221] K. C. Wheatstone and D. Gelsthorpe, *Analyst*, 1982, **107**, 731.
[222] K. H. Mancy and W. J. Weber, in I. M. Kolthoff, P. J. Elving, and F. H. Stross, ed., 'Treatise on Analytical Chemistry', First Edition, Part III, Volume 2, Wiley-Interscience, London, 1971, pp. 413–562.
[223] P. W. Anderson, J. J. Murphy, and S. D. Faust, in J. E. Kerrigan, ed., 'Proc. National Symposium on Data and Instrumentation for Water Quality Management', University of Wisconsin, Madison, 1970, pp. 261–281.
[224] E. M. Thain and P. H. H. Trendler, *Water SA*, 1978, **4**, 49.
[225] J. A. Cherry and P. E. Johnson, *Ground Water Monitoring Rev.*, 1982, **2**, 41.

226 M. J. L. Robin, D. J. Dytynshyn, and S. J. Sweeney, *Ground Water Monitoring Rev.*, 1982, **2**, 63.
227 R. Dawson, J. P. Riley, and R. H. Tennant, *Mar. Chem.*, 1976, **4**, 83.
228 R. E. Baier, H. B. Mark, and J. S. Mattson, ed., 'Water Quality Measurement', Dekker, New York, 1981, pp. 335–373.
229 A. H. Little, *Water Pollut. Contr. (London)*, 1973, **72**, 606.
230 D. M. Grant, *Water Engng Manag.*, 1982, June, 28.
231 R. L. Beschta, *Water Resour. Bull.*, 1980, **16**, 137.
232 D. L. Clark, R. Asplund, J. Ferguson, and B. W. Mar, *J. Environ. Engng Div., Am. Soc. Civ. Engrs.*, 1981, **107**, 1067.
233 P. D. Fisher and J. E. Siebert, *J. Environ. Engng Div., Am. Soc. Civ. Engrs*, 1977, **103**, 725.
234 V. D. Nguyen, P. Valenta, and H. W. Nurnberg, *Sci. Total Environ.*, 1979, **12**, 151.
235 L. F. Bliven, F. J. Humenik, F. A. Koehler, and M. A. Overcash, *J. Environ. Engng Div., Am. Soc. Civ. Engrs.*, 1979, **105**, 101.
235a Commission of the European Communities, 'Analysis of Organic Micropollutants in Water.' Report on the Meeting of Working Party 1, 'Sampling and Sample Treatment', Rome, 2 October 1982, C.E.C., Brussels, 1983.
236 H. L. Aichinger, *Sci. Total Environ.*, 1980, **16**, 279.
237 M. R. Coscio, G. C. Pratt, and S. V. Krupa, *Atmos. Environ.*, 1982, **16**, 1939.
238 E. E. Lewin and U. Torp, *Atmos. Environ.*, 1982, **16**, 795.
239 D. C. Tigwell, D. J. Schaeffer, and L. Landon, *Anal. Chem.*, 1981, **53**, 1199.
240 J. J. Westrick and M. D. Cummins, *J. Water Pollut. Contr. Fed.*, 1979, **51**, 2948.
241 G. A. Major, *Water Res.*, 1977, **11**, 525.
241a B. J. A. Haring, *Trib. Cebedeau*, 1978, **31**, 349.
242 J. D. Nelson and R. C. Ward, *Groundwater*, 1978, **16**, 617.
243 E. Wallick, *Bull. Alberta Res. Counc.*, 1977, **35**, 19.
244 G. Scarponi, *et al.*, *Anal. Chim. Acta*, 1982, **135**, 263.
245 R. Massee, F. J. M. J. Maessen, and J. J. M. de Goeij, *Anal. Chim. Acta*, 1981, **127**, 181.
246 P. Tschöpel, *et al.*, *Fresenius' Z. Anal. Chem.*, 1980, **302**, 1.
247 K. S. Subramanian, C. L. Chakrabarti, J. E. Sueiras, and I. S. Maines, *Anal. Chem.*, 1978, **50**, 444.
248 D. Maier, H. W. Sinemus, and E. Wiedeking, *Fresenius' Z. Anal. Chem.*, 1979, **296**, 114.
249 J. R. Moody and R. M. Lindstrom, *Anal. Chem.*, 1977, **49**, 2264.
250 V. T. Bowen, *et al.*, in Y. Nishiwaki and R. Fukai, ed., 'Reference Methods for Marine Radioactivity Studies', International Atomic Energy Agency, Vienna, 1970.
251 P. D. La Fleur, ed., 'Accuracy in Trace Analysis: Sampling, Sample Handling, Analysis', Volumes 1 and 2, (NBS Spec. Publ. 422), National Bureau of Standards, Washington, 1976.
252 P. Tschöpel and G. Tölg, *J. Trace Microprobe Techniques*, 1982, **1**, 1.
253 *Talanta*, 1982, **29**, 963–1055.
254 D. W. Denney, F. W. Karasek, and W. D. Bowers, *J. Chromatogr.*, 1978, **151**, 75.
255 D. P. H. Laxen and R. M. Harrison, *Anal. Chem.*, 1981, **53**, 345.
256 L. Mart, *Fresenius' Z. Anal. Chem.*, 1979, **296**, 350.
257 B. Kinsella and R. L. Willix, *Anal. Chem.*, 1982, **54**, 2614.
258 L. Mart, *Talanta*, 1982, **29**, 1035.
259 J. R. W. Kerr, D. I. Coomber, and D. T. Lewis, *Analyst*, 1962, **87**, 944.
260 J. H. Cragin, *Anal. Chim. Acta*, 1979, **110**, 313.
261 N. R. Schock and S. C. Schock, *Water Res.*, 1982, **16**, 1455.
262 S. P. Levine, *et al.*, *Environ. Sci. Technol.*, 1983, **17**, 125.
263 K. C. Thompson, 'Atomic Absorption Spectrophotometry, 1979 Version', H.M.S.O., London, 1980.
264 H. Scheuermann and H. Hartkamp, *Fresenius' Z. Anal. Chem.*, 1983, **315**, 430.
265 V. Cheam and H. Agemian, *Analyst*, 1980, **105**, 737.
266 V. Cheam and H. Agemian, *Anal. Chim. Acta*, 1980, **113**, 237.
267 G.-P. Husson, *J. Franc. Hydrologie*, 1976, **7**, 33.
268 J. C. Ryden, J. K. Syers, and R. F. Harris, *Analyst*, 1972, **97**, 903.
269 J. J. Latterell, D. R. Timmons, R. F. Holt, and E. M. Sherstad, *Water Resourc. Res.*, 1974, **10**, 865.
270 G. E. Carlberg and K. Martinsen, *Sci. Total Environ.*, 1982, **25**, 245.
271 G. E. Carlberg and K. Martinsen, in A. Bjørseth and G. Angeletti, ed., 'Analysis of Organic Micropollutants in Water', Reidel, London, 1982, pp. 42–44.
272 K. F. Sullivan, E. L. Atlas, and C.-S. Giam, *Anal. Chem.*, 1981, **53**, 1719.
273 H. Zawadzka, J. Zerbe, and D. Baralkiewicz, *Chem. Anal. (Warsaw)*, 1980, **25**, 469.
274 L. Weil and K.-E. Quentin, *Gas-Wasserfach, Wasser–Abwasser*, 1970, **111**, 26.
275 Standing Committee of Analysts, 'Organochlorine Insecticides and Polychlorinated Biphenyls in Waters. 1978', H.M.S.O., London, 1979.
276 C. W. Carter and I. H. Suffet, *Environ. Sci. Technol.*, 1982, **16**, 735.
277 J. P. Hassett and M. A. Anderson, *Environ. Sci. Technol.*, 1979, **13**, 1526.
278 N. D. Bedding, A. E. McIntyre, R. Perry, and J. N. Lester, *Sci. Total Environ.*, 1983, **26**, 255.

124   *Chemical Analysis of Water*

279 I. Ogawa, G. A. Junk, and H. J. Svec, *Talanta*, 1981, **28**, 725.
280 C. L. Chakrabarti, K. S. Subramanian, J. E. Sueiras, and D. J. Young, *J. Am. Water Wks Assoc.*, 1978, **70**, 560.
281 P. H. Whitfield and J. W. McKinley, *Water Resourc. Bull.*, 1981, **17**, 381.
282 J. P. Giesy and L. A. Briese, *Water Resourc. Res.*, 1978, **14**, 542.
283 J. Ranchet, F. Pescheux, and F. Menissier, *Tech. Sci. Municipales L'Eau*, 1981, **76**, 547.
284 U. Neis, Z. *Wasser Abwasser-Forsch.*, 1978, **11**, 3.
285 F. J. Agardy and M. L. Kidds, in 'Proc. 21st Industrial Waste Conference, 1966', Purdue University, Lafayette, undated, pp. 226–233.
286 J. D. Burton, T. M. Leatherland, and P. S. Liss, *Limnol. Oceanogr.*, 1970, **15**, 473.
287 P. Pichet, K. Jamati, and P. D. Goulden, *Water Res.*, 1979, **13**, 1187.
288 S. W. Hager, *et al.*, *Limnol. Oceanogr.*, 1972, **17**, 931.
289 D. W. Nelson and M. J. M. Romkens, *J. Environ. Qual.*, 1972, **1**, 323.
290 A. H. Johnson, D. R. Bouldin, and G. W., Hergert, *Water Resourc. Res.*, 1975, **11**, 559.
291 D. J. Hydes and P. S. Liss, *Analyst*, 1976, **101**, 922.
292 R. W. Macdonald and F. A. McLaughlin, *Water Res.*, 1982, **16**, 95.
293 S. H. Omang and O. D. Vellar, *Clin. Chim. Acta*, 1973, **49**, 125.
294 A. Otsuki and K. Fuwa, *Talanta*, 1977, **24**, 584.
295 C. P. Hwang, T. H. Lackie, and R. R. Munch, *Environ. Sci. Technol.* 1979, **13**, 71.
296 H.-H. Schierup and B. Riemann, *Arch. Hydrobiol.*, 1979, **86**, 204.
297 A. D. Eaton and V. Grant, *Limnol. Oceanogr.*, 1979, **24**, 397.
298 Standing Committee of Analysts, 'Dissolved Oxygen in Natural and Waste Waters. 1979', H.M.S.O., London, 1980.
299 A. Miyazaki, A. Kimura, and Y. Umezaki, *Anal. Chim. Acta*, 1981, **127**, 93.
300 M. I. Liddicoat, S. Tibbitts, and E. I. Butler, *Water Res.*, 1976, **10**, 567.
301 F. H. Rainwater and L. L. Thatcher, 'Methods for Collection and Analysis of Water Samples', U.S. Government Printing Office, Washington, 1960.
302 R. B. Barnes, *Chem. Geol.*, 1975, **15**, 177.
303 L. Ögren, *Anal. Chim. Acta*, 1981, **125**, 45.
304 G. E. Batley, *Anal. Chim. Acta*, 1981, **124**, 121.
305 G. Torsi, E. Desimoni, F. Palmisano, and L. Sabbatini, *Anal. Chim. Acta*, 1981, **124**, 143.
306 E. Michnowsky, L. M. Churchland, P. A. Thomson, and P. H. Whitfield, *Water Reourc. Bull.*, 1982, **18**, 129.
307 A. Henriksen and K. Balmer, *Vatten*, 1977, **33**, 33.
308 J. W. Owens, E. S. Gladney, and W. D. Purtymun, *Anal. Lett.*, 1980, **13**, 253.
309 W. C. Hoyle and A. Atkinson, *Appl. Spectrosc.*, 1979, **33**, 37.
310 G. P. Fitzgerald and S. L. Faust, *Limnol. Oceanogr.*, 1967, **12**, 332.
311 E. D. Klingaman and D. W. Nelson, *J. Environ. Qual.*, 1976, **5**, 42.
312 J. R. Majer and S. E. A. Khalil, *Anal. Chim. Acta*, 1981, **126**, 175.
313 Standing Committee of Analysts, 'Magnesium in Waters and Sewage Effluents by Atomic Absorption Spectrophotometry. 1977', H.M.S.O., London, 1978.
313a Standing Committee of Analysts, 'Calcium in Waters and Sewage Effluents by Atomic Absorption Spectrophotometry. 1977', H.M.S.O., London, 1978.
314 T. T. Chao, E. A. Jenne, and L. N. Heppting, in 'United States Geological Survey Professional Paper 600-D', U.S. Government Printing Office, Washington, 1968, pp. D13–D15.
314a J. Owerbach, *J. Appl. Photographic Engng*, 1978, **4**, 22.
315 J. M. Lo and C. M. Wai, *Anal. Chem.*, 1975, **47**, 1869.
316 E. A. Jenne and P. Avotins, *J. Environ. Qual.*, 1975, **4**, 427.
317 E. A. Jenne and P. Avotins, *J. Environ Qual.*, 1975, **4**, 515.
318 C. Feldman, *Anal. Chem.*, 1974, **46**, 99.
319 M. R. Christman and J. D. Ingle, *Anal. Chim. Acta*, 1976, **86**, 53.
320 Standing Committee of Analysts, 'Mercury in Waters, Effluents and Sludges by Flameless Atomic Absorption Spectrophotometry. 1978', H.M.S.O., London, 1978.
321 S. P. Piccolino, *J. Chem. Ed.*, 1983, **60**, 235.
322 J. Carron and H. Agemian, *Anal. Chim. Acta*, 1977, **92**, 61.
323 P. D. Goulden and D. H. J. Anthony, *Anal. Chim. Acta*, 1980, **120**, 129.
324 I. Sanemasa, *et al.*, *Anal. Chim. Acta*, 1977, **94**, 421.
325 K. Matsunaga, S. Konishi, and M. Nishimura, *Environ. Sci. Technol.*, 1979, **13**, 63.
326 J. R. Knechtel, *Analyst*, 1980, **105**, 826.
327 B. J. Farey, L. A. Nelson, and M. G. Rolph, *Analyst*, 1978, **103**, 656.
328 G. Kaiser, D. Götz, P. Schoch, and G. Tölg, *Talanta*, 1975, **22**, 889.
329 R. W. Heiden, and D. A. Aikens, *Anal. Chem.*, 1979, **51**, 151.
330 C. E. Oda and J. D. Ingle, *Anal. Chem.*, 1981, **53**, 2305.
331 J. Heron, *Limnol. Oceanogr.*, 1962, **7**, 316.

Sampling 125

332 D. C. Bowditch, C. R. Edmond, P. J. Dunstan, and J. A. McGlynn, 'Suitability of Containers for Storage of Water Samples', (Australian Water Resources Council Tech. Paper No. 16), Australian Government Printing Services, Canberra, 1976.

333 T. J. Williams, *J. Am. Water Wks Assoc.*, 1979, **71**, 42.

334 B. Riemann and H.-H. Schierup, *Water Res.*, 1978, **12**, 849.

335 Standing Committee of Analysts, 'Phosphorus in Waters, Effluents and Sewages. 1980', H.M.S.O., London, 1981.

336 V. Krivan, *Talanta*, 1982, **29**, 1041.

337 A. M. Gunn, Technical Report TR 169, Water Research Centre, Medmenham, Bucks., 1981.

338 J. M. Long and C. H. Anderson, 'Preservation of Water Samples for Asbestos Determination', Paper read at 12th Annual Symposium on the Analytical Chemistry of Pollutants, Amsterdam, 14–16 April, 1982.

339 E. S. Gladney and W. E. Goode, *Anal. Lett.*, 1977, **10**, 619.

340 A. D. Shendrikar, V. Dharmoivajan, H. Walker-Merrick, and P. W. West, *Anal. Chim. Acta*, 1976, **84**, 409.

341 German Chemists Association, *Water Res.*, 1981, **15**, 233.

342 D. S. Lee, *Anal. Chem.* 1982, **54**, 1682.

343 T. Korenaga, S. Motomizu, and K. Tôei, *Analyst*, 1980, **105**, 955.

344 R. A. van Steenderen, *Water SA*, 1976, **2**, 156.

345 R. Otson, *et al.*, *Bull. Environ. Contam. Toxicol.*, 1979, **23**, 311.

346 M. O. Andreae and P. N. Froelich, *Anal. Chem.*, 1981, **53**, 287.

347 S. de J. Alevato and A. de L. Rebello, *Talanta*, 1981, **28**, 909.

348 R. E. Pellenberg and T. M. Church, *Anal. Chim. Acta*, 1978, **97**, 81.

349 K. I. Mahan and S. E. Mahan, *Anal. Chem.*, 1977, **49**, 662.

350 A. A. El-Awady, R. B. Miller, and M. J. Carter, *Anal. Chem.* 1976, **48**, 110.

351 M. Stoeppler and W. Matthes, *Anal. Chim. Acta*, 1978, **98**, 389.

352 F. D. Schaumberg, *J. Water Pollut. Contr. Fed.*, 1971, **43**, 1671.

353 L. C. Michael, *et al.*, in L. H. Keith, ed., 'Advances in the Identification and Analysis of Organic Pollutants in Water', Volume 1, Ann Arbor Science Publishers, Ann Arbor, Michigan, 1981, pp. 87–114.

354 A. Dombi, Z. Galbaes, and D. Kiraly, *Hidrol. Kozl.*, 1980, **60**, 519 (*Anal. Abstr.*, 1982, **42**, 817, No. 6H63.)

355 B. Josefsson, P. Lindroth, and G. Östling, *Anal. Chim. Acta*, 1977, **89**, 21.

356 T. A. Bellar and J. J. Lichtenberg, in C. E. van Hall, ed., 'Measurement of Organic Pollutants in Water and Wastewater', (ASTM Spec. Tech. Publ. 686), A.S.T.M., Philadelphia, 1979, pp. 108–129.

357 D. J. Schultz, *et al.*, *J. Res. U.S. Geol. Survey*, 1976, **4**, 247.

358 H. Hirayama, *Bunseki Kagaku*, 1978, **27**, 252 (*Anal. Abstr.*, 1978, **35**, 432, No. 4H39.)

359 G. J. Piet, *et al.*, *Anal. Lett.*, 1978, **11**, 437.

360 R. L. Tate and M. Alexander, *J. Nat. Cancer. Inst.*, 1975, **54**, 327.

361 J. D. Miller, R. E. Thomas, and H. J. Schattenberg, *Anal. Chem.*, 1981, **53**, 214.

362 S. Bourne, *J. Environ. Sci. Health*, 1978, **13B**, 75.

363 M. J. Carter and M. T. Huston, *Environ. Sci. Technol.*, 1978, **12**, 309.

364 B. K. Afghan, P. E. Belliveau, R. H. Larose, and J. F. Ryan, *Anal. Chim. Acta*, 1974, **71**, 355.

365 K. Higuchi, Y. Shimoishi, H. Miyata, and K. Tôei, *Analyst*, 1980, **105**, 768.

366 W. T. Sullivan, *J. Am. Water Wks Assoc.*, 1975, **67**, 146.

367 R. J. Wilcock, C. D. Stevenson, and C. A. Roberts, *Water Res.*, 1981, **15**, 321.

368 W. Davison, *Freshwater Biol.*, 1977, **7**, 393.

369 A. C. Rossin and J. N. Lester, *Environ. Technol. Lett.*, 1980, **1**, 9.

370 G. A. Cutter, *Anal. Chim. Acta*, 1978, **98**, 59.

371 C. I. Measures and J. D. Burton, *Anal. Chim. Acta*, 1980, **120**, 177.

372 R. W. Hayes, M. W. Wharmby, R. W. C. Broadbank, and K. W. Morcom, *Talanta*, 1982, **29**, 149.

373 E. S. Gladney and W. E. Goode, *Anal. Chim. Acta*, 1977, **91**, 411.

374 A. W. Garrison and E. D. Pellizzari, in J. Albaiges, ed., 'Analytical Techniques in Environmental Chemistry. 2', Pergamon, Oxford, 1982, pp. 87–99.

375 J. E. Longbottom and J. J. Lichtenberg, eds., 'Methods for Organic Chemical Analyses of Municipal and Industrial Wastewater', (Report EPA-600/4-82-057), U.S. Environmental Protection Agency, Cincinnati, 1982.

376 R. D. Kleopfer, J. R. Dias, and B. J. Fairless, in W. J. Cooper, ed., 'Chemistry in Water Reuse', Volume 1, Ann Arbor Science Publishers, Ann Arbor, Michigan, 1981, pp. 207–227.

377 C. Boutron, *Anal. Chim. Acta*, 1979, **106**, 127.

378 R. M. Hirsch, J. R. Slack, and R. A. Smith, *Water Resourc. Res.*, 1982, **18**, 107.

379 A. Monteil, J. Carre, J. Devoucoux, and G. Bousquet, *Tech. Sci. Municipales*, 1981, **76**, 285.

380 R. W. Crabtree and S. T. Trudgill, *J. Hydrol.*, 1981, **53**, 361.

381 T. Kempf and M. Sonneborn, *Mikrochim. Acta*, 1981, **I**, 207.

382 J. M. Carter and C. S. Short, in 'Second Report of the Standing Committee of Analysts, February 1977—April 1979', (DOE/NWC Standing Technical Committee reports No. 23), National Water Council, London, 1981, pp. 45–50.
383 S. J. Nacht, *Ground Water Monitoring Rev.*, 1983, **3**, 23.
384 J.S.-Y. Ho, *J. Am. Water Wks Assoc.*, 1983, **75**, 583.
385 W. M. J. Strachan and H. Huneault, *Environ. Sci. Technol.*, 1984, **18,** 127.
386 Standing Committee of Analysts, 'Silver in Waters, Sewages and Effluents, 1982', H.M.S.O., London, 1983.

4 The Nature and Importance of Errors in Analytical Results

In this section we stress the fact that analytical results are subject to errors which cause the results to differ from the true concentrations of determinands, and which can, therefore, affect the validity of any decisions made on the basis of the results (Section 4.1). In Section 4.2 we define 'error' and the most common means for expressing its magnitude. Two main types of error – random and systematic – are usefully distinguished, and their natures and properties are considered in Sections 4.3 and 4.4, respectively; brief reference is made to their resultant – total error – in Section 4.5. In Sections 4.3 and 4.4 we also introduce statistical ideas and techniques essential in quantifying the sizes and effects of errors. Some simple examples of the effects of systematic and random errors on decision making are discussed in Section 4.6. These examples show the general need for every laboratory to ensure that the magnitudes of its errors are such that they do not invalidate the intended uses of analytical results. Finally, we argue in Section 4.7 that the need to control errors implies a general need for each laboratory to estimate the sizes of its errors. This estimation is taken up in detail in Sections 5 to 7.

In this and subsequent sections we shall repeatedly use several extremely common terms that are, however, used with different shades of meaning by different authors. To avoid ambiguity, the meanings throughout this book are as follows.

(1) The term *analytical method* denotes a set of written instructions specifying an *analytical procedure* to be followed by an analyst in order to obtain a numerical estimate of the concentration of a determinand in each of one or more samples.

(2) The term *analytical system* denotes a combination of analyst, analytical method, equipment, reagents, standards, laboratory facilities, and any other components involved in carrying out an analytical procedure. We do not include a sample or samples in 'analytical system'.

(3) The term *analytical result* denotes a numerical estimate of the concentration of a determinand in a sample, and is obtained by carrying out once the procedure specified in an analytical method. Note that a method may specify analysis of more than one portion of a sample in order to produce one analytical result.

(4) All analytical systems include a measurement sub-system (*e.g.*, spectrophotometric measurement of the absorbance of a solution) whose function – on presentation to it of a portion of a sample or treated sample – is to produce a numerical observation whose magnitude is related to the amount or concentration of the determinand in that portion. This numerical observation is termed an *analytical response*. One or more analytical responses (as specified by a method) are used, in conjunction with a calibration curve or factor, to produce an analytical result. Note that an analytical response may not be given directly by a measuring instrument, *e.g.*, when, in chromatographic procedures, peak areas must be calculated manually from traces on a recorder chart.

The 'analytical' in all the above terms will often be omitted in the text below when its presence is clearly implied. Given these meanings, two points are worth stressing. First, since analytical methods consist of writing on paper, they are not subject to error in the sense noted above and defined quantitatively in Section 4.2. Of course, methods may contain mistakes, ambiguities, and omissions, and they may be mis-applied; in such events, increased errors in analytical results may well ensue. Second, it is analytical systems that produce results and, therefore, the errors in these results*. Although analytical methods are sometimes the components of analytical systems with greatest effect on the errors of result, it is very important to note that this is by no means necessarily so.

4.1 THE IMPORTANCE OF ERRORS IN ANALYTICAL RESULTS

Analyses of waters are required for many different purposes. No matter what the purpose, the general aim of routine analysis is to obtain one or more analytical results that will then be compared with one or more other results (*e.g.*, when comparing the qualities of two or more waters) or with a specified numerical value (*e.g.*, when checking if a given water meets a specified standard of quality). In some applications, a function of the results – rather than the results themselves – may be used, but this does not affect the general arguments below. This comparison together with consideration of any other relevant factors (*e.g.*, economic, political, aesthetic) then forms the basis for deciding whether or not any action is required in relation to water quality. It is clearly essential to reach the correct conclusion as often as possible, and factors affecting the likelihood of making an incorrect decision should, therefore, always be carefully considered. The accuracy of analytical results is evidently of key importance in this situation.

Ideally, every analytical result would always be equal to the true concentration of the determinand, but it is impossible to achieve this desirable situation. For chemical analysis, just as for any other measurement process, it must be accepted that analytical results are subject to error and, therefore, generally differ from the true values. Account must be taken of the possible effects of such errors on the decision making processes. This implies immediately that the magnitude of the error of any result must be controlled and known to within limits sufficiently small to ensure that the validity of the results for their intended purpose is not destroyed. For example, analyses to check if a determinand concentration meets a water quality standard are essentially useless without knowledge of the errors of the results. Users of analytical results should, therefore, define the maximum tolerable magnitude of the error of results for each purpose, but bearing in mind that the difficulty and cost of analysis generally increase as the required accuracy is made more stringent [see Section 2.6.2(*a*)].

The above considerations thus identify two crucial topics:

(1) consideration of the effects of errors on decision making so that the maximum tolerable magnitude of error can be defined;

(2) estimation and, when necessary, reduction of the size of errors.

We use the term *analytical quality control* for (2) throughout this book, but other

* We are ignoring for the present any errors arising in the sampling process; see Sections 3.5.1(*b*) and 5.5.

terms are used by some authors, *e.g.*, analytical quality assurance. In the past, far too little attention has been paid to these two topics by both users and analysts. Within the last decade or so, however, there have been substantial increases in many countries in the attention paid to such matters (see Section 4.7). It seems likely that both topics will have a growing impact on the whole field of water quality measurement and control.

We believe that analytical quality control should be as central to the analyst's task as matters such as important developments in analytical techniques and procedures. For this reason alone, discussion of this subject forms an essential part of this edition just as it was in the first. There is, however, an additional reason, namely, that it seems that many analysts find some difficulty in dealing with the topic. We suspect that this arises partly because, for many, some unfamiliar statistical concepts and techniques are involved, but also because the central role of errors in any scientifically based work appears often to be insufficiently stressed in the training of analysts*. Accordingly, the remainder of Section 4 is devoted to what seems to us some of the more important aspects of analytical errors. Whole books have been written on such topics, and we have not thought the present section the place for a comprehensive treatment. Our aim has been to emphasise those situations in which it is important to consider errors, and to describe the underlying ideas and techniques that are of value. Nevertheless, the section has been markedly expanded in relation to the first edition in line with the growing interest in analytical errors. The text of Sections 4 to 7 together with the references and suggestions for further reading should allow any analyst to deal effectively with the estimation and control of analytical errors.

4.2 DEFINITION AND CLASSIFICATION OF ERRORS

The error, ε, of a result, R, is defined as the difference between the result and the true value, τ:

$$\varepsilon = R - \tau. \tag{4.1}$$

The error may thus be positive or negative, and it is common to refer to ε as the absolute error, implying that the units in which ε is expressed are those used for τ and R.

One also speaks of the relative error, ε_r, of a result, where

$$\varepsilon_r = (R - \tau)/\tau,$$

and the relative percentage error, ε_p, is even more frequently used, where

$$\varepsilon_p = 100(R - \tau)/\tau.$$

It is very important in speaking of errors always to make clear which of these three ways of expressing them is being adopted; otherwise, much confusion can easily result. Throughout this book, absolute error, ε, is used except where stated otherwise.

It should be noted that, in speaking of analytical error, there is no implication

---

* Our experience is that it is by no means only analysts who meet this problem; users appear often to find similar difficulty.

that the analyst has made a mistake in the analysis of a sample though, of course, such a mistake may well cause error. Two types of error are usefully distinguished, namely, *random* and *systematic*. The natures of these two types of error are quite different so that they may have markedly different effects on the validity of results; they are considered in Sections 4.3 and 4.4, respectively.

Throughout these and subsequent sections it is generally assumed that the errors of analytical results arise entirely during the analysis of samples. Of course, it is possible for the sampling process also to introduce error. If desired, such errors could be included with those from analysis, but on grounds of simplicity we think it preferable not to do this. Of course, the potential sources of error arising in the sampling process (sample collection and instability of samples after collection) should be considered, and the aim should be to use, whenever possible, procedures leading to negligibly small errors [see Section 3.5.1(*b*)]. If sampling errors cannot be essentially eliminated, their effect on the total error of analytical results should certainly be considered; see also Section 5.5.

4.3 RANDOM ERROR

4.3.1 The Nature of Random Error

In the following we use the term *discrimination* with the meaning given below.

> The term *discrimination* denotes the smallest interval by which differences in analytical responses (expressed in concentration units*) can be reliably distinguished. For example, if the scale of a spectrophotometer can be read to the nearest 0.001 absorbance unit, the discrimination would be 0.001*f*, where *f* is the factor to convert absorbance to concentration units. Note that the discrimination may depend on the concentration of determinand in samples.

To come now to random error, we begin by noting that it is general experience that, with one exception considered below, repeated analysis of a stable, homogeneous sample does not give a series of identical results. The results obtained differ among themselves and are more or less scattered about some central value. This scatter is attributed to random errors which are so named because the sign and magnitude of the error varies at random from one result to another. A consequence is that the error of a particular result cannot be exactly predicted. Random errors are caused by uncontrolled and uncontrollable random variations in the many factors that may affect analytical results, *e.g.*, slight variations in the volumes of reagents added to samples, variations in the times allowed for chemical reactions, contamination effects, instrumental fluctuations, temperature variations, *etc.* Several terms have been, and are, used to denote this scatter of results, but the term *precision* appears to be in most widespread use in the English language, and is employed with this meaning throughout this book. Precision is said to improve as the scatter of repeated results decreases. Note that some authors use the term imprecision with the same sense as our precision.

---

* Of course, some analytical systems produce results that are not in concentration units. To avoid continually noting this point, we speak here only of concentrations, but the appropriate units should always be understood.

In the preceding paragraph we mentioned an exception for which repeated results do not show scatter. This exception arises when the discrimination of an analytical system is coarse in relation to the size of the random errors of results. The apparently perfect agreement of repeated analyses, however, is then illusory for, by the same token, samples differing in the concentration of the determinand may also give identical results. We may, therefore, disregard this exception, noting that valid estimates of random error cannot be obtained if the discrimination is not sufficiently fine. An approximate indication of this effect of the discrimination can be obtained from the equation

$$\sigma^2 = \sigma_e^2 - d^2/12 \,,$$

where σ = the true standard deviation (a measure of the scatter of repeated results – see Section 4.3.2) of analytical results in concentration units, and σ_e = the standard deviation of analytical results observed experimentally (also in concentration units) when the discrimination is equal to d.

Thus, when $d = 0.5\sigma_e$, the error of σ_e is approximately 1 %, and when $d = \sigma_e$ the error is approximately 4 %. Such errors, as will be seen below, are usually negligible in comparison with other uncertainties in estimating and using the standard deviations of analytical results. As a general rule, therefore, the effect of the discrimination may be ignored provided that it is not greater than the observed standard deviation.

The random error of an analytical result clearly means that the result can be regarded as only an estimate of the true value. Such an estimate will be worthless in any decision making process if the magnitude of the error cannot be quantified. Statistical techniques are important in that they allow this quantification to be made in an explicit, efficient, and reproducible manner which is also – to a very large extent – objective, though see Section 4.3.3(*b*). These techniques are based on the concepts that: (1) there is a definite probability* of obtaining a result of a given value in the repeated analysis of a sample and (2) the relationship between that probability and the value of an analytical result may be reasonably approximated by a given mathematical equation. It is usual for the equation used as the model for this relationship to be referred to as *the distribution of results*. A number of models and their corresponding equations are used in statistical techniques; throughout this book we shall assume, except where stated otherwise, that analytical results follow a particular distribution usually known as the *normal distribution* (sometimes also called the Gaussian distribution). This distribution is described in the next section.

4.3.2 The Normal Distribution

In this and later sections in which statistical ideas and techniques are discussed, we

---

* Throughout this discussion, the term *probability* is to be understood as the relative frequency with which a particular value is observed over a very large number of independent measurements. Thus, if the probability of a particular value is 0.5, this implies that this value will be observed, on average, for 50 % of all measurements. Note, however, that in a limited number of measurements (such as could be made in practice), the proportion of results with this value will vary from one set of measurements to another as a result of the random nature of the errors. Probability is expressed on a fractional scale where the probability of an event sure to happen is 1.0, and the probability of an event sure not to happen is 0.

have not attempted to be mathematically or statistically rigorous. We have tried to stress what seem to us the most important ideas involved in analytical situations, and have not tried to make a detailed presentation of particular statistical techniques nor to quote long strings of computational formulae. The intention has been primarily to show the vital role of statistical ideas in quantifying random errors and their effects. We hope that the discussion will encourage analysts to delve more deeply and often into texts on statistics (see Section 4.8 for references) and to consult statisticians whenever possible.

(a) Nature of the Normal Distribution

The normal distribution is defined by the following equation:

$$p(R) = (1/\sigma\sqrt{2\pi})\exp(-[R-\mu]^2/2\sigma^2) \qquad (4.2)$$

where the symbols have the following meanings in the context of practical analytical systems:

$p(R)$ = the probability of obtaining a result of value R in the analysis of a stable, homogeneous sample,*

μ = arithmetic mean of the infinite number of results that could be conceptually obtained for this sample; this set of all relevant results is referred to as *the population* of interest, and μ is called *the population mean*,

σ = *the population standard deviation* – see below.

From equation (4.2) it can be seen that the relationship between the numerical value of a result, R, and the probability, $p(R)$, of obtaining such a value is governed by only two parameters, μ and σ, which are referred to as *the parameters of the distribution*. The shape of the normal distribution is shown in Figure 4.1, and inspection of equation (4.2) shows that the peak of the distribution occurs at $R = \mu$. The width of the distribution is determined solely by the value of σ; as this parameter increases so too does the width. Evidently, this distribution has general properties that one might expect (on a commonsense basis) random errors to follow. For example, the most probable result is equal to the mean, μ, and the probability of obtaining a result different from μ decreases as the difference increases.

The width of the normal distribution reflects the scatter of results, and since this width is determined by σ, it follows that the standard deviation is an appropriate parameter for the quantitative characterisation of precision. Thus, precision improves as σ decreases. The term '*relative standard deviation*' is commonly used to denote either σ/μ or $100\sigma/\mu$, and the term '*coefficient of variation*' is also often used to denote $100\sigma/\mu$. Note also that the square of the standard deviation, σ^2, is called *the variance of the distribution*; this parameter plays a role of fundamental importance in statistical techniques.

More detailed analysis of equation (4.2) allows other important properties of the normal distribution to be derived. In particular, the proportions of results falling

---

* Strictly, $p(R)\mathrm{d}R$ is the probability that the value of R will fall in the interval between R and $R+\mathrm{d}R$. The probability that R will fall between two given values, R_1 and R_2, i.e., $R_1 \leqslant R \leqslant R_2$, is given by $\int_{R_1}^{R_2} p(R)\mathrm{d}R$.

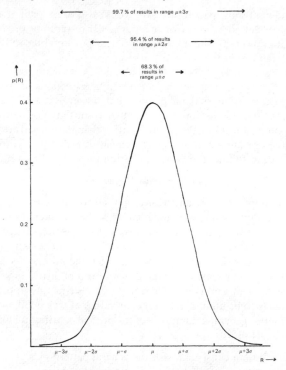

Figure 4.1 *The normal distribution*

within a given range from the mean are of special interest, and some of these proportions are shown in Figure 4.1 and in Table 4.1. Much more complete tabulations of these proportions are given in virtually all statistical texts.

TABLE 4.1
The Normal Distribution

Range of results/$\mu\pm$	1.00σ	1.50σ	1.64σ	1.96σ	2.17σ	2.58σ	2.97σ
Proportion of results within the range, %	68.3	86.6	89.9	95.0	97.0	99.0	99.7

Table 4.1 shows that the range may be written generally as $\mu\pm u_\alpha\sigma$, where the numerical value of u_α depends only on the proportion of results to be included within the range. Various symbols and conventions are used to denote this factor by which the standard deviation is multiplied. Here, we have used u_α, where $\alpha = 1 - P/100$, and P = the proportion (%) of results within the range defined by the two limits, $\mu\pm u_\alpha\sigma$. Thus, $u_{0.05}$ denotes the factor appropriate for the range that includes 95 % of the results from a given population. The numerical value of α is the probability that a result will fall outside the range, $\mu\pm u_\alpha\sigma$.

It is important to note that for the application of the normal distribution to be valid, each result must be independent of all other results in the population. The need for this condition can be seen from equation (4.2) since $p(R)$ is determined only by μ and σ and not by any past results. We would also stress again that the normal distribution does not allow the error of a particular result to be predicted exactly. All that it allows one to say is that, on average, a proportion equal to $100(1-\alpha)\%$ of all results will fall within the range $\mu \pm u_\alpha\sigma$.

Several other properties of normally distributed results are of very frequent use in analytical situations and are used repeatedly in later sections of this book; these properties are described in (b) to (d) below. First, however, it will be useful to say a little more about the assumption that analytical results are normally distributed.

The first point to note is that distributions other than the normal are possible and, indeed, occur. For example, in the measurement of radioactivity, the number of counts recorded, R, follows the Poisson distribution:

$$p(R) = \exp(-\mu)\mu R/R!$$

where μ = the population mean. Many of these other distributions, however, approximate the normal quite well under commonly encountered situations. Thus, except when very small counts are involved, the Poisson distribution closely approaches the normal. There is, in addition, the very important point that the means of n results from *any* distribution approach the normal distribution more and more closely as n increases. For the general form of distribution expected for analytical results from well controlled analytical systems, the approximation to normality is likely to be quite satisfactory for values of n as small as two or three. The reasons for our general assumption of the normal distribution may, therefore, be summarised as follows.

(1) It is a distribution consistent with the general pattern of scatter of analytical results expected from well controlled analytical systems.
(2) Many other distributions commonly approximate the normal in practice.
(3) Statistical procedures for dealing with normally distributed results are well established and described in many texts and publications readily available to, and written specially for, scientists.
(4) In our view, the greatest value of the statistical approach to be described below seldom if ever hangs on fine distinctions between different distributions. In practice, the true distribution of results will seldom if ever be known exactly, and experience indicates that the assumption of normality provides a valuable model for dealing with the errors arising in most of the analytical procedures considered in this monograph and commonly applied in water analysis. See also ref. 1 for a useful discussion of the validity of the assumption of normality, and see also the following paragraph.

The more advanced statistical texts (see Section 4.8) may be consulted for details of other distributions, and the advice of statisticians should be sought whenever there is doubt on the distribution appropriate to a particular application. It is also worth noting that there is a group of statistical techniques that makes essentially no assumption about the detailed nature of the distribution followed by results. Such techniques are usually referred to as 'distribution-free' procedures or 'non-para-metric' techniques (see Section 4.8 for references). They appear to have found relatively little use in analysis so far, and indeed it has been suggested that they will rarely have any significant advantage in analytical quality control.[2] Two results are, however, worth mention here as an indication of the reasonableness of the assumption of the normal distribution.[3,4] First, for *any symmetrical, single-peak distribution*, the probability that R differs from μ by more than $e\sigma$ is less than $(2/3e)^2$. Second, for *any distribution*, the probability that R differs from μ by more than $e\sigma$ is less than $1/e^2$. These theorems allow the proportions of results within a certain range from μ to be calculated with the conclusions shown in Table 4.2 where the corresponding values for the normal distribution are also shown for comparison. It can be seen that the proportions for a symmetrical, single-peak distribution (such as one would certainly expect from well controlled analytical systems) and for the normal distribution do not differ greatly.

TABLE 4.2
Proportion of Results within a Given Range from the Population Mean for Different Distributions

Type of distribution	Proportion of results within the range $\mu \pm e\sigma$, %		
	$e = 2.0$	$e = 2.5$	$e = 3.0$
Normal distribution	95.5	98.8	99.7
Any symmetrical, single-peak distribution	88.9	92.9	95.1
Any distribution	75.0	84.0	88.9

(b) Mean and Standard Deviation of Sums and Differences of Independent Random Variables

Let z be a random variable formed, in general, by a constant and the addition and subtraction of a number of independent, random variables; that is:

$$z = a + \sum_i b_i x_i \qquad (4.3)$$

where $x_i = $ the i th random variable ($i = 1$ to n) with population mean $= \mu_i$ and population standard deviation $= \sigma_i$,

$a\ = $ a constant,

$b_i = $ a constant whose sign and value may depend on i,

and $\sum_i$ denotes the sum of the individual values of $b_i x_i$ for all values of i, e.g.,

$$\sum_i b_i x_i = b_1 x_1 + b_2 x_2 + \ldots\ldots\ldots + b_n x_n .$$

The term 'independent' implies that the random error of any individual variable is not correlated with the random error of any of the other variables. That is, there is no tendency if the random error of one variable becomes, for example, more positive, for the random error of any other variable also to become more positive.

Given the above conditions, the population mean of z, μ_z, is given by

$$\mu_z = a + \sum_i b_i \mu_i \qquad (4.4)$$

and the population standard deviation, σ_z, is given by

$$\sigma_z = \sqrt{\sum_i b_i^2 \sigma_i^2} \qquad (4.5)$$

Equations (4.4) and (4.5) are valid whatever the distributions of the x_i, but when each of the x_i is normally distributed, so too is z. Note that the addition to, or subtraction from, a random variable of one or more constants does not affect the standard deviation of the sum or difference.

From the general results in equations (4.4) and (4.5), a number of particular results of common importance in analysis are easily deduced. For example, when $z = bx$, $\mu_z = b\mu_x$ and $\sigma_z = b\sigma_x$. Another common situation is where the difference between two analytical results is of interest, *i.e.*, $z = R_1 - R_2$. Using equation (4.5) and noting that in equation (4.3), $a = 0$, $b_1 = 1$, and $b_2 = -1$, we obtain $\sigma_z = \sqrt{\sigma_1^2 + \sigma_2^2}$, where σ_1 and σ_2 are the standard deviations of the populations corresponding to R_1 and R_2, respectively. If R_1 and R_2 both come from normal distributions, then z will also be normally distributed. Note that though $z = R_1 - R_2$, it is the *variances* that are *summed*.

There is sometimes interest in the standard deviations of functions of random variables that are not simple sums or differences, *e.g.*, when $z = x_1/x_2$. Equation (4.5) does not apply to such functions, but an approximate equation of common value is given in Section 4.6.2(*c*).

(c) Population Mean and Standard Deviation of the Mean of n Independent Results
The distribution of the mean, $\bar{R}$, of n independent analytical results, R_i ($i = 1$ to n), for the same sample is of common interest, and can be deduced from the results in (*b*) above. Let the population mean and standard deviation of individual analytical results be μ and σ, respectively. Now $\bar{R} = \sum_i R_i/n$, so that

$$\bar{R} = \frac{R_1}{n} + \frac{R_2}{n} + \cdots\cdots\cdots + \frac{R_n}{n}$$

$\bar{R}$ is thus the sum of n random variables, each with a population mean and standard deviation of μ/n and σ/n, respectively [from the results in (*b*) above]. Thus, using equations (4.4) and (4.5) we can obtain the mean, μ_m, and standard deviation, σ_m, of the distribution of $\bar{R}$:

$$\mu_m = \sum_i (1/n)\mu = (1/n)n\mu = \mu \tag{4.6}$$

and

$$\sigma_m = \sqrt{\sum_i (1/n^2)\sigma^2} = \sqrt{(1/n^2)n\sigma^2} = \sigma/\sqrt{n} \tag{4.7}$$

In other words, the population mean of the means of n results is the same as the population mean of individual results; the population standard deviation of the means of n results is smaller by a factor $1/\sqrt{n}$ than the population standard deviation of individual results. Thus, as would be expected from commonsense considerations, the scatter of the means of a number of results is smaller than the scatter of individual results. Finally, if the R_i are normally distributed, then so too are the means of a number of results, though this will be approximately true no matter what the distribution of the R_i [see Section 4.3.2(*a*)].

(d) Components of Variance of a Random Variable

When a number of independent sources of random error contribute to the total random error of a variable, it is often of value to use the property that the total variance, σ^2, is given by the sum of the contributing variances:

$$\sigma^2 = \sum_i \sigma_i^2 \tag{4.8}$$

where σ_i^2 is the variance attributable to the i th source of random error. This may be regarded as a special case of equation (4.5). For example, if the standard deviation of analytical results, σ, is made up of contributions from random errors arising in sampling, σ_S, and analysis, σ_A, then $\sigma^2 = \sigma_S^2 + \sigma_A^2$. Note again that it is the variances and not the standard deviations that are additive.

4.3.3 Quantifying the Uncertainty Caused by Random Error

(a) Distinction between Population Parameters and Their Estimates

In analysing a sample, the aim is generally, of course, to determine the true concentration. In the absence of systematic error (see Section 4.4), this aim is equivalent to determining the population mean for a sample, μ. We have seen, however, that analytical results generally differ from μ as a result of random error. Consequently, a result, R, must be regarded as only an *estimate* of the population mean. This distinction between population parameters – whose values we should like to know but cannot measure exactly – and their estimates – whose values we know exactly but which are not generally equal to the population values – is very important, and to emphasise the distinction it is usual to denote the former by Greek letters and the latter by English letters. For example, we have already used μ and σ for the population mean and standard deviation while here we denote their estimates by R (or $\bar{R}$) and s, respectively. One of the most valuable characteristics of statistical techniques in the present context is that, given an experimental estimate, they allow one to make a quantitative statement on the corresponding uncertainty in the value of the population parameter. To illustrate this approach, we show in (b) below how this uncertainty is derived for a population mean when the population standard deviation is known. This then leads to the problem that, in general, the population standard deviation is not known, and we consider, therefore, in Section 4.3.4 the estimation and interpretation of standard deviations. The results derived allow us to deal in Section 4.3.5 with the estimation and interpretation of means when only an estimate of the population standard deviation is available.

(b) Confidence Limits for the Population Mean: Population Standard Deviation Known

We saw in Section 4.3.2 that the range $\mu \pm u_\alpha \sigma$ includes $100(1 - \alpha)\%$ of results from a normal distribution. Thus, given a result, R, for a sample with population mean, μ, we can make the statement

$$\mu - u_\alpha \sigma \leqslant R \leqslant \mu + u_\alpha \sigma \tag{4.9}$$

and know that $100(1 - \alpha)\%$ of such statements will be correct on average. This statement can be simply re-arranged algebraically to

$$R - u_\alpha\sigma \leqslant \mu \leqslant R + u_\alpha\sigma \qquad\qquad (4.10)$$

which implies that the value of μ may lie anywhere between $R - u_\alpha\sigma$ and $R + u_\alpha\sigma$. The two values, $R \pm u_\alpha\sigma$, are usually called *the 100(1 − α) % confidence limits for μ*, the range between them is usually referred to as *the 100(1 − α) % confidence interval for μ*, and the numerical value of $100(1 - \alpha)$ is the (percentage) confidence level. Such confidence intervals are an excellent means of quantifying the uncertainty in the population mean, μ. Thus, $\pm u_\alpha\sigma$ may be regarded as the maximum possible random error at the confidence level corresponding to the value of α. For example, the 95 % confidence limits for μ are $R \pm 1.96\sigma$ (see Table 4.1) so that the maximum possible random error is $\pm 1.96\sigma$ (95 % confidence level). It must be emphasised that μ is equal to the true concentration in a sample only in the absence of systematic error (see Section 4.4).

It is also important to note that such confidence intervals imply that on 100α % of occasions, the value of μ will not lie within the confidence interval. Since the normal distribution is symmetric about μ [see Section 4.3.2(a)], of this 100α % of occasions, μ will be less than $R - u_\alpha\sigma$ on $100\alpha/2$ %, and greater than $R + u_\alpha\sigma$ on the other $100\alpha/2$ %. For example, if 95 % confidence intervals are used, they will fail to include μ on 5 % of occasions; on 2.5 % of occasions μ will be less than $R - 1.96\sigma$, and on the other 2.5 % μ will be greater than $R + 1.96\sigma$. If the proportion of occasions for which the confidence intervals for samples do not include the respective population means is greater than desired, the value of α must be decreased. For example, if $\alpha = 0.01$, the 99 % confidence limits become $R \pm 2.58\sigma$ (see Table 4.1), but note that this greater confidence that μ will lie within the confidence interval is obtained at the expense of increasing the uncertainty in the value of μ because this uncertainty now becomes $\pm 2.58\sigma$ as compared with $\pm 1.96\sigma$ when 95 % confidence limits are used. Two other aspects of these ideas require consideration.

First, if the uncertainty for a chosen confidence level is greater than desired, it can be reduced in only two ways.

(1)   Steps are taken to reduce the numerical value of σ, e.g., by using a different analytical system which allows better precision to be achieved.

(2)   A number, n, of replicate analyses is made on each sample, and the mean of the individual results is used as the final result to be reported. As seen in Section 4.3.2(c), the standard deviation of the mean is $\sigma/\sqrt{n}$, where σ is the standard deviation of individual results. Thus, the $100(1 - \alpha)$ % confidence limits for μ now become $\bar{R} \pm u_\alpha\sigma/\sqrt{n}$, where $\bar{R}$ is the mean of the n analytical results. Once more, we see how such simple statistical ideas allow proper quantification of commonsense ideas.

Which, if not both, of these approaches is appropriate for a particular situation clearly depends on many factors that need not be considered here, but see Section 9.2.

The second point concerns the choice of confidence level. In some circumstances, aspects such as the economic penalty associated with an incorrect decision may allow quantitative derivation of the most appropriate confidence level. This approach, however, seldom seems feasible in most water analysis situations, and some element of arbitrariness and subjectivity is involved in selecting the confidence interval to be used. In practice, it is essentially impossible to know with certainty the exact shape of the distribution followed by analytical results. Accordingly, it should not be thought that choice of, for example, the 95 % level necessarily ensures that one will, on average, be incorrect on exactly 5 % of occasions. It seems to the authors because of these and other uncertainties that experience of the practical value of a particular confidence level is an important factor. A level of 95 % is commonly regarded as reasonable for all but the most crucial applications of analytical results, and this level will be used throughout this book except where stated otherwise. It is emphasised though that the confidence level should be explicitly considered for each application, and the level selected should be made clear in any statements on the errors of analytical results.

4.3.4 Estimation and Interpretation of Standard Deviations

(a) Identifying the Population of Interest
We saw in Section 4.3.2(a) that the standard deviation is a parameter characterising the random variability of repeated analytical results for a particular sample. The magnitude of this variability is affected, in general, by a number of factors. For example, analyses repeated within an hour òr so commonly show less scatter than analyses made over a period of several weeks. This is equivalent to saying that, for a given sample, there are at least two populations, each having its own characteristic standard deviation. Clearly, if we were interested in the comparison of routine analytical results obtained over many weeks, the population of primary interest is that corresponding to such time periods and not to much shorter time periods. Thus, an essential first step in the estimation of any standard deviation is to identify the population of interest.

Two stages are generally involved in making this identification. First, one must identify the analytical results whose random variability one wishes to characterise. Second, the sources of random error affecting the scatter among these results must be identified. The population of interest is then composed of results each of which is subject independently to all those sources of random error. These considerations are discussed further in Sections 6.2 and 6.3 where we consider in detail practical approaches to the estimation of precision.

(b) Estimation of Standard Deviations
In essence, to estimate the standard deviation, σ, of a population of interest, a series of independent analyses of a stable, homogeneous sample are made under conditions such that each result is subject to all the sources of random error that determine the numerical value of σ. (The requirements of stability and homogeneity may cause problems in practice. These and procedures for overcoming them are considered in Sections 6.2 and 6.3; here, we wish to deal with the purely statistical aspects.)

Given a set of n results R_i ($i = 1$ to n), for the concentration of the determinand in a sample, the estimate, s, of the population standard deviation, σ, is given by one of two equations. When the population mean, μ, is known, we use

$$s = \sqrt{\sum_i (R_i - \mu)^2 / n} \qquad (4.11)$$

When the population mean is not known, we use

$$s = \sqrt{\sum_i (R_i - \bar{R})^2/(n-1)} \qquad (4.12)$$

where $\bar{R} = \sum_i R_i/n$. In practice, even when the true concentration, τ, of a sample used for these tests is known with negligibly small error (*e.g.*, when an accurately prepared standard solution is used), we recommend general use of equation (4.12). This allows for the possibility that systematic error in the measurements causes μ to differ from τ (see Section 4.4). Throughout this book, therefore, we use only equation (4.12). An algebraically equivalent form of this equation may be more convenient:

$$s = \sqrt{[\sum_i R_i^2 - (\sum_i R_i)^2/n]/(n-1)} \qquad (4.13)$$

The estimate, s, is said to have n *degrees of freedom* [when equation (4.11) is used] or $(n-1)$ *degrees of freedom* when equations (4.12) or (4.13) are used. The number of degrees of freedom of an estimate may be regarded as indicating the worth of the estimate. That is, as the degrees of freedom increase so the random error of the estimate, s, tends to decrease [see (d) below]. Since we are interested in the value of σ, it is clearly desirable to make the degrees of freedom of an estimate as large as possible. In practice, of course, this aspect must be balanced against the additional effort and time involved as n is made larger; see Sections 6.2 and 6.3.

Finally, it should be mentioned that equations (4.11), (4.12), and (4.13) are derived from the corresponding equations for estimates of the population variance, σ^2; for example,

$$s^2 = \sum_i (R_i - \bar{R})^2/(n-1) \qquad (4.14)$$

corresponds to equations (4.12) and (4.13). Equation (4.14) gives an unbiased estimate, s^2, of σ^2, *i.e.*, if many such estimates of s^2 were obtained for the same population, their mean tends to σ^2 as the number of estimates is increased. Equations (4.12) and (4.13) are obtained by taking the square root of both sides of equation (4.14), but this operation leads to the estimate, s, having a slight bias in the sense noted above.[5] For example, when $n = 5$, 10, 20, and infinity, the mean values of s are 0.940σ, 0.973σ, 0.987σ, and σ, respectively. Compared with the random error of an individual estimate, s, [see (d) below], these biases are negligible, and it is usual to ignore them; this practice is followed here.

(c) Combining Estimates of Standard Deviations
The situation frequently arises in analysis where one has a number of estimates, s_j ($j = 1$ to m), of the *same* population standard deviation, the j th estimate having v_j degrees of freedom, and it is required to combine them in order to obtain an estimate, s, with a larger number of degrees of freedom and, therefore, with a smaller random error. This operation is commonly referred to as pooling the estimates, and s is calculated from the equation:

$$s = \sqrt{\sum_j v_j s_j^2 / \sum_j v_j} \qquad (4.15)$$

The pooled estimate, s, has $\sum_j v_j$ degrees of freedom.

A common and useful application of equation (4.15) arises when precision is estimated by making a number, m, of duplicate analyses [see Section 6.3.2(c)], and pooling the estimates of standard deviation obtained from each pair of results. This leads to a very simple equation for the pooled estimate, s. Let the results for the j th pair of analyses ($j=1$ to m) be R_{ij} ($i=1$ and 2). Then the estimated variance, s_j^2, for the j th pair is calculated from equation (4.14):

$$s_j^2 = \sum_i (R_{ij} - \bar{R}_{\cdot j})^2 ,$$

since $n=2$; s_j has 1 degree of freedom. Now,

$$\sum_i (R_{ij} - \bar{R}_{\cdot j})^2 = (R_{ij} - [R_{1j} + R_{2j}]/2)^2 + (R_{2j} - [R_{1j} + R_{2j}]/2)^2$$

$$= \left(\frac{R_{1j} - R_{2j}}{2}\right)^2 + \left(\frac{R_{2j} - R_{1j}}{2}\right)^2$$

$$= (d_j/2)^2 + (-d_j/2)^2 ,$$

where $d_j = R_{1j} - R_{2j}$.

Thus, $\qquad\qquad s_j^2 = d_j^2/2.$

Substituting the value of s_j^2 in equation (4.15), and noting that $\sum_j v_j = m$, we obtain the simple equation for the pooled standard deviation, s:

$$s = \sqrt{\sum_j d_j^2 / 2m} \qquad (4.16)$$

(d) Confidence Limits for the Population Standard Deviation
We saw in Section 4.3.3(b) that, given an estimate, R, of the population mean, μ, the uncertainty in the value for μ can be defined by forming a confidence interval for μ. A similar approach can be used for standard deviations. Thus, given an estimate, s, with v degrees of freedom, the lower $100(1-\alpha)\%$ confidence limit for the population standard deviation, σ, (i.e., the probability that σ will be less than this limit is α) is given by:

$$s\sqrt{v/\chi_\alpha^2} \qquad (4.17)$$

where χ_α^2 is the numerical value of another statistic – χ^2 (chi squared) – for the α probability level and v degrees of freedom. Numerical values of χ^2 are tabulated in statistical texts, and some values are given for illustrative purposes only in Table 4.3.

Similarly, the upper $100(1-\alpha)\%$ confidence limit is given by:

$$s\sqrt{v/\chi_{1-\alpha}^2} \qquad (4.18)$$

These two limits form a confidence interval for the population standard

TABLE 4.3

Some Values of χ^2

Degrees of freedom	$\chi^2_{0.975}$	$\chi^2_{0.950}$	$\chi^2_{0.050}$	$\chi^2_{0.025}$
5	0·83	1.15	11.1	12.8
10	3.25	3.94	18.3	20.5
20	9.59	10.9	31.4	34.2
30	16.8	18.5	43.8	47.0
60	40.5	43.2	79.1	83.3

deviation, but note that this is a $100(1-2\alpha)\%$ confidence interval as there is a probability α that σ is less than the lower limit *and* a probability α that σ is greater than the upper limit. In general, therefore, to form a $100(1-\alpha)\%$ confidence interval for σ, one calculates the lower and upper confidence limits as:

Lower limit $s\sqrt{v/\chi^2_{\alpha/2}}$

Upper limit $s\sqrt{v/\chi^2_{1-\alpha/2}}$.

Some books tabulate the values of $\sqrt{v/\chi^2}$ directly.   (4.19)

From the values in Table 4.3 we can calculate that when we have an estimate, s, with 10 degrees of freedom, the 95 % confidence interval for σ is given by $0.57s$ to $1.43s$. It can be seen, therefore, that estimates of standard deviations are subject to rather large relative errors.

(e) Comparing Standard Deviations
The need to compare two standard deviations commonly arises in analysis in two situations. In the first, one has to decide if the precision achieved with one analytical system differs from that obtained with another. This is equivalent to deciding whether or not the population standard deviation of system 1 differs from that of system 2. In the second situation, the need is to decide whether or not the population standard deviation of one system is greater than that of the other. In both situations, there is the problem that the population standard deviation achieved with any analytical system is not known *a priori*, and must be estimated experimentally. The estimate obtained is subject to random error as we saw in (*d*) above so that, for example, a difference between s_1 and s_2 does not necessarily mean that σ_1 differs from σ_2. In order justifiably to claim that the precision of system 1 is worse that that of system 2, it is necessary, therefore, that the ratio s_1/s_2 exceed not unity but a greater value (often called the critical value). There is a simple statistical *significance test* – the F-test (sometimes also called the variance-ratio test) – which is used for these purposes, and whose applications and underlying principles are described in statistical texts together with tabulations of the appropriate critical values. The use of this test is required in subsequent sections of this book, and we thought it useful to summarise in (*i*) and (*ii*) below its application to the two situations mentioned above.

We should also note here that statistical tests are available for comparing more than two standard deviations; statistical texts should be consulted for details.

(i) Testing whether two population standard deviations differ Suppose we have estimates, s_1 and s_2 with v_1 and v_2 degrees of freedom, respectively, of two population standard deviations, and we wish to decide if these two latter standard deviations, σ_1 and σ_2, differ. The procedure is to calculate the ratio, F, of the larger to smaller variance estimates, *i.e.*, if $s_1 > s_2$, one calculates (s_1^2/s_2^2). The value of F obtained is then compared with $F_{\alpha/2}$, where $F_{\alpha/2}$ is the tabulated, critical value of the F-statistic for the $\alpha/2$ probability level and v_1 and v_2 degrees of freedom for s_1 and s_2. If $F > F_{\alpha/2}$, we conclude at the α significance level that σ_1 and σ_2 differ. The significance level, α, may be regarded as the probability of falsely concluding that $\sigma_1 \neq \sigma_2$. If $F \leqslant F_{\alpha/2}$, the evidence is insufficient to justify the conclusion that $\sigma_1 \neq \sigma_2$, but it is very important to note that this is not equivalent to a demonstration that $\sigma_1 = \sigma_2$. As an example, suppose that the values obtained for s_1, s_2, v_1 and v_2 were 1.0 μg l^{-1}, 2.0 μg l^{-1}, 5, and 5, respectively, and that α were chosen to be 0.05. the value of F is then 4.0, and the value of $F_{0.025}$ (see illustrative values in Table 4.4) is 7.15. Thus, $F < F_{0.025}$, and we cannot conclude (at the 0.05 significance level) that σ_1 differs from σ_2.

TABLE 4.4
Some Values of $F_{0.025}$

Degrees of freedom of smaller standard deviation	Values of $F_{0.025}$ Degrees of freedom of larger standard deviation						
	1	2	5	10	20	60	α
1	648.8	799.5	921.9	968.6	993.1	1009.8	1018.3
2	38.51	39.0	39.30	39.40	39.45	39.48	39.50
5	10.01	8.43	7.15	6.62	6.33	6.12	6.02
10	6.94	5.46	4.24	3.72	3.42	3.20	3.08
20	5.87	4.46	3.29	2.77	2.46	2.22	2.09
60	5.29	3.93	2.79	2.27	1.94	1.67	1.48
α	5.02	3.69	2.57	2.05	1.71	1.39	1.00

(ii) Testing whether one population standard deviation is greater than another In this situation, our interest is not whether σ_1 differs from σ_2, but whether σ_1 is greater than σ_2. Again, suppose that we have estimates, s_1 and s_2 with v_1 and v_2 degrees of freedom, of the two population standard deviations. If $s_1 \leqslant s_2$ there is clearly no evidence that $\sigma_1 > \sigma_2$, and no test is required. If $s_1 > s_2$, one calculates $F = (s_1^2/s_2^2)$, and compares the value with F_α, where F_α is as defined in *(i)* above. If $F > F_\alpha$, we may conclude at the α significance level [note the difference from the test in *(i)* above] that $\sigma_1 > \sigma_2$. If $F \leqslant F_\alpha$, the evidence is insufficient to justify the conclusion that $\sigma_1 > \sigma_2$, but as before this is not equivalent to a demonstration that $\sigma_1 = \sigma_2$.

A particular application of this type of F-test arises when σ_2 represents a required or target precision, and σ_1 represents the precision of an analytical system. In this situation, σ_2 is known exactly, and it is appropriate, therefore, to assign an infinite number of degrees of freedom to it. The test is then made as described in the preceding paragraph. For example, suppose that the target value corresponds to $\sigma_2 = 1.0 \mu$g l^{-1}, that the numerical value of s_1 is 1.4 μg l^{-1} with 10 degrees of freedom, and that α is chosen to be 0.05. Since $F = 1.96$ and $F_{0.05} = 1.83$ (see Table 4.5), we conclude at the 0.05 significance level that σ_1 is greater than the target value.

TABLE 4.5

Some Values of F_α *when* σ_2 *is Known*

Degrees of freedom of s_1	Values of F_α^*		
	$\alpha = 0.10$	$\alpha = 0.05$	$\alpha = 0.025$
5	1.85	2.21	2.57
10	1.60	1.83	2.05
20	1.42	1.57	1.71
60	1.24	1.32	1.39
α	1.00	1.00	1.00

* The values given are only for the case where an estimated standard deviation, s_1, is compared with a target value, σ_2, and $s_1 > \sigma_2$; see text

4.3.5 Estimation and Interpretation of Means

(a) Confidence Limits on the Mean of n Results: Population Standard Deviation Unknown

In Section 4.3.3(b) a procedure was given for calculating a confidence interval for a population mean when the population standard deviation was known. Here, we describe the appropriate procedure when the population standard deviation is not known and only an estimate of it is available.

Suppose that a simple series of n independent analyses has been made on identical portions of a stable, homogeneous sample under conditions such that all the results obtained are from the population of interest [see Section 4.3.4(a)]. Let the i th analysis ($i = 1$ to n) give a result, R_i, for the concentration of the determinand. The first step is to calculate an estimate, s, of the standard deviation exactly as described in Section 4.3.4(b); s has $(n-1)$ degrees of freedom. The $100(1-\alpha)$ confidence limits for the population mean are then obtained from the formula:

$$\bar{R} \pm t_\alpha s/\sqrt{n} \qquad (4.20)$$

where $\bar{R}$ is the mean of the n results and t_α is the t-statistic for the α probability level and $(n-1)$ degrees of freedom. Numerical values of t are tabulated in statistical texts which should also be consulted for a full discussion of the t-distribution and its applications. For present purposes it suffices to note that the numerical value of t_α depends on α and the degrees of freedom of the estimate, s; some illustrative values are given in Table 4.6.

Since the estimate, s, is itself a random variable, we can see from equation (4.20) that the width of the confidence interval for a given population will vary from one set of measurements to another. In a proportion $100(1-\alpha)\%$ of such sets on average, however, the calculated confidence intervals will include the population mean.

We saw in Section 4.3.3(b) that when the population standard deviation is known, $\pm u_\alpha \sigma/\sqrt{n}$ may be regarded as the maximum possible random error of $\bar{R}$ at the $100(1-\alpha)\%$ confidence level. The term $\pm t_\alpha s/\sqrt{n}$ has the same meaning when σ is not known, but in this case the maximum possible random error will vary from one set of measurements to another. If, however, we ignore the small bias in s [see Section 4.3.4(b)], the mean value of $\pm t_\alpha s/\sqrt{n}$ over a large number of confidence

TABLE 4.6

Some Values of the t-Statistic

	Values of t_α		
Degrees of freedom	$\alpha = 0.10$	$\alpha = 0.05$	$\alpha = 0.01$
1	6.31	12.71	63.66
2	2.92	4.30	9.92
3	2.35	3.18	5.84
4	2.13	2.78	4.60
5	2.02	2.57	4.03
6	1.94	2.45	3.71
10	1.81	2.23	3.17
20	1.72	2.09	2.85
30	1.70	2.04	2.75
60	1.67	2.00	2.66
α	1.64	1.96	2.58

intervals is $\pm t_\alpha \sigma / \sqrt{n}$. Thus, for given values of α and n, it is the values of t_α and u_α that govern the mean maximum possible random error of $\bar{R}$. Tables 4.1 and 4.6 show that t_α differs little from u_α for large values of n, but as n decreases t_α becomes progressively larger than u_α, the difference being very large when n is two or three. Clearly, therefore, to reduce the random error of $\bar{R}$, n should be as large as possible, but note that the decrease in the random error of $\bar{R}$ becomes smaller and smaller as n is made larger and larger. There is a further advantage in forming a mean from a number of results in that the assumption of a normal distribution becomes almost certainly valid [see Section 4.3.2(a)].

(b) Confidence Limits for Individual Analytical Results

The procedure given in (a) above for calculating confidence intervals can be used for any value of $n \geqslant 2$. It cannot, however, be used when $n = 1$ for it is then impossible to obtain an estimate of the standard deviation. Table 4.6 also shows that the large values of t will lead to relatively wide confidence intervals when $n = 2$ or 3. Notwithstanding the clear advantage, from the standpoint of error reduction, in analysing each sample many times and applying the procedure in (a) to calculate the confidence interval for each sample, most if not all laboratories find such an approach impracticable. Most often, each routine sample is analysed once or at most twice. The approach to this situation that has been recommended by many authors (including ourselves) is simple. By one or more special tests involving replicate analyses of specially selected samples, an estimate, s, is obtained with a reasonable number of degrees of freedom – say, ten or more. This estimate is then used to form the confidence intervals for subsequent singly analysed samples. That is, the confidence interval for any sample giving an analytical result, R, is $R \pm t_\alpha s$, and the value of t_α is that corresponding to the degrees of freedom of s and the chosen confidence level, $100(1 - \alpha)\%$. This account of the approach is simplified, but it suffices here for the statistical points to be made; the approach is considered in more detail in Section 6.

It should first be stressed that the approach suggested above lacks theoretical justification because the use of the t-statistic should strictly be reserved for situations where, each time a confidence interval is formed, one has an independent estimate of the population standard deviation. The repeated use of the same estimate suggested above clearly violates this condition and leads to the error that the proportion of samples for which the confidence intervals so formed (*i.e.*, using the one value of *s*) include the respective population means does not, in general, equal the proportion corresponding to the chosen value of α*. For example, if the confidence intervals are calculated from the value of *t* corresponding to the 95 % confidence level (*i.e.*, $t_{0.05}$), the proportion of samples for which the confidence intervals include the respective means is not, in general, 95 %. The reasons for, magnitude of, and ways of reducing this error are considered in the Appendix (p. 171) to this section, and we conclude that the simplest approach is to follow the procedure suggested in the preceding paragraph, accepting that this type of error is present. If this is done, it is essential to do everything possible to ensure that the precision of analytical results remains constant and that the estimate, *s*, has as many degrees of freedom as possible; both points will tend to minimise the error noted above and both are discussed in detail in Sections 6.1 to 6.3.

If one is prepared to accept the error involved in this procedure, the same approach may be used when $n = 2$ or 3 in order to reduce the value of t_α used to form the confidence intervals. If this is done, the confidence limits are still given by expression (4.20), but the value used for t_α is that corresponding to the degrees of freedom of the preliminary estimate of standard deviation.

(c) Comparing Estimates of Population Means

Caution is required in comparing estimates of population means for the same reasons as noted in Section 4.3.4(*e*); procedures for this purpose are discussed in Section 4.6.2.

4.3.6 Concluding Remarks

It is worth emphasising the explicit, quantitative, easily reproduced and largely objective nature of the statistical approach to the quantification of random errors and their effects. This is in marked contrast to the formerly prevalent (but, happily, decreasing) tendency for analysts either to ignore this source of error altogether or to quote the error on the basis of some undefined, semi-quantitative, and almost totally subjective approach that varied from one individual to another. There is no doubt that random error needs to be quantified in analytical work, and there is equally no doubt that statistical ideas and techniques should be used for this purpose. The general procedures described in this monograph are all based on the assumption that analytical results follow the normal distribution sufficiently well for it to be of primary interest in the present context [see Section 4.3.2(*a*)], but it is no valid argument against statistical procedures if instances are identified or suspected where other distributions are followed. The basic statistical concepts apply to any distribution, and it is only the detailed tests and computations that depend on the distribution. Statistical advice should be sought if it is thought that the appropriate distribution for a particular application needs more detailed consideration.

---

* This error was pointed out to the authors by Mr. J.C. Ellis (Water Research Centre) to whom we are most grateful. The conclusions on the practical approach to the problem are, however, those of the authors.

Experience, however, suggests that, for vast numbers of analytical applications, the assumption of the normal distribution is a fruitful approach in the estimation and control of analytical errors. In any event, quite apart from the validity of any particular distribution, the adoption of statistical procedures also has the great advantage that it forces the explicit and careful consideration of sources of error and their effects; this perhaps is the most compelling reason of all for their adoption.

We should not close this section without mentioning that statistical techniques find many other valuable applications apart from those noted in this book.

4.4 SYSTEMATIC ERROR

4.4.1 The Nature of Systematic Error

Systematic error is present when there is a consistent tendency for results to be either greater or smaller than the true value. One example of this type of error occurs when interfering substances are present in a sample. The term *'bias'* is used synonymously with systematic error by many authors in the English language, and this usage is followed throughout this book; bias is said to decrease as the magnitude of the systematic error decreases.

Some care is needed in defining bias quantitatively because any analytical results obtained to determine the magnitude of this error are also subject to random error. In Section 4.3.5(a) we saw that the $100(1-\alpha)$ % confidence limits for the population mean, μ, are $\bar{R} \pm t_\alpha s/\sqrt{n}$. As n increases, the interval between these limits tends to become smaller and smaller, and in the limit as n approaches infinity, the width of the interval approaches zero and $\bar{R}$ approaches the population mean, μ. When μ differs from the true concentration, τ, for a particular sample, bias is said to be present of magnitude, β, where

$$\beta = \mu - \tau \qquad (4.21)$$

The above definition of bias may appear somewhat abstruse and remote from reality, but it has been recommended by a number of authors and organisations. It must be noted, however, that others have adopted different definitions; see ref. 6. We believe that the definition of equation (4.21) is essential in order to avoid confusion of random and systematic errors. It will be seen that the definition is simply applied in practice (Section 5).

One very important conclusion stems directly from this definition. Since it is impossible to make the infinite number of analyses required to obtain the value of μ, it follows that in any experimental attempts to measure the magnitude of bias, only an estimate, $\bar{R}$, which is subject to random error, will be available to use in place of μ in equation (4.21). Thus, it is possible to obtain only an estimate, B, of the bias, where

$$B = \bar{R} - \tau \qquad (4.22)$$

so that B will also be subject to random error. Statistical ideas and techniques are, therefore, equally essential in the estimation of bias. This simple, and yet once more commonsense, observation has a number of important consequences that will be considered in later sections.

When a result is subject to a number of independent sources of bias, the resultant total bias is the sum of the individual components with due regard for their signs.

4.4.2 Differences and Inter-relationships between Systematic and Random Errors

(a) Differences between Systematic and Random Errors

The different natures of systematic and random errors lead to two important consequences in their effects.

First, the effective size of random error can be reduced in using the mean of a number of results in place of a single result [Section 4.3.2(*c*)]. This is not so for systematic error. The definition of equation (4.21) shows that β is independent of the number of results because it is defined by two parameters, μ and τ, that are error-free constants. Replication of analytical results, therefore, does not reduce the bias of the mean.

Second, because bias is a constant and the same for each result *from a given population*, it is possible, in principle, to make a correction to results to eliminate or decrease their bias. For example, suppose that an analytical method required a separation procedure that led to the loss of 20 % of the determinand. Estimation of the magnitude of this loss would allow correction of the analytical results by a factor 1.25 so that the bias of the corrected results was much decreased. In practice, however, this approach faces several important problems. For example, the bias may vary from time to time and from sample to sample, In this situation, a prohibitive amount of effort may well be needed to estimate the bias for each sample, and the approach becomes impracticable. Such a situation may occur, for example, when samples may contain a number of interfering substances or when the processes leading to interference effects are inherently irreproducible so that the size of the bias varies from time to time. The problems involved in the estimation of bias are considered in detail in Section 5.

Note also that the way in which individual sources of error contribute to the total error is different for systematic and random errors [see Sections 4.3.2(*b*) and 4.4.1].

(b) Inter-relationships between Systematic and Random Errors

It is important to note that the detailed procedures adopted in analysing samples and producing analytical results may sometimes lead to random errors appearing as bias, and *vice versa*. For example, when a linear calibration curve is prepared, the random errors of the individual measurements made to define the curve will lead to a random error in the slope of the curve, *i.e.*, independently prepared calibrations will show a random scatter of values for the slope. If a fresh calibration curve were prepared for each sample analysed, the random error of analytical results would include a contribution from the random errors of the slope values. On the other hand, if a calibration, once prepared, were used for many subsequent samples, the random error of this particular slope value would no longer be a source of variation among the results of those samples but would become a constant source of bias affecting all results. Currie[4] discussed this 'inter-conversion' of random and systematic errors in more detail. It suffices here to note the possibility which has two main consequences.

First, the nature of experimental designs – and the interpretation of their results – to estimate the size of errors depend, in general, on whether the effects of the errors are to cause random or systematic variations in analytical results. Careful consideration of such points is, therefore, essential if the experimental estimates are properly to reflect the actual errors of the analytical results of interest. Various aspects of this point are discussed in practical contexts in Sections 6.2 and 6.3.

Second, in subsequent sections we shall stress: (1) the importance of doing everything possible to minimise the bias of analytical results, and (2) the considerable difficulties involved in estimating and controlling this bias. Currie[4] makes the point that such difficulties may, in general, be decreased by designing routine analytical procedures so that as many sources of bias as possible are 'converted' to random effects. He gives a number of examples, one of which is to prepare a fresh calibration for each sample (see above). Of course, not all such possibilities may be considered practicable, and this may well often be so for the suggestion concerning calibration. Nevertheless, Currie's general point is important, and should always be borne in mind.

4.5 ACCURACY

Some authors use the term '*accuracy*' to denote what we call here bias – or rather its absence. Throughout this book, however, we follow the practice of many other authors and use accuracy to denote total error, ε_T, where

$$\varepsilon_T = \varepsilon_R + \varepsilon_S \tag{4.23}$$

and ε_R and ε_S are the random and systematic errors, respectively. Accuracy is said to improve as ε_T decreases.

Though useful in some contexts (see below), accuracy is not a suitable concept for detailed consideration of the errors of analytical results and their effects. For example, as we saw in the preceding sections, the effect of replicate analyses on the errors of the mean results can be determined only by separate consideration of random and systematic errors.

We saw also in Section 4.3.3(*b*) that the maximum possible random error [at the $100(1-\alpha)\%$ confidence level] may be equated with $\pm u_\alpha\sigma$. Thus, if the bias is β, the maximum possible total error (at the same confidence level) may, from equation (4.23), be regarded as $u_\alpha\sigma + |\beta|$*. It is sometimes useful to remember this expression when considering the relative contributions of random and systematic errors to the total errors of results; see, for example, Section 2.6.2(*b*).

4.6 EFFECTS OF ERRORS ON DECISION MAKING

We have stressed in previous sections that the errors of analytical results need to be controlled within limits such that the results satisfactorily fulfil the purposes for which they were obtained. These limits will, in general, vary from one application to another, and need to be decided individually for each application through consideration of how the errors will affect any decisions based on the analytical results. The many possible applications and types of decisions required prevent a comprehensive discussion, and we are, in any event, not qualified to attempt such a

* The symbol $|x|$ denotes the numerical value of x regardless of its sign.

task. Nevertheless, it seems to us that many applications in the field of water quality
can reasonably be assigned to one of relatively few classes for which rather simple
and straightforward considerations lead to some useful conclusions. These
considerations are of value also in indicating the usefulness of the simple statistical
ideas described in Sections 4.3 and 4.4. It is hoped also that the discussion will show
through simple examples how an analysis of the decision making processes allows
the effects of errors to be assessed so that one can derive values for the maximum
tolerable errors of analytical results. It is also worth noting that ref. 7 gives a useful
introduction to formal statistical decision theory.

In Section 4.6.1 we consider the effect of systematic error alone, in Section 4.6.2
the effects of random error alone, and in Section 4.6.3 we discuss the still more
problematic situation where both systematic and random errors are present.

4.6.1 Systematic Error

(a) Comparison of an Analytical Result with a Fixed Numerical Value
Whenever a result, R, is to be used to decide whether or not the true concentration,
τ, of the determinand is greater than some fixed value, λ, the bias, β, of the result
must be considered. Such situations arise, for example, when the aim is to control
water quality so that τ does not exceed a specified maximum acceptable
concentration. There are, of course, other similar situations where it is required that
the true concentration is never less than a specified value or where the true
concentration must be maintained between two specified values. The principles
involved in these two latter cases are exactly the same as those for the first, and for
simplicity only the situation where τ must not exceed λ is discussed below.

When error is present of size, ε, the result, R, is given by: $R = \tau + \varepsilon$. In general, ε is
composed of random as well as systematic errors, but the effect of the latter is most
clearly seen and discussed by assuming for the moment (see Section 4.6.3) that
random error is absent; this assumption does not affect the validity of the
conclusions. Given this assumption, $R = \tau + \beta$.

The simplest decision rule and one that seems often to be used (though it is not the
only possible approach) is to accept the water quality as satisfactory when the
analytical result is less than or equal to the specified value, *i.e.*, when $R \leqslant \lambda$. By
substituting $\tau + \beta$ for R, we obtain

$$\tau + \beta \leqslant \lambda \qquad (4.24)$$

as the condition for which the water quality is taken as acceptable. It can be seen
from inequality (4.24) that when $\tau \leqslant \lambda$ (*i.e.*, the quality is truly satisfactory), the
decision that quality is unsatisfactory will be made incorrectly whenever $\beta > \lambda - \tau$.
Similarly, when $\tau > \lambda$, the decision that quality is satisfactory will be made
incorrectly whenever β is negative and less than $\tau - \lambda$. In general, therefore, if
incorrect decisions are to be completely avoided, it is necessary to ensure that
$|\beta| \leqslant |\lambda - \tau|$. Clearly, if information on the likely concentrations of the determinand
in the water is available, more detailed analysis of the effects of bias is possible. For
example, if it were known that the values of τ varied with time in such a way that τ
followed, at least approximately, a normal distribution with known mean and

standard deviation, the effect of β on the proportion of incorrect decisions could be assessed.

Thus, and as would be expected, bias may, but does not necessarily, lead to difficulty in the proper implementation of water quality standards. In terms of the maximum bias tolerable by the user of analytical results, the above simple analysis shows that there is no unique value for this bias because it depends on the determined concentration actually present in the water of interest. When $\tau = \lambda$, the bias must be zero to avoid incorrect decisions, and ideally one would ensure $\beta = 0$ for all values of τ. This, however, is not generally possible (see Section 5), and is, therefore, not a general solution to the problem of bias in practice. Similarly, it is not generally possible to eliminate the effect by measuring bias and then making a correction to R (Sections 4.4.2 and 5). In this position, it seems to us that there is only one generally effective approach, and this requires changing the objectives of the control scheme so that the aim is to control τ at or below a concentration λ', where

$$\lambda' = \lambda - \beta_c \qquad (4.25)$$

and the value of β_c is governed by the range of possible bias of analytical results (see Section 5) as described in the three following cases.

Case 1: $0 \leqslant \beta \leqslant \beta_p$ In this case, β is always positive so that $R > \lambda$ whenever $\tau > \lambda$, i.e., unacceptable water quality will be detected efficiently. To avoid the possibility that $R > \lambda$ when $\tau \leqslant \lambda$, one aims to ensure that $\tau \leqslant \lambda - \beta_p$, and takes action whenever $R > \lambda$. Thus, in this situation, β_c of equation (4.25) is equal to β_p.

Case 2: $\beta_n \leqslant \beta \leqslant 0$ The possible values of β (never positive) ensure that $R \leqslant \lambda$ whenever $\tau \leqslant \lambda$, i.e., needless action will never be taken. To avoid the possibility that $R \leqslant \lambda$ when $\tau > \lambda$, one aims to ensure that $\tau \leqslant \lambda - |\beta_n|$, and takes action whenever $R > \lambda - |\beta_n|$. Thus, β_c of equation (4.25) is equal to β_n.

Case 3: $\beta_n \leqslant \beta \leqslant \beta_p$ In this situation where some results may have positive bias up to β_p and others may have negative bias as large as β_n, the aim should be to control $\tau \leqslant \lambda - \beta_p - |\beta_n|$, and to take action whenever $R > \lambda - |\beta_n|$. Thus, $\beta_c = \beta_p + |\beta_n|$.

The above approach does not, of course, eliminate the bias of analytical results, but it does ensure explicit recognition of the importance of bias in the design of the total scheme for the control of water quality. The approach calls for detailed discussion between the analyst and the user of analytical results because of the reciprocal influences of λ' and β_c. If there is some value, τ_o, below which it is impossible to control the value of τ, the maximum tolerable bias of analytical results can be derived. Clearly, however, there is value, other things being equal, in making the strongest possible attempts to use analytical systems with the smallest achievable bias. The estimation of bias is, therefore, of great and growing importance in the field of water quality as also is the choice of analytical systems subject to the least possible bias; these topics are discussed in Sections 5 and 9, respectively.

(b) Comparison of One Analytical Result with Another
Interest may centre on the difference between two analytical results. For example, the qualities of two different waters or of the same water at two different times or places may have to be compared. Decisions will then be based, for example, on the

difference, D, between two results, R_1 and R_2. Following the same approach as in (*a*) above, we obtain

$$D = (\tau_1 - \tau_2) + \beta_1 - \beta_2 \qquad (4.26)$$

and the correctness of decisions is, therefore, governed by the difference between the biases of R_1 and R_2.

Correct decisions are clearly ensured when $\beta_1 = \beta_2 = 0$, but we have noted in (*a*) above the problem of achieving this ideal solution. Neither is it generally possible, as also seen in (*a*), to eliminate the problem by measuring β_1 and β_2, and then making corrections to R_1 and R_2. Of course, selection and use of analytical systems such that bias is as small as practicable will always be worthwhile. Finally, by considering in more detail how decisions are governed by the value of D, it will also be possible to assess the effect of the two biases. As a simple example, suppose that the aim were to take corrective action whenever $\tau_1 - \tau_2$ exceeded a fixed value, δ, and the decision rule were to act whenever $D > \delta$. Putting $\Delta\tau = \tau_1 - \tau_2$ and $\Delta\beta = \beta_1 - \beta_2$, equation (4.26) becomes $D = \Delta\tau + \Delta\beta$, so that the situation will be acceptable whenever

$$\Delta\tau + \Delta\beta \leqslant \delta \qquad (4.27)$$

Comparison of inequalities (4.24) and (4.27) shows that we have essentially the same situation as that discussed in (*a*) above, and essentially the same considerations apply. The main point to note is that the $\Delta\beta$ term of (4.27) may tend to be larger than the corresponding β term in (4.24) because the former is the difference between two biases which may, in general, be of opposite signs. Thus, other things being equal, the present type of situation may be more stringent with respect to tolerable bias than that considered in (*a*).

Before leaving this section, it is worth considering another approach sometimes suggested for eliminating the effect of bias in situations characterised by inequality (4.27). It can be seen that decisions will be correct whenever $\beta_1 = \beta_2$, though neither need then be zero. On this basis it is sometimes argued that bias is relatively unimportant, and that equality of β_1 and β_2 is simply achieved by ensuring that R_1 and R_2 are both obtained by the same analytical method. Such situations may occur, but the unimportance of bias here is, in our view, often falsely emphasised for several reasons. First, the magnitude of bias depends, in general, on properties of the sample, *e.g.*, the concentrations of the determinand and interfering substances. The two waters under comparison will rarely if ever be identical in all respects that may affect bias. Second, when two laboratories each produce one of the results, their stated use of the same analytical method does not necessarily guarantee that each laboratory really does follow exactly the same detailed analytical procedure in practice. Laboratories tend to introduce their own individual variants of procedures, both consciously and unconsciously, and such variations may well lead to different liabilities to bias. Finally, other components of the two analytical systems may differ in the two laboratories and cause differences in bias. We feel, therefore, that this approach should, in general, only be used after considerable care has been taken to ensure the equality of β_1 and β_2. There is, however, one particular situation where the use of the same method by all laboratories is essential if the effects of bias are to be minimised, *i.e.*, when the determination of non-specific determinands is involved (see Sections 2.4.2, 7.2.5 and 9.3.1).

(c) Concluding Remarks

From the discussion above of two decision making situations commonly met in the analysis of waters, it is clear that such decisions are, in general, affected by bias in analytical results. Whether or not this effect is important depends on a number of factors that vary from one situation to another. As a general rule, we conclude that it is always preferable, and in most cases essential, to attempt to achieve negligibly small bias. Many authors have, of course, reached the same conclusion; see, for example, particularly useful discussions by Currie.[4,8,9] The same view has recently been expressed in an authoritative paper on analytical quality control in clinical chemistry.[2] Quite apart from the particular needs of an application, it is always worth remembering that analytical results may well be used for subsequent purposes other than that for which they were originally intended; biased results will much reduce their value for such subsequent uses, *e.g.*, in formulating and revising water quality standards. The problems involved in attempting to achieve negligibly small bias from analytical systems are considered in detail in Sections 5, 9.1, and 9.2.3(*a*).

4.6.2 Random Error

We saw in Sections 4.3 and 4.4 that statistical ideas and techniques are essential when random errors are present, and some basic concepts and procedures were described there. Rather than immediately consider here the impact of random errors on decision making in water quality situations, it will be useful first to derive some statistical results that are of such frequent relevance to analysts that their separate discussion is justified. Accordingly, in (*a*) below we consider the general situation where the difference between an analytical result and a fixed numerical value is of interest. In (*b*), the difference between two analytical results is discussed, and in (*c*) the general case is considered where some function of analytical results is involved. In (*d*) to (*f*) we consider decision making in a few, typical water quality applications. Throughout these six sections bias is assumed to be absent. The combined effects of random error and bias are examined in Section 4.6.3.

(a) Difference between an Analytical Result and a Fixed, Numerical Value

(i) Population standard deviation known Suppose we wish to determine whether or not the true concentration, τ, of a determinand differs from some fixed, numerical value, λ; depending on the magnitude of any difference that may exist between τ and λ, a decision is required on whether or not any action is necessary. Accordingly, a measurement of the determinand is made to give a result, R, and the difference, D, calculated where $D = R - \lambda$. Now, $R = \tau + \varepsilon$, where ε is the random error of the result, R (bias being assumed absent). It follows that:

$$D = (\tau - \lambda) + \varepsilon \tag{4.28}$$

Clearly, the random error of D is the same as that of R. Thus, a non-zero value for D may well be obtained experimentally even when $\tau = \lambda$, so that such a value for D does not alone justify the conclusion that $\tau \neq \lambda$, and any consequent actions. The commonsense approach to this difficulty is, of course, to note that as D becomes larger in relation to ε, so it becomes increasingly likely that $\tau \neq \lambda$. In this situation, what is needed is some quantitative criterion for deciding when the value of D justifies the conclusion that $\tau \neq \lambda$. The statistical ideas presented in Section 4.3 allow such a criterion to be derived as follows. Note that the random errors are assumed to follow a normal distribution and that bias is assumed absent.

In the absence of bias ($\beta = 0$), the population mean for the distribution of R is τ (Section 4.4.1). Thus, by considering the simple equation:

$$D = R - \lambda \qquad (4.29)$$

and using the results in Section 4.3.2(*b*), it follows that D is normally distributed with population mean, $\mu_D = \tau - \lambda$, and population standard deviation, $\sigma_D = \sigma$, where σ is the population standard deviation of R. The distribution of D is, therefore, completely defined, and we can make use of the results in Section 4.3.3(*b*).

Consider the situation where $\tau = \lambda$ so that $\mu_D = 0$. The results in Section 4.3.3(*b*) show that there is then only a probability, α, that a result for D will exceed the value $\pm u_\alpha \sigma_D$. If, therefore, we make α small (*e.g.*, 0.05), and find that $|D| > u_\alpha \sigma_D$ (*e.g.*, $u_\alpha \sigma_D = 1.96 \sigma_D$ for $\alpha = 0.05$), we may conclude that the probability of obtaining such a value when $\tau = \lambda$ is so small that it is justifiable to reject the possibility that $\tau = \lambda$, and, therefore, to conclude that $\tau \neq \lambda$. Thus, the value $u_\alpha \sigma_D$ is the quantitative criterion required in interpreting the observed value of D. Two points are particularly to be stressed in this procedure. First, if we conclude that $\tau \neq \lambda$ whenever $|D| > u_\alpha \sigma_D$, we shall be wrong on $100\alpha \%$ of occasions. Second, if $|D| \leqslant u_\alpha \sigma_D$, this does not necessarily imply that $\tau = \lambda$; it simply means that the evidence is not strong enough (given the chosen value of α) to support the conclusion that $\tau \neq \lambda$.

This procedure for assessing the value of D is usually called a '*test of significance*', and the value of D is said to be *statistically significant at the α significance level* when $|D| > u_\alpha \sigma_D$. It is very important to note that the term 'statistically significant' does not necessarily imply 'practically important'. For example, when the value of σ_D is very small in relation to $\tau - \lambda$, we may well be able to detect statistically significant differences between τ and λ much smaller than those of practical importance. Similarly, if σ_D is very large in relation to $\tau - \lambda$, we may well be unable to detect important differences between τ and λ because they fail to achieve statistical significance and thereby fail to justify the conclusion that $\tau \neq \lambda$. In such tests of significance, a value for α of 0.05 is very commonly used, but if one wishes to reduce the probability of falsely concluding that $\tau \neq \lambda$, a smaller value of α must be used. Readers should consult statistical texts for a full discussion of the concepts underlying tests of significance.

We considered in the preceding paragraph the situation where we wished to decide if τ differed from λ and positive and negative differences were both of interest. Situations arise where differences of only one sign are of interest: for example, when τ cannot be less than λ or when action would be taken only if τ were greater than λ. In these latter situations, we are, therefore, interested in only positive or only negative values of D, and thus in only the positive or negative half of the normal distribution of D when $\tau = \lambda$. This requires some modification of the value of the criterion against which D is compared; see Section 4.3.3(b). For a significance level of α in these '*single-sided*' tests, D is compared with $u_{2\alpha}\sigma_D$. For example, given that the significance level of the test is to be 0.05, we compare D with $1.96\sigma_D$ and $1.64\sigma_D$ for the double-sided and single-sided situations, respectively [see Section 4.3.3(b)].

Returning now to the double-sided situation considered earlier in this section, another very important point concerns the effect of the value of τ on the probability of obtaining a statistically significant difference. We saw above that, at the 0.05 significance level for example, the statistically significant value was $1.96\sigma_D = 1.96\sigma$ since $\sigma_D = \sigma$. Consider now what would happen if $\tau - \lambda = 1.96\sigma$. In this case, the population mean for D would be 1.96σ, and from the symmetry of the normal distribution it follows that 50 % of the observed values for D would be less than 1.96σ. In other words, there is only a probability of 0.5 of obtaining a statistically significant value for an observed value of D. Clearly, as $\tau - \lambda$ increases beyond 1.96σ, so the chance of obtaining a statistically significant value of D also increases. Suppose that we required the probability of not observing a significant difference to be no greater than 0.05. The value of $\tau - \lambda$ would then have to be such that the probability of obtaining a result *less* than 1.96σ from the corresponding population of D was not greater than 0.05. Section 4.3.3(b) shows that this is achieved when $\tau - \lambda = 1.96\sigma + 1.64\sigma$, i.e., when $\tau - \lambda = 3.6\sigma$. Different values of $\tau - \lambda$ can easily be calculated for other probability levels for the detection of a statistically significant difference. We shall use the 0.05 level for present purposes. For single-sided significance testing, it follows that the corresponding value of $\tau - \lambda$ at the 0.05 probability level is $1.64\sigma + 1.64\sigma$, i.e., 3.28σ.

To summarise, we have shown that $\tau - \lambda$ must be equal to or exceed 3.6σ if there is to be a probability of at least 0.95 that a measured value of D will be statistically significant (0.05 significance level). Thus, the smallest value of $\tau - \lambda$ that can be expected (0.95 probability) to give a statistically significant (0.05 significance level) difference is 3.6σ. The experimental procedure would, therefore, be unsatisfactory if values of $\tau - \lambda$ less than 3.6σ were of importance. This is a very important result because it focuses attention on the ability of analytical results to meet their intended purposes, and also provides a quantitative criterion for deciding if the precision of analytical results is adequate. For example, suppose that the smallest value of $\tau - \lambda$ of interest were δ. Then σ must be not greater than $\delta/3.6$. If the standard deviation were greater than $\delta/3.6$, the test could not achieve its purpose. As mentioned in Section 4.3.3(b), there are only two ways of improving precision: (1) by using a more precise analytical system, and (2) by making replicate analyses and using the mean result, the standard deviation of the mean of n results being smaller than standard deviation of individual results by a factor $1/\sqrt{n}$ [Section 4.3.2(c)]. If this latter approach were adopted, the appropriate value of n may be calculated by equating $3.6\sigma/\sqrt{n}$ to δ, i.e.,

$$n \approx 13(\sigma/\delta)^2 \tag{4.30}$$

A corresponding equation for single-sided testing can also be easily derived. This aspect of choosing n so that a test is capable of detecting a difference when the population mean differs from the fixed value by a specified amount is commonly known as 'ensuring adequate *power* of the test'. It is a very important aspect of the design of experiments, and is discussed in the more advanced statistical texts mentioned in Section 4.8.

(ii) Population standard deviation unknown   In (*i*) we showed how simple statistical ideas allow values for the maximum tolerable standard deviation to be derived. For this purpose it is valid to use the population standard deviation, σ. The discussion, however, introduced the idea of significance testing which has many applications even when only an estimate, *s*, of the population standard deviation is available. Although the aim here is not to give a rigorous description of significance testing, two situations recur so often in analysis that they are briefly described below. It is stressed though that readers should also consult statistical texts (see Section 4.8).

As might be expected from Section 4.3.5(*a*), when we have only an estimate, *s*, the criterion for statistical significance uses the *t*-statistic (see Table 4.6) in place of u_α, and the significance test is then often called a *t*-test. The general procedure for testing, at the α significance level, whether a population mean differs from a fixed numerical value is to calculate the quantity $|D|/s_D$, where D is the experimentally determined value of the difference of interest, and s_D is the estimate of the standard deviation of D. The value of $|D|/s_D$ is then compared with the value of t_α corresponding to the number of degrees of freedom of the estimate, s_D. If $|D|/s_D > t_\alpha$, it is concluded (at the α significance level) that the population mean of D differs from zero, *i.e.*, that the population mean of interest differs from the fixed numerical value.

In the present context, suppose that *n* replicate results, R_i ($i = 1$ to *n*), have been obtained from which the mean, $\bar{R}$, and standard deviation, *s*, are calculated exactly as in equation (4.13). Statistical significance is achieved if

$$\frac{|\bar{R}-\lambda|}{s/\sqrt{n}} > t_\alpha \tag{4.31}$$

where t_α is the *t*-statistic for $(n-1)$ degrees of freedom and the α probability level (see Table 4.6). If it is required to check not whether μ_R differs from λ but whether μ_R is either greater or smaller than λ (*i.e.*, a single-sided test is to be made), the same calculations are made but the value of *t* used for the α significance level is $t_{2\alpha}$. Thus, for double-sided and single-sided tests at the 0.05 significance level, the value of *t* for 10 degrees of freedom is 2.23 and 1.81, respectively (see Table 4.6).

Equation (4.31) indicates that the statistical significance of a single result ($n = 1$) may also be assessed provided that an appropriate estimate, *s*, is available from other tests. Difficulties arise, however, when the same estimate, *s*, is used repeatedly for a number of significance tests. The nature of this problem is noted in Section 4.3.5(*b*) and discussed in the Appendix (p. 171); its severity decreases as the degrees of freedom of *s* increase.

(b)   Difference between Two Analytical Results

(i)   Population standard deviation known   Suppose we wish to determine if the

true concentrations, τ_1 and τ_2, of a determinand in two samples differ. Accordingly, the determinand is measured in each of the two samples to give the results, R_1 and R_2, whose difference, D, is calculated:

$$D = R_1 - R_2 \tag{4.32}$$

Applying the results in Section 4.3.2(*b*) to equation (4.32) shows that D is normally distributed with population mean $\mu_1 - \mu_2$ and standard deviation $\sigma_D = \sqrt{\sigma_1^2 + \sigma_2^2}$. The distribution of D is, therefore, completely defined, and we can again make use of the results in Section 4.3.3(*b*) in the same way as in (*a*) (*i*) above.

There is no need to go through the derivation of the various results of interest for, with one exception, they are readily obtained directly from the analogous results in (*a*) (*i*) simply by using the appropriate value for σ_D, i.e., $\sqrt{\sigma_1^2 + \sigma_2^2}$. [The exception concerns significance testing when only an estimate, s_D, is available; for this, see (*ii*) below.] Thus, D will be statistically significant (0.05 significance level, double-sided test) when its value exceeds $1.96\sigma_D$. Similarly, the smallest difference between τ_1 and τ_2 for which there is a probability of 0.95 that a statistically significant (0.05 significance level) value of D will be obtained is $3.6\sigma_D$. Similar points apply for the power of tests. Given values for σ_1 and σ_2, the numerical values of such parameters may be calculated. As a simple example, it will sometimes be true that $\sigma_1 = \sigma_2$, e.g., when the same laboratory makes both analyses at the same time and the two waters are of similar general composition with τ_1 not very different from τ_2. Thus, when $\sigma_1 = \sigma_2 = \sigma$, $\delta_D = \sigma\sqrt{2}$, so that $3.6\sigma_D = 3.1\sigma$, i.e., the smallest detectable difference (0.95 probability level) is 3.1 times the population standard deviation of individual analytical results. This value may be compared with that of 3.6σ derived in (*a*) (*i*); as would be expected, it is easier to detect small differences between a population mean and a fixed value than between two population means because, in the latter case, the effects of two random errors are involved (those of R_1 and R_2).

(ii) Population standard deviation unknown   In general, significance testing of the difference between two results when population standard deviations are not known follows the same general procedure as that described in (*a*) (*ii*). That is, the statistic $|D|/s_D$ is calculated and compared with the appropriate t value. In the present context, suppose D is the difference between the means, $\bar{R}_1$ and $\bar{R}_2$, of two sets of results with n_1 results (R_{1i}, $i = 1$ to n_1) in set 1 and n_2 results (R_{2i}, $i = 1$ to n_2) in set 2; set 1 could correspond, for example, to repeated analyses on a sample of one water, and set 2 to repeated analyses on a sample of another water, but there are clearly many other possibilities. Thus, $D = \bar{R}_1 - \bar{R}_2$, so that from the results in Section 4.3.2(*b*) we obtain

$$\sigma_D^2 = \sigma_{\bar{R}_1}^2 + \sigma_{\bar{R}_2}^2 \tag{4.33}$$

From Section 4.3.2(*c*), $\sigma_{\bar{R}_1} = \sigma_1/\sqrt{n_1}$ and $\sigma_{\bar{R}_2} = \sigma_2/\sqrt{n_2}$, where σ_1 and σ_2 are the population standard deviations of individual analytical results for sets 1 and 2, respectively. Substituting these values in equation (4.33) gives

$$\sigma_D^2 = (\sigma_1^2/n_1) + (\alpha_2^2/n_2) \tag{4.34}$$

When it cannot be assumed that $\sigma_1 = \sigma_2$, the t distribution is strictly not applicable though a reasonable approximation is available and described in statistical texts (see Section 4.8). We shall deal here only with the case where $\sigma_1 = \sigma_2 = \sigma$ so that equation (4.34) simplifies to

$$\sigma_D = \sigma \sqrt{\frac{n_1 + n_2}{n_1 n_2}} \qquad (4.35)$$

Finally, we may replace the population standard deviations by their estimates to obtain

$$s_D = s \sqrt{\frac{n_1 + n_2}{n_1 n_2}} \qquad (4.36)$$

Thus, the statistic to be calculated is

$$\frac{|D|}{s_D} = \frac{|\bar{R}_1 - \bar{R}_2|}{s} \sqrt{\frac{n_1 n_2}{n_1 + n_2}} \qquad (4.37)$$

To obtain the value for s in equation (4.37), note that we can obtain an estimate, s_1, of σ from set 1 with $n_1 - 1$ degrees of freedom, and another estimate, s_2, of σ from set 2 with $n_2 - 1$ degrees of freedom [see Section 4.3.4(b)]. These two estimates can be pooled [see Section 4.3.4(c)] to give the best possible estimate of σ. Thus,

$$s = \sqrt{\frac{(n_1 - 1)s_1^2 + (n_2 - 1)s_2^2}{n_1 + n_2 - 2}} \qquad (4.38)$$

where

$$s_1^2 = \sum_i (R_{1i} - \bar{R}_1)^2 / (n_1 - 1)$$

and

$$s_2^2 = \sum_i (R_{2i} - \bar{R}_2)^2 / (n_2 - 1) \, .$$

The estimate, s, has $n_1 + n_2 - 2$ degrees of freedom.

For a double-sided significance test at the α significance level, the value of $|D|/s_D$ calculated using equations (4.37) and (4.38) is compared with the value of t_α corresponding to $n_1 + n_2 - 2$ degrees of freedom. All other details are as described in (*a*) (*ii*) above. Again, it is stressed that failure of $|D|/s_D$ to achieve statistical significance does not necessarily mean that the two population means, μ_1 and μ_2, are equal.

Equation (4.37) indicates that the statistical significance of the difference between two single results ($n_1 = n_2 = 1$) may also be assessed if an appropriate estimate, s, is available from other tests. Difficulties arise, however, if the same estimate, s, is used repeatedly for a number of significance tests. The nature of this problem is noted in Section 4.3.5(*b*) and discussed in the Appendix (p. 171); its severity decreases as the degrees of freedom of s increase.

(c) General Function of Analytical Results

In (a) and (b) above we considered situations where decisions were based on an experimental value for D that was a very simple function of the analytical result(s). Sometimes the decision may be based on a more complicated function of results such that it is much less easy to see how to calculate the effects of random errors in the analytical results. If a value of D is calculated that is a general function of a number of random and *independent* variables, *i.e.*, $D = f(R_1, R_2, \ldots \ldots \ldots R_n)$, it is often a reasonable approximation to obtain the standard deviation of D from the following equation:

$$\sigma_D \approx \sqrt{\sigma_1^2\left(\frac{\partial f}{\partial R_1}\right)^2 + \sigma_2^2\left(\frac{\partial f}{\delta R_2}\right)^2 + \ldots\ldots\ldots + \sigma_n^2\left(\frac{\partial f}{\partial R_n}\right)^2} \qquad (4.39)$$

where σ_i ($i = 1$ to n) denotes the standard deviation of the ith random variable, and $(\partial f/(\partial R_i))$ is the partial differential coefficient of $f(R_1,R_2, \ldots \ldots R_n)$ with respect to R_i. It is essential to the validity of this equation for all the variables to be independent of each other. Apart from this, equation (4.39) will be reasonably valid provided that the relative standard deviations, σ_i/μ_i, of each of the n variables do not exceed approximately 0.2.[10] The distribution of D is not necessarily normal even if each of the individual variables is normally distributed, but it will usually be reasonable to assume normality in initial assessment of the precision required of analytical results.

Equation (4.39) may, for example, be used to derive the standard deviation of D for products and quotients of random variables. Consider the function

$$D = b_1x_1/b_2x_2 \, ,$$

where b_1 and b_2 are constants, and x_1 and x_2 are two random variables with standard deviations of σ_1 and σ_2, respectively. Applying equation (4.39) leads to

$$\sigma_D^2 = \left(\frac{b_1}{b_2x_2}\right)^2\sigma_1^2 + \left(\frac{-b_1x_1}{b_2x_2^2}\right)^2\sigma_2^2$$

$$= \left(\frac{b_1x_1}{b_2x_2}\right)^2\left(\frac{\sigma_1^2}{x_1^2} + \frac{\sigma_2^2}{x_2^2}\right)$$

so that

$$\sigma_D^2/D^2 = (\sigma_1^2/x_1^2) + (\sigma_2^2/x_2^2) \, .$$

Thus,

$$\sigma_D/D = \sqrt{(\sigma_1/x_1)^2 + (\sigma_2/x_2)^2},$$

so that, putting r = relative standard deviation:

$$r_D = \sqrt{r_1^2 + r_2^2} \qquad (4.40)$$

Exactly the same equation applies when $D = b_1b_2x_1x_2$.

An application of equation (4.39) to a practical situation is given in Section 4.6.2(*f*).

(d) Decisions Based on the Comparison of a Result with a Fixed Numerical Value
This situation arises whenever an analytical result, R, or a mean, $\bar{R}$, of a number of results is used to decide whether or not the true concentration, τ, of a determinand is greater or smaller than some specified value, e.g., a water quality standard. This type of application corresponds exactly to the considerations presented in (a) above, and the procedures described there are directly applicable. Two additional practical points are considered below, and note that, at this stage, bias in the analytical results is still assumed to be absent; the combined effects of bias and random error are considered in Section 4.6.3.

(i) Effect of random errors not arising from analysis In (a) to (c) above there was the tacit assumption that the only source of random error was in the measurement process, and this will be true for many applications in the present category. In some circumstances, however, other types of random variation may be relevant. For example, suppose a given water were analysed on a number of occasions in order to check whether the mean concentration of a determinand were less than a given value. In this situation, the individual analytical results (and hence the mean) are subject not only to random errors of analysis, but the water itself may well show temporal variability that increases the overall uncertainty of the mean. Suppose further that this temporal variability were random and characterised by a standard deviation, σ_s, and that the standard deviation due to analysis were σ_A (here assumed independent of the concentration of the determinand). Then the total standard deviation, σ, of any one result is given by [see Section 4.3.2(d)]

$$\sigma = \sqrt{\sigma_S^2 + \sigma_A^2}$$

Thus, the standard deviation, σ_m, of the mean result is given by

$$\sigma_m = \sigma/\sqrt{n} = \sqrt{(\sigma_S^2/n) + (\sigma_A^2/n)} \qquad (4.41)$$

In many situations the (σ_S^2/n) term may well be the major contribution to σ_m. It can be calculated from equation (4.41) that if $\sigma_A \leqslant \sigma_S/2$, the analytical errors will cause σ_m to be increased by no more than 10 % of its value if $\sigma_A = 0$. Such considerations may sometimes allow relaxation of the required analytical precision, but this is not, of course, possible when individual results are to be used for deciding on the suitability of water quality [see (ii) below]; see also Section 2.6.2(b)(v).

(ii) Effect of random errors on short-term control of water quality As would be expected, random errors arising in analysis have a direct impact on the operational aspects of any scheme for the control of water quality, and this is illustrated for a threshold-control situation (see Section 3.3.3) in Figure 4.2. For example, it can be seen from Figure 4.2(a) that when the true concentration of the determinand, τ, is slightly less than the maximum acceptable concentration, λ, random errors will quite frequently lead to a result, R, greater than λ. Similarly, Figure 4.2(b) shows that when τ is a little greater than λ, there is a substantial probability that a result less than λ will be obtained. Thus, if the decision rule were, for example, to take corrective action whenever $R > \lambda$, there are finite probabilities that action will be

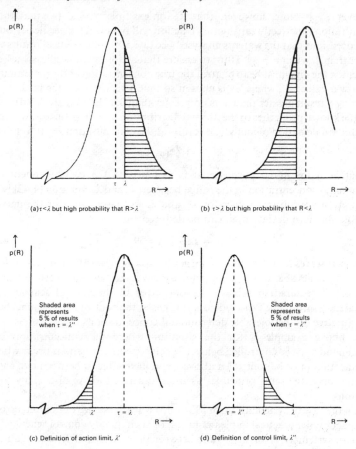

(a) $\tau < \lambda$ but high probability that $R > \lambda$

(b) $\tau > \lambda$ but high probability that $R < \lambda$

(c) Definition of action limit, λ'

(d) Definition of control limit, λ''

Figure 4.2 *Effect of random errors on control of water quality*

taken when it is not necessary (when $R > \lambda$ but $\tau \leqslant \lambda$), and also that action will not be taken when it is required (when $R \leqslant \lambda$ but $\tau > \lambda$). Suppose now that it is desired to ensure that the probability of each of these undesirable occurrences is not greater than some specified value (say, 0.05) so that the control scheme can be considered reasonably efficient. (Other values could be used for these probabilities, *e.g.*, 0.1 or 0.01, and it is not necessary that the same probability be used for each of the two occurrences; such differences do not affect the principles of the following argument.) To achieve the desired control system, a new action limit, λ', must be defined and action taken whenever $R > \lambda'$ [see Fig. 4.2(c)]. The value of λ' is chosen so that, when $\tau = \lambda$, the probability of obtaining a result less than λ' is no greater than 0.05. From the properties of the normal distribution, λ' equals $\lambda - 1.64\sigma_\lambda$, where σ_λ is the standard deviation of analytical results at the concentration λ [see Section 4.3.2(a)].

This value of λ' will ensure with the desired probability that action will be taken whenever $\tau > \lambda$. Note, however, that if, for example, $\tau = \lambda'$ (so that the true concentration is perfectly satisfactory), action will be called for needlessly on 50 % of the occasions that the water is analysed because 50 % of analytical results will be greater than λ' when $\tau = \lambda'$. Thus, to ensure that action is not needlessly taken too frequently, the aim must be to control the true concentration of the determinand at or below a value, λ'', where λ'' is chosen so that, when $\tau = \lambda''$, the probability of obtaining a result greater than λ' is no greater than 0.05 [see Figure 4.2(d)]. Again, from the known properties of the normal distribution, $\lambda'' = \lambda' - 1.64\sigma_{\lambda''}$, where $\sigma_{\lambda''}$ is the standard deviation of analytical results at the concentration λ''. It follows that

$$\lambda'' = \lambda - 1.64(\sigma_\lambda + \sigma_{\lambda''}) \tag{4.42}$$

If, at least for initial purposes, it is assumed that σ is independent of the determinand concentration in the range between λ'' and λ, equation (4.42) is easily solved given appropriate values for λ'' and λ, so that a value for the maximum tolerable standard deviation, σ, can be deduced:

$$\sigma = (\lambda - \lambda'')/3.29 \tag{4.43}$$

which then serves as a quantitative target for the analyst. Alternatively, if for any reason the achievable value of σ cannot be decreased beyond a certain value, the value of λ'' required to achieve the objectives of the control scheme can be calculated. As would be expected, as σ decreases, the control requirements become less restrictive on the tolerable determinand concentrations.

The above example is for the situation where the concentration of the determinand must be controlled below a specified value. Situations occur where the concentration must be controlled above a specified value or between two specified values; exactly the same principles as those described above also apply to these situations.

As in the case of systematic error [Section 4.6.1(a)], the size of random errors has an impact on the practical implementation of such quality control schemes. Thus, just as for systematic error, discussion between the analyst and the user of analytical results is desirable in relation to random errors.

(e) Decisions Based on the Comparison of Two Analytical Results
This situation arises whenever the true difference in the qualities of two waters is of interest, and a decision on whether or not action is required is based on the difference between corresponding analytical results. The considerations and procedures in (b) above are directly applicable, but note that the reservation made in (d)(i) above applies here also.

These procedures may be used when appropriate to derive numerical values for the maximum tolerable standard deviation of analytical results. For example, we saw that the smallest difference in true concentrations for which there is a probability of 0.95 that a statistically significant (0.05 significance level) difference between two individual results will be obtained is 5.1σ, where σ is the standard deviation of analytical results (assumed here to be the same for both waters). If it is required, therefore, to be able to detect (at the above probability levels) a true difference in concentration of δ, σ must be not greater than $\delta/5.1$. If the required value of σ cannot be achieved, replicate analysis of each of the two samples may help by reducing the standard deviation of the mean results [see Section 4.3.2(c)]; otherwise, it may be necessary through joint discussions between the analyst and user of results to increase, at least temporarily, the numerical value of δ.

(f) Decisions Based on a Function of Analytical Results

In general, decisions may be based on some more or less complicated function of analytical results, and knowledge of the random error of the function is needed to assess the effect of the analytical errors. For this purpose, equation (4.39) is often a useful approximation, and an example of its use is given below. It is again assumed that the random analytical errors follow the normal distribution and are independent.

Suppose that the percentage removal, E, of a determinand by a treatment plant were of interest, that an approximately 80 % removal is achieved, and that the percentage removal is to be measured such that the width of the 95 % confidence interval is no greater than 20 %. If R_i and R_e are the analytical results for the influent to, and effluent from, the plant, respectively, then

$$E = 100(R_i - R_e)/R_i \qquad (4.44)$$

Equation (4.44) gives the functional relation between E, the quantity of interest, and the two analytical results. We can, therefore, obtain an approximate value for the standard deviation, σ_E, of E by applying equation (4.39), and this leads to:

$$\sigma_E \approx \frac{100 R_e}{R_i} \sqrt{\frac{\sigma_i^2}{R_i^2} + \frac{\sigma_e^2}{R_e^2}} \qquad (4.45)$$

where σ_i and σ_e are the standard deviations of R_i and R_e, respectively. If we let r_i and r_e denote the relative standard deviations (expressed as percentages) of R_i and R_e, respectively, equation (4.45) may be rewritten as

$$\sigma_E \approx (R_e/R_i)\sqrt{r_i^2 + r_e^2}$$

and if it is now assumed that $r_i = r_e = r$, we obtain

$$\sigma_E \approx 0.2 r \sqrt{2} \qquad (4.46)$$

since $R_e/R_i \approx 0.2$. Now 95 % confidence limits for the true percentage removal are given approximately by $E \pm 1.96 \sigma_E$, so that the width of the confidence interval is $3.92 \sigma_E$. Accordingly, from the requirement on the maximum tolerable width of the confidence interval,

$$3.92 \sigma_E \not> 20$$

and substituting the above value for σ_E gives

$$r \not> 10/0.392\sqrt{2} \qquad (4.47)$$

Thus, from equation (4.47), r must be not greater than 18 % if the objectives of the measurements are to be achieved.

Of course, a number of approximations are made in such an approach, but they allow a reasonable target value for the analytical standard deviation to be simply derived, and the approach, therefore, is commonly useful.

4.6.3 Decisions Made when Both Bias and Random Error are Present

If the bias of analytical results were constant and known, its effect would be trivial because it could be eliminated by simple correction. In general, however, the bias may vary from time to time and/or from sample to sample, and is not known exactly for each analytical result (see Sections 4.4.1 and 5). In this situation and with the added complication of the effect of the inevitable random errors, the authors know of no simple and general approach that overcomes the problems involved in the interpretation of results that are both biased and imprecise. Given explicit formulation of any decision rules and estimates of the bias and precision of analytical results, the types of procedures outlined in Sections 4.6.1 and 4.6.2 can be applied to determine the effects of errors on the correctness of decisions, but in general this will not be simple or rapid. We know of no alternative. Consequently, the rest of this section is intended merely to note some general points that seem to us helpful in this rather complicated situation.

First, we believe that it is usually possible to obtain a reasonable estimate of the random error of a particular result, but much more difficult to obtain an experimental estimate of the bias of a particular result (see Section 5). For this reason we feel that the effect of random errors on the interpretation of analytical results is easier to deal with than in the case of bias. Consequently, whenever there is need to simplify such interpretative problems, it seems to us that it is usually better to place greater emphasis on the estimation and control of bias.

Given the problems in estimating bias experimentally, emphasis should, in turn, be placed on ensuring that its magnitude is as small as possible, and preferably negligible in relation to random error [see Section 4.6.1(c)]. To this end, it is crucial to attempt to select and properly apply analytical systems subject to as little bias as possible. This is much more easily said than done, but some detailed suggestions are made in Sections 5.5 and 9. Of course, the use of such systems may sometimes involve penalties such as greater analytical time, more expensive equipment, *etc.* Such aspects must be balanced against the advantages gained from greater confidence in the accuracy and interpretation of results.

Finally, joint consideration by the analyst and the user of results may sometimes allow solution of problems posed by inaccurate results by means of appropriate control of water quality. For example, we saw in Sections 4.6.1(*a*) and 4.6.2(*d*) that such problems could be avoided in certain control schemes by ensuring that the true concentration of a determinand was maintained at a value appreciably below that specified in a water quality standard. Thus, considering bias alone, if the standard specified that the concentration should not exceed a value, λ, the aim should be to control the true concentration below a value $\lambda - \beta_c$, where β_c is governed by the bias of results [Section 4.6.1(*a*)]. The effect of random errors could then be controlled as described in Section 4.6.2(*d*), *i.e.*, by maintaining the true concentration below a value $\lambda - \beta_c - 3.29\sigma$, where σ is the standard deviation of analytical results (assumed independent of the concentration of the determinand). Again, this approach may entail financial or other penalties which must be judged against the problems of interpretation that would otherwise arise. As mentioned previously, we believe that such explicit recognition and joint discussion of the consequences of errors can only benefit both analysts and users of analytical results. Such discussions will usually benefit from the presence of a statistician well versed in such practical problems.

4.7 NEED TO ESTIMATE ANALYTICAL ERRORS

Section 4.6 has stressed the important consequences of analytical errors, and the need to ensure that these errors are sufficiently small for their intended uses. Of course, one of the primary responsibilities of the analyst is to ensure that his results are of adequate accuracy, and this task can be accomplished only by knowing the magnitude of the errors. Provided that the user states his requirements for bias and precision, the task may seem to some relatively easy to carry out. For example, it seems sometimes to be thought that the use of 'Standard' or 'Approved' analytical methods (*i.e.*, methods approved and issued by one or more organisations with expertise in the field of water analysis) will alone ensure satisfactorily small errors. Such an approach is not recommended here for the following reasons.

The accuracy of analytical results clearly depends on a number of factors over which the analytical method has normally no direct control. For example, each of the following factors is well known as capable of markedly affecting accuracy: the skill and conscientiousness of the analyst, the type of instrumentation and the care with which it is maintained, the purity of reagents and accuracy of standards, the type of apparatus, the magnitude of contamination effects, the variability of environmental conditions in a laboratory, and the extent to which modifications to the prescribed procedure come to be used (knowingly or unknowingly). All these factors may vary from one laboratory to another as well as from time to time in any one laboratory. There is no need to give detailed accounts of the effects of such factors as any analyst will easily be able to draw on examples from his own experience.

A few additional words about one of the above factors – instrumentation – are worthwhile because within the last few years there has been a very marked growth in the numbers and types of relatively complicated instruments applied routinely in water analysis laboratories. Examples of general techniques for which a wide choice of various instruments is available and which are being increasingly applied are: atomic absorption and flame emission spectrometry, gas chromatography, anodic-stripping voltammetry, ion-selective electrodes, and the automation of a wide range of analytical techniques. There are also examples of similar types of instrumentation but for more specialised purposes, *e.g.*, instruments for determining total organic carbon. In addition, other techniques though not at present so widely used routinely are finding increasing application, *e.g.*, plasma emission spectrometry, high-performance liquid chromatography, mass spectrometry. The ingenuity of instrument manufacturers and the technological advances of recent years are now providing instruments that, with the minimal attention from the analyst, allow demanding analyses to be made. This analytical development is one of the most striking of recent years, but as always the benefits are not achieved without some cost; two particular points are, therefore, worth noting in the present context though the topic is dealt with in more detail in Section 10.4 and throughout the sub-sections of Sections 11 and 12.

(1) As a result of differences in design and function, the accuracy achieved with one particular manufacturer's equipment may well differ from that achieved with another manufacturer's nominally equivalent instrument. Indeed, differences in performance between two instruments of the same model from the same manufacturer may be observed. Further, superior performance may be achieved from the more expensive and sophisticated instruments. Laboratories possessing such equipment will not generally forsake it if an analytical method specifies the use of simpler equipment, and laboratories having only the simpler instruments may well be unable, or choose not, to obtain the more expensive types. Such problems seem not to be generally allowed for by the written procedures of 'Standard' or 'Approved' methods.

(2) The apparent simplicity of use of many of these instruments may well tend to conceal the often complicated series of processes taking place during an analysis, and the many possibilities of malfunction, some of which may not be easily recognised but which may nonetheless cause important errors in results. Analysts may, therefore, unless particularly vigilant, tend to be lulled into a false sense of security concerning the accuracy of their results.

For all the above reasons, there is clearly no fundamental reason why all laboratories using 'the same method' should necessarily obtain errors of the same magnitude. Indeed, with modern analytical techniques, we feel that it would be rather remarkable if equality of errors were regularly observed in different laboratories. There is, of course, much evidence to support the validity of this viewpoint; see the many published papers on inter-laboratory studies. Many examples will be found in a very useful summary of such studies in the field of water analysis in the United States.[11] Thus, the assumption that, because one or more laboratories have obtained errors of a certain size when a particular method was used, any other laboratory will also obtain errors of the same magnitude, seems to us quite unjustified. In this situation, the only practicable approach to ensure that errors are satisfactorily small is for each laboratory to estimate the magnitude of its own errors for each of the analytical systems it uses. If the errors are found to be intolerably large, corrective action is, of course, required. This is the approach recommended here, and its consequences are considered in detail in Sections 5 to 7 and 9.

The need for this measurement and control of errors has been increasingly recognised within the last decade by many countries and organisations. Many examples could be quoted to support this statement, but the following points will suffice to show this widespread recognition by both the users of analytical results and also the analysts. The United Nations Environment Programme/World Health Organization project on a Global Environmental Monitoring Scheme for fresh surface waters specifies analytical quality control as an integral part of the scheme.[12,13] Similar requirements are included in the United Kingdom's Scheme for the Harmonised Monitoring of River Water,[14,15] and in the United States in the implementation of water quality legislation[16,17] and in environmental monitoring.[18] Maximum tolerable errors (thereby implying the need for analytical quality control) are also specified in a Directive of the European Community concerned with the quality of surface waters.[19] It is interesting to note that such developments have not been restricted to the field of water analysis. For example, the field of clinical chemistry has shown great interest in analytical quality control, and many valuable publications on the topic appear within that field of analysis; see ref. 2 for an authoritative introduction to this literature. Further, within the last few years, many countries have developed or are developing national schemes for the accreditation of laboratories undertaking a wide variety of chemical analyses (and other types of measurements). The United Kingdom's scheme (NATLAS – National Laboratory Accreditation Scheme) is described in refs. 20 and 21, and requires analytical quality control in laboratories. The international co-ordination of such schemes is also being developed.[20,23] Several standards organisations have issued guides on the assessment of the technical competence of testing laboratories.[22a,b] Ref. 22 deals solely with laboratories in which waters and wastewaters are analysed, and it too requires analytical quality control. Finally, almost every authoritative publication on the analysis of waters recommends that analytical quality control should form an integral part of routine analysis.

Substantial amounts of effort may well be required for an adequate level of analytical quality control; estimates by different authors[22,24–29] range between 5 and 30% of the routine workload (or resources) of a laboratory. Of course, what constitutes an adequate level of control is likely to vary with the accuracy required by the user of analytical results, and also from one determinand to another. Further, as experience accumulates of the continuing, satisfactory performance of an analytical system, the amount of quality control for that system can probably be reduced. Clearly, each laboratory needs to give critical consideration to such points, and there will usually be considerable room for manoeuvre. Nevertheless, it seems to us unlikely that any water analysis laboratory will decide that, on average, less than one member in every four or five of the laboratory's staff will suffice for the minimal level of control. If a laboratory has insufficient effort for that level of control, there seems to us only two options:

(1) the laboratory informs the user of results that it cannot supply reasonable estimates of the errors of those results (*cf.* Section 8.2), or
(2) the number of routine samples is decreased so that some effort is freed.

These possibilities should be jointly discussed by the analyst and the user.

To close this section it is useful to add a few words about another practice that is increasing, *i.e.*, the inclusion of statements on accuracy, bias, and precision in published analytical methods. This is an excellent development because such information is of great value in indicating the size of errors achieved by particular laboratories. It can, however, be misleading if it is intended, or is taken, to imply that the reported errors are a fixed property of the method (see the introduction to this section). Care should be taken, therefore, to avoid any statement or action encouraging the assumption that the use of a particular method will automatically ensure specified magnitudes of errors. Such an attitude is the very antithesis of the critical approach so vital to the analyst.

The estimations of bias and precision are considered in detail in Sections 5 to 7.

4.8 FURTHER READING

For readers wishing to pursue the application of statistical concepts and techniques, there is a very large number of books from which to choose. We have reasonably detailed knowledge of only a small proportion of these books, and the suggestions made below should, therefore, not be regarded as a comprehensive list of references. We feel that readers would be well advised to examine as many books as possible in order to find those most suited to their needs. The books quoted below should, however, provide analysts with useful introductions to the subject. We would also stress once more the value of discussing problems and applications with statisticians whenever possible.

The nature of errors is discussed in all statistical texts, but Eckschlager[30] has written a book on this topic purely for chemical analysis. Three other, particularly good discussions of errors are also worth noting: one by Eisenhart[31] on measurement processes in general, and those by Currie[4] and Liteanu and Rîcă.[32]

Good introductory texts for those unfamiliar with statistical ideas are provided by Moroney,[33] Cormack,[34] Langley,[35] Wynne,[36] and Youden;[37] the last of these is written especially for chemists. All these books and those noted below have the useful feature of worked examples of various statistical calculations. Of the more advanced statistical texts for scientists, we mention those by Steel and Torrie,[38] Bennett and Franklin,[5] Johnson and Leone,[39] Brownlee,[40] Mandel,[41] Davies and Goldsmith,[10] and Caulcutt.[42] Of these, the first five tend to devote more attention to the underlying mathematical and statistical bases of practical techniques than do the last two. Many of these books include a chapter or some discussion of non-parametric statistical techniques [see Section 4.3.2(a)], but a number of books dealing solely with this topic are available; see, for example, those by Conover[43] and Tukey.[44]

There are also a number of books on the application of statistical techniques in analytical chemistry; see, for example, those by Nalimov,[3] Massart *et al.*,[6] Gottschalk,[45] Doerffel,[46] Liteanu and Rîcă,[32] and most recently Caulcutt and Boddy.[47] At least one analytical treatise includes a long section on the use of statistics in analytical chemistry.[48] The book by Eckschlager[30] also includes a short section on statistical techniques. Some of these tend to concentrate on particular applications, and it is probably better to read one or more of the more general texts before consulting these analytically oriented books. A review of the basic statistics required for analytical quality control in clinical chemistry is available,[49] and a volume of papers edited by Ku[50] contains many contributions of interest to analysts. A select bibliography of publications on accuracy in measurement has recently been published.[51]

Detailed aspects of the estimation and control of errors are considered, with references, in Sections 5 to 7. Useful publications on overall approaches to analytical quality control include refs. 2, 4, 18, 22, 27–29, and 52.

4.9 REFERENCES

1 M. Thompson and R. J. Howarth, *Analyst*, 1980, **105**, 1188.
2 J. Büttner, *et al.*, *J. Clin. Chem. Clin. Biochem.*, 1980, **18**, 69.
3 V. V. Nalimov, 'The Application of Mathematical Statistics to Chemical Analysis', Pergamon, Oxford, 1963.

[4] L. A. Currie, in I. M. Kolthoff and P. J. Elving, ed., 'Treatise on Analytical Chemistry', 2nd Ed., Part 1, Vol. 1, Wiley, London, 1978, pp. 95–242.

[5] C. A. Bennett and N. L. Franklin, 'Statistical Analysis in Chemistry and the Chemical Industry', Chapman and Hall, London, 1954.

[6] D. L. Massart, A. Dijkstra, and L. Kaufman, 'Evaluation and Optimization of Laboratory Methods and Analytical Procedures', Elsevier, Amsterdam, 1978.

[7] I. E. Frank, E. Pungor, and G. E. Veress, *Anal. Chim. Acta*, 1981, **133**, 433 and 443.

[8] L. A. Currie, *Anal. Lett.*, 1980, **13**, 1.

[9] L. A. Currie and J. R. de Voe, in J. R. de Voe, ed., 'Validation of the Measurement Process', American Chemical Society, Washington, 1977, pp. 114–139.

[10] O. L. Davies and P. L. Goldsmith, ed., 'Statistical Methods in Research and Production', Fourth Revised Ed., Oliver and Boyd, 1980.

[11] E. F. McFarren, R. J. Lishka, and J. H. Parker, *Anal. Chem.*, 1970, **42**, 358.

[12] R. Helmer, *Nat. Resourc.*, 1981, **17**, 7.

[13] S. Barabas, in L. H. Keith, ed., 'Advances in the Identification and Analysis of Organic Pollutants in Water, Volume 2', Ann Arbor Science Publishers, Ann Arbor, Michigan, 1981, pp. 481–494.

[14] E. A. Simpson, *J. Inst. Water Eng. Scient.*, 1978, **32**, 45.

[15] A. L. Wilson, *Analyst*, 1979, **104**, 273.

[16] J. A. Winter, in W. J. Cooper, ed., 'Chemistry in Water Re-use, Volume 1', Ann Arbor Science Publishers, Ann Arbor, Michigan, 1981, pp. 187–205.

[17] D. G. Ballinger, *Environ. Sci. Technol.*, 1979, **13**, 1362.

[18] American Chemical Society, *Anal. Chem.*, 1980, **52**, 2242.

[19] Council of the European Communities, *Off. J. Eur. Communities*, 1979, **22**, No. L-271, 44.

[20] NATLAS Executive, *NATLAS News*. 1983, No. 1, National Physical Laboratory, Teddington.

[21] D. Simpson, *Trends Anal. Chem.*, 1982, **1**, IV.

[22] American Society for Testing and Materials, '1983 Annual Book of ASTM Standards. Volume 11.01 Water (I)', A.S.T.M., Philadelphia, 1983.

[22a] International Organization for Standardization, ISO Guide 25, I.S.O., Geneva, 1982.

[22b] British Standards Institution, BS 6460:Part 1:1983, B.S.I., Milton Keynes, 1983.

[23] J. D. Campbell, *Trends Anal. Chem.*, 1982, **1**, VI.

[24] R. J. Pickering and J. F. Ficke, *J. Am. Water Works Assoc.*, 1976, **68**, 82.

[25] J. Büttner, R. Borth, P. M. G. Broughton, and R. C. Bowyer, *J. Clin. Chem. Clin. Biochem.*, 1980, **18**, 535.

[26] P. E. Mills, in W. J. Cooper, ed., 'Chemistry in Water Re-use', Volume 1, Ann Arbor Science Publishers, Ann Arbor, Michigan, 1981, pp. 229–244.

[27] R. L. Booth, ed., 'Handbook for Analytical Quality Control in Water and Wastewater Laboratories', (Report EPA-600/4-79-019), U.S. Environment Protection Agency, Cincinnati, 1979.

[28] J. K. Taylor, *Anal. Chem.*, 1981, **53**, 1588A.

[29] R. V. Cheeseman and A. L. Wilson, Technical Report TR 66, Water Research Centre, Medmenham, Bucks., 1978.

[30] K. Eckschlager, 'Errors, Measurement and Results in Chemical Analysis', Van Nostrand-Reinhold, London, 1969.

[31] C. Eisenhart, *J. Res. Natl. Bur. Stand.*, 1963, **67C**, 161.

[32] C. Liteanu and I. Rîcǎ, 'Statistical Theory and Methodology of Trace Analysis', Ellis Horwood, Chichester, 1980.

[33] M. J. Moroney, 'Facts from Figures', Fourth Ed., Penguin, Harmondsworth, 1969.

[34] R. M. Cormack, 'The Statistical Argument', Oliver and Boyd, Edinburgh, 1971.

[35] R. Langley, 'Practical Statistics Simply Explained', Second Ed., Pan Books, London, 1979.

[36] J. D. Wynne, 'Learning Statistics', Collier-Macmillan, London, 1982.

[37] W. J. Youden, 'Statistical Methods for Chemists', Wiley, London, 1964.

[38] R. G. D. Steel and J. H. Torrie, 'Principles and Procedures of Statistics', Second Ed., McGraw-Hill, London, 1982.

[39] N. L. Johnson and F. C. Leone, 'Statistics and Experimental Design in Engineering and Physical Sciences', Vols. 1 and 2, Wiley, London, 1964.

[40] K. A. Brownlee, 'Statistical Theory and Methodology in Science and Engineering', Wiley, London, 1965.

[41] J. Mandel, 'The Statistical Analysis of Experimental Data', Interscience, London, 1964.

[42] R. Caulcutt, 'Statistics in Research and Development', Chapman and Hall, London, 1983.

[43] W. J. Conover, 'Practical Non-parametric Statistics', Second Ed., Wiley, Chichester, 1980.

[44] J. Tukey, 'Exploratory Data Analysis', Addison-Wesley, Reading, Massachusetts, 1977.

[45] G. Gottschalk, 'Statistik in der Quantitativen Chemischen Analyse', Enke, Stuttgart, 1962.

[46] K. Doerffel, 'Statistik in der Analytischen Chemie', 3rd revised Ed., Verlag Chemie, Weinheim, 1984.

[47] R. Caulcutt and R. Boddy, 'Statistics for Analytical Chemists', Chapman and Hall, London, 1983.

[48] P. Moritz, in G. Svehla, ed., 'Wilson and Wilson's Comprehensive Analytical Chemistry', Volume XI, Elsevier, Amsterdam, 1981, pp. 1–169.

[49] G. W. Williams and M. A. Schork, *CRC Crit. Rev. Clin. Lab. Sci.*, 1982, **17,** 171.

[50] H. H. Ku, ed., 'Precision Measurement and Calibration – Statistical Concepts and Procedures', (U.S. Natl. Bur. Stand. Spec. Publ. 300), U.S. Government Printing Office, Washington, 1969.

[51] J. P. Cali and K. N. Marsh, *Pure Appl. Chem.*, 1983, **55,** 908.

[52] E. L. Berg, ed., 'Handbook for Sampling and Sample Preservation of Water and Wastewater', (Report PB 83–124503), U.S. National Technical Information Service, Springfield, Virginia, 1982.

APPENDIX: CONFIDENCE LIMITS ON SINGLE ANALYTICAL RESULTS

We assume throughout that analytical results are unbiased (*i.e.*, $\mu = \tau$, see Section 4.4), and normally distributed with population standard deviation, σ.

Given the mean, $\bar{R}$, of n results, we saw in Section 4.3.3(b) that the $100(1-\alpha)\%$ confidence limits for μ are given by $\bar{R} \pm u_\alpha \sigma / \sqrt{n}$. In practice, however, the exact value of σ is never known, and we can have only an estimate, s, with v degrees of freedom (Section 4.3.4). Section 4.3.5 showed that this problem is, in principle, simply overcome since the $100(1-\alpha)\%$ confidence limits for μ, when $\bar{R}$ and s are estimated from the results of n replicate analyses on each sample, are given exactly by $\bar{R} \pm t_\alpha s / \sqrt{n}$. Provided, therefore, that each sample is analysed at least twice [though preferably more times – see Section 4.3.5(b)], the uncertainties in the true concentrations of samples can be properly quantified.

Section 4.3.5(b) noted a difficulty in this approach using t, namely, that it is common in practice for each sample to be analysed only once since the resources available to most routine laboratories usually make replicate analysis impracticable. When each sample is analysed only once, a value for s cannot be obtained for each sample so that an independent value of $\pm t_\alpha s$ cannot be derived for each sample. In this situation the question arises – how can the uncertainty associated with a single analytical result be estimated when the population standard deviation is not known? This seems to us an extremely important question with which many of the statistical texts used by scientists do not deal, and it is to this question that this Appendix is addressed.

As noted in Section 4.3.5(b), the approach often adopted by analysts (including ourselves) is to obtain an estimate, s, with v degrees of freedom from special tests, and to calculate the $100(1-\alpha)\%$ confidence limits for subsequent singly analysed samples as $R_i \pm t_\alpha s$, where R_i is the analytical result for the ith subsequent sample with population mean, μ_i, and the value of t_α is that corresponding to v degrees of freedom. Now, if a proportion, $100(1-\alpha)\%$, of confidence intervals given by $R_i \pm t_\alpha s$ are to include the corresponding values of μ_i, it is necessary that an independent value of s is obtained on each occasion that a confidence interval is formed. The approach just described for singly analysed samples repeatedly uses the same estimate, s, and is, therefore, not theoretically valid; the result is that the actual proportion of confidence intervals including μ_i does not, in general, equal $100(1-\alpha)\%$. This is the error incurred by this simple approach, and we are very grateful to Mr. J.C. Ellis of the Water Research Centre for pointing it out to us. The source of this error and a procedure for calculating its size are best illustrated by an example.

Suppose the estimate, s, to be used repeatedly has 10 degrees of freedom, and that its numerical value is given by $s = 0.8\sigma$; such a value for s is quite likely to be obtained (see below). Suppose also that it has been decided that the 95 % confidence level is appropriate for expressing the uncertainty in μ. The 95 % confidence limits would then be calculated for each sample by the (theoretically incorrect) approach as $R_i \pm t_{0.05} s = R_i \pm 2.23 s = R_i \pm 1.78\sigma$, since $t_{0.05}$ for 10 degrees of freedom is 2.23 (see Table 4.6) and we have assumed that $s = 0.8\sigma$. Now, since we are also assuming that analytical results for each sample follow a normal distribution with standard

172   *Chemical Analysis of Water*

deviation, σ, we can immediately obtain the proportion of samples whose μ_i values will be included by the intervals, $R_i \pm 1.78\sigma$, from the known properties of the normal distribution (see Table 4.1); this proportion will be, on average, 92.5 %. Thus, although the value of t corresponding to the 95 % confidence level was used, the actual confidence level achieved in this example is 92.5 %. This error arises primarily because the value of s is substantially smaller than σ. Similar calculations for the case where $s = 1.2\sigma$ and $v = 10$ show that the actual confidence level then achieved is approximately 99.3 %. Thus, the error is now in the opposite direction because s is substantially larger than σ.

The sign and the magnitude of the difference between desired and actual confidence levels depends on the relative magnitudes of s and σ, and on v. To indicate the practical importance of this difference when $t_{0.05}$ is used (*i.e.*, the desired confidence level is 95 %), we have calculated the confidence levels achieved for various values of s and v with the results shown in Table 4.7. The values of s used for these calculations were those corresponding to the upper and lower limits expected (at a given probability) for s for a given value of σ.* For example, the lowest value of s (0.05 probability) means that there is a probability of 0.05 that lower values will be

TABLE 4.7

Errors in using $R_i \pm t_{0.05}s$ and $R_i \pm u_{0.05}s$ as Confidence Limits on Single Analytical Results, R_i

Value of s	Percentage of samples for which $R_i \pm t_{0.05}s$ or $R_i \pm u_{0.05}s$ include the population means of singly analysed samples†					
	$v = 10$	$v = 20$	$v = 30$	$v = 60$	$v = 120$	$v = \infty$
Smallest s expected	**83.8**	**87.6**	**89.1**	**91.0**	**92.3**	**95.0**
	79.2	*85.2*	*87.6*	*90.3*	*92.1*	*95.0*
(0.05 probability level)	(0.63σ)	(0.74σ)	(0.79σ)	(0.85σ)	(0.89σ)	(σ)
Smallest s expected	**88.0**	**90.0**	**91.0**	**92.2**	**93.0**	**95.0**
	82.9	*87.8*	*89.6*	*91.6*	*92.7*	*95.0*
(0.10 probability level)	(0.70σ)	(0.79σ)	(0.83σ)	(0.88σ)	(0.92σ)	(σ)
Smallest s expected	**93.3**	**93.4**	**93.5**	**93.8**	**94.1**	**95.0**
	89.2	*91.5*	*92.3*	*93.3*	*93.9*	*95.0*
(0.25 probability level)	(0.82σ)	(0.88σ)	(0.90σ)	(0.93σ)	(0.95σ)	(σ)
$s = \sigma$	**97.4**	**96.3**	**95.9**	**95.4**	**95.2**	**95.0**
	95.0	*95.0*	*95.0*	*95.0*	*95.0*	*95.0*
Largest s expected	**98.7**	**97.7**	**97.2**	**96.6**	**96.1**	**95.0**
	97.2	*96.8*	*96.5*	*96.2*	*95.9*	*95.0*
(0.25 probability level)	(1.12σ)	(1.09σ)	(1.08σ)	(1.06σ)	(1.04σ)	(σ)
Largest s expected	**99.5**	**98.7**	**98.2**	**97.4**	**96.8**	**95.0**
	98.7	*98.1*	*97.7*	*97.1*	*96.6*	*95.0*
(0.10 probability level)	(1.26σ)	(1.19σ)	(1.16σ)	(1.11σ)	(1.08σ)	(σ)
Largest s expected	**99.7**	**99.1**	**98.6**	**97.8**	**97.1**	**95.0**
	99.2	*98.6*	*98.2*	*97.6*	*97.0*	*95.0*
(0.05 probability level)	(1.35σ)	(1.25σ)	(1.21σ)	(1.15σ)	(1.10σ)	(σ)

† Each cell of the table contains two or three entries; the number in bold type is the percentage corresponding to $R_i \pm t_{0.05}s$ and the italicised number is the percentage corresponding to $R_i \pm u_{0.05}s$. The third entry (in all cells except those where $s = \sigma$) gives the numerical value of s in terms of σ, the population standard deviation

* These values of s were calculated by 'inverting' the approach described in Section 4.3.4(*d*). Thus, the lower and upper values for s (probability α) are given by $\sigma\sqrt{\chi^2_{1-\alpha}/v}$ and $\sigma\sqrt{\chi^2_{\alpha}/v}$, respectively.

obtained. There is, therefore, a probability of 0.1 that, for a given value of σ, the value obtained for s will lie outside the interval between the lowest and highest values (0.05 probability). To put these values of s in more practical terms, it may be helpful to think of a laboratory measuring ten determinands and obtaining an estimate of the standard deviation for each. On average, only one of these estimates will fall outside the interval between the lowest and highest values of s (0.05 probability) quoted in the table, and many of the estimates will differ from σ by substantially smaller amounts than these lowest and highest values.

Several points emerge immediately from the table, and in considering them we shall refer to the difference between the desired (95 %) and achieved confidence levels as 'the error'. Thus, it can be seen that the magnitude of the error decreases as v increases, but the error remains even when v is as large as 60 or 120, values that most analysts will usually find impracticable to achieve in the short-term. In practice, therefore, the error cannot be eliminated, though its size can be somewhat decreased by practicable increases in v. Table 4.7 also shows that the sign and magnitude of the error depend on the ratio, s/σ. When $s \geqslant \sigma$, the error is such that the intervals, $R_i \pm t_{0.05}s$, fail to include μ_i less frequently than 5 %; this also remains true for values of s smaller than σ so long as $s > 1.96\sigma/t_{0.05}$. For still smaller values of s, the intervals fail to include μ_i more frequently than 5 %, the greatest error occurring for $v = 10$ and the smallest value of s shown in the table when the confidence level achieved is approximately 84 %. This then is the nature of the error, and we turn now to what should be done about it.

We can begin by supposing that a user of analytical results has decided that results for a given determinand will serve his purposes provided that their uncertainty is not greater than ε at the 95 % confidence level. Suppose also that an analyst then obtains an estimate, s, such that $t_{0.05}s \leqslant \varepsilon$. Under these conditions, it seems to us that the user is most unlikely to experience any difficulty if the value of s were, in fact, greater than σ so that the confidence level achieved turned out to be greater than 95 %. On the other hand, the user may well feel more concern about the situation where s is substantially smaller than σ, because the confidence level achieved is then smaller than that desired. On this basis, we believe that the main practical problem concerns those instances where s is substantially smaller than σ, and we consider only these in the following. For such values of s, Table 4.7 also shows that, as would be expected, the use of $u_{0.05}$ in place of $t_{0.05}$ leads to rather greater errors,* so that we prefer the use of t rather than u.

It can be seen that, if it is desired (at some chosen probability level) that the confidence level is not less than 95 %, any statistical procedure for this purpose must involve an increase in the width of the confidence intervals calculated for each singly analysed sample. This will ensure that the desired proportion of confidence intervals including μ_i will be obtained when s is substantially smaller than σ. Various procedures for calculating such increased confidence intervals can be devised; one is described, for example, by Bennett and Franklin[5]. When, however, v is small and the number of singly analysed samples will be large (*i.e.*, conditions likely to occur at the beginning of a routine analytical programme), such procedures lead to so much greater widths of the confidence intervals that they are very likely

---

* The intervals, $R_i \pm u_{0.05}s$, will be correct in the (rather unlikely) event that $s = \sigma$. Note also that the use of u leads to smaller errors when $s \geqslant \sigma$.

not to meet the user's condition that the uncertainty of analytical results be not greater than ε. Further, this would occur for all determinands of interest to a laboratory even though many of them would, in fact, not need such extreme modification of the confidence intervals. The implications of the use of such 'correction' procedures are that the task of the analyst would be made even more difficult in that he would have to achieve sufficiently small values of σ that the estimates, s, would still give tolerably small uncertainties, ε. Bearing in mind the remarks made in Section 4.3.2(*a*) on the assumption of the normal distribution, in Section 4.3.3(*b*) on the choice of confidence level, and the various practical problems involved in obtaining valid estimates of standard deviations (see Section 6), it seems to us that any such increase in the difficulties confronting the analyst is not generally justified. Of course, this subject is very appropriately a matter for discussion between the user of analytical results, the analyst, and the statistician during the planning stage of any measurement programme. In general, however, we conclude that the best approach (subject to the conditions noted below) is simply to calculate the confidence intervals for single analytical results as $\bar{R}_i \pm t_\alpha s$, but noting the possible existence of the error in the confidence level achieved.

If this approach were adopted, several aspects which are under at least partial control by the analyst should be noted.

(1) It will be advantageous to adopt and properly apply analytical systems for which the standard deviation, σ, is expected to be smaller than the value implied by the maximum tolerable random error specified by the user of analytical results.

(2) The degrees of freedom, v, of the initial estimate, s, of standard deviation should be as large as possible.

(3) During routine analysis, special tests should be made to allow refinement (*i.e.*, to increase the degrees of freedom) of the initial estimate, s. As the number and frequency of such tests increase so the period, during which there may be a large difference between the desired and the achieved confidence levels, decreases.

These and related points are discussed in detail in Sections 6, 7, and 9.

5 Estimation and Control of the Bias of Analytical Results

5.1 INTRODUCTION

In Section 4.7 we stressed the need for each laboratory to obtain estimates of the errors of its analytical results so that it can be decided whether or not these errors are adequately small for the intended use(s) of the results. The estimations of bias and precision are considered in detail in the present section and Section 6, respectively. In addition, Section 7 describes a detailed approach to a problem of growing importance, namely, the achievement of comparable results meeting defined standards of accuracy by all laboratories of a group. Finally, in Section 8 we discuss several important questions that arise when considering exactly how analytical results should be reported, and that require consideration of the errors of the results.

We would stress at the outset that, in our view, the control and estimation of bias is one of the key problems of modern water analysis. Currie[1,2] and Liteanu and Rîcă[3] have made the same point for the whole of analytical chemistry, and the need to aim for essentially zero bias has been recommended in clinical chemistry.[4]

The magnitude of bias for a given sample is determined by the analytical system used. Of this system, the analytical method is often, but by no means necessarily, the component with greatest influence on the bias. The choice of method is, therefore, a crucial aspect in the control of bias, particularly when the smallest possible bias is desired [see Section 4.6.1(c)]. The factors involved in considering the suitability of analytical methods from this pont of view are very closely interconnected with the various sources of bias and their control and estimation. It is best, therefore, to consider these latter topics in the present section and to defer discussion of the choice of methods to Section 9.

The ability to detect, and the estimation of, bias both improve as the precision of measurements improves. In practice, therefore, one always faces the question – given a particular analytical method, is it better to estimate precision before bias or *vice versa*? Given the effect of precision on the estimation of bias, it would be natural to select the first alternative, but this runs the risk that considerable effort may be devoted to estimating and controlling precision only then to find that bias is intolerably large. As Currie[2] has noted, such problems may lead in practice to an iterative-type of approach, and each instance requires individual consideration. On balance, the strategy that seems to us of most general value proceeds in the following stages:

(1) great care is taken in selecting an analytical method so that the bias of results is expected to be minimal,

(2) the method selected is modified to remove any obvious sources of bias [see, for example, Sections 5.4.4(c) and (d), and 9.2.4(b)],

(3) an estimate of the precision of analytical results is obtained (see Section 6), and if this is satisfactory,
(4) experimental tests are made, if necessary, to estimate the magnitude of bias from any sources that could not be adequately assessed or eliminated in (1) and (2).

With this approach in mind, we thought it useful to deal with bias before turning, in Section 6, to precision.

The present section is primarily concerned with bias from the true value as defined in Section 4.4. In some circumstances, bias among the results from different laboratories (rather than bias from the true value) may be the overriding concern, and the control and estimation of interlaboratory bias is considered in Section 7. Even in this situation, however, we feel it usually desirable to follow one of the approaches described below for estimating bias from the true value in addition to any tests specifically to estimate interlaboratory bias.

There are three possible approaches to the experimental estimation of bias, and these are considered in Sections 5.2, 5.3, and 5.4.

5.2 USE OF SAMPLES OF KNOWN CONCENTRATIONS

If one or more *homogeneous and stable* samples are available whose determinand concentrations are known with negligible inaccuracy for a particular application, replicate analysis of each sample provides a simple and direct approach to the estimation of bias for those samples. If n portions of a sample (true concentration $= C_S$) are analysed to give results, R_i ($i = 1$ to n), with mean, $\bar{R}$, and standard deviation, s, bias is shown to be present (at the α significance level) if [see Section 4.6.2(a)(ii)]

$$|\bar{R} - C_S| \cdot \sqrt{n}/s > t_\alpha \qquad (5.1)$$

In this approach, four points should be noted.

(1) It is generally desirable to make the n analyses as one analysis on each of n separate analytical occasions rather than as n analyses all at essentially the same time. This reduces the possibility of apparent bias arising from the random errors of the calibration parameters used to calculate the analytical results (see Section 6.2.1). The approach also makes some allowance for the possibility that the size of a given bias may vary from time to time, and the estimate obtained will be a better representation of the average bias of routine analytical results.
(2) The best estimate of the magnitude of the bias is $\bar{R} - C_S$, but this is subject to uncertainty caused by the random errors of the analytical results. The $100(1-\alpha)\%$ confidence limits for the true bias are given by $(\bar{R} - C_S) \pm t_\alpha s/\sqrt{n}$ [see Section 4.3.5(a)]. If it is required to detect a statistically significant difference (thus demonstrating the presence of bias) between $\bar{R}$ and C_S when the true bias is equal to or greater than a given value, β_0, the value of n must be sufficiently large [see Section 4.6.2(a)(i)] that the uncertainty of $\bar{R}$ is adequately reduced.
(3) The magnitude of bias may vary from sample to sample depending on the

concentration of the determinand, the chemical and physical forms of the determinand present (see Section 5.4.5), and the natures and concentrations of other substances in the sample (see Section 5.4.6). Thus, unless a fully representative selection of relevant sample types is available with known determinand concentrations, caution is necessary in drawing general conclusions on the magnitude of bias from the results for only a few such samples, and recourse to the approaches in Sections 5.3 and/or 5.4 will be necessary.

(4) The value of this approach to the estimation of bias has been discussed by many authors – see, for example, refs. 2 and 5 to 11 – and should always be borne in mind. To the best of our knowledge, however, no such freshwater samples are available. For such sample types, this lack is not surprising given the essential requirements that the samples be homogeneous, stable, and available in large volume, and that the determinand concentrations be known with negligibly small error. If samples containing negligibly small determinand concentrations can be obtained, an alternative test that may be useful is described in Section 6.4.4. The use of standard solutions (prepared to simulate natural waters) in place of true samples may also be of value in detecting certain sources of bias, but such solutions do not usually allow adequate assessment of other types of bias [see (3) above]. Note also in this connection that the addition of chemical stabilising reagents [see Section 3.6.3(*d*)] to ensure the stability of such solutions may invalidate their use by laboratories whose analytical procedures employ other stabilisers [see Section 7.2.9(*b*)]. The U.S. National Bureau of Standards supplies several such solutions for a number of determinands,[5] and a much wider range of solutions (covering all determinands of common interest in water analysis) is prepared and used extensively by the U.S. Environmental Protection Agency.[12]

On the basis of points (3) and (4), we conclude that this approach is at present seldom useful for estimating the *total* bias of analytical results, and recourse to the approaches in Sections 5.3 and 5.4 will usually be necessary.

5.3 USE OF REFERENCE METHODS OF ANALYSIS

In this approach, which is closely related to that in Section 5.2, one or more samples are analysed by the method under consideration and by a second (reference) method known to give results of negligible bias. The value of this approach has been discussed by a number of authors.[2,6,7,9,11,95]

Suppose that, for a given sample, n results, R_{ri} ($i = 1$ to n), with mean, $\bar{R}_r$, are obtained by the reference method, and n results, R_{ti} ($i = 1$ to n), with mean, $\bar{R}_t$, are obtained by the method under test. Bias is shown to be present (at the α significance level) if

$$\frac{|\bar{R}_t - \bar{R}_r|}{s}\sqrt{\frac{n}{2}} > t_\alpha \qquad (5.2)$$

where s and t_α have the meanings described in Section 4.6.2(*b*)(*ii*).

Again, in this approach several points should be noted.

(1) For the reasons stated in point (1) of Section 5.2, it is better to make the n

analyses, R_{ti}, as one analysis on each of n occasions. If this is done, it is then also better to make one analysis by the reference method on each occasion.

(2) Inequality 5.2 assumes that the population standard deviations of results by the two methods are the same. When this is not so more involved calculations are required as described in refs. 3 and 13.

(3) The best estimate of the magnitude of the bias is $\bar{R}_t - \bar{R}_r$, but this is subject to the uncertainty caused by the random errors of analytical results. The $100(1-\alpha)\%$ confidence limits for the true bias are $(\bar{R}_t - \bar{R}_r) \pm t_\alpha s \sqrt{(2/n)}$ [see Sections 4.3.5(a) and 4.6.2(b)(ii)]. If it is required to detect a statistically significant difference (thus demonstrating the presence of bias) between $\bar{R}_t$ and $\bar{R}_r$ when the bias is equal to or greater than a given value, β_0, the value of n must be sufficiently large [see Section 4.6.2(b)(i)] that the uncertainties of $\bar{R}_t$ and $\bar{R}_r$ are adequately reduced.

(4) In application of this approach, it is common to make $n=1$, and to combine the results from all the samples tested to obtain some overall estimate of bias. This procedure certainly allows substantial economy of effort, but is not generally recommended here for two reasons. First, the magnitude of bias may vary from sample to sample [see (5) below]. Second, the statistical assessment of the overall bias will be complicated when, as is common [see Section 6.2.1(f)], the standard deviation of analytical results depends on the concentration of the determinand; see, for example, refs. 14 and 15.

(5) As in point (3) of Section 5.2, unless a fully representative series of relevant sample types is studied, caution is necessary in drawing general conclusions on the magnitude of bias from the results on relatively few samples.

(6) In practice, though this approach should always be borne in mind, it is seldom if ever that normal water analysis laboratories will be able to establish and use such a reference method even if – and this is very rare in water analysis – one were available. In this position, there seems to us little or no point in using in place of the reference method a method for which the sources and magnitude of bias are not accurately known.

On the basis of point (6) and the conclusion of Section 5.2, we conclude that neither this approach nor that of Section 5.2 is a generally useful means of estimating the *total* bias of analytical results for waters. The only other possibility is considered in detail in Section 5.4.

5.4 DETAILED CONSIDERATION AND ASSESSMENT OF INDIVIDUAL SOURCES OF BIAS

5.4.1 Introduction

This approach consists simply of the identification of individual sources of bias in the application of a method under consideration, the elimination of those that can readily be avoided, and the assessment of the magnitudes of those remaining. In the present context there are five sources of bias to which every analytical system is generally subject:

(1) a biased calibration curve or factor,

(2) a biased blank correction,
(3) an inability to determine all the chemical and physical forms of the determinand that may occur in samples,
(4) effects of other substances present in samples, *i.e.*, interference,
(5) mistakes by the analyst.

Of these sources, (5) may, of course, be very important and should never be overlooked. The relevant factors involved in the organisation, training and control of analysts, however, are generally well understood, and are not further considered here, but see Section 10.3.

It should also be noted that biased results may be caused by inappropriate procedures for collecting and storing samples in the period between sampling and analysis. Such problems are discussed in Sections 3.5 and 3.6; see also Section 5.5.

If the evidence available for a particular method shows that sources of bias (1) to (4) are negligible, the method may be adopted and used with some confidence – provided, of course, that all the other components of the analytical system are also suitably chosen, and that the analytical system is properly applied. If, on the other hand, one or more of these sources of bias is so large that the maximum bias tolerable by the user of analytical results is exceeded, then alternative methods should be considered. In the event that no suitable methods are available, discussion between the analyst and the user is necessary to decide if analysis of samples is worthwhile before a suitable method is established.

The two situations considered in the preceding paragraph are, as it were, extremes. In practice, analysts commonly face the situation where either there is insufficient evidence of the performance achievable when a method is used to decide the likely magnitude of bias or the evidence indicates the bias to be neither clearly negligible nor clearly intolerably large. In both these situations, detailed estimation of the bias is required.

A disadvantage of this third approach to the control and assessment of bias is that not all possible sources of bias – particularly for effects (3) and (4) – may be foreseen. It may be concluded, therefore, that the bias of results is tolerable when, in fact, it is not. This problem can be especially marked in the analysis of natural waters which contain so many different substances, often in a variety of chemical and physical forms. Notwithstanding the difficulty, the authors see no alternative to this approach as the two other approaches (Sections 5.2 and 5.3) are not generally applicable. The approach in this section accordingly places great importance on the analyst's detailed knowledge of the principles and mechanisms involved in each of the analytical procedures he uses. Such knowledge is indispensable in considering individual sources of bias. This observation is particularly worth emphasis in the light of modern developments in analytical instrumentation (see Sections 4.7 and 10.4.3) which allow very complicated analytical procedures to be carried out apparently satisfactorily by relatively inexpert and inexperienced analysts. Such instrumentation is often just that requiring the most intimate knowledge of principles and mechanisms if bias is to be properly estimated and controlled. Other authors have stressed the importance of this detailed knowledge of the basis of analytical methods; see, for example, refs. 2, 3, 6, 16, and 96.

The sources of bias, (1) to (4) above, are considered in Sections 5.4.3 to 5.4.6; the discussion on calibration and blank correction is aided by some preliminary considerations in Sections 5.4.2.

5.4.2 Preliminary Considerations on Calibration and Blank-Correction

In (a) to (d) below, we consider the theoretical basis of calibration and blank-correction procedures, and in (e) we summarise some useful results concerning a statistical technique called linear regression analysis. This technique is of value in defining the position of a calibration curve when the experimental points on which it is based are subject to appreciable random errors.

(a) Analytical Response Proportional to Determinand Concentration
We restrict attention here and in (b) and (c) to analytical systems for which the analytical response is proportional to the determinand concentration; other types of relationship are considered in (d). To simplify the discussion of bias, we assume here that random errors are absent; this does not affect the validity of the conclusions, and the additional effects of random errors are considered in Sections 5.4.3, 5.4.4, and 6.

We begin with analytical systems for which the analytical response, $\mathscr{A}_S$, caused by the determinand from the original sample is related to the concentration of the determinand C_M, in the processed sample at the measurement stage by an expression of the form

$$\mathscr{A}_S = \pi C_M \qquad (5.3)$$

where π is a parameter which is constant for given experimental conditions. We use the symbol $\mathscr{A}$ to denote the response in the absence of random errors; the symbol C (with appropriate subscripts) is used throughout to denote true concentration. We now assume that the relationship between C_M and the concentration of determinand in the original sample, C_S, is simply $C_M = \xi C_S$, where ξ is another parameter; this assumption is commonly but not necessarily true. Thus, equation 5.3 can be re-written as

$$\mathscr{A}_S = \chi C_S \qquad (5.4)$$

where $\chi = \pi \xi$. There is nothing new in this, but in comparing the values of χ for different analytical systems, it should be remembered that χ depends on both π and ξ.

Several other factors may affect the analytical response. With one exception [see (c) below], we ignore interference caused by substances in the sample; this is considered in Section 5.4.6. Similarly, discussion of the dependence of χ on the nature of the determinand in the original sample is deferred to Section 5.4.5. For the present, we consider only two factors.

(1) The determinand concentration in the processed sample may be increased through contamination from reagents, apparatus, *etc.* We assume that the contaminating determinand produces an analytical response also given by equation (5.3), and that the total concentration of determinand in the processed sample does not exceed any upper limit to the validity of that

equation. The amount of this contamination is expressed in terms of the corresponding concentration in the original sample, and denoted by C_C. Contamination is discussed in detail in Section 10.8.

(2) The analytical response may contain a component which is due neither to an increase in the determinand concentration nor to any other substance present in the sample. For example, in a spectrophotometric procedure, the inherent colour of a reagent may contribute to the absorbance. Another possibility is contamination by interfering substances (causing positive bias independent of C_S) during analysis of samples. This *additional* response is denoted by η, which is regarded as another parameter.

On this basis, the total analytical response, $\mathscr{A}_T$, for a sample is given by

$$\mathscr{A}_T = \eta + \chi(C_S + C_C) \tag{5.5}$$

so that, putting $\eta + \chi C_C = \lambda$,

$$\mathscr{A}_T = \lambda + \chi C_S \tag{5.6}$$

Thus, for this model, the relationship beteen $\mathscr{A}_T$ and C_S is a straight line of slope, χ, and intercept, λ. Parameters such as χ and λ that define the relationship between $\mathscr{A}_T$ and C_S are called here *calibration parameters*. We refer to the equation defining the relationship as the *calibration function*, and to a graphical plot of this function as the *calibration curve*. There is, of course, nothing new in the derivation of equation (5.6), but it is useful, and is employed extensively throughout this book. Currie[2] has described essentially the same model.

One value of the above model is that it helps to make clear the various assumptions involved in its use. For example, we have implicitly assumed above that any interfering substance entering the processed sample through contamination causes only an increase in η. That is to say, we have ignored the possibility of contamination by interfering substances whose effects are to decrease the response obtained from a given concentration of the determinand. Clearly, however, such effects can be included in the model if desired. For this purpose, one would have to postulate a relationship between the effect and the concentrations of the determinand and an interfering substance. For example, suppose that the effect were that a concentration, C_I, of an interfering substance prevented a concentration, γC_I, of the determinand from producing an analytical response. It is then easily shown for example that – provided $C_C \geqslant \gamma C_I$ – the relationship between $\mathscr{A}_T$ and C_S remains a straight line of slope χ, but that the intercept becomes $\eta + \chi(C_C - \gamma C_I)$. When, however, $C_C < \gamma C_I$, the analytical response is equal to η for all values of C_S between zero and $(\gamma C_I - C_C)$. Other types of effect are also possible, *e.g.*, the contaminant may affect η, and the mathematical relationships between the effects and the contaminant and determinand concentrations may be more complicated than in the simple example above. That such possibilities are not purely theoretical is shown, for example, by Reust and Meyer[97] who note that contamination by organic substances may lead to negatively-biased results in the electroanalytical determination of trace metals. Nonetheless, we have, for simplicity, ignored such effects below since they appear not to be a common feature of the routine analysis of waters. The possibility of difficulties of this nature should, however, be borne in mind, and the general approach followed here may be applied to assess their consequences.

Some analytical methods specify the addition of an internal standard to samples, and the measurement of the responses of both the determinand and the internal standard. The ratio of these two responses then takes the place of $\mathscr{A}_T$ in an equation of the same form as equation (5.6). All the points made below are equally applicable to such internal standard procedures.

Given the response, $\mathscr{A}_T$, for a sample of unknown concentration, the estimate of that concentration, *i.e.*, the analytical result, R, is obtained from equation (5.6) as

$$R = (\mathscr{A}_T - \lambda)/\chi \qquad (5.7)$$

Thus, to obtain an unbiased value for R, we must have unbiased values for λ and χ. (We ignore here the possibility that R may be unbiased as a fortuitous consequence of exactly counterbalancing biases in λ and χ.) In practice, the true values of calibration parameters are not known, in general, *a priori*, and must, therefore, be estimated experimentally. This estimation is here referred to as *calibration of an analytical system* or often and more simply as *calibration*. A number of generally important practical aspects of procedures commonly employed for calibration are considered in detail in Sections 5.4.3 and 5.4.4. Many approaches have been devised and are specified in analytical methods for calibration; Kaiser[17] provided a useful classification of the various approaches. For analytical systems for which equations (5.6) and (5.7) are valid, the relative advantages and disadvantages of different procedures for calibration and calculation of results can be deduced from these equations. In doing this, it must be borne in mind that the numerical values of λ and χ generally depend on experimental conditions and may, therefore, vary from time to time. In addition, the values of η and χ may also depend on the nature of the solution being analysed. Clearly, a large number of different circumstances may arise, and we shall consider in (*b*) and (*c*) below only a few corresponding to procedures commonly applied in routine water analysis.

(b) η and χ Independent of the Nature of the Solution Analysed
In this situation, since η and χ have the same values for samples and standard solutions, calibration is usually most conveniently based on the analysis of two or more standard solutions whose concentrations cover the range of interest. Such calibration standards must, in general, be analysed by *exactly* the same procedure as that used for samples so that all measurements are equally subject to all the many possible effects determining the values of η, χ, and C_C. Two situations commonly arise in practice, depending on the temporal stability of χ.

(i) χ is essentially constant over long time-periods—In this situation it is economic to make an initial calibration, which is then repeatedly used in evaluating subsequent samples. Denoting the values of the calibration parameters for the initial calibration by the subscript 0, and writing $\eta + \chi C_C$ for λ, we have from equation (5.7):

$$R = (\mathscr{A}_T - \eta_0 - \chi C_{C0})/\chi \qquad (5.8)$$

since we are assuming that χ is constant. Suppose now that a sample is measured at some subsequent time when the values of η and C_C are η_S and C_{CS}, respectively. The response for this sample is obtained from equation (5.5) as

$$\mathscr{A}_T = \eta_S + \chi C_{CS} + \chi C_S \qquad (5.9)$$

and if this value for $\mathscr{A}_T$ is substituted in equation (5.8), we obtain

$$R = C_S + C_{CS} - C_{C0} + (\eta_S - \eta_0)/\chi \qquad (5.10)$$

Thus, when the sample is evaluated using the initial calibration curve, the bias of R is given by

$$R - C_S = (C_{CS} - C_{C0}) + (\eta_S - \eta_0)/\chi \qquad (5.11)$$

Equation (5.11) shows that this method of evaluation gives unbiased results only when $C_{CS} = C_{C0}$ and $\eta_S = \eta_0$, *i.e.*, when C_C and η do not vary from one analytical occasion to another. Given the origins of η and C_C, such constancy is generally to be expected only when both η and C_C are always zero. Such circumstances do arise, *e.g.*, often, though not necessarily, in classical analytical procedures using gravimetric and titrimetric techniques. In general, however, it is not valid to assume that η and C_C will not vary with time. Accordingly, some correction to the result, R, obtained from equation (5.8) is necessary to eliminate the bias shown by equation (5.11).

The bias can be eliminated if the appropriate values of η and C_C are used in deriving analytical results for samples. Thus, suppose a standard solution of known determinand concentration, C_K, is analysed by exactly the same procedure, and at the same time*, as the sample. The response, $\mathscr{A}_K$, for this standard is given [from equation (5.5)] by

$$\mathscr{A}_K = \eta_S + \chi C_{CS} + \chi C_K \qquad (5.12)$$

If $\mathscr{A}_K$ is subtracted from $\mathscr{A}_T$, we obtain from equations (5.9) and (5.12)

$$\mathscr{A}_T - \mathscr{A}_K = \chi(C_S - C_K) \qquad (5.13)$$

so that an unbiased result, R, is obtained from

$$R = C_K + (\mathscr{A}_T - \mathscr{A}_K)/\chi \qquad (5.14)$$

So far as the bias of R is concerned, the value of C_K is immaterial provided (1) C_K is known with negligible bias, and (2) any upper limit to the validity of equation (5.4) is not exceeded. Usually, it is advantageous to make $C_K = 0$ since the random error of R is then usually minimised [Section 6.2.1(*f*)] and this also aids the detection of small concentrations (Section 8.3); negative values of $\mathscr{A}_T - \mathscr{A}_K$ are also avoided. Of course, this is the basis of the 'blank correction' commonly specified in analytical methods whereby R is given by

$$R = (\mathscr{A}_T - \mathscr{A}_B)/\chi \qquad (5.15)$$

where $\mathscr{A}_B$ is the analytical response for a standard solution with $C_K = 0$. When a blank determination is made on each analytical occasion, only χ need be estimated

---

* It is not generally possible to analyse two solutions at exactly the same time by the same analytical system, but we assume that η_S and C_{CS} show no systematic variations over a period during which a number of solutions can be analysed; see Section 6.2.1(*b*).

in the initial calibration, and we call this procedure *fixed-slope calibration*. It is essential to note, however, that equation (5.15) gives unbiased results only if $C_K = 0$. Thus, the water used for the blank should contain a negligible concentration of the determinand. Many methods either fail to make, or do not sufficiently stress, this point.

Note also that in deriving equation (5.12), we have implicitly assumed that the value of η is the same for the standard as for the samples. This assumption will be incorrect when the water used to prepare the standard (or blank) contains positively interfering substances [see (*a*) above]. If the analytical response caused by such substances in the blank water is $\mathscr{A}_I$, the bias introduced can easily be shown to be $-\mathscr{A}_I/\chi$. The nature of this effect is exactly the same as that caused by the presence of determinand in the blank water, and is further considered in Section 5.4.4(*b*).

For simplicity again, we do not consider here the special types of blank-correction procedures specified by some methods. Similar approaches may be used if it is desired to consider such corrections theoretically.

Note finally that a second and – for the present case – equivalent method of blank correction is to obtain the apparent concentrations, R'_S and R'_B, corresponding to $\mathscr{A}_T$ and $\mathscr{A}_B$ by using equation (5.8), and then to subtract R'_B from R'_S. It is easily shown that $R = R'_S - R'_B = C_S$.

If it is required to plot this type of linear calibration curve when a blank correction is used, one may plot the initial calibration as $\mathscr{A}_T$ against C_S or as $(\mathscr{A}_T - \mathscr{A}_{B0})$ against C_S. On balance, we favour the latter since the calibration curve should then pass through the origin of the graph. This plot may also be simply used for both methods of blank correction if visual calculation of R is to be made whereas the plot of $\mathscr{A}_T$ against C_S may sometimes require negative apparent concentrations to be handled.

(ii) χ *may vary from batch to batch of analyses* In this situation, calibration is, in general, required for each batch of analyses, and we speak of *within-batch calibration*. The values of η, χ, and C_C for the calibration standards will all be appropriate for samples analysed in the same batch, and the analytical results for such samples from equation (5.7) will, therefore, be unbiased. Since $\eta + \chi C_C = \mathscr{A}_B$, equation (5.7) is equivalent to equation (5.15) and an explicit subtraction of $\mathscr{A}_B$ from each $\mathscr{A}_T$ value is not necessary if the calibration curve is plotted as $\mathscr{A}_T$ against C_S. If, however, this curve is to be plotted as $\mathscr{A}_T - \mathscr{A}_B$ against C_S, then $\mathscr{A}_B$ must clearly be subtracted from all standard and sample responses.

(c) η *but not* χ *Independent of the Nature of the Solution Analysed*
In this situation, the values of χ for a sample and for standard solutions may differ so that the calibration and blank-correction procedures in (*b*) above are no longer generally applicable. For example, if equation (5.15) were used to obtain analytical results, it can easily be shown that R will be biased by an amount $-(C_S + C_{CS})(1 - \chi_S/\chi)$, where χ_S is the value of χ for the sample. This situation is that in which calibration by the *method of standard addition* is commonly recommended and applied. [Note that this procedure does not correct for interference whose magnitude is independent of the determinand concentration; see Section 5.4.6(*j*)]

(i) Need for blank-correction In the simplest form of the standard-addition method, a number of portions of the sample are 'spiked' with different, known amounts of the determinand, and these and an 'unspiked' portion are then analysed by the specified analytical procedure. The analytical responses are plotted against the equivalent concentrations (in the original sample) of determinand added to the sample portions, and a straight line through the points is extrapolated to cut the concentration axis at a value read from the graph. If the value is $-C_0$, the result, R, for the concentration in the original sample is given by C_0. In practice, correction for the dilution of the sample portions by the added standard solution may be required. This and other important aspects of the procedure are considered in Section 5.4.3(d). For simplicity, we neglect the effect of dilution in the following treatment since it does not affect the points we wish to make.

Let C_i denote the equivalent concentration of determinand added to the i th portion of the sample. Then, following the same approach as in (a) above, we may write the analytical response, $\mathscr{A}_{Ti}$, for the i th portion of the sample as

$$\mathscr{A}_{Ti} = \eta + \chi_S(C_S + C_C + C_i) \qquad (5.16)$$

From equation (5.16), when $\mathscr{A}_{Ti} = 0$, $C_i = -(C_S + C_C + \eta/\chi_S)$, so that

$$R = C_S + C_C + \eta/\chi_S \qquad (5.17)$$

Thus, this procedure produces a result biased by an amount $C_C + \eta/\chi_S$. Except, therefore, when both C_C and η are zero, a correction to eliminate this bias is required.[18-21]

Before considering approaches to this correction that involve some modification of the simple standard-addition procedure, it is worth noting that if the determinand can be removed *completely* from a portion of the sample without affecting any of the sample characteristics that affect χ_S or η, analysis of this determinand-free sample by the usual analytical procedure will give a response, $\mathscr{A}_{TB} = \eta + \chi_S C_C$. Thus, subtraction of $\mathscr{A}_{TB}$ from $\mathscr{A}_{Ti}$ gives

$$\mathscr{A}_{Ti} - \mathscr{A}_{TB} = \chi_S(C_S + C_i),$$

so that when $\mathscr{A}_{Ti} - \mathscr{A}_{TB} = 0$, $C_i = -C_S$ and $R = C_S$, i.e., the result is free of bias. This approach is not generally feasible, but can be used with some methods and is worth bearing in mind.[19,22]

(ii) Either η or C_C is greater than zero Returning to the more usual standard-addition procedures, the correction procedure often recommended consists in making, in addition to the analyses described above, a blank determination at the same time as the sample portions, subtracting the blank response from the responses for each of the sample portions, plotting the corrected responses against C_i, and equating the intercept on the concentration axis to $-R$. Now the response, $\mathscr{A}_B$, for the blank is given by*

$$\mathscr{A}_B = \eta + \chi(C_C + C_W) \qquad (5.18)$$

where χ is the slope of the calibration curve for standard solutions, and C_W is the

---

* For simplicity, we assume the blank water contains no interfering substances.

concentration of determinand in the water used for the blank determination. From equations (5.16) and (5.18)

$$\mathscr{A}_{Ti} - \mathscr{A}_B = \chi_S(C_S + C_C + C_i) - \chi(C_C + C_W) \tag{5.19}$$

so that, when $\mathscr{A}_{Ti} - \mathscr{A}_B = 0$, $C_i = \chi(C_C + C_W)/\chi_S - (C_S + C_C)$, and

$$R = C_S + C_C(1 - \chi/\chi_S) - C_W\chi/\chi_S \tag{5.20}$$

Thus, the bias of R is given by

$$R - C_S = C_C(1 - \chi/\chi_S) - C_W\chi/\chi_S \tag{5.21}$$

Equation (5.21) shows that this procedure eliminates any effect due to η. Results, however, are biased when $C_C > 0$, and an additional bias due to C_W also arises unless this latter parameter can be measured (see Section 10.7), an estimate of χ obtained from the analysis of standard solutions, and an appropriate correction made to the result, R. These two sources of bias appear often to be ignored, though Cammann[20] has recently noted the problem caused by C_C.

A second method of blank correction[3,19,20] [see (b)(i) above] consists in obtaining the apparent concentrations of determinand for the sample and the blank (when both are analysed and evaluated identically), and subtracting the latter from the former. In the present context, this corresponds to analysing portions of the sample and of the blank water as described above. The responses for the sample and blank portions are plotted separately to give intercepts on the concentration axis of $-R'_S$ and $-R'_B$, respectively, where R'_S and R'_B are the apparent concentrations for the sample and blank. The result, $R = R'_S - R'_B$. Now, the responses for the blank portions can be written as

$$\mathscr{A}_{Bi} = \eta + \chi(C_W + C_C + C_i),$$

so that, when $\mathscr{A}_{Bi} = 0$, $C_i = -(C_W + C_C + \eta/\chi)$ and consequently

$$R'_B = -C_i = C_W + C_C + \eta/\chi \tag{5.22}$$

From equation (5.16), $R'_S = C_S + C_C + \eta/\chi_S$, so that

$$R = R'_S - R'_B = C_S - C_W + \eta(1/\chi_S - 1/\chi) \tag{5.23}$$

Equation (5.23) shows that this procedure eliminates any effect due to C_C. Results, however, are biased when $\eta > 0$, and an additional bias due to C_W also arises unless this latter parameter can be measured (see Section 10.7), and an appropriate correction made to the result, R. Of course, if the value of χ is essentially constant with time, the procedure may be simplified by making only one normal ('unspiked') blank determination and dividing the observed response, $\mathscr{A}_B$ [$= \eta + \chi(C_W + C_C)$], by a previously determined value for χ to give a value for R'_B.

Thus, to summarise the above, when η and C_C are both zero for all samples, no blank correction is required in standard addition procedures. This is always a desirable position, but ensuring it may well prove extremely difficult, if not impossible, in many instances, especially in trace analysis. In addition, when either $\eta = 0$, $C_C \neq 0$ or $\eta \neq 0$, $C_C = 0$ obtain, unbiased blank correction is possible provided that (1) the blank is appropriately analysed and evaluated, and (2) C_W is either zero or is determined and an appropriate correction made to the result.

(iii) Neither η nor C_C are zero When neither η nor C_C are zero, no generally appropriate, unbiased correction to, or modification of, the standard addition procedure is known to us. In principle, this difficulty may be overcome by using the special procedure noted in (*i*), but in practice this will not usually be possible as most analytical methods at present neither specify nor provide details of suitable removal procedures. In this position, the best advice that we can offer, and we realise that it may well be of little help, is to consider the particular analytical system in detail, to attempt to quantify as well as possible the magnitude of bias involved, and to decide whether η or χC_C may be regarded as the dominant contribution to $\mathscr{A}_B$. For example, from the value for $\mathscr{A}_B$, the equations above (or others derived similarly for other types of calibration procedure) may be used to calculate the bias for a given correction procedure if the whole of $\mathscr{A}_B$ were due to either η or χC_C. If the bias is considered negligible, one may dispense with a correction procedure. If, on the other hand, the bias would be of concern, detailed consideration of the analytical procedure may suggest that $\mathscr{A}_B$ is mainly due to η (or χC_C), and the appropriate correction may be used. Special investigations may also, in some instances, be able to give estimates of η and C_C individually. Note also that in practice this problem is made more difficult by the random errors of experimental measurements. Finally, it is worth stressing that, as shown above, this problem does not arise at all when the slope of the calibration is the same for samples and standard solutions. Thus, considering only this aspect, there is much to be said for avoiding whenever possible the use of analytical systems necessitating the use of standard additon calibration.

In view of the remarks in the preceding paragraph, we have not attempted detailed discussion of the situation where both η and χ depend on the nature of the solution analysed. The approach used above may be followed if it is wished to derive expressions for the bias associated with the various possible circumstances and correction procedures. None of the simple correction procedures noted above gives a result generally free from bias, and again the procedure involving complete removal of the determinand from a portion of sample is satisfactory in principle.

(d) Analytical Systems with Response not Proportional to Concentration

In (*a*) to (*c*) we obtained some general results concerning the bias associated with common calibration and blank correction procedures when the analytical response is proportional to the concentration of determinand in the solution analysed. We wish now to consider the consequences when the analytical response is *not* proportional to concentration. Many different relationships are possible, and we make no attempt at exhaustive coverage. Rather, we consider only two of the more common non-proportional types of relationship; the general approach followed here can always be applied to any other relationship that may be of interest. We do not consider calibration functions which are of the form of equation (5.6) up to a certain concentration above which there is a departure from proportionality. The better and usual approach in such situations is to limit the concentration range of the analytical system so that proportionality applies. In the following, symbols have the meanings defined in (*a*) to (*c*) above.

(i) $\mathscr{A}_T = \eta + \chi(C_S + C_C)^\gamma$ It is useful to start wth a more general form of equation (5.4), namely,

$$\mathscr{A}_S = \chi C_S^\gamma \tag{5.24}$$

where γ is a calibration parameter. Following exactly the same approach as in (*a*), we can derive

$$\mathscr{A}_T = \eta + \chi (C_S + C_C)^\gamma \tag{5.25}$$

for the relationship between $\mathscr{A}_T$ and C_S. Except when $\gamma = 1$, this calibration function is non-linear, and some points concerning calibration and blank-correction procedures are noted below.

(1)   The calibration curve may be plotted as $\mathscr{A}_T$ against C_S. In general, however, non-linear calibration curves are more difficult to define accurately, and $\log \mathscr{A}_T$ is often plotted against $\log C_S$ in an attempt to produce a linear calibration curve. Denoting the nominal concentration of the j th calibration standard by C_j, the corresponding response by $\mathscr{A}_{Tj}$, and the determinand concentration in the water used to prepare these standards by C_W, we derive* from equation (5.25)

$$\log \mathscr{A}_{Tj} = \log \{ \eta + \chi (C_j + C_W + C_C)^\gamma \} \tag{5.26}$$

A plot of $\log \mathscr{A}_{Tj}$ against $\log C_j$ is essentially linear, therefore, only when $C_C \ll C_j$, $\eta \ll \chi C_j^\gamma$, *and* either $C_W = 0$ or C_W is known and its value added to C_j. In the latter event, if we put $C_j' = C_j + C_W$, the plot of $\log \mathscr{A}_{Tj}$ against $\log C_j'$ is a straight line with the equation

$$\log \mathscr{A}_{Tj} = \log \chi + \gamma \log C_j' \tag{5.27}$$

Thus, if a plot of $\log \mathscr{A}_{Tj}$ against $\log C_j$ is a straight line, one can assume that $\eta = C_C = C_W = 0$. If a straight line is not obtained, knowledge of the values of η and $(C_C + C_W)$ is required if a correction to produce a linear calibration curve is to be made. Note that a standard with $C_j' = 0$ can neither be plotted nor evaluated when equations (5.26) or (5.27) are used as calibration functions. In addition, when η and C_C are not zero, equation (5.26) shows that the calibration curve has a progressively decreasing slope as C_j' decreases; it may be better, therefore, to plot $\mathscr{A}_{Tj}$ against C_j' at very small C_j' values.

(2)   Provided that calibration is made on each occasion when samples are analysed, and that the values of η, χ and γ are independent of the nature of the solution, then the results for samples analysed at the same time as the calibration standards are unbiased – however the calibration is plotted – provided also that the concentrations assigned to the standards include the concentration of determinand in the water used to prepare those standards. If the values of η, χ, and γ are independent of the solution analysed, and if χ and γ do not vary from batch to batch of analyses, an initial calibration may be used for subsequent batches provided that correction for any variations of η and C_C can be made. This correction is considered in (3) below.

(3)   Consider a series of calibration standards (including a blank for which $C_j = 0$) analysed at a time when the values of η, C_C, and C_W are η_0, C_{C0}, and C_{W0}. For the first type of blank correction in which $\mathscr{A}_B$ is subtracted from $\mathscr{A}_T$ [see (*b*) above], we can derive from equation (5.23)

---

* For simplicity, we assume the blank water contains no interfering substances.

$$\mathscr{A}_{Tj} - \mathscr{A}_{B0} = \chi(C'_j + C_{C0})^{\gamma} - \chi(C_{W0} + C_{C0})^{\gamma} \qquad (5.28)$$

It follows that analytical results would be obtained from the equation

$$R = \left\{ \frac{\mathscr{A}_T - \mathscr{A}_B + \chi(C_{W0} + C_{C0})^{\gamma}}{\chi} \right\}^{1/\gamma} - C_{C0} \qquad (5.29)$$

Consider now the analysis of a sample and a blank on a subsequent occasion when the values of η, C_C, and C_W are η_S, C_{CS}, and C_{WS}. For this sample, with a true concentration C_S, from equation (5.25)

$$\mathscr{A}_{TS} - \mathscr{A}_{BS} = \chi(C_S + C_{CS})^{\gamma} - \chi(C_{WS} + C_{CS})^{\gamma},$$

and if this value of $\mathscr{A}_{TS} - \mathscr{A}_{BS}$ is substituted in equation (5.29), we obtain

$$R = \{(C_S + C_{CS})^{\gamma} - (C_{WS} + C_{CS})^{\gamma} + (C_{W0} + C_{C0})^{\gamma}\}^{1/\gamma} - C_{C0} \qquad (5.30)$$

Equation (5.30) shows that the bias of R ($= R - C_S$) is zero only when $C_{WS} = C_{W0}$ and $C_{CS} = C_{C0}$, and that the bias is unaffected by variations in η. The only general way, therefore, of ensuring zero bias is to ensure that both C_W and C_C are always zero; achieving this may present formidable difficulties.

Similar problems arise for the other form of blank correction in which the apparent concentrations, R'_S and R'_B, are obtained from the initial calibration curve, and the analytical result, R, is equated with $R'_S - R'_B$. For this approach, the calibration curve is best plotted as $\mathscr{A}_{Tj}$ against C'_j. By reasoning similar to that above, it can be shown that the bias of R is now given by

$$R - C_S = \left\{ \frac{\eta_S - \eta_0 + \chi(C_S + C_{CS})^{\gamma}}{\chi} \right\}^{1/\gamma} - \left\{ \frac{\eta_S - \eta_0 + \chi(C_{WS} + C_{CS})^{\gamma}}{\chi} \right\}^{1/\gamma} - C_S \qquad (5.31)$$

Equation (5.31) shows that, when $\eta_S = \eta_0$, the bias of R is $-C_{WS}$. Since we have assumed that C_{W0} is known, we may also assume that C_{WS} is known or can be determined. Accordingly, the result can be corrected to eliminate this bias, *i.e.*, the analytical result should be calculated as $R + C_{WS}$. Thus, this method of blank correction is effective provided that C_W can be determined, and that η is constant. The only general way of ensuring the latter condition is to ensure that η is always zero.

When neither η nor C_C are always zero, no unbiased method of blank correction appears to be generally possible, and we are in the same position as noted in (c) above. If C_C is always zero, the blank correction should involve $\mathscr{A}_{TS} - \mathscr{A}_{BS}$, while if η is always zero, $R'_S - R'_B$ should be used. The problem is that of deciding the relative magnitudes of η and C_C so that the more appropriate method of blank correction can be chosen, and the bias associated with it considered. As in (c) above, we have no helpful advice to offer in this situation, but see (5) below.

(4) When either χ or γ or both depend on the nature of the solution analysed, we meet a similar situation to that in (c) above and again the only general approach to the problem is to use a standard-addition procedure. In this event, however, because a logarithmic plot ($\log \mathscr{A}_{Ti}$ against $\log C_i$) cannot be extrapolated to zero, $\mathscr{A}_{Ti}$ must be plotted against C_i. This then involves

extrapolation of a curve, clearly a much less desirable and accurate situation than when a straight line is extrapolated, though see reference 23. Given what seems to us the rather dubious assumption that such non-linear extrapolations can be made without bias, equation 5.25 and the approach in (c) above can be used to derive expressions for the bias of results when one or other of the procedures for blank correction are used. This leads to conclusions similar to those in (c).

(5) From (1) to (4) above, it can be seen that calibration functions of the form of equation 5.25 ($\gamma \neq 1$) show important disadvantages compared with linear calibration curves. Of course, any bias arising from calibration and blank correction when $\gamma \neq 1$ may be negligible in some applications, but we feel that the problems noted above will often be important in trace analysis, and are good grounds for attempting to use, whenever possible, analytical systems for which the calibration curves are linear.

(ii) $\mathscr{A}_T = \eta + \chi log(C_S + C_C + C_E)$ This type of relationship between $\mathscr{A}_T$ and C_S arises in water analysis only when ion-selective electrodes are used, but this application is sufficiently important to warrant brief mention here. For a detailed, fundamental discussion of calibration and blank correction for these electrodes, a series of papers[24-28] and a review[29] by Midgley should be consulted.

The Nernst equation (see Section 11.3.6) may be written as

$$\mathscr{A}_T = \eta + \chi log C_T \qquad (5.32)$$

where C_T is the total concentration of determinand in the solution in which the electrode is immersed. In making explicit allowance for contamination of the sample by the determinand, two types of contamination must be distinguished: (1) that arising from transfer of the determinand from the electrode during the measurement stage, and (2) that arising, during processing and measurement, from all other sources. The magnitudes of these two types (expressed as equivalent concentrations in the original sample) are denoted by C_E and C_C, respectively. It should be noted that, for a given electrode, C_E decreases as $C_S + C_C$ increases, and that C_E also depends on the nature of the electrode. On this basis, $C_T = C_S + C_C + C_E$, and the more general form of equation (5.32) becomes

$$\mathscr{A}_T = \eta + \chi log(C_S + C_C + C_E) \qquad (5.33)$$

Considerations such as those indicated in (*i*) above may be made on the basis of equation (5.33), and we shall mention only a few particular points.

(1) Positive values for C_C and C_E will cause increasingly large deviations from linearity ($\mathscr{A}_T$ against $log C_S$) as C_S decreases.

(2) Correction is generally required for the determinand content of the water used to prepare calibration standards.

(3) Midgley[24-29] discusses various procedures for estimating the values of C_C and C_E.

(4) Taking anti-logarithms of both sides of equation (5.33), we obtain after some re-arrangement

$$10^{\mathscr{A}_T/\chi} = 10^{\eta/\chi}(C_C + C_E) + 10^{\eta/\chi}C_S \qquad (5.34)$$

Thus, a plot of $10 \mathscr{A}_T/\chi$ against C_S is linear so long as the variation of C_E with C_S is negligible. This transformation is the basis of the commonly-used Gran plots.

(5) When measurements must be made at C_S values such that a plot of $\mathscr{A}_T$ against $\log C_S$ becomes curved, it is usually better to plot $\mathscr{A}_T$ against C_S since this plot tends to the form $\mathscr{A}_T = \lambda + \chi' C_S$. Midgley[24-29] has derived general expressions for the values of λ and χ' in terms of the relevant physicochemical constants.

(6) When the calibration parameters are estimated freshly in each batch of analyses, there is no need for blank corection.[30] A blank determination, however, is generally necessary[25-29] when there is interest in detecting very small concentrations (see Section 8.3).

(7) Standard addition procedures may be based on Gran plots or, at very small determinand concentrations, on plots of $\mathscr{A}_{Ti}$ against C_i [see (c) above]. Blank correction is generally necessary for such procedures. For very small determinand concentrations, any effects of the variation of C_E with C_i should be carefully considered.

(e) Estimation of Calibration Parameters by Regression Analysis

(i) General In general, the numerical values of the calibration parameters for a particular analytical system are not known *a priori*, and must be established experimentally. The statistical technique known as regression analysis can be of considerable value as an objective and explicit method of calculation, from the observed responses for a series of calibration standards, of estimates of calibration parameters. We shall not attempt here to describe the basis of this technique; this can be found in any of the statistical texts mentioned in Section 4.8. It is, however, useful to summarise some of the main results for calibration functions of the form $\mathscr{A}_T = \lambda + \chi C_S$, since we use them later. For such applications, the technique is known as linear regression analysis, but general regression analysis is applicable to non-linear calibration functions (see below).

We assume that a number, p, of different calibration standards of concentrations, C_j ($j = 1$ to p), are analysed and their analytical responses, A_j, measured. We write A now for response to emphasise that, unlike $\mathscr{A}_T$, it is subject to random error. Thus, $A_j = \mathscr{A}_j + \varepsilon$, where ε denotes a random error. In estimating λ and χ, linear regression analysis assumes that the analytical responses are subject to independent, random errors. The population mean, $\mathscr{A}_j$, of the responses for the j th standard is assumed to be equal to $\lambda + \chi C_j$. This implies that the measurements of responses are unbiased, and that the equation $\mathscr{A}_j = \lambda + \chi C_j$ holds exactly. The population standard deviation of the responses for the j th standard is denoted by σ_{Aj}. Given these assumptions, the technique allows the estimates, l and k, of λ and χ to be calculated from the p pairs of data (A_j, C_j). It is very important, however, to distinguish two situations:

(1) σ_{Aj} is independent of C_j, *i.e.*, the standard deviation of responses is the same for all concentrations, and is denoted by σ_A;

(2) σ_{Aj} varies with C_j.

Failure to make this distinction can and does lead to very erroneous conclusions.

Situation (1) requires the use of *unweighted linear regression*, and this is the technique described in many elementary statistical texts which often fail to stress the requirement noted in (1). When, however, σ_{Aj} varies with C_j, *weighted linear regression* should be used, and this technique is rarely described completely in statistical texts written for scientists.

For both unweighted and weighted regression, when the calibration standards include a blank determination ($C_j = 0$), this should be treated in the calculations in exactly the same way as all the other standards. Statistical complications arise if the blank response, A_B, is subtracted from the responses of all the other standards, and linear regression is then used to estimate χ in $\mathscr{A}_{T'} = \chi C_S$, where $\mathscr{A}_{T'} = \mathscr{A}_j - \mathscr{A}_B$; see, for example, ref. 31.

(ii) Unweighted linear regression Writing $\bar{A} = \Sigma A_j/p$ and $\bar{C} = \Sigma C_j/p$, the estimates, k and l, are given by

$$k = \frac{\Sigma A_j C_j - p\bar{A}\bar{C}}{\Sigma C_j^2 - p\bar{C}^2} \tag{5.35}$$

and
$$l = \bar{A} - k\bar{C} \tag{5.36}$$

If another set of such measurement were made, different values would generally be obtained for k and l as a result of the random errors of the measured responses. Thus, k and l are random variables, but they are unbiased since their population means can be shown to be χ and λ (see Section 4.4.1). The variances, σ_k^2 and σ_l^2, of k and l are given by

$$\sigma_k^2 = \sigma_A^2/(\Sigma C_j^2 - p\bar{C}^2) \tag{5.37}$$

$$\sigma_l^2 = \sigma_A^2 \left\{ \frac{1}{p} + \frac{\bar{C}_2}{\Sigma C_j^2 - p\bar{C}^2} \right\} \tag{5.38}$$

Note also that l and k are not statistically independent, and that if the A_j values for every j are normally distributed, then so too are l and k.

Suppose n replicate analyses (n can be 1) of a sample are made, and give a mean response, $\bar{A}_T$. The analytical result is calculated from

$$R = (\bar{A}_T - l)/k, \tag{5.39}$$

and the variance, σ_t^2, of R is given approximately by

$$\sigma_t^2 \approx \frac{\sigma_A^2}{k^2} \left\{ \frac{1}{n} + \frac{1}{p} + \frac{(\bar{A}_T - \bar{A})^2}{k^2(\Sigma C_j^2 - p\bar{C}^2)} \right\} \tag{5.40}$$

The degree of approximation becomes smaller as σ_k/k decreases. In addition, if the A_j values are normally distributed, the distribution of R will approximate more and more closely to the normal as σ_k/k decreases. Note that equation (5.40) assumes that the standard deviation, σ_{AT}, of responses for the sample equals that for the standards, σ_A. If this is not so, the $1/n$ term within the bracket of equation (5.40) should be replaced by $\sigma_{AT}^2/n\sigma_A^2$.

In practice, σ_A will not be known. An estimate, s_A, can be obtained from other experiments or, when $p \geqslant 3$, from the set of calibration standard results using the equation

$$s_A = \sqrt{\left\{ \frac{\Sigma A_j^2 - p\bar{A}^2 - k^2(\Sigma C_j^2 - p\bar{C}^2)}{p-2} \right\}} \tag{5.41}$$

s_A has $p-2$ degrees of freedom, and may be used in place of σ_A in equation (5.40) to give an estimate, s_t^2, of σ_t^2.

When q replicate analyses of each calibration standard are made, the above equations remain valid provided: (1) each one of the pq pairs of results are included in the summations indicated by Σ, and (2) p is replaced by pq throughout.

(iii) Weighted linear regression Let $z_j = 1/\sigma_{Aj}^2$, $\bar{z} = \Sigma z_j/p$, $w_j = z_j/\bar{z}$, $\bar{A}_w = \Sigma w_j A_j/p$, and $\bar{C}_w = \Sigma w_j C_j/p$. Note that this requires knowledge of how σ_{Aj} varies with C_j; in practice, values for z_j may be equated to $1/s_{Aj}^2$, where s_{Aj} is an estimate of σ_{Aj}.

Values for k and l are given by

$$k = \frac{\Sigma w_j A_j C_j - p\bar{A}_w\bar{C}_w}{\Sigma w_j C_j^2 - p\bar{C}_w^2} \tag{5.42}$$

and

$$l = \bar{A}_w - k\bar{C}_w. \tag{5.43}$$

Again, k and l are unbiased random variables, are not statistically independent, and are normally distributed if the A_j values for every j are normally distributed. Their variances are given by

$$\sigma_k^2 = 1/\bar{z}(\Sigma w_j C_j^2 - p\bar{C}_w^2) \tag{5.44}$$

and

$$\sigma_l^2 = \frac{1}{p\bar{z}} + \frac{\bar{C}_w^2}{\bar{z}(\Sigma w_j C_j^2 - p\bar{C}_w^2)} \tag{5.45}$$

As with unweighted regression, the result, R, is calculated from equation (5.39) using the values of l and k from equations (5.42) and (5.43). The variance, σ_t^2 of R is given approximately by

$$\sigma_t^2 \approx \frac{1}{k^2} \left\{ \frac{\sigma_{AT}^2}{n} + \frac{1}{p\bar{z}} + \left[\frac{\bar{A}_T - \bar{A}_w}{k} \right] / \bar{z}(\Sigma w_j C_j^2 - p\bar{C}_w^2) \right\} \tag{5.46}$$

If q replicate analyses of each calibration standard are made, the calculations and modifications to the equations are exactly as outlined in (*ii*) above.

It should be noted that the use of unweighted linear regression leads to unbiased estimates of l and k even when σ_{Aj} depends on C_j. The variances of l and k are smaller, however, when σ_{Aj} depends on C_j and weighted regression is used. The variances, σ_k^2, σ_l^2, and σ_t^2, given by equations (5.37), (5.38) and (5.40) may be seriously in error when unweighted regression is used, and σ_{Aj} depends on C_j.

An alternative to weighted linear regression analysis is to use mathematical transforms of A_j and C_j chosen so that the calibration curve remains linear but the standard deviation of the transform of A_j no longer varies with j. For example, $\log(A_j - A_B)$ could be plotted against $\log C_j$, though see (i) above for a complication that then arises. This type of approach has been reported by several workers,[32,33] but it also requires some knowledge of how σ_{Aj} varies with C_j in order that suitable transforms can be chosen. The procedure is convenient in that it allows the relatively simple unweighted regression, but it may be difficult to find transforms that provide adequate independence of standard deviation and C_j. On balance, we favour (see also ref. 34) weighted regression using A_j and C_j as described above.

(iv) Calibration functions other than $\mathscr{A}_T = \lambda + \chi C_S$ Linear regression analysis may also be used for calibration functions of the form $\mathscr{A}_T = \chi C_S$; details of calculations, *etc.* are provided in statistical texts. The procedures described above may also be used for calibration functions of the form $f(\mathscr{A}_T) = \lambda' + \chi'_g(C_S)$, where f() and g() represent defined, mathematical functions. In such instances, the values of $f(A_j)$ and $g(C_j)$ are used in place of A_j and C_j, respectively. General regression analysis may be used for non-linear functions of the form $\mathscr{A}_T = \lambda + \chi_1 C_S + \chi_2 C_S^2 + \ldots$, or for more complicated functions; see, for example, Schwarz.[35,36]

5.4.3 The Control of Bias arising from Practical Aspects of Calibration

We consider here various sources of bias arising in the practical procedures for estimating calibration parameters. Although blank-correction is best seen as estimation of a particular calibration parameter, we thought it more helpful to deal with this separately in Section 5.4.4. The book edited by Ku[37] contains many papers dealing with these topics.

(a) Standards and Standard Solutions
If either faulty or inadequately pure standard materials are used or the analyst prepares the standard solutions incorrectly, bias will clearly result. Great care is essential, therefore, in preparing calibration standards. Provided that this care is ensured, standards need not be a general source of appreciable bias in water analysis. This does not mean, of course, that such bias is never important, particularly when it is difficult or impossible to obtain standard materials of adequate purity and accurately known determinand content. For example, this problem has arisen in the past for a number of organic compounds of interest in water analysis, *e.g.*, polycyclic aromatic hydrocarbons and pesticides. In the United Kingdom, the National Physical Laboratory can now supply many such substances with certified purities;[38] see also refs. 39–42 and refs. 8, 43 and 44 for lists of suppliers of standard materials.

In general (see Section 5.4.2), the nominal concentrations of calibration standards should be corrected by adding to them the determinand concentration in the water used to prepare standard solutions. Ideally, this water should be prepared and stored so that this concentration is negligibly small, but if this is not possible the determinand concentration should be measured and the correction made. In many applications this source of bias is negligible, but it can be very important in some trace analyses. See Sections 5.4.4(b) and 10.7 for further consideration of this problem.

Even when the concentrations of calibration standards are known initially with negligible bias, another possibility of bias arises if these standards are stored and used over a period of time. Several factors may cause changes in the initial concentrations as discussed in Section 3.6. Any well described analytical method should contain detailed information on this aspect, and include recommendations on appropriate storage conditions and periods. In the absence of such information, the analyst is well advised to be cautious and to prepare fresh standard solutions wherever they are required – at least until reliable information on their stability is available. Note also that any tests to check stability are subject to the random errors of analytical results so that statistical interpretation of the results of stability tests is generally necessary (see Section 3.6.4).

(b) Analytical Procedure Used for Calibration
For methods specifying calibration by the analysis of standard solutions, Section 5.4.2 shows that, in general, the analytical procedure by which the standards are analysed must be identical with that used for normal samples; otherwise, the values obtained for the calibration parameters may differ from those relevant to samples. This obvious point is by no means always followed, and any such differences between the two procedures can be an important source of bias, particularly for methods including long and complex separation and concentration procedures before the measurement stage, *e.g.*, as in many methods for determining trace organic substances.

If the procedures for calibration and sample analysis specified by the method differ in any way (no matter how small), the analyst should satisfy himself – experimentally, if need be – that the difference causes no important bias. Considerations such as those in Section 5.4.2 are of value in indicating important factors. Alternatively – and in our view this is generally preferable – the two procedures can be made identical before testing and applying a method, thereby eliminating the possibility of this source of bias.

(c) Relationship between Analytical Response and Concentration
The mathematical function describing the true relationship between the analytical response and the concentration of the determinand may differ from the function assumed. For example, a straight-line calibration curve may be used though the true calibration may be slightly curved, and this will lead to bias [see (d) below]. It is again suggested that the emphasis should be on eliminating this source of bias, and to this end, several points are worth emphasis.

(1) Great value attaches to the use of analytical systems whose principles are such that the type of relationship between response and concentration is known even though its precise numerical form must be estimated. For example, many spectrophotometric systems are expected to follow equation (5.6), while ion-selective electrodes at sufficiently high determinand concentrations show a different type of relationship, *i.e.*, equation (5.32).

(2) Other things being equal, the use of analytical systems for which defined functions of response and of determinand concentration are related linearly is preferred to the use of curvilinear calibrations. Calibration of the former type

can be made with better accuracy for a given amount of effort, and is easier to check. Whenever possible, it is advantageous (see Section 5.4.2) to use systems for which the response or a defined function of the response is proportional to the determinand concentration.

(3) Of course, calibration curves or functions should not be extrapolated beyond the concentration range for which they are known to be valid. In this connection, it should be remembered that the upper or lower (or both) concentration limits of the validity of a particular calibration function may depend on the particular type of instrument used for measurement of the response even though the analytical procedure is otherwise identical.

(d) Experimental Estimation of the Calibration Parameters
The numerical values of calibration parameters are not generally known *a priori*, and must be estimated experimentally from the analysis of calibration standards. Even when the concentrations of these standards are known without appreciable error and the analytical responses for them are measured without bias, the inevitable random errors of these responses lead to errors in the estimates of calibration parameters. In this section, we consider how these errors may lead to bias in analytical results. For simplicity, we deal only with calibration functions of the form $\mathscr{A}_T = \lambda + \chi C_S$. Fixed-slope calibration, within-batch calibration, and standard-addition procedures are discussed in (*i*), (*ii*), and (*iii*), respectively.

(i) Fixed-slope calibration   We assume that an initial estimate, k, of the slope of the calibration curve is used over many subsequent batches, that an estimate, A_B, of λ is obtained from one or more blank determinations in each batch of analyses, and that analytical results, R, are calculated from

$$R = (A_T - A_B)/k \qquad (5.47)$$

where A_T is the analytical response for the sample. We also assume that the true determinand concentrations of calibration standards are known, and that the measurements of analytical responses are made without bias. On this bias, bias in R will arise only from bias in k; see also Section 6.2.1. Bias in k can arise from three sources.

(1) The procedure used to determine k from the experimental data may give biased values of k.
(2) k will certainly have a random error so that it will generally differ from χ. When many samples over a long time-period are evaluated using the same value of k, its random error becomes, in effect, a bias because all the samples are equally subject to it. To minimise this source of bias, the random error of k should, in general, be minimised.
(3) The use of equation (5.47) will lead to bias if, in fact, the true calibration function is non-linear.

Given these sources of bias, we may consider the advantages and disadvantages of the three main approaches to determining k, assuming that – as is usual and desirable – a number of calibration standards with different determinand concentrations are used for the initial calibration.

In principle, the simplest procedure is to plot the observed values of A_{Tj} against C_j, and to draw the best straight line by eye through the points; the slope of this line is equal to k. When the errors of all the A_{Tj} values are small (so that their scatter about the true position of the calibration curve is small), this procedure will give an essentially unbiased value for k. For example, Goode and Northington[45] have reported that, in flame atomic-emission spectrometry, this procedure is usually as good as the more complex, statistical procedures considered below. Equally, when the experimental points all fall on or very close to a straight line, it may usually also be taken that the random error of k is negligibly small provided at least 4 to 6 calibration standards have been used.[46] Finally, the critical, human eye is a good detector of any systematic pattern in the differences between the observed A_{Tj} values and the values given by the best straight line; such systematic patterns may well arise when the true calibration is non-linear, particularly when the random errors of the A_{Tj} values are small. In many instances, therefore, we believe that this simple, visual procedure is quite adequate. As, however, the random errors of the A_{Tj} values become larger, so the procedure becomes progressively less satisfactory. Thus, when the scatter of the points about the true calibration line is relatively large, different analysts may well produce different judgments on the slope of the best straight line through the points. This visual procedure is also unable to provide any objective quantification of the random error of k, and of the effect of the number and concentrations of calibration standards on that error. Further, the procedure allows no objective assessment of whether or not it is valid to regard the true calibration curve as linear. Thus, more objective and quantifiable procedures are generally necessary when the random errors of the A_T values are appreciable.

The second approach to determining k involves calculation rather than visual determination of its value. For example, if A_{Tj} is the observed response for the j th calibration standard of concentration, C_j ($j = 1$ to p), and A_B is the response for a blank determination, a value for k can be obtained from

$$k = \frac{\Sigma(A_{Tj} - A_B)/C_j}{p} \qquad (5.47)$$

Such a procedure overcomes problems of subjectivity and gives unbiased estimates of k, but it suffers from the other disadvantages of the visual procedure when the random errors of the A_{Tj} values are appreciable. In addition, the random error of k may be greater than that when the appropriate statistical procedures described below are used.

The third approach, which has found greatly increasing use in recent years, involves the use of the statistical technique known as linear regression analysis, the relevant assumptions and calculations for which are summarised in Section 5.4.2(e). This technique allows the objective calculation of an unbiased value for k whatever the size of the random errors of responses. It also has the valuable feature of providing estimates of the standard deviation of k so that the random error of k can be quantified; the effects on this random error of factors such as the number and concentrations of calibration standards can also be calculated. Further, provided that at least duplicate analyses are made of each calibration standard, one can also assess whether or not there is any evidence that the true calibration curve is non-linear. Linear regression analysis, therefore, has all the capabilities ideally required. The calculations involved are rather cumbersome – in some applications, very cumbersome – if made manually, but this disadvantage is less important with the growing availability of cheap calculators programmed for regression analysis and of computing facilities associated with analytical instruments. The technique is, therefore, of general applicability and value.

We have already stressed a number of points concerning the use of the technique in Section 5.4.2(*e*) where the equations of principal interest are given, and we briefly summarise below several points relevant to controlling the random error of k.

(1)   In addition to statistical texts, applications of linear regression analysis to calibration are described in many analytical papers; see, for example, refs. 34 and 47 to 52.

(2)   Many analytical systems show a dependence of σ_{Aj} on C_j, and it is essential to distinguish, and use appropriately, unweighted and weighted linear regression.

(3)   When the random errors of responses are normally distributed, k is normally distributed. Its random error may, therefore, be quantified as described in Sections 4.3.3 and 4.3.5.

(4)   Equations (5.37) and (5.44) may be used to calculate the effects of p, q, and the C_j values on σ_k. Examples of such calculations may be found in refs. 31, 46, 48, 49 and 53. The choice of p and the C_j values should, however, also take into account the need to check for deviations of the true calibration function from linearity [see (5) below]. For this latter purpose, the calibration standards should be uniformly spread over the concentration range of interest. The common analytical procedure of analysing five or six standards in duplicate should usually ensure that the random error of k is negligible compared with other sources of bias.

(5)   Given that a number of different calibration standards have been analysed twice or more in one batch of analyses, a suitable statistical procedure for checking for deviations from linearity is described, for example, in ref. 54. Of course, as the values of p and q are increased, so it is possible to detect smaller deviations from linearity.

(6)   When linear regression is used to calculate k, we would nevertheless recommend visual examination of the derived calibration curve showing the individual points. This may reveal some abnormality, *e.g.*, a result deviating from the curve by an unexpectedly large amount, that may affect the accuracy of k and require further investigation.

(7)   Regression analysis is not restricted to calibration functions of the form $\mathscr{A}_T = \lambda + \chi C_S$; see Section 5.4.2(*e*) (*iv*).

(8)   The estimation of the precision of analytical results using linear regression is considered in Sections 6.2.1(*c*) and 6.3.1.

(9)   Many modern instruments have facilities which provide automatic computation of the calibration curve when appropriate standards are analysed. Such facilities may also include the capability for automatic correction of non-linear calibration curves to linear forms. Such computing systems may be of great value, but their speed, accuracy of calculation, and convenience in no way decrease the importance of any of the points mentioned above. Indeed, the

analyst generally needs to be even more critical because the routine use of such devices may well lead to the situation where the procedures involved in the calculation of calibration parameters are unknown, and the fit of the responses of calibration standards to the calculated calibration curve is never seen. In general, we recommend that such devices should be used only when the analyst knows the principles involved in the calculations performed, and is satisfied that the calculations themselves will lead to negligible bias in calibration parameters; see also Section 10.4.3. Note also that pocket calculators programmed for regression analysis usually provide, in fact, only unweighted calculations.

(ii) Within-batch calibration Fixed-slope calibration cannot be used for many analytical systems because of temporal variability in the value of χ. The required frequency for determining k is governed by the nature and size of such changes. When manual analyses are made, it often suffices to calibrate once on each occasion when samples are analysed, *i.e.*, once per batch [see Section 6.2.1(*b*)]. When semi- or fully-automatic systems are used, gradual drift in the values of λ and χ may occur and necessitate re-calibration at intervals during one analytical occasion. The analytical method for such systems may specify the frequency of such re-calibration, but if this is not so experimental investigation of the rate of drift may be required. Many papers deal with this topic; see, for example, refs. 3, 55 to 61*a*, 98 and Section 6.2.1(*e*)(*ii*).

Within-batch calibration is usually made by essentially the same procedures as those used for fixed-slope calibration. Of course, when one is confident that the calibration function will always be of the form $\mathscr{A}_T = \lambda + \chi C_S$ so that it is necessary only to estimate λ and χ, some simplification of the calibration procedure may well be possible. For example, analysis of a blank and one standard solution may suffice, though see Section 6.2.1(*c*) concerning the random errors of the estimates, *l* and *k*. For calibrations of this general type, the points in (*i*) above all apply with one exception. Although the random error of a particular value of *k* will equally affect all samples evaluated using that value, we regard the resulting errors in analytical results as random errors whose size varies from calibration to calibration [see Section 6.2.1(*b*)].

A completely different type of procedure for within-batch calibration has very recently been described in detail by Tyson and Appleton.[99] We know of no applications at present to water analysis, but the novel nature of the approach justifies brief mention here. In essence, a single standard solution is continuously pumped into a mixing-chamber containing a blank solution, and the solution emerging from this chamber is then passed continuously into the nebuliser of a flame atomic-absorption spectrometer. By suitable choice of equipment and control of experimental conditions, the determinand concentration in the solution emerging from the mixing-chamber and the time of its emergence can be accurately related by a simple, mathematical equation. This relationship is, in effect, plotted by passing the instrument response to a chart-recorder, and the analytical results for samples may then be calculated by comparison of their responses with the calibration curve on the recorder chart. The authors claim several advantages for this approach over the more traditional calibration procedures: (1) only one standard solution need be prepared, (2) the procedure is easily automated, (3) the entire calibration curve rather than a limited number of points is obtained, and (4) the procedure is directly applicable to any type of calibration function. We have no personal experience of this procedure, and it is too soon to decide its potential role in the routine analysis of waters. Readers wishing to pursue the topic will find full details of the theory, equipment and operation in the reference cited.

(iii) Calibration by standard-addition procedures   In recent years there has been a marked increase in the use of analytical systems for which calibration using standard solutions is inaccurate because χ is different for standards and samples. The calibration of such systems is usually made by standard-addition procedures whose principles are considered in Section 5.4.2(*c*).* These procedures may be subject to certain sources of bias that are difficult or impossible to eliminate as noted in that section. Here, we stress that it is generally necessary[62] that the value of χ_S for a sample be independent of the determinand concentration in that sample (at least within the concentration range of interest). This requirement may, for example, not be met when χ_S is governed by the proportion of the determinand complexed by a negatively interfering substance since the latter will be able to complex only a certain concentration of the determinand. Under these conditions, as the amount of added determinand is increased, the proportion of determinand complexed may show a marked decrease as the complexing capacity of the interfering substance is approached and exceeded. Considerable care is also required, in general, in deciding how best to make a blank-correction. A number of the practical aspects of analysis may have important effects on bias, and these are summarised below.

(1)   Incorrect standard solutions may, of course, introduce bias.

(2)   The volumes of standard solutions added to the portions of a sample should be such that any effect of sample dilution on the analytical response is either negligible or corrected.

(3)   The procedure assumes that the analytical response for a given concentration of determinand is the same for the determinand originally present in the sample and for the determinand added to the sample. This condition may not be met if the chemical/physical forms of the determinand in the sample differ from those of the added determinand; see, for example, refs. 17, 63 and Section 5.4.5. When this possibility occurs, it will generally be useful to attempt to ensure that the samples are treated so that the determinand from both sources is converted to the same form or forms.

(4)   The procedure generally relies on linear extrapolation – though see ref. 23. It is important, therefore, to ensure that the greatest determinand concentration in 'spiked' portions of samples does not exceed any upper limit to linearity of the calibration function. If it is certain that this function is linear, one 'unspiked' and one 'spiked' portion of a sample represents the smallest possible number of measurements on the sample. Otherwise, one 'unspiked' and two 'spiked' portions are the minimal requirement to provide some check on the validity of linear extrapolation.

* For a novel approach related to that described in (*ii*), see refs. 100 and 101; we know of no applications at present to the routine analysis of waters.

(5) The same three procedures as discussed in (*i*) may be used for the extrapolation, and the points made there apply equally here. Again, linear regression (weighted or unweighted as appropriate) is generally applicable, and provides unbiased estimates.

(6) Regression theory also allows calculation of the random errors of analytical results; this aspect is dealt with in Section 6.2.2.

The above points concern standard-addition procedures as normally practised in the routine analysis of waters. Recently, however, a much more mathematically complex calibration procedure has been developed – the Generalised Standard Addition Method (GSAM)[61,64–68] – and warrants mention here. In fact, the standard-addition component of GSAM is exactly as described above though the problems of blank-correction seem not yet to have been considered in detail. The more interesting component of GSAM is its ability to correct for a number of individual interference effects when each interfering substance may itself also be subject to a matrix-type interference. Accordingly, we consider this technique further in Section 5.4.6(*j*)(*iv*).

5.4.4 The Control of Bias arising from Blank Determinations

As discussed in Section 5.4.2(*b*), for analytical systems for which χ is independent of the nature of the solution, one purpose of blank determinations is to estimate the value of the calibration parameter, λ, in the calibration function, $\mathscr{A}_T = \lambda + \chi C_S$. Thus, analytical systems for which $\lambda = 0$ do not require blank determinations or corrections.* We would suggest, however, that it is a sound precaution to assume that $\lambda \neq 0$ until compelling evidence to the contrary is available. We assume that high-purity water is used as the 'blank' sample. When χ differs for samples and standards so that standard-addition procedures are used, we saw in Section 5.4.2(*c*) that difficulties may arise because λ is affected by χ, and, therefore, is different for samples and standards. The simple analysis of a single portion of high-purity water may not, therefore, provide an unbiased estimate of λ_S, and possible means of dealing with this problem are considered in Section 5.4.2(*c*). For simplicity, we shall consider below those sources of bias associated with the use of a simple blank determination to estimate λ. Note also that such blank determinations are generally essential when one is interested in detecting very small concentrations of a determinand; see Section 8.3.

(a) Analytical Procedure Used for the Blank Determination

Clearly, the blank determination must be subject to all the factors affecting λ. Thus, in general, the analytical procedure for the blank determination must be identical with that for normal samples. This obvious point is by no means always observed, and any such differences between the two procedures can be an important source of bias, particularly for procedures involving extensive treatment of samples before the measurement stage. Of course, there are circumstances where the blank procedure may validly be simpler than that for samples, but we would recommend

---

* We exclude from this, systems for which $\lambda = 0$ because of the effects of negatively interfering substances arising either from the water used for blanks and standards or from contamination of solutions during analysis [see Sections 5.4.2(*a*) and (*b*)].

that all such differences (no matter how small) be eliminated, at least until there is firm evidence that the differences cause no important bias in the blank correction.

(b) Water Used for the Blank Determination
It was shown in Section 5.4.2(b) that, if the water used for the blank contains the determinand, analytical results for samples will be biased if no allowance is made for this concentration, C_W. The magnitude of this bias is often, but not necessarily, equal to $-C_W$. Many analytical methods pay scant or no attention to this source of bias which can be important in many trace analyses, particularly those for ubiquitous determinands such as iron, sodium, silicon, ammonia, and organic carbon. Instructions in methods simply to use 'high-purity water' or 'determinand-free water' do little to ensure control of this source of bias. Clearly, the maximum tolerable value of C_W depends on the maximum tolerable bias of analytical results and on the sizes of other types of bias. As a general rule, it is very desirable to make C_W very much smaller than the maximum tolerable bias so that this source of bias becomes negligible (see Section 10.7.2).

If the desired purity of the blank-water cannot be achieved, it is essential to determine the concentration of the determinand in this water, and then to make an appropriate correction. It should be noted that the analytical procedure used for samples is generally incapable of providing an estimate of C_W; for this purpose a special procedure is required. This aspect is discussed in detail in Section 10.7.2. An additional caution to note is that though the value of C_W may be known when the water is first prepared, storage of the water may lead to increases or decreases of C_W depending on the determinand (see Sections 3.6, 10.7.1, and 10.8).

We saw in Section 5.4.2(b) that bias also arises if the blank-water contains interfering substances. Usually, if the determinand content of the water is satisfactorily small, then the effect of interfering substances may be ignored. This is, however, not necessarily true, and circumstances arise where it is interfering substances, rather than the determinand, whose concentration is the dominant contribution to this source of bias. This possibility should, therefore, be borne in mind, and appropriately purified water used whenever possible. If the concentrations of interfering substances cannot be reduced to negligible values, some procedure is required for estimating the bias caused by such substances in the blank-water so that a suitable correction can be made to analytical results (see Section 10.7.2).

The importance of this source of bias cannot be over-emphasised since it is so often apparently ignored. We recommend, therefore, that preference be given, in general, to those analytical methods that give explicit instructions on suitable procedures for preparing high-purity water and for measuring its determinand content (see Section 10.7).

(c) Need to Minimise the Analytical Response of the Blank Determination
For analytical systems with calibration functions of the form $\mathscr{A}_T = \lambda + \chi C_S$, where χ is independent of the nature of the solution, the results derived in Section 5.4.2(b) show that blank-correction provides an unbiased analytical result whatever the numerical value of λ. When, however, standard-addition procedures are used, it is usually advantageous, and may well be essential, to do everything possible to

minimise λ. Since $\lambda = \eta + \chi C_C$ [Section 5.4.2(a)], it follows that stringent precautions against contamination (see Section 10.8) will reduce C_C and, therefore, λ; appropriate selection of experimental conditions may also help to reduce η. Similar points apply to systems with non-linear calibration functions.

The precision of analytical results may also be improved by minimising λ; see Sections 6.2.1(e)(v) and 10.8.1.

(d) Effect of the Random Errors of Blank Determinations
Given that one or more blank determinations are made in each batch of analyses, we regard the random error of the blank value in a given batch as leading to random errors in analytical results; see Section 6.2.1(b).

5.4.5 Inability to Determine all Forms of the Determinand

When the determinand may occur in samples in a number of different physical and/or chemical forms, it is essential that the analytical response for a given concentration of the determinand be independent of the form of the determinand in the original sample, *i.e.*, the value of χ [Section 5.4.2(a)] must be independent of the form of the determinand. It may well not be easy to ensure this, but failure to do so can result in considerable bias. For example, suppose that the determinand were total iron, and that a method specified simply that the sample were collected into dilute hydrochloric acid and then analysed spectrophotometrically using a specified chromogenic reagent. For most such reagents, the simple treatment of the sample with dilute hydrochloric acid is unlikely to dissolve completely several forms of iron that may be present, *e.g.*, hydrated ferric oxide, magnetite; the iron in such forms would then not be readily available for reaction with the chromogenic reagent. Consequently, negatively biased results would be obtained whenever samples contained such forms of iron.

A number of common and important determinands in water analysis may often occur in different physical and/or chemical forms, *e.g.*, total phosphorus (where inorganic and organic as well as dissolved and undissolved forms may be present), total silicon (where various undissolved minerals and biological detritus in addition to monomeric and polymeric silicic acids may occur), total organic carbon (an enormous possible range of different compounds), total metals (where in addition to dissolved and undissolved forms there is also the possibility of different oxidation states and organometallic compounds), *etc.* Furthermore, virtually all measurement techniques commonly employed in analytical systems for waters show some dependence on the form of the determinand. The problem is, therefore, both widespread and important.

Accordingly, both the information available on this aspect of an analytical system's performance *and* also the forms of the determinand expected to occur in samples need very careful consideration in attempting to judge whether or not this source of bias is likely to be important in a particular application. Some analytical methods and/or associated publications provide evidence on this problem, but, unfortunately, many do not. In addition, knowledge of the forms that may occur is meagre for some determinands. In this far from ideal position, three main positive suggestions are made below for attempting to control this source of bias.

First, other things being equal, preference should be given to the choice and use of methods giving satisfactory or reasonable evidence on the magnitude of the problem.

Second, when there is doubt, it is safer to apply more rather than less vigorous pre-treatment procedures to samples in order to convert the various forms of determinand present to forms that are measurable without bias. Note, however, that the precision of analytical results may deteriorate – as a result, for example, of increased contamination – when more vigorous procedures are used. In practice, therefore, a compromise between bias and precision may be necessary.

Third, tests may be made to determine the suitability of an analytical method from this point of view. We would stress, however, that the required tests are seldom quick and simple to perform, and that it is sometimes difficult to obtain unambiguous results. Two main approaches to such tests can usefully be distinguished.

(1) Known amounts of pure, standard substances corresponding to those that may be present in samples are analysed as specified by the method of interest so as to provide estimates of the efficiency with which they are determined. In some instances this approach may be quite feasible even if time-consuming, *e.g.*, when a range of different organic compounds is to be tested. In other situations, however, it may be difficult or impossible to obtain standard substances corresponding to those present in samples, *e.g.*, when particulate forms of metals are of interest. Of course, even when this is so, the testing of model substances may still give useful information. It should also be borne in mind that the results of such tests will be subject to random errors so that bias may not be detected even when it is present. For the statistical aspects of this, see the account of significance testing in Section 4.6.2(*a*).

(2) Pre-treatments of increasing severity are applied to portions of a sample, and this is repeated for a number of samples. If the results indicate that pre-treatments of greater severity than that specified by the analytical method of interest do not lead to appreciably greater results, it may usually be reasonably assumed that that method is adequate for the forms present in the samples of interest. Again, as for (1) above, the statistical aspects of the tests and the interpretation of their results must be considered [see Section 4.6.2(*b*)].

In practice, our experience is that routine analytical laboratories can seldom if ever find the time and effort required for the proper conduct of tests such as those in (1) and (2). Great responsibility, therefore, rests on those developing and investigating analytical procedures to ensure that adequate attention is paid to this problem, and that the evidence on this source of bias is fully published.

From the above considerations it can be seen that it will generally be difficult, if possible at all, to estimate the bias due to this potential source of error when it is thought that it may be present. The only satisfactory approach, and one that we strongly recommend whenever practicable, is to make strenuous efforts to use analytical methods for which the bias is expected to be essentially zero. When this cannot be done, there seems to be no alternative to joint discussions between the analyst and the user of analytical results to decide the best approach. It may be, for example, that the user would prefer to change the determinand in order to have results of better known accuracy.

A final point should be made concerning the relationship between the ability of an analytical system to determine all forms of the determinand and the same system's liability to interference effects. Since the chemical and/or physical forms of a determinand may be affected by other substances in a sample, such effects may lead to biased analytical results. For example, aluminium may complex fluoride, and the aluminium fluoride complex produces no response with a fluoride ion-selective electrode. When the determinand is 'total fluoride', *i.e.*, the total concentration of fluoride no matter the form in which it is present, aluminium may, therefore, cause bias. The question arises whether this bias should be attributed to: (1) inability to measure all forms of the determinand, or (2) interference caused by aluminium. Of these two, we favour the latter; see our definition of interference in Section 5.4.6(*a*). Accordingly, the concept of 'the ability to measure all forms of the determinand' should be restricted to include only those effects not included in whatever definition of interference is adopted. This clear separation of effects seems to us essential since otherwise there may be the danger that the same effect may be counted twice when attempting to assess the total bias of analytical results through summation of individual effects.

5.4.6 The Effects of Interfering Substances

(a) Introduction

This source of bias is commonly one of the most important in the analysis of waters, and analysts need to be particularly careful in attempting to assess its magnitude when a given analytical method is used. An important preliminary to consideration of the problems involved is to define what is meant in this book by the term *interference*. This is all the more necessary since there is no universally agreed definition.

Many authors have discussed and proposed definitions of terms such as *interference*, *selectivity*, and *specificity*, and we do not wish to review the topic here; refs. 17 and 69 to 72 (and the references cited therein) provide a selection of relevant papers. For reasons already noted,[70] we consider the following definition essential, and it is used throughout this book.

> For a given analytical system, a substance is said to cause interference if its presence in the original sample for analysis and/or in the sample during analysis leads to systematic error in the analytical result, whatever the sign and magnitude of this error.

With the exception of the phrase 'and/or .... analysis', this definition is essentially the same as that suggested earlier.[70] We have included that phrase because it seems to us useful to include as interference any bias caused by the contamination of samples during analysis by substances other than the determinand. An IUPAC Commission has recently adopted and recommended for general use essentially the same definition,[72] and the U.K. Standing Committee of Analysts has also adopted what amounts to the above definition.[73] Note, however, that at least one authoritative analytical body[4] has recommended a definition that restricts 'interference' to the bias caused by substances that do not by themselves produce an analytical response, but see ref. 72 on this last definition. Our definition makes no such restriction, and its implications are considered below.

A special case of interference may arise when it is required to determine one or more forms of a substance in the presence of other forms of the same substance. Examples are the determination of ferrous iron in the presence of ferric, the determination of orthophosphate in the presence of condensed inorganic phosphates and/or organophosphorus compounds, and in general all those analyses involved in investigations of the speciation of substances in waters. Difficulties in achieving the desired freedom from interference may be particularly pronounced for such requirements, and there are also major problems in assessing the bias caused by those forms of a substance that are required not to be measured. This topic of analytical procedures for the study of speciation is discussed in detail in Section 11.5.

It is important to note that differences in the calibration procedures employed by otherwise identical analytical systems can have a marked effect on the magnitude of interference effects. An obvious example is the use of a standard-addition procedure in place of calibration by standard solutions, but there are many other possibilities. Approaches to minimising the size of interference effects by using special calibration and/or correction procedures are discussed in (*j*) below.

Most analytical methods for waters, however, specify no special calibration procedure for minimising interference. For such methods, analysts must, therefore, assess the likely bias that will result for their samples when a particular analytical method is used. Unfortunately, although published methods usually provide some information on interference effects, it is not uncommon for this information to be incomplete or ambiguous or both. For this reason, and also in view of the potential importance of the problem, we have given below a much fuller discussion of the assessment of the bias caused by interference than that in the first edition of this book. Part of this treatment is based on an earlier paper,[71] and we lay no claim to originality. We believe, however, that the integrated discussion is helpful in making clear both the various problems that arise and also the possible approaches to their solution. To a very large extent, identical considerations are involved in: (1) assessing the likely bias for a given sample and analytical system, and (2) deciding the tests necessary properly to characterise interference effects when a given analytical system is used. We have, therefore, attempted to deal with both aspects in the one discussion.

(b) Nature of Interference Effects
The bias caused by interference in the analysis of a given sample is the resultant of the effects of all those substances individually causing interference, and of any other substances that may affect the concentrations of such interfering substances. The sizes of the individual effects depend, in general, on the concentrations of the interfering substances, and may also depend on the concentration of the determinand. Interference may lead to positive or negative bias. Furthermore, the interference caused by one substance may depend on the concentration of another substance other than the determinand. For example, the effect of phosphate on the

spectrophotometric determination of 'reactive' silicon depends on the concentration of ferrous iron;[74] similarly, the effect of silicate on the determination of manganese by atomic absorption spectrometry depends on the concentration of calcium.[75] Indeed, there seems to be no theoretical reason why the effect of one substance should not depend on the concentrations of several other substances. It is apparent, therefore, that the assessment of the magnitude of bias for a particular sample containing, as is common for many waters, many substances other than the determinand is not, in general, simple if it is made on the basis of the effects of each of the individual interfering substances. This difficulty is compounded if, again as is common for many waters, the concentrations of determinand and interfering substances in samples are subject to marked variations from sample to sample. There is, therefore, an important question of how best to estimate the overall bias caused by interference.

If a series of samples of accurately known, homogeneous and stable determinand concentrations were available corresponding to all sample compositions of interest, and if it were known that no bias arose through inability to measure all forms of the determinand, the above difficulty could, in principle, be overcome by replicate analysis of these special samples and estimation of the bias for each as described in Section 5.2. Such samples are, however, not generally available. Similarly, the approach described in Section 5.3 is not generally feasible. There is, therefore, only one practicable means of assessing the bias caused by interfering substances, *i.e.*, one must consider the effects of individual substances or combinations of substances; see also (*i*) below.

(c) Choice of Analytical Method
Given that the magnitude of the interference effects for the samples of interest is to be assessed from the effects of individual substances (or combinations of substances), and given also the difficulties noted above in this approach, two general and important points follow. These points are part of what is often considered 'good analytical practice', but this lessens neither their importance nor the value of re-stating them here.

(1)   There is very great advantage in using analytical systems subject to interference from as few as possible of the substances that may be present in samples. Equally, the smaller the size of a particular interference effect the better.

(2)   There is very great advantage in using analytical procedures whose principles and mechanisms are known sufficiently well that some theoretical assessment of the likely nature and magnitude of interference effects is possible. Analytical methods specifying such procedures are frequently likely to offer the further advantage that the size of interference effects varies little, if at all, with time; see also Section 10.4.3.

Of course, as always in analysis, the final choice of analytical method generally depends on many other factors which may outweight the above points. When, however, the primary aim is to control and minimise the bias of analytical results, the importance of (1) and (2) cannot be over-emphasised. It is also essential to remember that other components of an analytical system may affect the magnitude of interference effects; see (*h*) below and Section 10.

(d) Quantitative Estimation of Interference Effects

Suppose that it is wished to estimate the bias caused by a concentration, C_I, of a substance when the determinand concentration is C_S. The simplest approach is to analyse n portions of a solution containing only this substance and the determinand at the concentrations of interest, to calculate the mean analytical result, $\bar{R}$, and to obtain the estimate, B_1, of the bias from

$$B_1 = \bar{R} - C_S \qquad (5.48)$$

The n portions of this solution should, in general, be treated, the analytical responses measured, and the analytical results calculated by exactly the same procedures as those specified for samples in the analytical method under consideration. To ensure the required composition of the solution, the latter must generally be an accurately prepared standard solution.

The above approach is satisfactory when the calibration parameters [see Section 5.4.2(a)] are known with negligible inaccuracy, and when they do not vary appreciably over the several (possibly many) batches of analyses likely to be required in a thorough investigation of interference effects. When these conditions do not apply, a second approach to estimating interference effects is of very common application; see, for example, refs. 69 and 70. For this, n portions of a second solution containing only the determinand at a concentration C_S are also analysed in the same way and at the same time as the first solution mentioned above. If the mean analytical results for the solutions with and without the interfering substance are denoted by $\bar{R}_1$ and $\bar{R}_S$, respectively, the estimate of the bias, B_2, given by this approach is obtained from

$$B_2 = \bar{R}_1 - \bar{R}_S \qquad (5.49)$$

The second of these two approaches has the merit that the estimate, B_2, is derived from two sets of measurements, the difference between which should be caused only by the interfering substance. The inaccuracy of calibration parameters is, therefore, of less importance than when the first approach is adopted. In general, we prefer this second procedure, and do not further consider the first. We stress, however, that, given the conditions noted above, the first procedure is simpler and its estimate of bias, B_1, has a smaller random error than B_2. The points made subsequently can easily be adapted to this first approach if desired.

It should be noted that a number of more complex experimental designs have been used with a variety of more complex, statistical procedures for estimating the magnitudes of interference effects. Such approaches usually have the potential advantages that, for a given number of analyses, they may allow more precise estimation of: (1) the effect of one interfering substance on the bias caused by another, and/or (2) the effect of an interfering substance on the result for the determinand. Approaches such as these also commonly form the basis of a number of mathematical procedures for correcting for the error caused by interference; see (*j*) below. Examples of such procedures will be found in refs. 21 and 76 to 83. All these procedures, however, gain their advantages at the expense of the need to assume that certain rather simple, mathematical relationships validly describe the dependence of the magnitude of an interference effect on the concentrations of determinand and interfering substances. One should be careful, therefore, to check the validity of such assumptions, and note that any checking is likely to involve a substantial amount of effort. In addition, to apply these more complex procedures it is usually necessary to know the substances that cause appreciable interference. On the other hand, the simple approach using equation (5.49) makes no assumptions on the mathematical relationships involved, and generally provides unbiased estimates of interference effects. It seems to us, therefore, that this approach is ideally suited for an initial survey of such effects for a particular analytical system. The results obtained may provide all the information required or may be used as a basis for further, more complex experiments if these are considered desirable.

Equations (5.48) and (5.49) show that the interpretation of experimental estimates of bias must take into account the random errors of the experimental values, *e.g.*, of $\bar{R}_I$ and $\bar{R}_S$ in equation (5.49). For example, a non-zero value of B_2 does not necessarily mean that the true bias, β, does not equal zero. To decide that bias is present, therefore, we must show that the numerical value of B_2 is statistically significant at a chosen significance level. If it is, then we can claim to have detected bias, and the best estimate of the true bias is B_2. The statistical aspects of this are dealt with in Section 4.6.2(*b*).

An important consequence of our definition of interference is that it is impossible to show experimentally that $\beta = 0$ because the random errors of $\bar{R}_I$ and $\bar{R}_S$ can never be made zero in practice. Accordingly, it is not possible to claim, from results of experimental tests, that a given substance does not interfere. The most that can ever be said is that the bias caused by a particular interference effect does not exceed a stated value which is determined by the size of the random errors of $\bar{R}_I$ and $\bar{R}_S$.

Before considering further the estimation of interference effects, some warning on the interpretation of published information is desirable. Many analytical methods contain statements such as 'substance X does not interfere' or 'the concentrations of substance X normally present in samples do not cause appreciable interference'. Although such statements may at first sight appear to be very satisfactory, a moment's thought shows that they are of little or no value in deciding the magnitude of the bias expected for particular concentrations of the determinand and substance X. Analysts intent on the control and quantitative estimation of bias should, therefore, be very wary of such vague descriptions of interference. To allow proper, quantitative assessment, the analyst needs to know the observed effect of a stated concentration of a substance (or combination of substances) on the determination of a stated concentration of the determinand. Such data should, therefore, be quoted in published methods together with information on the random errors of the quoted results [see (*g*) below]. Alternatively, methods should refer to where all the relevant information can be found, but the former approach seems to us preferable as the method is then more 'self-contained'. The information required for this purpose is considered in the next section.

(e) Minimum Amount of Information Required on Interference Effects

What is the minimum amount of information required on interference effects in order that an analyst may adequately assess the bias to which his samples will be subject from this source? This important question applies equally to analysts concerned with providing this information, *i.e.*, with estimating the magnitudes of interference effects for a particular analytical system. Given the potential complexity of such effects, it seems to us that there is, unfortunately, no definite answer that can be given that is both universally valid and also practicable. To some extent, each situation needs individual consideration, and the answer to the question may be markedly dependent on the analytical system of interest, on existing knowledge of its tendency to suffer from interference, and on the analyst's knowledge of the principles and mechanisms involved in the analytical procedure and the natures of the samples to be analysed. In this situation, the best that we feel able to do is to offer the following suggestions as a basis for deciding the minimum amount of information required on interference effects.

(1) Since interference effects may depend on the concentration of the determinand, the effect of any substance chosen for test should be estimated for at least two concentrations of the determinand. The lower and upper limits of the concentration range of interest are usually best if only two determinand concentrations are tested. If the results from these tests indicate that the effects of a substance differ appreciably at these two determinand concentrations, further tests at intermediate concentrations will usually be necessary adequately to define the quantitative nature of this particular interference.

(2) Any substance that may occur in samples at the same concentration as, or greater concentration than, the determinand should be considered initially as a potential interfering substance. All information available on the analytical system/method is then reviewed to decide if the effects of any of these substances can be taken to be negligible, and such substances are deleted from the list of substances whose effects are to be estimated. Of course, it is realised that substances present in smaller concentrations than the determinand may sometimes cause important interference. It is essentially impossible, however, to check the effects of all substances that may occur in samples such as natural waters, and some means of restricting the number of substances to be tested is necessary. Hopefully, the review of the literature and consideration of the basis of the analytical procedure will also reveal any such strongly interfering substances which may then be included among those to be tested; see also (*i*) below.

(3) The effects of each of the substances chosen in (2) should be estimated experimentally, each substance being tested separately when its concentration is slightly greater than the maximum expected in samples. Substances found to cause appreciable bias should then be tested at other concentrations in order quantitatively to define the nature of the particular interference.

The organisation and execution of these tests is considered in (*f*) below, but some difficulties caused by points (1) to (3) are first considered. First, the effects of greater concentrations of the substances tested are not determined so that the validity of extending the use of a method to other types of sample may not be known. This is not, however, a crucial disadvantage because it can, in principle, be easily remedied at any time simply by making additional tests. A second problem is that the suggested tests may fail to identify substances whose effects are greater when they are present at a smaller concentration than that used in the tests. This possibility probably arises only for substances that actually cause two different effects – one giving positive bias and the other negative; an example is the effect of phosphate in the determination of 'reactive' silicon.[74] Knowledge of the physical and chemical mechanisms of interference should help to decide if additional tests at smaller concentrations of a substance are necessary. Of course, there is no problem, in principle, in testing each substance at a number of concentrations, and this is clearly a safer approach; the consequences in terms of the effort needed for the tests, however, will usually be alarming. The final difficulty to be mentioned here is that the suggested tests take no account of the possibility that the effect of one interfering substance may depend on the concentration of another substance. Again, detailed knowledge of the processes involved in the analytical procedure will help to decide when such interactions are likely, and thus suggest combinations of substances for additional tests; see also (*d*) above. It is also often useful to test the effects of a few combinations of at least the major components of samples so that typical compositions of samples are simulated.

(f) Organisation and Execution of Interference Tests
The suggestions in (*e*) require that the concentrations of the determinand and other substances can be chosen at will, *i.e.*, by the preparation of appropriate standard solutions, and this is usually possible in the field of water analysis. In this approach, it is clearly essential that the substances whose effects are to be estimated contain negligible amounts of the determinand; otherwise, a correction for their determinand content is necessary.

As noted in (*d*) above, the ability to detect, and the accuracy of estimation of, interference effects improve as the random errors of measurements decrease. Further, uncertainties arising in the use of estimates of interference effects – see point (5) in (*h*) below – decrease as the accuracy of the estimates improves. Two means of reducing random errors can be employed.

First, since results are generally subject to within- and between-batch random errors [see Section 6.2.1(*d*)], the best experimental design is to estimate interference effects from tests within one batch of analyses.* A typical experimental design for one batch of analyses would, therefore, be as indicated in Table 5.1. Other batches

TABLE 5.1
Example of Experimental Design for One Batch of Interference Tests

Other substance	*Concentration of* other substance	Concentration of determinand	
		C_0	C_1
– (*j* = 0)	–	S_{00}	S_{10}
X (*j* = 1)	C_X	S_{01}	S_{11}
Y (*j* = 2)	C_Y	S_{02}	S_{12}
Z (*j* = 3)	C_Z	S_{03}	S_{13}
W (*j* = 4)	C_W	S_{04}	S_{14}

In this Table, S_{ij} denotes a standard solution with a determinand concentration C_i ($i = 0$ or 1) and the *j* th other substance present at a defined concentration; $j = 0$ corresponds to no added other substance.
Reprinted with permission from *Talanta*, **21**, A.L. Wilson, Performance Character-istics of Analytical Methods – IV, 1974, Pergamon Press Ltd.

* This assumes that interference effects do not vary with time; if they do, other considerations apply.

of similar design would be analysed until all the substances at all concentrations and combinations of interest had been tested.

The second means of reducing the effects of random errors is to make more than one determination for each solution, S_{ij}. Of course, the greater the number of replicates for each solution, the smaller will be the random errors of the mean results [Section 4.3.2(c)]. If it is required to be able to detect a statistically significant effect of a particular substance if its true effect is equal to or greater than a given value, β_0, and if there is existing knowledge of the within-batch standard deviation, σ_w, of analytical results, then an estimate of the minimum number of replicates can be derived [see (h), point (5) below and Section 4.6.2(b)]. Often, however, and especially during initial interference tests, it may simply be decided to make a compromise between the better precision provided by replication and the effort required for such replication. As a general suggestion, we feel it wise always to make at least duplicate analyses for each solution; this gives some reduction of the effect of random errors, and also allows an estimate of the standard deviation of B_2 to be derived from the interference tests themselves [see (g) below].

These suggestions on replication apply to the estimation of bias caused by other substances. It is possible that the latter may also affect the precision with which the determinand is measured. Replicate measurements for each solution allow investigation of this possibility since an estimate, s_{ij}, of the standard deviation for each solution can then be obtained, and each value, s_{0j} or s_{1j} ($j \neq 0$), can be compared with s_{00} or s_{10}, respectively [see Section 4.3.4(e)]. For this purpose, however, duplicate measurements are essentially worthless because of the very large uncertainties of estimates of standard deviations with only one degree of freedom [see Section 4.3.4(d)]. If, for example, it was wished to have a probability of 0.95 that, for each substance tested, s_{0j} would be significantly greater (0·05 significance level) than s_{00} when $\sigma_{0j} = 3\sigma_{00}$, then approximately 16 portions of each solution should be analysed.[31] The number of analyses involved in such testing when many substances are of interest is, therefore, very large, and it will usually be impracticable. Our experience, however, is that it is very commonly valid to assume that precision is not affected by the other substances tested. Thus, the simpler experimental designs (considered in the previous paragraph) for estimating bias should usually suffice. We assume hereafter that the other substances do not affect precision, and the statistical procedures described are on this basis. If the assumption is known or suspected not to be valid, we suggest statistical advice be sought.

(g) Interpretation and Reporting of Results
The estimate, B_{ij}, of the effect of the j th substance ($j \neq 0$) at the i th concentration of the determinand is given by

$$B_{ij} = \bar{R}_{ij} - \bar{R}_{i0} \qquad (5.50)$$

where $\bar{R}_{ij}$ is the mean analytical result for the solution, S_{ij}. When B_{ij} is calculated from equation (5.50), blank-correction of the results is not needed.

The B_{ij} values are the primary experimental estimates of the bias caused by the other substances, and it is, therefore, strongly urged that the individual values of B_{ij} be reported. These values should be accompanied by their confidence limits at a defined confidence level (see Section 4.3.5) for each value of i so that the significance and practical importance of the results may be assessed independently by any analyst. The calculation of these confidence limits is described below, and Table 5.2 shows a suitable format for presenting the results.

The method of presenting the results shown in Table 5.2 allows rapid examination to identify those substances causing statistically significant effects, *i.e.*, those for which $B_{0j} \pm L_0$ and/or $B_{1j} \pm L_1$ do not include zero. The biases observed for such solutions are directly recorded, may easily be converted into relative biases if desired, and may also be easily assessed at other confidence levels if required. Further, the Table clearly shows the results for any substances whose apparent effects have not achieved statistical significance. Further transformations of these primary results may also be of interest, *e.g.*, in the calculation of selectivity or non-specificity coefficients [see (*j*) below], but we recommend that such transformed results should be given as well as, rather than instead of, the primary results.[84]

The calculation of L_0 and L_1 is made using the procedures in Sections 4.3.4 and 4.3.5, and may be summarised as follows, assuming that precision is not affected by the other substances, that m analyses are made for each solution, that n other substances are tested, and that the means for each solution are distributed normally.

(1) Estimate the variance, s_{ij}^2, for each solution from its m results using equation (4.12) or (4.13).
(2) Combine all estimates of variance with a given value of i for each value of i to obtain pooled estimates of variance, s_i^2, using equation (4.15). Note that when $m = 2$, stages (1) and (2) may be conveniently combined by using equation (4.16). The degrees of freedom of s_i are $(n+1)(m-1)$.
(3) The values of L_i are given by

$$L_i \approx t_\alpha \sqrt{2} . s_i / \sqrt{m} \qquad (5.51)$$

where t_α is the tabulated value of the *t*-statistic at the α probability level (see Section 4.3.5), *e.g.*, $t_\alpha = 2.45$ for $\alpha = 0.05$ and $n = 5, m = 2$. Note that in this

TABLE 5.2

General Format for Presenting the Results of Interference Tests

Substance	Concentration of substance	Estimated effect* (concentration units) for determinand concentrations of	
		C_0	C_1
X	C_X	B_{01}	B_{11}
Y	C_Y	B_{02}	B_{12}
Z	C_Z	B_{03}	B_{13}
W	C_W	B_{04}	B_{14}

* The results in the table have $100(1-\alpha)\%$ confidence limits equal to the result $\pm L_0$ and the result $\pm L_1$ for the determinand concentrations C_0 and C_1, respectively.

Reprinted with permission from *Talanta*, **21**, A.L. Wilson, Performance Characteristics of Analytical Methods – IV, 1974, Pergamon Press Ltd.

approach the same value, s_i, is used in forming the confidence limits for n different means, and the use of t_α, therefore, involves an approximation [see Section 4.3.5(b)]. When $m \geqslant 2$ and $n = 20$ to 30, the degree of approximation is rather small (see Appendix to Section 4).

An example of such a tabulation of interference tests, extracted from a published analytical method,[85] is shown in Table 5.3. This approach has been adopted within the U.K. by the Standing Committee of Analysts, and the methods published by that body furnish many other examples of such tables.

(h) Assessing the Overall Bias Caused by Interference for a Particular Sample
After the above discussion, we can return to the central question with which we began this section, namely, when considering the suitability of an analytical method, how does an analyst assess the bias caused by interference expected to arise for his samples. We saw in (b) above that the only generally practicable approach required knowledge of the magnitudes of the effects of individual substances (or combinations of substances), and procedures for obtaining such information have been described in (c) to (g) above. It follows immediately that without information like that shown in Tables 5.2 and 5.3 no quantitative assessment of bias can be made. Without such information, therefore, there are only two possible approaches that an analyst can take: (1) the analytical method under consideration is abandoned in favour of one for which such information is available, or (2) the analyst produces the necessary information by carrying out the appropriate interference tests. In our experience, this second alternative is normally impracticable for routine water-analysis laboratories, and this, in turn, places great responsibility on all organisations and authors concerned with the development, standardisation, publication, *etc.* of analytical procedures for routine use. It seems to us essential that vigorous attempts be made by all such bodies and persons to

TABLE 5.3
Extract from Table X of Analyst, 1974, **99,** *426*

Other substance	Concentration of other substance/ mg 1^{-1}	Effect* of other substance, mg 1^{-1} of Al, at an aluminium concentration of	
		0.000 mg 1^{-1}	0.300 mg 1^{-1}
Ca^{2+}	500	$+0.006$	$+0.003$
Ca^{2+}	100	$+0.003$	$+0.004$
Mg^{2+}	100	$+0.004$	$+0.006$
Na^+	100	$+0.002$	$+0.004$
PO_4^{3-}	8.2 (as P)	-0.005	-0.050
PO_4^{3-}	4.1 (as P)	-0.005	-0.038
PO_4^{3-}	1.7 (as P)	0.000	-0.002
Fe^{3+}	1	$+0.011$	$+0.014$

* If the other substances had no effect, results would be expected to lie (95% confidence limits) within the following ranges:

0.000 ± 0.003 for 0.000 mg 1^{-1} of Al
0.000 ± 0.006 for 0.300 mg 1^{-1} of Al

(Note that this footnote to the table is equivalent to that given in the examples of Table 5.2.)

ensure that every possible effort is devoted to this quantitative characterisation of interference effects because this is one of the major lacks of current analytical methodology for waters – and, indeed, for many other types of sample.

Given the above analysis of the situation, we feel it worth repeating that if, for a given determinand, no method is available with the quantitative information suggested in (e) to (g) above, then in our view quantitative assessment of the bias caused by interfering substances is, in general, impossible. In this position, the analyst has, of course, recourse to his experience and intuition, but these seem to us generally insufficient for the confident control of bias, especially for the increasingly complex analytical systems being applied to the analysis of waters.

Having considered the situation where quantitative information on interference effects is lacking, we can turn now to the situation where the analyst has been provided with such data – in the analytical method or in some associated publication. There are still important problems in assessing the bias expected for particular samples, but clearly the analyst is in a much stronger position than in the absence of such data. In essence, the approach suggested is as follows. From the information on interference effects the analyst assesses the bias expected from each of the substances (for which there is information) at the concentration(s) expected in his samples. The overall bias for a particular sample is then estimated by summing all these individual biases with due regard for their signs. It may suffice initially to do this only for the sample most affected by interference. This approach, though simple in principle, meets several important problems that are considered in (1) to (3) below.

(1) Substances other than those for which information is available may cause interference. If the analyst suspects any such effects, he could make the appropriate tests to estimate the effects. This may be quite practicable because he should not now have to test a large number of different substances. Otherwise, it is suggested that the analyst proceed to calculate the overall bias as described above on the assumption that he has included all interfering substances whose effects are appreciable. This assumption may prove incorrect, and the analyst should be alive to this possibility, but it represents the only possible way of proceeding, and is, of course, at the root of all scientific work.

(2) There may be interactions involving various substances such that their overall interference effect is not the sum of their individual effects [see (b) above]. The approach suggested here is exactly the same as in (1). That is, suspected interactions should be experimentally assessed if they are not included in the given data on interference; otherwise, the analyst should proceed as though there were no such interactions.

(3) The given information on interference, though obtained using the analytical method being considered by the analyst, may have been derived from an analytical system different from that available to the analyst. For example, the analyst may have a different measuring instrument to that used to obtain the data on interference, and this may affect the magnitude of interference effects. This difficulty has been reported when the analytical procedure involves the use of an atomic absorption spectrometer since the size of interference effects

may depend on which of the many commercially available instruments is used (see Section 11.3.5). This type of difficulty is probably mainly centred around the type of instrumentation, and can occur for techniques other than atomic absorption (see Section 10.4.3). Other components of an analytical system, however, may also be involved. For example, the purity of a chromogenic reagent may affect the interference caused by some substances when these can react with impurities in the reagent and the products of such reactions affect the analytical response.[87]

It seems that there has been little systematic investigation of this general problem whose importance is growing as laboratories adopt more complex and sophisticated instrumentation for routine analysis. We know of no sure way of theoretically assessing the extent to which users of one type of instrument can make valid use of data on interference obtained when another type of instrument was employed. In the more classical types of analytical technique such as gravimetry and titrimetry, the problem is much less pronounced (if it exists at all) because the chemical reaction(s) forming the basis of a procedure are not generally affected by the apparatus in which they are carried out, and the measuring equipment (burettes, pipettes, balances) is essentially equivalent in all laboratories even when it has been obtained from different manufacturers. This is true also to a large extent for many spectrophotometric procedures because the absorption bands of the chromophores commonly involved in the analysis of waters tend to be much broader than the spectral band-widths of the monochromators used in the available instruments. The same problem may, therefore, arise if an analyst with a filter-absorptiometer were attempting to use data on interference effects obtained with a spectrophotometer.

For analysts faced with this problem, detailed investigations are clearly of value, but this raises the difficulty of finding the effort required for such work. It is suggested that, as a minimum, one or other of two approaches may be useful. The first approach is to undertake recovery tests [see (*i*) below] on a representative range of sample types. This would be of greatest value for analytical systems showing complex interactions among the effects of interfering substances with the result that the overall bias is approximately directly proportional to the concentration of the determinand. Electrothermal atomisation in atomic absorption spectrometry is an example of a technique where this approach may be useful. Recovery tests, however, cannot detect interference effects whose size is independent of the determinand concentration. When such effects are the main components of the overall bias, the second approach becomes relevant. This consists simply in tests such as those described in (*c*) to (*g*) above, but restricting them to only those substances reported to cause important effects.

In passing, it is worth noting that this problem would very profitably be investigated in collaborative studies involving a number of laboratories.

(4)   The given information on interference may show that a given substance causes bias, but that the concentration of this substance and/or that of the determinand differ from those expected in samples. The bias expected for the sample concentrations may sometimes be interpolated from the given

information, *e.g.*, when the data show that the effect of a substance is directly proportional to its concentration. Alternatively, the analyst may make appropriate tests to estimate the size of the particular effect in which he is interested.

Of course, this problem does not arise if it has previously been shown that a given mathematical relationship validly describes the dependence of the size of an interference effect on the concentrations of the determinand and interfering substance; see (*j*) below.

(5) When the given data on interference show that the observed effect of a substance was found not to be statistically significant, there is a question of the value to be assigned to this particular bias when calculating the overall bias for a given sample. It must be remembered [see Section 4.6.2(*b*)] that the true value of a non-significant effect is not necessarily zero. To deal with this problem, it is suggested that the value to be assigned to a non-significant effect is the experimental value for the effect since this is, in general, the best estimate available of its magnitude. For example, suppose that the effect of a substance were reported as '0.13 ± 0.20 mg 1^{-1}', and were, therefore, not statistically significant; the suggestion above corresponds to assigning the value 0.13 mg 1^{-1} to the effect for purposes of computing the overall bias. Similarly, a reported effect of -0.07 ± 0.20 mg 1^{-1} would be assigned the value -0.07 mg 1^{-1}, and the bias caused by these two effects together would be equated to 0.06 mg 1^{-1}. Note that, if the true values of interference effects for a given analytical system are all smaller than the precision of results, then tests of interference should give a series of non-significant results for these effects. Choice of a method giving such data on interference is beneficial because the overall bias is then likely to be close to zero since it will be the sum of a number of small positive and negative components.

For a given analytical system, the uncertainty in the value assigned to an interference effect can be decreased only by making more replicate analyses on each solution tested. From the results in Section 4.6.2(*b*)(*i*), the smallest true bias, β_0, for which there is a probability of 0.95 that a statistically significant (0.05 significance level) effect will be observed is equal to $5.1\sigma/\sqrt{m}$, where σ is the within-batch standard deviation of individual analytical results at the appropriate determinand concentration, and m is the number of replicate analyses on each solution. Thus, the smallest value for m is approximately 26 $(\sigma/\beta_0)^2$ if one wishes to detect (with 0.95 probability) interference effects as small as β_0. For example, suppose that one had a good estimate of the numerical value of σ, and had also decided that one required to detect interference effects as small as σ, *i.e.*, $\beta_0 = \sigma$; m would then have to be not less than 26. Such considerations show the enormous amount of effort required for testing in this detail. If, for example, the effects of one concentration of each of twenty substances were to be checked at each of two determinand concentrations, and β_0 were required not to exceed 2σ, a total of about 560 analyses would be required. Clearly, there is great value in achieving as small a value as possible for σ when bias is to be estimated. These points also indicate the value of collaborative inter-laboratory investigation of interference effects in spreading the large amount of effort generally involved, but see point (3) above for an important proviso to this.

(i) Use of Recovery Tests for Detecting Interference Effects

The approach described above seems to us essential for any complete assessment of interference effects, but such an approach requires more effort than routine analytical laboratories commonly have available. A useful and less demanding means of checking the overall bias due to certain types of interference effect is the well known and frequently used *recovery test*. The basis of this test is to analyse at essentially the same time two portions of a sample to one of which has been added a known amount of the determinand (by addition, for example, of a suitable standard solution). Analysis of both portions is then made by the usual analytical procedure, and the difference between the two results gives the recovery of the added determinand. This test is useful, particularly for detecting unsuspected sources of interference, and is always worth carrying out, but it also suffers from some important disadvantages considered below.

Let δ = the amount of determinand added to the second portion of the sample, expressed in units of concentration in the sample,

R_S and R_A = the analytical results for the first and second portions of the sample, respectively,

and f = percentage recovery of the added determinand.

Then
$$f = 100(R_A - R_S)/\delta \tag{5.52}$$

(Note that, in general, a correction to R_A is required in order to allow for any effect of the dilution of the second portion of the sample by the volume of standard solution added. It is assumed here that R_A has been so corrected.)

Assuming for the moment that random errors are absent, and letting the bias caused by interference be β_S and β_A for R_S and R_A, respectively, we can write $R_S = C_S + \beta_S$ and $R_A = C_A + \beta_A$, where C_S and C_A are the true concentrations of the determinand. Substituting these values in equation (5.52) gives

$$f = 100\{(C_A - C_S) + (\beta_A - \beta_S)\}/\delta$$

which may be written as

$$f = 100 + 100(\beta_A - \beta_S)/\delta \tag{5.53}$$

since $C_A - C_S = \delta$. Equation (5.53) shows a major disadvantage of the recovery test, *i.e.*, it cannot detect interference whose magnitude is independent of the determinand concentration because then $\beta_S = \beta_A$ so that the expected value of f is 100 %. The equation also shows that interference can, in principle, be detected whenever $\beta_S \neq \beta_A$, but the observed recovery allows the individual biases, β_S and β_A, to be estimated only when it is known that both are a fixed proportion of the true determinand concentration. For example, if it were known that $\beta = \gamma C$, then the result, f, provides an estimate, g, of γ given by

$$g = (f - 100)/100 \tag{5.54}$$

The effect of the random errors of R_S and R_A can be assessed by applying the results of Section 4.3.2(*b*) to calculate the standard deviation of *f*. Thus, from equation (5.52) we obtain

$$\sigma_f = 100(\sigma_S^2 + \sigma_A^2)^{1/2}/\delta \qquad (5.55)$$

where σ_f, σ_S and σ_A are the standard deviations of *f*, R_S and R_A, respectively. Note that equation (5.55) assumes that the random error of the calibration curve used to obtain R_S and R_A from the corresponding analytical responses is negligibly small. Equation (5.55) is useful in that it allows the calculation of the variability expected for *f* in the absence of any interference effects. For example, suppose that $\sigma_S = \sigma_A = \sigma$, that $\delta/\sigma = 10$, and that analytical results are normally distributed; the value of σ_f is then 14 %. Thus, 95 % of the observed recoveries would lie within the relatively broad range, 72 to 128 % [see Section 4.3.3(*b*)]. That is, results within that range would not justify (at the 95 % confidence level) the conclusion that interference had been detected. It can be seen from equation (5.55) that this range is reduced as the ratio $(\sigma_S^2 + \sigma_A^2)^{1/2}/\delta$ is decreased. This ratio, however, cannot be reduced indefinitely because of the need to ensure that $C_A (= C_S + \delta)$ remains within the concentration range of the analytical system. In practice, this may often mean that one must accept the recovery test as a relatively insensitive means of detecting interference; Provost and Elder[102] have recently made similar points. A further problem in the interpretation of observed recoveries arises whenever $\sigma_S \neq \sigma_A$; see Section 4.6.2(*b*)(*ii*).

(j) Reduction of Bias by Corrrection Procedures

When, for a given analytical system, the interference effect of one or more substances is larger than can be tolerated, there are only two approaches to overcoming the problem experimentally. The first is to adopt another analytical system; this needs no further consideration since the discussion above applies. The second is to make a correction to reduce the bias caused by interference, and this approach is considered below. We strongly agree, however, with the view of Liteanu and Rica[3] that "It is best to try to eliminate systematic errors; applying corrections for them should be a last resort."

Clearly, any correction procedure requires, in addition to measurement of the analytical response for the determinand, knowledge of the concentrations of interfering substances. In the analysis of waters, this usually implies the need to determine these concentrations. Correction procedures may, therefore, have the important disadvantage of requiring additional analyses. Four approaches to correction are distinguished below, though all have the same basis, *i.e.*, the bias caused by the known concentration of an interfering substance is subtracted from the analytical result for the sample. In practice, the correction is usually made by appropriate correction of the calibration curve or function. With the exception of simple standard-addition procedures, none of the other three approaches appears to find common application in current water analysis. Our discussion, therefore, is brief, but with the references cited should help readers faced with the need to consider correction of interference effects.

(i) Simple standard-addition procedures These have been discussed in Sections 5.4.2 and 5.4.3. They are useful when the overall effect of interfering substances is

simply to change the slope of a linear calibration curve. For calibration functions of the form $\mathscr{A}_T = \lambda + \chi_S C_S$ [see Section 5.4.2(c)], unbiased results are obtained when $\lambda = 0$, but appropriate blank correction may sometimes allow unbiased results to be derived when $\lambda \neq 0$.

If, in addition to an overall effect on the slope of the calibration curve, there is also an overall interference effect, β_S, independent of the determinand concentration, then equation (5.16) becomes

$$\mathscr{A}_{Ti} = \eta + \chi_S \beta_S + \chi_S (C_S + C_C + C_i) \qquad (5.56)$$

By exactly the same procedure as that in Section 5.4.2(c) we can then derived the analytical result, R, as

$$R = C_S + C_C + \beta_S + \eta / \chi_S \qquad (5.57)$$

The bias, β_S, is not, therefore, eliminated by this procedure. As noted in Section 5.4.2(c), if it is possible to remove the determinand completely from a portion of the sample without affecting any other factor that affects the analytical response for the sample, then the bias in R can be eliminated.

(ii) Use of appropriate calibration curves When the analytical method specifies calibration by means of standard solutions, addition of all important interfering substances (at the concentrations present in samples) to the calibration standards should eliminate the bias caused by these substances. This assumes, of course, that the interference effects produced by given concentrations of these substances in the sample are the same as those from the forms of the substances added to standards; this assumption is by no means necessarily correct (see Section 5.4.5). This approach clearly becomes cumbersome when the concentrations of interfering substances vary markedly from sample to sample. The procedure has been recommended[87] for overcoming the effects of potassium and calcium on the determination of sodium in waters by flame emission spectrometry. Automatic, computer-controlled systems for this approach have also been used when only one or two interfering substances are of importance.[88,89]

When calibration parameters and the magnitudes of interference effects do not vary appreciably with time, another equivalent approach can be used, usually when only one interfering substance is important. This consists in the initial preparation of a family of calibration curves, each curve corresponding to a different concentration of the interfering substance. Analytical responses for samples are then evaluated from the appropriate calibration curve. This approach has been recommended[87,90] for overcoming the effect of fluoride on the spectrophotometric determination of aluminium in waters.

Such approaches make no assumptions on the mathematical nature of the relationship between the interference effect and the concentrations of determinand and interfering substance Their accuracy clearly depends on the accuracy with which the concentrations of interfering substances in samples and calibration standards can be matched. Thus, the accuracy with which the concentrations of interfering substances are determined should be considered and suitably controlled in the light of the magnitudes of the interference effects and the required accuracy of analytical results for the determinand.

(iii) Use of mathematical correction procedures The basis of this approach consists in expressing the analytical response for a sample as a function of the concentrations of the determinand and interfering substances. For example, when the calibration function in the absence of interference is of the form $\mathscr{A}_T = \lambda + \chi C_s$ (see Section 5.4.2), one expression that has been used is of the form[76]

$$\mathscr{A}_T = \lambda + \chi C_S + \Sigma \chi_q C_q + \Sigma \chi_{Sq} C_S C_q \qquad (5.58)$$

where the suffix S refers to the determinand and the suffix q (q = 1 to i) refers to i interfering substances. The terms $\chi_q C_q$ represent interference effects whose magnitudes are independent of the concentration of the determinand, C_S, while the terms $\chi_{Sq} C_S C_q$ represent effects whose sizes are a constant proportion of C_S for a given C_q. To suit the nature of particular effects, some of the terms of equation (5.58) may, of course, be deleted and/or new ones added. For example, in inductively coupled plasma spectrometry, additional terms of the form $\chi_{q'} C_q^2$ have been used,[82] and refs. 91 and 92 discuss the use of other non-linear functions of C_q. Given numerical values for λ, χ, C_q, the various coefficients defining the interference effects, and the observed value of A_T, for a sample, the analytical result for the sample can be calculated from equation (5.58). The coefficients, χ_q, are often referred to as selectivity coefficients (sometimes non-specificity coefficients). Many publications describe applications of this type of procedure; see, for example, refs. 3, 21, 30, 76 to 83 and 91 to 94.

In considering the possible applications of this approach, several points are worth noting.

(1) Much initial work may well be involved in establishing the validity of the equations used to represent interference effects, and in deriving estimates of the numerical values of the appropriate selectivity coefficients. If simple equations such as (5.58) are inadequate for describing interference effects, as may be the case, the approach becomes very cumbersome if not impossible.

(2) Use of equations such as (5.58) usually implies the assumption that the values of the selectivity coefficients do not change with time because of the points made in (1).

(3) Estimations of selectivity coefficients obtained in one laboratory may be inapplicable in other laboratories [see *(h)* above].

(4) The estimates of selectivity coefficients are subject to errors which will lead to errors in analytical results for samples.

Given the above points, considerable care seems to us essential in considering the possibility of adopting this approach.

(iv) Generalized Standard Addition Method This is a recently developed procedure[61,64-68] representing an amalgam of the simple standard-addition procedure and the procedure using mathematical corrections in order to correct for interferences which cause a change in slope, χ, of the calibration curve and/or which cause an effect independent of the determinand concentration. For a determinand and i interfering substances, the procedure requires* that the analytical responses

---

* For present purposes our account of the procedure is deliberately simplified; the references cited should be consulted for full details.

for the determinand and each interfering substance be measured before and after $i+1$ standard additions are made to the sample. The equations used to represent the analytical responses for each of the $i+1$ substances measured are of the form

$$\mathscr{A}_{qs} = \Sigma \chi_q C_{qs} \qquad (5.59)$$

where $\mathscr{A}_{qs}$ represents the analytical response for the q th substance ($q = 1$ to $i+1$) after the s th standard addition ($s = 1$ to $i+1$), χ_q is the slope of the relationship (assumed to be linear) between $\mathscr{A}$ and C for the q th substance, and C_{qs} is the concentration of the q th substance after the s th standard addition. In effect, equation (5.59) represents a set of simultaneous equations which, given the above conditions, can be solved for χ_q, and this then allows calculation of the original concentrations of each of the substances in the sample. Matrix algebra and a computer are required for these calculations.

This approach overcomes problems due to non-constancy of the numerical values of selectivity coefficients and of their variation from sample to sample, but at the expense of greater experimental and computational complexity. Problems (1) and (4) noted in the previous section also apply. In addition, the procedure appears to ignore the problem of blank correction in standard-addition procedures [see Section 5.4.2(c)] by assuming that the analytical response to solutions containing none of the determinand is always zero.

The full description of the original theory and more recent extensions of it are given in refs. 61 and 64 to 67. A still more powerful version of the procedure has recently been developed[68] which, among other useful features, is said to allow on-line computation of the validity of equation (5.59).

5.5 ASSESSMENT OF THE OVERALL BIAS OF ANALYTICAL RESULTS

In Section 5.4 we have considered a number of independent sources of bias in analytical results. From the viewpoint of the use to be made of the results, it is the total bias due to all these individual sources that is important (see Section 4.6.1). Thus, we need now to consider how this total bias can be assessed. Assuming that the approaches of Sections 5.2 and 5.3 are not feasible, this assessment can be made only through consideration and summation of the individual biases noted in Section 5.4. This approach, however, meets immediate problems because, as we saw in Section 5.4, it may be difficult or impracticable to obtain even reasonably precise estimates of the magnitudes of some of these biases. In this position, it is clearly advantageous to make as many of the sources of bias as possible of negligible size; this will not only tend to decrease the overall bias but will also simplify its assessment and decrease its uncertainty. A particular set of suggestions is made below based on the sources of bias described in Sections 5.4.3 to 5.4.6. Much of what follows is no more than a restatement of what may be considered 'good laboratory practice'. We believe, however, that this restatement is well worthwhile given the many and various developments that have so markedly changed many aspects of water analysis in recent years. The suggestions are as follows.

(1) *Bias from the calibration procedure.* The aim should be to eliminate all

    appreciable bias from this source. The various points to be considered are noted in Sections 5.4.2 and 5.4.3.

(2) *Bias from blank determinations.* As for (1), the aim should be to eliminate all appreciable bias; see Sections 5.4.2 and 5.4.4.

(3) *Bias caused by the forms of the determinand.* This presents quite different problems to the first two factors. In general, if it were known that a procedure did not determine all forms of interest of the determinand, we assume that the method would not be used. The difficulty, however, arises because one will often not have conclusive evidence on this aspect of the performance of an analytical system (see Section 5.4.5). Therefore, there seems to us only one possible course to follow, *i.e.*, to act as if the magnitude of this bias were zero until it is proved otherwise. As soon as any bias from this source is shown to be appreciable, a procedure better suited to the requirement should, in general, be adopted or devised. Of course, this approach will sometimes lead to incorrect decisions, *e.g.*, when bias, though present, has not been detected, but this difficulty is characteristic of the whole of 'scientific method'.

(4) *Bias from interference effects.* Given that the aims in (1) and (2) are achieved and that the approach in (3) is adopted, it follows that interference is the only source of bias that needs consideration in assessing the overall bias expected for analytical results. Section 5.4.6(*h*) suggests an approach for assessing the total bias caused by the various interfering substances that may be present in samples.

    The above suggestions emphasise both the vital role played by sources (3) and (4) in estimating the total bias of analytical results, and also the urgent need to fill the many gaps in knowledge of them, at least for the more commonly used analytical methods for the analysis of waters. The suggestions also imply the need for analysts to be continually seeking ways of demonstrating the validity (or not) of their estimates of total bias. Tests in individual laboratories – see, for example, Sections 5.4.5, 5.4.6, 6.3 and 6.4 – are of value; participation in inter-laboratory collaborative tests (see Section 7) may also reveal unsuspected sources of bias, and is, therefore, usually worthwhile.

    Finally, it should be noted that any bias stemming from the sampling process (see Sections 3.5 and 3.6) may, if desired be added to that arising in analysis. In general, however, we feel that the approach to bias from sampling procedures should be that suggested under (3) above; see also Section 3.5.1(*b*).

5.6 REFERENCES

[1] L. A. Currie, *Anal. Lett.*, 1980, **13**, 1.
[2] L. A. Currie, in I. M. Kolthoff and P. J. Elving, ed., 'Treatise on Analytical Chemistry', Second Edition, Part 1, Volume 1, Wiley, Chichester, 1978, pp. 95–242.
[3] C. Liteanu and I. Rîcă, 'Statistical Theory and Methodology of Trace Analysis', Ellis Horwood, Chichester, 1980.
[4] J. Büttner *et al.*, *J. Clin. Chem. Clin. Biochem.*, 1980, **18**, 69.
[5] R. Alvarez, S. D. Rasberry, and G. A. Uriano, *Anal. Chem.*, 1982, **54**, 1226A.
[6] J. K. Taylor, *Anal. Chem.*, 1981, **53**, 1588a.
[7] G. A. Uriano and J. P. Cali, in J. R. De Voe, ed., 'Validation of the Measurement Process', American Chemical Society, Washington, 1977, pp. 140–161.
[8] O. G. Koch, *Pure Appl. Chem.*, 1978, **50**, 1531.
[9] N. W. Tietz, *Clin. Chem. (Winston-Salem, NC)*, 1979, **25**, 833.

224 *Chemical Analysis of Water*

[10] T. J. Dols and B. H. Armbrecht, *J. Assoc. Off. Anal. Chem.*, 1977, **60**, 940.
[11] G. A. Uriano and C. C. Gravatt, *Crit. Rev. Anal. Chem.*, 1977, **4**, 361.
[12] D. G. Ballinger, *Environ. Sci. Technol.*, 1979, **13**, 1362.
[13] A. L. Wilson, *Talanta*, 1974, **21**, 1109.
[14] M. Thompson, *Analyst*, 1982, **107**, 1169.
[15] D. F. Davidson and J. Williamson, *Lab. Pract.*, 1982, **31**, 992.
[16] R. L. Grob, ed., 'Modern Practice of Gas Chromatography', Wiley-Interscience, London, 1977, p. 39.
[17] H. Kaiser, *Spectrochim. Acta, Part B*, 1978, **33**, 551.
[18] A. Shatkay, *Anal. Chem.*, 1968, **40**, 2097.
[19] K. Cammann, *Fresenius' Z. Anal. Chem.*, 1982, **312**, 515.
[20] I. L. Larsen, N. A. Hartmann, and J. J. Wagner, *Anal. Chem.*, 1973, **45**, 1511.
[21] A. Parczewski and A. Rokosz, *Chem. Anal. (Warsaw)*, 1978, **23**, 225.
[22] R. Guevremont, *Anal. Chem.*, 1981, **53**, 911.
[23] M. A. Müller and H. Oelschläger, *Fresenius' z. Anal. Chem.*, 1981, **307**, 109.
[24] D. Midgley, *Anal. Chem.*, 1977, **49**, 1211.
[25] D. Midgley, *Analyst*, 1979, **104**, 248.
[26] D. Midgley, *Analyst*, 1980, **105**, 417.
[27] D. Midgley, *Anal. Proc.*, 1980, **17**, 306.
[28] D. Midgley, *Analyst*, 1980, **105**, 1002.
[29] D. Midgley, *Ion-Selective Electrode Rev.*, 1981, **3**, 43.
[30] D. Midgley and K. Torrance, 'Potentiometric Water Analysis', Wiley, Chichester, 1978.
[31] R. Caulcutt and R. Boddy, 'Statistics for Analytical Chemists', Chapman and Hall, London, 1983.
[32] P. C. Jurs, *Anal. Chem.*, 1970, **42**, 747.
[33] D. A. Kurtz, *Anal. Chim. Acta*, 1983, **150**, 105.
[34] J. S. Garden, D. G. Mitchell, and W. N. Mills, *Anal. Chem.*, 1980, **52**, 2310.
[35] L. M. Schwartz, *Anal. Chem.*, 1977, **49**, 2062.
[36] L. M. Schwartz, *Anal. Chem.*, 1979, **51**, 723.
[37] H. H. Ku, ed., 'Precision Measurement and Calibration – Statistical Concepts and Procedures', (Natl. Bur. Stand. Spec. Publ. No. 300), U.S. Government Printing Office, Washington, 1969.
[38] National Physical Laboratory, 'Certified Reference Materials and Transfer Standards', N.P.L., Teddington, 1982.
[39] J. D. Cox, *Anal. Proc.*, 1982, **19**, 2.
[40] H. Marchandise, *Anal. Proc.*, 1982, **19**, 4.
[41] S. D. Rasberry, *Anal. Proc.*, 1982, **19**, 5.
[42] C. A. Watson, *Anal. Proc.*, 1982, **19**, 12.
[43] Community Bureau of Reference (BCR), 'Catalogue of BCR Reference Materials', B.C.R., Commission of the European Communities, Brussels, 1982.
[44] International Organisation for Standardisation, 'Directory of Certified Reference Materials', I.S.O., Geneva, 1982.
[45] S. R. Goode and J. W. Northington, *Appl. Spectrosc.*, 1979, **33**, 12.
[46] T. Aronsson, C.-H. de Verdier, and T. Groth, *Clin. Chem.*, 1974, **20**, 738.
[47] J. Agterdenbos, *Anal. Chim. Acta*, 1979, **108**, 315.
[48] J. Agterdenbos, F. J. M. J. Maessen, and J. Balke, *Anal. Chim. Acta*, 1981, **132**, 127.
[49] J. Hilton, I. B. Talling, and R. T. Clark, *Lab. Pract.*, 1982, **31**, 990.
[50] R. Klockenkämper and H. Bubert, *Spectrochim. Acta, Part B*, 1982, **37**, 127.
[51] A. G. C. Morris, *Analyst*, 1983, **108**, 546.
[52] J. S. Hunter, *J. Assoc. Off. Anal. Chem.*, 1981, **64**, 574.
[53] L. Aarons, *Analyst*, 1981, **106**, 1249.
[54] O. L. Davies and P. L. Goldsmith, ed., 'Statistical Methods in Research and Production', Fourth Revised Edition, Longman, London, 1977.
[55] W. J. Youden, *Science*, 1954, **120**, 627.
[56] C. Liteanu and I. I. Panovici, *Talanta*, 1977, **24**, 196.
[57] J. Inczédy, *J. Autom. Chem.*, 1980, **2**, 186.
[58] K. A. H. Hooton and M. L. Parsons, *Anal. Chem.*, 1973, **45**, 2218.
[59] G. Svehla and E. L. Dickson, *Anal. Chim. Acta*, 1982, **136**, 369.
[60] D. L. Massart, A. Dijkstra, and L. Kaufman, 'Evaluation and Optimization of Laboratory Methods and Analytical Procedures', Elsevier, Amsterdam, 1978.
[61] J. Kalivas and B. R. Kowalski, *Anal. Chem.*, 1982, **54**, 560.
[61a] K. W. Petts, 'Air Segmented Continuous Flow Automatic Analysis in the Laboratory, 1979', H.M.S.O., London, 1980.
[62] J. W. Hosking, K. R. Oliver, and B. T. Sturman, *Anal. Chem.*, 1979, **51**, 307.
[63] A. Corsini, C. C. Wan, and S. Chiang, *Talanta*, 1982, **29**, 857.
[64] R. E. Saxberg and B. R. Kowalski, *Anal. Chem.*, 1979, **51**, 1031.
[65] J. Kalivas and B. R. Kowalski, *Anal. Chem.*, 1981, **53**, 2207.

[66] C. Jochum, P. Jochum, and B. R. Kowalski, *Anal. Chem.*, 1981, **53**, 85.
[67] J. H. Kalivas, *Anal. Chem.*, 1983, **55**, 556.
[68] B. Vandeginste, J. Klaessens, and G. Kateman, *Anal. Chim. Acta*, 1983, **150**, 71.
[69] M. J. Maurice and K. Buijs, *Fresenius' Z. Anal. Chem.*, 1969, **244**, 18.
[70] A. L. Wilson, *Talanta*, 1974, **21**, 1109.
[71] J. Inczédy, *Talanta*, 1982, **29**, 595.
[72] G. den Boef and A. Hulanicki, *Pure Appl. Chem.*, 1983, **55**, 553.
[73] Standing Committee of Analysts, 'General Principles of Sampling and Accuracy of Results, 1980', H.M.S.O., London, 1980.
[74] H. M. Webber and A. L. Wilson, *Analyst*, 1964, **89**, 632.
[75] J. A. Platte and V. M. Marcy, *Proc. Amer. Power Conf.*, 1965, **27**, 851.
[76] L. Pszonicki, *Talanta*, 1977, **24**, 613.
[77] L. Pszonicki and A. Lukszo-Bienkowska, *Talanta*, 1977, **24** 617.
[78] J. Kragten and A. Parczewski, *Talanta*, 1981, **28**, 901.
[79] M. Feinberg and C. Ducauze, *Analusis*, 1980, **8**, 185.
[80] T. C. Woodis, G. B. Hunter, and F. J. Johnson, *Anal. Chim. Acta*, 1977, **90**, 127.
[81] A. Madej, A. Parczewski, A. Rokosz, and R. Stepak, *Chem Anal. (Warsaw)*, 1978, **23**, 231.
[82] P. L. Kempster, J. O. R. Malloch, and M. V. D. S. de Klerk, *Spectrochim. Acta, Part B*, 1983, **38**, 967.
[83] W. Wegscheider, G. Knapp, and H. Spitzy, *Talanta*, 1979, **26**, 25.
[84] D. Garvin, *J. Res. Natl. Bur. Stand.*, 1972, **76A**, 67.
[85] W. K. Dougan and A. L. Wilson, *Analyst*, 1974, **99**, 426.
[86] A. L. Wilson, *Analyst*, 1968, **93**, 83.
[87] American Public Health Association *et al.*, 'Standard Methods for the Examination of Water and Wastewater', Fifteenth Edition, A.P.H.A., Washington, 1981.
[88] P. D. Gaarenstroom, J. C. English, S. P. Perone, and J. W. Bixler, *Anal. Chem.*, 1978, **50**, 811.
[89] M. Thompson, M. H. Ramsey, and B. J. Coles, *Analyst*, 1982, **107**, 1286.
[90] Standing Committee of Analysts, 'Acid Soluble Aluminium in Raw and Potable Waters by Spectrophotometry (using pyrocatechol violet). 1979', H.M.S.O., London, 1980.
[91] P. Koscielniak and A. Parczewski, *Anal. Chim. Acta*, 1983, **153**, 103.
[92] P. Koscielniak and A. Parczewski, *Anal. Chim. Acta*, 1983, **153**, 111.
[93] G. J. Moody and J. D. R. Thomas, *Talanta*, 1972, **19**, 623.
[94] E. H. Hansen, J. Růžička, F. J. Krug, and E. A. G. Zagatto, *Anal. Chim. Acta*, 1983, **148**, 111.
[95] J. O. Westgard, *CRC Crit. Rev. Clin. Lab. Sci.*, 1981, **13**, 283.
[96] P. B. Stockwell, *Talanta*, 1980, **27**, 835.
[97] J. B. Reust and V. R. Meyer, *Analyst*, 1982, **107**, 673.
[98] P. C. Thijssen, S. M. Wolfrum, G. Kateman, and H. C. Smit, *Anal. Chim. Acta*, 1984, **156**, 87.
[99] J. F. Tyson and J. M. H. Appleton, *Talanta*, 1984, **31**, 9.
[100] J. F. Tyson, J. M. H. Appleton, and B. A. Idris, *Anal. Chim. Acta*, 1983, **145**, 159.
[101] S. Greenfield, *Spectrochim. Acta, Part B*, 1983, **38**, 93.

6 Estimation and Control of the Precision of Analytical Results

In Section 6.1 we stress several general concepts that are basic to any estimation of precision. In Section 6.2 a general model for the random errors of analytical results is developed. This model is used in Section 6.3 to guide the discussion of practical aspects of precision estimation before an analytical system is put into routine use. Finally, we deal in Section 6.4 with tests to check that satisfactory accuracy is maintained in routine analysis. Both bias and precision must be considered at this stage, and it is convenient to consider both aspects together even though this section is primarily concerned with precision.

6.1 PRELIMINARY CONSIDERATIONS

Random errors can never be eliminated completely; indeed, in some situations they may well be the dominant of the two types of error. We saw in Section 4.3 the general need to quantify the size of the random errors of analytical results, and that this could be done provided an estimate of the standard deviation of the results is available. In practice, it is impracticable if not impossible to obtain an independent estimate of the standard deviation for each sample analysed, and analysts must normally use an estimate derived from previous measurements. This approach clearly requires that the population standard deviation shows no appreciable changes with time [see Section 4.3.5(b)]. Such stability is desirable even if an estimate of precision were obtained for each sample; otherwise, the maximum tolerable standard deviation (see Section 2.6.2) is likely to be exceeded erratically from time to time.

There are many factors that may affect the precision of analytical results, and all these factors may change with time. Thus, if reliable quantification of adequately small errors is to be achieved, the analyst must do three things.

(1) All relevant components of an analytical system must be controlled so that the precision of results is expected to be both stable and adequately small.
(2) An adequately precise estimate of the relevant standard deviations must be obtained.
(3) If (2) shows inadequate or irreproducible precision, the control of the analytical system should be improved or a new one adopted, and (2) repeated until satisfactory precision is achieved or it is agreed with the user that his needs on precision be relaxed.

Condition (1) is commonly referred to as achieving 'a state of statistical control', and Eisenhart[1] has given a valuable analysis of all that this term implies. He states that " . . . until a measurement operation has been 'debugged' to the extent that it has attained a state of statistical control it cannot be regarded in any logical sense as measuring anything at all." Other authors[2,3,3a] stress the same point.

Eisenhart[1] discusses several problems in achieving *and* demonstrating statistical control, and we consider such aspects in Section 6.3. We wish to stress here that the time for work on points (1) to (3) above is before applying an analytical system to routine analysis. If an analytical system is used routinely without undertaking this stage, there is the danger that a considerable period of routine analysis may elapse before it is found that the precision is intolerably large and/or unstable; much wasted time and effort and many invalid analyses may then well result.

It is at this preliminary stage of error estimation that any experimental tests of the size of bias should also be made (see Section 5.1).

A second (routine) stage of error estimation should begin as soon as an analytical system is put into routine use. In general, these tests (Section 6.4) consist primarily of checks that the accuracy achieved in the preliminary stage is maintained during routine analysis. Of course, during this second stage, bias as well as precision of analytical results is of interest; it is convenient to consider them both together in Section 6.4. The optimal choice of tests to be used in this routine checking of errors is aided by the information from the preliminary stage.

Before turning to the practical aspects of precision estimation, we need to derive a model for random errors (Section 6.2) that can serve as a basis for planning their experimental estimation.

6.2 MODEL FOR RANDOM ERRORS

The discussion is restricted to random errors within laboratories since we prefer to regard errors between results from different laboratories as caused by systematic errors (see Section 7).

6.2.1 Calibration by Standard Solutions

(a) Need to Consider Errors Arising from Calibration
An analytical result, R, is generally produced from two components: (1) the analytical response for a sample, and (2) estimates of the calibration parameters [see Sections 5.4.2(a) and (b)]. Both components are subject to random errors which, in turn, affect the errors of results. We consider here only analytical systems that are calibrated by standard solutions, and for which equation (6.1) is valid

$$\mathscr{A}_T = \lambda + \chi C_S \tag{6.1}$$

see Section 5.4.2(a) for the meanings of the various symbols. Calibration by standard solutions implies that λ and χ are independent of the nature of the solution analysed. When, however, χ depends on the nature of the solution, standard-addition procedures [see Section 5.4.2(c)] are required, and are considered in (b) below. The general approach described below is applicable for calibration functions of different form to equation (6.1).

When equation (6.1) holds, analytical results, R, are generally obtained (explicitly or implicitly) from the equation

$$R = (A_T - l)/k \tag{6.2}$$

where A_T is the measured analytical response for a sample, and l and k are estimates of λ and χ, respectively. Clearly, the random errors of A_T, l, and k will all cause error in R, and it is essential, therefore, to consider the random errors of l and k arising from the calibration procedure. Many different procedures are recommended and used, and we cannot deal with all of them here. We concentrate, therefore, on a few common calibration procedures, but again the principles involved can be applied to any procedure.

(b) Distinguishing between Random and Systematic Errors

We begin by noting, with Currie,[2] that the consideration and estimation of the random errors of analytical results would be very simple if analytical methods specified that λ and χ be freshly and independently estimated for each portion of each sample analysed. For such methods, a valid estimate of the population standard deviation of results (for a given value of C_S) is very simply obtained by analysing (as prescribed by the method) n portions of a solution of concentration, C_S. These analyses can be made at any time – given that the standard deviation is stable – and the standard deviation is simply calculated using equation (4.12) or (4.13). In practice, of course, the vast majority of analytical methods do not specify such calibration, but rather specify or imply that the same estimates of calibration parameters be used for a number of samples. This approach is clearly desirable (and often essential) in reducing the effort required for analysis, and is usually capable of providing adequately precise results. A consequence, however, is that the consideration and experimental estimation of random errors becomes rather more complicated.

Suppose an analytical method specifies: (1) equation (6.2) for the calculation of analytical results, (2) procedures for determining values for l and k, a value for each to be determined once in each batch of analyses, and (3) analysis of one portion of each sample. Such methods are, of course, common. Before proceeding, a few words should be said about the term 'batch' because it is basic to the following discussion. The variety of analytical systems and practices makes it difficult to give a simple yet universally applicable definition of 'batch'; compare the definition of the equivalent term 'run' by Büttner *et al.*[4] For present purposes it suffices to define 'batch' as that set of sample analyses and calibration standards for which the same numerical value of the estimate of at least one calibration parameter is used to derive the analytical results for samples. Analytical systems for which this definition needs closer consideration are discussed in (e)(ii) below.

Returning to the analytical method described above, suppose that, to obtain an estimate of the precision of analytical results, n portions of a sample were analysed in one batch, all other details of the method also being followed. The n results obtained would be scattered about their mean because of the random errors of the A_T values; this scatter, however, would be totally unaffected by the random errors of l and k since the same values of l and k were used to calculate all n results. Of course, the effects of the random errors of l and k have not been eliminated, but in this situation they lead to an error affecting each of the n results equally. If, on the other hand, the n portions of sample were analysed one in each of n batches, the scatter of results would now fully include the effects of the random errors of l and k since different values would have been used in each batch. Thus, we may regard the analytical results from such a method as subject to: (1) a random error determined by the random errors of the A_T values, and (2) a systematic error determined by the random errors of l and k, and which varies in size, therefore, from batch to batch. Alternatively, the analytical results may be regarded as subject to two independent sources of random error: (1) that arising from the within-batch variations of A_T values, and (2) that arising from the within-batch random errors of l and k which leads to between-batch random errors of analytical results. This same choice will arise for any analytical method specifying analysis in batches as defined above. A criterion for choosing between these two alternatives is considered below.

We assume now that:

(1) the true values of λ and χ do not vary during any batch of analyses,
(2) measurements of analytical responses are unbiased, and their random errors are mutually independent,
(3) l and k are unbiased estimates of λ and χ for the batch of analyses in which l and k were determined.

The validity of these assumptions is considered in (*e*) below. Note also that we are concerned here with only random errors; sources of bias in analytical responses and in the values of l and k have already been discussed in Section 5. Denoting values for the i th batch by the subscript i, and writing ε_{Ai}, ε_{li}, and ε_{ki} for the random errors of A_{Ti}, l_i, and k_i, respectively, it is easily shown from equation (6.2) that the resulting error, ε_{Ri}, of R_i is given by

$$\varepsilon_{Ri} = (\varepsilon_{Ai} - \varepsilon_{li} - \varepsilon_{ki}C_S)/(\chi_i + \varepsilon_{ki}) \qquad (6.3)$$

We also write $\varepsilon'_{Ai} = \varepsilon_{Ai}/(\chi_i + \varepsilon_{ki})$, $\varepsilon'_{li} = \varepsilon_{li}/(\chi_i + \varepsilon_{ki})$, and $\varepsilon'_{ki} = \varepsilon_{ki}C_S/(\chi_i + \varepsilon_{ki})$. Now the population mean of the random errors of measurements of a random variable is zero.* It will be consistent, therefore, to regard a random error, ε_i, as leading to only random error in R if the population mean, $E(\varepsilon_i)$ of ε_i is zero. On this basis, we use the following critera for deciding whether random errors in A_{Ti}, l_i, and k_i are to be regarded as causing random or systematic errors in R.

(1) A random error, ε_i, is regarded as causing only random error in R when $E(\varepsilon_i) = 0$.
(2) When $E(\varepsilon_i) \neq 0$, this is regarded as causing, in general, both systematic and random errors in R. The systematic error equals $E(\varepsilon'_i)$, and the random error equals $\varepsilon'_i - E(\varepsilon'_i)$.

Given these criteria and the preceding assumptions, $E(\varepsilon_{Ai})$ is always zero, so that the random errors of A_T are always regarded as causing only within-batch random errors in R. The values of $E(\varepsilon_{li})$ and $E(\varepsilon_{ki})$ depend – as we shall see below – on the constancy of λ and χ from batch to batch, and on the detailed calibration procedure specified by the analytical method. In general, however, the random errors of l_i and k_i will act as a source of between-batch random error in R. On this basis we can

---

* Consider the mean error of n measurements, x_i ($i = 1$ to n), of a random variable with population mean, μ. The mean error is $\sum_{i=1}^{n}(x_i-\mu)/n$ which is equal to $(\sum_{i=1}^{n}x_i/n)-\mu$, and as n approaches infinity, the summation term approaches μ [see Section 4.3.2(*a*)]. Consequently, $\sum_{i=1}^{\infty}(x_i-\mu)/n = 0$. We denote the limiting value of $\sum_{i=1}^{n}\varepsilon_i/n$ as n approaches infinity by $E(\varepsilon_i)$.

formulate the general model that analytical results are subject to both within-batch and between-batch random errors, and that these two errors are independent. If the standard deviations characterising these two errors are denoted by σ_w and σ_b, respectively, then from Section 4.3.2(*d*) we can write

$$\sigma_t^2 = \sigma_w^2 + \sigma_b^2 \qquad (6.4)$$

where σ_t is the total standard deviation of analytical results, *i.e.*, σ_t characterises the total random error of R. Figure 6.1 illustrates the nature of the model represented by equation (6.4).

Of course, in some situations the errors of l_i and k_i may be so small that σ_b is essentially zero. This is, however, by no means necessarily so, and equation (6.4) provides a useful basis for the consideration and estimation of the precision of analytical results. The above approach may be applied to any calibration procedure using standard solutions, and we briefly illustrate below its use for two types of calibration common in water analysis.

Fixed-slope calibration This term is used when an initial estimate, k_0, of χ is used for many subsequent batches, but an estimate, l_i, of λ is obtained for each batch. If χ varies at random from batch to batch with population standard deviation, σ_χ, then $E(\varepsilon_{ki})$ is not generally zero. Thus, from criterion (2) above, the consequent systematic error in R is ε_{k0}, and the between-batch random error in R from this source is governed by σ_χ [see (*c*) below]. If the true value of χ is constant for all batches, *i.e.*, $\sigma_\chi = 0$, only bias arises in R as a result of the random error of k_0.

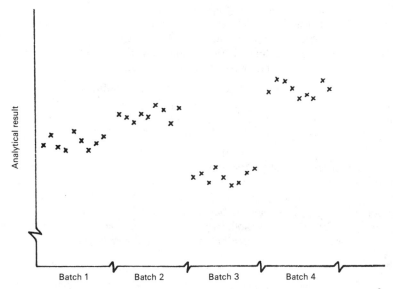

Figure 6.1 *Within- and between-batch variations (Each cross represents an analytical result for the same solution. The variability of results within each batch is the same, but the average value of results varies at random from batch to batch)*

Whether or not $\sigma_\chi = 0$, it is very desirable to ensure that ε_{k0}' is negligible in relation to the maximum tolerable bias of analytical results [see Section 5.4.3(d)].

For this type of calibration, l is usually equated to the mean response of one or more blank determinations in each batch [Section 5.4.2(b)]. In this event, $E(\varepsilon_{li}) = 0$, so that the random errors of such blanks lead to only between-batch random errors in R, but see also Sections 5.4.2(b) and 5.4.4.

Within-batch calibration This term is used when both l_i and k_i are determined freshly in each batch of analyses. In this event, $E(\varepsilon_{li}) = E(\varepsilon_{ki}) = 0$, and the random errors of l and k, therefore, lead to only between-batch random errors in R.

(c) Standard Deviation of Analytical Results
We consider now the derivation of quantitative expressions for the standard deviations of analytical results in terms of the standard deviations of A_T, l, and k. These expressions are useful both in planning tests to estimate precision and in considering means of controlling it. We make the further assumptions that:

(1) the standard deviation, σ_{AT}, of the analytical responses, A_T, for a given solution and population mean response, $\mathscr{A}_T$, is the same in all batches of analysis, and does not vary within batches;
(2) the water used to prepare blanks and other calibration standards contains a negligible concentration of the determinand;
(3) measurements to allow special correction procedures are not involved.

The derivation of equations for σ_t, σ_w, and σ_b for fixed-slope and within-batch calibration is described in (i) and (ii) below.

(i) Fixed-slope calibration We consider only the procedure where the analytical method specifies that l is the mean response, $\bar{A}_B$, of b blank determinations ($b \geqslant 1$) in each batch, that analytical results are calculated from

$$R = (\bar{A}_T - \bar{A}_B)/k_0 \qquad (6.5)$$

and that $\bar{A}_T$ is the mean response from the independent analyses of n portions of the sample ($n \geqslant 1$). Applying the procedure in Section 4.6.3 to equation (6.5) leads to

$$\sigma_t^2 \approx (\sigma_{\overline{AT}}^2 + \sigma_{\overline{AB}}^2 + \sigma_{k0}^2 C_S^2)/k_0^2 \qquad (6.6)$$

where σ_t, $\sigma_{\overline{AT}}$, $\sigma_{\overline{AB}}$, and σ_{k0} are the standard deviations characterising the random errors of R, $\bar{A}_T$, $\bar{A}_B$, and k_0. Now, $\sigma_{\overline{AT}}^2 = \sigma_{AT}^2/n$, and $\sigma_{\overline{AB}}^2 = \sigma_{AB}^2/b$, where σ_{AT} and σ_{AB} are the standard deviations for individual measurements of A_T and A_B [Section 4.3.2(c)]. We may also write σ_{AT}/k_0 and σ_{AB}/k_0 as σ_T and σ_B, respectively, where σ_T and σ_B are expressed in concentration units. If it is assumed that the true value of χ varies at random with standard deviation, σ_χ, then $\sigma_{k0} = \sigma_\chi$ [see (b) above]. Finally, the degree of approximation of equation (6.6) becomes smaller and smaller as σ_χ/k_0 decreases; we assume that this calibration procedure would be used only when σ_χ/k_0 is small, and the $\approx$ sign in equation (6.6) can, therefore, be replaced by the equality sign. Equation (6.6) may, therefore, be rewritten as

$$\sigma_t^2 = (\sigma_T^2/n) + (\sigma_B^2/b) + \sigma_\chi^2 C_S^2/k_0^2 \qquad (6.7)$$

As discussed in (b) above, the within-batch random errors of R are caused by the random error of $\bar{A}_T$, and the between-batch random errors of R are caused by the random errors of $\bar{A}_B$ and k_0. Comparing equation (6.4) with (6.7) shows that σ_w and σ_b are given by the following expressions*

$$\sigma_w^2 = \sigma_T^2/n \qquad (6.8)$$

and

$$\sigma_b^2 = (\sigma_B^2/b) + \sigma_\chi^2 C_S^2/k_0^2 \qquad (6.9)$$

Note the dependence of σ_b on C_S except when $\sigma_\chi^2 = 0$. Note also that, if $\bar{A}_T$ and $\bar{A}_B$ are normally distributed, then R will be approximately normally distributed provided that σ_χ/k_0 is small (Section 4.6.3).

(ii) Within-batch calibration   In this type of calibration, at least two standard solutions of different determinand concentration must be analysed in each batch to allow estimates of λ and χ to be obtained. More than two standards may be, and often are, used, and each or some of the standards may be replicated. Further, the procedures used to derive l_i and k_i from the responses of these standards vary [see Section 5.4.3(*d*)]. We shall consider only two somewhat extreme approaches since similar general conclusions follow for any approach. Except where noted otherwise, we make the same assumptions as in the previous sections.

Use of two calibration standards   These commonly consist of a blank and a standard whose concentration, C_D, is at or near the upper limit of the concentration range of the analytical system. Other concentrations for both standards, however, may be used if desired, and the treatment below is easily applied to such other conditions. Suppose the method prescribes the analysis of n portions of a sample, b of the blank, and d of the other standard, and that analytical results are calculated from

$$R = C_D(\bar{A}_T - \bar{A}_B)/(\bar{A}_D - \bar{A}_B) \qquad (6.10)$$

where $\bar{A}_D$ is the mean response for the higher concentration standard. Applying the procedure in Section 4.6.3 to equation (6.10), we obtain after some rearrangement

$$\sigma_t^2 \approx (\sigma_T^2/n) + r^2(\sigma_D^2/d) + (r-1)^2\sigma_B^2/b \qquad (6.11)$$

where $r = C_S/C_D$, $\sigma_D^2 = \sigma_{AD}^2/k_i^2$, and σ_{AD} is the standard deviation of individual responses for the higher concentration standard.

Again, we may write $\sigma_t^2 = \sigma_w^2 + \sigma_b^2$, where, as before, $\sigma_w^2 = \sigma_T^2/n$, but now

$$\sigma_b^2 \approx r^2(\sigma_D^2/d) + (r-1)^2\sigma_B^2/b \qquad (6.12)$$

Note that the dependence of σ_b on C_S differs from that for fixed-slope calibration. If $\bar{A}_T$, $\bar{A}_B$, and $\bar{A}_D$ are normally distributed, then R will be approximately normally-

---

* In a number of previous publications we have used the terms within-batch and between-batch standard deviation to denote, respectively, $\sqrt{\{(\sigma_t^2/n) + (\sigma_B^2/b)\}}$ and $\sigma_\chi C_S/k_0$. This grouping of errors can be useful in deciding if the variability of χ or the precision of analytical responses is the dominant contribution to σ_t^2; such information is of value when it is desired to decrease the value of σ_t^2. On balance, however, we prefer the definitions of σ_w and σ_b in equations (6.8) and (6.9) because they simplify consideration of the contributions of the various random errors to σ_t, particularly when within-batch calibration is used. We would stress also the fundamental importance of equation (6.7). This equation or its equivalents for within-batch calibration (see below) should always be borne in mind.

distributed provided that the standard deviation of $(\bar{A}_D - \bar{A}_T)$ (*i.e.*, $\sqrt{\{(\sigma_{AD}^2/d) + (\sigma_{AB}^2/b)\}}$) is small compared with $\bar{A}_D - \bar{A}_B$ (Section 4.6.3).

Use of more than two calibration standards Suppose that p ($p \geq 3$) different standards of concentration C_j ($j = 1$ to p) are used, and that q ($q \geq 1$) portions of each standard are analysed. The three main approaches to determining the values of l_i and k_i from the responses observed for the standards are discussed in Section 5.4.3(d); here we consider only the procedure using linear regression analysis. This technique is described in Section 5.4.2(e), and readers should consult that section before proceeding with this; we shall be very brief here. If the analytical method also prescribes the analysis of n portions of a sample in one batch, and calculation of the results from the equation

$$R = (\bar{A}_T - l_i)/k_i \qquad (6.14)$$

the standard deviation, σ_t, of R is given by the following expressions.

Unweighted regression
Some re-arrangement of equation (5.40) and inclusion of q [see Section 5.4.2(e)] leads to

$$\sigma_t^2 \approx (\sigma_T^2/n) + (\sigma^2/pq) + (v-1)^2\sigma^2\bar{C}^2/\Sigma(C_j - \bar{C})^2 \qquad (6.15)$$

where σ_T and σ are the standard deviations of the responses for the sample and standards, respectively (expressed in concentration units), and σ is independent of C_j, $v = C_S/\bar{C}$, and $\bar{C} = \Sigma C_j/p$.

Weighted regression
Some re-arrangement of equation (5.46) and inclusion of q [see Section 5.4.2(e)] leads to

$$\sigma_t^2 \approx (\sigma_T^2/n) + (1/\bar{z}'pq) + (g-1)^2(1/\bar{z}')\bar{C}_w^2/\Sigma w_j(C_j - \bar{C}_w)^2 \qquad (6.16)$$

where $\bar{z}'$ is $\bar{z}$ for standards expressed in concentration units and $1/\bar{z}'$ is equivalent to a variance, $g = C_S/\bar{C}_w$, $\bar{C}_w = \Sigma w_j C_j/p$, and w_j is a weighting factor [see Section 5.4.2(e)]. Thus, apart from details, equation (6.16) is of exactly the same form as equation (6.15), and the same remarks may, therefore, refer to both unweighted and weighted linear regression.

Both equations are of the same general form as those derived for other calibration procedures. We can again write $\sigma_t^2 = \sigma_w^2 + \sigma_b^2$, where as before $\sigma_w^2 = \sigma_T^2/n$, and σ_b^2 is given by the sum of the other terms in the equations for σ_t^2. Again, note the dependence of σ_b on C_S. As before also, when the random errors of analytical responses are normally distributed, equation (6.14) shows that results will be approximately normally distributed when σ_{ki}/k_i is small [see Sections 4.6.3 and 5.4.2(e)]. Equations (5.37) and (5.44) in Section 5.4.2(e) allow values of σ_{ki} to be calculated.

Equations (6.15) and (6.16) allow the calculation of the effects of the number and concentrations of calibration standards on σ_b and σ_t. This is a very useful feature of linear regression theory, and has been applied by many authors as a guide to the choice of p, q, and C_j values for standards; see, for example, refs. 5 to 9.

(d) General Model for the Standard Deviation of Analytical Results
Given all the assumptions made in (a) to (c) – their accord with reality is considered in (e) below – a very simple model can be formulated for the random errors of analytical results when discrete samples are analysed. The model is intended to apply only to the random errors of individual laboratories, and can be summarised as follows.

Analytical results are subject to within-batch and between-batch random errors (see Figure 6.1) characterised by within- and between-batch standard deviations, σ_w and σ_b, respectively. These two sources of random error are independent so that the total random error of a result is characterised by the total standard deviation, σ_t, where $\sigma_t = \sqrt{(\sigma_w^2 + \sigma_b^2)}$. Neither σ_w nor σ_b vary with time, and σ_t, therefore, is also constant. Since σ_t is the standard deviation characterising the random errors of individual analytical results, it is usually the fundamental parameter of interest in quantifying these random errors. Note that when r analytical results are obtained for a given sample in each of m batches of analyses, the standard deviation, $\sigma_{\bar{t}}$, of the mean result is given by

$$\sigma_{\bar{t}}^2 = (\sigma_w^2/rm) + (\sigma_b^2/m) \qquad (6.17)$$

The values of σ_w and σ_b depend on the random errors of the experimental measurements involved in the calculation of an analytical result. The nature of this dependence varies with the detailed procedures used to estimate the calibration parameters, and equations for σ_w, σ_b and σ_t can be derived for any calibration procedure as illustrated in (c) above. The validity of such equations depends, of course, on several assumptions, but they are generally of value in providing a theoretical framework for considering the estimation and control of the precision of analytical results. A few general points concerning σ_w and σ_b are worth noting.

(1) σ_w always equals $\sigma_T/\sqrt{n}$ $(n \geq 1)$, where σ_T is the standard deviation (expressed in concentration units) of the analytical responses for a sample, and n is the number of portions of a sample (as specified by the analytical method) independently analysed in order to provide an analytical result. Thus, σ_w is independent of the calibration procedure, and is governed by the degree of control over all the many factors affecting σ_T. Accordingly, for a given analytical system, σ_w can be decreased only by improving the degree of control of the system or by modifying the method and using a larger value of n. Alternatively, a different analytical method and system allowing more precise responses may be adopted.

(2)   Whenever the detection of small differences in concentration between two or more samples is of interest, it is generally advantageous [though the benefit may be small, see (3) below] to analyse all the samples in one batch. This is because the ability to detect differences improves as the random errors of results decrease; analysis of all samples in one batch eliminates the effect of the random errors of l_i and k_i. Thus, knowledge of σ_w may be of value in considering experimental designs for such special 'within-batch investigations'.

(3)   The value of σ_b represents the extent to which the random errors of estimates of calibration parameters have been controlled, and $\sigma_b = 0$ when these estimates

have zero random error. In practice, such a high degree of control, apart from being impossible, is not necessary. From equation (6.4) it follows that if $\sigma_b = 0.2\sigma_w$, then σ_t is only 2% greater than its value when $\sigma_b = 0$. For most purposes, such an effect is negligible. The equations given in (c) above for σ_b for different calibration procedures (or others equations derived similarly for other procedures) are approximations [see (c)(i) above], but they are useful in that they allow consideration of the factors affecting σ_b, and of means of reducing its size. Such equations also allow assessment of the merits of different calibration procedures in otherwise identical analytical systems. For example, within-batch calibration eliminates the possibility of errors arising from variability of χ in fixed-slope calibration, but at the expense of the greater effort needed to estimate χ_i for each batch. Whether or not this greater effort is justified depends on the relative values of σ_b for the two types of calibration, and on these values in relation to σ_w.

The concept of within- and between-batch random errors is used in the water industry of the U.K.[10,11] (though see the footnote on page 232), and in other fields of analysis.[4,12–14] Anticipating the arguments in (e), we conclude that this model is of general value whenever the concept of 'batches of analysis' applies (see also Section 6.2.2). We use the model constantly in Sections 6.3 and 6.4 where the experimental estimation of the precision of analytical results is discussed. Assumptions on which the validity of the model rests are considered in (e) below. A final, important aspect – the dependences of σ_w, σ_b, and σ_t on the concentration of the determinand in the sample – is discussed in (f) below.

(e) The Validity of the Model of Random Errors
Most models can be regarded as only approximations to reality. In the following we consider the various assumptions on which the model described in (d) is based, and the extent to which they are likely to hold in practice.

(i) Nature of the calibration function Only calibration functions of the form $\mathscr{A}_T = \lambda + \chi C_S$ were explicitly considered above. Such functions are observed for many current, routine water analyses, but others arise. The general model is equally applicable to any calibration function, but the expressions for σ_b must be derived for each particular function *and* calibration procedure.

For linear calibration functions with $\lambda = 0$ *and* when the analytical method does not specify obtaining an estimate, l_i, in each batch, note that the equations for σ_t and σ_b in (c) are incorrect since they assume such an estimate; see also (ii) below.

(ii) The definition of batch We noted in (b) the difficulty of finding a simple yet generally applicable definition of 'batch'. Three main situations need consideration.

(1) When fresh, independent estimates of *all* calibration parameters are obtained for each sample analysed, the definition of batch in (b) implies that each batch can contain only one sample and produce only one analytical result. In this situation, provided our other assumptions are correct, it is simpler to regard

results as simply subject to random errors characterised by σ_t; the standard deviation, $\sigma_{\bar{t}}$, of the mean of m results is then $\sigma_t/\sqrt{m}$ whenever the results are obtained. See also Section 6.2.2.

(2) Consider analytical systems with linear calibration functions but with $\lambda = 0$. If the method does not specify obtaining an estimate, l, the definition of 'batch' in (b) implies that *all* analytical results obtained using the same value of k are to be regarded as coming from the same batch. When k is freshly determined on each occasion when samples are analysed, our definition of batch is in accord with the common connotation of the term. It may, however, seem very strange if, for example, fixed-slope calibration were specified by the method because the one 'batch' of analytical results might then well correspond to many analytical occasions spread over a long period of time. We agree that this does appear to be strange, but it is a consequence of the calibration procedure specified by the method. Rather than calling for a change in the definition of 'batch', it suggests the need to check the validity of the calibration procedure when (or preferably before) estimating the precision of analytical results.

(3) The values of λ and χ for some analytical systems may gradually change during occasions when samples are analysed. Such behaviour is commonly called 'drift', and has already been mentioned as a source of bias in Section 5.4.3(d)(ii). In order to minimise errors caused by drift, sets of calibration standards are sometimes analysed at regular intervals during an analytical occasion, a set of samples being included between two sets of standards. Estimates of calibration parameters are derived separately from each set of standards, and then used to calculate appropriate values for each sample. Provided that values for l and k are calculated in this way for *each* sample, the definition of batch in (b) implies that each batch corresponds to one sample. It is important to note, however, that these values of l and k are not freshly-estimated and independent for each sample. Random errors in the l and k values from given sets of standards will tend to be perpetuated in all the l and k values used to calculate the analytical results for a given set of samples; we have, therefore, the same situation as that which led to the definition of 'batch' in (b). Accordingly, we regard a set of samples and the two 'bracketing' sets of standards as a batch. See also (iii) below.

(iii) Constancy of λ and χ in each batch   We assumed that λ and χ do not change during any batch of analyses, but as noted directly above this assumption is not always valid. The effect of such changes, if no correction is made for them, is clearly to cause greater variations in the analytical responses of samples and standards. The quantitative effect on σ_w, σ_b, and σ_t depends, however, on the precise manners in which λ and χ vary within batches, and we do not wish to discuss this problematic topic. In the worst case, it may be that random errors can no longer be reasonably quantified, while in the best σ_w, σ_b, and σ_t may simply be increased though their values may vary from batch to batch. In any event, the model described in (d) will no longer be valid. We have argued in Section 6.1 (see also Section 6.3.1), however, that an essential aim for the analyst is to achieve a state of statistical control in which the magnitudes of random errors are stable in time. This is equivalent to saying that the analyst must either control the analytical system so that appreciable

variations of λ and χ within batches do not occur or make suitable correction for such variations. Accordingly, we regard the model in (d) as in good accord with the aims of the analyst.

(iv) Constancy of precision of analytical responses We assumed that, for a given solution and analytical response, A, the standard deviation, σ_A, of A is the same in all batches of analysis. Again, this assumption may not be true, but it is subject to the same points concerning the aims of the analyst as in *(iii)*. One detailed point should be noted. We saw in Section 5.4.2(b) that variations in λ from batch to batch cause no bias in analytical results provided an estimate, l_i, is obtained in each batch and used to calculate results. If, however, the variations of λ from batch to batch are sufficiently large to cause appreciable changes in A_T values, *and* if σ_A depends on A, then σ_w and σ_b will vary from batch to batch. It is important, therefore, to control λ (*e.g.*, by controlling contamination) so that instability of precision from this effect is avoided.

(v) Dependence of precision on nature of solution analysed There are several reasons why the precision of analytical responses for samples may be worse than that for calibration standards of the same determinand concentration. Although none of the equations in (c) nor the general model of random errors assumes the equality of these two precisions, the reliable, experimental estimation of the precision of routine analytical results is greatly facilitated by such equality (see Sections 6.3 and 6.4). Whenever possible, therefore, analytical methods and systems should be selected and applied so that the responses of samples and standards of the same concentration have the same precision.

It is worth noting that this aspect is often insufficiently stressed in published papers. This is particularly true of those dealing with the use of unweighted regression analysis; these appear commonly to assume that the estimate of precision obtained from the standards is necessarily valid for samples.

(vi) Independence of random errors of analytical responses The general model of random errors and all the other statistical techniques described in this book assume that the random errors of the measured analytical responses are mutually independent [see Sections 4.3.2(a) and (b)]. Since any substantial degree of correlation between random errors will prevent any reasonably simple and accurate approach to the quantification of the random errors of analytical results, it is again essential to do everything possible to minimise the possibility of correlated random errors. In the present context, the main possibility of non-independent errors arises through systematic changes in the values of λ and χ from one sample to the next in a batch of analyses. This is most commonly associated with temporal variations, *e.g.*, drift, but spatial effects may also lead to this problem (see Section 10.8.1). In practice, complete elimination of such effects may not always be possible, and any residual problems may be further decreased by '*randomising*' the temporal and spatial arrangements for samples and standards in each batch of analyses; see also Currie[2] on this point. Procedures for randomising such factors are described in statistical texts (see Section 4.8).

(vii) Unbiased measurement of analytical responses This assumption requires the population mean of the analytical responses for a given solution to equal the true response for that solution (see Section 4.4). Factors such as interfering substances do not cause biased measurement of responses though they do, of course, cause bias in analytical results. The assumption will be sufficiently met provided:

(1) the discrimination of the analytical system does not appreciably affect experimental estimates of standard deviations (Section 4.3.1), and
(2) the numerical values of responses used in calculating results are those observed, *i.e.*, the values of responses should be recorded and used with the finest discrimination possible, and rounding-off should be avoided.

(viii) Unbiased estimation of λ and χ We assumed that the values of *l* and *k* used to calculate analytical results are unbiased estimates of λ and χ at the time when *l* and *k* were determined. Given the preceding assumptions and the discussion on procedures for determining *l* and *k* [Section 5.4.3(*d*)], this assumption should generally be reasonable.

(ix) Random variability of χ in fixed-slope calibration For this calibration procedure we assumed that any variations of χ from batch to batch of analyses were random. Circumstances are certainly possible in which χ varies systematically from batch to batch, *e.g.*, as a result of gradual deterioration of a reagent or gradual changes in the temperature of a laboratory. Such changes, when fixed-slope calibration is used, will clearly lead to systematically increasing bias in analytical results, and the aim must, therefore, be to control such variations to inappreciable levels. Accordingly, the assumption – though it may not be valid – corresponds to the degree of control of an analytical system for which the analyst should be striving.

(x) Normal distribution of analytical results We have already stated in Section 4.3.2(*a*) our general view on the question of whether or not analytical results are normally distributed. We wish here simply to note the point made frequently in (*c*) above, namely, analytical results are expected to be approximately normally distributed if the analytical responses are normally distributed. Since analytical responses are subject to so many different sources of random error, they will tend to be normally distributed, particularly when well controlled analytical systems are involved and the variations of each source of error are relatively small.[2,3]

(xi) Determinand content of water used for blanks and calibration standards We assumed that this water contained a negligible concentration of the determinand. If this is incorrect, analytical results will generally be biased unless a correction is made to results to allow for this concentration [see Sections 5.4.2 and 5.4.4(*b*)]. Such corrections will usually involve additional measurements and thereby introduce a further source of random error. The equations for σ_t in (*c*) may be modified to allow for this if desired, but it is simpler [see Section 10.7.2(*b*)] to ensure whenever possible that the determinand content of the water is negligible.

(xii) Use of special correction procedures In addition to the analysis of samples and calibration standards, some methods specify additional measurements to allow correction for certain sources of bias. For example, in spectrophotometric procedures the absorbances of samples without any added reagents may be measured to correct for turbidity in the sample. As in *(xi)*, such measurements introduce another source of random error whose contribution to the total random error should, in general, be considered. This can be done by writing the equation for the final corrected, analytical result, and following the procedures illustrated in *(c)* above.

(xiii) Conclusions We conclude that the general model for random errors described in *(d)* forms a reasonable basis for considering and estimating the precision of analytical results. It is in accord with good analytical practice when strenuous efforts are made to control the size and stability of the standard deviations of results. These aims are, in any event, essential in any approach to quantifying random errors. Of course, having these aims does not mean that they are necessarily achieved; many analytical problems may have to be overcome, and to the extent that this is not achieved, the model will lose validity.

(f) Dependence of Precision on Determinand Concentration

In quantifying the random errors of analytical results for samples whose determinand concentrations differ appreciably, it is important to know if, and how, the total standard deviation, σ_t, depends on concentration, C_S. The likely dependence of precision on concentration was stressed by Nalimov[15] and in the first edition of this book, and in recent years has been increasingly noted by many authors; see, for example, refs. 2,3,11 and 13–29. The discussion in *(c)* above provides no direct information on this point since it does not consider the most important governing factor, namely, the dependence of the standard deviation of analytical responses on concentration. Consequently, the primary evidence on this question must come from experimental estimates of the precision of results at different concentrations.

All the references just cited concur that the standard deviation of results very commonly increases as C_S increases. In many instances, the authors do not specify whether the standard deviation they quote is σ_t, σ_w, or neither; some papers appear to be concerned with σ_t while others apparently deal with σ_w. Our own experimental estimates of σ_t and σ_w for many analytical systems almost invariably increase with increasing C_S. There are many factors affecting analytical systems that can contribute to such a dependence of precision on C_S, and it is unnecessary to discuss them here. The essential point is that it is very common for σ_t and σ_w to increase as C_S increases. Accordingly, it seems to us prudent to assume such a relationship until there is evidence to the contrary. In general, therefore, the precision of results produced by a particular analytical system cannot be characterised by tests made at only one determinand concentration.

In deciding the number of values of C_S desirable in characterising precision, the nature of the variation of precision with C_S is important. A number of authors[15–17,22,25–27] have proposed and used the idea that precision is approximately proportional to concentration, *i.e.*, that $\sigma_w \approx \sigma_{w0} + \phi C_S$, where σ_w and σ_{w0} are the within-batch standard deviations at determinand concentrations of C_S and zero, respectively, and ϕ is a constant for a given analytical system. Other authors (see, for example, refs. 18 and 22, but compare refs. 20 and 24) have used what amounts to $\sigma_w = \phi C_S$, and this relationship is convenient when weighted linear regression analysis is used. It may be that the latter relationship is a sufficiently good approximation in the estimation of calibration parameters by linear regression techniques, but it implies that $\sigma_w = 0$ when $C_S = 0$. This is both theoretically impossible and in disagreement with a host of studies reporting that $\sigma_w \neq 0$ when $C_S = 0$. Thus, although the exact form of the relationship between σ_w and C_S need not be known for present purposes, we favour the assumption that

$$\sigma_w \approx \sigma_{w0} + \phi C_S \tag{6.18}$$

until evidence to the contrary is available. Since $\sigma_w = \sigma_T/\sqrt{n} \approx \sigma_{AT}/\chi\sqrt{n}$ [see (*c*) above], equation (6.18) implies that the standard deviation of analytical responses is also approximately proportional to C_S.

Whenever σ_b is small in relation to σ_w, σ_t will also be given approximately by equation (6.18). If, however, σ_b is not small, equation (6.18) may mis-represent the dependence of σ_t on C_S because σ_b may show a rather complex dependence on C_S when within-batch calibration is used [see (*c*) above].

In practice, one has only estimates of population standard deviations. In considering the conformity of, say, values of s_w (for different values of C_S) with equation (6.18), the uncertainties of these estimates [see Section 4.3.4(*d*)] must be borne in mind.

6.2.2 Calibration by Standard-Addition Procedures

These procedures have been considered with respect to bias in Section 5.4.2(*c*), and readers should consult that section before proceeding with the present one. We consider here only analytical systems with calibration functions of the form $\mathscr{A}_T = \lambda + \chi C_S$, where $\lambda = \eta + \chi C_C$. We assume that χ depends on the nature of the solution analysed, and is, therefore, estimated independently for each sample analysed. The parameters η and C_C are assumed to be independent of the nature of the solution.

When $\lambda = 0$, χ is the only calibration parameter. It follows, therefore, that there is no batch effect on the precision of analytical results [see Section 6.2.1(b)]. Such an effect can only arise – granted our assumptions in Section 6.2.1 – when $\lambda \neq 0$ *and* the one experimental estimate of λ is used for more than one sample. On this basis and with the same assumptions as those in Section 6.2.1, we can derive expressions for the standard deviations of analytical results. We use the following notation: p portions of each sample are analysed after adding to the j th ($j = 1$ to p) an amount of the determinand equivalent to a concentration C_j in the original sample; $C_j = 0$ for $j = 1$. We assume that the effects of dilution of the sample by these additions are negligible or that appropriate corrections for them are made to the measured analytical responses. The analytical response for the j th portion is denoted by A_{Tj}, and is assumed to come from a population with mean, $\mathscr{A}_{Tj}$, and standard deviation, σ_{Aj}; we write $\sigma_j = \sigma_{Aj}/k_S$, where k_S is the estimate of χ for a particular solution.

When $\lambda = 0$, the analytical result, R, is given by

$$R = h/k_S \qquad (6.19)$$

where h and k_S are the experimentally determined values of, respectively, the intercept on the response axis and the slope of the plot of A_{Tj} against C_j.

When $\lambda \neq 0$, there are two procedures that may be used for correcting for the bias caused by λ [see Section 5.4.2(c)]. One involves making a blank determination, and subtracting the response, A_B, from h. The second involves analysis of blank water in exactly the same way as the sample; the analytical result, R, is then given by $R = R'_S - R'_B$, where R'_S and R'_B are the results for sample and blank from equation (6.19), respectively. We assume that the water used for blank determinations contains none of the determinand [see Section 6.2.1(e)(xi)].

Standard-addition procedures vary in the values used for p and C_j, and in the approaches used to determine h and k_S from the experimental data. The standard deviation, σ_t, of R can be derived by the same procedures as those in Section 6.2.1(c), but note that h and k_S are not, in general, statistically independent. We consider only two variants of the simple standard-addition procedure in (a) and (b) below.

(a) $p = 2$
When $\lambda = 0$, R is most simply calculated from

$$R = C_2 A_{T1}/(A_{T2} - A_{T1}) \qquad (6.20)$$

An expression for σ_t may then be derived using the procedure in Section 4.6.3:

$$\sigma_t^2 = (r+1)^2 \sigma_1^2 + r^2 \sigma_2^2 \qquad (6.21)$$

where $r = C_S/C_1$.

When $\lambda \neq 0$ and blank correction is used, the expression for σ_t becomes

$$\sigma_t^2 = (r+1)^2 \sigma_1^2 + r^2 \sigma_2^2 + \sigma_B^2 \qquad (6.22)$$

and this applies for both methods of blank correction. If the same blank determination is used for more than one sample, $\sigma_b = \sigma_B$; otherwise, $\sigma_b = 0$.

(b) $p \geqslant 3$
We consider only the use of linear regression to determine h and k_S. In view of the conclusions in Section 6.2.1(f), we deal with only weighted linear regression; the corresponding equations for unweighted calculations are easily obtained from those below by putting $k_S^2 \bar{z} = 1/\sigma^2$, $k_B^2 \bar{z}_B = 1/\sigma_B^2$, and $w_j = w_{jB} = 1$ [see Section 5.4.2(e)].

When $\lambda = 0$,

$$\sigma_t^2 \approx \frac{1}{pk_S^2\bar{z}} + \frac{(C_S + \bar{C}_w)^2}{k_S^2\bar{z}\Sigma w_j(C_j - \bar{C}_w)^2} \qquad (6.23)$$

where $\bar{C}_w = \Sigma w_j C_j/p$, and $\bar{z}$ and w_j are as defined in Section 5.4.2(e)(iii).

When $\lambda \neq 0$ and a blank correction is made by subtracting A_B from h:

$$\sigma_t^2 \approx \frac{1}{pk_S^2\bar{z}} + \frac{(C_S + \bar{C}_w)^2}{k_S^2\bar{z}\Sigma w_j(C_j - \bar{C}_w)^2} + \sigma_B^2 \qquad (6.24)$$

σ_t = total A.d.

If the same blank determination is used for more than one sample, $\sigma_b = \sigma_B$; otherwise, $\sigma_b = 0$.

When $\lambda \neq 0$ and a blank correction is made by subtracting R'_B from R'_S:

$$\sigma_t^2 \approx \frac{1}{pk_S^2\bar{z}} + \frac{(C_S + \bar{C}_w)^2}{k_S^2\bar{z}\Sigma w_j(C_j - \bar{C}_w)^2} + \frac{1}{pk_B^2\bar{z}_B} + \frac{\bar{C}_w^2}{k_B^2\bar{z}_B\Sigma w_{jB}(C_j - \bar{C}_w)^2} \quad (6.25)$$

where k_B, $\bar{z}_B$ and w_{jB} are the appropriate values for the blank measurements. When the same value of R'_B is used for more than one sample, σ_b^2 is equal to the last two terms on the right-hand side of equation (6.25); otherwise, $\sigma_b = 0$.

(c) Optimisation of Precision

Equations such as (6.21) to (6.25) allow one to investigate the effects of the values of p and C_j on σ_t; examples of such studies for calibration functions of the form $\mathscr{A}_T = \lambda + \chi C_S$ will be found in refs. 18 and 30 to 32, and see ref. 33 for comments on errors in ref. 32. Many such studies have also been made for analytical systems using ion-selective electrodes; see, for example, refs. 31 and 34 to 37.

6.3 ESTIMATION OF PRECISION BEFORE BEGINNING ROUTINE ANALYSIS

In Section 6.3.1 we discuss and suggest a general approach to this stage of precision estimation. On this basis, we deal in detail with the estimation of the total standard deviation, σ_t, in Sections 6.3.2 to 6.3.4. The estimation of the within- and between-batch standard deviations is considered in Section 6.3.5. Although these sections are primarily concerned with the estimation of precision to ensure that it is adequately small for routine analysis, many of the considerations apply equally to tests to characterise the precision achievable when a particular analytical system is used.

6.3.1 General Approach

The primary objectives of the preliminary estimation of precision may be summarised as follows (see Section 6.1): to demonstrate that the standard deviations of interest are adequately small and do not vary appreciably in time. Since random errors arise through uncontrolled variations of the many factors that affect analytical responses, the achieved degree of control of precision and its temporal stability rests entirely on: (1) the extent to which the relevant factors are recognised, and (2) the degree to which the variations of these factors are consistently controlled in routine analysis. In other words, to ensure adequate control of the precision achieved with a given analytical system, all important factors must be identified and the appropriate degree of their control specified and reliably achieved. The analytical method should be the primary means through which this identification and specification are made explicit and quantitative. Many authors have stressed the importance of ensuring that methods describe all aspects of procedures as exactly as possible, including the tolerable variations of the important experimental factors; see, for example, refs. 1 and 19, and the discussion of these and related aspects in Section 9.2.4.

Of course, the method *and* all other components of the analytical system – analysts, instrumentation, reagents, apparatus, environmental conditions, *etc.* – should be carefully selected so that the system is potentially capable of the required precision. This is often easier said than done, but important aspects of selecting methods are discussed in Section 9, and the other components are considered in Section 10. We regard this critical approach to the selection, specification, and control of analytical methods and systems as an essential preliminary to any meaningful estimation of the precision of analytical results. This view is reinforced by the problem mentioned next.

For situations like that of our general model of random errors [Section 6.2.1(*d*)] where there are both within- and between-batch random errors, Eisenhart[1] has recommended the tests necessary to allow a reasonable check of whether or not a state of statistical control has been achieved. In the present context, for each determinand concentration of interest, at least 4 portions of a solution should be analysed in each of at least 25 batches of analyses. Eisenhart also notes the procedures appropriate for the analysis of the results obtained. On this basis, the minimal amount of testing for our purposes might be two or three times the above amount in view of the common dependence of precision on the concentration of the determinand [Section 6.2.1(*f*)]. We certainly do not argue with Eisenhart's proposal, but in our experience this amount of effort is rarely if ever available in routine analytical laboratories. It seems to us, therefore, that it will generally be impracticable in these preliminary tests to demonstrate a state of statistical control. If this is accepted, it follows that even greater importance attaches to the points noted in the preceding paragraphs, *i.e.*, everything possible should be done by choice, control and operation of the analytical system to ensure a state of statistical control.

Given the above conclusions, the main aim of the preliminary tests then becomes the estimation of the standard deviation(s) of interest. To the maximum extent practicable, however, the tests should be designed to maximise the likelihood that inadequate and/or inconsistent control of random errors will be revealed.

Two broad approaches to this more restricted preliminary estimation of precision can be distinguished. The first involves the derivation of equations for σ_t, σ_w and σ_b like those in Sections 6.2.1(*c*) and 6.2.2 but appropriate for the particular calibration procedure involved. Provided that one has experimental estimates of the standard deviations and other parameters appearing in such equations, estimates of the standard deviations of analytical results can be calculated. The second approach is simply to make replicate analyses of appropriate solutions and to calculate estimates of the standard deviations directly from the results obtained. The first approach may offer economy of experimental effort in some instances, but its validity depends on a number of assumptions that are by no means necessarily true [see Section 6.2.1(*e*)]. The second approach makes no such assumptions. Further, the economy of effort that may be offered by the first approach is such that the chance of detecting inadequate and/or unreliable control of precision is smaller than for the second approach. It seems to us, therefore, that the second approach is much to be preferred, and should be adopted whenever possible. Of course, if the results obtained show the validity of the first approach, it may well be used to advantage in later work. Currie[2] has also noted the value of comparing the estimates of standard deviations obtained by these two approaches; such comparisons are also a means of seeking evidence for lack of statistical control. For an example of the first approach, see refs. 23, 38 and 39.

Since we assume that a state of statistical control will not usually be demonstrated, we use the term 'well controlled analytical system' to denote a system that has been carefuly selected, assembled and operated as noted above, and that produces analytical results with adequately small standard deviations.

6.3.2 Factors to Consider in Planning the Precision Tests

Given that every standard deviation of interest will be estimated experimentally from replicate analytical results, one must decide the number and nature of the solutions to be analysed, and the way in which the replicate analyses are to be arranged in time. Five generally important factors are considered in (*a*) to (*e*) below. Note also that, whenever contamination may cause problems, the tests in Section 10.8.1(*b*) will be highly desirable as a preliminary to those considered here.

(a) Stability and Homogeneity of Solutions

As the precision will be estimated from the differences among the replicate results for each solution, it is essential that these differences are caused only by the imprecision of the analyses. Each solution, therefore, must show negligible instability of the determinand concentration over the duration of the tests (see Section 3.6). In general, also, the determinand should be distributed homogeneously throughout each solution so that each portion analysed contains the same concentrations of determinand. See also (*e*) below.

(b) Duration of the Tests

Routine analysis implies regular analysis of given types of samples over relatively long time-periods. The tests should, therefore, provide reasonable evidence that the factors capable of affecting σ_t over such periods have been sufficiently well controlled. Clearly, it is impossible to prove this point because routine analysis could then never begin; see also Section 6.3.1. Some compromise is necessary in deciding a suitable period over which the tests are made. As the length of this period is increased and an adequately small estimate of σ_t is obtained, so one's confidence increases that the analytical system is well controlled. The choice of period is to some extent arbitrary, but very short periods such as one day seem to us generally open to great risk of falsely concluding that the analytical system is well controlled when, in fact, it is not. We believe that a period of two to four weeks is usually suitable, but shorter or longer periods may sometimes be appropriate. Westgard[3a] notes some agreement in clinical chemistry that precision be estimated over a period of at least 20 days. This factor is also related to the next to be considered in (*c*) below.

(c) Number of Batches of Analysis

A high proportion of routine analyses are made in batches [as defined in Section 6.2.1(*b*)], and analytical results are then generally subject to within- and between-batch random errors [Section 6.2.1(*d*)] that determine the size of σ_t. Accordingly, any experimental estimation of σ_t for analytical systems involving batches of analyses must be based on replicate analyses made over a number of batches. When only σ_t is to be estimated, the simplest approach is to make one analysis of each solution in each batch, but see also Section 6.3.5. If m batches of analyses are made, the estimate, s_t, of σ_t has $m-1$ degrees of freedom, and the random error of s_t can be quantified as described in Sections 4.3.4(*b*) and (*d*). If one is aiming for a value of σ_t no greater than a specified value, this uncertainty in s_t causes difficulty in deciding if adequate precision has been achieved or not. Uncertain estimates of σ_t also lead to problems in quantifying the random errors in routine, analytical results [see Section 4.3.5(*b*)]. Since the random error of s_t decreases as its degrees of freedom increase, it is desirable to make m as large as possible. This is a factor to be decided for each application, but as a general guide we suggest that m should not be less than 10 whenever n, the number of replicate analyses of a given solution in a batch, is 1; m is preferably greater than 10. Values for m of 2 or 3 are particularly to be avoided because the upper 95% confidence limits for σ_t are then approximately $16s_t$ and $4.4s_t$, respectively; that is, the true value of σ_t may be 16 or 4.4 times greater than the estimate, s_t.

It may be argued that, when equations such as those in Section 6.2.1(*c*) indicate that σ_b is negligibly small, σ_t may then be reasonably estimated by replicate analysis of appropriate solutions in one batch. This, however, in addition to not meeting the points made in (*b*) above, is invalidated if the various assumptions (on which such equations are based) do not hold [see Section 6.2.1(*e*)]. We feel, therefore, that such 'one-batch' estimation of σ_t should be strongly avoided whenever possible. Circumstances arise where an analyst may consider that it is impracticable to make more than one batch of analyses to estimate precision, and that the results from one batch will be better than none. The analyst must judge the first point, but the second needs special care given the observations in Sections 6.1 and 6.3.1. When 'one-batch' estimation of σ_t is judged unavoidable, we recommend careful consideration of the appropriate equation in Sections 6.2.1(*c*) or 6.2.2 (or of other equations appropriate for different calibration procedures) so that the random errors affecting σ_b according to the model are identified. It may then be possible by a few additional analyses in the batch to obtain estimates of at least some of these random errors. Some confirmation of the assumption that σ_b is negligible and/or a better estimate of σ_t may thereby be obtained. For example, in fixed-slope calibration [see Section 6.2.1(*c*)(*i*)], σ_b depends on the within-batch standard deviation, σ_B, of blank determinations and the variability of χ. Clearly, tests in one batch can provide no information on the latter factor, but some replicate blank determinations in a batch can provide an estimate of the former.

For a given number of degrees of freedom of the estimate, s_t, the number of batches required may also be decreased by making a number of replicate analyses of each solution in each of fewer batches. This approach may be helpful when σ_b is expected to be small, but the points made in (*b*) above remain applicable. The calculations involved in obtaining s_t are more complex than when one analysis of each solution is made in each batch, and are described in Section 6.3.5.

The question of the number of batches does not arise when independent estimates of the calibration parameters are freshly determined for each sample (see Section 6.2.2). For such analytical systems, the required number of replicates can be analysed at any time subject to the points made in (*b*) above.

(d) Determinand Concentrations

The total standard deviation of analytical results may well depend on the concentration of the determinand [Section 6.2.1(f)]. Thus, to characterise the precision achieved by an analytical system over its concentration range, a minimum of two determinand concentrations is required, and they should be equal or close to the lower and upper concentration limits of the system. Given the uncertainties of the estimates and the approximate nature of equation (6.18), it is preferable to use more than these two concentrations. Of course, if the concentrations in routine samples are all essentially the same, only one determinand concentration need be tested.

(e) Standard Solutions and Samples

The requirements in *(a)* and *(d)* above can sometimes be met only by the use of standard solutions, and indeed we generally recommend their use. A number of effects, however, may cause the precision for samples to differ from that for standards. It should, therefore, always be the aim to check whether or not these two precisions differ markedly. Thus, it is generally essential to include both standard solutions and samples in precision tests; see also Section 6.4.

6.3.3 Experimental Design for Estimating Precision

The discussion in Section 6.3.2 virtually defines the general nature of the experimental design for the preliminary estimation of precision, and it is illustrated schematically in Table 6.1 and discussed further below. To ensure that the closest possible approach is made to a well controlled analytical system (see Section 6.3.1), we recommend that these tests be carried out by an experienced, conscientious analyst.

The following points should be noted concerning Table 6.1

(1) All analyses should be made exactly as described in the analytical method; both the method of calibration (including any blank correction) for each batch and also the calculation of the final analytical results for each solution must be exactly those specified by the method. Any departure from these conditions, no matter how small, should be avoided. This implies, of course, that the analytical method must be accurately and completely specified (see Section 6.3.1). In addition, the analytical responses and results should be recorded with the finest discrimination possible irrespective of any rounding-off that may be envisaged in routine analysis [see Section 6.2.1(e)(vii)].

(2) Table 6.1 shows the use of q standard solutions in each batch. In general, we would suggest that $q \geqslant 3$, and the concentrations suggested when $q = 3$ are: $C_1 = 0$, $C_2 = 0.1$ to $0.2C_U$, and $C_3 = 0.9$ to $1.0C_U$, where C_U is the upper concentration limit of the analytical system. For some methods, $C_1 = 0$ may be inappropriate, *e.g.*, if it is known either that the response is zero for all concentrations less than some limiting value as in gravimetric procedures or that the calibration function involves $\log C$.

(3) When the limit of detection (see Section 8.3) of the analytical system is of interest, it will usually be desirable to include at least two blank determinations

TABLE 6.1

Experimental Design for Estimating Total Standard Deviation

Solution	Batch of analyses								
	1	2	3	·	·	·	·	·	m
Standard solution, C_1	x	x	x						x
Standard solution, C_2	x	x	x						x
·									
·									
·									
Standard solution, C_q	x	x	x						x
Sample 1	x	x	x						x
Sample 2	x	x	x						x
·									
·									
·									
Sample r	x	x	x						x
Sample 1+	x	x	x						x
Sample 2+	x	x	x						x
·									
·									
·									
Sample r+	x	x	x						x

x represents the production of one analytical result for the indicted solution; see also the notes in the text.

in each batch. When $C_1 = 0$, the simplest approach is then to include one more independent analysis of the same solution in each batch.

(4) Each of the *m* portions of a standard solution must have essentially identical determinand concentrations. This can be ensured by the accurate preparation of fresh standard solutions just before each batch of analyses – if stability of the standard solutions over the whole period of the tests is not certain. For each batch of analyses, the *q* standard solutions should be prepared from one homogeneous batch of purified water (*e.g.*, distilled or de-ionised water), and this same batch of water is preferably also used for preparing calibration standards and blanks. If desired, different batches of purified water may be used to prepare the standard solutions for each batch of analyses.

(5) Table 6.1 shows the use of *r* different samples in each batch, but for unambiguous interpretation of the results it is essential that the determinand concentration in each sample be stable over the entire period of the tests. If this stability is not certain, the results should be interpreted with caution; see also Section 6.3.5. As with standard solutions, the greater the value of *r* the better, but we suggest that at least one sample should always be included unless this is precluded by their poor stability.

(6) The solutions 'Sample+' in Table 6.1 are intended to indicate the analysis of a second portion of the indicated sample after a known amount of the determinand (as a small volume of an appropriate standard solution) has been

added to it. Such 'spiked' portions of the samples may help to indicate how precision varies with concentration, and they also provide a useful check of certain sources of bias.

(7) As the tests involve repetitive analysis of the same solutions, the analyst's knowledge of this could lead to somewhat optimistic estimates of precision. This possibility may be decreased by other, more involved experimental designs. We feel, however, that such complication should not be necessary provided experienced and conscientious analysts undertake the work, and provided that the purpose of the tests is made clear to all analysts concerned (see Section 10.3).

(8) It is also generally desirable to make the analyses required in each batch in a random order that is selected freshly for each batch. This helps to ensure valid estimates of standard deviations [see Section 6.2.1(*e*)(*vi*)]. Procedures for selecting a random order for a number of tests are described in statistical texts (see Section 4.8). Note, however, that the instructions of the analytical method must have priority. If, for example, the method specified that samples must be analysed in order of increasing concentration, this should also be the order followed in precision tests provided that results for samples are only to be reported in routine analysis when they have been analysed in such an order.

(9) With values of $q = 3$, $r = 2$, $n = 1$, and $m = 10$, this experimental design requires 7 to 9 analyses (depending on whether or not the limit of detection is of interest) on each of 10 occasions spread over the chosen period. This number of analyses may be thought excessive, but in our view it represents the minimum capable of giving the information suggested above.

6.3.4 Calculation and Interpretation of Results

(a) Standard Deviations

Given the m analytical results for a given solution, the estimate, s_t, of the total standard deviation is obtained as described in Section 4.3.4(*b*) using equations (4.12) or (4.13). If desired, confidence limits for σ_t can be calculated as in Section 4.3.4(*d*), and a test to decide if σ_t is greater than some target value, σ_{t0}, can be made as in Section 4.3.4(*e*)(*ii*).

A problem arises in these calculations of s_t (and of other standard deviations) if one or more of the results appears to be quite different from the others for the same solution. This question of 'outliers' is discussed in Section 7.2.10(*d*)(*ii*). In the present context, however, such results can only be regarded as evidence of inadequate control of the analytical system. In this situation, no one standard deviation can represent the precision of analytical results, and we do not attempt, therefore, detailed discussion of the various statistical tests for detecting 'outliers'. It seems to us generally more realistic – if a standard deviation is to be calculated – to include all results whatever their values. The crucial point, however, is to seek and eliminate the source(s) of 'outliers'. We realise that this may present great problems, and if it cannot be achieved, the only recourse would appear to be the routine replication of the analysis of all samples.

A simple approach to deciding whether the precision achieved is adequate is to accept the precision as satisfactory if, for each of the solutions included in the tests, s_t is not statistically significantly greater than the appropriate value, σ_{t0}. Bearing in mind, however, the relatively large uncertainties likely for s_t values and the rather few solutions likely to have been tested, we would suggest rather more detailed consideration of the s_t values along the following lines.

Suppose that the value of σ_t for each solution were equal to the corresponding σ_{t0} value. Then, as a consequence of the random errors of the s_t values, we would expect few if any to be significantly smaller or greater than σ_{t0} when a significance level of, say, 0.05 is used. Some s_t values would, however, be smaller than the σ_{t0} values, and some would be greater. As the true precision of analytical results, σ_t, becomes greater than σ_{t0}, so the proportion of s_t values exceeding σ_{t0} values will tend to increase though the s_t values may still not be significantly greater than the corresponding σ_{t0} values. With this in mind, the following suggestions are made.

(1) If all solutions give ratios s_t/σ_{t0} less than 1, it may reasonably be taken that precision is adequate.

(2) If all s_t/σ_{t0} values are significantly greater than 1, the precision is inadequate.

(3) If all s_t/σ_{t0} values are greater than 1 but none is significantly so, it may usually be taken (for $q \geqslant 3$ and $r \geqslant 2$) that the target precision has not been achieved. Whether or not the precision is then taken as inadequate may depend on the s_t/σ_{t0} values. For example, if they were all only slightly greater than 1, it is probably reasonable to accept the precision provisionally, but efforts to improve it are almost certainly worthwhile.

(4) There are many other intermediate situations where decisions cannot be so clear-cut, and each should have individual consideration, perhaps in consultation with the user of analytical results. Such problems of interpretation become smaller as m is increased, and will also tend to be decreased if the analytical system of interest is chosen and applied with a view to obtaining rather better precision than the target value. This latter practice has other advantages [see Section 4.3.5(*b*)] in quantifying the random errors of analytical results, and in estimating bias (see Section 5), and we recommend it whenever practicable.

(5) If only the samples give unacceptable precision, the individual analytical results should be examined for any evidence of instability of the determinand concentration, *e.g.*, a systematic change in the results in passing from batch 1 to batch m. In the absence of instability, only two factors can cause poorer precision for samples than for standard solutions, *i.e.*, inhomogeneous distribution of the determinand in samples, and inherently less precise analysis of samples. Further investigations would be necessary to identify the nature of such effects and/or to reduce their size; alternatively, another analytical system should be chosen and tested in a similar fashion.

(6) In Section 6.2.1(*f*) we concluded that σ_t commonly increases with increasing determinand concentration. It is useful to examine the s_t values from this point of view. If, for example, these values suggest smaller σ_t values at the higher concentrations, it is very likely that the precision estimates are in some way unreliable and that further investigation is required.

By examining the results in such ways, a decision is finally reached on whether or not the precision achieved is adequate; the aim should be not to start routine analysis before satisfactory precision has been achieved.

(b) Recovery of Determinand Added to Samples
If, in the i th batch of analyses ($i = 1$ to m), the analytical results for the 'spiked' and 'unspiked' portions of a sample are denoted by R_{Ai} and R_{Si}, respectively, the percentage recovery, f_i, for that batch is given by

$$f_i = 100(R_{Ai} - R_{Si})/\delta \qquad (6.26)$$

where δ is the equivalent concentration of determinand added to the 'spiked' sample. (Note that a correction for the dilution of the 'spiked' sample by the standard solution used to 'spike' it may be necessary.) From the m values for f_i, the mean, $\bar{f}$, and standard deviation, s_f, can be calculated and the confidence interval for the population mean recovery derived as in Section 4.3.5(a). If the $100(1-\alpha)\%$ confidence interval includes the value 100 %, it may be taken that the mean recovery does not differ significantly (α significance level) from 100 %. Alternatively, a test of significance can be made as described in Section 4.6.2(a) to determine if the mean recovery differs from 100 % or some other fixed value. It is stressed that satisfactory recovery does not necessarily prove the absence of important bias in the analytical results for samples [see Section 5.4.6(i)].

(c) Within-batch Standard Deviation of Blank Determinations
In many situations, the limit of detection of an analytical system is determined by the within-batch standard deviation of blank determinations (see Section 8.3). This standard deviation can be estimated when each batch of the tests includes two or more blank determinations or their equivalent, *e.g.*, the standard solution with $C_1 = 0$. For this purpose, each of the blanks in a batch should have been prepared from the same homogeneous batch of purified water.

Provided that the calibration function is of the form $\mathscr{A}_T = \lambda + \chi C_S$, the blank responses in each batch are converted to concentrations using the estimated calibration function or curve for each batch. An estimate of within-batch standard deviation is calculated for each batch using equation (4.12) or (4.13), Section 4.3.4(b), and these m estimates are then pooled using equation (4.15), Section 4.3.4(c), to give the pooled estimate of within-batch standard deviation in concentration units. When two blanks were analysed in each batch, the calculations are simplified by using equation (4.16), Section 4.3.4(c). For other types of calibration function, it may be necessary to make these calculations using the responses rather than concentrations, *e.g.*, when the calibration function is of the form $\mathscr{A}_T = \lambda + \chi \log C_S$ [see Section 8.3.1(a)].

(d) Checks of the Degree of Control of the Analytical System
For the reasons stressed in Section 6.3.1, it is important also to examine the results from precision tests critically for any evidence that the analytical system cannot be regarded as well controlled. When $n = 1$ and $m = 10$, there will usually be rather little scope for such checks, and their nature is also affected by the calibration procedure for the analytical system under test. Some examples of these checks are given below, but the essential point is to examine the results for abnormalities

and/or inconsistencies; the considerations in Section 6.2.1(*e*) will also help to identify relevant aspects.

(1) For fixed-slope calibration, the results for standard solutions indicate the validity of the slope value used in calculating the results for samples; the variability of the true slope from batch to batch is also indicated [see Section 6.2.1(*c*)(*i*)]. The slope value used should also be compared with any value expected theoretically; discrepancy may indicate faulty standards that might otherwise not be detected.

(2) The linearity of the calibration curve may be checked if sufficient calibration standards were included in the tests.

(3) For within-batch calibration, there should be no systematic differences between the true and determined concentrations of standard solutions. Such differences indicate a source of bias.

(4) The results for each solution should be examined to decide if there is any evidence for non-randomness of errors. For example, if the first five results for a given solution were greater than the mean of all results, and the last five were smaller than the mean, there is a very strong suggestion that the variations of results from batch to batch are not solely due to random errors. Simple tests for examining results in this way are described in many statistical texts and by Liteanu and Rica.[3] Another feature of results to consider is gradual drift in their values over the period of the tests. Liteanu and Rica[3] describe statistical tests for this effect as do many other statistical texts; see also refs. 40 and 41.

If any such inconsistencies are found, it will generally be desirable to seek and eliminate their source – even if the s_t values are satisfactory – before applying the analytical system routinely.

6.3.5 Extension of the Experimental Design of Table 6.1

(a) Value and Nature of the Extended Experimental Design
The tests in Table 6.1 are intended only to provide estimates, s_t, of the total standard deviation for each of the solutions analysed; no information on the relative magnitudes of the within- and between-batch standard deviations, σ_w and σ_b, is obtained. For some purposes, lack of data on these two latter standard deviations is unimportant. When, however, unacceptably large values of s_t are obtained, estimates, s_w and s_b, may be helpful in indicating likely sources of dominant random error, and thereby the most fruitful approach to improving precision. For example, if s_b were much greater than s_w, some error arising in the calibration of blank-correction procedures would be indicated. Values for s_w may also be useful in designing tests of interference and other sources of bias (see Sections 5.4.5 and 5.4.6), and in establishing control-charts for the routine checking of precision (see Section 6.4). Finally, there is greater opportunity for detecting deviations from the general model of random errors [Section 6.2.1(*d*)] when $n > 1$ [see, for example, (*b*)(8) below]. The above points are, to some extent, matters of choice, but there is one situation where this extended design is essential; the estimation of σ_t for unstable samples will usually require estimates of σ_w and σ_b [see (*c*) below].

Thus, designs providing estimates of within- and between-batch standard deviations as well as the total standard deviation provide a greater amount of useful information, and are worth adopting whenever the additional effort required is available. Such designs require two or more analyses of each solution in each batch of Table 6.1. When $n = 2$ (*i.e.*, two portions of each solution are analysed in each batch), σ_w and σ_b are estimated with approximately equal efficiency since the estimates then have m and $m-1$ degrees of freedom, respectively. In general, therefore, we recommend $n = 2$. If, however, it is thought that σ_b is small compared with σ_w, it may be more convenient to increase n and decrease m; this reduces the number of batches required, and ensures that the larger standard deviation is estimated with more degrees of freedom. For example, if one used $n = 4$ and $m = 5$, s_w then has 15 degrees of freedom and s_b has 4 [see (*b*)(5 and (7)] below). All the other points noted in Section 6.3.3 on the conduct of the tests apply to this extended design.

(b)   Calculation of s_w, s_b, and s_t, and Interpretation of Results
The calculation of s_w and s_b with these experimental designs can be made by appropriate modifications of the procedures described in Section 4.3, but are much better made by a statistical technique known as the analysis of variance (ANOVA). Statistical texts should be consulted (see Section 4.8) for the basis and many applications of this powerful technique, and see also Westgard in the analytical context.[3a] We have simply summarised below the calculations involved for this particular experimental design where, in general, one has n analytical results from each of m batches, and estimates of the within-batch, between-batch, and total standard deviations of results are required. The description is based on that in ref. 42.

For a given solution, let the j th result ($j = 1$ to n) in the i th batch ($i = 1$ to m) be R_{ij}. The calculations for each solution then proceed in stages as follows.

(1)   Calculate $A = \Sigma R_{ij}^2$, *i.e.*, A is the sum of the squares of all the individual results.
(2)   Calculate $R_{i.} = \Sigma R_{ij}$ for each value of i, *i.e.*, calculate the sum, $R_{i.}$, of the results in each batch. Then calculate $B = \Sigma R_{i.}^2/n$, *i.e.*, square the batch sums, add these squares, and divide their sum by n.
(3)   Calculate $R.. = \Sigma R_{i.}$, *i.e.*, calculate the sum of the batch sums. Then calculate $C = R..^2/mn$, *i.e.*, square the sum of all the results, and divide this square by mn.
(4)   Note that since relatively small differences between large numbers may well be involved in subsequent calculations, rounding-off results should be avoided, and the calculations of A, B and C made with the greatest possible accuracy. A check on accuracy of calculation is provided by the fact that $(A-B)+(B-C)$ should equal $A-C$ (see below).
(5)   Calculate $M_w = (A-B)/m(n-1)$. M_w is called the within-batch mean square, and is an estimate of σ_w^2 with $m(n-1)$ degrees of freedom. Thus, M_w is our estimate, s_w^2.
(6)   Calculate $M_b = (B-C)/(m-1)$. M_b is called the between-batch mean square, and is an estimate of $n\sigma_b^2+\sigma_w^2$ with $m-1$ degrees of freedom. Thus, if $\sigma_b = 0$,

M_b is another estimate of σ_w^2. Accordingly, if M_b is statistically significantly greater than M_w, this is evidence of statistically significant between-batch variability, *i.e.*, that $\sigma_b > 0$. The statistical significance of M_b in relation to M_w is assessed by treating M_b and M_w as two variances, and applying the test described in Section 4.3.4(*e*)(*ii*); the testing proceeds as follows.

(7) When $M_b > M_w$, calculate M_b/M_w. Two cases arise:

If $M_b/M_w > F_\alpha$ (for the degrees of freedom of M_b and M_w), M_b is significantly greater (at the α significance level) than M_w, and an estimate, s_b^2, is obtained from $s_b^2 = (M_b - M_w)/n$; s_b has $m-1$ degrees of freedom. An estimate of σ_t^2 is then given by $s_t^2 = s_w^2 + s_b^2$, *i.e.*, $s_t^2 = (M_b + [n-1]M_w)/n$.

If $M_b/M_w \leqslant F_\alpha$, there is insufficient evidence (at the α significance level) to justify the conclusion that $\sigma_b > 0$. It is best then to regard the estimate s_b as 'not significant', but the estimate, s_t^2, is obtained from the same equation as that just given above.

(8) When $M_b < M_w$ some problems of interpretation arise. From the analytical viewpoint, the best approach is probably the following. Calculate M_w/M_b; again, two cases arise:

If $M_w/M_b \leqslant F_\alpha$, equate s_b to zero (note that this does not necessarily imply that $\sigma_b = 0$), and equate s_t^2 to M_w, *i.e.*, $s_t^2 = s_w^2$.

If $M_w/M_b > F_\alpha$, the results should not be regarded as providing valid estimates of random errors because the variability of the batch sums is smaller than could be expected from the variability of results within batches. The most likely explanation for this abnormality is that a systematic error has affected the differences between replicate results in the same batch by approximately the same amount in each batch [see Section 6.2.1(*e*)(*vi*)]. The experimental procedures and techniques should, therefore, be critically examined to discover and eliminate this source of error, and the tests repeated.

(9) It will often be necessary to know the number of degrees of freedom that can be assigned to the estimate s_t. The degrees of freedom, d, are given by:

When $M_b > M_w$,

$$d \approx \frac{m(m-1)[M_b + (n-1)\,M_w]^2}{mM_b^2 + (m-1)(n-1)M_w^2}.$$

This may give non-integral values for d, but use the nearest whole number.

When $M_b \leqslant M_w$, $d = m(n-1)$.

If the tests included 'spiked' and 'unspiked' samples, the simplest approach is to calculate the mean recovery, $\bar{f_i}$, for each batch, and then to follow the procedure in Section 6.3.4(*b*) using the $\bar{f_i}$ values in place of the f_i values. The within-batch standard deviation of blank determinations is calculated as described in Section 6.3.4(*c*); alternatively, the procedure in (1) to (5) above may be used. Comparison of estimates of standard deviations with target values and critical examination of results for abnormalities are described in Sections 6.3.4(*a*) and (*d*), respectively.

(c) Estimation of σ_t for unstable samples
An important application of this more involved experimental design arises when the determinand concentrations in samples are known not to be sufficiently stable to

allow the direct estimation of σ_t as described in Sections 6.3.3 and 6.3.4. In this event, values of s_w and s_b for standard solutions and of s_w for samples allow estimates to be derived for s_b and s_t for samples. Thus, since, σ_b depends only on the calibration procedure [see Sections 6.2.1(c) and (d)], which is independent of the nature of the analysed solution, the estimates, s_b, from standard solutions should be equally valid for samples. It must only be borne in mind that σ_b generally depends on the concentration of the determinand, and that the nature of this dependence is governed by the particular calibration procedure used [see Section 6.2.1(c)]. If the s_b values for standard solutions are in accord with the appropriate equation for σ_b, this equation may be used to calculate values of s_b for samples; s_t is then given by $s_t^2 = s_w^2 + s_b^2$. Alternatively, the values of s_b for standards may be used to obtain interpolated values for s_b at the determinand concentrations present in samples.

If this approach is followed, caution is needed on one point. If the same unstable sample is used for all batches of analyses, the value of σ_w may change during the tests if the instability is sufficient to change the determinand concentration substantially [see Section 6.2.1(f)]. An alternative is to use a fresh sample of the same type for each batch of analyses provided that the determinand concentrations do not change appreciably from sample to sample.

(d) Use of the Extended Experimental Design when Standard-Addition Procedures are Used

We saw in Section 6.2.2 that, according to the general model of random errors, analytical results produced by standard-addition procedures should not generally be affected by between-batch random errors. As noted, however, in Section 6.2.1(e), there are a number of reasons why this general model may be invalid. It is, therefore, useful to undertake this arrangement of precision tests in batches and the analysis of variance of the results even when standard-addition procedures are used. If, for example, statistically significant between-batch variability were found, this would be evidence of unsuspected sources of error (see Section 6.3.1).

6.4 ROUTINE CHECKING OF THE ACCURACY OF ANALYTICAL RESULTS

6.4.1 Need for Regular Programme of Tests

When the preliminary tests in Section 6.3 have shown an analytical system to be well controlled in the sense described in Section 6.3.1, *and* when the bias of analytical results is also adequately small (see Section 5.5), the system may then be put into routine use with some confidence that it is suitable for the task. Components of the system, however, will change with time, and the degree of control of the system may also show unintended variations. Such changes may affect the precision and/or bias of analytical results. It is essential, therefore, to maintain a continuing programme of regular tests, as an integral part of routine analysis, with the aim of detecting any important deterioration of accuracy as rapidly as possible so that the appropriate corrective action can be taken. Tests with this purpose are called here *control-tests*. When deterioration is detected, it will usually be necessary to suspend the analysis of samples until adequate control of the analytical system has been re-established.

In routine analysis, analysts often meet difficulties in finding the effort to analyse speedily all the samples presented to them. There is, therefore, a strong incentive to reduce to the maximum extent possible the amount of effort devoted to control-tests (see also Section 4.7). In this position, it is important to optimise the programme planned for control-tests in two respects:

(1) one must identify the control-test(s) most likely to reveal the errors to which analytical results for a given determinand are thought to be most subject, and
(2) one must ensure that the results from control-tests are used with maximal efficiency.

To aid with (1), we consider in Sections 6.4.3 to 6.4.7 several different types of control-test of common value, and in Section 6.4.2 we describe the use of quality-control charts which are useful in relation to (2).

We would stress that the information from preliminary tests on the sources and sizes of errors in analytical results is usually of considerable value in the selection of control-tests; it also allows control-charts to be established from the very beginning of routine analysis, and this is a very valuable feature. If no preliminary tests of precision or bias were made, a programme of control-tests is, of course, all the more necessary, but some time will pass before the optimal tests can be decided (see also Section 6.1), and control-charts established.

6.4.2 Principles and Use of Quality-control Charts

(a) Types of Control-Charts

Suppose that a control-test, *e.g.*, a standard solution, were analysed with each batch of routine analyses. The simplest procedure for using the results is merely to inspect each as it is obtained, and to decide whether or not it is sufficiently close to the expected value to indicate adequate control of the analytical system. This approach, though sometimes used, is subjective; the judgement is affected by the extent to which previous results for the control-test are remembered, and the results are not clearly and explicitly related to the accuracy achieved in preliminary tests or the accuracy required. An analytical quality-control chart largely overcomes these deficiencies by plotting all results of the control-test sequentially on a chart which includes means for statistical judgement [see (*b*) below].

There are several types of control-chart; that to be described below is known as the Shewhart-type whose use and interpretation are particularly straightforward. This type is discussed by Nalimov[15] in some detail and in many statistical texts (see Section 4.8). Another type of control-chart (the CUSUM-chart) was developed later (see, for example, refs. 43 to 43*d*), and has been suggested by a number of authors to be of value in analytical quality-control; see, for example, refs. 44 and 45. Investigation[45] of the relative advantages and disadvantages of the two types leads us to conclude that the CUSUM-charts may have some advantages in aiding the visual detection of increasing errors when the charts are examined. On balance, however, for laboratories establishing a quality-control programme, we would suggest the use of the Shewhart type because of its relative simplicity. Westgard and Groth[13] note that CUSUM-charts have been tried in clinical laboratories in the U.S.A. but that their use is not widespread, probably because of their greater complexity. In the following, therefore, we consider only Shewhart-type charts. For laboratories, however, in which an effective quality-control programme using Shewhart charts has already been established, CUSUM techniques can form a useful, additional check, and the references cited may be consulted for details. It should be noted that the use of control-charts in routine analysis is recommended by many authors; see, for example, refs. 4, 11, 15, 21, 26, 42, 46 and 47. We deal below with charts concerned with the errors of analytical results, but it can be of value also to use control-charts for important components of analytical systems, particularly the measuring instruments;[48,49] see also Section 10.4.2.

(b) Principles of Quality-control Charts and Control-tests

Consider a control-test whose results may be regarded as members of a normally distributed population with mean, μ, and standard deviation, σ. Suppose also that the result, τ, expected from this control-test in the absence of errors is known with negligible error, and that $\tau = \mu$. The properties of the normal distribution [see Section 4.3.2(a)] allow calculation of the proportion, α, of results from the control-test differing from $\tau(=\mu)$ by more than a given amount. In general, $100\alpha \%$ of results differ from τ by more than $\pm u_\alpha\sigma$. For example, only 0.3 % of all results differ from τ by more than $\pm 2.97\sigma$. If such a control-test were made weekly, we would expect, therefore, that only once every six years, on average, would such a result be obtained. Thus, it is so unlikely that such a result will be obtained from the population defined above that, given such a result, one may reasonably conclude that the population has changed. If this change is due to a change in the value of μ, this represents the detection of bias since μ will not now equal τ (see Section 4.4). If the change in the population is due to a change in σ, this represents detection of a worsening of precision. Of course, it is possible that the change in the population is due to changes in both μ and σ. In any event, however, a result from the control-test outside the limits, $\tau \pm 2.97\sigma$, may be taken as a strong indication that the accuracy achieved with the analytical system has deteriorated, and that corrective action is required.

The above considerations form the basis of the simple type of control-chart considered here. The results from a control-test satisfying the requirements noted above are plotted sequentially on a chart as they are obtained. The chart has lines drawn on it at values corresponding to τ and $\tau \pm 2.97\sigma$, and the need for corrective action is generally indicated whenever a result falls outside the two limits, $\tau \pm 2.97\sigma$. These two limits are, therefore, commonly referred to as the *action limits*. It is also common and useful to insert two other lines – the *warning limits* – at values corresponding to $\tau \pm 1.96\sigma$. On average, 5 % of results from the control-test would be expected to fall outside these limits. This frequency is such that a result falling outside the warning limits but within the action limits does not call for action so long as the next result falls within the warning limits. In general, therefore, the warning and action limits are set at $\tau \pm u_{\alpha w}\sigma$ and $\tau \pm u_{\alpha a}\sigma$, where α_w and α_a are chosen appropriately. It is, however, very common and practically useful in analysis to put $\alpha_w = 0.05$ and $\alpha_a = 0.003$. Note that there is a probability α_a of needlessly taking action when a result exceeds the action limits.

Some typical examples of such control-charts are illustrated in Figure 6.2 for the case where the control-test consists of the analysis of a standard solution of the

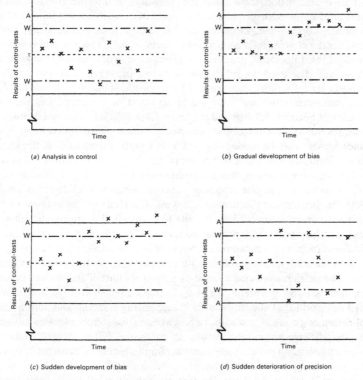

Figure 6.2 *Examples of control-charts using a control-standard.* $\tau = true\ concentration$; $W = warning\ limits$; $A = action\ limits$

determinand. As noted above, the results of control-tests of the type defined above are, in general, subject to changes in both precision and bias. Changes in bias are reflected by changes in the average level of results on a chart (Figure 6.2 b,c). Precision is reflected by the scatter of results; for example, if results begin to appear more frequently than expected outside both warning limits (Figure 6.2 d), deterioration of precision is clearly indicated. It should be stressed, however, that the precision and/or bias of the control-test may deteriorate without being detected immediately by an individual control-test. For example, suppose that between one batch of analyses and the next, σ increased by a factor of two. There would then still be a chance of 2 in 3 that the next result of the test would fall within the warning limits. One would only observe this deterioration of precision with some confidence after several batches of analyses. Much more detailed and quantitative considerations of this aspect of control-charts will be found in refs. 13, 44 and 45. It emphasises the value of running as many control-tests as possible since this will tend to increase the rapidity with which changes in the analytical system are detected. There will also be value in attempting to achieve rather better precision and, if possible, bias than those required by the user of analytical results because this will provide a 'safety-margin'.

The above description shows that the quantitative interpretation of control-charts requires knowledge of τ, σ, and the nature of the distribution followed by the results of the control-test. Deciding the value of τ does not generally present problems for the control-tests considered in Sections 6.4.3 to 6.4.6, and is discussed in those sections. Important, practical problems arise in determining the value of σ, and these are discussed in (c) below. Finally, provided that the results of control-tests are either analytical results for a special solution or linear combinations of such results, we feel that the assumption of a normal distribution is reasonable [see Sections 4.3.2(*a*) and (*b*), and 6.2.1(*e*)(*x*)]. If it is known or suspected that the results of a control-test are not normally distributed, appropriate allowance for this may be made if desired by calculating the values of the warning and action limits on the basis of the properties of the distribution followed by results. It is also worth noting that the assumption of normality becomes more and more likely to be valid as one replicates the control-test in each batch of analyses and plots the mean result [Sections 4.3.2(*a*) and 6.2.1(*e*)(*x*)]. The amount of effort available for control-tests is likely usually to preclude this approach, but if it is followed, careful consideration must be given to the effect of replication on the standard deviation of the mean result. Suppose, for example, that the control-test consisted of the analysis of a standard solution, two portions being analysed in each batch and the mean result being plotted on a control-chart. The standard deviation of such means is, in general, $\sqrt{(\sigma_b^2 + \sigma_w^2/2)}$ [see equation (6.17)], and it is this standard deviation that should be used in calculating warning and action limits.

Replication of a control-test within a batch of analyses offers a further advantage in that the differences among results in the same batch *directly* reflect the within-batch precision without any complicating influence of changes in bias. We consider one example of this approach in Section 6.4.5.

In Sections 6.4.3 to 6.4.7 we describe several types of control-test that are commonly useful for a wide variety of analytical systems and applications. The types described are:

(1) analysis of a standard solution (Section 6.4.3),
(2) analysis of a special sample (Section 6.4.4),
(3) analysis of two portions of a sample (Section 6.4.5),
(4) recovery tests (Section 6.4.6),
(5) blank determinations (Section 6.4.7).

None of these types is capable alone of indicating all the errors to which analytical results for routine samples are subject. Ideally, one would use all of them, but when this is impracticable the test(s) to be used should be very carefully selected for each analytical system of interest. We would also emphasise that any experimental test, whose results reflect in some way one or more errors to which routine analytical results are subject, may be used as a control-test provided that values for τ and σ are available, and that the nature of the distribution followed by the results can be reasonably assumed. For example, if an analytical system were particularly liable to an interference effect whose magnitude was difficult to control, a direct estimate of that effect could well be a more useful control-test than any of (1) to (5) above.

It is very desirable that at least one control test, and preferably more, be made for each batch of routine analyses – or analytical occasion if the concept of 'batch' does not apply [see Sections 6.2.1(*b*) and 6.2.2]. The results for samples in a batch should, in general, be reported only if the control-tests indicate no deterioration of accuracy. If, however, effort is limited, it may be acceptable to reduce the frequency of tests for those analytical systems that have in the past proved to show greatest stability and reliability.

Before considering commonly useful control-tests, we deal in (*c*) below with some practical aspects of the construction and use of control-charts.

(c) Practical Aspects of Control-Charts
In practice, the population standard deviation, σ, of the results of a control-test will not be known, and an estimate, s, of σ must be used. One of the valuable features of preliminary tests of precision (Section 6.3) is that a value for s will often be provided so that a control-chart may be constructed before routine analysis begins. When such an estimate is not available, the control-tests will have to be made routinely (without the benefit of a control-chart) until sufficient results (10 or more) become available to allow a reasonable estimate of σ to be calculated.

Given the error of the estimate, s, it is adequate to draw the warning and action limits at $2s$ and $3s$, respectively.

The use of s rather than σ involves the same difficulty discussed in Section 4.3.5(*b*). In the present context, however, we prefer the use of $\tau \pm u_\alpha s$ to $\tau \pm t_\alpha s$ because, though neither will generally include $100(1-\alpha)\,\%$ of results, the latter limits will include a greater proportion than the former. The chance of detecting worsening accuracy will, therefore, be greater when $\tau \pm u_\alpha s$ is used – though this will also gives a greater chance of falsely concluding that action is required when, in fact, it is not. The quantitative effect of this difficulty decreases as the degrees of freedom of s increase, and this is another important reason for making the degrees of freedom as large as possible in preliminary tests of precision (see Section 6.3.2).

The number of results from a control-test will gradually accumulate with time. Provided that no deterioration of accuracy has been detected, these results should be used to obtain an improved estimate of σ, and new warning and action limits inserted on the chart in place of the old. For example, suppose the chart were initially constructed using an estimate, s_1, with 10 degrees of freedom, and that after some weeks 20 control-tests had been made. The standard deviation, s_2, of the results from these control-tests should be calculated, the estimates, s_1 and s_2, pooled [Section 4.3.4(*c*)], and the pooled estimate, s (with 29 degrees of freedom) used to re-calculate and re-draw the warning and action limits. Such re-calculation is worthwhile again when a further 20 to 30 control-tests have been made.

Of course, given uncertainties in the estimate of σ and the precise nature of the distribution followed by the results, exact interpretation of the proportions of results falling within and without given limits is not possible. Many authors, however, have stressed – and we strongly agree – that this does not destroy the value of control-charts. They are useful in practice in ensuring that, as a result is plotted on a chart, it is interpreted in the light of previous results; in addition, the warning and action limits help to provide a reasonable and objective basis for making decisions.

To derive maximal value from control-charts, it is essential to plot the results from the control-tests in a given batch of analyses as soon as they have been obtained and before reporting the results for samples analysed in the same batch. Laboratory managers should ensure this prompt plotting of results.

6.4.3 Use of a Standard Solution for Control-Tests

For this type of test, one or more portions of a standard solution are analysed in each batch of analyses by exactly the same procedure as that used for samples. The value of τ is the true concentration of the standard, and it is clearly important that it be prepared so that its concentration is known with negligible error. If n ($n \geqslant 1$) portions of a standard are analysed in each batch, and the mean result is plotted on the control-chart, the relevant standard deviation, σ, is given by [see Section 6.2.1(*d*)]

$$\sigma = \sqrt{(\sigma_b^2 + \sigma_w^2/n)}$$

Thus, when $n = 1$, $\sigma = \sigma_t$.

The advantages and disadvantages of this type of control-test may be summarised as follows.

Advantages

(1) It is usually possible to use any determinand concentration of interest, and this concentration may be reproduced whenever required with excellent accuracy so that the value of τ is always known – except for non-specific determinands (see below) – and the same value can be used indefinitely.

(2) The results are subject to both within- and between-batch random errors as well as some sources of bias arising from the calibration and blank-correction procedures. Given the limited amount of effort usually available for control-tests, it is useful that one test responds to several sources of error. The scope of the test in this sense may be extended if desired by including potentially interfering substances as well as the determinand in the standard solution.

Disadvantages

(1) As results are subject to several sources of error, it may be difficult to decide which source of error has led to a detected deterioration of accuracy. The main function of control-tests, however, is to detect deterioration in the control of the analytical system; further investigations will usually be required in any event to identify and eliminate the cause. Thus, this disadvantage is considered relatively unimportant.

(2) The test does not directly reflect the accuracy of results for samples. This is a more important point, but it will be at least partly countered when the results from the preliminary precision tests (Section 6.3) have shown that the precisions for standards and samples were essentially the same. When this is so, it will often be reasonable to take continuing satisfactory precision for a standard solution as evidence of satisfactory precision for samples. In any event, it should be noted that control-tests giving direct estimates of the total standard deviation for samples are

impossible except when the determinand concentrations in samples are stable almost indefinitely (see Sections 6.4.4 and 6.4.5).

From the above we believe that a standard solution is usually the single most useful control-test, the main exception being when it is known that the precision for standard solutions does not reasonably represent the precision for samples. In addition, when non-specific determinands (see Section 2.4.2) are of interest, appropriate standards may be unavailable or the values of τ for standards may not be known, only experimental estimates subject to errors being available. In the latter event, it may be sufficient for a laboratory to use an experimental mean in place of τ, but the random error of this mean [Section 4.3.5(a)] will cause additional uncertainty in the construction and interpretation of the control-chart. Each situation requires individual consideration, but this type of control-chart may well be much less useful for non-specific determinands.

Several practical details concerning the use of control-standards are worth noting. First, there is the question of whether or not the analyst should know the concentration of the standard before it is analysed, and indeed whether or not he should know which of the samples he has to analyse are the controls. On general grounds, it seems desirable that he should know neither of these two points. To achieve this aim, however, may be extremely difficult, particularly in small laboratories, and time-consuming. We feel, therefore, that although this aim is useful, it is not essential provided that all members of a laboratory are properly aware of the need to ensure that results from the laboratory meet carefully defined standards of accuracy (see Section 10.3). It is worth noting, however, that if it is required that the analyst does not know the concentration of the control-standard before analysis, this involves a second analyst preparing the control-standard to have a concentration, varied from batch to batch, within a narrow (say, $\pm 5\%$) range of the chosen concentration, C. If this is done, the differences between the results and C should now be plotted on the control-chart. The value of τ is then zero, and the warning and action limits are drawn at $0 \pm 2s_t$ and $0 \pm 3s_t$, where s_t is the estimate of the total standard deviation of the results. It is usually necessary to make the range of concentrations rather narrow to ensure that the interpretation of the results is not complicated by any dependence of the standard deviation on the concentration of the determinand.

Many analytical systems involve within-batch calibration. With such systems, if the control-standard is prepared for each batch of analyses from the same stock standard solution as that used to prepare the calibration standards, deterioration of the stock solution – and hence inaccurate calibration – may well go undetected. Some defence against this may be obtained by preparing calibration and control standards from separate stock solutions and by recording any sensitivity settings of the measuring instruments for each batch of analyses. The better approach, of course, is to eliminate the possibility of any important deterioration in standard solutions.

The choice of the concentration(s) of determinand in the control-standard(s) involves several considerations. When the precision of analytical results varies with determinand concentration, and samples contain a wide range of concentrations, one would ideally analyse at least two control-standards with concentrations at or near the limits of the concentration range of interest. The effort required, however, may not be available, particularly if other types of control-test are also considered necessary. If only one control-standard is used in these circumstances, the best concentration will usually be that for which the relative standard deviation [Section 4.3.2(*a*)] is smallest because this will allow the smallest percentage changes in precision to be detected. The results from the preliminary tests (Section 6.3) are of value in indicating this concentration which is often at or close to the upper concentration limit of the analytical system. If the accuracy of the results at such a concentration remains satisfactory, it is very likely that the same will be true for standard solutions of other concentrations. Of course, the choice of concentration presents no problem if the precision of results is independent of concentration or if samples all contain essentially the same concentration of the determinand.

6.4.4 Use of a Sample for Control-Tests

In some circumstances it is possible to store a large volume of one or more samples of suitable determinand concentration, and to use them in place of, or as well as, a control-standard. This type of control-test has the advantage that its results provide direct evidence on the precision, σ_t, (and some sources of bias) for samples, but there are two very important limitations.

(1) The true concentration of determinand in a sample is not generally known. This introduces additional uncertainty on the correct positions of the lines on the chart for τ and the action and warning limits. If the sample has been analysed many times in the past, the mean result may have a sufficiently small uncertainty [see Section 4.3.5(*a*)] to allow this value to be used in place of τ.

(2) It is essential that the concentration of the determinand does not change over the long period of time for which control-tests are required. This may be difficult or impossible to ensure for many determinands in water analysis. Further, when a sample has all been used, it may well be difficult or impossible to replace it by another of the same concentration so that point (1) recurs.

In addition to these two points, non-homogeneous distribution of the determinand throughout a sample may prevent the assessment of the errors of the analytical system alone. For these reasons, we feel that this approach has only limited application. A variant, however, may be of considerable value, and is well worth noting. If a large volume of a sample containing a determinand concentration negligibly small in relation to the concentrations of interest is available, it can be 'spiked' with an accurately known concentration of the determinand so that the true concentration of the 'spiked' sample is known with negligible error. This procedure overcomes problem (1) above though the stability of the 'spiked' sample remains crucial. Of course, the chemical and/or physical forms of the determinand in the 'spiked' sample may differ from those present in samples naturally containing the determinand; whether or not this is an important factor can be judged only for each particular application (see Section 5.4.5). This approach can clearly also be followed if the determinand content of a sample containing a very small concentration can be measured with both random and systematic errors negligibly small in relation to the concentrations of interest in routine samples.

6.4.5 Analysis of Two or More Portions of a Sample as a Control-Test

Some direct evidence on the precision of sample analyses can be obtained by analysing two or more portions of one or more samples in each batch of analyses. The effort available will usually imply that only two portions of one sample need be considered here; clearly, however, when effort is available, analysis of more than two portions of a given sample will give less uncertain indications of the within-batch precision [Section 4.3.4(*d*)]. Two papers by Thompson and Howarth[17,50] on the estimation of precision in routine analysis by the use of duplicate analyses should be noted here. The choice of the samples from the viewpoint of determinand concentration is governed by similar considerations to those in Section 6.4.3. The construction of the control-chart involves slightly different ideas to those described so far.

Let the first and second results obtained in each batch for the selected sample be R_1 and R_2. The differences, $R_1 - R_2$, are then plotted on the chart, the second result always being subtracted from the first. Since both results are estimates of the same concentration (assuming homogeneous distribution of the determinand throughout the sample), the value of τ for this chart is zero. The main random errors affecting $R_1 - R_2$ are those arising from within-batch errors,* and are, therefore, characterised by the within-batch standard deviation, σ_w [see Section 6.2.1(*d*)]. Since both R_1 and R_2 have standard deviations of σ_w, the standard deviation of their difference is $\sqrt{2}.\sigma_w$ [see Section 4.3.2(*b*)]. Thus, the warning and action limits are drawn at values of $0 \pm 2\sqrt{2}.s_w$ and $0 \pm 3\sqrt{2}.s_w$. The chart may then be interpreted as described in Section 6.4.2, and its advantages and disadvantages may be summarised as follows.

Advantages

(1) Provided that no systematic effect operates within a batch of analyses so that R_1 tends to differ systematically from R_2, the results from this test can be interpreted purely with respect to precision without the added complication of bias.
(2) The results directly reflect the constancy of σ_w for samples. This is an important point if it is known or suspected that the precision of sample analyses is worse than that for standards.

Disadvantages

(1) The results, $R_1 - R_2$, are not affected by sources of between-batch random error. Such errors, however, do affect the results for control-standards so that if the latter remain in control and the differences also remain in control, it may reasonably be inferred that the total standard deviation for samples has remained in control.
(2) The test provides no information on bias, but compare advantage (1).
(3) If the experimental design of Table 6.1 were used in the preliminary tests of Section 6.3, no estimate is obtained of σ_w. The warning and action limits, therefore, cannot be drawn until a reasonable number (10 or more) of samples have been analysed in duplicate in routine analysis. If, however, the extended experimental

---

* This assumes negligible random error in the estimate of the slope of the calibration curve used to calculate the two results.

design of Section 6.3.5 were used, this provides an estimate of σ_w for samples, but see (4) below.

(4) If σ_w varies with determinand concentration and the determinand concentrations in samples vary appreciably, the results for this control-test will be governed by different standard deviations in different batches of analyses. Thus, no one set of warning and action limits will be generally applicable to all samples, and any reasonably quantitative interpretation of the control-chart becomes difficult or impossible. This problem may be decreased if it is possible to select samples for the test so that their concentrations do not differ appreciably. Alternatively, it may be feasible to establish a chart for each of a number of narrow ranges of concentration and results from the control-tests entered on the appropriate chart. Finally, this problem may be largely eliminated if, within the concentration range present in samples, the relative standard deviation is constant or essentially so. In this event, the differences between the two results can be plotted as a percentage of the mean result, *i.e.*, one plots $200 (R_1 - R_2)/(R_1 + R_2)$. The value of τ is then again zero, and the standard deviation to be used for calculating the warning and action limits is $100\sqrt{2}.s_w/C$, where C is the concentration of the determinand. This approach raises the problem that, even when R_1 and R_2 are normally distributed, $200 (R_1 - R_2)/(R_1 + R_2)$ may tend to depart from normality (see Section 4.6.3). Note also that results from preliminary tests (Section 6.3) are needed to decide at the beginning of routine analysis if σ_w varies with concentration, and if this variation is sufficiently eliminated when $100\sigma_w/C$ is considered.

From the above discussion, we conclude that it is well worthwhile to attempt to apply this type of control-test, particularly when it is known or suspected that the precision for standard solutions does not reasonably reflect that for samples. Very careful consideration must, however, be given to disadvantages (3) and (4) if valid interpretation of results is to be ensured. If the test is used, we also suggest that disadvantage (1) is a strong reason for also running a control-standard whenever possible.

6.4.6 Use of Recovery Tests

None of the above control-tests directly responds to certain sources of bias (*e.g.*, interference) in the analysis of samples. Recovery tests, however, can detect certain types of interference [see Section 5.4.6(*i*)] and may also be used as control-tests; indeed, they are strongly recommended as control-tests in refs. 21 and 26.

Suppose that a second portion of one of the samples in a batch of analyses is 'spiked' with an amount of the determinand equivalent to a concentration, δ, in the sample, and then analysed in the same batch as the 'unspiked' portion. The percentage recovery, f, may then be plotted on a control-chart. The value of τ for this chart should normally be 100 % since we assume that if it were known that the experimental mean recovery differed from 100 %, analytical results would be corrected accordingly. The warning and action limits are then drawn at values of $100 \pm 2s_f$ and $100 \pm 3s_f$. Again we meet the problem that if σ_w varies with the determinand concentration, then σ_f will vary with the concentration of the sample chosen for the test [see Section 5.4.6(*i*)]. This type of control-chart, therefore, will be of limited value except when one of the following conditions is met.

(1) σ_w is essentially independent of concentration. Note that the more extensive tests of Section 6.3.5 are necessary to obtain this information. If the design in Table 6.1 was used for the preliminary tests, it will be prudent to assume that this condition is not met [see Section 6.2.1(f)].

(2) The samples used for recovery tests can be selected so that their determinand concentrations are all essentially the same.

(3) Separate control-charts, with appropriate values of s_f, are kept for each of a number of narrow ranges of concentration in the 'unspiked' samples.

Provided that one of the above conditions is met, this type of control-chart is useful, but the relatively large value of σ_f and the inability of the test to detect bias independent of the determinand concentration [see Section 5.4.6(i)] should be borne in mind.

An estimate, s_f, of the standard deviation may be calculable from the results of the preliminary tests [Sections 6.3.4(b) and 6.3.5] if those tests included a sample of appropriate concentration and the 'spike' was that selected for routine use. Otherwise, s_f should be calculated as described in Section 6.3.4(b) when a reasonable number (10 or more) of recovery tests have been made during routine analysis.

6.4.7 Blank Determinations

The analytical response for blank determinations is a useful means of indicating the degree of control over factors affecting the calibration parameter, λ [see Section 5.4.2(b)]. For example, changes in the degree of contamination of samples during analysis are revealed by blank determinations. When such variations are of interest, *e.g.*, as in trace analysis for many determinands, it may also be helpful to maintain a sequential chart showing the analytical responses of blanks. In general, neither τ nor σ can be defined for this chart, but the mean and standard deviation estimated from previous measurements may be used if desired. The interpretation of the chart needs, however, to be approached with care because variations in the analytical response of the blank from batch to batch may have no effect on the precision or bias of analytical results for samples.

6.5 REFERENCES

[1] C. Eisenhart, *J. Res. Natl. Bur. Stand.*, 1963, **67C**, 161.
[2] L. A. Currie, in I. M. Kolthoff and P. J. Elving, ed., "Treatise on Analytical Chemistry", Second Edition, Part 1, Volume 1, Wiley, Chichester, 1978, pp. 95–242.
[3] C. Liteanu and I. Rica, "Statistical Theory and Methodology of Trace Analysis", Ellis Horwood, Chichester, 1980.
[3a] J. O. Westgard, *CRC Crit. Rev. Clin. Lab. Sci.*, 1981, **13**, 283.
[4] J. Büttner, *et al.*, *J. Clin. Chem. Clin. Biochem.*, 1980, **18**, 69.
[5] R. Caulcutt and R. Boddy, "Statistics for Analytical Chemists", Chapman and Hall, London, 1983.
[6] T. Aronsson, C.-H. de Verdier, and T. Groth, *Clin. Chem.*, 1974, **20**, 738.
[7] J. Agterdenbos, F. J. M. J. Maessen, and J. Balke, *Anal. Chim. Acta*, 1981, **132**, 127.
[8] J. Hilton, I.B. Talling, and R.T. Clark, *Lab. Pract.*, 1982, **31**, 990.
[9] L. Aarons, *Analyst*, 1981, **106**, 1249.
[10] A. L. Wilson, *Analyst*, 1979, **104**, 273.
[11] Standing Committee of Analysts, "General Principles of Sampling and Accuracy of Results. 1980", H.M.S.O., London, 1980.

[12] J. Büttner, et al., J. Clin. Chem. Clin. Biochem., 1980, 18, 78.

[13] J. O. Westgard and T. Groth, Clin. Chem. (Winston-Salem, NC), 1981, 27, 1536.

[14] D. Midgley and K. Torrance, "Potentiometric Water Analysis", Wiley, Chichester, 1978.

[15] V. V. Nalimov, "The Application of Mathematical Statistics to Chemical Analysis", Pergamon, London, 1963.

[16] A. L. Wilson, Talanta, 1970, 17, 21.

[17] M. Thompson and R. J. Howarth, Analyst, 1976, 101, 690.

[18] J. P. Franke, R. A. de Zeeuw, and R. Hakkert, Anal. Chem., 1978, 50, 1374.

[19] J. Mandel, in I. M. Kolthoff and P. J. Elving, ed. "Treatise on Analytical Chemistry", Second Edition, Part 1, Volume 1, Wiley, Chichester, 1978, pp. 243–298.

[20] J. Agterdenbos, Anal. Chim. Acta, 1979, 108, 315.

[21] R. L. Booth, ed., "Handbook for Analytical Quality Control in Water and Wastewater Laboratories", (Report EPA-600/4-79-019), U.S. Environmental Protection Agency, Cincinnati, 1979.

[22] British Standards Institution, B.S. 5497. Part 1: 1979, B.S.I., London, 1979.

[23] J. S. Garden, D. G. Mitchell, and W. N. Mills, Anal. Chem., 1980, 52, 2310.

[24] P. M. G. Broughton, Ann. Rev. Clin. Biochem., 1980, 1, 1.

[25] J. A. Glaser, et al., Environ. Sci. Technol., 1981, 15, 1426.

[26] American Society for Testing and Materials, "1983 Annual Book of ASTM Standards. Volume 11.01 Water (I)", A.S.T.M., Philadelphia, 1983.

[27] M. Thompson, Analyst, 1982, 107, 1169.

[28] D. A. Kurtz, Anal. Chim. Acta, 1983, 150, 105.

[29] L. Oppenheimer, T. P. Capizzi, R. M. Weppelman, and H. Mehta, Anal. Chem., 1983, 55, 638.

[30] I. L. Larsen, N. A. Hartmann, and J. J. Wagner, Anal. Chem., 1973, 45, 1511.

[31] K. L. Ratzlaff, Anal. Chem., 1979, 51, 232.

[32] J. Timm, H. Diehl, and D. Harbach, Fresenius' Z. Anal. Chem., 1980, 301, 199.

[33] J. P. Franke, Fresenius' Z. Anal. Chem, 1981, 308, 351.

[34] G. Horvai and E. Pungor, Anal. Chim. Acta, 1980, 113, 287.

[35] G. Horvai and E. Pungor, Anal. Chim. Acta, 1980, 113, 295.

[36] P. Longhi, T. Mussini, and S. Rondinini, Anal. Lett., 1982, 15, 1601.

[37] C. E. Efstathlou and T. P. Hadjiioannou, Anal. Chem., 1982, 54, 1525.

[38] D. G. Mitchell, W. N. Mills, J. S. Garden, and M. Zdeb, Anal. Chem., 1977, 49, 1655.

[39] D. G. Mitchell and J. S. Garden, Talanta, 1982, 29, 921.

[40] W. J. Youden, Science, 1954, 120, 627.

[41] K. A. H. Hooton and M. L. Parsons, Anal. Chem., 1973, 45, 2218.

[42] R. V. Cheeseman and A. L. Wilson, Technical Report TR 66, Water Research Centre, Medmenham, Bucks, 1978.

[43] R. H. Woodward and P. L. Goldsmith, "Cumulative Sum Techniques", Oliver and Boyd, Edinburgh, 1972.

[43a] British Standards Institution, BS 5703: Part 1: 1980, B.S.I., London, 1980.

[43b] British Standards Institution, BS 5703: Part 2: 1980, B.S.I., London, 1980.

[43c] British Standards Institution, BS 5703: Part 3: 1981, B.S.I., London, 1981.

[43d] British Standards Institution, BS 5703: Part 4: 1982, B.S.I., London, 1982.

[44] R. J. Rowlands, et al., Clin. Chim. Acta, 1980, 108, 393.

[45] D. J. Dewey and D. T. E. Hunt, Technical Report TR 174, Water Research Centre, Medmenham, Bucks, 1982.

[46] American Public Health Association, et al., "Standards Methods for the Examination of Water and Wastewater", Fifteenth Edition, A.P.H.A., Washington, 1981.

[47] G. Ekedahl, B. Rondell, and A. L. Wilson, in M. J. Suess, ed., "Examination of Water for Pollution Control", Volume 1, Pergamon, Oxford, 1982, pp. 265–315.

[48] W. Horwitz, Anal. Chem., 1978, 50, 521A.

[49] Research and Development Department, "Methods of Sampling and Analysis. Volume 1. Steam and Water", Central Electricity Research Laboratories, Leatherhead, Surrey, 1966.

[50] M. Thompson and R. J. Howarth, Analyst, 1976, 101, 699.

7 Achievement of Specified Accuracy by a Group of Laboratories

7.1 TYPES OF INTER-LABORATORY STUDIES

Inter-laboratory studies of one or more aspects of the accuracy of analytical results are undertaken for a number of purposes. Four main types of study can be distinguished, and their characteristics may be briefly summarised as follows.

Type 1 The main aim is to obtain a general picture of the magnitude of analytical errors in each of a group of laboratories. Such information may be required, for example, to identify those determinations presenting greatest problems. The objective of this type of study is usually achieved by relatively simple experimental designs involving laboratories in little additional work. A design devised by Youden[1] – the paired-sample technique – can be of special value for this type of study, and has been widely used in the field of water analysis; see, for example, refs. 2 and 3.

Type 2 The main aim is to investigate the natures, sources, and sizes of the errors in analytical results. Such information may be required for a number of purposes; one example arises in the collaborative development of analytical methods. More complex experimental designs and, consequently, a greater amount of effort in the participating laboratories are usually involved, and more complex statistical techniques for the interpretation of the results are generally employed; see, for example, refs. 4 to 6.

Type 3 The main aim is to establish one or more aspects of the accuracy of analytical results achievable when a particular analytical method is used by all laboratories. Such information is commonly required, for example, in characterising the precision of analytical results achieved when a method proposed as a 'Standard Method' is used. This type of study often has many similarities to Type 2. Detailed descriptions of particular approaches are given by many authors; see, for example, refs. 1 and 5 to 8a. The use of this type of study has recently been extended to deriving standards for judging the proficiency of analysts,[9] and the International Organization for Standardization is preparing a guide on the use of inter-laboratory tests for assessing the competence of laboratories.

Type 4 The main aim is to ensure that each participating laboratory achieves a specified accuracy required for some common purpose. This requirement has arisen with increasing frequency in recent years through national and international legislation and decisions concerning standards for water quality, and through the formation of regional organisations responsible for the management and provision of water resources, e.g., the Regional Water Authorities of England and Wales. Such developments have emphasised the need to ensure that all members of a group of laboratories obtain analytical results that can be compared with confidence; this usually requires that all laboratories produce results within specified limits from the true values.

Large numbers of inter-laboratory tests of various types have almost invariably shown that a substantial proportion of the participating laboratories obtain errors greater than they expect and usually greater than desirable; see, for example, refs. 3, 10 to 19. It may be concluded that there are important problems to be overcome whenever the errors of each member of a group of laboratories must be controlled within specified limits. Thus, although all the above types of inter-laboratory study are of general interest, Type 4 is of crucial importance in the context of this book because such studies are increasingly required as part of the normal operation of routine water-analysis laboratories. None of the other types, therefore, is further considered here, and Type 4 is discussed in detail in Section 7.2. For readers interested in these other types, apart from the references cited above, ref. 20 provides a short, comparative review of Types 1 to 3. Further a recent symposium[21] on inter-laboratory studies contains many papers of interest; this and a second symposium held on the same topic in October, 1984, the proceedings of which have yet to be published, reflect the importance attached to the subject. Ref. 22 is well worth consulting, and inter-laboratory test procedures used in the U.S.A. for waters and effluents have been reviewed in ref. 23. The effect of Type 1 studies on the performance of clinical laboratories within the U.K. National Health Service is discussed by Whitehead[24] who makes a number of points of general interest. Finally, we would also stress that participation in inter-laboratory studies of any type may be a useful means of detecting unsuspected sources of bias in analytical results.

7.2 ACHIEVING A SPECIFIED ACCURACY IN A GROUP OF LABORATORIES

7.2.1 Introduction

Previous sections of this book have made clear that any attempt to achieve a specified accuracy by each member of a group of laboratories involves a number of different but very closely inter-related considerations, *e.g.*, defining the determinand, defining the maximum tolerable errors, choosing and appropriately applying analytical methods and systems, estimating and controlling precision and bias. Surprisingly, very few publications dealing with integrated approaches to such work – in which all relevant factors are considered in detail – have appeared. Indeed, to the best of our knowledge, the only complete scheme that has been published in the usual analytical literature is that described by one of us.[25] This scheme was developed about 20 years ago, by members of the Research and Development Department of the Central Electricity Generating Board, for a survey of the quality of feed-water in power stations. During the last 10 years, this scheme (or modifications of it) has been widely applied within the U.K. water industry. For example, the Water Research Centre (under contract to the Department of the Environment and in collaboration with the Regional Water Authorities) has applied the scheme to river water analyses.[26-31] Similar schemes have been operated by individual Water Authorities.[31a] The approach has also been recommended by the Standing Committee of Analysts.[32] At the international level, the Water Research Centre has operated the scheme for the World Health Organization in a pilot project on the monitoring of lead and cadmium in drinking water, and the general approach has been recommended for water analyses in the United Nations

Environment Programme's Global Environmental Monitoring System. It is of interest that a very similar scheme in basic concepts and practical aspects has been recommended for the field of clinical chemistry.[33] Finally, the scheme is based on essentially the same concepts and strategy more recently suggested for analytical chemistry in general by Currie[34] in a penetrating analysis of errors in analytical results and their control.

The detailed account of this scheme given in Sections 7.2.2 to 7.2.10 represents an expanded version of an earlier paper.[25]

7.2.2 Summary of the Approach

The approach consists, for each determinand, of sequential execution of a number of closely-linked stages as summarised in Figure 7.1. The activities and their sequence shown in the figure are such that generally important problems and sources of error are progressively eliminated or controlled. In this way a sound basis for routine analytical quality-control is ensured with minimal chance of wasted effort. It is important, therefore, that no stage is started until the preceding stage has been satisfactorily completed. The reasons for, and activities involved in, the individual stages of Figure 7.1 are described in Sections 7.2.3 to 7.2.10.

The detailed considerations and work necessary in each stage are governed by the particular determinand and the required accuracy of results. The aim here, therefore, is to stress the principles involved; references 26 to 31 may be consulted for practical details and results obtained for particular determinands.

One aspect of the scheme is worth stressing at the outset because it underlies the order proposed for the experimental work (Stages 5 to 8), *i.e.*, everything possible is done to eliminate or adequately control sources of error before making inter-laboratory tests of the bias between the results from the various laboratories. There are three main reasons for this. First, it is often difficult in water analysis to obtain a direct, experimental estimate of bias for all the many types of sample commonly analysed by laboratories. Prime reliance for adequately small bias must, therefore, rest on previous elimination and/or control of sources of bias. Second, the ability to detect a systematic error is governed by the magnitude of random errors (see Section 4.6.2). Thus, tests of bias undertaken before adequate precision has been established may well prove incapable of detecting bias of the magnitude of interest; in this event, wasted effort may well result. Finally, if the tests in Stage 8 show important bias in the results of some laboratories, considerable difficulty may be experienced in eliminating it, particularly when the laboratories are widely dispersed geographically. In addition, Stage 8 would then, in general, need to be repeated with the additional work being required even in laboratories that previously achieved satisfactory results; the organisation of repeat tests may also not be simple when laboratories are widely dispersed geographically. The additional work called for by the laboratory preparing and distributing the samples for Stage 8 may also be considerable.

Many different circumstances and requirements arise in practice. We have attempted, therefore, to present the scheme in such a way that it is generally applicable. For a particular application, however, some simplification may be possible, and this should always be borne in mind. In our view, however, any such

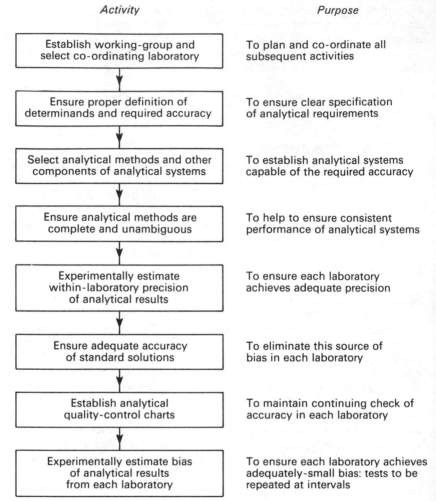

Activity	Purpose
Establish working-group and select co-ordinating laboratory	To plan and co-ordinate all subsequent activities
Ensure proper definition of determinands and required accuracy	To ensure clear specification of analytical requirements
Select analytical methods and other components of analytical systems	To establish analytical systems capable of the required accuracy
Ensure analytical methods are complete and unambiguous	To help to ensure consistent performance of analytical systems
Experimentally estimate within-laboratory precision of analytical results	To ensure each laboratory achieves adequate precision
Ensure adequate accuracy of standard solutions	To eliminate this source of bias in each laboratory
Establish analytical quality-control charts	To maintain continuing check of accuracy in each laboratory
Experimentally estimate bias of analytical results from each laboratory	To ensure each laboratory achieves adequately-small bias: tests to be repeated at intervals

Figure 7.1 *Sequence of activities for achieving a specified accuracy in a group of laboratories*

changes should be very carefully balanced against any risks of inadequate assessment of errors or wasted effort incurred by the changes. The guiding principle should be that the work undertaken must be capable of meeting its declared objectives. When the effort available in laboratories is insufficient to allow the objectives to be achieved, they should either be changed or the work not undertaken.

7.2.3 Establishment of a Working Group

The various activities shown in Figure 7.1 clearly require efficient and uniform execution in all laboratories. This objective is almost certain not to be achieved without very thorough planning and co-ordination of all the work. In our experience, it is of great value to establish a working group meeting regularly to agree and plan both the general approach and also the detailed procedures for each determinand. Whenever possible, this group should be composed of a representative from each of the laboratories involved. Such groups also act as a useful means of exchange of analytical information.

At least one laboratory is also required to act as a co-ordinating laboratory for the group. The tasks of this laboratory include the preparation and distribution of: (1) notes describing the tests required in each laboratory for particular determinands, (2) forms for entering the results of tests in a uniform manner, (3) standard solutions and samples required in Stages 6 and 8. This co-ordinating laboratory will usually also have to undertake the interpretation of results from Stages 6 and 8, the collation of all results, and the preparation of any reports on the work that may be required; a number of other tasks may well be involved. Detailed technical requirements will also commonly imply that the co-ordinating laboratory be capable of substantially better precision than that required for the routine analytical results of all laboratories. For example, it is generally necessary to ensure that the concentrations of all solutions distributed in Stage 8 do not differ by more than a small proportion of the maximum tolerable bias of routine analytical results. Whenever the ability to achieve this aim must, as is common, be investigated experimentally, the requirement on the precision of measurements by the co-ordinating laboratory becomes considerably more stringent than for the other laboratories. This special type of requirement calls for the use of special analytical systems and/or extensive replication of analyses; both require substantial additional effort. Prospective co-ordinating laboratories should be careful, therefore, not to underestimate the amount of work involved.

It is also highly desirable that the group has access to expert statistical advice on the design and interpretation of experiments.

The number of laboratories and/or their geographical distribution may sometimes be so large that it is impractical for one laboratory to act as co-ordinator for all the other laboratories involved. When this is so, a hierarchical scheme is useful, *i.e.*, the laboratories are divided into a number of groups each of which can be adequately serviced by its own co-ordinating laboratory. All the co-ordinating laboratories then form the first level in the hierarchy, and they follow the scheme in Figure 7.1. On satisfactory completion of all stages of the work for a particular determinand, each co-ordinating laboratory then initiates an identical approach within its own group of laboratories (the second level of the hierarchy). This hierarchical approach has been used for the work on river-water analysis in the U. K. mentioned above.

7.2.4 Ensuring Proper Definitions of Determinands and Required Accuracies

Unambiguous definition of the determinands of interest and numerical definition of the accuracy required of routine analytical results are essential for three main purposes: (1) to allow the choice of analytical methods and systems appropriate for the intended uses of the results, (2) to provide clear criteria for deciding whether or not the errors observed in particular laboratories warrant corrective action, and (3) to ensure that adequate numbers of tests are made to allow errors of the magnitudes of interest to be detected as statistically significant.

As stressed previously (Section 2.2), it is the responsibility of the user of analytical results to provide these definitions. The analytical working group's task is to ensure that appropriate definitions are available to provide a sound basis for its work. Detailed discussion between the user and the group will, in general, be required in formulating the initial definitions and in considering the need for subsequent revision in the light of changing requirements of the user and/or problems encountered in achieving the required accuracy. Ideally, the user or his representative should be a member of the working group.

The technical aspects involved in providing unambiguous definitions of determinands and accuracies have been considered in Sections 2.4 and 2.6, respectively. Four of the most important points are, however, worth repeating here.

(1) Non-specific determinands   These can, in general, be defined only by specifying the analytical method to be used by each laboratory; see Section 2.4.2.

(2) Maximum tolerable random and systematic errors   These two errors should be separately defined because their consequences, estimation, and control involve different considerations; see Sections 2.6, 4, 5, and 6. In general, the target for random error should be expressed as σ_{t0}, the maximum tolerable value for the total standard deviation of analytical results [see Section 6.2.1(d)]. It is also essential to bear in mind the likely dependence of analytical errors on the concentration of the determinand, and to consider carefully whether or not the maximum tolerable errors depend on determinand concentration.

(3) Errors arising during sampling   These may arise during the actual collection of samples (*e.g.*, contamination), and between sampling and analysis (*e.g.*, through instability of the determinand). Such errors may also lead to inadequately accurate analytical results; clearly, attention must be paid to such errors in addition to those arising during analysis if the required accuracy of analytical results is to be achieved. Control of the errors arising before analysis, however, usually involves different considerations to those relevant to analysis. In addition, analysts are often not directly involved with sampling. For these reasons we favour two separate but closely linked approaches to the control of sampling and analytical errors; a similar view has been expressed in the context of analytical quality-control in clinical chemistry.[33] On this basis, the scheme in Figure 7.1 is intended to deal primarily with the control of analytical errors. Note also that any experimental investigation of errors associated with sampling will generally require analyses of known precision and, possibly, bias. The nature and control of errors arising before analysis are considered in detail in Sections 3.5 and 3.6. Note, however, that procedures for ensuring constancy of determinand concentrations between sampling and analysis (*e.g.*, the addition of chemical stabilising reagents – see Section 3.6) can affect the performance of analytical systems. Close co-ordination of these two aspects of sample-stabilisation procedures is, therefore, essential by all laboratories and the working group [see Section 7.2.9(b)].

(4) Requirements of particular laboratories   When one or more laboratories of a group are normally concerned with the analysis of waters of generally different qualities to those of the other laboratories, certain difficulties may arise in planning

the practical work required. In particular, it may be difficult or impossible to distribute solutions in Stage 8 that represent the qualities of waters characteristic of routine analyses in all laboratories. Such problems should be considered by the working group during the planning stage.

7.2.5 Choice of Analytical Methods and Systems

Analytically, this is the most important stage of all because an inappropriate choice by one or more laboratories will not only prevent achievement of the required accuracy, but is also likely to cause much wasted time and effort, and this not only in laboratories making the unsuitable choice. The choice of analytical methods and other components of analytical systems is considered in detail in Sections 9 and 10. Some points of particular relevance in the present context are noted below.

(1) Criteria for judging the suitability of analytical methods It is highly desirable that each laboratory adopt a reasonably uniform approach to this choice. One of the first tasks of the working group should be, therefore, to reach agreement on the relevant criteria. The authors' suggestions for this purpose are summarised in Section 9.2.5. Consistent application of such criteria may well lead to the situation where only one analytical method is considered suitable; all laboratories should then adopt that method whenever possible.

(2) Choice of analytical methods With the exception of any non-specific determinands [see (3) below], we favour the approach that each laboratory may use any analytical method it wishes provided that it satisfies the criteria in (1) above. Generally to insist that all laboratories use the same method is not a means of necessarily ensuring specified standards of accuracy (see Sections 4.7 and 9.1).

(3) Non-specific determinands The methods to be used for these determinands may already have been defined or implied by the definitions of the determinands. When this is so, there is, in principle, no choice to be made, though it remains vital (and often difficult) to ensure that all laboratories actually follow the same analytical procedure (see Section 7.2.6). For such determinands, it is also possible, in principle, to allow the use of alternative methods for a particular determinand if they are first shown to give results in satisfactory agreement with the specified method. In practice, however, this approach can lead to severe experimental difficulties in proving satisfactory agreement and we would suggest that this alternative be avoided whenever possible (see Section 9.3.3).

(4) Lack of suitable methods Situations may well arise where no method meets the critera noted in (1). A similar situation may arise if the staff and/or equipment available in one or more laboratories necessitate the use of method(s) not meeting the group's criteria. In such situations it will usually be necessary for the group to raise the problem with the user of analytical results in order to decide if work on the particular determinand should be deferred until at least one suitable method can be made available or if the targets for accuracy should be relaxed to the extent that available methods become suitable. If the former possibility is followed, collaborative work by the group may considerably facilitate the provision of a suitable method, *e.g.*, by collaborative investigation of interference effects [see Section 5.4.6(*h*)].

7.2.6 Need for Completely Specified Analytical Methods

Virtually all publications on analytical quality-control stress the importance of completely and precisely specified analytical methods; see, for example, refs. 5, 34 to 37 and Section 6.3.1. Apart from the obvious need for such descriptions in defining the procedures to be followed by a laboratory, they are also essential in assessing likely sources of errors. Factors involved in ensuring adequate descriptions are considered in Section 9.2.4; here it need only be stressed that laboratories should not begin Stage 5 until they have completely specified analytical methods. The co-ordinating laboratory may sometimes usefully act as an independent reviewer of these methods.

Of course, it is also essential to do everything possible to ensure that the procedure specified by a method is followed as exactly as possible in all subsequent work – at least until any changes that may be contemplated have been validated by experimental tests. One of the aims of subsequent stages of Figure 7.1 is to characterise the accuracy obtained with a given analytical system; the achievement of this aim will be prevented if the analytical method is not faithfully and consistently followed. Should any changes to the method be considered, great caution is desirable; even apparently minor procedural details can sometimes have unexpectedly large effects, particularly in trace analysis.

7.2.7 Ensuring Adequate Within-laboratory Precision

This stage is the first of the experimental stages for two main reasons:

(1) To ensure that the analytical system of each laboratory is capable of reliable and adequate performance as indicated by the precision of analytical results; see Sections 6.1 and 6.3.1.
(2) To ensure that the ability to detect bias is adequate before any tests of bias are made; wasted effort caused by tests with inadequate precision (see Section 4.6.2) is thereby avoided.

A model for random errors and experimental designs for estimating precision are discussed in detail in Sections 6.2 and 6.3. The suggestions there form the basis of the tests of precision in each laboratory, but the working group must decide the exact experimental design appropriate for each determinand. Examples of the designs and results* obtained in the Harmonised Monitoring Scheme for River Water in the United Kingdom may be found in refs. 26 to 31. A few special points are noted below.

(1) Uniformity of tests and calculations It is essential that all laboratories carry out uniformly the tests and calculations agreed by the working group. To ensure this, the co-ordinating laboratory should prepare detailed written descriptions of all aspects of the tests for each determinand and suitable forms on which laboratories enter their results. These forms may also include notes on how to make any statistical calculations required of laboratories. If such aids are not provided to all laboratories, some may make inappropriate tests and calculations that may well invalidate their work. The co-ordinating laboratory is also likely to meet needless

* Note that the definition of σ_w and σ_b for these references differs from that used in this book; see page 230.

difficulties in checking and collating results if they are not calculated and presented uniformly. Of particular importance is the need to ensure that all experimental results are recorded to an agreed number of figures.

(2) Recovery tests If the tests include a sample and the 'spiked' sample [see Section 5.4.6(*i*)], the observed mean recovery of the determinand added to the 'spiked' sample provides an additional check on the internal consistency of results. This calculation and the assessment of whether or not the recovery differs significantly from a specified value are, therefore, usefully included by the working group in its tests. This procedure is often adopted in the Harmonised Monitoring Scheme for which the usual criterion is that the observed mean recovery, $\bar{f}$, is considered acceptable if it is either not significantly less than 95 % (when $\bar{f} < 95$ %) or not significantly greater than 105 % (when $\bar{f} > 105$ %).

(3) Comparison of observed and required precisions Problems may arise in deciding if the experimental estimates of standard deviation show that the required precision has been achieved. This is discussed in Section 6.3.4(*a*).

7.2.8 Accuracy of Standard Solutions

This stage is intended to ensure that the standard solutions (used for calibration purposes) of all laboratories are in satisfactory agreement, and hence do not cause important inter-laboratory bias. In principle, this aim could be achieved by the co-ordinating laboratory providing the necessary standard solutions for all laboratories. In practice, however, this approach is cumbersome, liable to practical difficulties, and very demanding of effort from the co-ordinating laboratory. A better approach is as follows.

All laboratories prepare their standards as prescribed by their analytical methods, and the co-ordinating laboratory distributes a portion of its own concentrated standard solution for the particular determinand to each laboratory. Great care is essential to ensure that the distributed standard's concentration is negligibly in error. In water analysis, special care is often required to ensure that the concentration of the determinand in the distributed solution is stable for an adequate time-period, and that the containers used for the standard cause neither contamination nor sorption of the determinand (see Section 3.6). Such problems are usually minimised by distributing relatively concentrated standard solutions; this also aids transport arrangements because small volumes of such solutions suffice for the subsequent tests. Each laboratory then compares the concentrations of its own and the distributed standard by replicate analysis of both in the same batch of analyses.

For these tests each solution is analysed n times, n being chosen so that a statistically significant difference between the two mean results is expected when the concentrations of the two standards differ by δ % or more [see Section 4.6.2(*b*)(*i*)]. The value of δ is decided by the working group before the tests on the basis of the maximum tolerable bias of analytical results, the other sources of bias, and the need to maintain good analytical standards. As δ is made smaller and smaller in relation to the within-batch standard deviation, σ_w, of analytical results (see Section 6.2), the required value of n rapidly becomes alarmingly large. In the Harmonised

Monitoring Scheme, δ is usually 2 %, and the significance level used in comparing the two means is 0.05.

Since the value of n decreases with decreasing σ_w, the analytical system used and the concentration at which the tests are made (*i.e.*, the extent of dilution of the stock standards before analysis) should be chosen to obtain the smallest possible σ_w. It may be advantageous, therefore, for these particular tests only, to use a different analytical method to that chosen for the normal routine analysis of samples. A randomised order of the total $2n$ tests will often be suitable, but systematic orders (*e.g.*, alternate analysis of the two diluted standards) may be preferable if the response of the analytical system to a given concentration of the determinand is likely to drift continuously in one direction.

If any laboratory finds that its standard solution differs significantly from that provided by the co-ordinating laboratory, and that the true difference could exceed δ %, the cause should be sought, reduced, and the tests repeated. When a laboratory's standard is in satisfactory agreement with the distributed standard, the next stage in Figure 7.1 is started.

7.2.9 Establishment of Analytical Quality-control Charts

At this stage in the overall procedure, the laboratories will have achieved adequate precision and ensured adequately accurate standard solutions. This is, therefore, a suitable time for each laboratory to establish appropriate analytical quality-control charts in order that a continuing check on precision and some sources of bias is maintained in all subsequent work. The basis and application of such charts has been considered in detail in Section 6.4 in the context of individual laboratories. For collaborative work among a group of laboratories, it is desirable that the working group decide and specify the type(s) of control-chart to be used for each determinand in order to ensure a uniform approach by all laboratories. For the reasons given in Section 6.4.3, we would suggest that it is frequently of value to specify use of a control-standard though other types of control-tests may also be required.

7.2.10 Assessment of Inter-laboratory Bias

(a) Summary of the Procedure
As soon as all laboratories have satisfactorily completed the previous stages, this final stage is begun. The approach suggested is very simple. Portions of one or more standard solutions and/or samples are distributed by the co-ordinating laboratory to each of the other laboratories. Each laboratory then makes sufficient replicate analyses on each solution to allow detection of a bias equal to the maximum tolerable value. The bias of the results obtained is assessed separately for each laboratory and solution because both bias and precision may, and often do, depend on the laboratory, on the determinand concentration, and on the nature and general composition of the solution. Other approaches to inter-laboratory tests have been described (see Section 7.1) that may allow some economy of effort, but this advantage is gained at the expense of having to make assumptions such as equal precision in all laboratories. For present purposes in obtaining unambiguous

estimates of bias, it is essential not to make such assumptions, and the following approach is suggested.

The co-ordinating laboratory prepares and distributes appropriate solutions to all laboratories [see (*b*) below]; only a broad indication of their determinand concentrations should be made known to the laboratories. Each laboratory then analyses *w* portions of each solution, the agreed control-tests also being included in each batch of analyses [see (*c*) below]. The results obtained are reported to the co-ordinating laboratory for assessment of bias [see (*d*) below]. Any laboratories with an unacceptably large bias are then informed so that they can seek and reduce the cause before undertaking further tests of bias [see (*e*) below].

(b) Distributed Solutions

The solutions to be used are governed by the following considerations. The main sources of bias needing to be checked are those arising in the analysis of samples rather than standard solutions (Sections 5.4.5 and 5.4.6). Emphasis, therefore, should whenever possible be given to the distribution of the former. Nevertheless, it is useful to include at least one standard solution so that the true concentration of at least one distributed solution is known. This may prove to be of value in locating possible sources of bias if unsatisfactory results are obtained for distributed samples. The distributed solutions should also cover the range of determinand concentrations of interest, and, in general, the more types of sample that can be included the better, provided that they are all relevant or of interest to all laboratories.

A useful scheme, and one commonly used in the Harmonised Monitoring Scheme, is to distribute one standard solution near the middle of the concentration range of interest, and two samples of different type with concentrations near the lower and upper limits of this range. Whenever possible (see below), these standard solutions are distributed at the above concentrations so that their dilution before analysis is not necessary; such dilution may prevent detection of certain sources of bias, *e.g.*, that due to inaccurate correction for the determinand concentration in the water used for blank determinations [see Section 5.4.4(*b*)]. The working group should decide the solutions to be used in the light of its requirements.

It is essential that the co-ordinating laboratory take great care to ensure that each laboratory receives essentially identical portions of each of the distributed solutions; otherwise, the basis for interpreting the results is clearly destroyed. Of particular and common importance in water analysis are the cleanliness of the containers in which the solutions are stored for distribution and the stability of the determinand (see Section 3.6). Heterogeneous distribution of undissolved determinands may sometimes also lead to problems if the bulk sample is not sub-divided very carefully. Such problems may depend on so many factors that preliminary tests in the co-ordinating laboratory are often desirable to ensure the necessary closeness of determinand concentrations in each of the distributed portions of a given solution. It is a useful precaution also for the co-ordinating laboratory to prepare and reserve spare portions of each solution in case any are lost or broken in transport or additional tests are required in any laboratories.

In water analysis, many determinands are so unstable that distribution of samples without special treatment may not only be pointless but also a source of wasted work by all laboratories. When this is so, the use of stabilising reagents [Section 3.6.3(d)] can be of value provided that tests have shown the reagent to be effective over the time-period required for the distribution and analysis of the solutions. Note also that the available information may relate to conditions not identical with those for the distributed solutions, *e.g.*, it may be necessary to distribute solutions in containers larger than those for which information on stability is available. Such differences may affect the stability, and this possibility should be borne in mind. The use of a stabilising reagent in the distributed solutions also strictly requires that all laboratories use this same reagent in their routine collection of samples. Otherwise, the analysis of distributed solutions containing a reagent not normally used by a laboratory will not duplicate normal sample analysis, and may lead to bias that would not normally affect routine analytical results. This aspect of the effect of stabilising reagents on the interpretation of the results of inter-laboratory tests of bias is potentially important for a number of determinands, and should be carefully considered by the working group when planning a programme of tests. Special investigations may be needed to check the effects of alternative stabilising reagents on the performances of the analytical systems involved.

One approach used to overcome such stabilisation problems is to distribute a stable standard solution with a sample, the former being used to 'spike' a portion of the latter just before analysis. The difference between the results for the 'spiked' and 'unspiked' portions is then used to assess bias. This procedure may be helpful, but it should, in general, be used with great caution because: (1) it does not detect those sources of bias whose magnitudes are independent of the concentration of the determinand [see Section 5.4.6(i)], (2) the behaviour of the added determinand may differ from that naturally present in samples (see Section 5.4.5), and (3) the precision of the recovery tests is usually worsened by the two random errors involved in analysing two portions of the sample. If, however, the distributed sample contains essentially none of the determinand, the 'spiked' sample will then have a determinand concentration known with sufficient accuracy to provide a much more powerful test of bias (see Section 6.4.4). In addition, there may then be no need to analyse the 'unspiked' sample so that the precision of results is governed by only the random error of the result for the 'spiked' sample.

Clearly, the best approach to these problems needs to be decided individually and with care for each determinand.

(c) Experimental Design
The working group must decide the experimental design to be used for the replicate analyses of the distributed solutions, and this design must be followed by every laboratory so that all results will be interpretable on the same basis. In general, to minimise the extent to which any apparent bias is caused by random errors stemming from the estimation of calibration parameters [see Section 6.2.1(d)], the simplest design – though not perhaps the most convenient – is to analyse one portion of each of the distributed solutions in each of w batches of analyses (see below for the value of w). The standard deviation of results obtained in this way is σ_t, and this approach, therefore, has the additional advantage of providing

estimates of σ_t for each of the distributed solutions. In each of these w batches, the agreed control-tests should also be made (Section 7.2.9).

If all w replicates for a given solution were analysed in only one batch of analyses, the standard deviation characterising the random errors among the results is the within-batch standard deviation, σ_w, which is, in general, smaller than σ_t. Such a design with the method of interpretation described in (d) below leads to a greater chance of falsely detecting bias between the results from different laboratories. This is because the random errors characterised by the between-batch standard deviation, σ_b, [see Section 6.2.1(d)] would appear in what is interpreted as bias. Clearly, this point becomes less important as σ_b decreases. Thus, analysis of the distributed solutions in just one batch of analyses will be adequate when σ_b is negligibly small in *every* laboratory. [Note that the extended design of Section 6.3.5 is required in the tests of precision (Section 7.2.7) if estimates of σ_w, σ_b, and σ_t are to be obtained, and that σ_b may show a rather complex dependence on determinand concentration; see Section 6.2.1(c)].

The value of w is governed by the required confidence level for assessing bias and the ratio, β_0/σ_t, where β_0 is the bias the tests are required to detect [see Section 4.6.2(a)(i)]. It is generally useful to equate β_0 with the maximum tolerable bias, and σ_t with the maximum tolerable total standard deviation; values for both these parameters should be available (see Section 7.2.4), but note that their values may depend on the concentration of the determinand. In the Harmonised Monitoring Scheme,[25] $\beta_0/\sigma_t = 2.0$ [see also Section 2.6.2(b)], and a confidence level of 95 % is used; the value of w should then be approximately 4, and a value of 5 has usually been used in that Scheme in order to provide a small safety margin. We would stress that w increases rapidly as β_0/σ_t decreases; for example, when this ratio equals 1.0, w should be approximately 14.

(d) Interpretation of Results

(i) Calculation of maximum possible bias For each laboratory and solution, the $100(1-2\alpha)$ % confidence interval for the population mean is calculated from the w results as described in Section 4.3.5(a). These calculations may, if desired, be made by the participating laboratories. The co-ordinating laboratory, however, must then calculate for each laboratory and solution the upper $100(1-\alpha)$ % confidence limit for the true bias as described below. (In the Harmonised Monitoring Scheme, a value of 0.05 is used for α.)

Suppose that for the i th ($i = 1$ to z) laboratory and a given solution, the $100(1-2\alpha)$ % confidence limits for the population mean are $\bar{R}_i \pm l_i$. The upper $100(1-\alpha)$ % confidence limit, B_{ui}, for the true bias of this laboratory and solution is calculated from [see Section 4.3.3(b)]:

$$
\left.
\begin{aligned}
B_{ui} &= \bar{R}_i + l_i - C_S && \text{when } \bar{R}_i \geqslant C_S \\
B_{ui} &= \bar{R}_i - l_i - C_S && \text{when } \bar{R}_i \leqslant C_S,
\end{aligned}
\right\} \qquad (7.1)
$$

or

where C_S is the determinand concentration in the distributed solution (see below). Equation 7.1 applies when C_S is such that the maximum tolerable bias is a fixed concentration. When the value of C_S is such that the maximum tolerable bias is a fixed percentage of C_S, the corresponding equations are:

$$B_{ui} = 100(\bar{R}_i + l_i - C_S)/C_S.\% \quad \text{when } \bar{R}_i \geqslant C_S$$

or

$$B_{ui} = 100(\bar{R}_i - l_i - C_S)/C_S\% \quad \text{when } \bar{R}_i \leqslant C_S.\qquad (7.2)$$

If B_{ui} is greater than the maximum tolerable bias, the laboratory has not demonstrated, for the particular solution, its ability to achieve an adequately small bias; further work by that laboratory is then, in general, required in order to improve the situation. Several points, however, concerning this assessment of results are worth noting.

First, it can be seen that B_{ui} depends on l_i which, in turn, derives from the variations among the replicate results for a given solution. Thus, if there has been appreciable deterioration of precision since Stage 5 of Figure 7.1, unacceptable estimates of bias may be caused by unsatisfactory precision rather than unsatisfactory bias; the results should be carefully examined from this point of view.

Second, the overall assessment of the bias of a laboratory should usually take into account the results obtained for all the distributed solutions. For example, a laboratory whose value for B_{ui} for one solution marginally exceeded the maximum tolerable value but whose values for all the other solutions were well within the target may usually be considered to have reasonably met the requirements for bias. On the other hand, a laboratory only just within the target for all solutions may well have need to check possible sources of bias.

The value to be used for C_S in equations (7.1) and (7.2) is considered in (*ii*) below.

(*ii*) *The numerical value of* C_S For both standard solutions and 'spiked' samples, the true value of C_S should be known with negligible error [see (*b*) above]. For samples, however, the true value of C_s is not generally known, and an estimate, $\hat{C}_S$, to be used in equation (7.1) or (7.2) must be derived from the z results, $\bar{R}_i$. This may present problems, and the following points should be noted.

(1)   Results for all solutions from one or more laboratories may be clearly biased with respect to the other laboratories. There is then a natural tendency to exclude the apparently aberrant laboratories, and to calculate the $\hat{C}_S$ values from the results of the other laboratories. In many circumstances this may well be the best approach, but, as has been noted elsewhere,[20,34] it is not impossible that the rejected laboratories have actually obtained results substantially less biased than the other laboratories. Clearly, a vital component in decisions to reject results must be a critical examination of sources of bias in the individual laboratories.

(2)   A variety of statistical procedures is often used to detect 'outliers' in the results from inter-laboratory tests, *i.e.*, results that deviate from the remaining results by amounts greater than would be expected on the basis of some hypothesis on the nature of the population from which all results are expected to derive. Dybczynski[38] has studied a number of such 'outlier' tests in the context of inter-laboratory studies, and has made some general recommendations on suitable tests for this purpose. Tests of this type are undoubtedly of value in providing a more objective approach to the identification of 'outliers', but all authors warn against the indiscriminate use of such tests. Before rejecting an

outlying result, it is highly desirable that some particular reason for its presence be identified [see (1) above].

(3) The probability of occurrence of 'outliers' in inter-laboratory studies will be decreased as analytical methods and systems are more critically chosen and applied to minimise bias, and as one devotes more effort to ensuring that the performance of the systems is reliably controlled. Essentially all the preceding stages of Figure 7.1 help to ensure that these two aspects are given detailed attention. Given that these preceding stages have been properly carried out with satisfactory results, one would expect that the importance of 'outliers' would be much decreased, and this, indeed, is our general experience. Given the above scheme, therefore, the simple, general approach to calculating $\hat{C}_S$ is to use the mean of all results retained after critical consideration of points (1) and (2) above, *i.e.*,

$$\hat{C}_S = \Sigma \bar{R}_i / z' \tag{7.3}$$

where z' is the number of laboratories whose results are retained. Of course, the random errors of the $\bar{R}_i$ values will lead to random error in $\hat{C}_S$, but the size of this error is progressively decreased as z' increases [see Section 4.3.2(c)]. For most purposes, the random error of $\hat{C}_S$ will be acceptably small in relation to the other errors affecting B_{ui} when five or more laboratories' results are used in equation (7.3). Finally, we should note that equation (7.3) assumes that all the $\bar{R}_i$ values come from populations with essentially equal standard deviations. Given the aims and preceding stages of the scheme, this is often likely to be sufficiently true to allow use of the equation. If, however, this approximate equality of precision in the different laboratories has not been achieved, it may be better to weight the $\bar{R}_i$ values differentially in calculating $\hat{C}_S$. The use of weighting factors in estimating parameters such as C_S has recently been considered in detail;[39] see also ref. 38.

(e) Further Experimental Work

If all laboratories achieve satisfactorily small bias, it may provisionally be taken that all laboratories have achieved the desired accuracy. If most of the laboratories did not achieve acceptable results, the test for bias should, in general, be repeated after the sources of bias and/or inadequate precision have been sought and controlled. If only a few of the laboratories did not achieve satisfactorily small bias, it has to be decided whether or not all laboratories should be asked to repeat these tests of bias or whether these repeats should be restricted to the laboratories that experienced problems in the first test. The latter has the advantage that it avoids probably unnecessary work in most of the laboratories, but it also means that direct experimental demonstration of the comparability of results from all laboratories is not achieved. In addition, the accuracy of $\hat{C}_S$ becomes much more problematic when only a few laboratories are involved. The better approach can only be decided for each individual situation, but in reaching a decision the tests described immediately below are worth bearing in mind.

On satisfactory completion of this final stage of Figure 7.1, it may provisionally be taken that laboratories have achieved the required accuracy. At this stage, however, it is unlikely that precision and bias will have been estimated for only a few sample types, and both parameters may also deteriorate subsequently. The use of control-charts will help to ensure that any deterioration in precision and some sources of bias is detected, but we also consider it essential to repeat the direct tests of interlaboratory bias at intervals. As always, the more frequent such tests and the greater the number of samples included in them, the better will be the assessment of bias. The effort involved is, however, substantial, and particularly when many determinands are of interest; compromise between what is desirable and what can be undertaken is, therefore, almost certainly necessary.

7.3 CONCLUDING REMARKS

Sections 5 and 6 stressed the many sources of error in analytical results. Consequently, close attention must be paid to many different aspects of analysis if the accuracy of routine analytical results is consistently to meet a specified value. We would emphasise that all the many points mentioned in discussing the procedure of Figure 7.1 represent problems actually encountered by groups of laboratories attempting to achieve a specified accuracy. To overcome the many difficulties, a systematic approach is essential, and the scheme described above represents one such approach that has been found useful in the field of water analysis. The principles of this approach are thought to be of general value and validity, but, of course, many detailed modifications are possible.

   The fact that the scheme is involved and unlikely to achieve rapid progress is a disadvantage, though the rate of progress can be markedly improved by thorough attention to the points involved in Stages 3 and 4. We believe, however, that problems in achieving progress arise primarily through the complexity of the analytical systems that one is attempting to control. This complexity, it seems to us, requires a scheme of the nature described above if progress is to be achieved on a sound and permanent basis. With the exception of the working group, most of the preliminary planning stages (*e.g.*, precise definition of objectives, critical choice of analytical methods and systems, exactly specified analytical methods, *etc.*) are increasingly called for by many authors and organisations as an integral part of the normal work of all laboratories. The same point also applies to ensuring that adequately accurate calibration standards and control-charts are maintained. Of course, the need for estimation of precision and bias is equally stressed by many authors, but the experimental designs proposed here for these purposes are more involved, lengthy, and demanding of effort than many others that have been suggested and used. We have explained our reasons for such designs here and in other sections of the book, but each particular application, clearly, requires detailed, individual consideration. Nonetheless, we would strongly suggest that simplifications of the above approach should be made only after very careful assessment of all their consequences; lack of success and wasted effort may well result if short-cuts are taken on inadequate grounds.

   The results obtained by this approach for a number of more commonly measured determinands have been published,[26-31] and reports on others are in preparation. For determinands that many might consider to present relatively few problems when concentrations are well above detection limits (*e.g.*, chloride, ammonia, nitrate, nitrite), these results suggest that an accuracy of approximately ± 20 % is somewhere near the limit of what can be achieved routinely by a group of laboratories without the expenditure of very much more effort than is usually available. The reliably achievable accuracy for many trace metals and organic substances may well be appreciably worse than this. Great caution is, therefore, essential in considering the feasibility of requirements calling for substantially better accuracy from all members of a group of laboratories in routine analysis.

7.4 REFERENCES

[1] W. J. Youden and E. H. Steiner, "Statistical Manual of the Association of Official Analytical Chemists", A.O.A.C., Washington, 1975.
[2] G. Ekedahl, B. Röndell, and A. L. Wilson, in M. J. Suess, ed., "Examination of Water for Pollution Control", Volume 1, Pergamon, Oxford, 1982, pp. 265–315.
[3] Anon., "Pilot-Scale Intercomparison Studies of Methods for Water Analysis", World Health Organisation, Regional Office for Europe, Copenhagen, 1980.
[4] J. Mandel, in I. M. Kolthoff and P. J. Elving, ed., "Treatise on Analytical Chemistry", Second Edition, Part 1, Volume 1, Wiley, Chichester, 1978, pp. 243–298.
[5] American Society for Testing and Materials, "1983 Annual Book of ASTM Standards. Volume 11.01 Water (I)", A.S.T.M., Philadelphia, 1983.
[6] J. Mandel, "The Statistical Analysis of Experimental Data", Interscience, London, 1964.
[7] British Standards Institution, B.S. 5497. Part 1: 1979, B.S.I., London, 1979.
[8] International Organisation for Standardisation, ISO 5725-1981, I.S.O., Geneva, 1981.
[8a] A.O.A.C. Committee on Interlaboratory Studies, *J. Assoc. Off. Anal. Chem.*, 1983, **66**, 455.
[9] A. J. Malanoski, *J. Assoc. Off. Anal. Chem.*, 1982, **65**, 1333.
[10] E. F. McFarren, R. J. Lishka, and J. H. Parker, *Anal. Chem.*, 1970, **42**, 358.
[11] P. W. Austin, *et al.*, Technical Report TR 65, Water Research Centre, Medmenham, Bucks, 1977.
[12] R. R. Edwards, T. L. Schilling, and T. L. Rossmiller, *J. Water Pollut. Contr. Fed.*, 1977, **49**, 1704.
[13] R. Dybczynski, A. Tugsaval, and O. Suschny, *Analyst*, 1978, **103**, 733.
[14] B. Griepink, *Anal. Proc.*, 1982, **19**, 564.
[15] P. G. W. Jones, *Anal. Proc.*, 1982, **19**, 565.
[16] A. Knöchel and W. Petersen, *Fresenius' Z. Anal. Chem.*, 1983, **314**, 105.
[17] H. S. Hertz, *et al.*, in C. E. van Hall, ed., "Measurement of Organic Pollutants in Water and Wastewater", (ASTM Spec. Tech. Publ. No. 686), American Society for Testing and Materials, Philadelphia, 1979, pp. 291–301.
[18] N. Andersen, *et al.*, Intergovernmental Oceanographic Commission Technical Series 22, U.N.E.S.C.O., Paris, 1982.
[19] R. Smith, *Water SA*, 1980, **6**, 37.
[20] R. V. Cheeseman and A. L. Wilson, Technical Report TR 66, Water Research Centre, Medmenham, Bucks, 1978.
[21] H. Egan and T. S. West, ed., "Collaborative Inter-laboratory Studies in Chemical Analysis", Pergamon, Oxford, 1982.
[22] H. H. Ku, ed., "Precision Measurement and Calibration – Statistical Concepts and Procedures", (Natl. Bur. Stand. Spec. Publ. No. 300), U.S. Government Printing Office, Washington, 1969.
[23] A. C. Green and R. Naegele, "Development of a System for Conducting Inter-laboratory Tests for Water Quality and Effluent Measurements", (National Technical Information Service Report PB-272 274), U.S. Department of Commerce, Springfield, Virginia, 1977.
[24] T. P. Whitehead, *Pure Appl. Chem.*, 1980, **52**, 2485.
[25] A. L. Wilson, *Analyst*, 1979, **104**, 273.
[26] Analytical Quality Control (Harmonised Monitoring) Committee, *Analyst*, 1979, **104**, 290.
[27] *Idem, Analyst*, 1982, **107**, 680.
[28] *Idem, Analyst*, 1982, **107**, 1407.
[29] *Idem, Analyst*, 1983, **108**, 1365.
[30] *Idem, Analyst*, 1984, **109**, 431.
[31] *Idem, Analyst*, 1985, **110**, 1.
[31a] The Severn Chemists' Sub-Committee, *Analyst*, 1984, **109**, 3.
[32] Standing Committee of Analysts, "General Principles of Sampling and Accuracy of Results", H.M.S.O., London, 1980.
[33] J. Büttner, *et al.*, *J. Clin. Chem. Clin. Biochem.*, 1980, **18**, 69.
[34] L. A. Currie, in I. M. Kolthoff and P. J. Elving, ed., "Treatise on Analytical Chemistry", Second Edition, Part 1, Volume 1, Wiley, Chichester, 1978, pp. 95–242.
[35] J. K. Taylor, *Anal. Chem.*, 1981, **53**, 1588A.

36 J. F. Lovett, in H. Egan and T. S. West, ed., "Collaborative Interlaboratory Studies in Chemical Analysis", Pergamon, Oxford, 1982, pp. 139–142.
37 Organization for Economic Co-operation and Development, "Good Laboratory Practice in the Testing of Chemicals", O.E.C.D., Paris, 1982.
38 R. Dybczynski, *Anal. Chim. Acta*, 1980, **117**, 53.
39 R. C. Paule and J. Mandel, *J. Res. Natl. Bur. Stand.*, 1982, **87**, 377.

8 Reporting Analytical Results

When the analysis of samples has been completed, the analyst's final task is to report the results to the user; several important questions arise in deciding exactly how this should be done. Three aspects involving consideration of the accuracy of results are discussed in Sections 8.1 to 8.3. The question of the units in which results are reported is mentioned in Section 2.7.1.

8.1 ROUNDING-OFF RESULTS

Analysts often round-off their results before reporting them with the intention that the number of figures in a reported result indicates its accuracy. This practice is discussed and recommended in many analytical texts; see, for example, ref. 1. Several points concerning problems of, or stemming from, this approach are worth making here.

8.1.1 Problems in Indicating Accuracy by Rounding-off Results

Rounding-off provides at best an ambiguous, semi-quantitative indication of accuracy. For example, suppose the analyst's initial results were 1.44; if the analytical error were ± 0.01, most texts would recommend reporting 1.44. But how then could an error two or three times larger be indicated, bearing in mind that the same texts would also recommend rounding to 1.4 when the error is ± 0.1? Clearly, rounding-off is an inadequate means of quantitatively indicating the error of a result. Further, this practice cannot distinguish between random and systematic errors though such distinction is urged by a number of authors (see, for example, Currie[2]), and strongly supported by ourselves. When it is required to indicate the accuracy of a result, much more direct and exact means should be used; see Section 8.2. Other authors[3,3a] have also urged that the number of figures in a result should not be used to indicate its accuracy.

8.1.2 Number of Figures in Analytical Results

Suppose the analytical response for a sample were 0.312 and the discrimination (see Section 4.3.1) of the analytical system were equivalent to 0.001 response units, *i.e.*, the value 0.312 can be reliably distinguished from 0.311 and 0.313. Of course, when this response is converted by the calibration function to a concentration, one can, in general, produce as many figures in the concentration result as desired. There seems to us little point in producing more figures than in the primary experimental measurement, the analytical response. On the other hand, if the concentration result is rounded to fewer figures than in the response, an error is introduced into the result. The size of this error will vary from result to result, and will not be known by the user of analytical results. We believe, therefore, that the simple and best approach is to ensure that analytical results *produced* by the analyst contain the same number of figures as the corresponding analytical responses for samples. This practice will, of course, sometimes lead to apparently different relative discriminations in results and responses. For example, the response of 0.312 could lead to results of 1.01 or 9.99 depending on the calibration function. The first result implies a relative discrimination corresponding to 1 % of the result while the second result implies 0.1 %; in fact, the relative discrimination in the response is 0.3 %. Such differences will, however, only exceptionally be of any importance in water analysis.

Most analytical systems used for the quantitative analysis of waters are capable of providing analytical responses that can be meaningfully read to three figures so long as the determinand concentration is well above the lower concentration limit of the analytical system. Often, however, as the concentration decreases, one reaches a point where the discrimination of the system no longer allows a third figure to be reliably read. A simple and clear-cut example is provided by any instrument with a digital read-out of three figures from 1 to 999; values below 100 can be read to only two figures and, indeed, below 10 to only one. This does not mean, in general, that such instruments are inadequate because it has also been suggested above [Section 2.6.2(*b*)] that the user's needs for the accuracy of results (expressed as a percentage of the result) will usually relax as the determinand concentration decreases. When this is so, the inaccuracy introduced by the coarser relative discrimination can usually be ignored provided that the discrimination at low concentrations is sufficiently small not to affect estimates of standard deviation by important amounts (see Section 4.3.1). If the discrimination near the lower concentration limit is 0.5 to 1.0 times the within-batch standard deviation of blank determinations, and if the discrimination remains constant over the entire concentration range, it may then easily be shown that the discrimination becomes equivalent to 1 % or less of the concentration at concentrations greater than 10 to 20 times the limit of detection (see Section 8.3).

8.1.3 Requirements of the User of Analytical Results

Reported results are intended for their user and not the analyst. It is the former, therefore, who should decide the number of figures in reported results, though this should be consistent with the discrimination of the analytical system. Of course, the user is well advised to consult with the analyst on such matters, and also bear in mind that if the results produced by the analyst are rounded before reporting: (1) real differences in the concentrations of samples from different locations or occasions may be needlessly concealed, (2) similarly, relatively small differences may sometimes be made to appear falsely large, (3) biased values for the mean, standard deviation, and/or other statistics of a set of results may be produced, and (4) quality-control procedures [see, for example, Section 4.6.2(*d*)(*ii*)] may be on a coarser basis than is desirable or necessary. Many quality measurement programmes concerned with the environment involve estimating statistics such as the annual average concentration (based on a number of individual analytical results), and points (1) to (3) are then especially relevant.

When results from more than one laboratory are to be compared, the need to ensure uniformity in rounding-off procedures is also important.

8.1.4 Suggested General Practice

Subject to any special requirements, the following suggestions are made. The analytical results *produced* by the analyst should have the same number of figures as the corresponding analytical responses. When the latter can be meaningfully read to three figures for concentrations well above the limit of detection, errors due to the discrimination of an analytical system may usually be ignored in routine water analysis. The degree of rounding (if any) in the reported results should be decided by the user in consultation with the analyst.

8.2 METHODS OF EXPRESSING THE ERRORS OF ANALYTICAL RESULTS

We saw in Sections 5 and 6 that analytical results are subject to both systematic and random errors, the latter always being present though the former may be negligible. Initially, therefore, we consider only random errors.

In Section 4.3 we also saw that one of the consequences of random errors is that each analytical result can be regarded as only an estimate of the true value, and which generally differs from that value. The resulting uncertainty in the true value can be quantified by calculating the confidence limits for the true value (Section 4.3.5). These confidence limits are a vital and integral part of any result for without them individual results cannot, in general, be used with confidence. It is of value, therefore, if confidence limits are whenever possible reported for each result to the user. Although this appears at present generally not to be done, we feel the procedure of sufficient importance that analysts should consider its adoption as standard practice. Similar recommendations have been made by other authors; see, for example, refs. 2 to 4.

It seems to us best to quote each reported result, R, as the $100(1-\alpha)$ % confidence limits for the population mean, *i.e.*, as $R \pm t_\alpha s$ [see Sections 4.3.5(*a*) and (*b*)], where the value of α is agreed in advance by the user and the analyst, and the estimate, s, is appropriate for the value of R (see Sections 6.1 to 6.3). We would include in this suggestion negative results so long as the confidence intervals included zero. If, however, a confidence interval did not include zero, there is an additional problem to consider. Our own view is that such a result is evidence that the analytical system has gone out of control, and that the result (and probably others obtained in the same batch of analyses) should not be reported; rather, sources of error should be sought, eliminated, and the analyses repeated. This interpretation rests, of course, on the assumption that bias is normally negligible. The reporting of negative results has recently also been suggested by the American Society for Testing and Materials.[33]

The reporting of results as $R \pm t_\alpha s$ would have the merit of continually bringing to the user's attention the fact that his ability to draw valid conclusions from the results is limited by their random errors. At the same time the maximum possible amount of useful information is provided on the true determinand concentrations in samples. As an alternative to quoting confidence limits with every result, the analyst could simply inform the user in advance that the confidence limits on results will have stated values unless the user is informed to the contrary. Such practices require reasonable estimates, as well as stability, of the precision of analytical results, but these are needed in any event [Sections 4.3.5(*b*) and 6.1]. The estimation of precision is discussed in detail in Sections 6.3 and 6.4.

If we now consider bias, we would stress first that it is almost always highly desirable and often essential to seek to use analytical systems giving results with negligible bias [Sections 4.6.1(*c*) and 5.5]. When this aim is thought to be achieved (see Section 5.5), only the confidence limits expressing the random error are necessary. We need, therefore, to consider only those analytical systems for which this aim is not achieved, and for these two points should be noted. First, the consequences of systematic errors generally differ from those of random errors (Section 4.4); it is, therefore, generally desirable clearly to distinguish the sizes of the two types of error.[2-4] An additional reason for maintaining this distinction is that, though the size of random error may usually be estimated with reasonable accuracy, the estimation of bias for analytical systems subject to this error may present considerable problems. Thus, we do not recommend as a general practice that results should be accompanied solely by an estimate of their total error, though circumstances may arise where this is useful. Second, the great problems commonly involved in estimating the bias of each analytical result lead us to suggest that the best general approach is for the analyst and the user to discuss what is known of the bias of results and, on this basis, to decide for each application how best to describe this error quantitatively.

8.3 REPORTING RESULTS CLOSE TO THE LOWER CONCENTRATION LIMIT OF AN ANALYTICAL SYSTEM

There is a long established concept that every analytical system has a lower concentration limit below which the determinand cannot be detected. There is, however, considerable variability in what precisely is meant by "cannot be detected", and in how the limit is defined numerically and estimated experimentally. In addition to such points, there is the important and apparently related problem of how analytical results should be reported when the determinand concentration is found to be close to or below this limit. Both aspects are of interest because the measurement of very small concentrations is often important in water analysis. In Section 8.3.1 we consider the definition and experimental estimation of several relevant lower concentration limits, and in Section 8.3.2 we discuss the reporting of results at small concentrations.

8.3.1 Lower Concentration Limits of Analytical Systems

At least three different types of lower limit have been proposed, and at least two different connotations applied to the word 'detected'. Even when there is underlying agreement between authors on fundamental concepts, their definitions of a given limit, and procedures proposed to estimate it experimentally, may vary substantially. A very large literature has developed on this topic, and we have made no attempt to review it here in view of our conclusions [see Section 8.3.1(*e*)]. Liteanu and Rîcă[5] provide an extremely detailed discussion of 'detection' and its statistical aspects, and Long and Winefordner[6] have recently reviewed a number of approaches to defining 'limit of detection'. In spite of the apparent complexity, it seems to us that the concepts basic to the consideration of detection are essentially very simple, and we have attempted to describe them in (*a*) to (*c*) below. In (*d*) we consider some important practical aspects arising from these ideas, and in (*e*) we draw some overall conclusions. Note also that the American Society for Testing and Materials has recently[33] discussed the concepts of Criterion and Limit of Detection in the context of reporting results at small concentrations.

(a) The Criterion of Detection
We begin by expressing our view that the phrase 'to have detected the determinand' is most usefully interpreted as meaning 'to have obtained experimental evidence that the determinand concentration is greater than zero'. Of course, there is nothing original in this view, but it forms the basis for subsequent discussion. Given this view, how can one obtain such evidence in routine analysis? Until recently, virtually all approaches have been based on the idea of comparing the sample and blank determinations, as recommended by IUPAC.[7] Within recent years, however, a number of authors have approached the problem on the basis of regression theory as applied to calibration curves. In our view, comparison of sample and blank is the better procedure, and we describe its basic principles and assumptions directly below. Brief consideration of the approach based on regression theory is deferred to (*d*).

Suppose that one makes one or more blank determinations [see Section 5.4.2(*b*)] at essentially the same time as a sample is analysed, *i.e.*, in the same batch of analyses. The water used for the blanks and the analytical procedure to which they are subjected are chosen so that, to the maximum extent possible, the analytical responses for the blank and sample are affected equally by all factors except the initial determinand concentrations in the blank water and the sample. Provided that this aim is achieved *and* that the concentration of determinand in the blank water is zero (or negligibly small in relation to the concentrations of interest in the sample), a positive difference between the analytical responses for the sample and the blank indicates the presence of the determinand in the sample. Of course, all experimental measurements are subject to random errors, and caution is necessary in interpreting the observed difference in sample and blank responses.

For generality, suppose that m independent blank determinations and n independent analyses of a sample are made in one batch of analyses; the values of m and n should normally be those prescribed by the analytical method. Denoting the mean responses for the sample and blank by $\bar{A}_T$ and $\bar{A}_B$, respectively, judgement on whether or not the determinand has been detected is governed by the value of D, where

$$D = \bar{A}_T - \bar{A}_B \qquad (8.1)$$

Clearly, a claim of detection is not justified solely by a positive value of D because the random errors of $\bar{A}_T$ and $\bar{A}_B$ will lead to such a value on 50 % of occasions when a sample containing none of the determinand is analysed. As D becomes larger in relation to the random errors of $\bar{A}_T$ and $\bar{A}_B$ so it becomes increasingly likely that the sample contains the determinand. What is needed, therefore, is a quantitative criterion for deciding when the value of D is large enough to justify a claim to have detected the determinand. This criterion is called here, following Roos,[8] the Criterion of Detection, and is denoted by L_{AC}, but note that other authors use different names and symbols for the same concept; see, for example, refs. 2, 5, and 9. The value of this criterion can be derived by arguments exactly the same as those in Section 4.6.2(b), but in view of the common interest in this topic, we give a derivation below.

Consider the situation where the concentration, C_S, of determinand in the sample is zero. The population mean for D is then also zero, and its variance, σ^2_{ADO}, is given by

$$\sigma^2_{ADO} = \sigma^2_{\overline{ATO}} + \sigma^2_{\overline{AB}} \qquad (8.2)$$

where $\sigma^2_{\overline{ATO}}$ and $\sigma^2_{\overline{AB}}$ are the within-batch variances of $\bar{A}_T$ and $\bar{A}_B$ [see Section 4.3.2(b)]. Given the assumption that the response of the blank is affected equally by all factors affecting the response for the sample, and since we are assuming that $C_S = 0$, it follows that $\sigma_{ATO} = \sigma_{AB}$, where these two standard deviations refer to individual measurements. Since $\sigma^2_{\overline{ATO}} = \sigma^2_{ATO}/n$ and $\sigma^2_{\overline{AB}} = \sigma^2_{AB}/m$ [Section 4.3.2(c)], equation (8.2) may be rewritten as

$$\sigma^2_{ADO} = (\sigma^2_{AB}/n) + (\sigma^2_{AB}/m)$$

so that

$$\sigma_{ADO} = \sigma_{AB}[(n+m)/nm]^{1/2} \qquad (8.3)$$

Equation (8.3) is true for any probability distribution of A_B. If, however, we assume that A_B follows a normal distribution [see Section 6.2.1(e)(x)], then so too does D. On this basis, when $C_S = 0$, D is normally distributed with population mean, $\mu_D = 0$, and population standard deviation, σ_{ADO}, given by equation (8.3). The known properties of the normal distribution [see Section 4.3.2(a)] can, therefore, be used to calculate the probability, when $C_S = 0$, of obtaining a value of D *greater* than any chosen value. In general, the probability that a value exceeding $+u_{2\alpha}\sigma_{ADO}$ will be obtained is α. Thus, if we put

$$L_{AC} = u_{2\alpha}\sigma_{ADO} \qquad (8.4)$$

and claim detection whenever $D > L_{AC}$, we shall falsely claim detection, on average, in a proportion, α, of instances. If we wish to decrease the probability of false detection, α must be decreased, and this causes $u_{2\alpha}$, and consequently L_{AC} to increase. In numerical terms, if we are prepared to accept a probability of 0.05 for false detection, $L_{AC} = u_{0.1}\sigma_{ADO} = 1.645\sigma_{ADO}$. In the common situation where $m = n = 1$, $\sigma_{ADO} = \sqrt{2}.\sigma_{AB}$ [equation (8.3)], so that $L_{AC} = 1.645/\sqrt{2}.\sigma_{AB} = 2.33\sigma_{AB}$. Of course, other values for α may be used if desired. Note that L_{AC} refers to the *difference* between $\bar{A}_T$ and $\bar{A}_B$. The observed response corresponding to L_{AC} is $\bar{A}_B + L_{AC}$.

There is general interest in expressing the Criterion of Detection as the corresponding value of C_S, and, following Currie,[9] we denote this expression of the Criterion of Detection by L_C. Problems arise[10] in converting L_{AC} to L_C when the calibration curve is markedly non-linear at small values of C_S. Midgley's discussion[11-15] of the criterion of detection when ion-selective electrodes are used provides a good example of the type of approach that can be followed; see also ref. 16 on ion-selective electrodes, and ref. 17 on non-linear calibration curves in general. We confine attention here to calibration functions of the form $\mathscr{A}_T = \lambda + \chi C_S$, and for these

$$L_C = u_{2\alpha}\sigma_{ADO}/\chi \qquad (8.5)$$

Note that σ_{ADO}/χ is the value of σ_{ADO} expressed in concentration units.

If the responses for blank determinations are always zero, then $\sigma_{ADO} = 0$ and $L_C = 0$. This situation may arise when the discrimination of the analytical system is too coarse to reveal the random errors of A_B or when the value of λ is zero. This latter condition appears to be implied by Glaser *et al.*[18] in their definition of 'method detection limit; see also ref. 19. It is important to avoid the former situation when 'detection' is of primary interest.

We should stress that L_C is *not* the smallest concentration that can be detected with probability 0.95.[8,9] Figure 8.1 shows that a sample for which $C_S = L_C$ will give

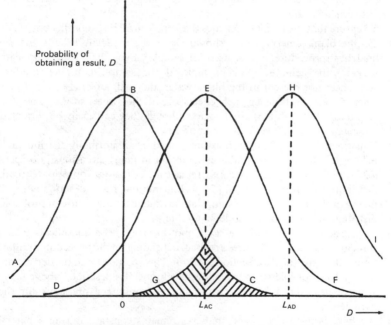

Figure 8.1 *Schematic illustration of the criterion and limit of detection (The shaded area of distribution ABC as a proportion of the whole area of the distribution represents the probability, α, of obtaining a value for D greater than L_{AC}. The shaded area of distribution GHI similarly represents the probability, β, of obtaining a value for D less than L_{AC}. For the figure, it has been assumed that $\alpha = \beta = 0.05$. For a distribution DEF with population mean, $\mu_D = L_{AC}$, the probability that D will exceed L_{AC} is clearly 0.5)*

analytical results greater than L_C on only 50 % of the occasions on which it is analysed; because of the symmetry of the normal distribution, 50 % of analytical results are greater than the true concentration and 50 % are smaller. The derivation of the smallest value of C_S for which there is a given probability of detection is described in (b) below. First, however, we consider the six main assumptions on which equations (8.4) and (8.5) rest.

(1) If the water used for the blank contains a concentration, C_W, of the determinand, the Criterion of Detection becomes

$$L_C = (u_{2\alpha}\sigma_{ADO}/\chi) + C_W \qquad (8.6)$$

Of course, as stressed already [Sections 5.4.2(b) and 5.4.4(b)], it is always highly desirable to ensure that C_W is essentially zero. When this cannot be achieved, it is generally necessary to estimate the value of C_W (see Section 10.7); this may cause practical difficulties but does not affect the validity of equation (8.6). See also Section 10.7.2(a) on interfering substances in the blank water.

(2) The approach is not valid for analytical systems for which $\mu_D = 0$ for all concentrations less than some limiting value, C_O, e.g., as may occur in gravimetric procedures. If, however, C_W is then made equal to C_O, equation (8.6) remains valid.

(3) To ensure that the blank duplicates the sample in all respects affecting A_T save C_S, the blank determination should clearly be made by exactly the same analytical procedure as that used for samples. Even then, exact duplication may not be achieved. For example, if the sample contains interfering substances not present in the blank water, they will affect A_T so that $\mu_D \neq 0$ when $C_S = 0$. Whenever such effects occur, the values of L_{AC} and L_C in equations (8.4) and (8.5) are incorrect, though they can easily be corrected if the magnitudes of the effects are known.

(4) Equations (8.4) and (8.5) assume that A_B and A_T are normally distributed. We believe this to be a reasonable assumption in many situations [see Section 6.2.1(e)(x)], but if desired one may replace $u_{2\alpha}$ by factors appropriate to other distributions[2,6] [see Section 4.3.2(a)]. Liteanu and Rîcă[5] describe non-parametric procedures (i.e., no assumptions are made on the nature of probability distributions) for estimating detection limits.

(5) The procedure for deriving L_{AC} and L_C may need modification when standard-addition procedures are used, depending on the procedure employed for blank-correction [see Section 5.4.2(c)]. This applies also to the derivation of L_{AD} and L_D considered in (b) below. The principles described above may be applied, however, to the derivation of appropriate expressions for these parameters.

(6) We have implicitly assumed that σ_{AB} remains constant during a batch of analyses. This is considered in (d) below.

(b) The Limit of Detection

This term was suggested by Roos[8] to denote the smallest value of C_S for which there is a desired probability of detection, i.e., that $\bar{A}_T - \bar{A}_B > L_{AC}$. We denote this value of C_S by L_D, and the corresponding value of $\bar{A}_T - \bar{A}_B$ by L_{AD}. Again, other terms and symbols are used by different authors for the same concepts. Figure 8.1 shows how a value for L_{AD} can be derived. To ensure a probability of $1-\beta$ for detection, the value of C_S must be such that there is only a probability, β, of obtaining a value for $D \leqslant L_{AC}$. Making the same assumptions as in (a), the standard deviation of D is given by

$$\sigma_{AD}^2 = (\sigma_{AT}^2/n) + (\sigma_{AB}^2/m) \tag{8.7}$$

where σ_{AT} is the within-batch standard deviation of A_T. Thus, when the population mean of D is $L_{AC} + u_{2\beta}\sigma_{AD}$, the probability of obtaining a value for D less than L_{AC} is β, so that

$$L_{AD} = L_{AC} + u_{2\beta}\sigma_{AD}$$

that is,
$$L_{AD} = u_{2\alpha}\sigma_{ADO} + u_{2\beta}\sigma_{AD} \tag{8.8}$$

In numerical terms, when $\alpha = \beta = 0.05$, and $n = m = 1$, $L_{AD} = 1.645\sigma_{ADO}$ $+1.645\sigma_{AD} = 1.645(\sqrt{2}.\sigma_{AB} + [\sigma_{AT}^2 + \sigma_{AB}^2]^{1/2})$. If we assume, as is often reasonable, that $\sigma_{AT} = \sigma_{AB}$ for very small values of C_S, then $L_{AD} = 3.29\sqrt{2}.\sigma_{AB} = 4.65\sigma_{AB}$. Of course, other values for α and β may be used if desired, and it is not necessary that $\alpha = \beta$. The observed response corresponding to L_{AD} is $\bar{A}_B + L_{AD}$.

As before, for linear calibration curves, $L_D = L_{AD}/\chi$, *i.e.,*

$$L_D = (u_{2\alpha}\sigma_{ADO} + u_{2\beta}\sigma_{AD})/\chi \tag{8.9}$$

It should be noted that the determinand may be detected in a sample whose concentration is less than L_D; the probability of detection of such concentrations, however, is less than $1-\beta$. For example, as noted above, when $C_S = L_C$, the probability of detection is 0.5.

(c) The Lower Limit of Determination

The relative percentage standard deviation of analytical results is rather large for samples for which $C_S = L_D$. For example, we saw in (b) that for $n = m = 1$, $\alpha = \beta = 0.05$, and $\sigma_{AT} = \sigma_{AB}$, the value of L_D is $4.65\sigma_{AB}/\chi$; the standard deviation of analytical results when $C_S = L_D$ is given by $\sigma_{AD}/\chi = \sqrt{2}.\sigma_{AB}/\chi$.* The relative percentage standard deviation, ρ, is, therefore, $(100\sqrt{2}.\sigma_{AB}/\chi)/(4.65\sigma_{AB}/\chi) \approx 30\%$; ρ is twice as large when $C_S = L_C$. Because of this rapid increase of ρ at low concentrations, a number of authors (see, for example, refs. 5, 9, 17) have suggested an additional concept – the Lower Limit of Determination – which is defined as that concentration above which a given value of ρ is achieved; $\rho = 10$ has been suggested as suitable. If the chosen value for ρ is denoted by ρ_Q, the Lower Limit of Determination, L_Q, is given by

$$L_Q = 100\sigma_t/\rho_Q \tag{8.10}$$

where σ_t is the total standard deviation of analytical results at a determinand concentration of L_Q [see Section 6.2.1(d)].

It is worth noting that many applications of analytical results may well be able to

---

* This assumes that the random error of the estimate of the slope of the calibration curve has negligible effect on the random errors of analytical results at very small concentrations [see Sections 6.2.1(c) and (d)].

tolerate relatively large values of $100\sigma_t/C_S$ at small determinand concentrations (see Section 2.6.2).

(d) Other Aspects of Criterion and Limit of Detection
A number of points arising in the application of L_C and L_D are considered in (*i*) to (*v*) below.

(i) Constancy of L_C and L_D   In Section 5.4.2(*a*) we derived equation (5.6) which, for linear calibration curves, relates analytical response and the determinand concentration in the solution analysed. If we put $C_S = 0$ in that equation, we obtain for the population mean response, $\mathscr{A}_B$, for the blank

$$\mathscr{A}_B = \eta + \chi C_C \tag{8.11}$$

where the meanings of the terms are defined in Section 5.4.2(*a*). We regard χ as constant within batches of analyses [Sections 6.2.1(*b*) and (*e*)(*iii*)], but the factors η and C_C are subject, in general, to random variations from sample to sample. Thus, a measured value of the analytical response, A_B, for a blank determination may be written as

$$A_B = (\eta + \varepsilon_\eta) + \chi(C_C + \varepsilon_C) + \varepsilon_M \tag{8.12}$$

where ε_η, ε_C, and ε_M are the random errors of the factors represented by η, C_C, and the measurement sub-system of the analytical system. These random errors are, in general, independent so that, using the results in Section 4.3.2(*d*),

$$\sigma_{AB}^2 = \sigma_\eta^2 + \chi^2\sigma_C^2 + \sigma_M^2 \tag{8.13}$$

where σ_η, σ_C, and σ_M are the within-batch standard deviations of the factors represented by η, C_C, and the measurement sub-system. For a given analytical system and value of α, L_C will be constant if σ_{AB}/χ is constant [see equations (8.5) and (8.3)]. It follows that each of the standard deviations, σ_η, σ_C, and σ_M as well as χ must be maintained constant if L_C is also to remain constant; we ignore the possibility of counter-balancing variations. This condition corresponds to what we have called a well controlled analytical system [see Sections 6.2.1(*e*) and 6.3.1].

A similar condition applies for the constancy of L_D, but in addition note that L_D also depends on σ_{AT}, the within-batch standard deviation of the analytical response for the sample. If σ_{AT} varies from sample to sample, so too will L_D, and equations (8.3), (8.7) and (8.9) allow calculation of the size of this effect for any ratio of σ_{AT}/σ_{AB}. Clearly, it is very desirable to avoid such variations whenever possible.

(ii) Use of L_C and/or L_D as performance-characteristics of analytical systems   The concept of performance-characteristics is considered in Section 9.4. In essence, the idea is quantitatively to characterise the performance achieved with a particular analytical system so that other analysts can better judge the likely suitability of that system for their own requirements. L_C and L_D (and similar parameters) are often recommended for inclusion in statements on performance-characteristics. It should be noted, however, that their values depend on α and β, respectively. Except, therefore, when the relevant analytical method prescribes the values to be used for α and β (and this is extremely rare), the values of L_C and L_D are,

in our view, better not regarded as fundamental characteristics of analytical systems. We believe, however, that the values for m and n should be considered an integral part of the procedure prescribed by an analytical method. Given this, we regard the parameter, σ_{ADO}/χ, as a suitable performance-characteristic of analytical systems. Note also that when $n = m = 1$, $\sigma_{ADO} = \sqrt{2}.\sigma_{AB}$, so that σ_{AB}/χ may then be regarded as a performance-characteristic.[20] Finally, we would stress that the various factors affecting the values of L_C and L_D are by no means necessarily the same in all laboratories applying the same analytical method.

In passing, we would also discount the use of L_Q as a performance-characteristic since it involves the arbitrary parameter, ρ_Q, which is not a property of analytical systems. Information on the standard deviation of analytical results and its dependence on the determinand concentration provides all the necessary information required on precision.

(iii) Using estimated rather than population standard deviations All the preceding discussion has been in terms of population standard deviations, σ. In practice, analysts only ever have estimates, s, which are subject to random error (see Section 4.3.4). If, therefore, the analyst wishes to calculate values of L_C and/or L_D, some allowance must be made for the uncertainties of s_{ADO} and s_{AD}. Many publications (including previous ones of ours) have suggested that this difficulty is overcome by replacing $u_{2\alpha}\sigma_{ADO}$ and $u_{2\beta}\sigma_{AD}$ in equations (8.5) and (8.9) by $t_{2\alpha}s_{ADO}$ and $t_{2\beta}s_{AD}$, respectively, where t is the t-statistic (see Section 4.3.5). This procedure is correct* when the values of s_{ADO} and s_{AD} are estimated from replicate measurements on a blank and a sample, and independent estimates of s_{ADO} and s_{AD} are obtained from independent analyses of blank and sample for every sample. Statistical complications arise, however, if one uses the estimates obtained from one set of measurements to estimate L_C and/or L_D, and then applies these estimated L_C and L_D values to assessing the results from a number of subsequently analysed samples and blanks. These complications arise from exactly the same causes as those discussed in detail in the Appendix to Section 4. In effect, the difficulty is that the use of $t_{2\alpha}$ and $t_{2\beta}$ does not lead to the required probabilities α (for false detection) and β (for failure to detect). The probability of false detection when $C_S = 0$ and $L_C = t_{2\alpha}s_{ADO}/\chi$, and the probability of failure to detect when $C_S = L_D = (t_{2\alpha}s_{ADO} + t_{2\beta}s_{AD})/\chi$ can be assessed by the procedure described in the mentioned Appendix. In view of our conclusions in (e) below, we do not further consider this problem here.

Liteanu and Rîcă[5] derive expressions for the confidence intervals of parameters such as L_C and L_D.

(iv) Estimation of L_C and L_D using regression theory As already mentioned, recent years have seen increasing application of regression theory to calibration [Sections 5.4.2(e) and 5.4.3(d)] and to the estimation of the precision of analytical results [Sections 6.2.1(c) and 6.2.2]. The same tendency is present in estimating L_C and L_D or similar parameters. The essential ideas underlying this approach are that, from the responses for a set of calibration standards and a sample, one can use

---

* We are assuming that $\sigma_{AT} = \sigma_{AB}$; if this is known or suspected to be untrue, statistical advice should be sought.

regression theory to calculate values for l, k and s_{AB}, which are estimates of λ and χ (the calibration parameters) and σ_{AB}. The estimates, k and s_{AB}, may then be used to obtain an estimate of L_C, and the uncertainty of this estimate can also be quantified.[5] Alternatively, regression theory has been used to calculate that concentration in a sample for which the confidence interval for the true concentration just includes zero; this concentration has been called the *minimum reportable concentration* or similar terms.[17,21-24] We shall not discuss these applications in detail; the references cited provide all necessary information. We wish rather to comment on the suitability of the general approach in the context of the detection of very small concentrations.

It should be noted that when regression theory is used in this way, the estimates, s_{AB} and l, are derived from calculations using some or all of the calibration standards whose concentrations are greater than zero. Since s_{AB} and l refer to zero concentration, the accuracy of these estimates depends on the accuracy with which the true relation between response and determinand concentration is known, and on the accuracy with which the true relation between the standard deviation of responses and concentration is known. In contrast, the procedure described in (*a*) above is less dependent on such assumptions because s_{AB} is determined from measurements of only blanks. Thus, the procedure in (*a*) seems to us more directly related to the particular problem of detection, and to require fewer assumptions for its validity. It is of interest that Mitchell and Garden[23] report that their minimum reportable concentration (determined by regression) is always greater than 'detection limits' calculated by procedures such as those in (*a*) and (*b*), and that the minimum reportable concentration decreased continuously as the concentrations of calibration standards were decreased.

(v) L_C and L_D when blank correction is not used   The calculation of L_C and L_D described in (*a*) and (*b*) above assumed that the analytical method specified blank correction. In water analysis there are two common situations where blank correction is not made even though the detection of traces of determinands is important.

(1) Instrumental zero adjustment   In some procedures, blank determinations are made but their responses are adjusted to zero by instrumental subtraction of a constant from all responses; the sample responses are then read with respect to this adjusted zero. Provided that these blank determinations are made by exactly the same procedure as that used for samples, this procedure is equivalent to that described above. The response for a sample becomes $A_T = A'_T - A'_B$, where A'_T and A'_B are the unadjusted responses. Thus, $D = A_T$ and $\sigma_D^2 = \sigma_{AT}^2 + \sigma_{AB}^2 = \sigma_{AT'}^2 + \sigma_{AB}^2$ since the subtraction of a constant from a measurement does not affect its standard deviation [Section 4.3.2(*b*)]. The only difference from the treatment in (*a*) and (*b*) above is then that the observed responses corresponding to L_{AC} and L_{AD} are L_{AC} and L_{AD}, respectively. Note also that this procedure conceals any systematic temporal variations of A_B from batch to batch.

(2) Procedures involving measurement of a peak superimposed on a background   In a number of analytical procedures, the determinand's presence is revealed by a peak rising above a background; chromatographic techniques are a familiar example but there are many others. In such procedures, a blank-correction is commonly made, in effect, by measuring the height or area of the determinand peak above the

base-line, and a blank determination and correction [as defined in (*a*) above] is not made. This approach is equivalent to that in (1) above, the base-line being used as the adjusted zero, but several points are worth making.

First, the procedure is valid as a blank-correction only if exhaustive testing has failed to detect an analytical response above the base-line when true blank determinations are made or if there are compelling reasons for expecting this response for such blanks always to be zero.

Second, the sensitivity of such analytical systems is sometimes adjusted so that no variations in the base-line (at and around the determinand peak) are visible when blanks are analysed, *i.e.*, σ_{AB} is apparently zero. Under these conditions, L_{AC} and L_C are both zero, though such conditions are undesirable when 'detection' is of primary interest.[25,26] When, however, the sensitivity is so adjusted, it follows that the discrimination, *d* (see Section 4.3.1), of the analytical system is large compared with the within-batch standard deviation of blank determinations. In this situation, it may well be that σ_{AT} (the apparent standard deviation of sample responses) is also zero for samples containing very small concentrations of the determinand. If this is so, statistical procedures for defining L_{AD} and L_D are inappropriate because random errors have been prevented from appearing. A practical limit of detection would then be that concentration producing an analytical response greater than *d*; this concentration can be deduced from the calibration function of the analytical system. It is also possible that σ_{AT} is not zero – when, for example, the responses for samples are much more variable than those for blanks. In this event, L_D is given approximately by [compare equation (8.9)]

$$L_D \approx d + u_{2\beta}\sigma_{AT''}/\chi\sqrt{n} \qquad (8.14)$$

where $\sigma_{AT''}$ is given approximately by (see Section 4.3.1)

$$\sigma_{AT''} \approx (\sigma_{AT}^2 - d^2/12)^{1/2} \qquad (8.15)$$

We would repeat, however, that whenever the detection of very small concentrations is of primary interest, it is better to adjust the sensitivity so that the discrimination has negligible effect on the apparent size of random errors.

Finally, when the sensitivity of analytical systems of the type under consideration is such that the random variations of the base-line are apparent, the principles described in (*a*) and (*b*) above may, in general, be applied. Thus, since measurement of the determinand response, A'_T, is made with reference to the base-line, A'_B, we have that $D = A_T = A'_T - A'_B$. The observed responses above the base-line corresponding to L_{AC} and L_{AD} are L_{AC} and L_{AD}, respectively. Similarly, $\sigma_D^2 = \sigma_{AT}^2 + \sigma_{AB}^2 = \sigma_{AT'}^2 + \sigma_{AB'}^2$ (but see also below). The new point that now emerges is the need to define σ_{AB}, the standard deviation characterising the random error of the base-line derived from the experimental record and used in measuring A_T.

The estimation of σ_{AB} is a rather complex topic, and is further complicated by the very strong likelihood that the random errors of A_T and A_B are correlated and not independent so that the simple equation in the preceding paragraph for σ_D^2 is unlikely to be correct. Numerous, and often mathematically complex, procedures have been described for dealing with these problems. Details of a variety of approaches may be found in refs. 25 to 30 and the publications cited therein; ref. 27 is of particular interest in view of the common use of automatic signal-processing and computing systems in the quantitative evaluation of chromatograms in water analysis. Given values for σ_{AT} and σ_{AB} and knowledge of the degree of correlation between the random errors of A_T and A_B, values for L_{AC}, L_C, L_{AD} and L_D may be derived as described above. Such complexity is, of course, avoided when A_B is obtained by measuring the response of a true blank determination.

(e) Conclusions

From the discussion in (*a*) and (*b*) above, we hope that we have made clear the concepts involved in quantifying 'detection'. On the basis of simple arguments and reasonable assumptions, one can derive expressions for, and estimate experimentally, parameters such as the Criterion and Limit of Detection. There is, however, the question of whether such parameters really serve a useful function. It seems to us that there are two main uses to which they may be put, though see also Section 2.6.2(*b*)(*v*). These uses are: (1) as performance-characteristics of analytical systems, and (2) in reporting routine analytical results. We argued in (*d*)(*ii*) that L_C, L_D and L_Q should not be regarded as performance-characteristics. Further, as will be seen in Section 8.3.2, it seems to us that the use of any or all of these three parameters in reporting results in also better avoided. Accordingly, notwithstanding the vast literature on this subject, we feel that parameters such as L_C, L_D and L_Q are of little value for either of the two above purposes.

8.3.2 Reporting Results at Small Concentrations

Having established the various concepts described in Section 8.3.1, we can now turn to the consideration of the reporting of analytical results at small concentrations of the determinand. It should be noted that our suggestions differ from those in the first edition of this book.

Suppose that an analytical result, R, is less than the Criterion of Detection, L_C, so that a claim to have detected the determinand is not justified at the chosen probability level, α. In this position, it is common to report the result as 'not detected' (ND). Such a result provides no information at all on the determinand concentration. It would, however, be unwise to report the result simply as 'less than L_C', the appropriate numerical value of L_C being given, because, when $R < L_C$ there may be a high probability that C_S (the true determinand concentration in the sample) is greater than L_C. To overcome this difficulty, the result could be reported as 'less than L_D'; the probability of this statement being incorrect is less than β [Section 8.3.1(*b*)]. If this were done, however, how would a result be reported when it was greater than L_C but less than L_D? If it were reported as 'less than L_D', valid information would be wasted (*i.e.*, that the determinand had been detected), and the probability of the statement being correct is also less than $1 - \beta$. If, on the other hand, it were reported as the result itself, the rather confusing situation would then arise where results on successive samples might be, say, 7, less than 10, 6, less than 10, *etc.* All such approaches – and others are possible – meet similar difficulties.

A second, and perhaps more important, point is that the result 'less than X' provides the user of analytical results with no information on the most likely value of the determinand concentration in the range 0 to X. Accordingly, it is difficult or impossible to make quantitative use of such results for many purposes, *e.g.*, in calculating the mean (or other statistics) of a set of results; see, for example, refs. 2, 31 and 33. Such problems may well lead to bias in the estimation of such statistics using the reported analytical results.

These problems seem to us easily overcome, in principle, simply by following the practice already suggested in Section 8.2, *i.e.*, to quote all results (explicitly or implicitly) as $R \pm t_\alpha s$. This approach gives all the useful information available from the sample analysis on the determinand concentration, and has been previously suggested in refs. 2 and 32. We see no reason, apart perhaps from unfamiliarity, why such reporting should cause difficulty for the user of results. For a somewhat different approach to the reporting of results at very small concentrations, see ref. 33.

Given the above suggestions, we see no value in the use of L_Q in the reporting of results.

8.4 REFERENCES

[1] American Public Health Association *et al.*, "Standard Methods for the Examination of Water and Wastewater", Fifteenth Edition, A.P.H.A., Washington, 1981.

[2] L. A. Currie, in I. M. Kolthoff and P. J. Elving, ed. "Treatise on Analytical Chemistry", Second Edition, Part 1, Volume 1, Wiley, Chichester, 1978, pp. 95–242.

[3] J. K. Taylor, *Anal. Chem.*, 1981, **53**, 1588A.

[3a] W. B. Crummett and J. K. Taylor, in H. Egan and T. S. West, ed., "Collaborative Inter-laboratory Studies in Chemical Analysis", Pergamon, Oxford, 1982.

[4] D. Garvin, *J. Res. Natl. Bur. Stand.*, 1972, **76A**, 67.

[5] C. Liteanu and I. Rîcă, "Statistical Theory and Methodology of Trace Analysis", Ellis Horwood, Chichester, 1980.

[6] G. L. Long and J. D. Winefordner, *Anal. Chem.*, 1983, **55**, 712A.

[7] I.U.P.A.C., *Spectrochim. Acta, Part B*, 1978, **33**, 242.

[8] J. B. Roos, *Analyst*, 1962, **87**, 832.

[9] L. A. Currie, *Anal. Chem.*, 1968, **40**, 586.

[10] J. D. Ingle and R. L. Wilson, *Anal. Chem.*, 1976, **48**, 1641.

[11] D. Midgley, *Analyst*, 1979, **104**, 248.

[12] D. Midgley, *Analyst*, 1980, **105**, 417.

[13] D. Midgley, *Anal. Proc.*, 1980, **17**, 306.

[14] D. Midgley, *Analyst*, 1980, **105**, 1002.

[15] D. Midgley, *Ion-Sel. Electrode Rev.*, 1981, **3**, 43.

[16] C. Liteanu, E. Hopirteau, and I. C. Popescu, *Anal. Chem.*, 1976, **48**, 2013.

[17] L. M. Schwarz, *Anal. Chem.*, 1983, **55**, 1424.

[18] J. A. Glaser, *et al.*, *Environ. Sci. Technol.*, 1981, **15**, 1426.

[19] C. J. Kirchmer, *Environ. Sci. Technol.*, 1982, **16**, 430A.

[20] A. L. Wilson, *Talanta*, 1973, **20**, 725.

[21] D. G. Mitchell, W. N. Mills, J. S. Garden, and M. Zdeb, *Anal. Chem.*, 1977, **49**, 1655.

[22] J. S. Garden, D. G. Mitchell, and W. N. Mills, *Anal. Chem.*, 1980, **52**, 2310.

[23] D. G. Mitchell and J. S. Garden, *Talanta*, 1982, **29**, 921.

[24] L. Oppenheimer, T. P. Capizzi, R. M. Weppelman, and H. Mehta, *Anal. Chem.*, 1983, **55**, 638.

[25] J. L. Sullivan, in R. L. Grob, ed., "Modern Practice of Gas Chromatography", Wiley-Interscience, London, 1977, pp. 213–288.

[26] H. W. Johnson and F. H. Stross, *Anal. Chem.*, 1956, **31**, 1206.

[27] G. F. Moler, R. R. Delongchamp, and R. K. Mitchum, *Anal. Chem.*, 1983, **55**, 842.

[28] L. R. Snyder and Sj. van der Wal, *Anal. Chem.*, 1981, **53**, 877.

[29] J. Ševčik, "Detectors in Gas Chromatography", Elsevier, Oxford, 1976.

[30] H. C. Smit and H. L. Waig, *Chromatographia*, 1975, **8**, 311.

[31] K. Heydorn and B. Wanscher, *Fresenius' Z. Anal. Chem.*, 1978, **292**, 34.

[32] R. V. Cheeseman and A. L. Wilson, Technical Report TR 66, Water Research Centre, Medmenham, Bucks, 1978.

[33] American Society for Testing and Materials, "1983 Annual Book of ASTM Standards. Volume 11.01 Water (I)", A.S.T.M., Philadelphia, 1983.

9 The Selection of Analytical Methods

9.1 IMPORTANCE OF SATISFYING THE USER'S REQUIREMENTS

9.1.1 General Considerations

Section 2 emphasised the need for the user of analytical results to formulate his requirements precisely so that the analyst may choose appropriate methods of analysis. Usually, several methods are available for each determinand; indeed, sometimes the number of potentially applicable methods is almost bewildering. The analyst may, therefore, have some difficulty in selecting a method. The growing interest in water quality and the increasing number of applicable analytical techniques and methods are intensifying this problem. Of course, analysts are accustomed to this situation, and no attempt is made here to discuss all the aspects relevant to the choice of analytical methods for routine use. Rather, certain points that seem to us of general importance with respect to the accuracy of analytical results are noted in Sections 9.2 to 9.4. Our approach is pragmatic, but we refer in Section 9.1.2 below to theoretical frameworks that either have been or might be applied to the selection of analytical methods. Finally, Section 9.5 gives details of useful sources of analytical methods for waters.

Throughout this section, the user's needs are taken to be of primary importance; in our view any other basis would be irrational. The user has, therefore, an extremely important responsibility to define these needs as carefully and precisely as possible (see Sections 2 and 4.6) since they form the basis of so much subsequent analytical activity and expenditure. In the present context of selecting analytical methods, this definition of needs should always include the following:

(1) the determinands, and then for each determinand,
(2) the concentration range of interest,
(3) the maximum tolerable bias and standard deviation of analytical results throughout the concentration range,
(4) the required frequency and/or times of sample collection (so that the number of samples to be analysed in a given period can be decided),
(5) the maximum tolerable period between sample collection and receipt by the user of the analytical result (so that the maximum allowable time for analysis can be deduced).

Given this information, the analyst has to decide, for each determinand, which method(s) is (are) potentially suitable; a suggested sequential approach to this decision is described in Section 9.2. From these methods the analyst can select those to be adopted for routine analysis on any grounds considered important. Sometimes, the user's requirements necessitate the use of a particular method for a given determinand, *e.g.*, when non-specific determinands must be measured (see Section 2.4.2). Such situations are discussed in Section 9.3.

We believe that the initial identification of potentially suitable methods should be made without regard for the analyst's current facilities; this will allow the possible benefits of improved facilities (*e.g.*, more staff, new instrumentation) to be defined. After these initial considerations, the user and the analyst can profitably discuss any necessary modifications of the user's requirements and/or the analyst's preliminary choice of methods so that the routine analytical programme may be agreed. For example, it may be that no available analytical method is capable of allowing the required precision for a particular determinand to be achieved. In this event, the user has to decide if it is worthwhile, in the short-term, including that determinand in the routine programme. Another situation that may arise is that the achievement of the user's requirements is possible only if the analyst is provided with better facilities; if such facilities cannot be made available, the user must again decide how to modify his needs. The aim throughout must clearly be that both the user and the analyst are confident that the methods chosen for routine analysis will allow the user's requirements to be met. It should also be remembered that many other components of analytical systems may markedly affect the accuracy achieved when a given analytical procedure is followed. Great care is, therefore, also necessary in selecting and controlling these other components. Thus, even when potentially suitable methods have been identified, experimental confirmation that the requirements for bias and precision can actually be met is necessary (see Sections 4.7, 5, 6, and 7).

It seems to us that the broad type of approach indicated above has the merits of logicality and ability to meet objectives. We have met, however, two common objections to these ideas, and these are worth considering in some detail.

First, analysts will know that the pressures and problems of the day-to-day work of laboratories often prevent or hinder the implementation of improved procedures. Many analysts may well believe, therefore, that the above approach is impracticable, particularly in a routine laboratory. The difficulty is exacerbated by the common problem of obtaining a carefully considered and precise statement of his needs from the user, and by the frequent situation that the period between the statement of new requirements and the beginning of the corresponding routine programme is too short to allow proper planning and provision of appropriate analytical facilities. Such difficulties will certainly make it impossible at present always to follow the suggested approach. Nevertheless, this clearly does not decrease the desirability of making the basic goal the establishment and implementation of the most suitable and efficient analytical systems for each measurement programme. The analyst, through the discussions proposed above with the user, has a vital role to play in attempting to reduce the severity and frequency of problems preventing achievement of this goal.

Second, there seems to be a rather common view that as long as an analyst uses methods from one or more of the various collections of 'Standard' or 'Recommended/Approved' methods, his results will necessarily be satisfactory. Such collections are of considerable value, and reliance on them may sometimes be the only practicable approach in the short-term. Complete reliance, however, is undesirable as many authors have stressed; see, for example, refs. 1 to 5 and one of the longest established collections itself.[6] The main points against unquestioning acceptance of the procedures in compilations of methods can be summarised as follows (see also Section 4.7).

(1) The methods are usually conceived and presented as general purpose and applicable in many situations. This, however, clearly takes no account of the possibility that the user's requirements may vary from one application to another – as, in fact, they do. For example, one analytical programme may require better precision than another, and a 'Standard' method may be incapable of the better precision. Similarly, the samples arising in some programmes may contain interfering substances that cause too great a bias when a particular method is used; this possibility is noted in most compilations of 'Standard' methods – see, for example, refs. 3 and 6.

(2) It is regrettable but nonetheless true that many 'Standard' methods give insufficient information to allow the likely magnitude of analytical errors in a particular application of a method to be decided. It is quite possible that an analyst may have much more detailed information concerning errors when a non-'Standard' method is followed.

(3) There are often a number of methods for a given determinand in a given collection. Thus, even if only such methods are considered, the question of choosing between them may still be involved. This choice is clearly made difficult when insufficient information is given on the accuracy that may be expected when a given procedure is followed.

(4) The revision and up-dating of collections of 'Standard' methods is almost invariably slow, and a number of years may well elapse between the development of a useful, new analytical procedure and its recommendation in the form of a 'Standard' method. Analysts dedicated solely to the use of the latter deny themselves the advantages of using better methods.

(5) The precision and bias of routine analytical results do not depend solely on the analytical method. In the past, the method was commonly the most important component of analytical systems, but increasingly today other components are becoming more and more important. For example, analysts, reagents, standards, laboratory conditions and facilities, apparatus, and instrumentation may all have very important effects on analytical accuracy, and, in some instances, may well be the dominant source of errors. Thus, use of a 'Standard' method in such analytical systems by no means ensures a given precision and bias. This point is being continually emphasised by the growing use of sophisticated instrumentation in routine analysis (see Sections 4.7 and 10.4), and by the smaller and smaller concentrations that analysts are required to measure. Indeed, such developments lead us to wonder about the real value of 'Standard' methods for such analytical systems if these methods cannot or do not specify in greater detail than is usual the required values and conditions for *all* important components of the corresponding analytical systems.

At present, therefore, although 'Standard' methods continue to play an important role in water analysis, they should not dissuade the analyst from attempting to choose and apply methods most appropriate to the user's needs. Essentially the same recommendation has been made in connection with the analytical methods published by the U.K. Standing Committee of Analysts.[3]

Requirements calling for the use of a particular method are considered in Section 9.3.

9.1.2 Theoretically Based Approaches to the Selection of Analytical Methods

As noted above, our main emphasis in Sections 9.2 and 9.3 concerns the selection of analytical methods when the primary interest is the accuracy of analytical results. This emphasis seems to us appropriate for a large proportion of the analyses required of natural and drinking waters. When, however, analyses are required *solely* for process-control purposes (see Section 3.3.3), the overall efficiency of this control is governed, in general, by both the frequency and the accuracy of analysis. Thus, the optimal analytical system and method can strictly only be decided by simultaneous consideration of both these factors. For example, in comparing the suitabilities of two or more methods, it is possible that the most suitable is not that allowing the most precise results to be achieved provided that poorer precision is more than compensated by a greater achievable rate of analysis.* Leemans[7] has described a quantitative procedure (derived from control-engineering theory) that allows this simultaneous consideration of the rate and precision of analyses, and has also illustrated its application to the control of a fertiliser-production process. The same theory has been described in rather more detail by Kateman and Pijpers,[8] and see also ref. 9 for an interesting extension of the approach. We are not aware of any applications of this theory in the field of water analysis, but it seems to be well worth exploration. Full details are given in the references cited, and we have not, therefore, further discussed the approach here.

Another theoretical approach,[10,11] combining statistical decision-theory and information theory, is also worth noting. This attempts to characterise the value of analytical information for the purposes for which it is obtained. For example, this value depends, in general, on the precision of analytical results. The approach may, in principle, be useful in aiding the selection of analytical methods. The theory requires, however, quantitative expression of the penalties arising from incorrect decisions based on analytical results, and we have noted the difficulty of such expression (see Section 3.4.3). Further, the theory is presented in a highly abstract form, and no examples are given of its use. Nonetheless, it represents a valuable attempt to integrate within one theoretical framework the aspects of analysis and use of analytical results. Again, we know of no applications of the theory, and do not consider it further here.

9.2 IMPORTANT FACTORS IN SELECTING ANALYTICAL METHODS

Many factors are involved, in general, in deciding the most suitable analytical method for a particular laboratory and measurement programme. Consideration of a number of these factors is straightforward and needs no discussion here. Our aim is rather to describe a sequential approach in which four aspects of primary

---

* This assumes that: (1) analysis rather than sampling is the rate-determining step in the delay between the beginning of collection of a sample and the reporting of the analytical result, and (2) any desired sampling frequency can be used.

importance with respect to the accuracy of analytical results are considered (Sections 9.2.1 to 9.2.4). Of course, this does not imply that the factors omitted are unimportant. For example, the speed and cost of analysis or the required man-power may be crucial aspects, but there is generally no difficulty in assessing such points from the description of an analytical procedure specified by a method. Furthermore, the emphasis on achieving the user's requirements also indicates the primacy of the accuracy of results; there is little point in making speedy analyses if the results are incapable of providing the required information. It is suggested, therefore, that the order of consideration of factors described below be followed; if, at the end of this process, more than one analytical method has been identified as potentially suitable, any other factors can be used by the analyst as a basis for the final selection of method.

9.2.1 The Determinand

It is essential that the user's definition of each determinand be unambiguous (see Section 2.4), and note that some determinands may be defined by specification of the analytical method to be used (Sections 2.4.2 and 9.3.1). When a particular method is not so specified, the analyst must consider the ability of the analytical procedures described in available methods to measure the specified determinand (see Section 5.4.5), and retain for further consideration only those for which satisfactory evidence for their suitability is available. If no such methods exist, this is a problem to be discussed immediately with the user. If one or more suitable methods are available, their suitability with respect to the concentration range of interest is next considered.

9.2.2 Concentration Range of Interest

In the analysis of waters it is generally possible to achieve any required upper limit of concentration. For example, even when this limit is greater than the upper limit of an analytical system, dilution of the samples is a simple means of overcoming the problem. Of course, such an approach may be undesirable from the standpoints of time and simplicity, but its use does not prevent analytical results of the required accuracy from being obtained – assuming that dilution or equivalent procedures are carried out with the good accuracy readily achievable. Thus, in the present context it is the lower concentration limit of interest, C_L, which is vital in selecting analytical methods. Of course, when a larger concentration range is of interest and two or more methods allow achievement of the required lower limit, preference will usually be given to the procedure involving minimal use of sample-dilution – provided that the required accuracy of results (see Section 9.2.3) is also achievable.

When the precision required of results is such that C_L can be equated with the limit of detection, L_D, of the analytical system [see Section 2.6.2(b)(v)], the candidate methods are simply examined to decide those for which the reported values of L_D are smaller than or equal to C_L. If no method meets this condition, this is a problem to be discussed immediately with the user. If one or more appropriate methods exist, their potential suitability with respect to achievable accuracy is next considered (Section 9.2.3). In considering reported limits of detection, several points should be noted.

(1) A variety of methods of defining and estimating the limit of detection of an analytical system are used. Such differences may well lead to apparent but false differences in the limits achievable when different analytical methods are followed. As suggested in Section 8.3.1, the limit of detection should normally be defined statistically, and that section may be consulted to see the various factors governing its numerical value. For example, this value depends on the chosen probability levels, α and β, so that care should be taken to ensure that the same values of α and β have been used for limits of detection that are being compared; otherwise, false conclusions on the potential suitabilities of different analytical methods will be drawn.

(2) The parameter of basic importance in governing the limit of detection is commonly the within-batch standard deviation of blank determinations. Thus, provided that values for this parameter are quoted or can be deduced, the relative abilities of analytical systems can be validly compared – though see the cautions concerning estimates of standard deviations in Section 9.2.3(b). When a value for this standard deviation is not available, great caution is necessary in drawing conclusions on the limit of detection achievable when a particular analytical method is used. Two approaches are possible in this situation. In the first, an estimate of the within-batch standard deviation of blank determinations is obtained experimentally (see Section 6.3). Alternatively, an estimate of the smallest possible value of this standard deviation can be obtained from the sensitivity of an analytical system and an estimate of the best possible standard deviation of the measurement sub-system [see equation (8.13), p. 294]. Such an estimate may indicate substantially better precision than would actually be achieved, *e.g.*, when the chemical processing of the blank is the dominant source of random error. The estimate, however, may well help to identify methods incapable of allowing achievement of the required limit of detection. For example, suppose that a spectrophotometric procedure were being considered and that the sensitivity were such that 10 mg l^{-1} of the determinand gave an absorbance of 0.5. By assuming that the best possible standard deviation of absorbance measurements at low concentrations is approximately 0.001 absorbance units, it follows that this corresponds to a standard deviation for the blank of approximately 0.02 mg l^{-1}.

(3) The numerical value of the limit of detection may be decreased by replicate analysis of the sample and/or blank. The size of this effect can be calculated using the equations in Section 8.3.1. For example, in many circumstances, duplicate analysis of the sample and the blank gives a 40 % reduction in the limit of detection compared with the value when only single analyses are made.

(4) In comparing the required value of C_L with values of L_D, some discretion may be necessary given that the exact value of C_L will seldom be crucial to the success of a measurement programme. For example, it could well be unwise to reject a method, satisfactory in all other respects, simply because the reported values of L_D are marginally greater than C_L. The numerical value of C_L is usually better regarded as a general target helping the process of selection of

analytical methods rather than as a fixed constant. Borderline cases may always be discussed with the user to resolve any doubts.

At the limit of detection of an analytical system, the relative standard deviation of analytical results is approximately 30 % [see Section 8.3.1(*c*)]. Thus, if the user requires better precision at the concentration, C_L, the limit of detection generally becomes irrelevant to the choice of analytical method. The dominant factor then is simply the achievable precision [see Section 9.2.3(*b*)]. Such a requirement will usually imply that the limit of detection will need to be substantially smaller than the value of C_L.

9.2.3 Accuracy

In principle, the required accuracy (as expressed by the maximum tolerable random and systematic errors – see Section 2.6.2) need only be compared with the published information on the accuracy achievable when the candidate methods are used. In practice, however, several problems may arise in interpreting such information, and these are considered in (*a*) and (*b*) below. When appropriate data are not available for a given method, it should be rejected in favour of methods for which such data are available – at least until either the accuracy achievable when the suspect method is used has been checked or its use has been agreed jointly by the analyst and the user of results. Again, this comparison of required and achievable accuracies should be made with discretion bearing in mind that the exact achievement of accuracy requirements will seldom be crucial to the success of a measurement programme. Thus, it may be unwise to reject a method, suitable in all other respects, simply because, for example, the reported precision is slightly worse than that required. Such situations can always be discussed with the user. Note, however, that the estimation of bias (Section 5) and the detection of deteriorating accuracy in routine analysis (Section 6.4) become more efficient as the precision of analytical results improves.

On the above basis, all methods reported to allow achievement of the required accuracy are to be regarded as potentially suitable. To choose from among such methods, one should also consider three other aspects of analytical systems which are often important in affecting the assessment and control of the accuracy of routine analytical results.

(1) There is great advantage in using robust analytical procedures; see Section 9.2.4.
(2) It is of value whenever possible to select analytical methods for which there is a well established theoretical understanding of the principles and mechanisms of the chemical and physical processes involved in the analytical procedures; see, for example, Sections 5.4.1 and 10.4.3.
(3) There is often advantage in using analytical methods which lead to proportionality between the analytical response and the concentration of determinand in the sample; see Sections 5.4.2 and 5.4.3(*c*).

As a final preliminary point we should also note that McFarren *et al.*[12] have proposed a numerical criterion for assessing the suitability of analytical methods with respect to accuracy. The criterion is based on the scatter of analytical results observed in an inter-laboratory study using the method in question. Eckschlager[13] and Midgley[14] have indicated problems in applying this criterion, and have also suggested the need for some modification. The main objection to all such proposals, however, seems to us that such criteria are based on judgements on what constitutes acceptable accuracy which are independent of the user's requirements.

(a) Bias

The assessment of the bias to which analytical results are potentially subject meets a number of important problems. These and the various sources of bias have already been discussed in detail in Section 5, and on this basis proposals for assessing the total bias of analytical results were made in Section 5.5. Readers should, therefore, consult the relevant parts of Section 5 before comparing the required and achievable biases. Special care is needed in considering bias caused by interfering substances because it is not uncommon for published methods to give insufficient and/or ambiguous information on this source of bias so that its reasonable assessment from published data may be difficult if not impossible. An additional point is that the magnitude of interference effects when a particular analytical method is employed may depend on the type of instrument used [see Section 5.4.6(*h*)].

It should also be borne in mind that sampling processes before analysis begins may also introduce bias; see Sections 3.5.1(*b*)(*v*), 5.5 and 9.2.4(*b*)(*viii*).

(b) Precision

It is becoming increasingly common for published methods to contain some statement concerning the achievable precision, and this is of great value. In using such information, however, to decide whether or not the required standard deviation is likely to be achieved when a particular analytical method is used, a number of points arise for which caution is necessary, and these are discussed below. See also Sections 6.2 to 6.4.

(i) Use of different measures of precision Some publications quote directly the standard deviation of individual analytical results; see, for example, the analytical methods published by the U.K. Standing Committee of Analysts[3] and the American Society for Testing and Materials.[15] This seems to us the best practice. A number of other practices, however, are followed, and unless proper account is taken of this, false conclusions may well be reached in comparing the precisions achievable by different analytical methods, and in deciding which methods are suitable for particular requirements. Three main points arise.

(1) Use of multiples of the standard deviation Instead of quoting the standard deviation, some publications quote multiples of this parameter as the precision, *e.g.*, refs. 16 and 17. So long as the multiple used is made clear, it is simple to derive the standard deviation from the quoted numerical values of precision.

(2) Use of terms other than standard deviation or precision Terms other than standard deviation or precision may be used, the two most common being *repeatability* and *reproducibility*. The use of these two terms is recommended by the International Organization for Standardization[16] and the British Standards

Institution.[17] As both these organisations issue methods for analysing waters, it is useful to cite the definitions given for these two terms. We quote the definitions given by the British Standards Institution; those of I.S.O. are essentially the same.

Repeatability is defined as "The value below which the absolute difference between two single test results obtained with the same method on identical test material under the same conditions (same operator, same apparatus, same laboratory and a short interval of time) may be expected to lie with a specified probability; in the absence of other indications, the probability is 95 %."

Reproducibility is defined as "The value below which the absolute difference between two single test results obtained with the same method on identical test material under different conditions (different operators, different apparatus, different laboratories and/or different time) may be expected to lie with a specified probability; in the absence of other indications, the probability is 95 %."

The import of these definitions in the light of the model of random errors in Section 6.2.1 is discussed in (3) below.

(3) Sources of random error included in reported values   The sources of random error included in quoted estimates of precision may vary from one publication to another. For example, some publications may quote values for the within-batch standard deviation while others may give the total standard deviation [see Section 6.2.1(d) for the definitions of these and related terms]. Since the former is, in general, smaller than the latter, care is needed in ensuring that the nature of a quoted standard deviation is both understood and appropriate to one's needs. In general, the total standard deviation is of primary interest [see Section 6.2.1(d)].

As a further example, it is useful to consider the interpretation of 'repeatability' in terms of a standard deviation. The numerical value of this parameter for 95 % probability is given by[17] $2\sqrt{2}.\sigma_r$, where σ_r is the repeatability standard deviation. The factor $\sqrt{2}$ arises because the difference between two results is being considered [see Section 4.3.2(b)], and the factor 2 is required for the 95 % probability level [see Section 4.3.2(a)]. With a slight complication that may be ignored here, σ_r is essentially the same as σ_w, the within-batch standard deviation of analytical results in our model of random errors [Section 6.2.1(d)]. Thus, by dividing a quoted repeatability (95 % probability) by $2\sqrt{2}$, an estimate of the within-batch standard deviation of individual analytical results may be obtained.

The inter-conversion of different measures of precision is, however, not always as simple as in the preceding paragraph. Let us consider 'reproducibility'. Tests in a number of laboratories are necessary to estimate this parameter, and the model of errors for a particular sample proposed by the British Standards Institution[17] is

$$R_{ij} = m + B_i + e_{ij} \tag{9.1}$$

where R_{ij} is the j th result of the i th laboratory, m is the overall average result, e_{ij} is the random error of the j th result of the i th laboratory, and B_i is the deviation of the results of the i th laboratory from m that cannot be accounted for by the random errors, e_{ij}. In this model, e_{ij} is characterised by what is termed the within-laboratory standard deviation, σ_w, and it is assumed that differences in the values of σ_w for different laboratories will be so small that an average value will be valid for all laboratories using the particular method. This average value is the repeatability standard deviation, σ_r.* The variability of B_i is characterised by the between-laboratory standard deviation, σ_L, which includes the longer-term random variabilities arising in single laboratories (due to factors such as operators, instruments, calibration, and environmental changes) as well as permanent systematic differences between laboratories, *i.e.*, B_i is equated to $B_{i0} + B_{is}$, where B_{i0} is a random component and B_{is} is a systematic component. On this basis, the reproducibility standard deviation, $\sigma_R = (\sigma_L^2 + \sigma_r^2)^{1/2}$, and the reproducibility is given by $2\sqrt{2}.\sigma_R$ (95% probability). It can be seen immediately that, even though σ_R can be readily derived from a quoted reproducibility, no simple interpretation of σ_R in terms of our model of random errors [Section 6.2.1(*d*)] is possible because σ_L represents, in general, the combined effect of random and systematic errors. All that can be said is that σ_L will not be less than the between-batch standard deviation, σ_b, of Section 6.2.1(*d*), and may often be substantially greater.

It seems to us that the model of Section 6.2.1(*d*) is much more relevant to any laboratory attempting to achieve a specified precision, and that reproducibility values are of little help to such laboratories. Exactly the same conclusion has been reached in the field of clinical chemistry,[18] and Mandel[19] has noted the general desirability of separating the longer term random errors within laboratories (B_{i0} which is approximately equivalent to our σ_b) from the systematic errors between laboratories (B_{is}). It should be mentioned that the problems of interpretation arising through this combination of random and systematic errors are discussed in the British Standard itself.[17] The definition of σ_R adopted by the International Organization for Standardization[16] is similar but not identical to that of the British Standard, and the interpretation of σ_R in terms of our model of random errors meets similar problems.

(ii) The concentration-dependence of precision In Section 6.2.1(*f*) we concluded that it is prudent to assume that σ_w and σ_t increase with increasing determinand concentration, at least until evidence to the contrary becomes available. This possible dependence of precision on concentration must be borne in mind in considering any published values for precision; otherwise, misleading conclusions may well be drawn on the abilities of methods to allow a specified precision to be achieved at one or more particular concentrations. Special care is needed for published methods giving only one precision estimate, especially if the concentration to which the estimate applies is not explicitly stated. The minimum amount of information regarded by us as acceptable is that there should be estimates of precision for two concentrations at or near the lower and upper limits of the concentration range of an analytical system; see also Sections 6.2.1(*f*) and 6.3.2(*d*).

(iii) Types of solution for which precision is quoted Data on precision may be quoted for standard solutions or for samples or both. Some care is necessary if only the former is quoted because several factors may cause worse precision for samples than for standards. The basic need, therefore, is that the published information should include data for typical samples. Note also that when data for standard

---

* Given the many factors affecting σ_w, the validity of this assumption for many methods – particularly for trace, instrumental analysis – seems to us very dubious.

solutions and samples are quoted and show essentially the same precision for both, this suggests either that samples are not liable to give worse precision than standards or that the analytical procedure has been so designed that any such tendency is nullified. In either event, the data allow greater confidence that the reported precision is achievable routinely by other laboratories.

(iv) Uncertainty of estimates of precision It must be remembered that published standard deviations are only estimates which are themselves subject to random error [see Sections 4.3.4, 6.3.2(c), and 6.3.5]. The magnitude of this random error depends on the number of degrees of freedom of the estimate of standard deviation, and published data should, therefore, include the degrees of freedom of each estimated standard deviation. When the degrees of freedom are not quoted, care is necessary in considering the data.

These uncertainties of estimated standard deviations may lead to the situation where the estimates from two different analytical systems differ though the population standard deviations do not. Similarly, an estimated standard deviation may be greater than the required standard deviation even though the population standard deviation is not. Such possibilities should be borne in mind when comparing the precisions achievable when different methods are used or when deciding if a given method is likely to allow a specified precision to be achieved; see also Section 4.3.4(e).

(v) Likelihood of achieving the reported precision Given that many factors may affect the precision achieved by a laboratory when a particular method is used (see Section 4.7), it is of value to examine the published data to check the number of laboratories that estimated precision and the range of standard deviations obtained – bearing in mind point (iv) above. One will naturally have greater confidence in selecting a method which, when used by many laboratories, led to results showing only a narrow range of standard deviations. Though we know of no way of quantifying this aspect, we feel that such methods are to be preferred, other things being equal, to those where a wide range of standard deviations was obtained or where only one laboratory estimated precision. In considering such points, care should be taken to ascertain the extent to which extreme values of the estimates of standard deviations have been deleted before the data were published. Note also that some approaches (*e.g.*, refs. 16 and 17) to the publication of data on precision quote only the average standard deviation, for a given solution, of the participating laboratories.

It is worth noting also that when many laboratories achieve similar precision when a given analytical method is used, there are the implications that: (1) the analytical procedure is robust, and (2) the analytical method is complete and unambiguous; see Section 9.2.4.

(vi) Effect of the instrument used for measurement Today, many different instruments for making a particular type of measurement (*e.g.*, spectrophotometers, atomic absorption spectrometers, gas chromatographs) are commercially available. It may well be, therefore, that the instrument in a laboratory considering the use of an analytical method differs from the instruments used by the

laboratories whose precision estimates are quoted for that method. The question then arises whether this will have an important effect on the precision that this particular laboratory can achieve. There is no sure way of answering this question without information on the precision of measurements made by the relevant instruments. In general, a laboratory in this position must make its own estimates of the precision of analytical results to answer the above question. It is worth noting that this effort may sometimes be avoided by some simple calculations. For example, suppose that a standard deviation of 0.01 mg 1^{-1} has been reported for a spectrophotometric procedure using a specified spectrophotometer, and that with this instrument the sensitivity of the analytical system was such that 1 mg 1^{-1} of the determinand gave an absorbance of 0.1. From these data it can be calculated directly that the standard deviation is equivalent to 0.001 absorbance units. Assuming that the sensitivity is little affected by the instrument used, it follows that to achieve this precision requires an instrument capable of reading to the nearest 0.001 absorbance units; instruments with appreciably coarser discrimination will not allow the reported precision to be achieved (see Section 4.3.1). Thus, any laboratory with an instrument of clearly inadequate discrimination may avoid the need for experimental tests. Of course, it does not follow that all instruments with adequate discrimination will necessarily be suitable with regard to their precision of measurement.

To be able to make calculations such as that above, it is necessary that the published data also includes the sensitivities of the relevant analytical systems as recommended in Section 9.4.

(vii) Concluding remarks Points *(i)* to *(vi)* above show that a reported value for precision must not be regarded as some precisely defined characteristic constant of an analytical system; such values must be interpreted and used critically. If the reported precision is worse than that required, care is also necessary in considering the effect of replication of analyses on the precision of analytical results. Although replication will generally improve precision, the magnitude of this effect depends on which measurements are replicated; see Sections 6.2.1(*c*) and (*d*).

9.2.4 Robustness and Description of Analytical Procedures

(a) Robust Analytical Procedures
The term '*robust analytical procedure*' is used here to denote a procedure which is so designed that the accuracy of analytical results is not appreciably affected by small deviations from the experimental conditions (*e.g.*, procedural, environmental) prescribed by the analytical method. Youden,[20] and recently Currie,[4] Sunderman *et al.*,[21] Winter,[22] and Massart *et al.*[23] use the term 'rugged' in the same sense as robust, and stress the value of such procedures in helping to achieve reliable results in routine analysis. Clearly, the adoption of robust procedures is always highly desirable, and especially so when analytical results are required to be of specified accuracy. Given two or more analytical methods that have met all the criteria discussed in Sections 9.2.1 to 9.2.3, preference should be given usually to that which represents the most robust procedure.

Unfortunately, there is at present no numerical definition of robustness which allows it to be adequately represented by a single numerical value. However, careful consideration of the effects and choice of the experimental conditions (as described in the published method and/or supporting publications) will help to provide a qualitative assessment of the degrees of robustness of different procedures [see (*b*)(*iii*) below]. Youden[20] has described simple experimental designs (and appropriate statistical procedures for interpreting the results obtained) for investigating the robustness of analytical procedures.

Currie,[14] Massart *et al.*,[23] Liteanu and Rîcă,[24] and many other authors have noted that it is to some extent possible, during the development of a new or improved analytical procedure, to arrange that its robustness is optimised. It may often be necessary to effect a compromise between robustness and other aspects of a procedure, but we feel that robustness should, whenever possible, be given priority for procedures intended to be applied routinely. In many circumstances, maximal robustness is achieved with respect to a given experimental factor (*e.g.*, the amount of a reagent) when the numerical value of that factor is that producing maximal analytical response for a given concentration of the determinand; see, for example, ref. 24. When only one or two such factors are involved, simple and intuitive experimental designs will usually suffice to determine the optimal values. More often, however, a number (sometimes many) of factors are relevant for analytical procedures, and more complex experimental designs and statistical interpretation of the results are both necessary and economic of effort. Such approaches to the optimisation of analytical procedures are reviewed in some detail in refs. 23 and 24.

(b) Need for Complete and Unambiguous Specification of Analytical Procedures
We have already stressed in Section 6.3.1 the need for, and vital role played by, a complete and unambiguous specification of any analytical procedure to be used routinely; that is to say, the analytical method should be complete and unambiguous. This point has been advocated by many authors; see, for example, refs. 3, 4, 15, 18, 19, and 25 to 30. This aspect of analytical methods is, therefore, important in considering their suitability for use in routine analysis. Before adopting any method, we would suggest that it be critically examined from this point of view, the aim being not to use a method until it is complete and unambiguous. This aim should, in our view, generally be regarded as mandatory for methods for non-specific determinands since such determinands can only be defined by the analytical procedures used to measure them; see also Section 9.3.1.

There is an International Standard[29] for the lay-out and content of analytical methods, and a British Standard deals with similar topics for gas-chromatographic methods.[30a] Burns[31] provides a useful collection of information on standards and recommendations concerning the nomenclature and publication of analytical methods and data. We feel, however, that a number of aspects are so important that brief discussion is useful here. Of course, particular responsibility falls on those who write analytical methods since they will usually be best qualified to judge when a particular method is complete and unambiguous. The various points involved are noted in (*i*) to (*ix*) below; they are addressed primarily to authors of methods, but may also be used as a set of criteria for judging the suitabilities of analytical methods from the present point of view. We draw some conclusions in (*x*).

(i) Completeness of analytical methods   Any method should specify all important, procedural details concerning every aspect of the analysis of samples, calibration, and calculation of analytical results. A number of particular points are noted in subsequent sections, but we wish here to stress, as Eisenhart[25] has observed, that this general aim is easier stated than achieved. It is generally impracticable to include *every* procedural detail, and the writer of a method must, therefore, decide which details are important and which are unimportant. This is often not simple for procedures as complex as those involved in analysis. Two devices seem to us of value in ensuring adequately complete descriptions.

(1)   Common analytical operations such as the use of volumetric glassware and measuring instruments need be described in analytical methods only when some particular and/or unusual manner of use is required. Organisations publishing analytical methods, however, should (and usually do) also publish recommended practices for the use of common equipment, and analytical methods may then refer to such practices. Each laboratory should ensure that it has, and follows, such a set of 'Standard Operating Procedures'[27,28,28a] for its analytical equipment.

(2)   An analytical method should include such detail that, if it were faithfully followed by an analyst with no previous experience of the procedure, he would reproduce all the operations believed by the author of the method to be essential to satisfactory execution of the procedure. A method may, therefore, be regarded as incomplete if, at one or more points in the method, an independent analyst is not sure of exactly what he is intended to do. Whenever authors are in doubt, the safe course is to include more rather than less detail.[18,30] The method should always specify how many portions of a sample and calibrating solutions (including blank determinations) should be analysed in order to obtain results of the accuracy quoted in the method.

(ii) Ambiguous instructions in analytical methods   A number of authors – see, for example, Eisenhart[25] – have noted the difficulty caused by instructions such as "Shake vigorously", "Add rapidly", "Clean thoroughly", *etc.* These and other similar examples are liable to widely-differing interpretations by analysts, with the potential problem that sources of error may be less well controlled than necessary. We believe it is always possible to replace such ambiguous instructions by specifying more clearly the procedures known by the author of the method to be suitable; see also *(iii)*.

(iii) Tolerances on experimental conditions   It is impossible to maintain any experimental condition completely constant for all analyses. The sizes of random errors and some types of bias depend, however, on the degree of control of experimental conditions. Thus, to aid the achievement of stable and adequate accuracy, the degree of control of all relevant experimental conditions needs to be specified and achieved in routine analysis (see Section 6.3.1). A simple means of specifying the required degree of control is to quote a tolerance for each prescribed experimental condition; this has been recommended in a number of publications.[19,25,29,32] For example, an addition of a reagent may be specified as "Add 5.0 ml

(± 0.2 ml)", and this seems to us much more helpful and informative than the bald "Add 5.0 ml". Appropriate tolerances for masses, volumes, times, temperatures, currents, voltages, flow-rates, *etc.*, will aid the accurate description of analytical procedures, and thereby help to ensure the minimum of difficulty in the routine use of analytical methods. Of course, the numerical values of tolerances should be carefully chosen; for example, specification of the smallest achievable tolerance for every experimental condition whether or not such a high degree of control is actually required is likely to be counter-productive. Such an approach conceals from the analyst using a method the more important conditions of a procedure for which special care is required, and calls for needless effort in unnecessarily close control of relatively unimportant factors. This approach will also lead analysts to the view that the procedure is not robust [see (*a*) above] because the size of tolerances is directly related to robustness. This use of tolerances has been generally adopted in the methods published by the U.K. Standing Committee of Analysts. See also (*ix*) below.

A related point concerns the specification of equipment appropriate for a particular procedure. For example, the particular material used for containers (beakers, flasks, *etc.*) in which samples are chemically processed may be crucial in reducing the extent of contamination in some procedures. Any such items, and special methods for cleaning them, should be specified in detail. A problem of growing importance concerns the specification of appropriate analytical instrumentation. Difficulties that may arise through this component of analytical systems are discussed in Sections 4.7, 5.4.6(*h*), and 10.4, and we would suggest that analytical methods should include detailed recommendations on the type of instrumentation and any associated precautions required to ensure that the accuracy quoted in a method may be achieved.

(iv) Measurement of analytical responses   For some analytical procedures there is little or no problem in deciding the numerical value of the analytical response to be recorded for each solution analysed, *e.g.*, when the measurement sub-system has a stable, digital read-out. In other procedures the discretion of the analyst may be required in judging the analytical response, *e.g.*, when a peak rising above a noisy base-line must be measured on a recorder-chart. The need for such discretion introduces a potential source of uncontrolled error, and, to minimise this, the method should specify as exactly as possible the procedure to be followed in determining the response for a solution.

(v) Specification of calibration procedure   To convert the analytical responses obtained for samples to concentrations of the determinand, estimates of the calibration parameters [see Section 5.4.2(*a*)] are required, and the errors of these estimates will contribute to the errors of analytical results [see Sections 5.4.3, 5.4.4, and 6.2.1(*d*)]. The calibration procedure(s) used to obtain the estimates of calibration parameters will govern their errors, and should, therefore, be specified completely and unambiguously. Several points are worth stressing.

(1)   In general, the analytical procedure used for calibration standards should be identical with that used for samples [Sections 5.4.3(*b*) and 5.4.4(*a*)].

(2) The concentrations, and number of analyses, of each calibration standard should be specified.

(3) The *exact* procedure used to determine the estimates of calibration parameters should be specified.

(4) Information should be given on the time-period (or number of samples) over which the numerical values of calibration parameters have been found not to vary appreciably. Such information is invaluable in deciding the frequency with which fresh estimates of calibration parameters should be obtained. Of course, an individual laboratory may find through use of an analytical method that this frequency should be increased, but a recommendation in a method provides a reasonable basis for laboratories to judge the amount of effort required for calibration and to carry out preliminary tests to estimate the precision of analytical results (see Section 6.3).

(5) In many procedures for trace analysis, the determinand content (and the concentrations of interfering substances) in the water used to prepare calibration standards (including blank determinations) is a source of bias if no correction is made for this content [see Sections 5.4.2, 5.4.4(*b*), and 10.7]. Whenever this content cannot confidently be assumed to be negligible in any laboratory, the analytical method should include detailed description of one or more procedures for producing water of minimal impurity content *and* a special procedure* for measuring the determinand content (see Section 10.7).

(vi) Calculation of analytical results The method must prescribe exactly how analytical results for samples are to be calculated from the one or more analytical responses for each sample and the estimates of calibration parameters; any special corrections should also be specified in detail. The question of the number of figures to be quoted in each reported result is considered in Section 8.1.

(vii) Stabilities of reagent and standard solutions Few such solutions are stable indefinitely, and some deteriorate rather rapidly. Such deterioration may well occur without visible signs in the solutions but with important effects on the accuracy of analytical results. Analytical methods should, therefore, give detailed instructions on all relevant aspects of the storage of such solutions and also provide information on their stability under the prescribed storage conditions. It seems to us invalid for authors to argue that information on the period of stability is not essential since any laboratory's general precautions and analytical quality control procedures will either prevent or rapidly detect deterioration. Reliance on such means of control alone may well cause unnecesary work in laboratories through needlessly frequent preparation of solutions or wasted work when deterioration is detected in solutions already used to analyse a number of samples. Further, quality control tests will not generally detect deterioration of solutions immediately it becomes appreciable [see Section 6.4.2(*b*)].

(viii) Sampling procedures† Several aspects of sampling procedures may have

---

* It will, in general, be impossible to measure the determinand content by the analytical procedure used for samples.
† See Section 3.5.1(*a*) for definition of what we include under the heading of 'sampling procedures'.

important consequences on analytical procedures and *vice versa*. We would suggest, therefore, that any relevant points should be specified in analytical methods. The various factors that may be involved are discussed in many parts of Sections 3.5 and 3.6, and common examples of important points are sample containers, sample stabilisation, the volume of sample required for analysis, and special sample-collection containers to prevent contamination of samples or losses of determinands.

(ix) Information on main sources of error   To aid analysts when they are considering the adoption of an analytical method, we feel that it is very helpful if methods mention the likely dominant sources of error together with the precautions known to be suitable for controlling the sizes of errors. This indicates the factors most needing to be carefully controlled, and may also help analysts to understand the reasons behind the procedures specified in a method. The methods published by the U.K. Standing Committee of Analysts generally include a section in which such information is given.

(x) Concluding remarks   The discussion in *(i)* to *(ix)* above identifies many aspects of analytical methods that analysts, considering the adoption of a particular method, should critically examine. In general, we would strongly urge that preference be given to methods that have clearly been very carefully written to provide the maximum amount of useful information. Whether or not this approach is followed, it is very important not to begin using any method until it is complete and unambiguous. The critical examination mentioned above will identify any inaccuracies, lacks and uncertainties in the description of a procedure, and these should generally be remedied before carrying out any preliminary tests of precision and/or bias.

9.2.5 Suggested Criteria for Judging the Suitability of Analytical Methods

In Section 7.2.5 we noted that, when a group of laboratories has common requirements for each laboratory, it is highly desirable that all laboratories adopt a reasonably uniform approach to the selection of analytical methods. To this end, it was also suggested that the group should agree a set of criteria to be observed by all laboratories in judging the potential suitability of analytical methods. Such a set of criteria, based on the considerations in Sections 9.2.1 to 9.2.4, is summarised below; methods not meeting these criteria should, in general, be rejected.

(a) Quantitative Information on Precision and Bias
These should be quantitative information on the following aspects of the precision and bias of analytical results.

(1)   The efficiency with which relevant chemical and physical forms of the determinand are measured (Section 9.2.1).
(2)   The limit of detection when this parameter is of interest (Section 9.2.2). If this parameters is of interest, the group of laboratories must also agree a definition of 'limit of detection' (see Section 8.3.1).
(3)   The effects of substances other than the determinand which are likely to be

present in samples at appreciable concentrations (Sections 9.2.3 and 5.4.6). As a minimum, this information should be available for at least two concentrations of the determinand at or near the lower and upper limits of the concentration range of interest, and for at least one concentration of each of the other substances equal to or slightly greater than the maximum concentration expected in samples [see Section 5.4.6(e)]. Under some circumstances [see Sections 5.4.6(h) and (*i*)], it may suffice to replace this criterion by one requiring evidence on the recovery of at least two different amounts of the determinand added to a series of relevant samples of different general composition.

(4) The standard deviation of analytical results at all concentrations of interest [Section 9.2.3(*b*)]. In general, it is the total standard deviation of analytical results which is of primary interest. As a minimum, the standard deviations for two concentrations of the determinand at or near the lower and upper limits of interest are required [Section 9.2.3(*b*)(*ii*)]. If not included in this information, there should also be some data on the precision achievable when real samples rather than standard solutions are analysed [Section 9.2.3(*b*)(*ii*)].

(b) Accuracy of the Specification of the Analytical Procedure
The analytical method should be complete and unambiguous [Section 9.2.4(*b*)], and the analytical procedures for blank and other calibration measurements should be exactly the same as that used for samples unless there is firm evidence that any differences between the procedures cause negligible errors [Section 9.2.4(*b*)(*v*)]. The value of using robust analytical procedures should also be noted [Section 9.2.4(*a*)].

(c) Effects of Sampling Procedures
There should be firm, quantitative evidence that any specified or proposed procedures concerning sample stabilisation and storage before analysis are adequately effective and cause no important worsening of the aspects of performance in (*a*) above [Sections 9.2.4(*b*)(*viii*) and 3.6]. In general, this implies that preference should be given to those methods for which those aspects were investigated when a specified sample-stabilisation procedure was followed. Otherwise, additional tests may well be necessary by the group of laboratories to assess the efficiency and other possible effects of sample-stabilisation procedures.

Any other sampling requirements dictated by the analytical method should be carefully considered for their feasibility and effectiveness [Section 9.2.4(*b*)(*viii*)].

9.3 APPLICATIONS REQUIRING THE USE OF A PARTICULAR ANALYTICAL METHOD

For many determinands and purposes, the particular analytical method used is not crucial provided that the user's requirements are met. There are, however, some instances where a particular method must be used, and these are briefly considered in Sections 9.3.1 and 9.3.2. Such requirements may well lead to practical difficulties, and a possible means of avoiding this problem is discussed in Section 9.3.3.

9.3.1 Measurement of Non-specific Determinands

As noted in Section 2.4.2, non-specific determinands can, in general, be defined only by specifying *very closely* the procedures to be used for measuring them. That is to say, the determinand is defined by a complete and unambiguous analytical method. Two examples are considered below to illustrate the types of problem that arise for such determinands, but similar points apply to all non-specific determinands.

Many techniques and different types of optical instruments have been used and are available for the measurement of turbidity. However, each technique and class of instrument (*e.g.*, absorptiometer, spectrophotometer, nephelometer) even if calibrated with identical suspensions may give different results for samples because of the factors affecting the transmission and scatter of light by samples.[6,48] For example, the wavelength of light used in the instrument, the path-length of the light in the sample, and the angle at which scattered light is measured may all affect the analytical results. Rigid standardisation of the analytical procedure is, therefore, necessary to obtain reliable results, and the use of different procedures and instruments in different laboratories is virtually certain to lead to inability to compare the results from different laboratories with confidence.

A second example is the measurement of filtrable ('dissolved') materials to the exclusion of non-filtrable ('undissolved') (see Section 11.2.1). This measurement may be required for particular substances (*e.g.*, iron, phosphorus) or for all substances in the sample when 'total dissolved solids' is of interest. The necessary separation is almost always attempted by filtration, but is, in general, incomplete. The extent to which non-filtrable materials are removed depends on the nature of the filter and filtration apparatus as well as other details of the filtration procedure (*e.g.*, rate of filtration, volume of sample filtered). Such factors may also affect the removal of filtrable materials (see Section 11.2.1). Again, therefore, rigid standardisation of the analytical procedure is essential.

Many other determinands of common interest in water analysis are non-specific (*e.g.*, colour, biochemical oxygen demand, suspended solids), and the problems of defining and measuring such determinands lead to two situations where all laboratories of a group should follow *exactly* the same analytical method, though see also Section 9.3.3.

(1) When the analytical results from each laboratory are required to judge the conformity of one or more waters with some specified value such as a mandatory standard for water quality. Strictly, of course, water quality standards for non-specific determinands ought, therefore, to specify in detail the analytical procedures to be used for non-specific determinands as a means of defining them. Otherwise, confusion in comparing the conformities of waters measured by different laboratories is virtually certain to arise. Insufficient attention seems to us to be paid to this aspect in some water quality standards, *e.g.*, those in some Directives of the European Communities[33–35] where the recommended analytical methods are usually incomplete and ambiguous.

(2) When the analytical results from two or more laboratories are to be compared with each other.

9.3.2 Measurement of Specific Determinands

If, for a particular user's requirements, only one analytical method will allow those requirements to be met, then it is clearly desirable that all laboratories concerned use that method until other methods are shown to be equally suitable (see Section 9.3.3).

Another situation arises if a user's requirements are such that, for a particular determinand, none of the available analytical procedures is sufficiently well characterised with respect to two sources of bias – efficiency of determination of different forms of the determinand (see Section 5.4.5) and interference (see Section 5.4.6) – to allow reasonable assessment of the likely bias of analytical results. Until this unhappy situation can be remedied – by further investigation of available procedures or development of better characterised procedures – the comparability of results from different laboratories may possibly be improved by all laboratories adopting the same analytical method. It should be noted, however, that any such beneficial effect is rather a hope than a firm, logical expectation. The postulated lack of knowledge on the above sources of bias and any variability of sample composition from one laboratory to another prevent, in general, any logical inference that the adoption of the same method by all laboratories will necessarily lead to more comparable results than when a variety of methods is used.

9.3.3 The Concept of 'Equivalent Analytical Methods'

The need for each laboratory involved in meeting a particular analytical requirement to use a particular analytical method for a given determinand may, of course, lead to practical difficulties. For example, not all laboratories may have the necessary instrumentation. Thus, when use of a particular analytical method is specified, it is also common to allow laboratories to use alternative methods provided that it is shown that the results obtained are 'equivalent' to those when the specified method is used. This concept of 'equivalent analytical methods' appears, for example, in the European Communities' Directives[33-35] on water quality; the United States Environmental Protection Agency employs the same idea in its quality assurance activities in support of water quality standards in the U.S.A.[22,49] The concept is also essentially the same as that stressed throughout this book (*i.e.*, that the accuracy of results in relation to the required accuracy is the prime factor determining the suitability of analytical methods) provided – and it is a very important proviso – a detailed and unambiguous definition is given of 'equivalent'. The lengthy discussion in Sections 5 and 6 on the experimental estimation of bias and precision emphasises that the assessment of accuracy is usually difficult and laborious. Thus, to demonstrate that the accuracy of results is not affected when an 'equivalent analytical method' is used, one is likely to face similar difficulty and require substantial effort. There may be an understandable tendency, therefore, to restrict the tests specified for the demonstration of 'equivalence' to such an extent that they no longer allow a proper assessment of the accuracy of results. It seems to us that any such restriction is likely to lead to problems concerning the accuracy and comparability of results, particularly for non-specific determinands. In these situations, therefore, we would suggest great caution in defining 'equivalence' in such a way that less than a complete assessment of the accuracy of results is required. Detailed proposals for equivalence testing[49] and an example of their application[50] have been reported.

9.4 PERFORMANCE CHARACTERISTICS OF ANALYTICAL SYSTEMS

From the above discussion it can be seen that the task of analysts in choosing analytical methods appropriate for given requirements is much facilitated when published methods contain (or give references to) detailed and unambiguous information on factors such as precision, bias, limit of detection, concentration range, *etc.* Thus, those concerned with the publication of analytical methods should strive always to include such information in addition to the description of the analytical procedure. Factors such as these, relating to analytical performance, may be termed *performance characteristics*,[36] though other terms are used by other authors, *e.g.*, 'figures of merit'.[26] Performance characteristics relate to the performance achieved by analytical systems of which analytical methods are but one component. Although the method is sometimes the component governing the numerical values of performance characteristics, this is by no means necessarily so. We think, therefore, it is better to speak only of the '*performance characteristics of analytical systems*'; such usage has the merit of explicitly recognising the potentially important role of components of systems other than methods.*

Since the first edition of this book there has been a welcome tendency for more and better defined information on performance characteristics to be included in published methods, though it seems to us that a number of organisations issuing collections of recommended methods have some way yet to go. A variety of practices is followed at present by organisations and individual authors in respect of the performance characteristics quoted and the manner in which the information is presented. A lengthy discussion of the relative advantages and disadvantages of the various approaches is inappropriate here, but the preceding sections of this book show the value of quoting at least the following performance characteristics.

(1) *Calibration function* This shows both the nature of this function and also the sensitivity† at any determinand concentration. Any concentration limits for the validity of the function should be noted.

(2) *Total standard deviation of analytical results* The quoted estimates should be accompanied by their degrees of freedom. When estimates for a given concentration of the determinand are available from more than one laboratory, we would suggest also that the range of these estimates be indicated, *e.g.*, by quoting the smallest and largest for a given concentration.

(3) *Within-batch standard deviation of analytical results at zero concentration of the determinand* This information is to characterise the ability to detect small concentrations [see Section 8.3.1(*d*)(*ii*)]. Again, the range and degrees of freedom of estimates should be quoted. Other or additional information may need to be given when the analytical response is not proportional to the determinand concentration [see Section 8.3.1(*a*)].

(4) *Sources of bias* Quantitative information on the size of bias from every relevant source of this type of error should be quoted.

---

* One of us (ALW) in reviewing a number of aspects of performance characteristics some years ago[31,37–39] spoke, rather illogically it is regretted, of the performance characteristics of analytical methods.

† By 'sensitivity' we mean the rate of change of analytical response with the concentration of the determinand; this usage is now almost universal.

Since performance characteristics refer to analytical systems, they should strictly be accompanied by information defining all the components (other than the method) used when the quoted values were determined. Provided, however, that the method is complete and unambiguous, and that it was faithfully followed by competent analysts, it will usually suffice to quote the instruments used for the measurement of analytical responses. The need to define other components should, however, be borne in mind.

It is also helpful to present such data and other relevant information so that an analyst may, at a glance, decide if a method is likely to be appropriate to his needs. Again, opinions differ on what constitutes other relevant information, but we would suggest that this should always include summary statements on: (1) the definition of the determinand, (2) the types of sample for which the method is suitable, and (3) the analytical procedure involved. It is also generally useful to include an indication of the total time required to analyse samples, and the proportion of this time for which the analyst's attention is required. All this information can be conveniently summarised in tabular form, and one method of tabulation was devised many years ago by one of us (ALW) and colleagues at the Central Electricity Research Laboratories. This approach was later adopted by the Water Research Centre and then by the U.K. Standing Committee of Analysts for its published methods. An example of such a table is reproduced in Figure 9.1.

The approach illustrated by Figure 9.1 seems to us to go a long way towards achieving its aims – the concise presentation of the nature of an analytical procedure and of the performance characteristics observed when that procedure was followed. We would, therefore, recommend this method of presentation. The approaches followed by other organisations issuing collections of analytical methods may be seen in their publications (see Section 9.5.2), and we feel that these approaches are generally less helpful than that in Figure 9.1. For readers wishing to read of other approaches to the definition and estimation of performance characteristics, refs. 4, 18, 23, 26, 29, 31, 36 to 47, 51 and 52 provide a useful introduction.

9.5 ANALYTICAL METHODS FOR THE ANALYSIS OF WATERS

9.5.1 Introduction

There is no shortage of published methods for the analysis of natural and drinking waters. Indeed, there is an almost embarassing richness of sources providing collections of 'Approved/Standard/Recommended' methods for a wide range of determinands – not to mention the very large numbers of individual methods published regularly and with growing frequency in the scientific literature. This super-abundance of methods makes it increasingly difficult critically to assess the relative values and roles of each of the many methods for a given determinand. This difficulty is, in turn, aggravated by the varying practices followed in presenting analytical methods and in defining, estimating, and reporting performance characteristics, and even more by the fact that a high proportion of published

Acid Soluble Aluminium in Raw and Potable Waters by Spectrophotometry

Tentative Method

(1979 version)

Note: Throughout this method aluminium is expressed as the element (Al).

1 Performance Characteristics of the Method (For further information on the determination and definition of performance characteristics see another publication in this series)	1.1 Substance determined	Those forms of aluminium reacting with pyrocatechol violet, α, α-bis (3, 4-dihydroxyphenyl) toluene-2, α-sultone. (See Sections 2 and 8).
	1.2 Type of sample	Raw and potable waters.
	1.3 Basis of method	The reaction of aluminium with pyrocatechol violet to form a blue-coloured complex the concentration of which is measured by spectrophotometry at 585 nm.
	1.4 Range of application	Up to 0·3 mg/l.
	1.5 Calibration curve (a)	Linear to at least 0·3 mg/l.

	1.6 Total Standard Deviation (a)	Aluminium Concentration (mg/l)	Standard Deviation (mg/l)	Degrees of Freedom
		0·060 (b)	0·005	18
		0·180 (b)	0·004	18
		0·300 (b)	0·005	17
		0·017 (c)	0·005	14
		0·318 (d)	0·008	12

	1.7 Limit of detection (a)	0·013 mg/l with 10 degrees of freedom.
	1.8 Sensitivity (a)	0·1 mg/l gives an absorbance of approximately 0·16.
	1.9 Bias (a)	No bias detected except when interferences occur (see Section 1.10).
	1.10 Interferences (a)	Certain substances are known to cause interference in this determination (see Section 3).
	1.11 Time required for analysis (a)	The total analytical and operator times are the same. Typical times for 1 and 10 samples are approximately 60 and 90 minutes excluding any pretreatment time.

(a) These data were obtained at the South-West Water Authority[1] using a spectrophotometer with 10-mm cells at 585 nm.
(b) Distilled water spiked with the stated concentration of aluminium.
(c) Tap water.
(d) Same tap water spiked with 0·3 mg/l aluminium.

Figure 9.1 *Example of a tabulation of performance characteristics* (The example is p. 4 of ref. 69 and is reproduced with the permission of the Controller of Her Majesty's Stationery Office)

methods do not meet all the criteria noted in Section 9.2.5. Such a critical assessment of the published methods of analysis for waters would be a monumental task if it took into account – as in our view it should – the various types of water, the widely differing qualities of certain types of water, the different purposes of analysis, the many relevant determinands, the different analytical resources available to laboratories, and the different weights attached to various features of analytical procedures by different analysts. In any event, the results of such a review would be subject to so many provisos that we doubt its value. Further, as noted repeatedly in preceding sections, we think it essential for analysts to attempt to select those methods most appropriate to their particular purposes and analytical resources. We hope that the main contribution of this book to such questions will be the discussion of, and suggestions on, the many factors that should be considered in selecting analytical methods, and in estimating and controlling the sizes of the errors affecting analytical results. It is also important to note that many countries, including the U.K., have organisations whose responsibilities include the recommendation and publication of analytical methods for waters. In addition to such national bodies, a number of international organisations also issue authoritative collections of recommended methods. Yet more lists of recommendations seem to us only likely to confuse still more an already rather confused situation. For all these reasons, we had no wish and have made no attempt to add here yet another list of recommended methods to those already available.

Analysts familiar with the field of water analysis will need no information on the available collections and other sources of analytical methods, but for those without this familiarity we have given some suggestions in Section 9.5.2 on useful publications. Further, to provide a general picture of commonly applied, routine analytical procedures, we have summarised in Section 9.5.3 the recommendations in two authoritative collections on the analytical methods for a number of determinands of common interest. As a further aid to the selection of methods, Section 11 deals in some detail with the principles and applications of many analytical techniques playing important roles in modern water analysis. In that section we try to indicate also the relative advantages and disadvantages of the various techniques as well as their general characteristics as means of providing routine analyses. Finally, in Sections 12 and 13 we discuss various aspects of the partial and complete automation of analytical procedures.

9.5.2 Useful Publications on Methods for Analysing Waters

A number of international organisations as well as various bodies in many countries produce comprehensive collections of methods for one or more types of water. Our detailed knowledge of such publications is mainly of those in the English language, and in the following we mention some of the more readily available and comprehensive.

In the United Kingdom, two national bodies – the Standing Committee of Analysts of the Department of the Environment, and the British Standards Institution – both issue analytical methods in a series of booklets, each of which usually deals with one determinand though a number of relevant methods may be included. The former body has at present issued some eighty booklets, and a list of those in print is provided in the current Sectional Publication List No. 5 of Her Majesty's Stationery Office in London. These publications cover a high proportion of determinands of common interest (see Section 9.5.3), and the Committee will be issuing both methods for other determinands and also new or revised methods for those determinands for which methods have already been published. The Committee's work is described in periodic reports, ref. 53 being the latest at the time of writing, and new publications are noted in several analytical journals, *e.g.*, *The Analyst* and *Analytical Abstracts*. The publications of the British Standards Institution appearing under the general heading BS 6068 are those stemming from the work of Technical Committee TC 147 of the International Organization for Standardization on which the Institution is represented. These British Standards are generally identical with the publications of I.S.O. (see below). Issue of these methods has only recently begun so that few are at present available, but increasing numbers should appear in the future. The methods that have been issued are listed in the annual Yearbook of B.S.I. See below also for other British Standards on methods of water analysis.

Several well known and widely quoted and used collections of methods are produced in the United States. Perhaps the two most comprehensive are those produced as large books by the American Public Health Association *et al.*[6] and the American Society for Testing and Materials.[15] The first is revised and issued at approximately 5-year intervals. The second is revised and published annually, and has a greater emphasis on industrial waters though many of the methods are of wide applicability. Valuable collections of methods are also produced by the United States Environmental Protection Agency[54,55] and the United States Geological Survey.[56,57] Note that ref. 57 does not, in general, give detailed analytical methods, but it remains a valuable source which summarises the analytical procedures and cites references to the detailed methods.

Of the international organisations working in this field, the World Health Organization has recently provided a very comprehensive collection of methods;[58] this is based to a large extent on refs. 59 and 60 both of which have been superseded by later publications though many of the analytical procedures remain essentially the same or very similar. As noted above, Technical Committee TC 147 of the International Organization for Standardization has been working for some time on the provision of methods for all determinands of general importance. These methods have now begun to appear as a series of separate booklets dealing with particular determinands, and are listed in the annual Yearbook of the Organization. Finally, we should mention a useful collection of methods for the analysis of freshwaters that was originally produced as part of the International Biological Programme; a second edition has now been published.[2]

The analysis of sea and saline waters can pose a number of problems that are not commonly met in the analysis of freshwaters. Although some of the methods in the above publications are relevant to such waters, two collections of methods for sea water analysis should be noted.[61,62] Riley[63] also gives a very detailed review – with many references – of analytical procedures for sea water, but this does not include detailed methods.

Industrial waters (*e.g.*, boiler and boiler-feed waters) are dealt with particularly in ref. 15 and two British Standards. The first[64] is intended for rapid control analyses where the accuracy required of results is not stringent; the second[65] provides methods for more accurate measurements. The Central Electricity Generating Board has produced a manual[66] giving a number of good and well written methods for power-station waters. Finally, a review[67] on the analysis of industrial waters though somewhat out-dated remains of value.

In addition to these collections of methods, individual methods for particular purposes appear in the scientific literature with growing frequency. Not only journals dealing solely with analysis but also many others (*e.g.*, those concerned with the environment, water resources, *etc.*) contain relevant papers. Fortunately, a number of abstracting journals (*e.g.*, *Chemical Abstracts*, *Analytical Abstracts*) contain sections on waters and their analyses, and are of great value in attempting to keep abreast of the many and varied developments. Further, two journals – *Analytical Chemistry* and *Journal of the Water Pollution Control Federation* – include regular reviews of the literature of water analysis. The former appears every two years (odd-numbered years), and is particularly comprehensive; the latter review appears annually.

9.5.3 Typical Analytical Techniques and Methods in Common Use

We have summarised in Table 9.1 the methods recommended for determinands of common interest in two authoritative and well established collections of methods in the English language, namely, those of the Standing Committee of Analysts in the United Kingdom, and those of the American Public Health Association, the American Water Works Association, and the Water Pollution Control Federation in the United States. Our aim is simply to try to show a broad picture of the types of analytical technique and method that are recommended at present for routine water analysis. We would stress, however, three points.

(1) We do not imply that our choice of these two collections of methods means that the methods noted in Table 9.1 are necessarily to be preferred to those in other publications.

(2) With two exceptions [see (3) below], the picture presented by Table 9.1 seems to us reasonably representative of the procedures and techniques used in the many laboratories concerned with fresh, natural and treated waters. Laboratories dealing routinely with special types of water (*e.g.*, sea water, high-purity industrial waters) may be less well represented because of their special analytical needs.

(3) The table does not indicate the rather common use of electrothermal atomisation/atomic-absorption spectrometry for the determination of trace metals. Such procedures are to be published by the Standing Committee of Analysts, and though no particular method is given in ref. 6, this does provide some general guidance on the use of the technique. See also Section 11.3.5(*c*) for a description of electrothermal atomisation and a discussion of the problems that its use may incur. Finally, Table 9.1 does not indicate sufficiently clearly the great use of semi- and fully-automatic procedures (see Section 12.2) in many laboratories.

TABLE 9.1
Methods Recommended by the U.K. Standing Committee of Analysts and the American Public Health Association/American Water Works Association/Water Pollution Control Federation

Recommended analytical procedure
(see notes at end of table for abbreviations and other information)

Determinand	U.K. Standing Committee of Analysts*	American Public Health Association et al.[6]
Alkalinity[68]	TIT: using acid, to indicator or potentiometric end-point	TIT: as ref. 68 – see opposite

Aluminium[69]	SPEC: using pyrocatechol violet	SPEC: using eriochrome cyanine R
		AAS (N_2O/C_2H_2 flame):
		  (a) direct aspiration of sample
		  (b) solvent-extraction of Al oxinate, aspiration of extract

Antimony[70]	SPEC: coprecipitation of Sb on $Zr(OH)_4$, filtration and dissolution of precipitate, solvent-extraction of Sb/crystal violet ion-pair, measurement of extract	AAS (air/C_2H_2 flame): direct aspiration of sample
Arsenic[71,72]	AAS†: formation of AsH_3, decomposition and measurement in a heated tube SPEC: formation of AsH_3, absorption in a solution, conversion to reduced arsenomolybdate and measurement	AAS: (a) formation of AsH_3, measurement in a hydrogen flame; the preferred method (b) electrothermal atomisation when interferences absent SPEC: formation of AsH_3, reaction with silver diethyldithiocarbamate and measurement
Barium	Method not yet published	AAS (N_2O/C_2H_2 flame): direct aspiration of sample

Beryllium	Method not yet published	AAS: (a) aspiration of sample into N_2O/C_2H_2 flame
		  (b) solvent-extraction of Be oxinate, and aspiration of extract into N_2O/C_2H_2 flame
		  (c) electrothermal atomisation for low concentrations.
		SPEC: using aluminon; use when AAS not available

Boron[73]	SPEC: (a) using curcumin (b) using azomethine-H (c) using azomethine-H TIT: using alkali after adding mannitol	SPEC: (a) as (a) ref. 73 – see opposite (b) using carmine
Bromide[74]	TIT: oxidation of Br^- to BrO_3^-, iodometric titration of BrO_3^- (Only for high concentrations as in sea water)	SPEC: using phenol red and chloramine-T
Cadmium[75]	AAS (air/C_2H_2 flame): solvent-extraction of Cd pyrrolidine dithiocarbamate, aspiration of extract	AAS (air/C_2H_2 flame): (a) direct aspiration of sample (b) as ref. 75 – see opposite SPEC: solvent-extraction of Cd dithizonate; use when AAS not available
Calcium[76]	AAS (air/C_2H_2 flame): direct aspiration of sample	AAS (air/C_2H_2 flame): as ref. 76 – see opposite TIT: (a) pptn of Ca oxalate, titration with $KMnO_4$ (b) using EDTA; this is the preferred procedure for general use
Carbon dioxide, free	Method not yet published.	TIT: using alkali CALCULATION: from pH and alkalinity
Chloride[77]	TIT: (a) using $AgNO_3$ (K_2CrO_4 indicator) (b) using $Hg(NO_3)_2$ (diphenylcarbazone indicator) (c) using $AgNO_3$, potentiometric end-point SPEC†: using $Hg(SCN)_2$ and Fe^{3+}	TIT: (a) (b) } as (a), (b) and (c) ref. 77 – see opposite (c) SPEC†: as ref. 77 – see opposite
Chlorine[78]	TIT: (a) iodometric (starch indicator) for concentrations > 1 mg l⁻¹ (b) reaction with diethyl p-phenylenediamine and titration of product with $Fe(NH_4)_2(SO_4)_2$, visual end-point SPEC: using diethyl p-phenylenediamine	TIT: (a) as (a) ref. 78 – see opposite (b) as (b) ref. 78 – see opposite (c) amperometric titration using phenylarsine oxide SPEC: (a) as ref. 78 – see opposite (b) using leuco crystal violet

TABLE 9.1 (cont.)

Recommended analytical procedure
(see notes at end of table for abbreviations and other information)

Determinand	U.K. Standing Committee of Analysis*	American Public Health Association et al.[6]
Chromium (III, VI)[79]	SPEC: oxidation of Cr^{III} to Cr^{VI}, solvent-extraction, reaction with diphenylcarbazide AAS (air/C$_2$H$_2$ flame): sample concentrated (5x) by evaporation before aspiration; background correction probably necessary	(c) using syringaldoxine SPEC: using diphenylcarbazide after oxidation to Cr^{VI} AAS (air/C$_2$H$_2$ flame): (a) direct aspiration of sample (b) oxidation to Cr^{VI}, solvent-extraction of Cr pyrrolidine dithiocarbamate, aspiration of extract
Cobalt[80]	AAS (air/C$_2$H$_2$ flame): (a) sample concentrated (10x) by evaporation before aspiration; background correction essential (b) solvent-extraction of Co pyrrolidine dithiocarbamate, aspiration of extract	AAS (air/C$_2$H$_2$ flame): (a) direct aspiration of sample (b) as (b) ref. 80 – see opposite
Colour	Method not yet published	(a) Sample filtered, colour compared visually with standard solutions of K$_2$PtCl$_6$/CoCl$_2$ (b) Spectrophotometric measurement of absorbance at several wavelengths and calculation of the tristimulus value
Copper[81]	AAS (air/C$_2$H$_2$ flame): sample concentrated (10x) by evaporation before aspiration; background correction may be necessary	AAS (air/C$_2$H$_2$ flame): (a) direct aspiration of sample (b) solvent-extraction of Cu pyrrolidine dithiocarbamate, aspiration of extract SPEC: (a) using bathocuproine (b) solvent-extraction of Cu^I-neocuproine complex, measurement of extract
Cyanide	Method not yet published	SPEC: using chloramine-T, pyridine and barbituric acid ISE: CN$^-$ distilled into absorbing solution, measured with CN$^-$ ion-selective electrode
Fluoride[82]	ISE: direct measurement after adding buffer	ISE: as ref. 82 – see opposite

Hardness, total[83]	SPEC: (a) using La chelate of alizarin-fluorine blue complexone (b) using eriochrome cyanine R–zirconium complex (colour bleached by F⁻) Ion-chromatography [see Section 11.3.8(g)] TIT: using EDTA (eriochrome black T indicator) SPEC: (a) using zirconium-SPADNS complex (colour bleached by F⁻) (b)† as (a) ref. 82 but semi-automatic TIT: as ref. 83 (indicator may also be calmagite) CALCULATION: from separately measured concentrations of Ca and Mg; this is the preferred method
Iodide	Method not yet published SPEC: (a) oxidation of I⁻ to I₂, reaction with leuco crystal violet. (b) catalytic effect of I⁻ on reduction of Ce^{IV} by arsenious acid, residual Ce^{IV} measured by adding Fe(NH₄)₂(SO₄)₂ and measuring Fe³⁺ with thiocyanate
Iron (II, III)[84]	SPEC: using 2,4,6-tripyridyl-1,3,5-triazine SPEC: using *o*-phenanthroline AAS (air/C₂H₂ flame): (a) direct aspiration of sample (b) solvent-extraction of Fe pyrrolidine dithiocarbamate, aspiration of extract
Lead[85]	AAS (air/C₂H₂ flame): solvent-extraction of Pb pyrrolidine dithiocarbamate, aspiration of extract AAS (air/C₂H₂ flame): (a) direct aspiration of sample (b) as ref. 85 – see opposite SPEC: solvent-extraction of Pb dithizonate
Lithium	Method not yet published AAS (air/C₂H₂ flame): direct aspiration of sample FAES: direct aspiration of sample
Magnesium[86]	AAS (air/C₂H₂ flame): direct aspiration of sample AAS (air/C₂H₂ flame): as ref. 86 – see opposite GRAV: removal of Ca, precipitation of MgP₂O₇ CALCULATION: as difference between total hardness and Ca

TABLE 9.1 (*cont.*)

Recommended analytical procedure
(see notes at end of table for abbreviations and other information)

Determinand	U.K. Standing Committee of Analysts*	American Public Health Association et al.[6]
Manganese[87]	SPEC: using formaldoxime	SPEC: oxidation of MnO_4^- whose absorbances then measured AAS (air/C_2H_2 flame): (a) direct aspiration of sample (preferred method) (b) solvent-extraction of Mn pyrrolidine dithiocarbamate, aspiration of extract
Mercury[88]	AAS (cold-vapour technique): (a) direct (b) solvent-extraction of Hg dithizonate, back extraction of Hg, and then cold-vapour technique; for saline waters	AAS (cold-vapour technique): preferred method SPEC: solvent-extraction of Hg dithizonate, measurement of extract
Nickel[89]	AAS (air/C_2H_2 flame): sample concentrated (10x) before aspiration; background correction essential	AAS (air/C_2H_2 flame): preferred technique (a) direct aspiration of sample (b) solvent-extraction of Ni pyrrolidine dithiocarbamate, aspiration of extract SPEC: (a) solvent-extraction of Ni-heptoxime complex back extraction, measurement of the complex (b) as above but using dimethylglyoxime to measure Ni in the back extract
Nitrogen: Ammoniacal[90]	TIT: distillation of NH_3, titration of distillate with acid ISE: direct measurement after adding buffer SPEC (formation of an indophenol blue): (a) using dichloroisocyanurate, salicylate, nitroprusside; for non-saline waters (b) using dichloroisocyanurate, phenol, ferrocyanide; for saline waters (c)† as (a) above but semi-automatic (d)† as (b) above but semi-automatic	TIT: as ref. 90 – see opposite ISE: as ref. 90 – see opposite SPEC: (a) using Nessler's reagent before or after distillation of NH_3 (b) formation of an indophenol blue using hypochlorite, phenol, Mn^{2+} (c)† formation of an indophenol blue using hypochlorite, phenol, nitroprusside

Determinand	Method	
Nitrate[91]	SPEC: (a) using sulphosalicylic acid (b) measurement of absorbance at 210 nm; correction for effects of organic materials usually necessary ISE: direct measurement after adding buffer; special care needed to check accuracy CALCULATION: as difference between TON (see below) and nitrite	SPEC: using chromotropic acid CALCULATION: as ref. 91 – see opposite
Nitrite[91]	SPEC: using sulphanilamide and N-1-naphthylethylenediamine to form azo dye (See also total oxidised nitrogen)	SPEC: as ref. 91 – see opposite
Organic	Method not yet published	SPEC: Kjeldahl digestion, distillation and measurement of NH_3; correction for original NH_3 content of sample may be necessary
Oxidised, total[91]	SPEC: (a) reduction by Davarda's alloy, distillation of NH_3 and measurement by TIT or SPEC (Applied to residual solution after distillation of NH_3 in original sample) (b)† reduction of NO_3^- to NO_2^- using Cu^{2+}/N_2H_4, measurement of NO_2^- as above (c)† as (b) but metallic Cd as reductant; preferred for sea water (Note that both (b) and (c) may be used for nitrite alone by omitting reductant)	SPEC: (a) as (a) ref. 91 – see opposite (b) as (b) and (c) ref. 91 – see opposite but manual method and Cd/Hg reductant (c)† as (c) ref. 91 – see opposite (Note that both (b) and (c) may be used for nitrite alone by omitting reductant)
Organic determinands: Biochemical oxygen demand[92]	Measurement of dissolved oxygen in sample before and after incubation for 5 days ± 2 h at 20 ± 0.5 °C in the dark	As ref. 92 – see opposite

TABLE 9.1 (*cont.*)

Recommended analytical procedure
(see notes at end of table for abbreviations and other information)

Determinand	U.K. Standing Committee of Analysis*	American Public Health Association et al.[6]
Carbon, organic[93]	Commercially available instruments reviewed, and general guidance given on their use	As ref. 93 – see opposite
Chemical oxygen demand[94]	TIT: sample refluxed with H_2SO_4/$K_2Cr_2O_7$/$HgSO_4$/Ag_2SO_4, and amount of $K_2Cr_2O_7$ consumed measured by titration with $Fe(NH_4)_2(SO_4)_2$ SPEC: as above but direct measurement of the absorbance of residual $K_2Cr_2O_7$	TIT: as ref. 94 – see opposite
Chlorinated phenoxyacid herbicides	Method not yet published	GC (electron-capture detector): solvent-extraction and hydrolysis of herbicides in extract, conversion of herbicide acids to methyl esters, injection onto column of gas chromatograph
Halogenated methanes[95]	GC (electron-capture detector): solvent-extraction, injection of concentrated extract onto column of gas chromatograph	GC (halogen-specific detector): 'purge-and-trap' procedure [see Section 11.3.8(c)]. (This method is said to be suitable for a larger range of compounds than the method in ref. 95)
Organochlorine insecticides[96]	GC (electron-capture detector): solvent-extraction, purification of extract (Al_2O_3/$AgNO_3$ column), separation from PCB compounds (silica gel column), injection onto column of gas chromatograph	GC (electron-capture detector): solvent-extraction, purification of extract (MgO/silica gel column), injection onto column of gas chromatograph
Organophosphorus pesticides[97]	GC (flame-thermionic or flame-photometric detector): solvent-extraction into two different solvents, concentration of both extracts, injection onto column of gas chromatograph	No method included
Phenolic compounds[98]	SPEC: (a) distillation of phenols, reaction of distillate with 4-aminoantipyrine (AAP), solvent-extraction of product, measurement of extract	SPEC: (a) as (a) ref. 98 – see opposite (b) as (c) reference 98 – see opposite, but no solvent-extraction GC (flame-ionisation detector): direct injection of

	(b) as (a) but using 3-methyl-2-benzothiazoline hydrazone in place of AAP	sample onto column of gas chromatograph
	(c) as (a) but reaction with AAP at a different pH (This method is particularly for chlorinated phenols)	
	GC (flame-ionisation detector): solvent-extraction, formation of trimethylsilyl ethers of phenols, injection onto column of gas chromatograph	
Polynuclear aromatic hydrocarbons[99]	Method not yet published	No method included
Surfactants: Anionic[99]	SPEC: (a) solvent-extraction of methylene blue complex, measurement of extract	SPEC: as (a) ref. 99 – see opposite.
	(b)† as (a) but semi-automatic	IRS: adsorption of surfactants on activated carbon, solvent-extraction of carbon, purification of extract, evaporation to dryness, dissolution of residue in CS_2, measurement by IRS
Non-ionic[99]	TLC: solvent-extraction, evaporation of extract, measurement by TLC using visual estimation after spraying with Dragendorff reagent	
	TIT/SPEC/AAS: solvent-extraction (by sublation), non-ionic surfactants in extract precipitated with barium tetraiodobismuthate, the precipitate separated, dissolved, and the liberated bismuth measured by TIT or SPEC (in the U.V) or AAS	
	(Two other tentative methods are also given)	
Oxygen, dissolved[100]	TIT: using the Winkler titration.	TIT: as ref. 100 – see opposite
	Commercially available electrodes for measuring dissolved oxygen	As ref. 100 – see opposite

TABLE 9.1 (*cont.*)

Recommended analytical procedure

(see notes at end of table for abbreviations and other information)

Determinand	U.K. Standing Committee of Analysis*	American Public Health Association et al.[6]
pH value[101]	ISE: using glass electrode	ISE: as ref. 101 – see opposite
Phosphorus, molybdate-reactive[102]	SPEC: (*a*) formation of reduced phosphomolybdate (*b*) as (*a*) but solvent-extraction and measurement of reduced phosphomolybdate in extract	SPEC: (*a*) formation of vanadophosphomolybdate (*b*) as (*a*) ref. 102 – see opposite; two different methods given (*c*) as (*b*) ref. 102 – see opposite; but different experimental conditions (*d*)† as (*a*) ref. 102 – see opposite; but semi-automatic
Phosphorus, total[102]	Both refs. 6 and 102 describe a number of oxidative procedures for converting organic and inorganic compounds of phosphorus to orthophosphate which is then measured by a procedure for molybdate-reactive phosphorus	
Potassium[103]	AAS (air/C₂H₂ flame): direct aspiration of sample FAES: direct aspiration of sample	AAS (air/C₂H₂ flame): as ref. 103 – see opposite FAES: as ref. 103 – see opposite
Selenium	Method not yet published	AAS (hydrogen flame): evolution of H₂Se, transfer into flame SPEC: oxidation to Se^VI, reduction to Se^IV, reaction with diaminobenzidine, solvent-extraction of product, measurement of extract
Silicon, molybdate-reactive[104]	SPEC: (formation of reduced silicomolybdate): (*a*) ascorbic acid as reductant (*b*) 1-amino-2-naphthol-4-sulphonic acid as reductant	SPEC: (formation of silicomolybdate): (*a*) no reductant used. (*b*) as (*b*) ref. 104 – see opposite (*c*)† as (*b*) above but semi-automatic
Silicon, total[104]	Both refs. 6 and 104 describe procedures for converting non-reactive forms of silicon to reactive forms before measurement by the above methods	
Silver[105]	AAS (air/C₂H₂ flame): sample treated with CNI	AAS (air/C₂H₂ flame):

under alkaline conditions, concentrated (10x) by evaporation before aspiration into flame

(a) direct aspiration of sample
(b) solvent-extraction of Ag pyrrolidine dithiocarbamate, aspiration of extract (The possible need for treatment with CNI is also noted in this publication)
SPEC: solvent-extraction of Ag dithiozonate, measurement of extract. (The AAS methods are preferred)
AAS (air/C_2H_2 flame): as ref. 106 – see opposite
FAES: as ref. 106 – see opposite
AAS (air/C_2H_2 flame): direct aspiration of sample
FAES: direct aspiration of sample, preferably into N_2O/C_2H_2 flame

Sodium[106]
AAS (air/C_2H_2 flame): direct aspiration of sample

Strontium
FAES: direct aspiration of sample
Method not yet published

Sulphate[107]
GRAV: precipitation and weighing of $BaSO_4$
TIT: (a) precipitation of $BaSO_4$, excess Ba titrated with EDTA
(b) direct titration with Ba^{2+} (carboxyarsenazo indicator)
AAS (air/C_2H_2 flame): precipitation of $BaSO_4$, measurement of excess Ba by direct aspiration into flame
SPEC: precipitation of SO_4^{2-} with 2-aminoperimidine, measurement of excess reagent at 305 or 525 nm

GRAV: as ref. 107 – see opposite
SPEC†: precipitation of $BaSO_4$, reaction of excess Ba^{2+} with methylthymol blue, and measurement of the excess of the latter
TURBIDIMETRY: precipitation of $BaSO_4$, measurement of turbidity so produced

Sulphide[108]
TIT: iodometric
SPEC: reaction with *NN*-diethyl-*p*-phenylenediamine and $K_2Cr_2O_7$
Method not yet published

TIT: as ref. 108 – see opposite
SPEC: reaction with *NN*-dimethyl-*p*-phenylenediamine and $FeCl_3$
NEPHELOMETRY: measurement at ~90° to incident beam; formazin as standard suspension.
Visual procedures involving either simple comparison

Turbidity

TABLE 9.1 (*cont.*)

Recommended analytical procedure
(see notes at end of table for abbreviations and other information)

Determinand	U.K. Standing Committee of Analysts*	American Public Health Association et al.[6]
		with standard suspensions or the use of a candle turbidimeter are also given
Vanadium	Method not yet published	AAS (N$_2$O/C$_2$H$_2$ flame): direct aspiration of sample SPEC: the catalytic effect of V on the rate of oxidation of gallic acid by persulphate is measured

Zinc[109]	AAS (air/C$_2$H$_2$ flame): sample concentrated (10x) by evaporation before aspiration; background correction may be necessary	AAS (air/C$_2$H$_2$ flame): the preferred technique
		  (*a*) direct aspiration of sample
		  (*b*) solvent-extraction of Zn pyrrolidine dithiocarbamate, aspiration of extract
		SPEC: (*a*) solvent-extraction of Zn dithizonate, measurement of extract.
		  (*b*) using zincon.

Notes:

* The references to the individual booklets of this series of publications are noted against the determinand
The abbreviations used are as follows: (the symbol † denotes a semi-automatic, air-segmented, continuous-flow procedure)
  SPEC = spectrophotometry, the measurements being made on the treated aqueous sample at wavelengths greater than 400 nm unless stated otherwise
  AAS   = atomic-absorption spectrometry
  TIT   = titrimetry
  GRAV = gravimetry
  ISE   = potentiometry using ion-selective electrodes
  FAES = flame atomic-emission spectrometry (flame photometry)
  TLC   = thin-layer chromatography
  GC   = gas chromatography
  IRS   = infrared spectrophotometry
The very brief descriptions of the procedures given in the table are, of course, intended only to show the general nature of the methods. Thus, methods that the table shows to be identical may differ in a number of details, and various steps in the procedures may not be mentioned. The alternative methods given for many determinands are often intended to cover different concentration ranges and/or to have different susceptibilities to interference. The publications cited must, of course, be consulted for such information.

9.6 REFERENCES

[1] K. H. Mancy, ed., "Instrumental Analysis for Water Pollution Control", Ann Arbor Science Publishers, Ann Arbor, Michigan, 1971, pp. 17–18.

[2] H. L. Golterman, R. S. Clymo, and M. A. M. Ohnstad, "Methods for Physical and Chemical Analysis of Fresh Waters", Second Edition, Blackwell Scientific Publications, Oxford, 1978.

[3] Standing Committee of Analysts, "General Principles of Sampling and Accuracy of Results, 1980", H.M.S.O., London, 1980.

[4] L. A. Currie, in I. M. Kolthoff and P. J. Elving, ed., "Treatise on Analytical Chemistry', Second Edition, Part 1, Volume 1, Wiley, Chichester, 1978, pp. 95–242.

[5] G. F. Lee and R. A. Jones, *J. Water Pollut. Contr. Fed.*, 1983, **55**, 92.

[6] American Public Health Association, American Water Works Association, and Water Pollution Control Federation, "Standard Methods for the Examination of Water and Wastewater", Fifteenth Edition, A.P.H.A., Washington, 1981.

[7] F. A. Leemans, *Anal. Chem.*, 1971, **43**, 36A.

[8] G. Kateman and F. W. Pijpers, "Quality Control in Analytical Chemistry", Wiley, Chichester, 1981.

[9] T. A. H. M. Janse and G. Kateman, *Anal. Chim. Acta*, 1983, **150**, 219.

[10] I. E. Frank, E. Pungor, and G. E. Veress, *Anal. Chim. Acta*, 1981, **133**, 433.

[11] I. E. Frank, E. Pungor, and G. E. Veress, *Anal. Chim. Acta*, 1981, **133**, 443.

[12] E. F. McFarren, R. J. Lishka, and J. H. Parker, *Anal. Chem.*, 1970, **42**, 358.

[13] K. Eckschlager, *Anal. Chem.*, 1972, **44**, 878.

[14] D. Midgley, *Anal. Chem.*, 1977, **49**, 510.

[15] American Society for Testing and Materials, "1983 Annual Book of ASTM Standards. Volumes 11.01 and 11.02 Water (I and II)", A.S.T.M., Philadelphia. 1983.

[16] International Organization for Standardization, ISO 5725-1981, I.S.O., Geneva, 1981.

[17] British Standards Institution, BS 5497: Part 1: 1979, B.S.I., Milton Keynes, Bucks, 1979.

[18] J. Büttner, *et al.*, *J. Clin. Chem. Clin. Biochem.*, 1980, **18**, 69.

[19] J. Mandel, in I. M. Kolthoff and P. J. Elving, ed., "Treatise on Analytical Chemistry", Second Edition, Part 1, Volume 1, Wiley, Chichester, 1978, pp. 243–298.

[20] W. J. Youden and E. H. Steiner, "Statistical Manual of the Association of Official Analytical Chemists", Association of Official Analytical Chemists, Washington, 1975.

[21] F. W. Sunderman, S. S. Brown, M. Stoeppler, and D. B. Tonks, in H. Egan and T. S. West, ed., "Collaborative Studies in Chemical Analysis", Pergamon, Oxford, 1982, pp. 25–35.

[22] J. A. Winter, in W. J. Cooper, ed. "Chemistry in Water Reuse, Volume 1", Ann Arbor Science Publishers, Ann Arbor, Michigan, 1981, pp. 187–205.

[23] D. L. Massart, A. Dijkstrat, and L. Kaufman, "Evaluation and Optimisation of Laboratory Methods and Analytical Procedures", Elsevier, Amsterdam, 1978.

[24] C. Liteanu and I. Rîcă, "Statistical Theory and Methodology of Trace Analysis", Ellis Horwood, Chichester, 1980.

[25] C. Eisenhart, *J. Res. Natl. Bur. Stand.*, 1963, **67C**, 161.

[26] H. Kaiser, *Spectrochim. Acta, Part B*, 1978, **33**, 551.

[27] J. K. Taylor, *Anal. Chem.*, 1981, **53**, 1588A.

[28] Organization for Economic Cooperation and Development, "Good Laboratory Practice in the Testing of Chemicals", O.E.C.D., Paris, 1982.

[28a] Health and Safety Commission, "Principles of Good Laboratory Practice", H.M.S.O., London, 1982.

[29] International Organization for Standardization, ISO Guide 18-1978, I.S.O., Geneva, 1978.

[30] J. F. Lovett, in H. Egan and T. S. West, ed., "Collaborative Studies in Chemical Analysis", Pergamon, Oxford, 1982, pp. 139–142.

[30a] British Standards Institution, BS 5443: 1977, B.S.I., Milton Keynes, Bucks., 1977.

[31] D. T. Burns, *Lab. Pract.*, 1979, **28**, 247.

[32] A. L. Wilson, *Talanta*, 1970, **17**, 21.

[33] Council of the European Communities, *Off. J. Eur. Communities*, 1980, **23**, No. L-229, 11.

[34] *Idem, Off. J. Eur. Communities*, 1978, **21**, No. L-222, 1.

[35] *Idem, Off. J. Eur. Communities*, 1979, **22**, No. L-271, 44.

[36] F. W. J. Garton, W. Ramsden, R. Taylor, and R. J. Webb, *Spectrochim. Acta*, 1956, **8**, 94.

[37] A. L. Wilson, *Talanta*, 1970, **17**, 31.

[38] A. L. Wilson, *Talanta*, 1973, **20**, 725.

[39] A. L. Wilson, *Talanta*, 1974, **21**, 1109.

[40] J. Büttner, *et al.*, *J. Clin. Chem. Clin. Biochem.*, 1980, **18**, 78.

[41] D. G. Mitchell and J. S. Garden, *Talanta*, 1982, **29**, 921.

[42] G. Gottschalk, *Fresenius' Z. Anal. Chem.*, 1975, **275**, 1.

[43] G. Gottschalk, *Fresenius' Z. Anal. Chem.*, 1975, **276**, 81.

[44] G. Gottschalk, *Fresenius' Z. Anal. Chem.*, 1975, **276**, 257.

[45] G. Gottschalk, *Fresenius' Z. Anal. Chem.*, 1976, **278**, 1.

46 G. Gottschalk, *Fresenius' Z. Anal. Chem.*, 1976, **280**, 205.
47 G. Gottschalk, *Fresenius' Z. Anal. Chem.*, 1976, **282**, 1.
48 R. D. Vanous and R. E. Heston, *J. Ind. Water Wks Assoc.*, 1980, **12**, 95.
49 P. H. Gibbs, W. J. Cooper, and E. M. Ott, *J. Am. Water Wks Assoc.*, 1983, **75**, 578.
50 W. J. Cooper, P. H. Gibbs, E. M. Ott, and P. Patel, *J. Am. Water Wks Assoc.*, 1983, **75**, 625.
51 J. O. Westgard, *et al.*, *Am. J. Med. Technol.*, 1978, **44**, 290.
52 J. O. Westgard, *CRC Crit. Rev. Clin. Lab. Sci.*, 1981, **13**, 283.
53 Standing Committee of Analysts, "Second Report, February 1977–April 1979", (DOE/NWC Standing Technical Committee reports No. 23), National Water Council, London, 1981. (Future reports will be published by H.M.S.O.)
54 J. F. Kopp and G. D. McKee, ed., "Methods for Chemical Analysis of Water and Wastes", (Report EPA-600/4-79-020), U.S. Environmental Protection Agency, Cincinatti, 1979.
55 J. E. Longbottom and J. J. Lichtenberg, ed., "Methods for Organic Chemical Analysis of Municipal and Industrial Wastewater", (Report EPA-600/4-82-057), U.S. Environmental Protection Agency, Cincinnati, 1982.
56 E. Brown, M. W. Skougstad, and M. J. Fishman, "Methods for Collection and Analysis of Water Samples for Dissolved Minerals and Gases", U.S. Government Printing Office, Washington, 1970.
57 U.S. Geological Survey, "National Handbook of Recommended Methods for Water-Data Acquisition", U.S.G.S., Reston, Virginia, 1977.
58 M. J. Suess, ed., "Examination of Water for Pollution Control, Volume 2", Pergamon, Oxford, 1982.
59 Department of the Environment, "Analysis of Raw, Potable and Waste Waters", H.M.S.O., London, 1972.
60 American Public Health Association, American Water Works Association, Water Pollution Control Federation, "Standard Methods for the Examination of Water and Wastewater", Fourteenth Edition, A.P.H.A., Washington, 1976.
61 J. D. H. Strickland and T. R. Parsons, "A Practical Handbook of Seawater Analysis", Fisheries Research Board of Canada, Ottawa, 1968.
62 K. Grasshof, M. Ehrhardt, and K. Kremling, "Methods of Seawater Analysis", Second Edition, Verlag Chemie, Weinheim, 1983.
63 J. P. Riley, in J. P. Riley and G. Skirrow, ed., "Chemical Oceanography, Volume 3", Second Edition, Academic, London, pp. 193–514.
64 British Standards Institution, BS 1472: 1962, B.S.I., Milton Keynes, Bucks, 1962.
65 British Standards Institution, BS 2690: Parts 1–15 and 100–125, B.S.I., Milton Keynes, Bucks, parts published between 1964 and 1983.
66 Research and Development Department, "Methods of Sampling and Analysis. Volume 1. Steam and Water', Research and Development Department, Central Electricity Research Laboratories, Leatherhead, Surrey, 1966.
67 K. H. Mancy and W. J. Weber, in I. M. Kolthoff, P. J. Elving, and F. H. Stross, ed., "Treatise on Analytical Chemistry", First Edition, Part III, Volume 2, Wiley-Interscience, London, 1971, pp. 413–562.
68 Standing Committee of Analysts, "The Determination of Alkalinity and Acidity in Water, 1981", H.M.S.O., London, 1982.
69 *Idem*, "Acid-Soluble Aluminium in Raw and Potable Waters by Spectrophotometry (using pyrocatechol violet), 1979", H.M.S.O., London, 1980.
70 *Idem*, "Antimony in Effluents and Raw, Potable and Sea Waters by Spectrophotometry (using crystal violet), 1982", H.M.S.O., London, 1983.
71 *Idem*, "Arsenic in Potable Waters by Atomic Absorption Spectrophotometry (semi-automatic method), 1982", H.M.S.O., London, 1983.
72 *Idem*, "Arsenic in Potable and Sea Water by Spectrophotometry (arsenomolybdenum blue procedure), 1978", H.M.S.O., London, 1980.
73 *Idem*, "Boron in Waters, Effluents, Sewage and Some Solids, 1980", H.M.S.O., London, 1981.
74 *Idem*, "Bromide in Waters. High Level Titrimetric Method, 1981", H.M.S.O., London, 1981.
75 *Idem*, "Cadmium in Potable Waters by Atomic Absorption Spectrophotometry, 1976", H.M.S.O., London, 1977.
76 *Idem*, "Calcium in Waters and Sewage Effluents by Atomic Absorption Spectrophotometry, 1977", H.M.S.O., London, 1978.
77 *Idem*, "Chloride in Waters, Sewage and Effluents, 1981", H.M.S.O., London, 1982.
78 *Idem*, "Chemical Disinfecting Agents in Water and Effluents, and Chlorine Demand, 1980", H.M.S.O., London, 1980.
79 *Idem*, "Chromium in Raw and Potable Waters and Sewage Effluents, 1980', H.M.S.O., London, 1981.
80 *Idem*, "Cobalt in Potable Waters, 1981", H.M.S.O., London, 1982.
81 *Idem*, 'Copper in Potable Waters by Atomic Absorption Spectrophotometry, 1980", H.M.S.O., London, 1981.
82 *Idem*, "Fluoride in Waters, Effluents, Sludges, Plants and Soils, 1982", H.M.S.O., London, 1983.

[83] *Idem*, "Total Hardness, Calcium Hardness and Magnesium Hardness in Raw and Potable Waters by EDTA-Titrimetry, 1981", H.M.S.O., London, 1982.

[84] *Idem*, "Iron in Raw and Potable Waters by Spectrophotometry (using 2,4,6-tripyridyl-1,3,5-triazine), 1977", H.M.S.O., London, 1978.

[85] *Idem*, "Lead in Potable Waters by Atomic Absorption Spectrophotometry, 1976", H.M.S.O., London, 1977.

[86] *Idem*, "Magnesium in Waters and Sewage Effluents by Atomic Absorption Spectrophotometry, 1977", H.M.S.O., London, 1978.

[87] *Idem*, "Manganese in Raw and Potable Waters by Spectrophotometry (using formaldoxime), 1977", H.M.S.O., London, 1978.

[88] *Idem*, "Mercury in Waters, Effluents and Sludge by Flameless Atomic Absorption Spectrophotometry, 1978", H.M.S.O., London, 1978.

[89] *Idem*, "Nickel in Potable Waters, 1981", H.M.S.O., London, 1982.

[90] *Idem*, "Ammonia in Waters, 1981", H.M.S.O., London, 1982.

[91] *Idem*, "Oxidised Nitrogen in Waters, 1981", H.M.S.O., London, 1982.

[92] *Idem*, "Biochemical Oxygen Demand, 1981", H.M.S.O., London, 1983.

[93] *Idem*, "The Instrumental Determination of Total Organic Carbon, Total Oxygen Demand and Related Determinands, 1979", H.M.S.O., London, 1980.

[94] *Idem*, "Chemical Oxygen Demand (Dichromate Value) of Polluted and Waste Waters, 1977", H.M.S.O., London, 1978.

[95] *Idem*, "Chloro- and Bromo-trihalogenated Methanes in Water, 1980", H.M.S.O., London, 1981.

[96] *Idem*, "Organochlorine Insecticides and Polychlorinated Biphenyls in Waters, 1978", H.M.S.O., London, 1979.

[97] *Idem*, "Organo-Phosphorus Pesticides in River and Drinking Water, 1980", H.M.S.O., London, 1983.

[98] *Idem*, "Phenols in Waters by Gas-Liquid Chromatography, 4-aminoantipyrine or 3-methyl-2-benzothiazolinone, 1981", H.M.S.O., London, 1982.

[99] *Idem*, "Analysis of Surfactants in Waters, Wastewaters and Sludges, 1981", H.M.S.O., London, 1982.

[100] *Idem*, "Dissolved Oxygen in Natural and Waste Waters, 1979", H.M.S.O., London, 1980.

[101] *Idem*, "The Measurement of Electrical Conductivity and the Laboratory Determination of the pH Value of Natural, Treated and Waste Waters, 1978", H.M.S.O., London, 1979.

[102] *Idem*, "Phosphorus in Waters, Effluents and Sewages, 1980", H.M.S.O., London, 1981.

[103] *Idem*, "Dissolved Potassium in Raw and Potable Waters, 1980", H.M.S.O., London, 1981.

[104] *Idem*, "Silicon in Waters and Effluents, 1980", H.M.S.O., London, 1981.

[105] *Idem*, "Silver in Waters, Sewages and Effluents by Atomic Absorption Spectrophotometry, 1982", H.M.S.O., London, 1983.

[106] *Idem*, "Dissolved Sodium in Raw and Potable Waters, 1980", H.M.S.O., London, 1981.

[107] *Idem*, "Sulphate in Waters, Effluents and Solids, 1979", H.M.S.O., London, 1980.

[108] *Idem*, "Sulphide in Waters and Effluents, 1983", H.M.S.O., London, 1983.

[109] *Idem*, "Zinc in Potable Waters by Atomic Absorption Spectrophotometry, 1980", H.M.S.O., London, 1981.

10 General Precautions in Water-analysis Laboratories

10.1 INTRODUCTION

Even when suitable analytical methods are available and selected, other components of analytical systems (*e.g.*, analysts, reagents, apparatus, instrumentation, laboratory facilities) may prevent achievement of the required accuracy. The aim of the present section is to note a number of general precautions of value in controlling the effects of these other components. Emphasis is placed on trace analysis because the precautions are then usually most important. Nonetheless, the principles involved are equally applicable to any analyses, and analysts should consider the appropriate precautions for all determinands whatever their concentrations.

Many of the collections of analytical methods cited in Section 9.5 as well as many other texts on analytical chemistry make at least some mention of some or many of the points discussed below. Ref. 1 is particularly helpful and comprehensive on many aspects.

10.2 DESIGN OF LABORATORIES

The design of laboratories for water analysis is subject to essentially the same considerations as any other laboratory, and needs no detailed discussion here. General information on laboratory design is provided in many texts, *e.g.*, ref. 2, and various aspects of the design of water-analysis laboratories are dealt with in refs. 3 to 6. The design can be markedly influenced by safety considerations, and this aspect is discussed in detail for water analysis in ref. 6. Ref. 1 provides much useful information on laboratory services.

Three aspects of special relevance in water analysis are worth brief mention.

(1) Much trace analysis may well be involved, and special care is then necessary to ensure that the design and operation of a laboratory are such as to minimise problems of contamination. This aspect is discussed in Section 10.8; see especially Section 10.8.4.

(2) Many determinands tend to be unstable in samples of waters, and adequate sample storage space at a temperature of approximately 4 °C is generally required (see Section 3.6). Special 'cold-rooms' may well be necessary for laboratories handling many samples.

(3) Waters commonly require a rather wide range of different types of analysis, and this is likely to involve extensive movement of samples between different parts of a laboratory. This is worth bearing in mind when planning the spatial distribution of different analytical facilities within a laboratory.

10.3 ANALYTICAL PERSONNEL

All analyses made in a laboratory require some attention from an analyst, and some analytical procedures call for considerable manual dexterity and experience for their proper execution. Of course, the role of such abilities has tended to become less important with the growth in use of automated procedures, but even then analysts may still have a profound effect on the accuracy of results, *e.g.*, by inaccurate preparation of reagents and calibration standards. It remains essential, therefore, to ensure that all staff of a laboratory receive appropriate training and education in the relevant analytical procedures and techniques. In Sections 10.3.1 to 10.3.4 we note a number of points of special relevance to the achievement of specified precision and bias of analytical results.

10.3.1 Nomination of an Analyst to be Responsible for Analytical Quality Control

Previous sections of this book have shown the complexity of tasks arising in any laboratory aiming to control the precision and bias of analytical results. To ensure a consistent and adequate approach to the various problems, it is desirable that an analyst be made responsible for: (1) defining a laboratory's policy and detailed approach to analytical quality control, and (2) ensuring that this approach is properly implemented and regularly reviewed. Refs. 4 and 7 to 9 also recommend the need for a quality control officer (or similar title). The manager of a laboratory may often play this role, but note that refs. 8 and 9 recommend that this function should be independent of the purely analytical function.

10.3.2 Education and Training of Laboratory Staff

Naturally, as other authors have noted,[4,9] initiation of the quality control policy may lead some members of the staff of a laboratory to feel resentment that their competence and/or conscientiousness are apparently being questioned. It is very important, therefore, that strenuous efforts be made to ensure that all staff understand and agree with the need for analytical quality control. The achievement of these aims will be facilitated when staff know how their analytical results will be used and the deleterious effects of inadequate accuracy on such use(s). In this respect, the dialogue between analysts and users of results – as recommended at many points in this book – is of value in helping to demonstrate that the accuracy of results is important to users. It will be much more difficult to secure the willing co-operation of staff if they feel that the accuracy of their results has little practical importance. This willing co-operation is also more likely to be obtained if staff have the opportunity to contribute their ideas on one or more elements of a total programme of analytical quality control.

Two aspects of the training of staff are especially important. First, all staff should be given a clear explanation of a laboratory's policy on analytical quality control and of the detailed consequences of this in terms of the work and special tests that must be done.[4] This explanation, we feel, should include the fact that analysts *are* components of analytical systems and *can* affect the accuracy of results; see also Section 10.3.3. Consequently, the second point is that staff must receive sufficient training to ensure that they are capable of the required accuracy, and they must also demonstrate experimentally this ability.[4] Participation in inter-laboratory studies (see Section 7.1) is often useful in providing analysts with an additional interest in the accuracy of their results. Refs. 1 and 7 deal with other general aspects of staff training.

10.3.3 Need to Follow Analytical Methods Exactly

The need to follow exactly the procedure specified by an analytical method adopted for routine use may seem an obvious point, and it certainly has been recommended by many authors for many years. Nevertheless, our experience is that it is not uncommon for laboratories to introduce – consciously or unconsciously – variations to the prescribed procedure. Indeed, different analysts in the same laboratory may use their own individual and unexpressed modifications of a method. We have already stressed the need for analytical methods to be complete and unambiguous, and for the prescribed procedures to be followed as exactly as possible [Sections 6.3.1 and 9.2.4(*b*)]. Even small departures from the prescribed procedure may very markedly – and unsuspectedly – affect the accuracy of results (see below). Thus, any such departures are generally to be avoided, particularly when, as is often the case, they have only marginal or negligible effects on aspects such as the speed, simplicity, and cost of analyses. Taylor[9] has observed that the attitude of laboratory staff is a key factor in achieving the strict following of analytical methods, and the points in Section 10.3.2 are relevant to this.

We do not suggest, of course, that analysts should slavishly and unthinkingly follow published methods [see, for example, Section 9.2.4(*b*)(*x*)]. It is suggested, however, that they should avoid unconscious adoption of modified procedures, and that changes to methods be made only when there are compelling reasons and the effects of the changes can be properly assessed. Many examples could be quoted of the problems caused by inadequately checked modifications; the following example should suffice to illustrate the need for caution.

The use of the chromogenic reagent, *sym*-tripyridyltriazine, for the spectrophotometric determination of iron can lead to negative bias when certain other metals (*e.g.*, copper and nickel) are present. This effect is caused by the competition of copper and nickel with iron for the reagent; the problem is easily overcome by adding a sufficient excess of the reagent to samples.[10] Suppose now that an analyst had decided to use a method in which a suitable amount of this reagent was specified, but, as a means of economising in the use of a comparatively expensive reagent, he also decided to reduce the amount added to samples. Following usual procedures, he could then check the calibration curve, sensitivity, and precision using standard solutions of iron, and find that all these performance characteristics were completely unaffected by his modification of the procedure. On this basis, he might well feel reasonably confident that the modified method was satisfactory, but the accuracy of his results could well be quite unacceptable for samples containing appreciable concentrations of copper and nickel. This simple example also demonstrates the importance of knowledge of the processes involved in analytical procedures (see Section 9.2.3).

Although we believe that the need for caution in modifying well defined analytical methods cannot be over-emphasised, modifications (or closer specification of the desired procedure) may well be needed for methods that are incomplete and/or ambiguous and that specify inappropriate procedures, *e.g.*, for calibration. Such exceptions are discussed in Section 9.2.4(*b*).

10.3.4 Use of Complex Analytical Instrumentation

The growing tendency for routine analysis to use relatively complex instrumentation requiring minimal attention during the analysis of samples should not lull analysts into the view that, for such equipment, analytical personnel are no longer a vital concern.[11-16] Such a view is perhaps not difficult to form when many procedures now consist basically in the operator presenting a series of samples to an instrument which then automatically provides a print-out of the determinand concentrations in the samples. This type of equipment, however, almost invariably involves a complex string of physical and chemical processes whose natures and mechanisms should be well understood if its continuing and successful use is to be ensured. Thus, for each such item of instrumentation we would suggest that at least one member of the laboratory should be expert not only in its use but also in the theory and mechanisms of the processes involved. All analysts using such equipment for routine analysis should also be encouraged to examine critically all aspects of its performance during operation with the aim of detecting any malfunction as rapidly as possible. These and other aspects of modern analytical instrumentation are discussed in Section 10.4; see especially Section 10.4.3.

10.4 ANALYTICAL INSTRUMENTS

Any modern laboratory is likely to possess a number of instruments of varying degrees of complexity, *e.g.*, balance, spectrophotometer, atomic absorption spectrometer, gas chromatograph, and one or more of these may be automated to various degrees. The principles and applications of these are discussed in Sections 11 to 13, and it goes without saying that they should be handled and used carefully. Several general points concerning such equipment are worth brief mention in Sections 10.4.1 and 10.4.2. We have also added Section 10.4.3, which was not present in the first edition of this book, in order to discuss some important problems arising from the increasing use of relatively complex instrumentation in routine analysis.

10.4.1 Location and Installation

Sensitive instruments should be installed in suitable locations, and their suppliers can give detailed advice. Points to consider are given below.

(a) Vibration
Anti-vibration mountings can be used for some instruments when vibration is a problem.

(b) Power Supplies
The performance of many instruments can be adversely affected by excessively fluctuating mains voltages. When this is a problem, special voltage-stabilising equipment can be obtained, and is of considerable value.

(c) Laboratory Atmosphere

The atmosphere around an instrument can affect its performance either by causing deterioration of one or more of its components (*e.g.*, deposition of ammonium chloride on optical components) or by causing contamination of samples during analysis. It will often be useful to instal delicate equipment in special instrument laboratories in which chemical work is excluded or restricted (see also Section 10.8).

(d) Fume Extraction

Some instruments require fume-extraction facilities, *e.g.*, atomic absorption spectrometers. Such instruments must, therefore, be sited where such facilities can be provided.

(e) Temperature and Humidity

Variations in ambient temperature may affect the performance of many instruments, but the importance of this effect depends markedly on the instrument, on its intended uses, and on the degree of temperature variation. It is generally desirable, therefore, to seek the advice of the instrument supplier on the need for, and nature of, any temperature-control system. The same points apply to humidity, though control of this factor will seldom be essential in temperate climates.

10.4.2 Maintenance and Control of Performance

All instruments are likely to develop faults and show worsening performance from time to time, and such problems can be particularly embarassing when they occur just when equipment is urgently required to analyse samples. This latter problem is further aggravated when the instrument is of such size and/or cost that only one is installed in a laboratory. To minimise such difficulties it is well worthwhile to plan a programme of maintenance and performance-checking of instruments.[1,4,17,18]

Most instrument suppliers offer servicing facilities of their equipment, and it is usually beneficial to arrange for such maintenance to be carried out regularly when appropriate personnel are not directly available to a laboratory. Notwithstanding any such arrangements, it is also useful to make regular checks of the performance of instruments since this offers the possibility of detecting developing faults before they have become practically important. Such checks should be made in such a way that they directly reflect the performance of the instrument and are not influenced by errors arising in an analytical procedure. For example, regular measurements of the wavelength of maximum absorption and the absorbance of stable, reference solutions require little time and effort, and can be of value in checking the performance of spectrophotometers and absorptiometers. Such checks have been suggested in two manuals dealing with the analysis of water,[1,18] and are often advised in texts on spectrophotometric analysis. Similar tests may easily be arranged for other types of instrument provided that appropriate, reference solutions or other reference standards can be prepared or obtained.

10.4.3 Problems in the Use of Complex Analytical Instrumentation

(a) Introduction

We begin by stressing that the instrumentation considered below has many advantageous features; for example, it often provides lower detection limits and greater rates of analysis than more conventional procedures. Rightly, therefore, many complex analytical instruments have become or are becoming an everyday part of modern water analysis; such instruments are discussed in Sections 11 to 13. Examples are: atomic absorption spectrometers (flame and particularly electrothermal atomisation), instruments for stripping voltammetry and other electroanalytical techniques, chromatographs (for gas and liquid chromatography), instruments for the determination of total organic carbon and related determinands. Systems of this type are commonly equipped (and this tendency is increasing) with automatic devices for controlling and executing one or more stages of analytical procedures. All such instruments are also increasingly being provided with microprocessor/computing systems for automatic processing of the analytical signals so that all the calculations required to derive the calibration function and analytical results for samples are made by the instrument with automatic presentation of the results. We believe that the use of such instrumentation may, and not infrequently does, introduce certain problems which appear sometimes to be given relatively little consideration amid the more general concentration on the advantageous features. These problems concern several hitherto fundamentally important aspects of analytical practice, and it seems to us essential that these problems be recognised so that appropriate attention can be paid to them. We felt, therefore, that it would be useful to note these difficulties and some of their implications in the present section. The problems spring from two main causes.

First, it seems to us that there is a growing tendency for analytical systems, and especially those including complex instrumentation, to be regarded as '*black boxes*'.[12,19] That is to say, the detailed nature of the processes leading to the production of an analytical result from a sample are either not known or are not considered; the analytical system is viewed as a machine which, if provided with a sample at one end, supplies a result at the other end. We emphasise that we are speaking of a tendency to this approach, and the extent to which the tendency appears naturally varies greatly from one laboratory to another. Nevertheless, we believe that this tendency exists. Indeed, it would be surprising if it did not given the ever increasing number of samples laboratories are required to analyse, and the number of analyses which consist, in essence, of the operator pouring portions of samples into containers at one end of an instrument and observing analytical results at the other. In addition, knowledge of the intervening processes may require acquaintance with chemical, optical, electrical, electronic, and computing aspects. In some instances even, rather little may be known of the detailed processes involved in some commercial instruments, *e.g.*, in some electrothermal atomisers used in atomic absorption spectrometry [see Section 11.3.5(*c*)]. In the following discussion we shall, for brevity, speak of 'black boxes' in the sense described above. Of course, the simplest analytical system may be treated by an analyst as a 'black box' if he chooses. Thus, when considering the various points in (*b*) to (*g*) below, the reader should bear in mind that the extent to which our discussion applies to a particular instrument must be judged by each analyst in his own context.

The second factor is simply the growing complexity of modern instruments – in the sense of the number and nature of the components and processes involved – as these instruments undertake more and more operations previously carried out by analysts. Coupled with this is the fact that the detailed designs of instruments of the same type and function may differ sufficiently from one manufacturer to another that, for a given analytical method, the performance characteristics achieved may well vary with the instrument.

(b) Problems Caused by Inadequate Knowledge of the Processes Involved in Producing the Analytical Response for a Sample

For the successful control and estimation of the precision and bias of analytical results, it is highly desirable to have a detailed knowledge of the processes involved in the production of the analytical response for a sample (see Sections 5.1 and 6.3.1). Such knowledge is particularly important with respect to the bias caused by: (1) inability to measure all relevant forms of the determinand (Section 5.4.5), and (2) interfering substances (Section 5.4.6). One would expect, therefore, that the use of 'black-box' systems will tend to lead to larger random and systematic errors as well as greater variability of both types of error, though such problems will tend to be larger for systematic than for random errors. In many respects, these tendencies are well demonstrated by the various forms of electrothermal atomisers that have found widespread use for routine analysis in recent years [see Section 11.3.5(*c*)], but similar difficulties occur to varying extents with many other types of analytical system.

Generally speaking, this greater difficulty in controlling the size and constancy of errors will tend to increase as the number and complexity of the processes involved in a 'black box' increase. Again, this tendency may well be more marked for bias caused by interfering substances because the greater number of processes provides, in general, more opportunity for substances in, and other properties of, samples to affect the analytical response for a given concentration of the determinand. Finally, these problems are likely to be more marked in systems for which the analytical response is represented by a transient rather than a constant or equilibrium signal.

Although such problems are relevant to all types of error, a particular difficulty arises in connection with the control and estimation of bias caused by interfering substances. We noted in Section 5.4.6 the difficulty that laboratories usually find in estimating this bias for samples as complex and variable as natural waters. In principle, one approach for dealing with this problem is for one or more laboratories to make very detailed tests of interference effects when a particular analytical system is used. The results obtained may then be used by other laboratories using the same analytical system to assess the errors due to interference for their own samples. This approach is, of course, commonly applied to great advantage in the more classical analytical techniques such as gravimetry, titrimetry, and spectrophotometry. It meets, however, serious difficulties for many of the more modern instrumental systems under consideration because of the near impossibility of being sure that one analytical system is effectively the same as another. For example, instruments of the same general nature and performing similar functions but supplied by different manufacturers may show considerable differences in the size of interference effects (and other performance characteristics).[20,21] Indeed, important differences may occur between instruments of exactly the same type supplied by the same manufacturer. The use of such instruments, therefore, tends to lead to a very important problem, namely, the increasing isolation of laboratories in the sense that the only general and reasonably sure way for a laboratory to assess the bias of its analytical results is for it to make all the required tests itself. The problem is, however, compounded since a routine laboratory will often find it difficult or impossible to make sufficient effort available for the necessary tests. Yet further complication arises for 'black-box' analytical systems because of the difficulties of: (1) deciding the nature of the tests required, and (2) variability in the sizes of interference effects.[20,22]

Another feature of 'black-box' systems is that it is not generally possible accurately to predict the analytical response expected for a given concentration of the determinand even in the absence of sources of bias. 'Black boxes' must, therefore, be calibrated empirically, and the values obtained for the calibration parameters provide little or no check that the operation of the system has been satisfactory. Again, this is in contrast to more classical analytical systems whose calibration parameters can often be predicted with reasonable if not excellent accuracy. Various practical consequences follow from such considerations. For example, 'black boxes' generally require more frequent and detailed calibration, and *effective* analytical quality control is likely to be more complicated and effort-demanding.

All the above actual or potential problems are especially likely to arise for 'black boxes', but may also occur for any complex instrument even when it need not be regarded as a 'black box'. We feel that such difficulties are a fundamental problem in modern water analysis, and they have very important implications for anyone concerned with the provision of 'Standard/Recommended/Approved' methods of analysis.[21] Hitherto, a basic concept behind such methods seems to us to have been that the analytical method is the dominant factor controlling the performance of an analytical system. Thus, provided a method were truly complete and unambiguous and were faithfully followed by analysts using it, one might expect closely similar performance characteristics from any analytical system using a particular standard method. As noted at many points in this book, we doubt the general validity of this idea even for methods not specifying the use of complex instrumentation. The concept, however, seems to us completely invalid when complex instruments are involved unless methods also prescribe one or more procedures for ensuring equivalent performance of such instruments.

The final point to be made here is that as the complexity of an instrument increases, there will be, in general, a tendency for its liability to malfunction to increase. When malfunctions of 'black boxes' occur, ignorance of the processes involved will tend to make it more difficult to trace and eliminate the fault(s). This tendency will be particularly inconvenient for the more sophisticated and expensive instruments because there will often be only the one such instrument in a laboratory. This has important implications for staffing.

We would stress that though the above problems have been discussed, for brevity, in very general terms, each of these problems has arisen in laboratories with which we have been concerned. We believe that many other analysts will have experienced the same or related difficulties.

(c) Problems Caused by Incomplete Knowledge of the Processes Involved in Producing Analytical Results from the Analytical Signals

Instruments of the type under consideration are increasingly including automatic processing of the analytical signals from the measurement sub-system. A wide range of various computations is available, and typical examples are: calibration (including linearisation of non-linear calibration functions), drift correction (for the blank or a standard or the entire calibration), blank correction, and the calculation and presentation of analytical results for samples. Clearly, the procedures employed for such computations should be soundly chosen. When the analyst – in consultation, if necessary, with a statistician/computing expert – chooses the procedures appropriate for a particular analytical system *and* requirements for accuracy, there is, in principle, no problem. In many instances, however, this choice is now made by the instrument manufacturer, and problems may well arise if his choice is not appropriate. Special care is required in relation to the procedure used for calibration. These potential problems are aggravated if, as sometimes happens, the manufacturers refuse to disclose the exact details of the computations. We would strongly recommend that the computational procedures to be applied should always be known in detail by the analyst. Difficulties of this nature are beginning to be discussed with growing frequency in the literature; see, for example, refs. 12 to 15.

(d) Implications for Selection of Instruments
The selection of a particular instrument for routine analysis – often from a number of similar types available commercially – presents several problems that are well known to analysts. A series of related papers[23] provides a useful review of the subject, and refs. 1 and 24 give advice on the selection of instruments for water analysis; see also the recent suggestions on the evaluation of atomic-absorption spectrophotometers in ref. 67. We simply note below a few broad suggestions deriving from the above discussion of 'black boxes' and instrumentation in general.

(1) To the maximum extent possible, avoid 'black boxes' for both the analytical and also the signal-processing/calculation functions.

(2) To the maximum extent possible, use instruments for which the laboratory either has or can develop expertise in the underlying processes.

(3) Before purchase, undertake as full an evaluation as possible of the performance characteristics achievable for the types of sample of interest. This evaluation should, in general, involve both theoretical considerations and experimental tests, and special emphasis should be placed on sources of bias in analytical results. Whenever possible, secure guaranteed performance from the instrument supplier. See also ref. 67 on the comparison of instruments of the same type.

(4) Restrict purchase to well established instruments from suppliers with good records for the reliability, servicing, and 'trouble-shooting' of instruments.[25] Comprehensive and well written instruction manuals are another feature to consider.[68]

(e) Implications for Staffing
Some suggestions concerning analytical personnel have already been given in Section 10.3.

There is the additional need to consider staffing policy relevant to the routine maintenance and, as required, 'trouble-shooting' of complex instruments; such tasks increasingly require skills not all of which are commonly possessed by analysts. Large organisations may well have an Instrument Servicing Department or equivalent unit which can discharge this function. It may, however, present an awkward problem for smaller organisations and individual laboratories. In many instances, the manufacturers or suppliers of instruments can provide this service, but the analyst should satisfy himself that this approach will be feasible with respect to time delays and costs. Special problems may arise when instruments are in locations where the manufacturer cannot respond quickly to requests for help from a laboratory. In such circumstances, analysts should be careful not to put too great a reliance on complex instruments that cannot be rapidly cured of any malfunctions; consideration of alternative analytical procedures to be used in such an event is also important.

(f) Implications for Analytical Quality-control Procedures

All the principles discussed in Sections 5, 6 and 7 apply equally to simple manual analyses and the most complex, modern instruments. We saw in Section 6.4 that the ability to identify the most important factors affecting the bias and precision of analytical results was of great value in helping to decide the simplest feasible scheme for routine analytical quality control. It follows that the use of 'black-box' analytical systems calls for very careful consideration of quality-control procedures which will, in general, need to be more elaborate than for analytical systems whose processes are well understood. In particular, bearing in mind the remarks on interference in (b) above, special care is necessary to ensure that, as far as possible, any routine control tests will adequately reflect bias from this source. Little more can usefully and briefly be said here because the translation of the principles of Sections 5, 6 and 7 into a practical scheme needs to be considered individually for each analytical system.

One detail that is worth bearing in mind when considering routine quality-control procedures for semi- and fully-automated instruments is that these will often produce relatively large amounts of quality-control data. To ensure the timely use of these data as described in Section 6.4, their handling by computerised techniques is generally desirable.

(g) Concluding Remarks

From the above discussion it is clear that there are a number of potential dangers and problems in the use of complex, modern analytical instrumentation; such difficulties can be very markedly aggravated when analytical systems either are or are treated as 'black boxes'. Of course, we certainly do not advocate that such equipment should not be used. Rather, we would stress that analysts proposing to employ such analytical systems should very carefully consider the problems and ensure that they are well enough controlled for their purposes. It seems to us highly desirable to do everything possible to ensure that 'black boxes' play as small a role as possible in analysis. Progress towards this goal can be achieved by the analyst adopting a continually critical approach both to the analytical systems he uses and also to instrument manufacturers with whom he discussed his needs for instrumentation.

10.5 GLASSWARE AND OTHER APPARATUS

Complete and unambiguous analytical methods should specify the particular apparatus required; we note, therefore, only a few general points below.

One of the most important aspects to consider in choosing glassware and other analytical apparatus is their possible contaminating effect on samples and solutions; this is considered in Section 10.8. The selection of the best material for sample containers is also discussed in Section 3.6.2.

It is always useful to bear in mind the tolerances on the capacities or volumes delivered of volumetric equipment so that the most appropriate and convenient is used. There is little point in making a very accurate and relatively slow addition of a reagent with, for example, a Grade-A pipette when the volume of the reagent need be controlled to only a few percent. Note also that volumetric glassware does not necessarily meet the standards specified for Grade-A or Grade-B apparatus.[26] Of course, the use of semi-automatic pipettes and sample diluters (many are commercially available) can much increase the speed – and sometimes the precision – with which reagents can be dispensed and samples diluted. Such apparatus can be of great value in a busy laboratory, but care is necessary to choose types whose components cannot cause important contamination. The performance of such apparatus may also deteriorate as components age and wear.

Apparatus must, of course, be kept clean, especially in trace analysis. This often requires the use of special cleaning solutions and/or procedures, and details are then usually given in the individual analytical methods. The composition of cleaning solutions must be chosen so that their components cannot affect subsequent analyses, *e.g.*, avoid phosphate-containing detergents when phosphates are to be determined. Information on procedures for cleaning apparatus is given in most general analytical texts, and a good account in the context of water analysis is in ref. 1; see also Section 3.6.2(*b*).

The importance, and sometimes the difficulty, of ensuring adequately clean apparatus may often make it desirable in trace analysis to reserve a set of apparatus solely for a particular determinand. The storage of clean apparatus when not in use also needs to be considered with care since laboratory atmospheres may cause its contamination; storage in plastic bags in cupboards reserved solely for this purpose can be helpful in this respect, for inorganic determinands.

10.6 REAGENTS

Again, complete and unambiguous analytical methods should specify all relevant details on the preparation, storage and use of reagents. Therefore, we note below only a few general points.

Useful discussions of the preparation, storage and standardisation of commonly required reagents for water analysis are given in refs. 1, 4 and 27. Usually it is prudent to use the purest grade of reagents available, at least until the suitability of other grades has been assessed. The composition of some reagents, particularly organic compounds such as dyes, may vary appreciably from batch to batch. Thus, when different batches of reagents are brought into use, it is advisable to check whether aspects such as the nature of the calibration function have changed.

Some reagents and many of their solutions deteriorate on standing. Methods of storing such chemicals and their useful lives should be given in analytical methods, but, unfortunately, this is not always done. Analysts should, therefore, be on guard to ensure that reagents are freshly prepared at suitable intervals. Regular quality control of analytical results (see Section 6.4) will help in this respect, but may well not immediately reveal deterioration; alternatively, special tests to estimate the stability of reagents may be made (see Section 3.6.4).

It is always desirable, though often not essential, that reagents and solvents contain negligible amounts of the determinand, and their contents of interfering substances may also need to be controlled. The need for special purification procedures should be judged in the light of the accuracy required of results. In general, the need for, and methods of, reagent purification should be stated in analytical methods, but analysts should maintain a critical approach; it is always possible that they may be supplied with a batch of reagent of unusually poor quality. A solvent of special importance is water itself, but this is considered in Section 10.7.

Many publications deal with one or more aspects of reagent purification. The book by Zief and Mitchell[28] provides an extensive discussion, and more recent reviews will be found in refs. 1 and 29 to 31.

Many reagents, solvents, and their solutions are toxic, and suitable precautions against any harmful effects must be taken; see, for example, refs. 32 and 33.

10.7 HIGH-PURITY WATER

Water containing negligibly small concentrations of the determinand and interfering substances is generally required for the preparation of blank and other standard solutions (see Sections 5.4.2 to 5.4.4 and 8.3.1). We refer to this water as '*high-purity water*' or '*blank-water*'. Provided that the determinand content of this water is adequately small, then the concentrations of interfering substances will also usually be negligible, but this is certainly not necessarily so. For example, it is, in general, relatively easy to prepare blank-water containing negligible concentrations of organochlorine insecticides, but this water may well be contaminated by phthalate esters which interfere in the measurement of the insecticides.[4,27,33a] It is always important to keep in mind these two aspects of the purity of the blank-water.

In general, the purity of the water required to prepare reagent solutions need not be so high because blank-correction eliminates, in principle, the effect of determinand and interfering substances in reagents [see Sections 5.4.2(a) and (b)]. Of course, there is never any harm in using blank-water to prepare reagents, and we shall assume, for simplicity, that this is done so that we need speak only of blank-water in the following discussion. The preparation and testing of blank-water are considered in Sections 10.7.1 and 10.7.2, respectively. Note also that 'organic-free' water can be purchased from some suppliers of chemicals.

10.7.1 Preparation of Blank-Water

The purity required of this water depends, of course, on the application and the required accuracy of analytical results. It is virtually impossible, therefore, to state the most suitable means of producing this water for each of the very many different analytical situations. The appropriate procedures should normally be specified in analytical methods, and the following remarks are intended only to give some general guidance, and to stress the points involved. The preparation of blank-water is also discussed in refs. 1, 4 and 27.

Distilled water is quite suitable for many purposes, and any of the available stills can be satisfactory provided that appropriate precautions are taken to prevent contamination of the distillate by scale and corrosion products that may be present in the still. Deionised water (*i.e.*, water after passage through a mixed-bed deionisation unit) may also be used, but care is necessary to replace or regenerate the mixed-bed unit before the ion-exchange resins have become too exhausted. Both these purification techniques have their own technical advantages and disadvantages, and the choice between them may often come down to one based largely on cost and convenience. Deionisation can remove impurities (*e.g.*, ammonia and carbon dioxide) that distillation does not. On the other hand, non-ionic and colloidal materials may pass through mixed-bed units[34] but be removed by distillation.

A commonly useful approach is to use both purification techniques, the mixed-bed system being used to 'polish' the distilled water. This technique provides water suitable for almost all normal purposes (but see further below), though traces of organic materials are leached from ion-exchange resins,[4] and may render the water unsuitable when such substances are to be determined. In addition, many manufacturers can supply cartridge-filters of small pore-size when it is required to filter the water before use; this can be of value when particulate contamination of the water is of interest,[4] *e.g.*, for measurements of turbidity. Columns of activated carbon may also be useful in removing organic contaminants in the water, and irradiation of the water with ultraviolet light has been reported[35–37] to be an excellent means of reducing the concentrations of organic substances to very low levels. See also ref. 69 for a detailed study of the purification of water from organic compounds. A number of systems comprising various units for purification are commercially available for the preparation of high-purity water. For example, systems including activated carbon, deionisation, and filtration units have been reported by many workers to be excellent when trace metals are of interest; see, for example, refs. 38 and 39. Purification systems in which the water may be recirculated can be of special value.

The materials of construction used in purification systems may cause difficulty in preparing water of adequate purity for some determinands, and these materials may need to be specially chosen. For example, the use of all-glass or fused silica[40] distillation equipment may be advantageous when the determinands of interest are trace metals. Similarly, plastic materials should generally be avoided when trace organic compounds are to be measured since a variety of such compounds may be leached from a number of commonly used plastics.[28,33a,41,42,69] Whatever the purification system adopted, it is often useful to monitor the electrical conductivity of the purified water since this provides a convenient means of indicating certain changes in the operational efficiency of the system, *e.g.*, exhaustion of ion-exchange resins, priming in stills.

A number of determinands (*e.g.*, turbidity, organic materials, chlorine, dissolved gases) may require special water-purification procedures. These are normally specified in the appropriate analytical method, and no attempt is made to discuss them in detail here.

High-purity water can quickly become contaminated either by its containers or from the atmosphere; see, for example, refs. 43 and 69. Great care should be paid to both aspects, particularly since it is both desirable and convenient if the concentrations of any impurities in the water are held constant. Such constancy can markedly reduce the amount of analytical effort required in checking the purity of the blank water (see Section 10.7.2). Polyethylene or polypropylene containers are generally suitable for most inorganic determinands, and are preferable to glass when substances such as sodium and silicate are of interest (see also Section 3.6.2). For organic determinands, however, glass containers should generally be used.

Whatever the method employed to purify the water, its purity should be checked before it is used, and this check should be repeated as often as necessary (see Section 10.7.2). In this connection, it is convenient to be able to store large volumes of blank water having measured its determinand content, and then to be able to assume that this content remains constant for all subsequent analyses for which the same batch of water is used.

10.7.2 Determination of the Purity of Blank Water

(a) General Considerations
As discussed in Sections 5.4.2 to 5.4.4 and 8.3.1, the requirement that blank determinations and other calibration standards be analysed by exactly the same procedure as that used for samples implies that blanks and standards should be prepared from water whose concentrations of determinand and interfering substances are zero. If these concentrations are not zero, bias will generally arise in the analytical results for samples. The magnitude of this bias depends on the calibration function and the precise details of the procedures used to estimate calibration parameters and the concentrations of determinand in samples. For any procedure, the bias can be calculated by the approach described in Section 5.4.2. For simplicity here, we consider only calibration functions of the form $\mathscr{A}_T = \lambda + \chi C_S$ [the notation is that of Section 5.4.2(a)], the analytical results being calculated from

$$R = (A_T - A_B)/k \qquad (10.1)$$

where A_T and A_B are the measured analytical responses for sample and blank, respectively, and k is an estimate of the slope, χ, of the calibration curve. Such procedures are very common in water analysis.

Let the concentration of the determinand in the blank water be C_W, and let the effect of positively interfering substances [see Section 5.4.2(a)] in the blank water on the analytical response, $\mathscr{A}_B$, for the blank determination be $\mathscr{A}_I$. Then the analytical response for the blank in the absence of random errors may be written as

$$\mathscr{A}_B = \eta + \chi C_C + \chi C_W + \mathscr{A}_I \qquad (10.2)$$

The analytical response, $\mathscr{A}_T$, for the sample in the absence of random errors is given by (assuming that the sample contains no interfering substances)

$$\mathscr{A}_T = \eta + \chi C_C + \chi C_S \tag{10.3}$$

so that the analytical result in the absence of random errors is given by

$$R = (\mathscr{A}_T - \mathscr{A}_B)/\chi = C_S - C_W - \mathscr{A}_I/\chi \tag{10.4}$$

Thus, the bias of R is $-(C_W + \mathscr{A}_I/\chi) = -C'_W$, and the purity of the blank water should be such that C'_W is negligible in relation to the maximum tolerable bias of analytical results. Alternatively, a correction for this source of bias should be made by obtaining an estimate, $\hat{C}'_W$, of C'_W and adding $\hat{C}'_W$ to R. The former procedure is generally preferable since this avoids any increase in the random error of the corrected analytical result through the random error of $\hat{C}'_W$.

For many purposes, water prepared by procedures such as those outlined in Section 10.7.1 will have negligibly small values for C'_W. Experience with particular analytical systems may then validly allow one to assume that C'_W is negligible without the need to estimate it. Such an approach is, however, by no means necessarily valid for all determinands, and there is a general need, therefore, for experimental procedures for estimating C'_W. The importance of this will generally increase as C_S decreases. This source of bias is of potential importance in many trace analyses, and was discussed in the first edition of this book; the topic has recently been reviewed by Hunt and Morries.[44] We deal with it here at some length because it still appears frequently to be ignored. For generality in the following discussion, we shall continue to speak in terms of C'_W, but for many analytical systems the dominant contribution to C'_W is C_W, the determinand content of the blank water.

(b) Procedures for Estimating C'_W

Equation (10.2) shows that the procedure used for analysing samples cannot be used to estimate C'_W when $\eta + \chi C_C \neq 0$ since the contribution of $\eta + \chi C_C$ to $\mathscr{A}_B$ cannot be separated from that of $\chi C_W + \mathscr{A}_I$. Thus, a special procedure is generally required for estimating C'_W. Various approaches are possible, and the most appropriate should be devised for each combination – determinand/interfering substances/analytical method/accuracy required of analytical results. To this end, it is useful to bear in mind that special measurements to estimate C'_W need not necessarily follow exactly the same analytical procedure as that used for samples; it is only necessary that such measurements give unbiased values for $\hat{C}'_W$. For example, it may be essential that normal blank determinations be analysed by exactly the same procedure as samples when contamination by the determinand occurs during chemical processing of samples; otherwise, the value of C_C for blanks would differ from that for samples. When estimating C'_W, however, this chemical processing would preferably be avoided provided that it had no effect on the analytical response produced by the determinand and interfering substances in the blank water.

Various types of procedure for estimating C'_W are noted briefly in (1) to (6) below.

(1) If one is adequately confident that η and C_C are essentially zero or can be made so without affecting C'_W, $\hat{C}'_W$ is given directly by A_B/k. If these conditions cannot be achieved when the routine analytical system is used, it may, in principle, be possible to achieve them with another analytical system; for example, a different analytical method might be used. In practice, however,

this latter approach will seldom be practicable. It will also not be valid if $\mathscr{A}_1/\chi$ makes an appreciable contribution to C_W' *and* if the relative effects of the interfering substances are not the same when the routine and special procedures are used.

(2) When η and/or C_C are not zero, A_B/k gives an estimate of an upper limit for C_W'. If, therefore, A_B/k is negligible in relation to the maximum tolerable bias, the blank water may be concluded to be of adequate purity. Alternative analytical procedures may also be used to obtain estimates of upper limits for C_W'. For example, for some purposes the electrical conductivity of the blank water gives an upper limit for the concentrations of ionically dispersed substances. Similarly, the determination of total organic carbon provides an upper limit for organic compounds. Again, use of such alternative procedures is not valid if $\mathscr{A}_1/\chi$ is appreciable and the relative effects of the interfering substances differ from those for the routine analytical procedure.

(3) A portion of the blank water can be concentrated by evaporation, and both the concentrated and unconcentrated waters analysed. If the concentration factor is ϕ, and the analytical responses are A_{BC} and A_{BU}, respectively, $\hat{C}_W'$ is given by

$$\hat{C}_W' = (A_{BC} - A_{BU})/k(\phi - 1) \tag{10.5}$$

This approach assumes that: (1) neither the determinand nor the interfering substances are decomposed or lost through volatility during the evaporation, (2) no contamination of the water occurs during the evaporation, and (3) the relative effects of the interfering substances are the same in the concentrated and unconcentrated waters. Experimental tests will usually be necessary to show the validity of these assumptions. An example of the use of this approach – in a method for aluminium in water – is given in ref. 45.

A variant of the principle involved in this approach may be useful when the normal analytical procedure involves the quantitative extraction of the determinand from samples, *e.g.*, by solvent-extraction. It may then be possible to obtain $\hat{C}_W'$ from equation (10.5) by making two blank determinations, one using the normal volume, V_U, of blank water and the other a greater volume, ϕV_U. Again, special tests will generally be required to demonstrate the validity of the procedure, and correction for differences in the final volumes of the two blanks may be necessary. Examples of the procedure appear in methods for copper,[46] iron,[47] and nickel[48] in boiler-feed water.

(4) Two blank determinations can be made, the amount of each reagent added to one of them being ϕ times the normal amount. If the analytical responses for the normal and other blank are A_{BN} and A_{BS}, respectively, $\hat{C}_W'$ is given by

$$\hat{C}_W' = (\phi A_{BN} - A_{BS})/k \tag{10.6}$$

This approach assumes that: (1) the additional amounts of reagents do not affect the analytical responses for given concentrations of determinand and interfering substances in the blank water, and (2) no appreciable amounts of determinand or interfering substances enter the blanks during analysis from any source other than the reagents. In some procedures, assumption (1) will be invalidated if the two blanks are not made to the same final volume before measurement. This difficulty may, in general, be overcome by multiplying A_{BS}

by V_S/V_N, where V_S and V_N are the final volumes of the special and normal blanks, respectively. As before, experimental tests will generally be necessary to show whether or not the assumptions involved in this approach are valid; an example of its use is given in a method for iron in water.[49]

(5) If a means can be devised for preventing C_W and $\mathscr{A}_I$ from contributing to the analytical response for the blank water, analysis of this special blank (analytical response, A_{BO}) and a normal blank allows calculation of $\hat{C}'_W$ from

$$\hat{C}'_W = (A_{BU} - A_{BO})/k \qquad (10.7)$$

As before, special tests are generally necessary to show the validity of such an approach. Examples of its use are provided by methods for 'reactive' silicon in water,[50] and ammonia in water.[51]

(6) For some analytical systems it may be possible to make measurements not involving the analysis of any water but which provide direct or indirect estimates of $\eta + \chi C_C$. Given this estimate and a normal blank determination, $\hat{C}'_W$ can be obtained from equation (10.2). One example of this approach is given in a flame-photometric method for sodium in high-purity water.[52]

As noted above, it is very desirable always to attempt to ensure that C'_W in negligibly small since this avoids the complications involved in considering the effect of the error of $\hat{C}'_W$ on the errors of analytical results. When $\hat{C}'_W$ is not negligible, results for samples should be corrected by adding $\hat{C}'_W$ to them. Two points should be noted. First, assuming that the procedure for determining $\hat{C}'_W$ gives unbiased results, the random errors of the measurements involved in determining $\hat{C}'_W$ will lead to random error in $\hat{C}'_W$, and this will affect the random errors of analytical results for samples. The random error of $\hat{C}'_W$ may be reduced by replication of the measurements [see Section 4.3.2(c)]. Second, if an independent determination of $\hat{C}'_W$ is made for each value of A_B used in calculating analytical results for samples, the random error of $\hat{C}'_W$ will cause only the between-batch standard deviation of results to increase [see Sections 6.2.1(b), (c) and (d)]. If, however, the same estimate, $\hat{C}'_W$, is used for a number of batches of analyses and its random error is appreciable compared with other sources of within- and between-batch random errors, then a 'supra-batch' random error of analytical results will arise for the reasons discussed in Section 6.2.1(b).

10.8 CONTAMINATION

By *contamination* we mean the entry of determinand and/or interfering substances into a sample at any stage during its collection, preparation, storage and analysis [see Section 5.4.2(a)]. Note that contamination in this sense does not necessarily affect the analytical response. For example, if the contaminating forms of a determinand were solid particles of a substance that did not dissolve during the chemical treatment of samples, the analytical response from many analytical systems would be little if at all affected. Usually, however, contamination does affect responses, and is one of the greatest general problems of trace analysis.

No amount of writing can replace a critical appraisal by each analyst of potential sources of contamination and of procedures for eliminating or minimising them.

The importance of this problem in routine water analysis has been accentuated in recent years by the growing need for many trace analyses. We have made no attempt to gather from the literature all the very many individual sources of contamination that have been observed; rather, we have tried to state and discuss what seem to us the most important general aspects of the problem. In Section 10.8.1 we make a number of general observations concerning the effects, and assessment of the magnitude, of contamination. In Sections 10.8.2 to 10.8.7 we briefly consider some of the more important factors involved in controlling contamination.

Since the first edition of this book, a valuable text dealing solely with contamination in trace analysis has appeared.[28] Two collections of papers[53,54] deal with many aspects of the subject, and are valuable sources of additional references. Recent general reviews of the topic will also be found in refs. 11, 29 and 55 to 57. All authors stress the difficulty of controlling analytical accuracy when appreciable contamination occurs. Kosta[56] even states that reliable analytical results at concentrations much below 1 mg kg^{-1} can be expected only from specialised and experienced laboratories. The need for special tests to assess the magnitude of contamination is stressed in refs. 4, 11, 33a and 55.

10.8.1 Effects and Assessment of Contamination

(a) Effects of Contamination

When contamination occurs, the bias and/or precision of analytical results may well be worsened. Whether or not this effect is important depends on the magnitude of the contamination *and* the accuracy required of analytical results. Thus, a given source of contamination may be tolerable in some applications but completely unacceptable in others. It is useful, therefore, to have a clear view of the effects of various types of contamination on the bias and precision of analytical results. Tests to assess the size of bias caused by contamination during the sampling process have been considered in Section 3.5.5, and we concentrate here on the analytical process.

If contamination of a sample occurs during analysis, the analytical response for the sample will not reflect its original concentration of the determinand. Provided that the calibration of an analytical system is made using exactly the same analytical procedure as that by which normal samples are analysed, some correction for the effect of contamination is achieved (Section 5.4.2). Commonly, prime reliance is placed on blank determinations as a means of estimating the correction to be made to the analytical responses observed for samples, and we consider only this approach in the following.

Two broad types of contamination can be distinguished.

(1) The magnitude of contamination is essentially the same for all samples, blanks and calibration standards in the same batch of analyses; see Section 6.2.1 for the definition of 'batch'. This type commonly arises when the dominant source of contamination is reagent solutions. For such contamination, it follows (see Section 5.4.2) that blank correction prevents bias – from this source – in analytical results. This benefit is gained at the expense of the increased random error of analytical results since the within-batch random error of blank determinations leads to an increased between-batch standard deviation of

results (see Section 6.2.1). Notwithstanding this worsening of precision, the desirability of eliminating sources of bias (Section 5.5) requires that correction for this type of contamination should generally be made. In addition to this effect on precision, the within-batch standard deviation of results may also be increased if the size of the contamination is sufficient to affect the analytical response appreciably *and* if, as is common [see Section 6.2.1(*f*)], the standard deviation increases with increasing analytical response. An additional, undesirable effect of contamination is that the effective concentration range of an analytical system will tend to be reduced.

Of course, all such problems are eliminated by the complete prevention of contamination. This is, however, much easier written than achieved. A more realistic requirement is that this type of contamination be controlled to the extent that the required precision of analytical results and the required concentration range can be achieved.

(2) The magnitude of contamination may vary appreciably from one determination to another in the same batch of analyses. This type may arise, for example, when the dominant source of contamination is dust in the laboratory atmosphere. For this type, blank correction will not generally provide *complete* correction since the size of contamination for the blank will, in general, differ from that for a sample. To minimise bias, however, blank correction remains necessary, but, in addition to the two sources of increased random error noted for (1) above, the residual error arising from the incomplete correction of this type of contamination must also be considered.

When the size of contamination varies at random from one determination to another within a batch of analyses, the long-term average value over many samples of this residual error is zero. In this situation, we regard this error as leading to only an increase in the random errors of analytical results [see Section 6.2.1(*b*)]. The error appears – in our general model of random errors (Section 6.2.1) – as an increase in the values of both the within- and between-batch standard deviations of analytical results. We would stress, however, that the detailed procedure followed in making a batch of analyses may also lead to the situation where the size of contamination does not vary at random, but follows some systematic pattern. For example, if the laboratory atmosphere is the source of contamination, the size of the effect may increase systematically with the time during which a solution is in contact with the atmosphere. If then the blank were always analysed first, its contamination would always be less than that of samples. Thus, although the blank correction would eliminate some of the bias, the residual error for samples would always be positive so that analytical results would be biased. Similar effects may also arise through systematic spatial patterns of analysis, *e.g.*, if a blank were always placed in the position on a hot-plate least subject to contamination. Whenever such systematic patterns occur, the residual, uncorrected effect of contamination is to cause both bias and worsened precision of analytical results. To minimise this source of bias, therefore, one should either eliminate such sources of contamination and/or avoid relevant, systematic patterns of analysis [see Section 6.2.1(*e*)(*vi*)]. Given that one or other of these two approaches is practicable, it follows that the practical requirement for the control of this type of contamination becomes the same as for (1) above, *i.e.*, the degree of contamination should be controlled so that the required precision of analytical results and the required concentration range are achieved.

It should be noted that an extreme instance of this type of contamination sometimes occurs and can cause great difficulties, *i.e.*, when sporadic sources of contamination are present. In such situations, the size of contamination may be zero (or essentially so) for a high proportion of analysed solutions but positive for the remaining proportion. Analytical systems subject to this effect cannot be regarded as well controlled (see Section 6.3.1), and blank correction will be of little value for the degree of replication of analyses that is usually practicable in routine laboratories. The only suitable approach to such contamination is to do everything possible to eliminate it. If this cannot be achieved, it becomes virtually impossible to quantify the errors of analytical results.

On the basis of the above considerations, some suggestions are made in (*b*) below for tests that are useful in assessing the nature and importance of contamination for a given analytical system.

(b) Assessment of Nature and Importance of Contamination

(i) General approach Many types of investigation may be needed in the assessment and control of contamination. The detailed nature of the tests are governed by a number of factors that vary from one application to another, and no attempt is made here to consider the many possibilities. Our aim is rather to describe a simple and general approach that will almost always be of value as an initial stage in the establishment of an analytical system – potentially liable to contamination – for routine analysis. This approach consists of a detailed scrutiny of all components of the analytical system that could lead to contamination, followed by appropriate choice and/or control of all factors judged to be sources of appreciable contamination. This stage will generally be much facilitated by carefully written, complete and unambiguous analytical methods [see Section 9.2.4(*b*)]. The next stage involves making a series of blank determinations in one batch of analyses, the number of blanks being chosen on the basis of the practice to be used in routine analysis. For example, if it were proposed to make routine analyses in batches of ten determinations (*e.g.*, eight samples, one blank, and one control-test), then at least ten, and preferably more, blanks should be measured in this preliminary test, using all the equipment to be used routinely (see below). This implies that the equipment used for a particular determinand be reserved solely for that purpose; see Section 10.8.5. The results from these replicate blanks can then be used to provide some assessment of the nature and importance of contamination as described below.

(ii) Practical details for the tests It is useful to be able to identify which piece of apparatus was used for a particular blank determination. For this purpose, each piece can be numbered. When the analytical procedure requires the use of several different pieces of apparatus for each determination (*e.g.*, a beaker, filter funnel, and calibrated flask), use all items with the same number for a given determination.

This procedure will help in identifying any items of apparatus causing unacceptably large contamination. It will also help, should subsequent batches of tests be necessary, to keep a note of the order in which the numbered apparatus was used. In the first batch, the simplest approach is to use apparatus numbered "1" for the first blank, the numbered "2" for the second, and so on. It is essential, of course, that these replicate blanks be analysed by exactly the same procedure as that to be used for samples; otherwise, they may not be subject to all sources of contamination affecting samples. It is also very desirable that the water used for these blank determinations contain as small a concentration of the determinand as possible (see Section 10.7). It is necessary to be able to convert the analytical responses for the blanks to apparent concentrations of the determinand. For this purpose, an estimate, k, of the slope of the calibration curve [assumed here to be of the form $\mathscr{A}_T = \lambda + \chi C_S$ – see Section 5.4.2(*a*)] is required. This estimate need not be very precise, and we assume that such an estimate is available; if this is not so, two or three standards should be included in the same batch as the blanks to provide a value for k. Finally, the blank determinations should be made as carefully as possible so that the precision of the results is the best that can be achieved. The aim of the tests is not to estimate the routine precision of analytical results but to decide if contamination is a problem; this decision will be aided if the effects of the other factors affecting the accuracy of results are made as small as possible.

(iii) Interpretation of the results   The results should first be examined to decide if there is any evidence of either sporadic contamination (*i.e.*, one or two of the results much greater than the others) or a systematic change in the magnitude of contamination throughout the batch of analyses. For this purpose it may well be useful to plot the responses for the blanks sequentially in the temporal and/or spatial order of the analyses; see (*a*) above, item (2). Either or both of these two types of contamination may be so marked as to be revealed immediately by the plot. In general, however, statistical tests are required to decide if any apparent effects are statistically significant.

To decide if a response can be regarded as not a member of the population represented by the other responses, a number of 'outlier' tests may be used, and are described in many texts; one that will usually be suitable is given, for example, in ref. 58. Various procedures may also be used for testing the significance of any apparent trend in the blank responses. For example, an apparent linear trend may be tested using appropriately modified forms of the equations for unweighted linear regression given in Section 5.4.2(*e*)(*ii*). The slope, k', of this trend is significantly different from zero at the α significance level when $k'/s_{k'} > t_\alpha$, where $s_{k'}$ is the estimate of the standard deviation of k', $t\alpha$ is the *t*-statistic at the α probability level for $n-2$ degrees of freedom, and n is the total number of blanks used in calculating k' and $s_{k'}$. Any 'outliers' should not be included in these calculations.

If either or both of these two types of contamination are present, the magnitude of the effect (in concentration units) will show its practical importance in relation to the requirements on the bias of analytical results. In general, however, we suggest [see (*a*) above] that sources of bias be eliminated whenever possible. To eliminate gradual drift of the response of the blank, either the source of contamination should be eliminated or the systematic pattern of analysis [see (*a*) above] should be avoided.

If one or two outlying results were detected, they may be due either to sporadic contamination or to greater contamination from the apparatus used for these blanks. Which of these two possibilities is correct may be assessed by repeating the batch of analyses. If the blank from the apparatus that gave an outlying result is again greater than the other blanks, it may reasonably be concluded that this particular set of apparatus causes greater contamination. That apparatus may then either be rejected or subjected to special cleaning before a further test. If, on the other hand, the blank from the apparatus that gave the outlying result initially is now normal, one can conclude either that a sporadic source of contamination is present or that the process of making a blank determination can remove a source of contamination from the apparatus. The first explanation is the more likely if the second batch of determinations again shows a few results much greater than the others.

The second stage of interpretation consists in assessing the within-batch standard deviation of blank determinations after elimination of any systematic variations caused by the two types of contamination considered above. For this purpose, if any blanks were concluded to be 'outliers', reject those results, and use only the remaining blank responses. Two cases then arise.

(1) No systematic change in the blank responses was detected. In this event, an estimate of the within-batch standard deviation, s_B (in concentration units), of blank determinations can be obtained from the equation:

$$s_B = [\sum (A_{Bi} - \bar{A}_B)^2/(n-1)]^{1/2}/k \qquad (10.8)$$

where k is the estimated slope of the calibration curve, A_{Bi} is the analytical response of the ith blank determination ($i = 1$ to n), $\bar{A}_B$ is the mean response of the n blanks, and n is the total number of blank determinations after rejection of 'outliers'.

(2) A systematic change in the blank responses was detected. Provided that it is possible to eliminate systematic changes in subsequent tests, an estimate, s_B, of the within-batch standard deviation of blank determinations can be obtained from the scatter of the individual blank values around the line defining the trend. For example, if the trend were linear (or approximately so), linear regression theory may be used [see Section 5.4.2(e)(ii)] to obtain an estimate of the residual standard deviation, s_A, and an estimate of the within-batch standard deviation, s_B, of blank determinations is then given by

$$s_B = s_A/k \qquad (10.9)$$

The value obtained for s_B from equations (10.8) or (10.9) may then be used to estimate the precision of analytical results at or near zero concentration of the determinand. The calculations required for this step depend on the calibration procedure and method of calculation of analytical results, and may be derived from the theoretical considerations in Section 6.2.1(c). Of course, since at this stage of testing one has analysed only blanks, it is necessary to assume that samples will be analysed with the same precision as blanks, and that the imprecision of the estimate, k, of the slope of the calibration curve is negligible. Such assumptions are, however, reasonable in the present context of preliminary tests. Suppose, for example, that the analytical method specified single analyses of each sample and the blank water in each batch, subtraction of the blank response from that of each sample, and calculation of analytical results by division of the corrected responses for samples by k. An estimate, s_t, of the best possible total standard deviation of analytical results is then given simply by $s_t = \sqrt{2}.s_B$.

If the value of s_t is satisfactory in relation to the maximum tolerable value of σ_t for small concentrations of the determinand [see also Section 6.3.4(a)], it may be concluded with reasonable confidence that any random contamination has been adequately controlled. If, on the other hand, the value of s_t is greater than acceptable, one cannot conclude that contamination is necessarily the cause since inadequate control of other experimental conditions affecting the responses of blanks could also be responsible. When, however: (1) the analytical method is complete and unambiguous [see Section 9.2.4(b)], (2) the method has been faithfully followed, and (3) satisfactory precision has been achieved by other laboratories using the same method, then contamination is one of the most likely reasons for an unsatisfactory value of s_t. It will then usually be worthwhile to seek and minimise other sources of contamination, but the possible need to improve the degree of control of other experimental conditions should be kept in mind.

The final stage in the interpretation of results concerns the mean value, $\bar{A}_B$, of the blank responses; two aspects are involved. The first is simply that too great a value of $\bar{A}_B$ may restrict the concentration range in samples that can be measured with the required accuracy. The value of $\bar{A}_B$ is, therefore, another criterion for deciding whether or not contamination has been adequately controlled. The second aspect is that the value of $\bar{A}_B$ provides information on the average amount of contamination if there is some way in which the mean value of the blank in the absence of contamination can be estimated. Various possible approaches to this estimation exist depending on the detailed analytical procedure, and each situation needs individual consideration.

10.8.2 Determinand

Some determinands are much more liable to contamination by the determinand than are others. The measurement of trace concentrations of commonly occurring substances such as iron, sodium, silicon, and organic carbon is always likely to meet greater problems of contamination than when relatively uncommon substances such as beryllium, vanadium, and organochlorine insecticides are of interest. Indeed, the difficulties presented by the former type of determinand may be so great that special sampling and analytical equipment are required; the determination of small concentrations of dissolved oxygen in water is a familiar example. It is always of value to examine the laboratory and its environment, equipment and apparatus in order to identify, for each determinand, those materials that may contact samples and that contain appreciable amounts of the determinand. Generally speaking, it is desirable to eliminate as many such sources of potential contamination as practicable before beginning any analyses; see also Sections 10.8.3 to 10.8.6.

The above type of examination is particularly important in attempting to control sporadic types of contamination. For example, striking a match may contaminate a sample with phosphate, cosmetic products may be a source of boron, cigarette smoking may increase the alkali metal content of the atmosphere around a flame photometer, handling a coin may lead to contamination by copper, *etc.*

The importance of contamination by interfering substances rather than the determinand clearly depends markedly on the degree of contamination *and* the interference effect caused by a given concentration of a contaminant. In our experience, such contamination is not usually important in inorganic water analysis, but it can be of great importance when trace amounts of many organic compounds are of interest. For example, phthalate esters are a common contaminant in many laboratories, and may cause deterioration in the accuracy of results for organochlorine insecticides.[4,27,33a] Again, each analytical system requires individual consideration.

10.8.3 Analytical Method

Analytical methods ought to include detailed recommendations on all the precautions found necessary for adequate control of contamination for the purposes for which the method was devised [see Sections 9.2.4(*b*)(*iv*) and (*ix*)]. Such recommendations can be invaluable in helping to avoid many problems that laboratories would otherwise meet. Such advice is, however, not always given in analytical methods, and analysts should examine methods critically from this point of view, especially when choosing methods to be applied routinely. It must also be borne in mind that the degree of contamination caused by, for example, a piece of apparatus may depend on the manufacturer. Thus, tests of the precision of results in one laboratory may give perfectly satisfactory results while another laboratory using apparatus from another supplier may obtain unacceptable precision even though the instructions in an analytical method are followed exactly. It should also be remembered that when a method is to be used for a more rigorous purpose than that for which it was intended originally, then the precautions specified in the method may be insufficient for the new application. Equally, application of a method to less demanding requirements may allow some relaxation of precautions.

10.8.4 Laboratory

Many aspects of laboratory design and construction may affect the size of contamination effects, Detailed reviews of the problems and procedures for overcoming or minimising them are provided in refs. 2, 11, 28, 29, 39, 55 to 57, and 59 to 61.

Clearly, it is generally desirable to avoid materials of construction containing large amounts of substances of which trace amounts are to be measured in samples, but this approach is not always possible. Alternatively, such materials may be covered by non-contaminating materials such as paint and plastic coatings. The latter are often of special value in minimising contamination from inorganic substances, but may cause problems in trace organic analysis. Conflicting requirements of this type are usually best met by having separate laboratories for inorganic and organic analysis, and this approach is also of value in relation to several of the other points mentioned below. Such difficulties are best discussed with experts in the design, construction and furnishing of analytical laboratories.

The laboratory atmosphere is another potentially important source of contamination. Depending on its location, a laboratory may meet such difficulties from contamination by atmospheric dust that complete air-conditioning systems are essential; see, for example, refs. 28, 29, 39, 55, 57 and 59 to 62. Sometimes, local air conditioning (*e.g.*, laminar-flow hoods) may provide sufficient protection if installed at the locations of those analytical operations liable to important contamination.[28,29,39,55,57,59] Another type of atmospheric contamination is that generated by analyses within a laboratory; the vapours from solutions of ammonia and hydrochloric acid solutions and organic solvents are common examples, but many others are possible. Clean working, good housekeeping, and suitable isolation – in time and/or space – of particular analyses are all helpful and sometimes essential.

Of course, cleanliness in a laboratory is important. Regular wiping with a damp cloth of all surfaces that may act as sources of contamination (*e.g.*, bench-tops, interiors of fume-hoods) can be surprisingly effective in preventing difficulties from contamination. Such cleaning is aided by smooth working surfaces with rounded rather than angular contours.

10.8.5 Equipment and Apparatus

Clearly, those parts of equipment and apparatus coming into contact with samples should be of materials that cannot cause important contamination. Special care is required in identifying the materials of relevant components in instruments since analytical methods do not usually recommend the use of particular instruments and the materials of construction may vary from one manufacturer to another. Attention should also be paid to apparently minor components of equipment and apparatus, *e.g.*, rubber stoppers, springs for holding ground-glass joints together, bottle-caps and any liners they may have,[63] small pieces of plastic connection tubing,[41,42] marking ink,[64] micropipette tips,[65] *etc.* Such items may be rather easily overlooked, and yet can be important sources of contamination. In general, a sound approach is to regard every item as a potential source of contamination until evidence to the contrary has been obtained. Mart[39,59] provides a good example of the rigorous precautions taken for a particular purpose – the determination of trace metals by anodic stripping voltammetry.

Careful cleaning of glassware and other apparatus in which samples are handled in often essential, and suitable procedures are then generally described in analytical methods. No attempt is made, therefore, to deal with such procedures here, but three general points are worth noting.

(1) A particular piece of apparatus may cause greater contamination than apparently identical pieces. A specified cleaning procedure is, therefore, not necessarily a guarantee that all apparatus will be adequately clean. An approach to this problem has been described in Section 10.8.1(*b*)(*ii*).
(2) To ensure adequate cleanliness of apparatus, it is often preferable to reserve it solely for a particular determination.
(3) When not in use, apparatus should be stored in such a way as to minimise its contamination.

10.8.6 Reagents

Reagents and their solutions may also be important sources of contamination, but the blank determination corrects, in principle, for the contaminant content of the reagents no matter what its value – for one exception to this, see below. Thus, purification of reagents is, in general, necessary only when the degree of contamination is such that either the concentration range of an analytical system is too greatly reduced or the precision of analytical results is unacceptably worsened [see Section 10.8.1(*a*)]. The exception noted above arises when the contaminant is heterogeneously distributed throughout a reagent in the form in which the latter is added to samples. The size of contamination may then vary markedly from one determination to another, and reagent purification may well be essential. Such heterogeneous distribution is most likely to occur with solid reagents, but the problem has also been met with reagent solutions.[47] Of course, reagents and their solutions should be stored in containers and locations that do not lead to contamination.

Purification of reagents in order to minimise contamination has been discussed by many authors; see, for example, refs. 1, 28 to 31, 55 and 66.

10.8.7 Analysts

Analysts themselves may be important sources of contamination. Clothing, cosmetics, perspiration, dirt on shoes, *etc.*, may all cause problems. Appropriate protective clothing including overshoes and plastic gloves and other commonsense precautions are commonly recommended; see, for example, refs. 11, 28, 29, 39, 55 to 57 and 59.

10.9 REFERENCES

[1] R. L. Booth, ed., "Handbook for Analytical Quality Control in Water and Wastewater Laboratories", (Report EPA-600/4-79-019), U.S. Environmental Protection Agency, Cincinnati, 1979.

[2] K. Everett and D. Hughes, "A Guide to Laboratory Design", Butterworths, London, 1975.

[3] F. Malz, *et al.*, in M. J. Suess, ed., "Examination of Water for Pollution Control", Volume 1, Pergamon, Oxford, 1982, pp. 316–342.

[4] American Society for Testing and Materials, "1983 Annual Book of ASTM Standards. Volume 11.01 Water (I)", A.S.T.M., Philadelphia, 1983.

[5] D. W. Clark, *Public Wks*, 1979, **110**, 78.

[6] W. J. H. Gray, *J. Inst. Water Engrs Scientists*, 1981, **35**, 483.

[7] E. L. Berg, ed., "Handbook for Sampling and Sample Preservation of Water and Wastewater", (Report PB 83-124503), U.S. National Technical Information Service, Springfield, Virginia, 1982.

[8] Organization for Economic Cooperation and Development, "Good Laboratory Practice in the Testing of Chemicals", O.E.C.D., Paris, 1982.

[8a] Health and Safety Commission, "Principles of Good Laboratory Practice", H.M.S.O., London, 1982.

[9] J. K. Taylor, *Anal. Chem.*, 1981, **53**, 1588A.

[10] H. Diehl and G. F. Smith, "The Iron Reagents: Bathophenanthroline, Bathophenanthroline-Disulfonic Acid, 2,4,6-Tripyridyl-s-Triazine, Phenyl-2-Pyridyl Ketoxime", Second Edition, G. Frederick Smith Chemical Company, Columbus, Ohio, 1965.

[11] P. Tschöpel and G. Tölg, *J. Trace Microprobe Techniques*, 1982, **1**, 1.

[12] K. Cammann, *Fresenius' Z. Anal. Chem.*, 1982, **312**, 515.

[13] W. Kipiniak, *J. Chromatogr. Sci.*, 1981, **19**, 332.

[14] G. H. Morrison, *Anal. Chem.*, 1983, **55**, 1.

[15] S. A. Borman, *Anal. Chem.*, 1983, **55**, 519A.

[16] J. K. Foreman and P. B. Stockwell, "Automatic Chemical Analysis", Ellis Horwood, Chichester, 1975.

[17] W. Horwitz, *Anal. Chem.*, 1978, **50**, 512A.

[18] Research and Development Department, "Methods of Sampling and Analysis. Volume 1: Steam and Water", Central Electricity Research Laboratories, Leatherhead, Surrey, 1968.

[19] D. L. Massart, A. Dijkstra, and L. Kaufman, "Evaluation and Optimization of Laboratory Methods and Analytical Procedures", Elsevier, Amsterdam, 1978.

[20] M. S. Cresser, *Lab. Pract.*, 1977, **26**, 171.

[21] H. P. J. van Dalen and L. de Galan, *Analyst*, 1981, **106**, 695.

[22] R. I. Botto, *Anal. Chem.*, 1982, **54**, 1654.

[23] *J. Autom. Chem.*, 1980, **2**, 22–23.

[24] American Water Works Association, "Guidelines for the Selection of Laboratory Instruments. Manual No. M-15", A.W.W.A., Denver, 1979.

[25] F. L. Mitchell, *J. Autom. Chem.*, 1980, **2**, 23.

[26] J. D. Edmond, *Chem. Br.*, 1983, **19**, 386.

[27] American Public Health Association *et al.*, "Standard Methods for the Examination of Water and Wastewater", Fifteenth Edition, A.P.H.A., Washington, 1981.

[28] M. Zief and J. W. Mitchell, "Contamination Control in Trace Element Analysis", Wiley-Interscience, Chichester, 1976.

[29] A. Mizuike and M. Pinta, *Pure Appl. Chem.*, 1978, **50**, 1519.

[30] J. R. Moody and E. S. Beary, *Talanta*, 1982, **29**, 1003.

[31] J. W. Mitchell, *Talanta*, 1982, **29**, 993.

[32] N. I. Sax, "Dangerous Properties of Industrial Materials", Third Edition, Reinhold, New York, 1968.

[33] G. D. Muir, ed., "Hazards in the Chemical Laboratory", Second Edition, The Chemical Society, London, 1977.

[33a] J. E. Longbottom and J. J. Lichtenberg, ed., "Methods for Organic Chemical Analysis of Municipal and Industrial Wastewater", (Report EPA-600/4-82-057), U.S. Environmental Protection Agency, Cincinnati, 1982.

[34] W. Hoffmeister, *Fresenius' Z. Anal. Chem.*, 1978, **290**, 289.

[35] G. N. Peterson and J. R. Montgomery, *Anal. Chim. Acta*, 1980, **113**, 395.

[36] S. J. Poirier and P. M. Sienkiewicz, *Am. Lab.*, 1980, December, 69.

[37] C. G. B. Frischkorn and H. Schlimper, *Fresenius' Z. Anal. Chem.*, 1982, **312**, 541.

[38] M. Oehme and W. Lund, *Talanta*, 1980, **27**, 223.

[39] L. Mart, *Talanta*, 1982, **29**, 1035.

[40] H. J. Petrick, F. W. Schulze, and H. K. Cammenga, *Mikrochim. Acta*, 1981, **II**, 277.

[41] G. A. Junk, H. J. Svec, R. D. Vick, and M. J. Avery, *Environ. Sci. Technol.*, 1974, **8**, 1100.

[42] S. Mori, *Anal. Chim. Acta*, 1979, **108**, 325.

[43] E. Hamilton and M. J. Minski, *Environ. Lett.*, 1972, **3**, 53.

[44] D. T. E. Hunt and P. Morries, *Anal. Proc.*, 1982, **19**, 407.

[45] Standing Committee of Analysts, "Acid Soluble Aluminium in Raw and Potable Waters by Spectrophotometry. 1979", H.M.S.O., London, 1980.

[46] A. L. Wilson, *Analyst*, 1962, **87**, 884.

[47] J. A. Tetlow and A. L. Wilson, *Analyst*, 1964, **89**, 442.

[48] A. L. Wilson, *Analyst*, 1968, **93**, 83.

[49] W. K. Dougan and A. L. Wilson, *Water Treat. Exam.*, 1973, **22**, 100.

[50] H. M. Webber and A. L. Wilson, *Analyst*, 1964, **89**, 632.

[51] J. A. Tetlow and A. L. Wilson, *Analyst*, 1964, **89**, 453.

[52] H. M. Webber and A. L. Wilson, *Analyst*, 1969, **94**, 209.

[53] P. D. La Fleur, ed., "Accuracy in Trace Analysis: Sampling, Sample Handling, Analysis", Volumes 1 and 2, (N.B.S. Spec. Publ. No. 422), U.S. National Bureau of Standards, Washington, 1976.

[54] *Talanta*, 1982, **29** (No. 11B), 963–1055.

[55] P. Tschöpel, *et al.*, *Fresenius' Z. Anal. Chem.*, 1980, **302**, 1.

[56] L. Kosta, *Talanta*, 1982, **29**, 985.

[57] J. R. Moody, *Anal. Chem.*, 1982, **54**, 1358A.

[58] O. L. Davies and P. L. Goldsmith, ed., "Statistical Methods in Research and Production", Fourth Revised Edition, Oliver and Boyd, Edinburgh, 1980.

[59] L. Mart, *Fresenius' Z. Anal. Chem.*, 1979, **296**, 350.

[60] E. S. Boatman, *J. Am. Water Wks Assoc.*, 1982, **74**, 533.

[61] D. Gardner, *Lab. Pract.*, 1979, **28**, 1071.

[62] P. M. Baker and B. R. Farrant, *Analyst*, 1968, **93**, 732.

[63] D. W. Denney, F. W. Karasek, and W. D. Bowers, *J. Chromatogr.*, 1978, **151**, 75.

[64] E. J. Calabrese, R. W. Tuthill, T. L. Sieger, and J. M. Klar, *Bull. Environ. Contam. Toxicol.*, 1979, **23**, 107.

[65] S. Salmala and E. Vuori, *Talanta*, 1979, **26**, 175.

[66] R. P. Maas and S. A. Dressing, *Anal. Chem.*, 1983, **55**, 808.

[67] Analytical Methods Committee, *Anal. Proc.*, 1984, **21**, 45.

[68] IUPAC Commission on Automation, *IUPAC Inf. Bull.*, 1978, No. 3, 233.

[69] J. B. Reust and V. R. Meyer, *Analyst*, 1982, **107**, 673.

11 Analytical Techniques

11.1 INTRODUCTION

Virtually all analytical techniques* may be – indeed have been – used for the analysis of water, though many have not been widely used in routine work. It is beyond the scope of a monograph such as this to give detailed descriptions of all techniques and the approach, therefore, has been briefly to describe the principles, broad characteristics and applications of those techniques considered by the authors to be of interest. In this way, it is hoped to present a reasonably accurate and realistic picture which will be of help in assessing the value of different techniques.

Sample pre-treatment procedures (*e.g.* filtration, concentration) are of common importance in water analysis, and are considered in Section 11.2. Generally applied measurement techniques are discussed in Section 11.3 and other measurement techniques are more briefly mentioned in Section 11.4. (It should be noted that, for convenience, chromatographic techniques are discussed in Section 11.3 despite being, strictly, sample pre-treatment, rather than measurement procedures.) Dividing the measurement techniques in this manner involves rather subjective judgements on the part of the authors, and some laboratories and individual analysts would probably accord a different emphasis to the various techniques. For example, a research laboratory concerned with the identification of organic compounds in water will find mass spectrometry indispensible, but the technique is considered in Section 11.4 because it is normally found in such laboratories, or in large, central laboratories, Conversely, anodic stripping voltammetry has been considered in Section 11.3, although it is not widely applied in fresh water analysis, because it is often employed by laboratories concerned with seawater analysis.

A section on speciation analysis – the determination of particular forms, or groups of forms, of an element or substance – has been included in this second edition, because of the increased recognition of its importance and the greater attention paid to it, since the first edition.

Automatic and semi-automatic analytical procedures are finding ever-increasing use in water analysis, and such procedures are discussed in Section 12.

The discussions of the different techniques are necessarily brief and thus tend to be rather superficial. Readers requiring further details can consult the references cited in each section, many of which refer to books or articles dealing only with a particular technique. In addition to these publications, any of the comprehensive texts on analytical chemistry cover most if not all of the techniques of inter-

---

* The term 'analytical technique' is used here to signify a class of procedures all of which employ the same principle of sample treatment or measurement. Thus, sample concentration and filtration exemplify sample treatment techniques; titrimetry, atomic absorption spectrometry and mass spectrometry provide examples of measurement techniques.

est – those emphasising instrumental procedures are particularly useful, given the predominance of such techniques in modern water analysis. Refs. 1–5 are examples of such texts.

Books covering the whole field of water analysis are not common. Those edited by Mancy[6] and Ciaccio,[7] cited in the first edition, are now somewhat dated but remain valuable – the latter is particularly comprehensive. The manual on water analysis edited by Suess, published recently under the auspices of the World Health Organisation Regional Office for Europe, contains a review of analytical techniques[8] which gives valuable general information – particularly on those techniques most commonly applied in routine work. The book edited by Mark and Mattson,[9] 'Water Quality Measurement. The Modern Analytical Techniques', also contains much useful information, but its coverage is not entirely representative of routine analysis. The determination of organic compounds, as befits its growing importance, is the subject of a number of books, containing the proceedings of major conferences – those edited by Keith[10] and by Bjorseth and Angeletti[11,12] are particularly noteworthy. (The last of these was published just as this section on analytical techniques was being completed). Procedures and equipment for the field analysis of water are dealt with in a recent book[12a] by Hutton, and seawater analysis is the subject of a book edited by the late Klaus Grasshoff and his associates[13] and of a comprehensive chapter by Riley in the book 'Chemical Oceanography'.[14]

The reviews which appear in *Analytical Chemistry* are also of great value. Methods for analysing water are reviewed biennially in odd-numbered years, and these reviews give a good impression of the extent to which different techniques are being applied to water. The reviews are very comprehensive, as also are those published annually in the *Journal of the Water Pollution Control Federation*, and no attempt has, therefore, been made here to provide an exhaustive coverage of the literature. Rather, references are cited primarily to illustrate particular points in the text. For analysts faced with the need to undertake a new determination, perhaps not covered by collections of standard or recommended methods, these reviews represent a valuable starting point (together with *Chemical Abstracts* and *Analytical Abstracts*) for a search of the literature.

In even-numbered years, *Analytical Chemistry* provides another type of review – namely, of developments in particular analytical techniques rather than of their applications. Again, these represent a valuable route for becoming familiar with the recent developments in a technique of interest.

11.2 SAMPLE PRE-TREATMENT TECHNIQUES

Before the measurement stage of the analytical method is reached, it is often necessary to pre-treat the water sample in some way. This may be done for one or more of the following purposes:

(1) to *concentrate* the sample (*e.g.*, to obtain an adequately small limit of detection).

(2) to *convert* otherwise underterminable forms of the element or substance of interest into one or more forms which can subsequently be measured (or to *convert* potential interferents into innocuous forms).

(3) to *separate* the determinand from other constituents of the sample which would otherwise interfere with the measurement (the 'other constituents' may, of course, be forms of the same element or substance as the determinand, which it is not desired to include in the determination).

Sample preservation techniques are not considered explicitly in this context and are dealt with separately (see Section 11.3.6). However, it should be noted that a preservation procedure *may* also effect a desired pre-treatment (*e.g.*, the addition of acid to samples taken for trace metal determination).

The separation of one or more forms of an element or substance from other forms which it is not desired to determine (or which it is desired to determine separately) forms part of the general topic of 'Speciation' which is dealt with in detail in Section 11.5. However, the separation of dissolved and undissolved materials is widely undertaken as the sole speciation step in a method, and is therefore considered in this section, together with separation, concentration and conversion techniques used for purposes other than speciation.

11.2.1 Separation of Dissolved and Undissolved Materials

As noted above, the need to distinguish between the "dissolved" and "undissolved" forms of elements or substances arises quite frequently in water analysis. Although such differentiation can sometimes be achieved by choice of a suitable analytical technique*, it is usually necessary to separate the two forms physically before either or both is measured. A number of separation techniques is available, including dialysis and centrifugation but filtration is usually the technique of choice†. However, care is needed in the selection of filter, filtration apparatus and procedure if valid results are to be obtained. Moreover, the separation achieved will depend upon the type of filter used (as well as its nominal pore size), and, to emphasise the operational nature of the separation it is desirable to use the terms "filterable" and "non-filterable" in preference to "dissolved" and "undissolved", respectively. It is also, of course, essential to provide appropriate details of the separation performed (especially filter type and pore size) whenever results obtained after filtration are reported.

Filtration in water analysis is the subject of reviews by Riley,[14] by Hunt[15] and by de Mora and Harrison,[16] the latter two reviews dealing specifically with filtration in relation to trace metal determinations.

The criteria which should be satisfied by a filter for water analysis have been listed by Riley.[14] No single filter satisfies the requirements exactly, for all applications, but cellulose-based membrane filters (particularly of 0.45 μm nominal pore size) have been most extensively used[14] and are specified in the standard procedures of many organisations (*e.g.*, refs. 17–19). Other types of filter, such as glass-fibre filters (*e.g.*, Whatman GF/C and GF/F) and etched-track polycarbonate filters (*e.g.*, Nucle-

---

* For example, ion-selective electrodes do not respond to undissolved materials. However, even when such measurement techniques are available, their use will often require the application of other pre-treatment techniques which may cause undissolved forms of the element or substance of interest, if present, to enter solution (by desorption or dissolution) and be measured as dissolved. In such circumstances, only a prior physical separation of dissolved and undissolved forms will suffice.

† The use of some of the other techniques is considered in Section 11.5, which deals with speciation.

pore) may, however, be preferable in certain applications – see below. It should also be noted that cellulose and polycarbonate filters, in particular, are available in a wide range of pore sizes and the most commonly used sizes (0.4 to 0.5 μm) may not always be optimal, depending upon the nature of the particular problem under study.

Both cellulose-based membrane filters and glass-fibre filters have a complex, convoluted system of channels (of somewhat variable size) in the body of the filter; particles are trapped in these channels in addition to being retained on the surface. In consequence, the median pore size* may be much smaller than the nominal pore size, and the retention of small particles increases as the filter becomes more loaded.[20] By contrast, etched-track polycarbonate filters consist of a system of unbranched channels of very uniform diameter, produced by etching tracks left by nuclear irradiation of the filter material. They therefore offer a closer correspondence of median and nominal pore sizes, and the median pore size is relatively unaffected by filter loading, remaining essentially constant until blockage occurs; these characteristics permit the use of polycarbonate filters for investigating particle size distributions, for which cellulose filters are unsuitable.[20] Further advantages of polycarbonate filters are their small surface area (at the molecular level) which limits adsorption of dissolved metal ions during filtration (see below) and their low levels of trace metal contamination.[21] However, they are unable to sustain such high loadings of particulate matter as can cellulose membranes, and this – as well as their greater cost – may limit their application in water analysis.

Glass-fibre filters are sometimes recommended[22] because they block less readily than, for example, cellulose-based membranes, whilst giving similar filtration efficiency in terms of size retention. They may also be of value when organic materials are of interest.[14] They have also been used for mercury determination because they can be cleaned of that element by heating.[14]

Metallic filter membranes (*e.g.*, of silver) have been used in water analysis,[14,20] but have not found common application; they are reported to have a low adsorption capacity for humic acid and salts.[14] Paper filters meet few of the criteria listed by Riley[14] and are now rarely used in water analysis.

When filtration is to be performed, several potential problems must be considered:

(*i*) Samples should usually be filtered during or immediately after their collection to prevent – or at least minimise – changes in the concentrations of the filterable and non-filterable fractions of the substance(s) of interest. Simple techniques for filtering in the field have been described (see, for example, ref. 23).

(*ii*) Rupture of algal cells may cause erroneously high determinand concentrations in the filterable fraction, particularly in the case of trace metals which are often concentrated by such organisms, or erroneously low concentrations, *e.g.* of plant pigments,[14] in the non-filterable fraction. Additionally, rupture may introduce natural chelating agents into the sample, thereby altering the speciation of the dissolved metal.[24] Excessive pressure or vacuum should, therefore, be avoided.

(*iii*) Contamination may arise from both the filters and the filtration apparatus. For example, membrane filters have been reported to cause contamination by

---

* That size at which 50% by number of the particles are retained – see ref. 20.

nitrogen, phosphorus and organic compounds,[19] zinc[25] and copper.[26] The trace metal content of Millipore HA filters has been investigated by Spencer and Manheim,[27] who concluded that the determination of a number of metals in the non-filterable fraction of seawater might be subject to important contamination from that source. In a later paper[28] the results were used to calculate the maximum contamination of the filterable fraction which would arise if *all* the metal present in a 4.7 cm diameter Millipore HA filter were leached during filtration of a 10-litre seawater sample. Only for iron, chromium and lead did the potential contamination exceed 5% of the typical seawater concentration. For a volume of 1 litre, the potential copper contamination would also have exceeded this value. Extension of these calculations to 1 litre river water samples has indicated[15] that only for chromium, nickel and copper would the potential contamination exceed 5% of tabulated average river water concentrations.

It should be emphasised, however, that calculations of this kind can only provide estimates of likely *maximum* contamination. The results are encouraging, but also illustrate the need for vigilance.

The analysis of filter membranes has been extended by Watling and Watling to cover a number of different makes of membrane.[29] It was concluded that none of the types tested was markedly superior to the others, though "plain" filters were generally better than their "gridded" counterparts. The degree of contamination often varied considerably between individual filters of the same type.

Procedures for the control of contamination in particular cases have been described (*e.g.* refs. 19, 24, 25, 26, 28, 30, 31), but it is always necessary to check the magnitude of contamination effects for each application.[28] Moreover, it is possible that particular procedures for overcoming contamination may increase the risk of serious adsorption effects occurring (see below). For example, it has been suggested[24] that acid leaching to reduce trace metal contamination may activate surface sites upon which metal ions can be adsorbed*.

(*iv*) Losses of filterable material can occur by adsorption onto the filter and/or the filtration apparatus. Such effects may, in some instances, be more problematic than contamination because they are more difficult to measure and, if serious, to control. Trace metals are particularly prone to important adsorption effects, but other constituents – especially those present at low concentration – are also likely to be at risk in this respect.

Until recently, the problem of adsorption by filters and filtration equipment had received scant attention, and even now our understanding of it is by no means satisfactory. The available information has been reviewed[15] in 1979 and again[16] in 1983, and the following is a brief summary of the position.

The first comprehensive investigation of trace metal adsorption during filtration appears to have been that of Marvin *et al.*,[26] who studied copper in distilled water and sea-water matrices, using a variety of filters. Overall, membrane filters were found to be the least objectionable, but the design and execution of the experimental work are open to a number of criticisms[15] (*e.g.* unrealistically high spiked copper concentrations, the possibility of copper precipitation leading to apparent losses by

---

* The suggestion has been made with respect to sample containers for trace metal speciation work, in which unacidified water samples are to be kept. However, there is clearly a possibility that a similar problem could arise in respect of filters and filtration apparatus.

adsorption, lack of details of filtration procedure) and the results must be considered with caution. Large adsorption losses (up to about 90%) were encountered with some adsorption media, however.

Loss of cadmium from solution during filtration was observed by Gardiner,[32] using 0.45 μm membrane filters. Detailed data were not reported, but about 7% of the metal was lost from solution when 25 ml samples containing 2–50 μg Cd l^{-1} were filtered. Copper and mercury were also reported to suffer losses, but no figures were quoted.

Posselt[33] also reported summary data for cadmium adsorption, which was most pronounced in systems of low concentration (< 11 μg Cd l^{-1}) at pH values above 7. Adsorption was observed on a variety of plastics, glass and stainless steel but not, apparently, on PTFE (from which filtration apparatus was therefore constructed). Glass-fibre and paper filters adsorbed more cadmium than did membranes (cellulose-based), and the loss was finally reduced to negligible levels by using the larger pore sizes (0.45 μm) with faster rates of filtration, and by discarding substantial volumes of filtrate.

Batley and Gardner[24] have cited unpublished work by Gardner relating to mercury adsorption. Between 10 and 30% of the dissolved mercury was lost from seawater when filtered using untreated membranes, but silicone-treated glass-fibre filters were, apparently, successful in reducing losses to less than 7%.

Extensive studies of lead adsorption (and brief investigations of other trace metals) have been undertaken at the author's laboratory using synthetic solutions and spiked samples, and the results were reported in refs. 15, 30 and 31. The main conclusions of the detailed studies[30,31] were as follows:

(1) Trace metal adsorption during filtration of synthetic solutions may be very large; losses of up to 90% of the dissolved lead were encountered on borosilicate filtration apparatus, for example. Of the divalent trace metals examined, lead and copper were most prone to adsorption (in comparison with zinc and cadmium) but no such metal should be assumed free from the effect*.

(2) None of the materials tested (plastics, glass, stainless steel) showed markedly lower adsorption of lead than the others. The principal factor, it was suggested, which should be considered when selecting apparatus, is the exposed surface areas. (The large area of the sintered-glass filter support screen of the borosilicate apparatus was considered likely to be responsible for the large lead adsorption it produced, since glass itself did not give markedly higher losses compared with other materials, when equal areas were tested).

(3) Lead adsorption tended to be greater from spiked soft drinking water than from spiked hard water, and the presence of calcium in synthetic solutions was found to reduce adsorption. These effects were attributed to competition of calcium for adsorption sites. Factors other than hardness were, however, involved in determining the extent of adsorption from spiked samples. These were not specifically identified, but pH, for example, was likely to have been important. Percentage losses of lead during filtration of spiked potable waters

---

* This order of propensity to adsorption is consistent with the general order of adsorption affinity of trace metals with respect to oxide and mineral surfaces.[34] It reflects (in part at least) the tendency of the metals to hydrolyse – the more readily hydrolysable metals absorb more strongly.

ranged from about 5 to 18% (cellulose membranes, polycarbonate apparatus) and from about 15 to 40% (cellulose membranes, borosilicate apparatus).[31]

(4) The dependence of the extent of adsorption upon lead concentration was not investigated in detail. However, in one test,[31] the percentage loss of lead was similar from synthetic solutions containing about 20 and 70 μg Pb l^{-1}, consistent with a large adsorption capacity on the filtration apparatus.

(5) Etched-track polycarbonate filters gave lower adsorption losses than did cellulose membranes, but the filtration of successive portions of sample (without intermediate cleaning) was effective in conditioning membrane filters and apparatus to give lower adsorption losses.

(6) This process of conditioning, by discarding several portions of filtrate before retaining one for determination of filterable lead, was found to be capable of reducing adsorption losses to levels which might be considered acceptable in many applications (around 10%). Such conditioning does not, of course, aid determination of the non-filterable fraction directly. Conditioning with an unspiked hard water sample was not very effective in reducing lead loss from a synthetic solution, and was not recommended. (Conditioning using concentrated solutions of Ca/Mg has, however, been employed by other workers – see, for example, ref. 35).

Adsorption of copper and lead filtration has also been studied by Truitt and Weber,[36] who again observed smaller losses using polycarbonate filters and apparatus than with glass apparatus and cellulose filters.

By contrast with the above studies, Florence[37] has reported that glass filtration apparatus can be used for filtration of natural waters for cadmium, copper, zinc and lead determination, without adsorption losses. He has noted that several other studies (refs. 35 and 38) have also found no evidence of serious adsorption using real samples. It is possible, therefore, that the adsorption losses described above are a consequence of using spiked samples, in which the metal forms may have been different from those in unspiked samples, and less prone to adsorption. However, it seems unlikely that the dissolved metal speciation in spiked samples is wholly atypical of that in *all* waters, particularly those containing relatively high levels of trace metals (*e.g.*, drinking waters) and low levels of complexing organic matter. Moreover, the use of pre-filtered, unspiked samples in tests of adsorption is not without possible ambiguity – readily adsorbed metal species may be lost during the pre-filtration (necessary to remove any non-filtrable metal forms), and thus cause erroneously low estimates of loss to be obtained from the second, test filtration.[31]

Despite the further studies of adsorption during filtration which have been undertaken since the first edition of this book, there remains a need – identified in that edition – for more study of this source of error, which may occur with organic, as well as inorganic, determinands.[14,16] There are numerous pitfalls in the design and execution of tests to measure the extent of adsorption likely to occur with real samples, and great care is needed if a valid assessment is to be made. The problems, and procedures for dealing with them, have been considered in detail in refs. 15, 16, 30 and 31.

11.2.2 Concentration and Separation Techniques

Many determinands are present in such small concentrations that direct analysis of samples is either impossible or beset with many difficulties. When this is so, some means of increasing the concentration must be employed. Many techniques are available to effect such concentration, and some examples have already been given in Section 3.3.2. Elimination or reduction of the effects of interfering substances in the sample can also be accomplished when selective concentration techniques or separation techniques (not effecting concentration of the determinand) are used. However, the use of concentration and separation techniques usually necessitates greater analytical complexity, time and effort, and should be avoided whenever possible. The development and application of more sensitive and selective measurement techniques (*e.g.*, graphite furnace atomic absorption spectrometry, plasma emission spectrometry) has decreased the need for concentrating and/or separating certain determinands in particular applications (*e.g.*, trace metals in drinking water). However, concentration and separation techniques are, and are likely to remain, an indispensable part of the analyst's tools – particularly when organic compounds are of interest, when saline waters are to be analysed and, of course, when speciation is under investigation.

Many different techniques are in common use, for example, evaporation, freeze-concentration, solvent extraction, co-precipitation, ion-exchange, adsorption. Experimental details of the procedures employed are given in individual analytical methods and are not discussed here. For those wishing to read more about the theoretical bases and practical applications of common concentration techniques, the review by Andelman and Caruso[39] (though somewhat dated) is still of value, and the chapter[14] on analysis by Riley in the second edition of "Chemical Oceanography" discusses the topic in the context of seawater analysis. The comprehensive review of concentration techniques applicable to those trace elements included in quality standards for raw and potable waters, by Orpwood,[40] deals with the literature between 1970 and 1976, and includes sections on solvent extraction, freeze concentration, carbon adsorption, evaporation, precipitation, ion exchange and electrochemical concentration. Orpwood[40] concluded that no one technique was optimal, and a number of the techniques may be useful in particular circumstances – although chelating ion-exchange resins have many features appropriate to water analysis.

Since the first edition of this book was written, there has been a considerable growth of interest in the determination of organic substances present at very low concentrations (μg–ng l^{-1}) in both natural and treated waters. Such analysis involves very considerable sample pre-treatment, both for concentration and separation. Much work of this type remains in the realm of research, rather than of routine water quality monitoring, but it is by no means exclusively undertaken in support of research, and it is likely that a number of determinations currently undertaken only for that purpose will in time be undertaken on a routine basis, at least in the larger water analysis laboratories.

Among the most important concentration techniques employed in trace organic analysis are solvent extraction, resin adsorption (using non-ionic macroreticular resins), carbon adsorption, freeze-drying and the gas purging and direct headspace techniques which exploit the volatility of many of the organic compounds of interest in water. (The latter techniques have found particular application with gas chromatography, a powerful separation technique for volatile organic compounds, and are discussed further in Section 11.3.8 on gas chromatography.) Concentration techniques for organics are discussed in refs. 41, 42, and 10, and the last of these also contains many papers describing the application of the techniques in a complex scheme for the determination of trace organic compounds in waters (the 'Master Analytical Scheme' of the U.S. Environmental Protection Agency).

Although some degree of separation is achieved during the application of these concentration techniques, further purification and separation is usually required. The separation techniques most commonly employed are gas chromatography (GC), high performance liquid chromatography (HPLC) and thin-layer chromatography (TLC). Although strictly these are separation techniques rather than measurement techniques, the field of chromatography is conveniently considered under 'measurement techniques' – see Section 11.3.8. Additionally, the interfacing of chromatographic techniques with mass spectrometry is best considered under that particular measurement technique – see Section 11.4.1.

In recent years, the non-routine identification/determination of trace organic substances in waters has depended heavily upon the use of gas chromatography and combined gas chromatography and mass spectrometry (GC-MS). The data have, therefore, been heavily biased towards relatively volatile compounds. Increasing emphasis is now being placed upon the development of procedures for the characterisation of the 80–90% of the organic compounds in raw and drinking water (particular) which do not fall into this category.[43] High performance liquid chromatography is becoming widely used for this purpose.

When concentration or separation techniques are to be employed, certain general precautions should be observed:

(1) Many techniques are not completely quantitative and, while this may not present a serious source of error (provided, as is recommended, calibration standards are taken through the concentration/separation stage) it is always desirable to check the *absolute* recovery attained. Radioactive tracers, if available, can be particularly useful for such checks.

(2) The chemical and physical forms of the determinands may affect the efficiency of the concentration. For example, certain forms of trace metals (most notably colloidal and particulate forms) may not be efficiently retained by (or eluted from) an ion-exchange resin; this is considered further, with relevant references, in Section 11.5, which deals with speciation analysis.

(3) The concentration process may change the chemical and physical forms of determinands. This rarely presents a problem to the subsequent measurement, but can, in the limit, result in the loss of determinand. For example, evaporation can result in the loss of volatile materials, and can cause the destruction of heat-sensitive organic compounds.

(4) Non-selective techniques result in the simultaneous concentration of all components of the sample, including interfering substances, and this may markedly affect analytical accuracy. This situation often arises when trace organic compounds are being determined, and necessitates subsequent

separation ("clean-up") stages to remove interfering substances from the concentrate – see above. Another example is the use of evaporation to concentrate samples prior to atomic absorption spectrometry, which may be limited by the enhanced background (non-atomic) absorption of the concentrated matrix.

(5) The systems used for concentration may cause, or allow increased opportunity for, contamination of the sample or concentrate, either by the determinand or by interfering substances. For example, trace organic compounds from the degradation of ion-exchange resins may cause problems when such resins are used for concentration. It is usually advisable to purify such components of concentration systems, and to make a blank determination by following exactly the same procedure as that used for samples (see Sections 5.4.4, 10.7 and 10.8). Greater difficulty may be encountered as a result of the increased opportunity for contamination from the laboratory environment, resulting from the additional manipulation and exposure necessary to perform the concentration or separation. Such contamination is known to present particular difficulties in trace metal analysis. Although the development of more sensitive techniques, such as graphite furnace atomic absorption spectrometry may have reduced the need for concentration techniques in certain applications, they are still necessary in many others. For example, there has been an increase in the need for routine determination of trace metals at very low concentrations ($\mu g\ l^{-1}$ and sub-$\mu g\ l^{-1}$), and it is in such applications that concentration techniques often remain necessary – and contamination most likely to present problems. In such circumstances, special laboratory facilities (clean rooms or laminar-flow hoods fed with filtered air) may be needed to allow manipulation of the sample, for concentration or separation, to be performed without unacceptable contamination.

11.2.3 Dissolution and Conversion Techniques

Conversion techniques (including dissolution) are usually employed to convert different chemical and/or physical forms of the determinand to one form that can subsequently be measured, although conversion may also be employed to render innocuous substances which would otherwise interfere with the measurement. The latter is often known as 'masking', and an example is the use of chelating agents to suppress the interference of trace metals in the determinations of arsenic and selenium [see Section 11.3.5(d) and references therein].

Conversion techniques are often applied in order to release the determinand from particulate matter, as in the case of trace metals, or to release the determinand from organic compounds or from some other association with organic matter (*e.g.* inorganic colloidal material "sheathed" in adsorbed organic substances). Acid digestion (with or without heating) has been widely used to release trace metals from particulate matter, and is employed in national recommended methods (*e.g.*, refs. 44–46). Depending upon the nature of the sample, the action of acid added to prevent trace metal hydrolysis and loss by adsorption on container walls may be sufficient to effect release from particulate matter, or more vigorous digestion (involving heating, and often evaporation) may be necessary. Many methods (*e.g.*, refs. 44–46) recommend that representative samples can be analysed with and without the pre-treatment, and the results compared to assess whether or not it need be applied routinely. This approach is endorsed.

The use of ultraviolet (u.v.) irradiation as a technique for oxidative destruction of organic matter, noted in the first edition as a recent development having found relatively little application, has become well established, particularly for the determination of organic carbon (conversion to carbon dioxide with subsequent measurement) and for the conversion of organic metal or metalloid species into forms to which the measurement technique can respond. The use of u.v. irradiation for the determination of total organic carbon in waters has recently been reviewed[47] and instruments employing this approach are available commercially (see refs. 48 and 49 for details). In trace metal/metalloid analysis, the technique has found application in many speciation schemes (see Section 11.5.6) and it forms part of a national recommended method for the determination of total arsenic.[50]

The technique has the following advantages over conventional chemical pre-treatments:

(1) the need to add large amounts of reagents is avoided (and the risk of contamination thereby reduced),
(2) little attention is needed from the analyst, and
(3) the conversion efficiency for some compounds may be better.

When oxidative pre-treatment of relatively small sample volumes is required, u.v. irradiation may often be a very convenient and effective technique. However, when large volumes of sample are required for subsequent concentration, the use of u.v. irradiation may be incovenient because an adequate throughput of samples requires much expensive equipment. Interest has therefore arisen[51,52] in the use of ozonolysis as an alternative. Ozone may be generated from air or oxygen using relatively simple equipment, and could be used to treat a batch of samples simultaneously using a suitable manifold. The technique has not been tested as thoroughly as u.v. irradiation (with respect to conversion efficiency of a wide range of organic compounds), but the limited evaluation which has been made[52] suggests that it may indeed be a viable alternative.

Whenever conversion techniques are necessary, some general precautions should be taken to control errors which may be caused by the treatment:

(1) It is important to ensure that the particular method employed gives adequate conversion efficiency for the particular application; failure to check this point can lead to large errors (see Section 5.4.5).
(2) Contamination is often a problem, and the reagents, apparatus and procedure should be carefully selected to minimise this effect (see Section 10.8). The increased opportunity for contamination presented by more vigorous conversion procedures may result in a deterioration of precision (see, for example, ref. 53).
(3) A particular technique, chosen to effect a desired conversion, may cause loss of other forms of the determinand. For example, an evaporative acid digestion for the determination of arsenic in potable waters may cause loss of volatile arsenic compounds.[50] Although this problem may be encountered only rarely,

it is likely, when it does arise, to be difficult, if not impossible, to solve without impractical elaboration of the pre-treatment.

11.3 GENERALLY IMPORTANT MEASUREMENT TECHNIQUES

11.3.1 Titrimetry

The principles of titrimetric analysis are so well known that their discussion here is unnecessary. Books dealing with the technique in general are given in refs. 54–60, and texts dealing with specific aspects of the technique in refs. 61–65. The present section is included mainly to emphasise that the technique finds common application, and is one of the basic techniques for most laboratories concerned with routine water analysis. Thus titrimetric methods for the following determinands have been included in recent, authoritative collections of standard or recommended methods produced in the United States and the United Kingdom*: acidity,[66,67] alkalinity,[66,67] biochemical oxygen demand,[66,68] boron,[69] bromide,[70] calcium,[66] calcium hardness,[71] carbon dioxide,[66] total organic carbon,[48] chemical oxygen demand,[66,72] chloride,[66,73] chemical disinfecting agents and related determinands (including chlorine, chlorine dioxide, chlorine demand, ozone, bromine and iodine),[66,74] cyanide and cyanides amenable to chlorination,[66] dissolved oxygen,[66,75] (total) hardness,[66,71] magnesium hardness,[71] Kjeldahl-nitrogen,[66] ammonium-nitrogen,[66,76] nitrate-nitrogen,[77] nitrite-nitrogen,[77] organic acids,[66] salinity,[66] sulphate,[78] sulphide[66,79] and sulphite.[66]

The accuracies and limits of detection attainable with titrimetric methods are frequently governed by errors in the detection of end-points. Visual observation of a colour change is very commonly used for end-point detection in water analysis but other, more accurate methods are often useful. For example, potentiometric titration has been employed for the determination of acidity,[66,67] alkalinity,[66,67] boron,[69] carbon dioxide,[66] chemical oxygen demand,[80] chloride,[66,73,81] chlorine, chlorine dioxide and chlorite,[82] extractable organo-chlorine (see below),[83] cyanide,[84] hardness,[85] sulphate,[86,87] sulphite[88] and total sulphur.[89] Amperometric titration has been used for the determination of calcium,[90] chloride,[91] residual chlorine,[66] dissolved oxygen,[66,92] iodine,[66] magnesium[90] and nitrite-nitrogen.[77] Thermometric titration has been applied to the determination of sulphate, chlorinity and alkalinity in seawater[93] and photometric titration to the determination of alkalinity,[94] total carbonate,[94] calcium,[95–97] and sulphate.[98]

The analytical performance of titrimetric methods varies widely, according to the nature and concentration of the determinand, the reactions exploited and the exact procedure followed, especially with regard to end-point detection. Examination of the collections of methods cited above shows that relative standard deviations and systematic errors of 0.5–5% are commonly attained for many determinands (except at concentrations near the limit of detection) and much better performance can often be attained, especially with instrumental procedures. Thus, for example: Graneli and Anfalt[94] have reported relative standard deviations of about 0.1% for the determination of alkalinity and total carbonate in sea-water, with systematic

---

* Note that the U.K. method series has yet to be completed and further titrimetric methods may be produced.

errors of -0.15% (alkalinity) and $+0.4\%$ (total carbonate), using photometric titration; Millero *et al.*[93] reported relative precisions of 0.3%, 0.04% and 0.1% for sulphate, chlorinity and total alkalinity, respectively, in sea-water by thermometric titration; Jagner[96] reported a relative standard deviation of about 0.03% for the determination of calcium in sea-water by photometric titration; Sato and Mamoki[97] obtained relative standard deviations of less than 0.5% for calcium and magnesium by photometric titration; Anfalt and Graneli[95] reported relative precisions of 0.03% and 0.04% for calcium and magnesium, respectively, in sea-water by photometric titration; Montiel and Dupont[81] reported relative precisions of $0.3–0.5\%$ for chloride determinations at concentrations of $15–100$ mg 1^{-1}, using a microcoulometric method (see below) with potentiometric end-point detection. A useful monograph on high-precision titrimetry is available.[99]

(An interesting, if rather specialised, application of titrimetry in water analysis has been the measurement of the chlorinity of standard seawater – see reference[100] and literature cited therein. Strictly a gravimetric/titrimetric procedure, it involves the mixing of weighed portions of sea-water and concentrated silver nitrate solution, such that there is a slight excess of chloride remaining which is titrated against dilute silver nitrate using potentiometric end-point detection. A standard deviation of 3×10^{-4} parts per thousand, for chlorinities of about 19 parts per thousand – *i.e.*, a relative standard deviation of 0.0015% – has been reported for the results of the method.)

In titrimetry, titrants are usually added as conventional standard solutions. However, coulometric generation of titrants,[101] which avoids the need to prepare and standardise solutions, is often capable of better accuracy, and facilitates automation. Coulometric titrations have been used for the determination of ammonia/total nitrogen,[102,103] biochemical oxygen demand,[104] chloride,[81,92] dissolved oxygen,[105] mercury,[106] organic carbon,[107] phenols,[108] various sulphur determinands (including sulphate)[109–112] and, of particular current interest in water analysis, various organo-halogen determinands.[113–121] Commercial coulometric titration equipment is available, and has been used for many of these determinations, especially for ammonia/total nitrogen, sulphur compounds and organo-halogens.

The determination of the total concentration of organo-halogens in water has become of considerable interest because of their potential toxicity, both to man and other organisms. The subject has been reviewed recently by Wegman,[122] and is conveniently considered here because coulometric titration is most frequently employed as the final measurement technique. No single method exists which is capable of determining the true total concentration of organo-halogens; various techniques are used to extract and concentrate the compounds of interest. These include solvent extraction (*e.g.*, refs. 113, 116–118, 121), carbon adsorption (*e.g.*, refs. 119 and 120), resin adsorption (*e.g.*, refs. 115 and 123) and, for the volatile compounds, gas purging (*e.g.*, refs. 118–120). The organo-halogen compounds are then converted into inorganic halide, commonly by pyrolysis in a flame or furnace, the halide then being trapped in a suitable liquid and determined. As noted above, coulometric titration with silver ions has been very commonly employed for this determination, but other measurement techniques have also been used, notably potentiometric titration without coulometric generation of the titrant,[83,123] direct potentiometry with an ion-selective electrode [123-125] and ion chromatography.[126]

A variety of names has been given to the various determinands – Wegman[122] gives a comprehensive list. They include total organic chlorine (TOCl), extractable organic chlorine (EOCl), carbon adsorbable organohalides (CAOX), total organo halides (TOX or OX) and the purgeable (POX) and non-purgeable (NPOX) fractions of OX. There now seems to be a movement, which the present authors support, towards the use of more accurate terminology which reflects the operational nature of the concentration process (adsorption, extraction) and the non-specific nature of many of the halide measuring systems (*e.g.*, coulometric titration by silver, responding to halides rather than to chloride alone).

It is beyond the scope of this monograph to consider the determination of organo-halogen 'sum parameters'[122] in detail – the reader is referred to the above references, especially the review by Wegman.[122] It will be noted, however, that methods involving coulometric titration have permitted the detection of very low levels of these compounds in water. Thus, for methods employing solvent extraction and microcoulometry, detection limits in the range $0.3-10 \ \mu g \ Cl \ l^{-1}$ have been reported,[122] for methods employing absorption (carbon or resin) and microcoulometry, detection limits in the range $2-35 \ \mu g \ Cl \ l^{-1}$ have been reported[122] and, for methods employing gas purging and microcoulometry, the corresponding figures[122] are $0.1-1 \ \mu g \ Cl \ l^{-1}$.

Microcolouometric titration systems of the type used for the 'total' organo-halogen determinations described above are also used for detection of eluted compounds in GC. An example of such use is the determination of vinyl chloride by GC/microcoulometric titrator detection, by Dressman and McFarren.[127]

11.3.2 Gravimetry

As with titrimetry, detailed discussion of gravimetric analysis is unnecessary here; the subject is dealt with comprehensively in many text books – see, for example, refs. 128, 129 and 130-132. Gravimetric methods have been very largely replaced by more rapid techniques and few determinands are now measured gravimetrically.

Thus a recent, authoritative compilation of methods for the analysis (almost exclusively) of fresh waters[133] gives gravimetric methods for the determination of only eight determinands (or group of determinands, in the case of 'residues' – see below). Moreover, for many of those eight, alternative, more rapid methods are also given. Similarly, examination of a comprehensive review[134] of methods for sea-water analysis suggests that only for potassium (and, by difference, sodium) are gravimetric procedures not obsolete.*

Use of the technique remains essential for the determination of filterable and non-filterable (*i.e.*, 'suspended') solids or 'residues',[135] and oils, greases and other organic substances are sometimes measured by gravimetry after adsorption or extraction and evaporation of the solvent.[136,137]

Other determinands for which the above-quoted compilation[133] gives gravimetric methods are magnesium,[138] silica,[139] sulphate[140] and floatables.[141]

---

* Leaving aside 'suspended solids'

11.3.3 Ultra-Violet and Visible Solution Spectrophotometry and Colorimetry

(a) General Basis of the Technique
In this technique the absorption of light by one or more components of a solution is used to measure their concentrations. Several different approaches to measurement can be used,[142,143] and the terminology used varies from author to author. The first edition of this book used the term 'Colorimetry' to refer to procedures in which measurements of solution colour intensity are made by eye (for which the absorption must occur in the visible region of the spectrum, *i.e.*, approximately 380–750 nm), and reserved the term 'Spectrophotometry' for procedures involving photoelectric measurement of the light absorption. In a recent review of the applications of the technique in water analysis,[143] however, the term 'Colorimetry' was used in reference to measurements in which " . . . the radiation passing through the absorption cell comprises a relatively broad band of wavelengths isolated with a simple gelatin, coloured glass or an interference filter". The term 'Spectrophotometry' was used to denote measurements using "essentially monochromatic" radiation (from a monochromator or, exceptionally, from a good quality interference filter).

In the following, the policy of the first edition will be continued, but the reader should bear in mind that this particular use of the terminology is not universal.

A schematic representation of a typical photometer is shown in Figure 11.1. In simple filter instruments (often called 'colorimeters' – see above) the wavelength selector is a gelatin, glass or interference filter, whereas in more elaborate instruments a monochromator is employed, invariably now of the diffraction grating type.[143] The detector in filter systems is usually a photocell, but photomultipliers are used in monochromator instruments. The readout on modern spectrophotometers normally consists of a digital display, but many also have an associated pen recorder (which is valuable for plotting absorption spectra).

Both single and double beam instruments have been developed. In the latter, the light beam is split into two, one beam being passed through the sample cuvette while

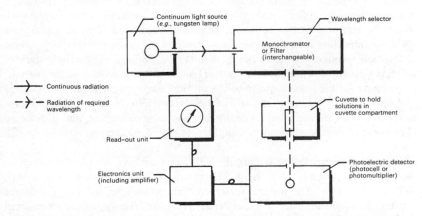

Figure 11.1 *Schematic arrangement of typical instrumentation for spectrophotometry*

the other is passed through a reference cuvette containing the pure solvent. In commercial instruments, a single detector is used and a rotating "chopper" system permits the amplifier to resolve the signals from the two beams, the ratio of which is then independent of source intensity fluctuations.[143]

Thompson[143] has compared the two types of system, and tabulated the advantages and disadvantages of each, noting that most commercial scanning u.v. spectrophotometers employ double beam optics to allow automatic plotting of absorption spectra.

In some cases, the determinand itself absorbs light sufficiently strongly for measurement to be possible without chemical treatment of the sample. Examples in water analysis include the determination of nitrate,[144,145] fulvic acids[146] and, of course, so called "colour".[147] Similarly, organic carbon can be estimated by spectrophotometry[148] and instrumentation for continuous monitoring of natural waters for organic pollution has been devised.[149] Additionally, it has been suggested[143] that certain organic substances, having a characteristic absorption spectrum and capable of being extracted from water into a suitable solvent, could be determined spectrophotometrically after simple extraction and separation. (Strychnine and pentachlorophenol were identified[143] as two such determinands, for which rapid screening procedures might be developed).

However, it is usually necessary in spectrophotometry to add one or more reagents to the sample to convert the determinand into a strongly absorbing substance. Familiar examples include the determination of nitrite by formation of a pink azo dye[144] and the determination of chlorine using DPD (diethyl-*p*-phenylene-diamine).[74] The ability to manipulate chemical reactions to produce a suitably absorbing substance allows the possibility of indirect procedures. The determination of fluoride by means of its ability to destroy coloured metal-dye chelates (through formation of colourless metal/fluoride complexes) is one example.[150,151] Indirect spectrophotometric procedures based on catalysis have also been used in water analysis (see Section 11.4.4).

Because so many chemical reagents and reactions can be exploited in solution spectrophotometry, the technique has a wide applicability – to both inorganic and organic substances. Examination of any compilation of water analysis methods will reveal the vital role of the technique, despite its continued displacement from a formerly important area of application – trace metal determination – by atomic absorption spectrophotometry (see Section 11.3.5). Solution spectrophotometry remains virtually essential to laboratories concerned with general water analysis, particularly for the determination of the nutrient elements (nitrogen, phosphorus and silicon).[144,152,153] Even in the case of metals, well tested solution spectrophotometric methods are available for many determinands of interest, and provide an alternative to atomic absorption using flames (see, for example, refs. 45, 46, 53, 154).

Apart from books dealing specifically with water analysis, useful information on the technique is given in ref. 143 and in the many text books devoted to it (*e.g.*, refs. 142 and 155–160). The latter usually contain procedures for many different substances, albeit not often in great detail. The biennial reviews (even-numbered years) on light absorption spectrometry in *Analytical Chemistry*[161] offer an excellent summary of chromogenic reagents and analytical procedures. Colorimetric procedures are also, of course, used for qualitative analysis, and ref. 162 gives a comprehensive account of spot tests.

(b) Comparison of Colorimetry and Spectrophotometry
A more detailed account of the equipment and experimental procedures for these two approaches is given in ref. 143, and in texts (such as refs. 156 and 157). The remarks made above concerning definitions should be borne in mind, however, when such works are consulted.

The principal advantage of colorimetry is that the equipment required is simpler, cheaper and more robust. It is also convenient to use under field conditions, where spectrophotometry cannot be used. Less analytical skill and experience are usually required of the operator. When permanent comparison standards can be used, colorimetry is more rapid, but the saving in time may often be negligible, particularly when lengthy chemical treatment of samples is necessary.

One of the principal disadvantages of colorimetry is its tendency to produce results of relatively poor accuracy. Thus, the technique is subjective and differences in the visual acuity of observers can lead to problems of bias. Again, colorimetric measurement can be seriously affected if samples contain coloured materials or other substances affecting the hue of the colours of concern – such effects can often be minimised in spectrophotometry by suitable choice of the wavelength at which measurements of intensity are made. Similarly, the use of coloured chromogenic reagents is often precluded in colorimetry, whereas by suitable choice of the wavelength of measurement they provide the bases of very satisfactory spectrophotometric methods. The precision of colorimetric methods is also generally poorer; Thompson (in ref. 143) has noted that the average observer can only just detect a $\pm 3\%$ change of intensity in the yellow-green region (where the human eye is most sensitive), compared with $\pm 0.3\%$ or better for a typical photoelectric detector.

Other disadvantages of colorimetry include the restriction to the visible region of the spectrum (many useful procedures produce materials whose maximum absorption lies in the u.v. region, *i.e.*, at wavelengths below about 400 nm) and the problem of observer fatigue with large sample numbers.

In summary, the main applications of colorimetry are either under field conditions where spectrophotometry cannot be used, or in circumstances where relaxed requirements for analytical accuracy allow advantage to be taken of its simplicity and low cost. A collection of colorimetric methods for many different substances is available.[163] Spectrophotometry is generally a more powerful technique, and is further considered in the next section.

(c) Principal Analytical Characteristics of Spectrophotometry
The characteristics of the different types of commercially available spectrophotometer have been considered in (a) above. In general, the differences hardly affect the principles of measurement but the narrower wavelength band provided by monochromators (or interference filters) compared with simple glass or gelatin filters is advantageous in some applications. These advantages are discussed elsewhere (*e.g.*, refs. 142 and 143), and include improved sensitivity and more linear calibrations, especially at high absorbances. However, for many methods of water analysis both types of instrument are satisfactory and those using filters are cheaper

(except in the case of interference filters, a set of which to cover adequately the 325–1000 nm range can cost more than a typical monochromator[143]). For simplicity, the two types of instrument (simple filter systems having large spectral bandpasses, and interference filter or monchromator systems having much lower spectral bandpasses) are not distinguished in the following discussion. It should be borne in mind, however, that monochromator instruments will usually be more and more preferable as the wavelength of maximum absorption of the solution departs further and further from the wavelengths transmitted by the most suitable filter. In addition, most commercial filter instruments cannot be used below about 325–350 nm, whereas most monochromator-based instruments cover the range 190–900 nm, using a tungsten or tungsten halide lamp for the range 325–900 nm and a hydrogen/deuterium arc lamp for the 190–325 nm range.

In addition to the method of wavelength selection, considerable diffences in the natures of the other components of Figure 11.1 exist in commercial instrumentation, and such differences have important effects on both capabilities and cost. One particular area in which differences may be marked is the use of microprocessors, both for rapid setting of instrumental conditions and for data processing.[143]

These points notwithstanding, an attempt is made in the following to summarise the main analytical features of the technique.

(i) Calibration   The fundamental equation of spectrophotometry combines the Lambert-Bouguer Law, and Beer's Law*, and is:

$$A = \log_{10}(I_O/I_T) = \varepsilon.c.l \tag{11.1}$$

where A = absorbance or optical density = $\log_{10}(100/T)$
T = percentage of light of wavelength λ transmitted by the solution = $100\ I_T/I_O$
I_O = intensity of incident radiation
I_T = intensity of transmitted radiation
ε = molar extinction coefficient of absorbing species at wavelength λ ($1\ mol^{-1}\ cm^{-1}$)
c = concentration of absorbing species ($mol\ l^{-1}$)
l = length of light path through solution (cm)

Calibration curves are prepared by measuring values of A for standard solutions of known concentrations, and plotting the values against concentration. Straight lines are usually obtained, but causes of deviation are discussed elsewhere (see, for example, refs. 142 and 143) and include – as noted previously – large spectral bandpass. In contrast to many other (more recent) instrumental techniques, calibration curves can often be used without change over many weeks (so called 'fixed calibration').

---

* These Laws may be summarised[143] as follows:
Lambert–Bouguer Law: The fraction of radiation absorbed by a given amount of substance under fixed conditions is independent of the intensity of the incident radiation.
Beer's Law: Successive increments in the number of identical absorbing molecules in a given path length of a beam of monochromatic radiation absorb equal fractions (dI/I) of the intensity, I, upon it. The equation above is often called the Beer–Lambert Law, but is more properly known as the Beer–Lambert–Bouguer Relationship.[143]

(ii) Limit of detection The approach to defining the limit of detection, using the within-batch standard deviation of the results of blank determinations, has been described in Section 8.3.1. However, order-of-magnitude calculations for detection limits [using equation (11.1) above] are useful for demonstrating the potential capabilities of the technique. Taking reasonable values for A (0.01) and l (4 cm), the concentration corresponding to the detection limit is given by $2.5 \times 10^{-3}/\varepsilon$. For many determinands, light-absorbing compounds having molar extinction coefficients (ε) between 10^4 and 10^5 litre mol^{-1} cm^{-1} can be produced, so that (provided the sample is not greatly diluted by added reagents) the detection limit is in the range 2.5×10^{-7}–2.5×10^{-8}M. Such limits have been achieved for many determinands.

Smaller limits of detection can often be achieved simply by the simultaneous formation and extraction of the coloured compounds into an organic solvent, the optical density of which is then measured. Such an approach commonly provides a concentration factor between 5 and 20, and is often used in water analysis. It should be emphasised that some procedures which employ solvent extraction in this way can be more rapid and simple than many direct spectrophotometric procedures. (Pre-concentration of the determinand itself, before formation of the coloured compound, may, of course, also be employed to allow lower detection limits to be attained).

Some examples of reported limits of detection (defined as in Section 8.3.1 and therefore comparable) by solution spectrophotometry are given in Table 11.1.

(iii) Precision Relative standard deviations in the range 1–5% are commonly reported, though routine analysis is likely to give performance nearer the latter figure. Special techniques (*e.g.*, ref. 171) allow standard deviations around 0.1% of determinand concentration. Of course, relative standard deviations become much worse as the limit of detection is approached (see Section 8.3.1). The publications cited in Table 11.1 include estimates of precision at several determinand concentrations, usually for both standard solutions and real samples, and often obtained by several laboratories using the particular method. Readers wishing to obtain further information on precision, particularly for determinations commonly undertaken by solution spectrophotometry, are therefore advised to consult the references given in Table 11.1.

(iv) Selectivity The selectivity of spectrophotometric procedures varies considerably. Interference effects can arise in a number of ways: other components of the sample can absorb at the wavelength used, or can react with the added reagent(s) [either to form a species that absorbs, or to cause an interference indirectly by reducing the concentration of reagent(s) available for reaction with the determinand], and turbidity in the sample itself can cause problems.

By suitable selection of chromogenic reagent, pH, temperature, reaction time, *etc.*, many interferences can be eliminated or minimised. Additionally, masking agents may be employed to prevent an interferent reacting with the chromogenic reagent, and it may also be possible to reduce interferences by using a wavelength displaced from the absorption maximum of the species of interest (though care should be taken that this practice does not lead to unacceptable deterioration of precision and sensitivity). If none of these approaches is effective, recourse must be

TABLE 11.1

Examples of Limits of Detection Achieved using Spectrophotometry

Determinand	Procedure*	Limit of Detection (μg l^{-1})	Ref.
Aluminium	D	13	46
Ammonia	D	3–8 (as N)	76
Arsenic	HGT	0.2	50
Boron	D	4	69
Chloride	I	3400	73
Chlorine (Free/Total Available)	D	4	74
Chromium	PCSE	5	154
Copper	SE	1	164
Fluoride	DB	200	169
Fluoride	D	50	169
Hydrazine	D	0.4–1.0	165
Iron	D	3–15	53
Iron	SE	2	166
Manganese	D	5	45
Nickel	D	0.28–1.75	167
Total Oxidised Nitrogen	D	1.3 (as N)	144
Nitrate	D	3–13 (as N)	144
Nitrate	DNC	30 (as N)	144
Nitrite	D	0.26 (as N)	144
Phenols	D	9 (as phenol)	168
Phenols	SE	3 (as phenol)	168
Reactive Phosphorus	D	3–6 (as P)	153
Reactive Phosphorus	SE	0.5–0.7 (as P)	153
Reactive Silica	D	2 (as SiO$_2$)	152
Sulphate	IER(A)	~2000 (as SO$_4$)	78
Sulphate	IER(B)	~4000 (as SO$_4$)	78
Surfactants	SE/BE	20 (as Manoxol OT)	170

* *Key to Procedures:*

D	Direct
DB	Direct, bleaching of colour
DNC	Direct, no chromogenic reagent
HGT	After generation and collection of arsine
I	Indirect
IER	Indirect; spectrophotometric measurement of excess reagent, (A), without, and (B), with, chromogenic reagent
PCSE	After prior concentration by solvent extraction
SE	After solvent extraction of coloured complex
SE/BE	As SE, but including back extraction of coloured complex

NOTES: (1) Where a range of values is given, a number of estimates of the Limit of Detection were obtained by different laboratories.
(2) The Limits of Detection were calculated as in Section 8.3.1 of this book.
(3) The Limits of Detection quoted should not be regarded as necessarily the lowest attainable for that determinand by solution spectrophotometry. They are indicative, however, of the capabilities of the technique in water analysis, particularly in respect of determinations for which it is commonly employed.

made to pre-treatment to separate the determinand from the interfering substance(s), with concomitant increase in the time and cost of analysis.

Correction for the turbidity and/or colour of the original sample is often made by carrying out the procedure in the usual manner, but omitting a reagent (usually the chromogenic reagent itself) essential for the formation of the absorbing species to be measured (see, for example, ref. 53), or by reversing the order of reagent addition, if this prevents production of the absorbing species (see, for example, ref. 45).

Because it relies upon chemical reactions in solution, the technique has some selectivity with respect to different forms of the same element. This has been exploited in trace metal speciation, for example (see Section 11.5.) and in the selective determination of different nutrient species (*e.g.*, nitrate/nitrite, see ref. 144). However, it also means that some form of pre-treatment is required if, for example, total metal concentrations are of interest. Acidification, for sample preservation, may be adequate in some instances, but in other cases a more vigorous digestion may be necessary (see, for example, refs. 45, 46, 53, 154, and Section 11.2.3).

For practical purposes, therefore, adequate selectivity can be attained for many inorganic determinands. However, chromogenic reagents for organic substances tend to react with groups of compounds (*e.g.*, primary amines, aldehydes, phenols) rather than with individual homologues. This important limitation should always be borne in mind when considering spectrophotometric methods for organic determinands. The substitution of different groups into an organic molecule can also markedly affect its reactivity with a chromogenic reagent. This is often a particular problem when phenolic compounds are to be determined by solution spectrophotometry.[168]

(d) Application to High Performance Liquid Chromatography (HPLC)
Many substances determined by HPLC absorb in the u.v. region at wavelengths around 250 nm, and fixed-wavelength HPLC detectors are available. These are similar to a flow-through filter spectrophotometer for conventional automatic analysis by spectrophotometry [see (*e*) below], but they use a mercury source lamp and an interference filter able to transmit the 253.7 nm mercury line, and have small volume flow cells with silica windows.[143] For the large number of substances absorbing at wavelengths other than 253.7 nm, dedicated variable wavelength detectors are available to cover the range 190–350 nm. For further details, Section 11.3.8 should be consulted.

(e) Automated Spectrophotometry
It has been remarked[143] that most laboratories involved in routine water analysis probably make the greatest use of solution spectrophotometry in automatic analysis systems. Most instruments of this type use interference filters, and are double beam for long-term stability and low baseline drift. Further details are given in ref. 143 and in a similar review of air-segmented continuous flow automatic analysis.[172] See also Section 12.

(f) Summary

The instrumentation and experimental procedures for solution spectrophotometry are well established, and a wide range of reliable instrumentation is available so that the technique can be put into service at relatively little cost. The technique is not generally applicable to the determination of several determinands in the same portion of sample, and both the total analytical time and the operator time tend often to be lengthier than with other techniques requiring less chemical treatment of samples. If performed manually, such treatment often requires relatively greater experience and skill on the part of the analyst. The trend towards lower limits of detection being required in trace metal determination, and the greater operational simplicity of flame atomic absorption spectrometry [see Section 11.3.5(b)] have largely displaced solution spectrophotometry from that field of water analysis. However, it remains an important technique in routine analysis, particularly in the guise of automatic analysis for nutrient determination, and has a new role in the technique of HPLC.

11.3.4 Flame Emission Spectrometry

(a) General Basis of the Technique

In this technique, the sample is aspirated as a mist of fine droplets into a flame in which the solvent is evaporated and the residual solutes then vapourised and partially converted to their constituent atoms. As a result of the high temperature of the flame, these atoms are excited and emit light at wavelengths characteristic of the element, *e.g.*, the yellow and red colours observed when sodium and lithium, respectively, are present in a flame. Radiation of the wavelength of interest is isolated by optical filters or a monochromator, and its intensity measured photoelectrically and displayed on an appropriate read-out device.* The intensity of the emitted radiation is a measure of the concentration of the determinand in the sample. A schematic diagram of a flame emission spectrometer is shown in Figure 11.2.

A number of texts dealing with the technique are available (refs. 173–176 for example) and books on atomic absorption spectrometry frequently contain information on atomic emission also (see Section 11.3.5). The recently published text 'Metal Vapours in Flames', by Alkemade *et al.*,[177] contains a wealth of information relevant to all forms of flame spectrometry, including flame emission. The Annual Reports on Analytical Atomic Spectroscopy[178] and the biennial reviews of techniques published in *Analytical Chemistry*† give details of developments in the technique, and the former also review its applications in various fields, including water analysis. Readers requiring more details of the theory and practice of flame emission should consult the above, or similar, publications.

The technique normally finds its greatest application in determining metallic

---

* Some monochromator instruments have a facility for scanning a range of wavelengths, such that the entire emission spectrum may be recorded. Instruments employing filters, rather than a monochromator, for wavelength selection are frequently called 'flame photometers'.

† In the 1982 set of such reviews, both that on 'Emission Spectrometry'[179] and that on 'Atomic Absorption, Atomic Fluorescence and Flame Spectrometry'[180] contained information relevant to the flame emission technique.

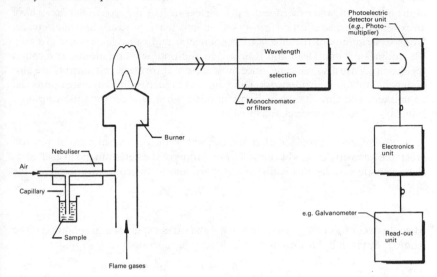

Figure 11.2 *Schematic of a flame emission spectrometer*

elements, because the sensitivities for non-metals are typically rather poor. Indirect methods for anions have been described, however (*e.g.*, refs. 181 and 182), and the use of radiation emitted by molecular, rather than atomic, species may also be used to determine some non-metallic elements. The use of phosphorus and sulphur flame photometric detectors in GC represents the most important current application of such molecular emission measurement in water analysis [see Section 11.3.8(*a*)].

Many metallic elements can be determined by flame emission, though certain elements (*e.g.*, Cd, Hg, Zn, As, Se), having their main resonance lines below 270 nm, cannot be dealt with, because flames provide insufficient energy for their excitation, and others (*e.g.*, La, Ce) cannot be atomised satisfactorily in flames.[176] Despite the wide scope of flame emission, its application in water analysis was for many years overshadowed by atomic absorption spectrometry. A discussion of the relative advantages and disadvantages of these two techniques is given in Section 11.3.5(*b*). More recently, the development of plasma source optical emission spectrometry (see Section 11.4.2) has probably detracted attention from the flame technique.

The response of a flame emission spectrometer typically stabilises within a few seconds from the time when a sample is first sprayed into the flame. Analysis, therefore, involves merely placing the capillary tube of the nebuliser* in the sample, and reading the result a short time later. Thus, one of the main characteristics of the technique is great simplicity and rapidity. All compilations of analytical methods recommend flame emission for the determination of sodium and potassium, and lithium is also readily determined by the technique.

(b) Principal Analytical Characteristics
The analytical performance of flame emission spectrometers depends on the design

---
* Nebulisation is the process of forming a fine spray from the sample.

of the instrument, and experimental parameters such as the type of flame can have an extremely marked effect on factors such as sensitivity, precision and interference. Good descriptions of the effects of instrumental and other factors are given by Mavrodineau[173] and Dean and Rains.[175] In addition, texts on atomic absorption spectrometry are also relevant, since most atomic absorption instruments are also designed for flame emission work. Differences in the capabilities of instruments can be marked, and this should be borne in mind when reading the following brief summary of performance characteristics.

(i) Calibration   Provided that the concentration of determinand is not too great, the intensity, I, of light emitted by the atoms of the determinand at each of its characteristic wavelengths is directly proportional to the concentration, c, in the sample:

$$I = kc$$

where k is a constant for given instrumental and other experimental conditions. The reading, R, given by the instrument is directly proportional to I, so that:

$$R = k'c$$

Many factors determine the numerical value of k' and empirical calibration curves are always prepared from the values of R obtained for standard solutions of different concentrations. In practice, it is not uncommon for calibration curves to show slight – and relatively unimportant – deviations from linearity. As it is virtually impossible to control accurately all factors affecting k' over more than very short periods of time, the calibration must be determined for each batch of analyses and checked at intervals during each batch.

(ii) Limits of detection   Thompson and Reynolds[176] have tabulated figures (effectively Criteria – rather than Limits – of Detection; see Section 8.3.1) for some twenty elements, for the nitrous oxide–acetylene flame, and using an instrument with a 0.2 nm spectral bandpass. For the great majority of elements, the values quoted were less than 100 μg l^{-1}, and for many, values of less than 10 μg l^{-1} were reported. Even with simple flame photometers, limits of detection for sodium and potassium of 30 and 80 μg l^{-1}, respectively, have recently been reported for two national recommended methods of analysis,[183,184] and more refined instrumentation can provide detection limits of $< 1 \mu$g l^{-1} for these elements. An example of the use of such refined instrumentation is the determination of sodium in high-purity waters in power stations.[185,186] The detection limits of flame emission and atomic absorption are considered further in Section 11.3.5(*b*).

(iii) Precision   Flames are relatively stable sources for exciting elements to emit radiaton and good precision is usually readily achieved. Provided that concentrations not too close to the detection limit are measured, relative standard deviations of 1–2% can be attained for sodium, potassium and lithium (see, for example, refs. 183, 184 and 187). Values around 0.5% can be achieved with greater care. At very low sodium concentrations (1–10 μg l^{-1}), standard deviations of about 1 μg l^{-1} have been achieved.[186]

(iv) Selectivity   The apparent simplicity of the technique is deceptive, in that many chemical and physical processes are involved and these lead to the possibility of several different types of interference. These include interferences resulting from differences in aspiration rate of samples and standards (caused by differences in viscosity), interferences resulting from compound formation and spectral interferences. The technique is more prone to spectral interference than is atomic absorption spectrometry, but both techniques are affected by the other two types of interference.

Interference effects in emission spectrometry are too complicated for detailed discussion here; the reader is referred to texts on the technique – and to that by Dean and Rains[175] in particular – for a more comprehensive treatment. Potential users of the technique should ensure that they are familiar with the various effects.

For sodium and potassium, correction for interference is usually readily achieved by calibrating the spectrometer with standard solutions whose composition approximates that of samples, but even this may be unnecessary for the analysis of raw and potable waters.[183,184] It is emphasised, however, that much more complex effects can occur with other determinands; calcium, for example, is particularly prone to various interferences (but is now more usually determined by atomic absorption). Moreover, the magnitude of interference can depend markedly upon factors such as the type of flame and burner, and the optical system used for isolating the emitted radiation.

Provided that the determinand is present in dissolved forms the result obtained seldom depends on the chemical forms of the determinand in the sample. In general, however, undissolved forms are not efficiently determined and if they are present, and must be determined, they should be dissolved before the sample is analysed.

(v) General   Instrumentation of widely ranging sophistication and cost is commercially available, from simple filter instruments (commonly known as 'flame photometers') to atomic absorption spectrometers with flame emission capability (including wavelength-scanning facilities). The possibility exists of simultaneous determination of more than one element – thus, for example, commercial instruments for simultaneous determination of sodium and potassium have been produced, and the capability of many modern atomic absorption spectrometers to permit rapid, sequential qualitative or 'semi-quantitative' determination of a number of elements has been remarked.[176]

Despite the inherent capabilities of the technique, its use in water analysis now seems likely to be restricted to the determination of a relatively small number of elements, for which its simplicity is matched by good analytical performance (*i.e.*, sodium, potassium, lithium). Thus, for example, Basson and von Staden[187] have combined flow injection analysis (see Section 12.3.2) with flame emission spectrometry to provide a rapid procedure (up to 128 samples per hour) for the determination of sodium and potassium in water. For other elements amenable to the technique, atomic absorption spectrometry offers better overall performance and, for multi-element work, plasma emission systems have eclipsed flame instruments.

As noted previously, plasma emission spectrometry is treated in a separate section of this text (Section 11.4.2), constituting as it does a well developed, though not as yet generally important, water analysis technique. Emission spectrometry may also be carried out using an electrothermal atomiser*, rather than a flame or plasma, as the atom cell. Though not a generally important technique in water analysis, the use of electrothermal atomisation atomic emission spectrometry may be advantageous in certain cases and is conveniently considered here. It appears to offer potential advantages over flame atomic absorption spectrometry [see Section 11.3.5(b)] for the determination of barium – namely, improved detection power and lower calcium interference. Thus, Ebdon et al.[188] and Suzuki et al.[189] have reported detection limits of 0.8 μg Ba l^{-1} and ca. 0.6 μg Ba l^{-1}, respectively, using electrothermal atomisation atomic emission, compared with a detection limit of 20 μg Ba l^{-1} obtained using flame atomic absorption.

It should also be noted that, like other forms of emission spectrometry, electrothermal atomisation atomic emission has the potential for simultaneous multi-element analysis – see, for example, ref. 190. As in the case of electrothermal atomisation atomic absorption [see Section 11.3.5 (c)], the use of a platform within the furnace atomiser has been examined; it reduces interference effects and enhances sensitivity through the attainment of higher temperatures during the lifetime of the atomic population.[191,192] Thus Bezur et al.[191] have reported detection limits in the range 0.05–50 μg l^{-1} for a number of elements; most of the limits were below 5 μg l^{-1} and many were less than 1 μg l^{-1}. Although it is thought unlikely to be exploited commercially in the near future, electrothermal atomisation atomic emission spectrometry has obvious potential as a multi-element technique offering better detection limits than does plasma emission spectrometry for a range of elements.[192]

In another variant of the approach, electrothermal atomisation has been combined with a glow discharge to produce the excited species.[193,194] Detection of picogram quantities of determinand in small samples (1–80 μl) can be attained. As Ottaway[192] has remarked, such an approach, though more complicated than the use of electrothermal atomisation alone, may allow extension of the electrothermal atomisation atomic emission technique to elements (such as arsenic and selenium) whose high excitation potentials preclude the use of electrothermal atomisation alone for both atom production *and* excitation.

11.3.5 Atomic Absorption Spectrometry (AAS)

(a) General Basis of the Technique

A schematic diagram of an atomic absorption spectrometer is shown in Figure 11.3. It will be seen that the basic instrumentation has many similarities to that for flame emission spectrometry, and it is usual for commercial atomic absorption spectrometers to have facilities for flame emission work.

The principle of AAS is that determinand atoms in a non-emitting ground state will absorb light of a characteristic resonance wavelength. The extent of the absorption will increase as the number of atoms increases.

The light source is selected so that it emits only the atomic spectrum of the chosen

---

* For a description of electrothermal atomisation, see Section 11.3.5(c).

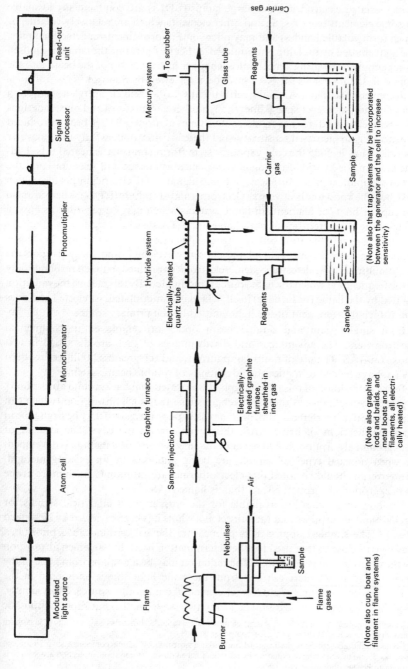

Figure 11.3 *Schematic of Atomic Absorption Spectrometer Systems*

element*. Hollow-cathode lamps (HCLs) are usually employed, though radiofre-quency-excited electrodeless discharge lamps (EDLs) are commercially available for some elements (*e.g.*, As, Se and other elements which are relatively volatile, or which form volatile halides) and may offer some improvements in detection limits. The light emitted by the lamp is modulated, either by operating the lamp on a pulsed power supply or by using a mechanical chopper prior to entry of the beam into the atom cell. Light sources for AAS have been reviewed by Sullivan.[195]

In conventional flame AAS, the cell is usually a 10–12 cm air/acetylene flame or a 5 cm nitrous oxide/acetylene flame, though other gas mixtures and burner designs have been employed. However, alternative types of atom cell have been introduced in order to gain increased sensitivity, and these are now quite widely used in water analysis. They include the cold vapour system (for mercury), the heated tube/hyd-ride system (for hydride-forming elements such as arsenic and selenium) and the various graphite or metal furnaces, rods and filaments (all electrically heated) which have become known collectively as electrothermal atomisers (ETAs). There are also the cup (or boat, or filament)-in-flame systems which may be regarded as having some of the characteristics of both flame and ETA systems.

After traversing the atom cell, the light beam enters the monochromator. This is adjusted so as to isolate light of the chosen wavelength and focus it upon the photomultiplier. The detector system's electronics are tuned so that it responds only to the frequency at which the light source is modulated. By this means, the radiation emitted by the flame (or furnace) itself, which is unmodulated, is rejected and does not disturb measurement of the light emitted by the primary source.

Both single beam and double beam optics are employed in commercial spectrometers. The advantages and disadvantages of each are discussed in text books on AAS. The use of multi-lamp turrets, and the greater stability of modern HCLs, have tended to reduce the advantages of double beam systems.[196]

In the first edition of this monograph, it was stated that background absorption† is generally negligible. Although it remains true that such absorption (which can lead to positive interference and, in some cases, poor precision) may be negligible in many applications of flame AAS, this is not always the case. The absence of background absorption should never be assumed when a new method is developed, or when unusual types of sample are being analysed by an existing method. Moreover, it should be noted that electrothermal atomisation is much more prone to background absorption effects than is flame AAS.[197]

Various approaches are available for the correction of analytical signals for background absorption, and the subject as a whole has been reviewed by Newstead *et al.*[198] The simplest approach is to measure the background absorption at a wavelength close to that used for the determination itself, but at which absorption by the determinand is negligible. The radiation may be a non-absorbing line of the determinand or an inert gas line from the same light source as is used in the determination, or a line of a different element from another source. Most simply, sequential measurements are made at the two wavelengths, but rapid, automatic

---

* Lamps emitting the spectra of several elements are available, but they may not allow maximum sensitivity to be achieved for any of their components.

† Background absorption is the result of molecular absorption by sample components which are not dissociated, and of scattering of radiation by particles formed during atomisation; molecular absorption is thought to be the more important.[196]

correction is possible if a dual channel spectrometer is employed. Such correction is essential in graphite furnace AAS, because the background absorption is likely to show important differences between successive firings with the same sample.[198]

A second, and more widely used approach, is to measure the background absorption using continuum radiation from a suitable source – commonly a deuterium arc or HCL. Because the atomic absorption profile is narrow in relation to the monochromator bandpass, only a negligible part of the continuum radiation is absorbed by determinand atoms and the attenuation of this radiation therefore provides a measure of the background absorption. Automatic, simultaneous correction is available with commercial spectrometers, the instruments being able to switch rapidly between measurements of the radiation from the two sources during individual determinations.

In recent years, two other approaches to automatic simultaneous background correction have become commercially available (on particular instruments) which are reported to provide rather more accurate correction of high background absorption.

The first of these – Zeeman background correction – exploits the splitting and polarisation of spectral lines (the direct Zeeman effect) or of atomic absorption profiles (the inverse Zeeman effect) when the source or atom cell (respectively) are placed in a magnetic field. Reviews of Zeeman background correction have been provided by Grassam and Dawson,[199] Yasuda *et al.*[200] and De Galan;[201] the theory of the various Zeeman techniques has been described by De Loos-Vollebregt and De Galan,[202] who have concluded that the direct Zeeman effect (source placed in magnetic field) offers rather similar performance to that of continuum source systems, in that for both to be entirely effective, the background needs to be virtually continuous near the analytical wavelength, whereas the inverse Zeeman effect (atom cell placed in magnetic field) allows accurate correction for structured* background, because correction is made at the same wavelength as is used for the determination.

The second new approach (Smith–Hieftje background correction) exploits the fact that, when a hollow cathode lamp is run at high current, self-reversal† occurs and the atomic absorption is greatly diminished while background absorption remains essentially unaffected. By alternately operating the lamp at normal and high current, automatic simultaneous background correction is achieved. Smith–Hieftje background correction is described by Sotera and Kahn.[203]

An assessment of the relative advantages and disadvantages, in relation to water analysis, of these various new approaches to background correction is not yet possible, but it is clear that they are likely to find particular application in the direct analysis of waters of high dissolved solids content. Indeed, application of inverse Zeeman background correction to the determination of manganese[204] and cadmium[205,206] in sea-water by graphite furnace AAS has been reported.

Checking that background correction systems are functioning correctly can be undertaken by various means – the review by Newstead *et al.*[198] gives details. Insertion of a gauze or filter into the light beam is simple, but may not reflect the

---

* Background absorption which is strongly dependent on wavelength.

† Absorption of the emitted radiation by absorbing atoms within the source itself, preferentially at the centre of the emission profile such that the latter becomes double-peaked.

conditions of normal analysis (especially with electrothermal atomisation, in which transient signals occur). Tests under analytical conditions using specially prepared solutions are therefore also recommended.[198]

The rapid growth in the application of AAS to water analysis, noted in the first edition of this book, has continued apace. The use of the technique has become virtually as general as solution spectrophotometry, at least in water analysis laboratories in western Europe and the U.S.A., and indeed AAS has displaced solution spectrophotometry from many of its former areas of application. Moreover, atomic absorption has (by means of preconcentration techniques and electrothermal atomisation) kept pace with (and, it must be admitted, been a factor in the establishment of) requirements for lower detection limits in trace metal analysis. Although plasma emission spectrometry (Section 11.4.2) and anodic stripping voltammetry (Section 11.3.7) may offer some advantages over AAS in certain circumstances, the flexibility, robustness and relatively low cost (compared with plasma emission) of atomic absorption have combined to make it the most important single technique for routine metal determinations in water analysis. The space afforded to the technique in this edition reflects that status, and also the continuing rapid development, particularly of non-flame atomisation systems.

Many text books devoted to AAS are available, and a selection is given in refs. 196, 197, 207–213. These often contain a section devoted to water analysis. Given the rapid developments in electrothermal atomisation, it is inevitable that the older texts in particular tend to give only a limited coverage of that subject. The existence of a monograph[197] dealing specifically with electrothermal atomisation is, therefore, of considerable value. In addition to general reviews of water analysis published on a regular basis, which give much information on the application of AAS in that field, a comprehensive annual review of the technique in general is provided (and has been since 1971) by the Royal Society of Chemistry's 'Annual Report on Analytical Atomic Spectroscopy'.[214] There are also a number of specialist journals which frequently contain important papers and reviews on AAS; these include *Spectrochimica Acta, Applied Spectroscopy, Spectroscopy Letters* and *Progress in Analytical Atomic Spectroscopy*. Some of the instrument manufacturers publish useful house journals on AAS, and one, *Atomic Spectroscopy* (formerly *Atomic Absorption Newsletter*) has continued to provide extensive bibliographies. Reviews of atomic spectroscopy are also published (in even-numbered years) in *Analytical Chemistry*.[215]

A review dealing specifically with the application of atomic absorption spectrometry in water analysis has been published as part of a series of U.K. recommended methods.[216] In addition to providing a review of the literature, this publication also contains useful advice on instrumentation, routine maintenance and fault finding. A recently published book[216a] deals with the use of atomic absorption for environmental monitoring for regulatory purposes and includes a chapter on procedures for water analysis.

AAS has been most widely applied to the determination of metals, because the sensitivity for non-metals is generally poor. However, indirect methods for anionic and other impurities have been reported. A review of such methods has been provided by Kirkbright and Johnson.[217] An example of the technique in water analysis is the determination of sulphate by barium sulphate precipitation, in which residual barium is determined by flame AAS.[78] Very recently, a text[218] on the determination of organic compounds by AAS has been published; it contains a review of the determination of non-ionic and anionic surfactants, an example of the indirect use of AAS for the determination of organics which has found application in water analysis.

As would be expected for such a widely used technique, methods employing AAS are frequently to be found in collections of standard or recommended methods for water analysis (*e.g.*, refs. 17, 18, 216, 216*a*). In general, flame methods are most often encountered in such compilations but methods based upon the hydride generation, cold vapour and electrothermal atomisation techniques have also appeared. Particular difficulty attends the production of such methods when electrothermal atomisation is involved, because the performance and exact methodology are often heavily dependent upon the characteristics of the individual furnaces, which differ more from manufacturer to manufacturer than do flame systems.* These differences have also tended to make more difficult the task of selecting the appropriate system from the wide range of AAS instrumentation which is commercially available.

Regardless of the type of atom cell, measurements are based upon the absorption of light and the fundamental equation of the technique is

$$A = kc$$

where A = optical density = $\log(100/T)$
$\quad T$ = percentage transmission of light by the atom cell
$\quad k$ = a constant for a particular determinand, instrument and operating conditions
$\quad c$ = the concentration of determinand

In all variants of the technique, k is dependent upon so many factors that calibration curves (of absorbance, A, or a related measurement, against concentration, c) are always prepared using solutions of known determinand concentrations. Such curves should be prepared for (and checked during) each batch of analyses. Calibration curves in flame AAS are often linear over the concentration range of interest, but this is not always the case – and is often not with electrothermal atomisation and hydride generation systems. Most modern atomic absorption spectrometers are equipped with curve linearisation facilities and, in microprocessor-based systems, these may be mathematically quite sophisticated. While such facilities are useful in certain applications, it must be borne in mind that the fundamental disadvantages of a grossly non-linear calibration cannot be eliminated by mere signal processing.

The subject of microprocessors in analytical instrumentation is considered in general terms in Section 13. With particular regard to AAS, the reader's attention is drawn to an article by Koirtyohann,[220] who asserts that "a tendency, especially in advertising literature, to oversimplify the entire process of AA measurements", always present, has "been enhanced by microprocessor technology". He cites examples to support his thesis that manufacturers have tended often to emphasise

* It must be emphasised, however, that the magnitudes of interference effects in flame atomic absorption spectrometry are by no means always independent of the type of instrumentation employed.[219]

unsound applications of the new technology rather than the sound benefits which can derive from its proper use. He emphasises that "there is no way for a microprocessor to improve AA data that were improperly collected", and the present authors endorse this warning, both with respect to AAS and to analytical techniques generally.

(b) Flame Atomic Absorption Spectrometry
Flame atomisation, by virtue of being the earliest mode of atomisation employed in AAS, is the best understood and the most widely applied.

(i) Limits of Detection   The scope of the technique is demonstrated by Table 11.2. which lists typical detection limits which can be achieved using direct flame atomisation.

TABLE 11.2
Detection Limits by Direct Aspiration into Flames

Element	Flame*	Detection† Limit $\mu g\, l^{-1}$	Element	Flame*	Detection† Limit $\mu g\, l^{-1}$
Ag	A	0.9–3	Li	A	0.5–2
Al	N	25–30	Mg	A	0.1–0.3
As	A, N or H	90–300	Mn	A	1–3
Ba	N	8–20	Mo	N	20–30
Be	N	1	Na	A	0.1–0.5
Ca	A or N	1–2	Ni	A	4–10
Cd	A	0.5–1.5	Pb	A	5–15
Co	A	4–6	Se	A, N or H	60–500
Cr	A	2–6	Si	N	60–250
Cu	A	1–3	Sn	N or H	25–110
Fe	A	3–6	V	N	25–70
Hg	A	140–170	Zn	A or N	0.8–1.2
K	A	1–3			

* A = air/acetylene, N = nitrous oxide/acetylene, H = air/hydrogen or argon/ hydrogen.
† The ranges of values given are based on data supplied by four major manufacturers of AAS equipment: Instrumentation Laboratory (UK) Ltd., Perkin-Elmer Ltd., Pye-Unicam Ltd. and Varian Associates Ltd. These companies were asked to supply data for a 'top-of-the-range' instrument but several reported that the figures given applied to all instruments in their range or did not identify a particular instrument in their replies. It is understood that the detection limits were all defined in broad conformity with the definition of 'Criterion of Detection' as used in this book (see Section 8.3.1). In cases where a manufacturer gave more than one value (for example, for different nebulisation or flame conditions) the lowest was always used in constructing the Table.

The detection capability of flame AAS is limited by the inefficiency of the flame as an atom source and cell. No more than 10% of the sample reaches the flame when conventional pneumatic nebulisation is employed,[208] and the high velocity of flame gases allows the determinand atoms a very limited period of residence in the spectrometer beam. Impact-bead nebulisation, in which the larger droplets are broken up as they impact upon a bead near the nozzle of the nebuliser, can increase the fraction of sample reaching the flame by 50–100%,[208] but this can only improve detection limits to a modest degree. The use of impact-bead systems is now widespread, however, and may account, in part, for the generally lower detection limits quoted in Table 11.2 compared with those given in the first edition of this book some 12 years ago. The effects of impact beads and other aids to efficient nebulisation on droplet distributions have been examined by Cresser and Browner.[221]

Useful improvements in flame AAS detection limits, particularly for the more volatile metals, have been reported by Watling,[222,223] who used a slotted quartz tube above a conventional, slotted burner to reduce the effective flame speed and thereby increase the residence time of atoms in the optical path. This approach has now been exploited commercially, and detection limit improvements of two- to more than ten-fold have been reported.[224]

Other means of improving the detection limits achievable with flame AAS include the so called "sampling cup and boat" procedure.[225–227] In these, the sample is placed in a cup (or boat) and evaporated to dryness, following which the cup (or boat) is placed in the flame of an atomic absorption spectrometer. There appears to have been little application of such techniques to water analysis, probably as a result of the growth of interest in electrothermal atomisation, although the use of the "Delves Cup" for drinking water analysis has been reported.[228] The method of sample introduction, involving volatilisation of solid material from a surface, would seem likely to be prone to matrix interferences – especially as the residence time in flames is relatively short, so that variations in atomisation rate are likely to have considerable impact upon peak absorbances. Thus signal integration was employed in the Delves Cup method cited above.[228] More recently, Berndt and Messerschmidt[229] have described the use of an electrically heated platinum loop to introduce dried samples into a flame. Although the detection limits achieved were such as to permit the direct determination of lead and cadmium in drinking water, comprehensive interference tests were not reported and the value of the technique in water analysis remains uncertain.

Because the detection limits achievable using conventional flame AAS are inadequate for many applications in water analysis, preliminary preconcentration techniques are often used in conjunction with a flame AAS finish. Such techniques have recently been reviewed, with particular reference to water analysis, by Orpwood.[232] Many techniques are available (see also Section 11.2.2), including solvent extraction, evaporation, ion exchange, freeze concentration, coprecipitation, carbon adsorption and electrodeposition. However, simple evaporation (by heating) solvent extraction and the use of chelating ion-exchange resins have been most popular.

The simplicity inherent in evaporating the acidified sample is offset by a number of potential disadvantages. These include: (1) simultaneous concentration of the matrix, with the concomitant danger of interference effects, (2) the risk of contamination and (3) the need for close control of the extent of evaporation in order to avoid irreversible adsorptive losses. Although such problems may not prevent the technique being applied in many circumstances in water analysis, they do limit its scope and control of (2) and (3) may render it unacceptably slow and labour-intensive. It should be noted, however, that evaporation (unlike solvent extraction and ion exchange) will concentrate all trace metal forms – except those that are volatile. Some examples of methods for water analysis involving evaporation followed by flame AAS are given in refs. 167 and 231–234.

Chelating ion-exchange resins have the advantage of providing a high degree of matrix rejection, being less prone to contamination and requiring less operator attention than evaporation and solvent extraction, and have been widely applied in water analysis (see, for example, refs. 230 and 235–237). However, they are not completely selective for trace metals, and variable recoveries resulting from differing concentrations of major cations (especially calcium) may prove to be a limitation if waters of a wide range of hardness are to be analysed.[237] Moreover, it has been reported[236] that chelating ion-exchange is less efficient than solvent extraction in recovering all forms of trace metals from water samples.

Solvent extraction has been most widely used with flame AAS for water analysis. In addition to the review by Orpwood[230] the monograph by Cresser[238] contains many references to water analysis.

Many organic reagents and extraction solvents have been used. The solvent must be such that, in addition to providing satisfactory extraction, its combustion characteristics are suitable and do not lead to hazards (if direct aspiration of extracts is to be performed). Additionally, the need to handle the solvent must not give rise to risks to operator health – most of the available extraction solvents are potentially injurious, at least to some degree. The book by Cresser[238] contains much valuable advice on solvent selection and handling.

Efficient extraction systems, which may be necessary to reduce operator exposure to the solvent, may introduce contamination problems by causing large volumes of laboratory air to be drawn over the work. This can be overcome by working in a clean room, but this is an expensive facility not available in many laboratories. A combination of a laminar flow filtered air system with a (flow-balanced) extraction system may provide an acceptable solution at much lower cost. For many applications, however, the use of an ordinary fume cupboard will be perfectly adequate.

Of the many solvents which have been used 4-methylpentan-2-one (less properly, but more commonly, called methyl isobutyl ketone or MIBK) is without doubt the most common. It is relatively inexpensive, not particularly toxic and less unpleasant to handle than many others, and it has excellent characteristics for extraction from aqueous media and for aspiration in flame AAS.[238] Although less unpleasant to use than many alternatives, it can (in common with other ketones) cause headaches, and appropriate extraction must be arranged if it is to be used.

Amongst the organic chelating reagents which have been often used are sodium diethyldithiocarbamate (DDC),[239–241] 8-hydroxyquinoline (oxine, 8-quino-linol)[241–243] and ammonium tetramethylenedithiocarbamate (more commonly known as ammonium pyrrolidinedithiocarbamate, APDC).[241,244–247] Mixtures of chelating agents have also been employed.[248,249] However, APDC (usually in conjunction with MIBK as solvent) has been by far the most commonly used chelating agent, because it forms chelates with many trace metals and gives excellent recoveries from both fresh and saline waters. Back-extraction into acid may be necessary in some applications, however, because of the limited stability of the complexes.

When chelating ion-exchange resins or solvent extraction systems are employed with flame AAS, detection limits in the $100 \, \text{ng} \, \text{l}^{-1}$ to $10 \, \mu\text{g} \, \text{l}^{-1}$ range are commonly achieved. However, the use of such preconcentration techniques brings certain disadvantages:

(1) Flame AAS tends to give the same analytical response to different dissolved trace metal forms, but this is not true of the chelating ion-exchange resins and solvent extraction systems, which may fail to recover all the metal present in naturally occurring chelates, for example.[230,236] A potential advantage in speciation studies (see Section 11.5), this property is a distinct disadvantage when total metal concentrations are of interest. A digestion procedure (*e.g.*, u.v. photolysis, heating with acid) may therefore be needed in addition to the preconcentration step. This need for a digestion step represents an important potential limitation upon pre-concentration techniques such as resin chelation, which are otherwise readily applicable to large sample volumes. Chemical digestion of large sample volumes (many litres) may be impractical, though continuous-flow u.v. irradiation remains a possible approach.

(2) The preconcentration procedures are time-consuming and relatively labour-intensive. Although automated solvent extraction systems have been produced[238,250] they are complicated to construct and operate, and may not allow the desired concentration factor to be achieved in all applications.

(ii) Precision In direct flame atomic absorption spectrometry, excellent precision is usually readily achieved (even with high sample throughputs) except when concentrations are close to detection limits. Relative standard deviations of 1% or better are not uncommon, and values greater than 5% are exceptional. The standard deviation tends to be approximately proportional to concentration.

(iii) Selectivity The erroneous impression that flame AAS is virtually interference-free, which seemed to be commonplace when the technique was first introduced, appears to have disappeared as the use of the technique has become better established, and the interferences more widely publicised. Many authors have described both simple and complex interference effects, and the topic is discussed in modern textbooks and has been reviewed (from a mechanistic standpoint) by Rubeška and Musil.[251] Interferences in flame AAS (apart from non-specific background adsorption, discussed above) fall into three main categories:

(1) 'Transport' effects occur when the surface tension and viscosity of samples and standards differ, so that the aspiration rates (and hence the rates of determinand transport to the flame) also differ.

(2) Suppression effects due to the formation of refractory compounds of the determinand which are not dissociated during their residence in the flame. A well known case is the suppressive effect of phosphate on calcium determination in the air–acetylene flame.

(3) In hot flames, the alkali and alkaline earth elements (in particular) undergo a considerable degree of ionisation. Thus, for example, calcium may be nearly 50% ionised in nitrous oxide–acetylene flames.[208] The sensitivity of atomic absorption is thereby reduced, and, because the proportion of atoms ionised reduces as the determinand concentration increases, the calibration tends to

curve away from the concentration axis. Moreover, the phenomenon can also result in interference (an enhancement of signal) when samples contain a more easily ionised element, whose presence reduces the ionisation of the determinand (*e.g.*, potassium causing enhancement in calcium determination).

Users of the technique should familiarise themselves with the types of interference which may arise in their applications, so that appropriate measures can be taken. However, interference effects in water analysis by flame AAS often present no problem and, where they are a problem, the means to overcome them are often simple and well established. Thus, for example, the acid strength of samples and standards is kept constant to avoid transport effects which would otherwise occur and lanthanum is often used to overcome phosphate interference in calcium determinations (by combining preferentially with the phosphate). Again, the deleterious effects of ionisation can be overcome by adding an 'ionisation buffer' – that is, a readily ionised element such as caesium[184,182] – in such an excess that its enhancement effect becomes constant, as ionisation of the determinand is effectively eliminated. These and other approaches to overcoming the various types of interference effects are discussed in modern texts on AAS.

Although, therefore, the technique of flame AAS can be said to possess very good selectivity, especially in water analysis applications, care is needed, particularly as the magnitude of interference effects can depend on factors which may vary from instrument to instrument.[219]

As noted previously, flame AAS tends to give the same analytical response to different dissolved metal forms. This is not always the case, however, as the work of Thompson and others on chromium and iron has so clearly demonstrated (see refs. 252–254 and literature cited therein). Also, the response for undissolved metal forms is usually less than that for dissolved forms, so that if determination of the former is required it is generally necessary to dissolve them (or at least to leach adsorbed metal from the surface of particles, for example by acid digestion) prior to measurement.

(iv) General aspects of performance   Many instruments of differing sophistication (both in terms of optical systems and data processing) are commercially available, and offer robustness and reliability of performance. Little manipulative skill or experience is demanded of the operator, numerous well proven methods for water analysis are available and the literature on such applications is extensive. All these factors, together with the wide scope of the technique, have secured for flame AAS a key role in water analysis laboratories.

A continuing limitation of the technique is its lack of multi-element capability. Although multi-element atomic absorption systems continue to be constructed by research groups (see, for example, literature cited by Horlick in a review[255] of flame spectrometry), the authors know of no such system being available commercially. The most that is available on a commercial instrument, as at the time of the first edition of this book, is a dual-channel (*i.e.*, simultaneous two-element) capability. Again, the rapid development of plasma emission spectrometry (see Section 11.4.2) may have diverted effort from the pursuit of multi-element AAS.

(v) Comparison of flame atomic absorption and emission The conclusion reached in the first edition, namely that flame atomic absorption and flame emission are complementary, remains unaltered. Emission offers lower detection limits for a number of elements (most notably, the alkalis and the alkaline earths – except magnesium) for which the most sensitive spectral lines are at wavelengths about 300–350 nm, while flame AAS is generally more sensitive than flame emission for lines at smaller wavelengths. However, the lower detection limits achievable by flame emission for the alkalis and alkaline earths may be unimportant in many applications.

Flame emission is more prone to spectral interferences than is flame AAS, and is also more seriously affected by flame temperature fluctuations. It does not, on the other hand, require hollow cathode lamps and, unlike atomic absorption, it lends itself to rapid qualitative analysis by wavelength scanning. Overall, however, the scope and versatility of flame atomic absorption spectrometry are far greater.

Of greater current concern than the relative merits of flame AAS and flame emission is the comparison of AAS and plasma emission spectrometry; this is dealt with in Section 11.4.2(*b*).

(c) Electrothermal Atomisation Atomic Absorption Spectrometry

(i) General basis of the technique It has been noted previously that the detection limits attainable using direct flame AAS are often too high for many applications in water analysis. The use of preconcentration procedures with flame AAS to overcome such limitations detracts considerably from the inherent speed and simplicity of the technique, and is therefore far from ideal. The reasons for the inherent inefficiency of the flame as a means of producing free atoms of the determinand and maintaining them in the beam – namely, the very limited efficiency of nebulisation and the limited residence time engendered by the flame speed – have also been discussed.

Electrothermal atomisers (ETAs) offer markedly improved sensitivity, partly because the total amount of determinand present in a small volume of sample is atomised, and partly because the residence time of the atoms in the beam is markedly increased, particularly in the case of the 'enclosed' or furnace atomisers. Detection limits are typically about two orders of magnitude lower with graphite furnaces than with flames, for common furnace sample sizes (10–50 μl).

It was therefore to be expected that the development, some ten years ago, of commercially available ETAs would be of considerable interest to those engaged in water analysis. Moreover, the promise of very low detection limits by a variant of atomic absorption spectrophotometry – a technique noted for its reliability and relative freedom from interferences – was particularly attractive. Unfortunately, it soon became apparent that commercial ETA systems, though indeed permitting very low detection limits to be achieved, were subject to serious problems from matrix interference. The first edition of this book noted that the ETA technique, then in its infancy, seemed likely to be " . . . more subject to interference than normal flame atomic absorption". The validity of that warning has been amply demonstrated by subsequent events, at least in respect of commercial atomisers, and much of the remainder of this section will be devoted to an examination of the interference problem.

The subject is complicated, and the treatment of it is necessarily lengthy, not least because the problem of matrix interferences is intimately connected with atomiser design and operation. However, the authors believe that the space devoted to electrothermal atomisation AAS is justified by the enormous potential of the technique for direct determination of many elements of interest in natural and treated waters, with the low limits of detection which are increasingly necessary to meet legislative and other requirements.

Because the subject of atomiser design is of major importance to an understanding of the current status and limitations of the technique, a brief summary of ETA development will be given before the principal analytical characteristics are discussed. Attention will be focused upon the 'closed' or furnace system (rather than the 'open' systems – rods, filaments and braids) because of the far greater importance of the former to current practice.

(ii) The development of electrothermal atomisers   The monograph on electrothermal atomisation by Fuller[197] gives a valuable account of the evolution of ETA designs, and the following brief summary owes much to that source.

The forerunner of all modern atomisers was the graphite cuvette developed in the 1950s and 1960s by L'vov. In the final versions of this system, the body of the atomiser consisted of a graphite tube, 3–5 cm in length, maintained at *constant* high temperature by resistive heating and sheathed in argon. Samples were pipetted onto a separate graphite pedestal and dried thereon, and the pedestal was then raised to fit into a matching aperture in the cuvette, the determinand being atomised into a constant temperature environment. The properties of this system, including its freedom from serious interferences, led L'vov to believe that there were "good prospects for developing an absolute method of analysis, based on cuvettes" (ref. 213, p. 221).

However, L'vov's atomiser was, understandably, considered less suitable for commercial development than the simpler and smaller Massmann design, introduced in 1967. The latter became the basis of most commercially available graphite furnaces (the exceptions being the small Varian furnaces, which in any case shared the most important feature of the Massmann furnace – pulsed heating). In the Massmann system, the liquid sample is pipetted directly into the cold tubular furnace, which may be heated resistively to temperatures of 2500 °C or more in seconds. Such a system clearly has operational advantages, and the entire analytical process, including sample introduction, has been easily automated. Moreover, although the absolute sensitivities attained using such atomisers have not generally reached those achieved by the L'vov design, the performance of Massmann systems has been more than adequate for very many applications, without the need for sample preconcentration [see (iii) below].

Other types of electrothermal atomiser have also been developed. For example, West and Williams produced a carbon rod atomiser, and other workers have produced a variety of metal filament and boat systems. None of these designs appears to offer important advantages over Massmann-type atomisers, and they have been used much more rarely in water analysis. Consequently, they will not be considered in the remainder of this section. Woodriff has introduced another type of constant temperature furnace (CTF)*, having some similarities to the L'vov cuvette, but of greater length (25 cm) to produce long residence times of atomic vapour. The physical dimensions of the Woodriff furnace have undoubtedly militated against its commercial exploitation by major manufacturers of atomic absorption equipment, but its performance – like that of the L'vov cuvette – provides an interesting and instructive contrast to that of Massmann-type systems.

(iii) Principal analytical characteristics Although the detailed performance of the technique depends markedly on the individual instrumentation and its mode of operation, particularly with respect to interference effects, modern commercial atomisers exhibit broadly similar performance in many respects. In the following sub-sections, therefore, attention will be directed towards major features, such as the differences between the interference effects between commercial, pulse-heated furnaces and constant temperature furnaces. The reader should bear in mind, however, that this approach does not imply that identical performance will necessarily be achieved by different commercial atomisers, or even (with respect to interferences, particularly) by the same atomiser on different occasions.

(1) *Limits of Detection*

Table 11.3 (see p. 415) lists the detection limits and sensitivities for a number of elements, reported by a number of manufacturers of atomic absorption instrumentation. It is evident that extremely small limits are generally achievable, and comparison with Table 11.2 shows the marked improvement over the detection capabilities of the flame technique.

Further enhancement of detection capability can, of course, be achieved by applying a preconcentration procedure. However, in contrast to flame AAS, the limits of detection attainable with furnace atomisation are adequate for many applications in water analysis, without such preconcentration. Indeed, for many determinands, only anodic stripping voltammetry (see Section 11.3.7) and inductively coupled plasma/mass spectrometry [Section 11.4.1(*d*)] offer comparable or better performance by direct analysis.

(2) *Precision*

Before the widespread availability of commercial automatic sampling systems, the precision attainable using Massman-type furnaces was relatively poor, being limited by the ability of the operator to pipette small sample volumes reproducibly† into the confined space of the furnace. Thus, Ranson[256] concluded that relative standard deviations of 5–10% were typical for manual operation.‡

The situation has improved markedly with the development of automatic sampling systems, and relative standard deviations of about 2% or better

---

* Although the Woodriff furnace can be purchased from a specialist supplier, it is not marketed by major manufacturers of AAS equipment, is too large to be accommodated in modern spectrophotometers, and will not be regarded here as a 'commercial' system.

† With respect to location within the furnace, as well as to volume.

‡ At concentrations distant from the detection limit.

should be readily attainable with commercial equipment, in simple cases.* However, precision may be degraded if special procedures (*e.g.*, standard additions calibration) are performed to reduce interference effects. Overall, however, relative standard deviations of better than 5% should usually be achievable, except at concentrations close to the limit of detection. As with flame AAS, standard deviations typically increase with increasing determinand concentration.

(3) *Selectivity, and the search for ways to reduce interferences*
With respect to different dissolved forms of the elements, it is to be expected that electrothermal atomisation AAS, like flame AAS, will tend to give the same analytical response. However, the experience with flame AAS suggests that particular caution should be exercised when different oxidation states may be present. The response to undissolved forms is likely often to be smaller than to dissolved forms, and some form of digestion may be necessary if their determination is required. (Even if the response to undissolved forms were the same as to dissolved forms in a particular case, the small sample volumes typically required – say 10–50 μl – could make representative sub-sampling of suspended matter extremely difficult).

It is the propensity of the technique to severe matrix interferences, however, which has represented a serious limitation to its usefulness in water analysis. Numerous studies (see, for example refs. 257–260) have demonstrated that the determination of trace metals in waters by direct electrothermal atomisation is subject to severe interference from major constituents. Particular attention has been paid to the determination of cadmium and lead, because of the general importance of those metals in environmental pollution and the concern about their effects upon human health, but other trace metals are also subject to similar interferences.[197] Although interference tests with individual salts have shown that both suppression and enhancement of the signal can occur,[260] the overall effect of the matrix in water samples is generally suppressive.[257-260] The magnitude of the interferences varies considerably with both the determinand and the sample type. For lead and cadmium, suppressions of the order of 50% of determinand concentration or more have been frequently encountered and very severe effects (> 70% suppression) have been reported for lead determination by several groups (see ref. 259 and literature cited therein). Clearly, such effects, if uncontrolled, would render the technique valueless in all likely applications in water analysis.

The effects of the individual ions present together in real samples are dependent upon the overall composition of the matrix, and the extent of interference can vary markedly from one analytical occasion to another, even when the same instrumentation and conditions are employed. For these reasons, it is not possible to predict the interference effect for a particular sample from a knowledge of the major ion composition, and thereby correct the result obtained. Standard additions calibration has been employed to overcome the interferences, but it adds considerably to the time required for analysis. Moreover, experience at the authors' laboratory has shown that it

---
* At concentrations distant from the detection limit.

does not necessarily produce wholly unbiased results – probably because the percentage suppression produced by a particular sample is not completely invariant with changing determinand concentration.[261]

Given the serious interference effects to which the technique is subject, and the clear potential advantages of electrothermal atomisation in respect of detection limits, it is not surprising that considerable effort has been devoted throughout the last decade to the search for ways to overcome the interferences. In the early years, attempts to overcome interferences were hampered by the lack of understanding of their origin and nature. More recently, however, a reasonably comprehensive and rational – though still imperfect – theoretical approach to the problem has been developed. In part, this has involved an appreciation that commercial atomiser designs have departed from theoretical requirements for the sake of speed and convenience of operation, and that the propensity to interference effects has increased in consequence. The history, and present status, of the search for means to overcome the matrix interference problem is summarised below.

Interference effects in electrothermal atomisation have been classified by Matousek.[262] Leaving aside spectral interference from background absorption and scattering (discussed above), the interferences can be categorised as follows:

(a) *Solute volatilisation (condensed phase) effects*:
Volatilisation losses due to volatile compound formation; incomplete volatilisation resulting from occlusion of the determinand within matrix particles, or from refractory compound formation; effects resulting from variations in the rate of atomisation from different matrices.

(b) *Vapour phase effects*:
Shifts in dissociation equilibria (due to volatile compound formation); shifts in ionisation equilibria; changes in the rate of loss of atoms from the atomiser.

In the absence of a clear theoretical framework which took account of atomiser characteristics, early work concentrated upon the use of chemicals – either added to samples, or used to treat the graphite furnace – which have been described as "releasing agents"[263] or (in the case of substances added to samples) as "matrix modifiers".[264] The substances which have been added to samples include ascorbic acid,[258,265] EDTA,[257,266,267] orthophosphoric acid,[268] thiourea[269] and lanthanum.[263] The last of these has also been used to coat furnace tubes prior to analysis[259,263] as also has molybdenum (in conjunction with orthophosphoric acid).[268] The rationale underlying the use of these materials has not always been entirely clear, but the basic idea has often been to ensure that, regardless of the detailed composition of the matrix, the determinand has been atomised from one paticular chemical form. In this way, differences in atomisation rate arising from differences in the form of the determinand prior to the atomisation step* are eliminated. Thus, for example,

---

* With commercial atomisers, a three-step temperature programme is commonly applied. The aqueous sample is first dried at low temperature; next, a pyrolysis (charring or ashing) stage is performed at intermediate temperature – the aim here is to raise the temperature as high as possible without loss of

Regan and Warren[258] added ascorbic acid (and other organic compounds) to samples to try to produce a "molecular mixture" of carbon and sample – the carbon then assisting the reduction of metal oxide to metal. Similarly, thiourea has been used in an attempt to overcome interferences in lead determination by reacting to form the sulphide[269] although it has also been considered to eliminate chloride interference by lowering the lead atomisation temperature.[262,269] This latter effect illustrates another aspect of chemical treatment of samples – separation of the determinand and a matrix component of similar volatility. This is exemplified also by the use of nitric acid (*e.g.*, 270) or ammonium nitrate (*e.g.*, 264) to covert sodium chloride to sodium nitrate and volatile hydrogen chloride or ammonium chloride which will then be removed from the atomiser at relatively low temperatures prior to determinand atomisation. In this way, both the problem of non-specific background absorption due to sodium chloride, and specific interferences due to chloride, may be reduced.

Chakrabarti *et al.*[271] have pointed out that, if a chemical is added to the sample to form a stable compound of the determinand, with higher appearance temperature (*e.g.*, phosphate in lead determination[262]), the term "analyte modification" (*i.e.*, "determinand modification" in the terminology recommended here) is more appropriate than "matrix modification". However, Matousek,[262] in his comprehensive review of interferences in the ETA technique, has remarked that, since both determinand and matrix may be changed in some cases, the term "chemical treatment" is better than either.

Despite the considerable efforts which have been devoted to them, simple chemical treatments alone have not proved very successful in overcoming the serious interference effects on, for example, lead and cadmium determination in fresh waters. A variety of such treatments has been investigated by Thompson, Wagstaff and Wheatstone,[259] and at the authors' laboratory (see ref. 272 for a summary), including EDTA, APDC, ascorbic acid, oxalic acid, orthophosphate, tantalum (tube-coating) and lanthanum. Of these procedures, only the use of lanthanum (either as a furnace coating, or added to samples) produced worthwhile reductions in interference on a consistent basis. The use of lanthanum in electrothermal atomisation AAS to overcome sulphate interference in the determination of lead in fertilisers was reported by Andersson,[273] and its use in water analysis has been pioneered by the Severn-Trent Water Authority in England.[259,263,274]. Although it may not reduce interference effects to acceptable levels for all applications, it has been shown[259,263,272,274,275] to reduce suppression in lead and cadmium determinations to about 10% or less in many fresh water samples – both natural and treated – and has also been found to be effective in reducing interference effects on these determinands in saline samples.[276]

The generally disappointing results obtained by most chemical treatment procedures are not, in retrospect, particularly surprising. It has become

---

determinand, to break down and/or vaporise the matrix (or component of it) to reduce interference effects. Finally, the temperature is raised further to atomise the determinand. More complex temperature programmes may be used for particular matrices, but the three-stage programme described is the most typical.

increasingly clear over recent years that matrix interferences in electrothermal atomisation AAS are extremely complex, and most chemical treatments appear to have been proposed on the basis of their ability to deal with certain specific problems. Thus, the effectiveness of lanthanum treatment is perhaps a greater surprise than the inefficiency of other procedures. Its action in preventing sulphate interference by formation of a thermally stable lanthanum compound with sulphur seems reasonably well established[273,277] but its effectiveness in reducing interferences in water analysis generally might seem to imply that prevention of sulphate interference is not its only action. Indeed, it has been suggested[277] that metal carbide coatings may affect atomisation processes by "blocking" certain active graphite sites, and lanthanum does form a carbide (albeit one which is decomposed by water). However, it must also be noted that lanthanum appears always to have been used in water analysis with nitric acid, and the latter may have assisted the reduction of certain types of interference, notably chloride effects, as discussed previously.

Although interest in chemical treatments continues, attention has been directed increasingly towards the fundamental problems of commercial atomiser design in recent years. This trend is welcomed by the authors because, if there are basic deficiencies in atomiser design which can be remedied, this would seem to be a sounder approach than to attempt to minimise the consequences of poor design by employing a variety of chemical treatments (sometimes of uncertain mechanism), specific to particular determinand/sample combinations. Again, if atomiser modifications do not eliminate the need for chemical treatments, they can at least increase their effectiveness.

The limitations of commercial atomisers of the Massmann type, in relation to the theoretical requirements for efficient atomisation, appear to have been first reviewed comprehensively by L'vov,[278] although other workers had previously noted that commercial atomisers do not fulfil the necessary conditions for achievement of the theoretical maximum sensitivity (see ref. 279 and literature cited therein). L'vov emphasised the effects of atomiser design upon the susceptibility of the technique to interferences.[278] Space does not permit a complete examination of his treatment in this section, but the basic reasoning will be summarised because it is fundamental to many of the recent attempts to minimise interferences by modification of Massmann-type atomisers. Because it is somewhat simpler than that given by L'vov[278] the treatment given by Grégoire and Chakrabarti[279] will be used. (The reader interested in the detailed theoretical background should consult L'vov's review[278].)

Commercial atomisers produce transient absorption pulses, and signal measurement may involve either peak absorbance or integrated absorbance. If the atomisation time, t_1, is much smaller than the residence time of the atoms in the analysis volume, t_2, then $N_{peak} = N_o$ (where N_{peak} is the number of atoms at the peak of the absorption pulse, and N_o is the total number of determinand atoms in the sample). If the condition is met (*i.e.*, $t_1/t_2 \ll 1$), interference due to changes in atomisation rate will not be observed. However, as L'vov[278] has noted, this condition is difficult to meet because of the relatively short mean residence times (t_2) of atoms in furnaces. Pressurisation of the furnace is one means of increasing t_2, but sample introduction is then rendered more difficult. In the Woodriff CTF atomiser, the length of the furnace (some 25 cm) gives long residence times (residence time is proportional to the square of furnace length[278]) but the size of the furnace again poses practical problems. In commercial atomisers, the ideal condition ($t_1 \ll t_2$) for the use of peak height measurement is certainly not met, and therefore effects of sample matrix upon atomisation rate will be manifested as changes in peak height (*i.e.*, as interferences). However, the use of integrated absorbance (peak area) measurement with commercial atomisers will not, on its own, completely solve interference problems because such atomisers do not meet the theoretical conditions for correct use of this mode of measurement either, as L'vov[278] has demonstrated.

The integrated absorbance will be directly proportional to the function Q_N, which is given by the equation:

$$Q_N = \int_0^{t_3} N(t)\, dt = N_0 t_2$$

(t_3 is the time over which the signal is measured, $N(t)$ is the number of atoms inside the furnace at time t, and N_0 and t_2 have the same significance as before).

It can be seen that Q_N (and hence the integrated absorbance) is a function of residence time t_2, but is independent of the atomisation time, t_1. Thus matrix effects upon atomisation kinetics should not influence peak areas. However, for the equation $Q_N = N_0 t_2$ to be valid, two conditions must be fulfilled: the determinand must be completely atomised in the time t_3, and t_2 – the residence time – must be constant. The first of these conditions, as L'vov has remarked,[278] is the easier to fulfil because (with the exception of elements of the lowest volatilities) virtually complete atomisation can be ensured by increasing atomisation temperature and/or the period of furnace heating. The second condition, constancy of residence time t_2, is not fulfilled by Massmann-type atomisers because atomisation frequently occurs when the furnace tempera- ture is still rising – and constancy of t_2 demands that the furnace is isothermal (spatially and temporally) during the absorption pulse. Spatial non-isotherma- lity is to some extent found with the L'vov cuvette, but it is more serious with Massmann-type furnaces held at the ends.[278] Lack of isothermal conditions during the absorption pulse is probably a more serious problem, and is characteristic of Massmann-type furnaces (though of course absent in constant temperature atomisers such as those of L'vov and Woodriff).

The theoretical supremacy of constant temperature furnace systems over commercial atomisers with respect to interferences has been confirmed by experimental observation. It has already been noted that the freedom from interferences of the graphite cuvette led L'vov to believe that absolute analysis (*i.e.*, analysis without standardisation) might ultimately be possible. Again, the relative freedom from interferences of the Woodriff CTF atomiser has been demonstrated both with respect to manganese[280] and lead[281,282] determination. [When all the major ionic components of fresh water were tested in a mixture, at concentrations greater than those expected in drinking water samples, the *maximum possible* bias (95% confidence) was less than 6% when the Woodriff

CTF was used to determine lead at 45 μg l^{-1} – see ref. 282]. Moreover, all the above studies employed peak height measurement successfully, and this is presumably a consequence of the fact that the condition $t_2 \gg t_1$ (see above) is more closely fulfilled in the Woodriff furnace than in the other types.

Recognising the superiority of CTF atomisers over pulse-heated systems, and the crucial role of non-isothermal conditions in the latter, many groups of workers have attempted to ameliorate interference effects in commercial furnaces by modifications designed to attain more nearly isothermal status during atomisation. Two basic approaches have been followed: in the first, an attempt is made to delay sample atomisation until the furnace has attained a more nearly isothermal state, in the second, an attempt is made to heat the furnace to equilibrium temperature in a shorter time than the duration of the adsorption pulse.[278] This second approach is equivalent, as L'vov has remarked,[278] to making $t_2/t_1 > 1$, and can be achieved by increasing the length of the furnace (to increase t_2) and/or by decreasing t_1 by increasing furnace heating rates.

Lundgren and co-workers[283,284] developed the first "temperature-controlled" furnace. In this system, a photodiode monitored the furnace temperature and, when the desired temperature was reached, a feedback circuit sharply decreased the input power so that a constant temperature could be maintained (to within ± 10 °C). By this means, the final temperature and heating rate could be chosen independently, and high heating rates achieved with little "overshoot" of the final desired temperature. With a large input power (15 kw), only relatively modest heating rates were obtained (up to 900 °C s^{-1}). The furnace was longer than commercial Massmann-based systems (10 cm rather than 2–3 cm), and thus the residence time, t_2, was relatively long.[278] Frech[285,286] showed that the use of a temperature-controlled furnace gave a reduction in matrix effects in the determination of lead and antimony in steels, compared with commercial atomisers, but the high power requirements of the Lundgren furnace (together with the relatively large size) would appear to have discouraged its commercial exploitation. Temperature control by a feedback system has, however, been incorporated in commercial atomisers (see, for example, ref. 287).

Much more rapid heating can be achieved using capacitative discharge systems, and this approach has been exploited both by L'vov[278] and by Chakrabarti and his colleages.[288-290] L'vov[278] noted that the idea was patented in 1974, and surmised that the early workers in the field had been hampered by the low resistance of conventional graphite furnaces. Both L'vov's[278] and Chakrabarti's[288] systems have dealt with the problem by employing high voltage capacitors, so reducing the physical size of the capacitor bank needed. Additionally, both systems exploit the properties of anisotropic pyrolytic graphite to ensure spatial isothermality. Heating rates of 40 000[278] to 75 000 °C s^{-1}[288] were obtained. Chakrabarti's group have reported both enhanced sensitivity[288] and much reduced matrix interferences[289,290] in comparison with commercial atomisers, though it was noted that, at its present state of development, the system did not allow direct determination of trace elements in sea-water.[290] At the time of writing (Spring 1984), the authors are not aware of any commercial capacitative discharge atomisers being available.

Turning next to attempts to delay atomisation, the most important development has been the use of graphite platforms within commercial Massmann-type atomisers – an idea first investigated by L'vov and his colleagues.[291] In this approach, a small platform of anisotropic pyrolytic graphite is inserted into the conventional Massmann-type furnace, and the sample pipetted onto it. The lamellae of the graphite are oriented parallel to the base of the platform, and the contact between the platform and the furnace itself is very small. In this way, direct heating of the platform (and sample) by conduction is minimised, and atomisation is delayed until the furnace is more nearly isothermal. L'vov's group reported[291] increased peak height sensitivity for determinands of high (lead and cadmium) and medium (copper and nickel) volatility, and a reduction of interference effects, considered by L'vov to be consistent with theoretical expectations. The platform concept has been taken up by the Perkin-Elmer Corporation (under the name "L'vov Platform", a trademark of the Corporation[292]), and a further study on lead determination was reported by Slavin and Manning[292]).

That study confirmed the finding of L'vov's group that chloride interference was much reduced when the platform was used, and showed that sulphate interference was also greatly diminished. However, even when the platform was used with temperature-controlled heating and lanthanum treatment, potentially serious interference effects were reported for a natural water sample. Subsequently, use of the L'vov platform has been further rationalised by Slavin and his associates[293] in the 'Stabilised Temperature Platform Furnace' (STPF) concept. These authors have noted[293] that, to obtain accurate results, the analyst *may* need to employ (in addition to the L'vov platform): temperature-controlled heating, a spectrometer with suitably fast signal processing (to avoid distortion of peaks*), signal integration, pyrolytically coated furnaces *and* chemical treatment. By incorporating some or all of these features, a number of studies have demonstrated that interference effects in water analysis can be markedly reduced by the platform technique; see, for example, refs. 204 (manganese in sea-water), 205 and 206 (cadmium in sea-water), 295 (aluminium in sea-water) and 296 (lead and cadmium in potable waters). Recently studies by Koirtyohann *et al.*[297] and Chakrabarti *et al.*[298,299] have confirmed that the use of platforms in commercial atomisers does indeed allow a close approach to isothermal atomisation conditions.

Another approach to achieving constant temperature atomisation in pulse-heated furnaces is the 'second surface atomiser' developed by Holcombe and his associates – see ref. 300 for details of the latest version, in which a gas-cooled tantalum plug is inserted through the wall of a conventional pulse-heated furnace. This system involves the use of a high-temperature 'ashing' stage, in which the determinand atoms are recondensed onto the cooled plug, followed by atomisation when the coolant flow is stopped. Background absorption is much reduced, because the matrix salts tend to be

---

* See also ref. 294 for a discussion of this topic

removed during the high temperature ashing, and atomisation again takes place into a constant temperature environment.

Perhaps the simplest approach to truly constant temperature atomisation with a conventional Massmann-type furnace is the introduction of a dried sample into a furnace pre-heated to atomisation temperature; *i.e.*, the use of the pulse-heated furnace as a constant temperature furnace! Early attempts to achieve this[301,302] used a tungsten wire probe, onto which the sample was dried, which could be inserted into the hot furnace. As predicted by theory, interference effects were reduced compared with conventional pulsed atomisation;[301,302] the system could not be readily automated, however, and (unlike the platform technique) has not been widely used. Recently, a more readily automated probe atomisation technique has been developed by Ottaway and his associates.[303,306] In their system, a probe is inserted through an additional orifice cut into the front face of the furnace, and the sample can be injected onto it using a conventional automatic sampler system. Drying and ashing can be performed in the usual manner, and the probe then removed while the furnace is heated to atomisation temperature, after attainment of which the probe is re-introduced. These operations can be automated using a solenoid system to drive the probe in and out of the furnace,[305] and interferences on lead determination in chloride media[303] and whole blood[304] are reported to be substantially reduced in comparison with conventional atomisation. The technique appears not to have been studied in relation to water analysis as yet, but would seem to offer considerable promise.

Certain commercially available atomisers employ aerosol deposition, rather than introduction of a discrete droplet, as the means of introducing the sample into a Massmann-type furnace. It is reported[307,308] that the more even distribution of sample so obtained decreases the size of matrix salt crystals formed on drying, and thereby reduces suppressive interference resulting from loss of determinand occluded within such crystals when they are ejected from the furnace without thermal degradation (as suggested by Copeland and co-workers[309,310]). Moreover, aerosol deposition has also been claimed[307,308] to reduce two other potential sources of interference, namely spreading of the sample within the furnace, and soaking into the graphite.

It is not possible, in a book of this kind, to provide a complete review of electrothermal atomisation in all its facets. The reader wishing to pursue the subject in greater detail is referred to the comprehensive reviews by Sturgeon and Chakrabarti[311] and Matousek,[262] to the influential analysis by L'vov,[278] to the recent issue of *Spectrochimica Acta*[312] devoted to graphite furnace technology and atomic absorption (which includes a history of the technique by L'vov[313]) and to the individual references cited above. The aim of the present section has been to present the non-specialist with a succinct and considered appraisal of the current position, with particular emphasis upon the role of theories of atomiser function, which have exerted a considerable influence upon recent developments to reduce matrix interferences.

Given the major benefits of electrothermal atomisation – in particular, the capacity for direct trace metal determinations in natural and treated waters – it is to be expected that future years will bring further major improvements to the technique. It seems likely that, although interest in chemical treatments will continue, the main thrust of future work will continue to be concerned with more comprehensive approaches involving further changes in commercial atomiser operation and design, and with improvements to the cost-effectiveness of the technique (such as the use of longer-lived glassy carbon furnaces[314]).

Although the above treatment has emphasised direct analysis, it must also be noted that preliminary separation of the determinand from the salt matrix can be used to overcome interference effects. Although it might at first sight seem equally disadvantageous to perform solvent extraction (for preconcentration) with flame AAS and solvent extraction (for matrix removal) with graphite furnace AAS, the latter approach may allow a more convenient scale of extraction. This benefit is well illustrated by the work of Mitcham[315] in which a small scale APDC/MIBK extraction (based upon a nationally recommended flame method for the determination of lead in drinking water[244]) is combined with a graphite furnace finish.

The combined use of solvent extraction and electrothermal atomisation AAS is also of particular importance in sea water analysis. In this case, the extraction serves the dual purpose of separating the determinand from the salt matrix and providing a suitable preconcentration. Chelating ion-exchange resins may also be used. References to such methods are given in Table 6.4 of the monograph by Fuller[197] and in ref. 214.

In summary, electrothermal atomisation AAS using graphite furnaces has the capability for direct determination of trace elements at very low concentrations in many types of water, but the full potential has yet to be realised because of the propensity of the technique to serious matrix interferences. A better understanding has now been gained of the causes of these problems (which derive from the pulse-heated nature of commercial furnaces), much progress has been made and further improvements can be expected.

Commercial atomisers currently available are of very similar design and construction, and no particular type can be recommended as having major advantages. This position may change, however, and the reader who finds himself faced with making a choice is advised to consider, in particular, the extent to which a particular system is likely to allow isothermal atomisation conditions to be met. In the case of pulse-heated furnaces, the heating rates attainable are likely to be of particular importance, but other measures may be introduced to aid the attainment of more nearly isothermal conditions during atomisation. Constant temperature furnaces have proved very successful in regard to interferences, but there seems little likelihood of their being introduced commercially by major manufacturers in the immediate future.

Recovery tests using determinands and water types of interest, particularly those likely to give rise to serious interferences (if such information is available), are strongly recommended as a means of assessing furnace systems prior to purchase.

(d) Hydride Generation AAS

(i) General basis and instrumentation The detection limits attained for arsenic and selenium by direct aspiration flame AAS are inadequate in relation to most

Analytical Techniques

potential applications in water analysis, being about 1–2 orders of magnitude higher than would be required, for example, in drinking water monitoring. Although the detection limits attainable by graphite furnace AAS are markedly lower, as shown in Table 11.3, such that direct analysis would be possible in many water monitoring applications, matrix modification is necessary to stabilise the determinands, which would otherwise suffer serious losses during ashing (see, for example, refs. 316–318). Moreover, matrix interferences have been observed, and standard additions calibration may be necessary.[317,318]

The generation, and subsequent atomisation prior to AAS measurement, of the gaseous hydrides of these elements, and those of certain others (antimony, bismuth, germanium, tellurium, tin and lead) allows greatly improved detection limits to be achieved compared with those which can be attained by direct flame AAS.[216,319–323] Furthermore, the detection limits are usually lower than those attainable by direct graphite furnace AAS, though larger sample volumes are, of course, required). Although the hydride/AAS technique is not without interferences and other problems, these appear to have been less serious and/or more readily solved than those affecting the graphite furnace determination, at least for arsenic.

There has been a considerable growth of interest in the technique since the first

TABLE 11.3
*Sensitivities and Detection Limits for Graphite Furnace AAS**

Element	Sensitivity† pg	Detection‡ Limits ng l^{-1}	Element	Sensitivity† pg	Detection‡ Limits ng l^{-1}
Ag	0.5–2	1,5	Li	2.4–4	10,300
Al	4–10	10	Mg	0.07–0.3	0.2,4
As	8–12	80,200	Mn	0.7–1.6	0.5,10
Ba	4–22	40	Mo	8 12	20,30
Be	0.6–1	3	Na	0.2–1	4
Ca	0.6–1.3	10.50	Ni	6.0–20	50,2000
Cd	0.2–3	0.2,3	Pb	1.8–4	7,50
Co	3–11	8,20	Se	8–26	50,500
Cr	1–4	4,10	Si	10–73	100,600
Cu	2–6.4	5,20	Sn	7–22	30,2000
Fe	2–3	10,20	V	20–40	100,200
Hg	40–360	200,2000	Zn	0.2–0.3	1
K	0.4–0.8	4			

* Data for the construction of this Table were sought from four major manufacturers of AAS equipment: Instrumentation Laboratory (UK) Ltd., Perkin-Elmer, Ltd., Pye-Unicam Ltd. and Varian Associates Ltd. These companies were asked to supply detection limits for a 'top-of-the-range' instrument but several reported that the data given applied to all the instruments in their range or did not identify a particular instrument in their replies. Where a manufacturer reported more than one value (for different atomisation conditions, for example) the lowest was always used in constructing the Table.
† Sensitivities are in picograms for 1% absorption, in peak height mode. [Sensitivity data, rather than detection limits, were supplied by two companies and both sensitivity and detection limit data were given by a third. For completeness, sensitivity data in peak height terms were obtained from the fourth company, though the latter pointed out that increasing use of peak area measurement (see text) has, in their view, reduced the relevance of peak height sensitivity values.]
‡ It is understood that the detection limits were all defined in broad conformity with the definition of 'Criterion of Detection' as used in this book (see Section 8.3.1).

edition of this book and the range of procedures which has been employed – particularly for the determination of arsenic and selenium – is vast. Thus, for example, a comprehensive review of the determination of these two elements in waters, by hydride generation/AAS, produced recently by Gunn[319] contains over 180 references. A wide variety of approaches, both to the generation of the hydrides and to their subsequent measurement, have been employed. The reader is referred to the aforementioned review, and literature cited therein, for full details; the following description summarises the main points, with particular reference to arsenic and selenium, which are the elements most commonly determined by this technique.

Generation of the hydride is now often achieved using sodium borohydride, which has numerous advantages over alternative systems (among which are zinc/acid and aluminium/acid). These advantages include its ability to be used in solution form (allowing ready purification and convenient use in automated systems) and its suitability for generating the hydrides of elements other than arsenic and selenium.

The generation of arsine is slower from As^V than from As^{III}, so that a pre-reduction to convert any As^V present to As^{III} is usually incorporated. Potassium or sodium iodide is normally employed for this purpose. Alternatively, advantage can be taken of the oxidation of As^{III} to As^V by a reagent (*e.g.*, persulphate) added to ensure conversion of organic arsenic species to inorganic forms (see below). Although reduced sensitivity may result, because the arsine is then generated from As^V, the performance may still be adequate for many purposes. This approach has been exploited successfully in a recently developed automated procedure for the determination of total arsenic in drinking water.[326]

In the case of selenium, prior reduction of Se^{VI} to Se^{IV} is essential, since a virtually negligible yield of the hydride will be obtained from any Se^{VI} present. Heating the sample in the presence of hydrochloric acid has usually been employed for this purpose, though care has to be taken to avoid loss of selenium during this process.[323] Similar behaviour has been reported in the case of tellurium, and heating with hydrochloric acid again proved effective.[323,324] In the case of antimony, the sensitivity of the hydride technique has been reported to be less for Sb^V than for Sb^{III}, by a factor of about two, but pre-reduction of the former with potassium iodide in acid solution was readily achieved;[323] in another study,[325] only Sb^{III} was reduced at near-neutral pH but both Sb^{III} and Sb^V could be reduced together under highly acidic conditions in the presence of iodide. Lead and bismuth are expected to occur in water to an important extent only as Bi^{III} and Pb^{II}.

Although certain alkylated arsenicals can be determined directly by generation of the corresponding substituted arsines, other organo-arsenic species are unlikely to be recovered without prior digestion. A variety of procedures has been employed, including acid/permanganate and u.v. photolysis, but it has been concluded[319] that acid/persulphate is most generally suitable, both for arsenic and selenium.

After generation, the hydride is swept out of solution and carried to the atom cell (either directly, or *via* some form of collection system) by a stream of inert gas. Collection systems (*e.g.*, cold-trapping) offer increased peak-height sensitivity, but complicate the procedure and may be unnecessary in many applications. Collection of arsine in a reagent solution (*e.g.* aqueous potassium iodide/iodine), with subsequent determination of arsenic by graphite furnace, has also been used[327] but passage of arsine, directly or indirectly, to the atom cell is far more common.

The usual flames for AAS are not well suited to As and Se determination, because of their strong absorption of radiation in the relevant, far-u.v., region. Consequently, argon–hydrogen or nitrogen–hydrogen diffusion flames (much more transparent at such wavelengths) or heated quartz tubes are normally employed. Gunn[319] has reviewed the use of such cells, and has concluded that quartz tubes are preferable to hydrogen diffusion flames, having even lower background absorption and longer determinand residence times (and hence improved sensitivity). Quartz tubes may be heated by flame (usually the normal air–acetylene flame), or electrically. Tube replacement may be less simple with electrical systems, but the heating is more uniform than with flames. Temperatures between 600 and 1200 °C are usually employed, and the same temperature has been reported to be satisfactory for both As and Se.[319] It has been recently reported by Welz and Melcher[328] that the atomisation of gaseous hydrides in quartz tube atom cells proceeds not by direct thermal decomposition, but by collision with free hydrogen radicals formed by reaction with oxygen at temperatures in excess of 600 °C. The addition of oxygen to the gas mixture entering the tube may, therefore, produce enhanced sensitivity. Again, this mechanism of atomisation explains the depression of signals which can occur when dust and dirt are present in the tube – such materials catalyse radical recombination reactions and thereby reduce the concentration of the hydrogen radicals necessary for atomisation. Cleaning the tube with hydrofluoric acid can help to overcome this problem.[328] Conventional graphite furnaces have also been employed as atom cells in hydride systems – see, for example, ref. 329 – but are not commonly used.

Most, if not all, manufacturers of AAS equipment offer hydride generation accessories based on quartz tube cells. Such systems usually incorporate apparatus for the generation of the hydrides and permit, to varying degrees, partial automation of the analytical procedures. Fully automated hydride generation systems (using continuous flow technology) have been developed by many workers (*e.g.*, refs. 326, 330–333).

(ii) Principal analytical characteristics   As has been noted, a very wide range of instrumentation and methodology have been employed for the determination of As and Se by hydride generation. This makes difficult the task of summarising performance characteristics, but an attempt to do so is made below. The summary given relies heavily on the wealth of data collated by Gunn[319] who has provided a detailed assessment of the capabilities of published procedures for As and Se. Much less information is available concerning other determinands amenable to the hydride/AAS technique, but some data are presented as a guide.

(1) *Limits of detection*
   For arsenic and selenium, detection limits of about 0.1–1 μg l^{-1} have been commonly attained using both hydrogen diffusion flames and quartz tubes, directly without intermediate hydride collection. With collection systems, such as cold-traps, much lower limits (down to about 1 ng l^{-1}) have been achieved.[319]

For both determinands, the use of fully automated systems (without hydride collection) would appear to have permitted lower limits of detection than those attainable with (direct) manual operation. Thus, values of about 0.01 μg l^{-1} (As) and 0.02 μg l^{-1} (Se) have been reported[330] for an automated system without hydride collection, and a recently developed system[326] for the determination of total arsenic in raw and potable water, using similar automated instrumentation, has allowed a limit of detection (as defined in Section 8.3.1) of 0.07 μg l^{-1} to be attained.

Reported detection limits for other elements determined by hydride generation/AAS include the following: antimony, 0.5 μg l^{-1} (automated procedure[331] adopted by the World Health Organisation in a recent compilation of methods[333]), *ca.* 0.5 ng l^{-1} (cold-trapping procedure[325]); bismuth, 20 ng l^{-1};[323] germanium, 0.5 ng l^{-1} (cold-trapping procedure[329]); lead, 100 ng l^{-1};[320] tellerium, 20 ng l^{-1};[323] tin, *ca.* 10 ng l^{-1} (cold-trapping procedure[334]).

(2) *Precision*

Examination of collected data[319] indicates that, for arsenic and selenium, relative standard deviations of 1–5% are readily attainable at concentrations greater than ten times the detection limit. Similar precision can be expected for other elements amenable to determination by the hydride/AAS technique. Thus, for example, a relative standard deviation of 3.7% at ten times the detection limit has been reported for antimony determination by a fully automated method[331] and one of 2.5% for lead determination by a manual method at 100 times the detection limit.[320]

It has been concluded,[319] for arsenic and selenium determination, that precision tends to be better with automated systems than with manual analysis, but only marginally so.

As with other AAS variants, standard deviation tends to increase with increasing concentration of the determinand.

(3) *Selectivity*

Detailed data on interferences in the hydride/AAS determination of arsenic and selenium have again been collated by Gunn.[319] Suppressive interferences can arise from three main sources: metal ions, other hydride-forming elements and oxyanions/acids.

Metal ion effects of widely differing severity have been reported by various studies of arsenic and selenium determination, but nickel and copper appear to present the most serious difficulties.[319] Interference by nickel and copper present simultaneously[335] and by copper alone[336] has also been reported in the hydride/AAS determination of lead. It has been proposed that the suppressive effects of metal ions result from their preferential reduction to a lower oxidation state, or the free metal itself, with consequent precipitation. The precipitate may interfere by co-precipitation of the determinand, or by absorbing, decomposing or retarding the generation of, the hydride of interest.[337] Effects of metal ions on selenium determination are generally more severe than in the case of arsenic, possibly because of the formation of highly stable selenides under reducing conditions.[338,339]

Masking procedures have been employed to reduce or eliminate interferences from metals. Thus, for example, 1,10-phenanthroline has proved efficacious in reducing Ni interference in the determination of As,[340] Te (IV) in reducing interferences from a number of heavy metals (and hydride-forming elements) in the determination of selenium[338,339,341] and potassium cyanide in reducing copper interference in the determination of lead.[336] However, careful attention to the conditions of hydride generation may obviate the need for masking in some cases – thus, for example, it has been reported that increasing borohydride concentration can reduce the suppressive interference of Cu in the determination of As.[326]

As in the case of metals, interference effects on As and Se determination from other hydride-forming elements have varied widely in severity, and the mechanism(s) of interference are poorly understood.[319] Again, as with metal ion interferences, the efficiency of hydride generation may be an important influence upon the extent of interference, and the nature and concentration of the reducing agents used a major factor in explaining the wide variety of reported effects.[319] Thus, for example, Yamamoto and Kumamaru[342] have reported that the presence of potassium iodide eliminated interference from Bi, Se and Te in the determination of As, when borohydride was used as the principal reductant.

Various oxyanions and acids (including nitrate, nitric acid, sulphuric acid and perchloric acid) have been reported to interfere in the determination of Se and/or As (see ref. 319 and literature cited therein). Although such effects may have an important influence on the choice of reagent or preserving acid(s), such that nitric acid is usually avoided, for example, little difficulty should arise from oxyanions present naturally in most types of water.

Apart from the interferences noted above, the hydride/AAS technique appears to be largely unaffected by the major ionic components of natural waters; indeed, the technique has been applied to the direct analysis of saline waters [see, for example, refs. 343 (As) and 344 (Se)].

With respect to recovery of different forms of the determinands, it has already been noted that a preliminary digestion will usually be necessary to convert organic species to inorganic forms prior to hydride generation, if total element concentrations are required. Persulphate has been recommended for this purpose, not only for arsenic and selenium[319] but also for antimony.[333] Although the rate and extent of hydride generation from inorganic As and Se, for example, is dependent upon the oxidation state of the determinand, it has been remarked that appropriate pre-treatment (usually reduction to As[III] or Se[IV]) can overcome this potential source of bias in the determination of total concentrations.

The dependence of hydride generation, under certain conditions, upon the oxidation state of the element of interest, and the ability to form substituted hydrides directly from certain organic forms of the elements, have been exploited in speciation studies to determine the concentrations of different forms of hydride-forming elements in waters. This aspect of the hydride-generation technique is considered further in Section 11.5 which deals with speciation analysis.

In summary, the hydride generation/AAS technique offers very low limits of detection, and excellent general precision, for a number of elements which cannot be determined with adequate performance for many applications by flame AAS. The technique is, however, subject to a number of potential difficulties in respect of selectivity – although the data presented here have referred mainly to arsenic and selenium, broadly similar problems (in respect of interferences, for example) are to be expected for other determinands amenable to the production of volatile hydrides. Means of reducing or eliminating these difficulties are often available, however, and hydride generation/AAS is now widely used in water analysis. The equipment required is readily available, and recommended methods employing the technique have been published by various national and international organisations (*e.g.* refs. 332, 333, 345). The reader is cautioned, however, that the interference effects to which the technique is subject are of a complex nature, and published methods often lack detailed information about them.

(e) Cold Vapour AAS for Mercury Determination

(i) General basis and instrumentation   The detection limits attained for mercury by direct aspiration into the air–acetylene flame and by direct graphite furnace AAS are inadequate in relation to most potential applications in water analysis (for which concentrations of about 1 μg l^{-1} or less are typically of interest) – see Tables 11.2 and 11.3. However, much lower detection limits can be attained by using the cold vapour technique, which exploits the fact that mercury has an appreciable vapour pressure at room temperature, and exists as a monatomic gas. In this technique, ionic mercury in solution is reduced to elemental mercury which is swept* by a carrier gas (simple compressed air is adequate) to the absorption cell (a glass or plastic tube with quartz end windows). The transfer may be direct, or an intermediate trap may be used to increase sensitivity.

The early work on the cold-vapour technique by Hatch and Ott[347] is well known. Developments have been reviewed by Ure[348] (in 1975) and – with particular reference to natural water analysis – by the Watlings.[349]

Stannous chloride is usually employed as the reductant, though the use of sodium borohydride has also been reported.[350] In the simplest systems, a 'one-pass' procedure is used, in which the mercury vapour is passed through the atom cell once only. Recirculation of the vapour has been reported[351] to give an improved detection limit, but for work at very low levels some form of concentration technique is required. Although preconcentration by means of dithizone[352] and chelating resin[353] has been employed, collection of the generated mercury prior to AAS measurement represents a particularly attractive approach to improving detection capability. Trapping has been achieved using liquid nitrogen,[354] gold amalgam formation[355] and solutions of permanganate[356] or brominating reagent[357] (albeit in the latter case for an atomic fluorescence finish).

All methods involving reduction with stannous chloride face the problem that organomercurials (such as methylmercury) are not converted to the free metal by that reagent, and a digestion step must be incorporated to effect their conversion to inorganic mercury first. Methods have been developed which employ a simple

* Water displacement of the mercury-laden air has also been used.[346]

acid/permanganate digestion, but it has been reported that such digestion gives poor recovery of organomercurials, which can be readily recovered if persulphate is included in the digestion (see, for example, ref. 157 and literature cited therein). It has been concluded[352] that the essential factor is digestion period; the use of permanganate alone requires a longer digestion than when permanganate *and* persulphate are employed. Other procedures for digestion of organic mercury forms which have proved effective include u.v. photolysis[358] and bromination.[359]

As in the case of hydride generation, many methods for the determination of mercury in waters by cold vapour AAS have been published, and some automated systems have been developed (see ref. 358). Again, kits for cold vapour mercury work are marketed by manufacturers of atomic absorption spectrometers.

(ii) Principal analytical characteristics

(1) *Limits of detection*

With direct methods not involving collection of generated mercury, typical detection limits are 40–200 ng l^{-1} for sample volumes of 50–100 ml.[216,360] Substantially lower detection limits, down to about 1–10 ng l^{-1}, can be attained using intermediate mercury collection (see, for example, ref. 361).

(2) *Precision*

Relative standard deviations of about 2–5% should be attainable, using all forms of the technique, provided that the determinand concentrations are not near to the detection limit and that (in the case of low concentration work particularly) contamination can be adequately controlled. Again, standard deviation will tend to increase with increasing determinand concentration.

(3) *Selectivity*

Reference has already been made to the need for a digestion step to convert organomercurial forms to inorganic mercury prior to reduction. Potential interferents in the AAS measurement stage are volatile organic compounds and free chlorine, generated by oxidation of chloride ion, both of which could absorb at the mercury resonance line (253.7 nm). Sufficiently vigorous oxidation to destroy organic matter, and the use of background correction, should deal with these problems in most applications, but u.v. photolysis has been developed as an alternative to permanganate/persulphate oxidation for saline waters – not only because of absorption by free chlorine, but also because generation of the latter may exhaust oxidative capacity and hence impair organomercurial recovery.[358] Intermediate collection of generated mercury has the benefit of separating the latter from other volatile materials, and the technique of gold amalgam trapping has been applied to water analysis.[361]

11.3.6 Ion-selective Electrodes

The glass electrode, widely used for the determination of pH, remains the most

422 · Chemical Analysis of Water

familiar example of an ion-selective electrode*. The measurement of pH is so well established that it is not considered in detail here. Texts dealing with the subject in great depth are available (see refs. 362–364), but the need for particular care to control several sources of error when determining the pH of high-purity and poorly buffered natural waters is again emphasised.[362,365] The choice of reference electrode may be of particular importance in such determinations;[365] Midgley and Torrance have examined[366] a number of such electrodes for the determination of pH in ammonia-dosed boiler feedwater (of low conductivity). The same authors have also reported the results of a similar study of glass electrodes.[367]

The development of electrodes responding to ions other than hydrogen has produced a number which have found application in water analysis [including on-line analysis (see Section 12) to which the technique is well suited]. However, only a limited number of electrodes have proved suitable for widespread use in routine water analysis, though a number of others have found application in research work. Ion-selective electrodes remain, therefore, more limited in range of application than (for example) solution spectrophotometry or atomic absorption spectrometry.

(a) General Basis of the Technique

Ion-selective electrodes can be regarded as electrochemical half-cells consisting of a suitable reference electrode/reference solution system, separated from the sample by a membrane (see Figure 11.4)†. The nature of the membrane determines the ions to which an electrode responds – when the electrode is placed in a solution, an electric potential is developed across the membrane, the magnitude of which is

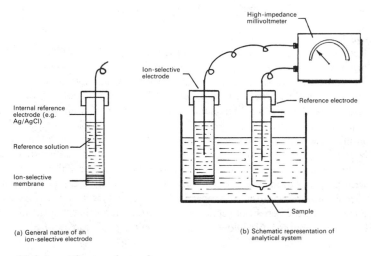

Figure 11.4 *Ion-selective electrodes*

* The term 'specific-ion electrode', criticised in the first edition of this book, has, fortunately, disappeared from the literature.
† Membranes containing silver salts may be connected directly to a metallic conductor, avoiding the use of the reference electrode and solution.

dependent upon the thermodynamic activity (and therefore, at constant ionic strength, upon the concentration) of the ion to which the electrode responds. Measurement of this potential is accomplished by completing the electrochemical cell by means of a suitable reference electrode, and connecting the electrodes to a high-impedance millivoltmeter (see Figure 11.4). Most laboratory pH meters have millivolt or pIon scales and are suitable for use with all ion-selective electrodes, and many portable meters have similar facilities.

A number of texts on the ion-selective electrode technique are available,[368–372] and one[370] deals specifically with its application to water analysis, giving methods (and performance characteristics) for many determinands. Reviews of the application of the technique to water analysis have been produced recently by Hulanicki and Trojanowicz[373] and by Herman.[374] Reference electrodes are the subject of a text by Ives and Janz,[375] and have also been reviewed by Covington and Rebelo.[376] They are also discussed in texts on ion-selective electrodes.

One of the most attractive features of the ion-selective electrode technique is its simplicity and rapidity. Thus, in principle, it is necessary only to immerse the two electrodes in the sample and then to read the result either directly from the meter or from a calibration curve. In practice, most applications to water analysis require the addition of a reagent to the sample before measurement of the potential (see below); however, this addition does not usually have to be made with great accuracy so the simplicity of the technique is preserved. In certain cases, the 'standard addition' method of calibration (see Section 6.2.2) may be necessary and this will, of course, reduce the speed of analysis.

Although the number of ions for which suitable electrodes are available has increased since the first edition of this book, the range of determinands covered remains limited, particularly with respect to water analysis, in comparison with certain other common techniques. The use of electrodes as end-point detectors in potentiometric titrations allows the range of determinands to be extended considerably. For example, aluminium can be titrated with fluoride, using a fluoride-selective electrode;[377,378] sulphate can be titrated with lead[379,380] or barium[382] using lead- or barium-selective electrodes, and phosphate can be titrated with lead using a lead-selective electrode.[383] Although potentiometric titration may often allow better accuracy than does potentiometry[370] (as well as increasing the range of determinands amenable to the ion-selective electrode technique), the associated disadvantages – especially the increased time for analysis, even with autotitration equipment – are such that the direct potentiometric approach will be emphasised here. For the reader interested in potentiometric titrations using ion-selective electrodes, chapter 6 of the monograph by Midgley and Torrance,[370] chapter 9 of the monograph by Bailey[371] and the review by Mascini[384] should be consulted.

(b) Available Ion-selective Electrodes and Their Applications

The selectivity of an electrode is conferred by the membrane employed. In addition to the glass electrodes (for hydrogen, sodium and – more rarely – potassium), there are two basic types of electrode: solid state electrodes having inorganic salt membranes and ion-exchange electrodes, in which the membrane contains an organic material having a selective ion-transport capability. (The latter type include the so called neutral-carrier electrodes, see below). Gas-sensing membrane 'electrodes' are not strictly electrodes at all, but complete electrochemical cells in which an ion-selective electrode is used to monitor changes in cell chemistry caused by determinand transfer across the cell membrane separating the internal medium from the sample. The term 'gas-sensing membrane probe' would seem preferable, but the similarities between such probes and true ion-selective electrodes are such that they will be considered here along with the latter.

Electrodes based on inorganic salts are of various types. The term 'homogeneous membrane' is commonly applied to electrodes in which the membrane material is in the form of a single crystal (*e.g.*, of lanthanum fluoride in the fluoride-selective electrode) or a compressed pellet (*e.g.*, of silver bromide and silver sulphide in a bromide-selective electrode). The term 'heterogeneous membrane' is applied to electrodes in which the active material is contained within an inert matrix composed of, for example, PVC or silicone rubber.

In both types of inorganic salt electrode, the current may be carried by the determinand or by another ion whose concentration is controlled by the determinand, *via* a solubility equilibrium involving the membrane.

Certain commercial inorganic salt electrodes allow the user to impregnate a suitable conducting matrix with active material and (by shaving away the existing electrode tip and re-impregnating) to change, subsequently, the active material to produce an electrode selective for another ion. Such systems may offer advantages to the analyst making only occasional determinations of the ions concerned. Ref. 370 lists the determinands covered by several commercially available systems of this type, but also warns that the electrodes usually require conditioning before use. Certain types of electrode having ready prepared membranes have interchangeable tips and permit the same electrode body to be used for different determinations.

Ion-exchange electrodes have a membrane composed of an organic material, immiscible with water, having a selective ion-transfer capability. The first such electrodes to be introduced had the ion-exchanger dissolved in an organic solvent, separated from the sample by a hydrophobic, porous membrane (such as a cellulose acetate filter). The name 'liquid ion-exchange electrode' was applied to such devices. It is now more common for the ion-exchanger to be incorporated in a suitable inert matrix (such as PVC or silicone rubber) from which a solid membrane can be produced. The term 'ion-exchange electrode', now used widely, reflects the predominance of this type of construction.

Bailey[371] has provided a summary of the advantages and disadvantages of the two types. He concludes that the liquid type, which usually have longer lives, are better suited for frequent use and for continuous monitoring purposes, whereas the solid type are better for less frequent analysis, for potentiometric titrations and for situations where there is a danger of brief exposure to strong interferents (from which the solid electrodes recover more quickly). He also concludes that, despite the predominance of the solid type in commercial electrodes, the advantage of one type over the other may frequently be marginal.

In some 'ion-exchange' electrodes the active material is not a large organic ion (as in true ion-exchange electrodes) but a large organic molecule (a 'neutral carrier') which forms a complex with the determinand ion by an ion-dipole mechanism.

As in the case of inorganic salt electrodes, interchangeable electrode tips are a feature of many commercial ion-exchange electrodes.

Gas-sensing probes, as noted previously, consist of an electrochemical cell – a component of which is an ion-selective electrode – separated from the sample by a membrane permitting diffusion of the determinand gas from sample to cell, but impervious to ions. Gas sensing probes for ammonia and acidic gases (carbon dioxide, sulphur dioxide and oxides of nitrogen) employ pH electrodes in the cell, but other ion-selective electrodes have been used for other determinands. Thus a sulphide electrode is used in a probe for H_2S determination, and a fluoride electrode for HF determination.

Of the true ion-selective electrodes (other than glass electrodes), inorganic salt electrodes include those for chloride, bromide, iodide, fluoride, sulphide, thiocyanate, silver, copper, lead and cadmium; ion-exchange electrodes include those for calcium, 'water hardness'*, sodium, potassium, ammonium (ion), lithium, barium, lead, nitrate, perchlorate, tetrafluoroborate and chloride.

Fuller details of electrode (and gas-sensing probe) construction, and of the solids and ion-exchange systems used, are given in texts on the technique; the book by Bailey[371] is particularly helpful in this respect.

As an indication of the applicability of the technique, some of the commercially available electrodes, of greatest interest in general water analysis, are described in summary form in Table 11.4. The list is not exhaustive, and the reader considering use of the technique should obtain up-to-date information from the numerous manufacturers of electrodes (ref. 9 gives a useful list of manufacturers and, where appropriate, their U.K. and U.S. agents). The different types and makes of electrode often available for the same determinand may offer important differences

TABLE 11.4
Some Ion-selective Electrodes used for Water Analysis

Determinand†	Main interfering species	Approximate lower limit of concentration M‡	References to some applications
Ammonia (P)	Amines	10^{-6}	385–396
Bromide (S)	S^{2-}, CN^-	10^{-6}	397
Calcium (I)	Zn^{2+}, Fe^{2+}, Pb^{2+}	10^{-5}	398–401
Chloride (S)	S^{2-}, I^-, Br^-, CNS^-, CN^-	10^{-6}	402–409
Cyanide (S)	S^{2-}, I^-	10^{-6}	410–412
Fluoride (S)	OH^-	10^{-6}	413–419
Hardness (I)	Zn^{2+}, Fe^{2+}, Cn^{2+}, Ni^{2+}	$< 10^{-5}$	420–422
Iodide (S)	S^{2-}, CN^-	10^{-7}	423–424
Nitrate (I)	HCO_3^-, Cl^-	10^{-5}	425–427
Sodium (G)	Ag^+, H^+	10^{-7}	431–435
Sulphide (S)	—	10^{-6}	428–430

† The type of electrode is shown in parentheses:
    P   Gas-sensing membrane probe
    S   Inorganic salt
    I   Ion-exchange
    G   Glass
   ‡ See text for a discussion of these data

* An electrode of essentially equivalent sensitivity to calcium and magnesium.

in performance (*e.g.*, selectivity), and one may be more suitable than another in particular circumstances.

Of the applications in Table 11.4, the determination of fluoride has been most widely adopted because the ion-selective electrode technique has advantages of speed, convenience and selectivity over the alternative approach, solution spectrophotometry. The fluoride electrode is included in collections of standard or recommended methods for water analysis.[169,436] The nitrate electrode is also quite widely used, and is included in collections of standard methods.[144,437] The ammonia gas-sensing probe, recently developed at the time the first edition of this monograph was published, has also found wide application and has again been included in collections of standard methods.[76,438]

The development of sulphate-selective electrodes, noted in the first edition, does not appear to have led to the introduction of commercial systems because of poor selectivity and limited concentration range.[370] Sulphate, if determined by ion-selective electrode methodology, is usually titrated using a lead-selective electrode to determine the end-point, as described previously.

Enzyme electrode systems, another recent development at the time of the first edition, do not appear to have had an important impact upon routine water analysis; they may, however, find application as wide-spectrum pollution monitors.[439] Enzyme electrode development has been reviewed recently by Guilbault.[440]

Related to ion-selective electrodes are devices known as ion-sensitive field effect transistors (ISFETs). These are basically field effect transistors (FETs) upon which are mounted ion-selective membranes. Development of these devices – which offer the possibility of very small sensor systems – has been reviewed by Janata and Huber,[441] and is also briefly discussed in refs. 371 and 372. ISFETs do not, however, appear to have had any impact on water analysis to date.

The capabilities of the ion-selective electrode technique can be further enhanced by the combined use of different electrodes. Thus, for example, Guterman *et al.*[430] have combined a sulphide electrode and a pH electrode, with suitable electronic signal processing, to form a system for the determination of *total* dissolved sulphide at concentrations of 10^{-5}–10^{-1}M, in the pH range 7.5–11.5. Again, by combining a platinum redox element with an iodide electrode, a probe for total residual chlorine determination has been produced. When iodide is added to the sample solution, all the forms of residual chlorine (free and combined) are reduced, with concomitant oxidation of iodide to iodine. The platinum electrode provides a measure of the iodine/iodide ratio, and the iodide electrode responds only to iodide. By taking the signals from the two electrodes, a response to iodine formed by the action of residual chlorine on iodide can be obtained, thus allowing measurement of the total residual chlorine. The system, which is available commercially, has been found capable of determining very low concentrations (to *ca.* 10^{-7}M) of total residual chlorine in both fresh[442,443] and saline[443] waters. By replacing the iodide ion-selective electrode with a bromide ion-selective electrode, and adding bromide to samples in place of iodide, a similar system for determining free residual chlorine has also been devised.[444] However, the determination of free residual chlorine by this means is negatively biased when combined residual chlorine is also present, so practical use of the system is thought likely to be limited. (If combined residual chlorine is absent, the probe would have no advantage over that for total residual chlorine which would, in such circumstances, provide a measure of free residual chlorine[444]).

The scope of the ion-selective electrode technique can also be extended by exploiting responses of electrodes to species other than that which they are designed to determine. A recent example is the use of a mercury(I) chloride/mercury(II) sulphide electrode (primarily responding to chloride) for the determination of sulphite or dissolved sulphur dioxide in water.[445] This use exploits a reaction between sulphite and mercury(I) chloride occurring at the electrode surface, in which chloride ions are produced. The electrode responds to the latter and hence, indirectly, to sulphite. The electrode is incorporated in a flow injection system, in which the sample is injected into a flowing stream of nitric acid. The sulphur dioxide so produced passes across a PTFE membrane into a continuously flowing stream of less concentrated nitric acid, passing over the electrode. By adopting this transfer approach, chloride concentration changes from sample to sample do not cause interference.[445]

(c) Principal Analytical Characteristics

(i) Calibration The potential, E, of the cell formed by the ion-selective and reference electrodes under normal conditions of use is (ideally) given by the Nernst equation:

$$E = E° + (2.303RT/zF) \log a \qquad (11.1)$$

where $E°$ = a constant for given electrodes (selective and reference) and temperature,

R = the gas constant (8.31432 J K^{-1} mol^{-1}),

T = the absolute temperature (K),

F = the Faraday (9.64845×10^4 C mol^{-1}),

z = the charge of ion to which electrode responds,

a = the activity, in the sample, of the ion to which the electrode responds.

The activity, a, is equal to $f.c$ where f and c are the activity coefficient of the ion, and its solution concentration, respectively. For a given ion and temperature, the term $2.303 RT/zF$ is a constant – for a monovalent cation, at 25 °C, it has the value of 59.16 mV per decade change of activity. Under a given set of experimental conditions (including constancy of temperature and ionic strength, such that the activity coefficient, f, is constant) equation (1) above reduces to the form:

$$E = k + m \log c \qquad (11.2)$$

where k and m are constants. Therefore, in preparing calibration curves it is usual to plot E against $\log c$, when a straight line should be obtained. In practice, several effects may cause departures from linearity. These include variation of f with the ionic composition of the sample (if constancy of ionic strength is not maintained), the presence of the determinand in the water used to prepare calibration standards or reagents, the solubility of the electrode membrane itself (affecting the determinand concentration in solution) and the action of interfering substances.

The first of these effects (variation of f) is often allowed for by adding sufficient inert electrolyte to samples and standards to 'swamp' variations in ionic strength.

The subject of ion-selective electrode response at low determinand concentrations, below the limit of Nernstian behaviour, has been extensively treated and rationalised since the first edition of this text. Midgley[446,447] has derived a number of equations, from theoretical principles, to describe the response of various types of electrode in terms of the last three causes of non-Nernstian behaviour noted above – *i.e.*, determinand in reagents, solubility of the electrode material and interferences. Relatively simple equations are adequate to describe electrode response under certain conditions – for example, if the only cause of non-Nernstian behaviour is electrode solubility, and if the resulting determinand concentration greatly exceeds that initially present in samples and standards, the following equation applies:

$$E = q + r.c \qquad (11.3)$$

(q and r being constants). In this case, a simple plot of E *versus* c (rather than log c, as in the Nernst equation) provides a linear calibration. When electrode solubility is not limiting, or when several causes of non-Nernstian behaviour are simultaneously important, the equations describing electrode response are much more complicated, and their use has been questioned by Jain and Schultz[448] who have applied a non-linear regression instead. The latter authors have also stated that equation (11.3) has only very limited applicability, but it may be noted that it has been successfully applied to the determination of low levels of chloride ion.[406,409]

It is apparent from the above that use of ion-selective electrodes below the concentration range of Nernstian response may pose a number of problems, and the present authors do not generally favour such use unless the calibration can be linearised – for example, because equation (11.3) holds or because a simple correction can be applied for the determinand content of added reagent.

Even under the conditions of Nernstian response, both the slope and the intercept of the calibration curve may differ from the theoretical values, and may change with time. The calibration should, therefore, be checked whenever a batch of analyses is performed. Other methods of evaluation are used, especially the 'standard-addition' procedure, and can be very useful in circumstances where sample compositions vary markedly. Texts on ion-selective electrodes, and the review by Mascini,[384] should be consulted for details of these alternative methods.

(ii) Limit of detection   In the absence of interfering substances, limits of detection of inorganic salt and ion-exchange membrane electrodes are governed largely by the solubility of the ion-selective membrane in the sample (strictly, in the case of ion-exchange electrodes, by the partition equilibrium of the determinand between the exchanger and the sample[370]). Typical, approximate values for the lower concentration limits of a number of electrodes are given in Table 11.4.

For reporting the detection limits attained using particular ion-selective electrodes, the International Union of Pure and Applied Chemistry (IUPAC) has recommended[449] definition of the detection limit as the concentration of determinand at the intersection point of the extrapolated linear portions of a plot of potential against logarithm of concentration (*i.e.*, the point of intersection of the Nernstian response line and the limiting 'flattened' curve at the extreme of the non-Nernstian region). Midgley[450] has pointed out that such a definition (and other, similar ones also relating to some measure of the limit of Nernstian response) fails to take account of the random errors associated with the measurement and is inconsistent with the fact that ion-selective electrodes have been used routinely to determine chloride at concentrations well below the 'detection limits' which would be derived from it. He proposed[450] instead the use of statistically based criteria and limits of detection as described in Section 8.3.1 of this text, and illustrates the results of applying the two different approaches to data obtained using a number of different electrodes. For sodium- and ammonium-selective glass electrodes, the differences between the values obtained were relatively small, but for two chloride electrodes the IUPAC definition gave detection limits about 35 and 60 times greater than those obtained on a statistical basis.[450]

It should also be pointed out that, in solutions in which the determinand ion is well buffered (*e.g.*, of copper in the presence of an excess of a strong complexing agent), Nernstian response may be observed to extremely low ion activities – less than 10^{-10} M in some cases. This behaviour has been exploited in studies of trace metal speciation (see Section 11.5).

The lower concentration limits quoted in Table 11.4 are, as noted above, intended to give an indication of the approximate detection limits attainable under routine analysis conditions. However, if special precautions are taken, and special procedures adopted (*e.g.*, mounting the electrodes in a flow cell through which the sample is passed), lower limits than those given in the table may be achieved. Thus, for example, Warner and Bressan[416] have measured fluoride down to 1 μg l^{-1}, Eckfeldt and Proctor[432] have determined 0.1 μg l^{-1} Na and a number of studies have reported[406,408,409] determination of chloride at μg l^{-1} levels. Such methods are usually appreciably slower than those for use at higher concentrations, not least because the response of ion-selective electrodes to changes in determinand concentration tends to be slower at lower concentrations. Response times can be improved by special techniques, such as polishing the membrane surface.

(iii) Precision If equation (11.1) is written in the form:

$$E = E^\circ + (RT/zF) \ln c + (RT/zF) \ln f$$

and then differentiated with respect to *c*, we obtain:

$$dE = (RT/zF).dc/c$$

$$\therefore 100 \, dc/c = 100 \, (F/RT).z.dE$$

The term F/RT is approximately equal to 0.039 at 25 °C. Thus, an error of 1 mV in measuring *E* is equivalent to a relative error of 3.9% for a univalent ion and 7.8% for a divalent ion. In routine analysis, standard deviations of about 1 mV in potential are commonplace, so that relative standard deviations of analytical results often lie in the range of 5–10%. With special care, however, such errors may be reduced and relative standard deviations of about 1–2% have been reported.[76,144,169,370]

(iv) Selectivity Ion-selective electrodes are never completely specific for one ion and, indeed, lack of adequate selectivity is often a problem in the application of electrodes, though some (*e.g.*, the fluoride electrode) are highly selective.

Equations of the following form have been used to describe the effects of other ions:

$$E = E^\circ + (2.303RT/z_iF) \log (a_i + \sum_j k_{ij}a_j^{z_i/z_j}) \qquad (11.4)$$

Where the suffixes 'i' and 'j' refer to the ion being determined and interfering ions, respectively and k_{ij} is a 'selectivity coefficient' describing the effect of the *j*th interfering ion on the determination of the ion of interest. The values of selectivity coefficients may depend upon the concentrations of the two ions and upon solution composition and temperature, as well as upon the exact procedure used to obtain them;[371] and it is therefore essential to employ caution when using them to calculate likely interference effects. It is prudent to regard selectivity coefficients as no more than ' . . . an approximate guide to the possible interference effects'[370] and, if there is any doubt, to check the magnitude of interference effects under one's own intended conditions of electrode use. This frequent lack of specificity should be borne in mind when use of the technique is under consideration, and Table 11.4 identifies some of the main interfering species for a number of electrodes of interest in water analysis. Some interferences may be readily overcome, *e.g.*, by adding a buffer to reduce the concentration of hydroxyl ions when determining fluoride, or to reduce the concentration of hydrogen ions when determining sodium,[370,371] or by adding a complexing agent to 'mask' interference from copper on the determination of cadmium.[452]

Equation (11.4) shows that interfering substances will tend to cause departures from a linear relationship between E and $\log c$, an effect which becomes more pronounced as c becomes smaller. Reference has already been made [in (*i*) above] to this effect, and to the work of Midgley[446] on the interpretation of non-Nernstian electrode responses resulting from, *inter alia*, interferences.

Ion-selective electrodes respond only to the free, uncomplexed form of the ion. This is clearly disadvantageous if it is desired to determine the total concentration and an important fraction of the determinand is not present as the free ion. An important example of this problem arises in the determination of total dissolved fluoride by means of a fluoride ion-selective electrode. Polyvalent cations such as Al^{3+}, Fe^{3+} and Ca^{2+} form complexes with fluoride, to which the electrode will not respond. The degree of interference will depend, *inter alia*, upon the concentrations of such cations in the sample, but it is normally necessary to add a suitable reagent capable of preferentially complexing the cations and thereby releasing the bound fluoride for measurement. It is common practice to incorporate this complexing agent in a combined reagent which also buffers the pH to the optimum range (pH 5–6[169,371]) – thus avoiding hydroxyl ion interference – and maintains a high and essentially constant ionic strength. Many different formulations for this combined reagent have been employed,[371] but two recently published, authoritative collections of recommended methods for water analysis[169,437] specify an acetate buffer with sodium chloride for ionic strength adjustment and CDTA (*trans*-1,2-diamino-cyclohexane-*NNN'N'*-tetra-acetic acid) for complexing the interfering cations. It should be noted that the action of the complexing agent is not instantaneous – at least 10 minutes standing of the sample/reagent mixture has been advised prior to potential measurement, to allow time for the bound fluoride to be released[169,453]).

By contrast, a response to only the free ion may be advantageous in speciation studies – indeed, no other technique appears to be available which is capable of responding to a single, defined solution species. Section 11.5, dealing with speciation analysis, should be consulted for further details of this application of the technique.

(v) General The technique is attractive in that the equipment required is simple, usually reliable and relatively cheap, and experimental procedures can, in favourable circumstances, be straightforward and rapid; in some cases an analysis can be completed in a few minutes. Battery operated meters (though not capable of the highest precision) can, in some types of determination, permit measurements to be made in the field.

The simplicity and rapidity of analysis is often lost, however, when measurements are made at low concentrations because many electrodes then take a much longer time to reach an equilibrium potential, and because special procedures may be required.

Rather close temperature control (typically ± 1–2 °C, but ± 0.1 °C in some cases[370]) is usually necessary, the most important aspect being equality of the temperatures of calibrating standards and samples.

The technique is basically suitable for determining only one ion at a time, although arrays of several electrodes can be used to determine a number of ions, using more complex potential-measuring apparatus (see, for example, ref. 454).

It would appear that the ion-selective electrode technique has remained rather limited in application to routine water analysis since the first edition of this monograph was published. Now, as then, the most common applications are to the determination of pH and fluoride, though nitrate electrodes and ammonia probes are also fairly widely used. For the analysis of natural and potable waters, it seems unlikely that laboratory applications of the technique will expand greatly in the near future; possible areas of development are speciation analysis and the use of enzyme monitors for protective monitoring of potable water abstractions.

11.3.7 Voltammetric and Related Techniques

Many analytical techniques have their basis in electrochemical processes. For example, conductimetric, amperometric and coulometric titrations have been mentioned in Section 11.3.1, ion-selective electrodes have been dealt with in Section 11.3.6 and conductivity measurement and amperometry are considered in Section 11.3.9.

In the first edition of this work, two related electroanalytical procedures – polarography and stripping voltammetry – were dealt with under the general heading 'Other Measurement Techniques', that is, as techniques not widely used in routine water analysis. (They were classified as 'voltammetric' procedures, but the IUPAC recommendations on electroanalytical nomenclature published since – see ref. 455 and also ref. 456 – are not consonant with that usage, and will be used here whenever possible).

Since the first edition, anodic stripping voltammetry (ASV) has become sufficiently widely used in water analysis (particularly in the analysis of saline waters) to merit inclusion as a 'Generally-Important Measurement Technique', though it remains much less widely applied than, say, atomic absorption spectrometry. Two other techniques – polarography and potentiometric stripping analysis (PSA) – are not widely used, but are so closely related to stripping voltammetry that they are also dealt with here, rather than in Section 11.4. There are, of course, other electroanalytical techniques which have little or no application to water analysis, details of which can be obtained from specialist texts (*e.g.*, refs. 457–462) The biennial reviews (even-numbered years) on analytical techniques that appear in *Analytical Chemistry* are also a useful source of information on electroanalysis generally.

With respect to the three techniques considered here, a number of textbooks and reviews are available dealing with polarography and stripping voltammetry, and their application to water analysis. Potentiometric stripping analysis is a much newer technique, and only one review of it is known to the authors. Table 11.5 gives details of some relevant texts and reviews.

(a) Bases of the Techniques

Polarography and stripping voltammetry have many variants; potentiometric stripping, as expected for a recently developed technique, has not. For ease of understanding, the basis of the simplest variant in each case is first considered, and the relevant variations are briefly described subsequently. References to the variants are given only when they are not dealt with in appropriate detail in texts or reviews in Table 11.5.

(i) *Conventional d.c. polarography* The term 'polarography' is used here to describe the group of related variants in which one of the two electrodes of an electrochemical cell consists of mercury flowing in a capillary, the tip of which is immersed in the solution which is to be analysed, forming a series of drops successively growing and detaching from the capillary. This is the so called 'Dropping Mercury Electrode' or 'DME'.

The equipment and principles of classical d.c. polarography are shown

TABLE 11.5

Some general references dealing with polarography, stripping voltammetry and potentiometric stripping analysis

Technique	Type of reference	Ref.
Polarography	Text-books	457–462
	General reviews	466–469, 480
	Reviews of applications in (or including) water analysis	470–473, 477, 480
Stripping voltammetry	Text-books	460–463
	General reviews	464–469
	Reviews of applications in (or including) water analysis	470–474, 476, 477, 479, 480
Potentiometric stripping analysis		475

schematically in Figure 11.5. The voltage applied across the two electrodes is slowly increased at a constant rate and the current flowing through the system is measured and recorded. A typical record is shown in Figure 11.5(b) and is explained qualitatively as follows:

As the potential of the DME reaches a value at which a reaction involving electron transfer can occur at the surface of the mercury drop, the current increases

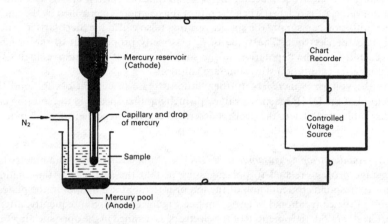

(a) Schematic arrangement for d.c. polarography

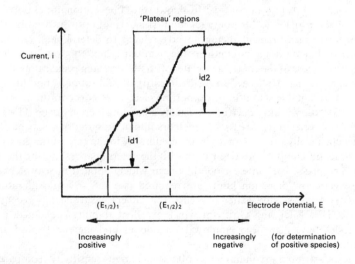

(b) Example of a d.c. polarogram

Figure 11.5 *D.c. polarography*

as a result of the increased rate of chanre transfer. An example of such a reaction is reduction of a metal ion:

$$Zn^{2+} + 2e^- \rightleftharpoons Zn(Hg)$$

[where Zn(Hg) represents zinc amalgamated with the mercury].

The electrode reaction tends to deplete the concentration of the reacting species in the vicinity of the electrode, but this is counterbalanced by diffusion of that species from the bulk solution to the electrode. As the voltage is further increased, the rate of the electrode reaction increases so that the current also grows until a limiting value, i_{d1}, is reached when the rate of the reaction is controlled by the rate at which diffusion can supply the species reacting (also called the 'depolariser' – here Zn^{2+}) to the electrode. The value of i_{d1} is directly proportional to the original concentration of the depolariser in the solution, so that quantitative analysis is possible after calibration with standard solutions.

As the voltage is increased further, little increase in current occurs until the potential of the DME reaches a value at which another species in the solution can undergo reaction. Then, the process described above is repeated to yield another polarographic 'wave' with a limiting value of i_{d2}.

(ii) Anodic stripping voltammetery (ASV) Anodic stripping voltammetry is a two-stage process, related to polarography in that the first stage – the 'plating' step – corresponds to reduction of the ion of interest at a potential in the plateau region of the conventional polarogram [see Figure 11.5(b)]. Subsequently, after a short 'rest' period, the potential of the electrode is scanned in the opposite direction to polarography (*i.e.* in the positive direction for the determination of positively charged species by ASV*). In this 'stripping' step, the metal is electro-oxidised back to the ionic, solution form; the potential–current graph has the form of a current peak (see Figure 11.6 in contrast to the wave of conventional d.c. polarography [Figure 11.5(b)]. The stripping current peak is almost invariably used for quantitative measurement, being linearly related to determinand concentration provided that the ionic strength of the medium is not too low;[476] at low ionic strengths the use of the charge associated with the stripping peak, rather than of the peak current, may be needed to ensure linearity of calibration[476] but this is usually avoided by suitable selection of supporting electrolyte concentration.

In ASV, the working electrode is usually a hanging mercury drop (HMDE) or a mercury film electrode (MFE). The mercury film of the latter type may be deposited electrochemically prior to analysis, or, commonly in the case of the glassy carbon electrode, by simultaneous deposition with the determinand(s) during the 'plating' step. The glassy carbon electrode has been widely applied in seawater analysis, where very low detection limits are required (see ref. 477 and literature cited therein). A physically coated mercury (on silver) electrode has recently been described by Riley and Gu[481] for which excellent stability is claimed, and gold electrodes have also been employed for copper and mercury[482] and arsenic[483] determinations.

Anodic stripping voltammetry offers increased sensitivity compared with

---

* Cathodic stripping voltammetry, the inverse of ASV, is applicable to the determination of anionic species but is much less widely used.

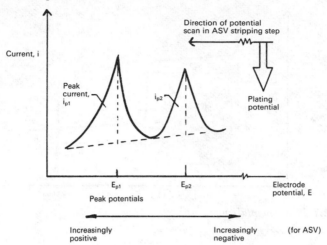

Figure 11.6 *ASV current–potential curve*

conventional polarography by virtue of the preconcentration effected by amalgam formation at the 'plating' stage. Mercury film electrodes, having a much higher surface area to volume ratio than hanging mercury drops, permit higher amalgam concentrations to be achieved in a given plating time, and thus higher sensitivities. [However, the chance of intermetallic compound formation within the mercury – leading to interference effects – is also increased; see (*c*)(*iii*) below].

In ASV, 'plating' times may vary from a few minutes to an hour or more, depending upon the degree of preconcentration needed. Typically, times shorter than 10 minutes are employed, and film electrodes used in place of the hanging mercury drop electrode to attain higher sensitivity, in preference to extending the 'plating' time unduly. Stripping scan rates may also be varied over a considerable range according to the nature of the determination but the stripping step is typically accomplished in a period of a few minutes.

The technique has the advantage of permitting low detection limits to be attained without the need for large sample volumes. Sample sizes of 25–50 ml are usually employed, although even smaller volumes may be adequate in some cases.

(iii) Potentiometric stripping analysis (PSA) Potentiometric stripping analysis (PSA) is an electroanalytical technique of recent development, closely related to ASV. In this technique, the plating step proceeds in a similar manner to ASV, but subsequent 'stripping' is effected not by the imposition of a potential ramp or scan, but by merely disconnecting the potentiostatic deposition circuitry, thus allowing chemical re-oxidation of the amalgamated metal to occur. This chemical re-oxidation may be accomplished by $Hg(\text{II})$ ions (in deaerated solutions) or by oxygen itself (in non-deaerated solutions).

The potential of the mercury working electrode (a mercury film on carbon electrode) is monitored as a function of time during the chemical re-oxidation or 'stripping' step. Potential–time curves of the form shown in Figure 11.7 are

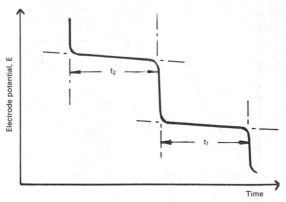

Figure 11.7 *PSA potential–time curve*

obtained, and the time of duration of the plateau is used for quantitative measurement, being proportional to determinand concentration.

(b) Variants of the Techniques
In order to improve the performance of the basic analytical techniques described above (usually to obtain higher sensitivities and lower detection limits), many variants have been described. The range – and indeed nomenclature – of such variants is bewildering to the non-specialist, and it is beyond the scope of this monograph to consider them all. However, the more important variants with respect to water analysis are briefly described, together with their advantages, in the following sections.

(i) Variants of polarography When the potential, E, of an electrode is varied, both a Faradaic current and a capacitive current flow in the circuit. The latter arises because the electrode/solution interface acts as a condenser, but it is the Faradaic current which is of interest in polarography. Many techniques have, therefore, been devised to increase the ratio of the Faradaic to capacitive current components, and thereby increase sensitivities and detection limits.

After imposition of a sudden change in potential, the resulting capacitive current decays more rapidly than any Faradaic current produced. By replacing the simple linear potential sweep of conventional d.c. polarography by a sweep incorporating potential 'pulses' or 'steps', and by sampling the current at an appropriate time after application of the pulse or step, the ratio of the Faradaic to capacitive current components can be increased. This is the basis of such polarographic variants as staircase polarography, and (normal) pulse polarography.

By measuring not the current flowing itself, but the difference of current flowing after and before the application of the potential pulse, the normal polarographic wave-form is replaced by a current peak [above an (ideally) flat baseline], with consequent improvement in measurement accuracy and resolution of successive reactions. This is the basis of differential pulse polarography in which the two currents are both taken during the lifetime of a single drop*.

Another means of increasing the sensitivity is to employ a linear potential ramp with a superimposed, *alternating*, component of potential. By recording the current at zero phase angle, the capacitive component is (ideally) eliminated. Again, this variant (a.c. polarography) produces a current peak in the current–potential plot, as also does square-wave polarography, in which a square-wave potential ramp is applied and the resulting square-wave current measured.

A new development in the field of polarography is the 'Static Mercury Drop Electrode' (SMDE).[484] With this system, mercury flows from a reservoir through a capillary to form drops as in the HMDE, but the use of a wide-bore capillary and a valve to stop the mercury flow at the desired intervals allows drops to be grown rapidly and then held at a constant size until mechanically dislodged. This feature permits virtual elimination of the charging current associated with growth of the drop, since, although a large charging current flows during the phase of rapid drop growth, this fades rapidly after growth of the drop has ceased. By using current-sampled d.c. polarography, with current measurement at the end of the stationary phase of drop life, the Faradaic current can be sampled accompanied only by a very small capacitive component. It has been found[485] that detection limits attained using current-sampled d.c. polarography are improved by at least an order of magnitude when the SMDE, rather than the DME, is used; even greater advantage occurs when very fast scan rates are used to achieve short analysis times. The same authors claim[485] that the technique of current-sampled d.c. polarography is so improved by the use of the SMDE that it rivals differential pulse techniques in many situations.

(ii) Variants of anodic stripping voltammetry   As in polarography, the need also arises in ASV to measure a Faradaic current in the presence of a capacitive or 'charging' current. Here, the Faradaic current is that due to stripping of the determinand and the capacitive current is that arising from charging the double layer at the electrode/solution interface, as the applied potential is varied during the stripping step†. In order to increase the sensitivity and speed of analysis, and to improve detection limits, d.c. stripping voltammetry has given way to differential pulse anodic stripping voltammetry (DPASV). In this variant, a square-wave pulse is superimposed on the linear ramp of conventional d.c. stripping, and the current is sampled just before, and at the end of, each pulse. The difference between the second and the first of these current measurements is displayed on a recorder; the first measurement is made at essentially constant potential whereas the second reflects (mainly) the Faradaic current due to the application of the potential pulse. (The capacitive current component arising from the application of the pulse dies

---

* Measurement of the difference between the current flowing after application of a pulse and that flowing after application of the previous pulse, *when each pulse occurs after formation of a new drop*, is the basis of derivative pulse polarography. This variant gives a similar current–potential curve to that of differential pulse polarography.

† In conventional polarography at the DME, the charging current arises both from the changing applied potential and from the changing drop size (and hence changing double layer area). It has already been seen how the use of the SMDE virtually eliminates the contribution of the latter. In ASV, the area of the HMDE (or the MFE) does not alter during the measurement, so that the charging current arises only in response to the changing applied potential during stripping.

away more rapidly, and is again discriminated against by this measuring procedure).

Although the differential pulse mode is the commonest of the variants in water analysis, other potential wave-forms have been used, *e.g.*, in staircase voltammetry and phase sensitive a.c. voltammetry, in a manner analogous to their use in polarography.

Another approach to improving discrimination of the Faradaic current component involves the use of 'Ring-Disc Electrodes' (RDE). These consist of a disc electrode, separated from an annular ring electrode by a non-conducting material. Normal anodic stripping voltammetry is performed on the disc, but during stripping the ring electrode is held at the reduction potential of the depolariser. Thus, the depolariser ions stripped from the disc are reduced at the ring. The high Faradaic current resulting is measured, with little or no capacitive component since the potential of the ring is constant.

Another approach in improving the performance of ASV is the use of multiple scanning with background subtraction.[486] In this, the initial plating step is followed by a very rapid (*e.g.* less than 1 s) potential scan (stripping) step. Immediately after the scan, the electrode potential is reduced to the value required for plating, held at that value for a short time, and a second scan (stripping) then initiated. This sequence of rapid scans, with intervening short plating periods, is repeated many times and the current readings obtained are summed by the computer. Because the stripped metal ions remain within the diffusion layer of the electrode (because the stripping is effected in a short time) some will be plated again in the intervening plating periods. Thus, sensitivity is increased for a given initial plating time. The electrode surface changes continually, however, so that background (capacitive) current is not reproducible and cannot be calculated or corrected electronically. It can be determined experimentally for each sample, however, by stripping all the metals except mercury from the electrode after an analytical run, stirring to remove the stripped ions from the neighbourhood of the electrode, and then scanning under the same conditions as before. The summed current value so obtained is then used to correct the summed current recorded in the analytical step. More recently, Brown and Kowalski[487] have developed a background-substracted d.c. ASV system similar to that of Kryger and Jagner[486] but providing enhanced performance over conventional ASV without the need for multiple scanning. This is achieved by using multipoint-averaging by computer in a single scan, rather than multiple scanning, to improve signal-to-noise ratios.

(iii) Variants of potentiometric stripping analysis Potentiometric stripping analysis is of comparatively recent introduction, and therefore few variants have been reported. However, improved sensitivity for a given plating time has been achieved using the technique of differential potentiometric stripping analysis.[488] This is similar in its principles to the multiple scanning ASV approach, since it involves successive re-plating and re-stripping of the depolariser.

(c) Principal Analytical Characteristics
The large number of variants of polarography and voltammetry, together with the very wide choice of operating conditions available with these techniques, makes it

difficult to provide a comprehensive, yet succinct, summary of the main analytical characteristics. However, in the following sections an attempt is made to do so, for the principal variants used in water analysis.

(i) Calibration – all techniques Before considering other analytical characteristics, by technique, it is convenient to deal with calibration, for all techniques together.

The detailed equations relating the measured response to determinand concentration in the sample depend upon the particular technique and variant. In all cases, however, the fundamental relationship is of the form:

$$R = KC$$

where R = measured response,
C = solution concentration of determinand,
K = a constant under given analytical conditions.

Because K depends on so many factors, in all the techniques, calibration curves are prepared for each batch of analyses, or standard additions calibration employed. The measured response, R, is different for different techniques. For example, in d.c. polarography it is the diffusion current measured at the plateau of the polarogram, in ASV it is the peak stripping current* and in PSA it is the duration of the plateau in the potential–time curve. The factor K, as noted above, also varies according to the technique and variant in question, and is composed of a number of different terms. The appropriate texts should be consulted for the exact form of K for a particular variant of polarography or ASV, and the review of Jagner[475] for PSA.

Standard additions calibration may be necessary with any of the techniques, because of sample matrix effects. It has also been employed with ASV using mercury film electrodes because of difficulties in maintaining adequately consistent film thickness.[489] It has been claimed[490] that PSA is better suited to calibration curve use than ASV, being less subject to surface active materials,[475,490] variations in the ionic matrix,[490] and variations in film thickness.[491] It is recommended that the need or otherwise for standard additions calibration be assessed experimentally whenever a polarographic, ASV or PSA method is developed, or when an existing method is applied to new types of sample.

It should be noted that the temperature-dependence of the response is also relatively large for these techniques, and close control of temperature may be needed if results of the highest accuracy are required and standard additions calibration is not employed.

Consistency of stirring is also important in the achievement of good precision in ASV and PSA. For greatest precision, rotation of the working electrode, rather than the use of a magnetic stirrer, is employed.

(ii) Polarography

(1) *Limits of detection*
Limits of detection with conventional d.c. polarography at the DME are

---

* Less frequently, the charge under the stripping peak is employed.

typically in the range 10^{-5} to 10^{-6}M, whereas differential pulse polarography
and square-wave polarography often allow detection limits in the range 10^{-7}
to 10^{-8}M to be attained. Detection limits intermediate to those of conventional
d.c. polarography and the very sensitive variants noted above are commonly
obtained with (normal) pulse polarography and (phase-sensitive) a.c. polarography. As noted previously, the detection limits attainable using the recently
developed SMDE with current-sampled d.c. polarography can rival those
achieved with the differential pulse technique at the DME.

(2) *Precision*

Detailed information on the precision achievable – using the more refined
variants, in particular – is scarce, but relative standard deviations of 2–5%
would appear to be typical, except at concentrations close to the detection
limit.

(3) *Selectivity*

Both the measured current (i_d) and the half-wave potential ($E_\frac{1}{2}$) may vary, for a
given determinand concentration, according to the form in which the
determinand is present. Although it has been suggested[492] that the dependence
of measured current upon determinand form may be useful in allowing
separation of 'soluble' and 'insoluble' lead, the separation of forms achieved by
the electrochemical techniques considered here is not clear-cut (see Section
11.5 on speciation analysis). Therefore, the dependence of measured current
upon determinand form is likely to be a hindrance to the measurement of total
concentrations, without providing an ideal separation to be exploited in
speciation work. The dependence of half-wave potential upon determinand
form is a somewhat different matter – this feature of the technique may be used
to measure the stability constants of complexes. Although not an analytical
application, this is of importance in allowing *calculations* of trace metal
speciation to be made (see Section 11.5).

If total metal concentrations are of interest, therefore, some kind of
transformation technique will usually be necessary when polarographic (or
stripping) analysis is employed. Thus acidification, u.v. photolysis and ozone
oxidation have all been used to convert so called "non-labile" complexes prior
to electrochemical measurement.

The presence of organic materials often causes interferences by virtue of
their adsorption on the mercury electrode. Again, wet-oxidation pretreatment
may be necessary for removal of such materials, as well as to convert
electro-inactive complexes to a determinable form.

The presence of dissolved oxygen in samples causes problems, not only
because of its reduction at small potentials but also because of the changes in
solution chemistry (*e.g.*, pH changes) which accompany that reduction.[493]
Although the removal of oxygen prior to analysis is not always required
(particularly when phase-sensitive a.c. polarography is used[493]) the performance – especially limit of detection – will often be degraded,[493] at least with
current-sampled variants (d.c. and pulse).

If the sample contains two substances reduced at potentials such that incomplete resolution is achieved, interference results. In conventional d.c. polarography, a separation of $E_{\frac{1}{2}}$ values of *ca.* 200 mV is required for adequate resolution, but the use of such variants as differential pulse polarography permits reactions having even closer $E_{\frac{1}{2}}$ values to be resolved satisfactorily. Additionally, some control of the problem may also be possible by varying the nature of the medium in which the polarogram is obtained, so that the separation of the $E_{\frac{1}{2}}$ values for the determinand and interferent is increased (see, for example ref. 458).

(4) *General*

Even in its more sensitive variants, polarography is not widely used in water analysis, in comparison with (say) atomic absorption spectrometry. Although the range of determinands accessible to polarography is smaller than with AAS (if organic species are not considered), polarography does have the ability to determine several metal ions sequentially in the same sample, and costs substantially less to establish*. Moreover, the limits of detection attainable by differential pulse polarography are at least similar to those attainable by flame AAS for many determinands of interest (*e.g.* lead, cadmium), though inferior to those attainable with graphite furnace AAS. (In some cases – *e.g.*, nickel with dimethyglyoxime enhancement[495] and arsenic[496] – the limits of detection attainable by DPP are much *superior* to those attainable by flame AAS).

Despite these potential advantages, polarography does not compete with atomic absorption, in most circumstances, because of the inherent complexity, and increase in analysis time, associated with the need usually to deoxygenate samples, and remove organic materials (in order to avoid interferences and incomplete recovery of the total determinand concentration, when electro-inactive complexes are present). Polarographic techniques have been used for trace metal determination in sea-water, after pre-concentration using a chelating resin;[497] in such applications, the need for pre-concentration (and therefore pre-treatment to ensure recovery of complexed metal by the resin) applies to both AAS and polarography, so the disadvantages of the latter in comparison to AAS essentially disappear. However, in such work, where very low trace metal concentrations are of interest, anodic stripping voltammetry and potentiometric stripping have advantages over both polarography and atomic absorption.

Applications of polarography in routine water analysis seem likely, therefore, to be restricted mainly to determinations of trace elements at relatively high concentrations (*e.g.*, in heavily polluted waters such as metallurgical plant effluent[498]) and to certain determinations where, in contrast to flame AAS, direct determination is possible (*e.g.*, arsenic in drinking water, though hydride generation/AAS would be an effective competitor), or where polarography possesses particular advantage (*e.g.*, the determination of uncomplexed sulphide ions[499]).

Other determinands to which polarography has been applied, or developed

---

* A report by the U.S. Environmental Protection Agency concluded[494] that the cost of establishing differential pulse anodic stripping voltammetry (DPASV) was less than half that of establishing *flame* atomic absorption spectrometry. The DPASV system would, of course, be suitable for polarography also, with perhaps some low-cost items added.

for water analysis include: bromide,[500] chlorite,[501] carbon disulphide,[502] chromium,[503] ammonia and primary amines,[504] iodate (and iodide after transformation to iodide under u.v. irradiation,[505] iron,[503] manganese[503] and dissolved oxygen (see ref. 506 and literature therein). Indirect methods are also possible; for example, the determination of sulphate by titration with lead nitrate and polarographic determination of excess lead ions.[507]

The use of a.c. polarography at a hanging mercury drop electrode (rather than a dropping mercury electrode) has been reported[508] for the determination of surface-active substances in water. In this case the capacitive (rather than the Faradaic) current was measured.

(iii) Anodic stripping voltammetry

(1) *Limits of detection*

These again vary widely according to the electrode type used and variant employed. Nürnberg[509] quotes the following as typical "determination limits at a precision of $< \pm 20\%$ RSD":

Conventional (linear-sweep) ASV at the HMDE*	10^{-9}M
Conventional (linear-sweep) ASV at the MFE† ⎱	
Differential pulse ASV at the HMDE ⎰	10^{-10}M
Differential pulse ASV at the MFE	10^{-11}M

The same author has also quoted, in a different paper,[489] the following "determination limits" for lead and cadmium specifically:

	Pb	*Cd*
Conventional (linear-sweep) ASV at the HMDE	2.5 μg l^{-1}	10 μg l^{-1}
Conventional (linear-sweep) ASV at the GCE‡	—	0.01 μg l^{-1}
Differential pulse ASV at the HMDE	0.01 μg l^{-1}	0.01 μg l^{-1}
Differential pulse ASV at the GCE	—	< 0.001 μg l^{-1}

The deposition time, in the case of lead, was 5–10 min and the HMDE area was 3.58×10^{-2} cm^2.

Using a much longer plating time of 60 min., Gillain et al.[510] reported the following detection limits for simultaneous determination of the quoted elements in sea-water, by DPASV at the HMDE:
Zn, 0.1 μg l^{-1}; Cu, 0.1 μg l^{-1}; Cd, 0.01 μg l^{-1}; Pb, 0.01 μg l^{-1}; Sb, 0.05 μg l^{-1}; Bi, 0.05 μg l^{-1}.

In a report of snow analysis, using DPASV at the rotating GCE, Landy[511] reported concentrations as low as 0.005 μg l^{-1} for Cd, 0.02 μg l^{-1} for Cu and 0.05 μg l^{-1} for Pb, using 5–10 min. plating periods.

Using DPASV at the HMDE, with 3 min deposition period, Davison[512]

---

* HMDE – Hanging mercury drop electrode. † MFE – Mercury film electrode. ‡ GCE – Glassy carbon electrode (mercury film plated *in-situ*).

reported Criteria of Detection (defined as in Section 8.3.1) of 0.128 μg l^{-1} (Zn), 0.03 μg l^{-1} (Cd) and 0.06 μg l^{-1} (Pb).

Recently, O'Halloran[513] has reported a detection limit of 0.01 μg l^{-1} for the determination of Mn^{II} in sea-water by DPASV at a glassy carbon electrode with 5 min plating time.

Using a glassy carbon rotating ring-disc electrode, detection limits of approximately 5 ng l^{-1} have been attainable for Cd, Pb and Cu in sea-water by ASV using a plating time of 1 h.[514]

Florence[515] has reported a detection limit of 5 ng l^{-1} for the determination of bismuth in sea-water using d.c. ASV at a glassy carbon electrode.

Brown and Kowalski,[487] using the multipoint-averaging single-scan background subtraction d.c. ASV technique at a rotated glassy carbon electrode, reported detection limits of 0.54 ng l^{-1} (Cd), 0.71 ng l^{-1} (Pb) and 2.1 ng l^{-1} (Cu), with only very short plating times (5 min).

It is apparent from the above selection of detection limit data that differential pulse anodic stripping voltammetry and multipoint-averaging background subtraction d.c. ASV, at mercury film electrodes, are capable of extremely low limits of detection, unrivalled by atomic absorption spectrometry, and suitable for the direct determination of the trace metals in question at the low concentrations prevailing in many unpolluted natural waters (sea-water, snowmelt, clean fresh waters). The same may be true of DPASV at the HMDE, though with the requirement of longer deposition (plating) periods. If such low detection limits are not required, less sensitive systems and/or shorter deposition periods can be selected. In this way, virtually any desired detection limit may be attained that is likely to be necessary, at least in routine water analysis.

(2) *Precision*

Detailed data on the precision of results attainable by ASV is again scarce. At concentrations greater than ten times the "determination limit", Nürnberg quoted[489] relative standard deviations of 2–3%. Data collected by Edwards[476] also suggests that relative standard deviations of this order should often be attainable at concentrations greater than about five times the detection limit. In a recent paper on the use of ASV at a rotating ring-disc electrode for the determination of Cd, Pb and Cu in sea-water, relative standard deviations of about 25% and 6% were reported at concentrations of 10 ng l^{-1} (twice the detection limit) and 1 μg l^{-1} respectively.[514] A relative standard deviation of 7% has been reported by Florence[515] for the determination of bismuth in seawater (by d.c. ASV with a GCE) at a concentration of about 40 ng l^{-1} (eight times the detection limit). At the low concentrations attainable using the more sensitive variants of ASV, the precision might well be limited by contamination effects rather than by the inherent reproducibility of the technique itself. The series of papers by Mart and his colleagues[516–518] illustrates the care which has to be taken when working at such concentrations. (The last paper of this series describes a multiple cell ASV system specially designed for ultra-trace analysis). The reader is also referred to Section 10.8 of the present work, which deals with the subject of contamination.

(3) *Selectivity*

The problems previously noted in respect of polarographic techniques apply also to ASV. Thus, peak currents and peak potentials may be affected by the forms of the determinand present (though the analytical utility of the dependence of the peak current is again likely to be limited), and the variation of peak potential has been used (in a manner analogous to the variation of half-wave potential in polarography) to estimate complex stability constants – see Section 11.5. Again, pre-treatment to release trace metals present in "non-labile" or electro-inactive complexes has usually been applied when total concentrations are of interest – usually u.v. oxidation and/or acidification, but also ozonolysis. Acidification may not be adequate to release all "bound" metal, and may in fact exacerbate the effects of surface active organic compounds which, as in polarography, can cause interference effects in ASV.[519]

As with polarography, the need to deoxygenate samples is also a disadvantage in anodic stripping voltammetry. Additionally, standard additions calibration has often to be employed, as noted previously.

ASV is prone to interference effects arising from the formation of intermetallic compounds (other than mercury amalgams) within the electrode. Examples of metals forming such compounds are Au/Cd, Au/Zn, Sn/Ni, Zn/Cu and Cd/Cu (see refs. 476 and 477 and literature cited therein). The formation of such compounds may change the stripping peak potentials, reduce peak currents and produce a new stripping peak attributable to the compound.[476] The seriousness or otherwise of the problem depends, *inter alia*, upon the relative concentrations of the two elements concerned (see, for example, ref. 520).

Formation of an intermetallic compound (or compounds) between copper and zinc is likely to be the most important manifestation of the problem in the application of ASV to water analysis, though the possibility of practically-important effects with other element combinations should always be borne in mind. The Cu/Zn effect results in a negative interference on both determinands, more marked for zinc than for copper*, but the mechanism of the interference is not fully understood (see ref. 521 and literature cited therein). The use of a simple standard additions calibration does not overcome the problem.[520,521] Thus, for example, Gerlach and Kowalski[521] observed suppressions of about 3 μg l^{-1} for zinc and about 0.5 μg l^{-1} for copper when these elements were present together at about 13 μg l^{-1} (zinc) and 5 μg l^{-1} (copper), using this mode of calibration and a rotating glassy carbon electrode. They found, however, that a more complex procedure – the so called 'Generalised Standard Addition Method', involving a number of standard additions of each determinand and solution of the resulting set of simultaneous equations – was able to reduce the errors to about 0.1 μg l^{-1} (Zn) and 0.3 μg l^{-1} (Cu). This calibration procedure would, however, greatly increase the overall analysis time.

---

* Because the intermetallic compound is oxidised at a potential close to that at which the copper itself is stripped.

The dependence of the effects of intermetallic compound formation upon the exact conditions of analysis and upon the nature of the samples precludes any simple assessment of the likely importance of the phenomenon in the application of ASV to water analysis. In many circumstances, the effects may be so small as to be of no practical importance, but this will not always be the case and special tests are recommended when methods are developed or applied to new types of sample. If problems are disclosed, various approaches are available to reduce their magnitude. These include the following: reduction of amalgam concentrations by changing from an MFE to the HMDE or by reducing plating time by employing, if necessary, a more sensitive variant of ASV,[522] the use of different plating potentials for the elements involved (*i.e.*, determining them separately),[477] the addition of a third element which forms a more stable intermetallic with one of the affected determinands (thus acting as a releasing agent for the other)[513,520,523] and, finally, the adoption of a complex standard additions procedure[521] as described above.

Another source of interference in ASV is inadequate resolution of the stripping peaks of determinands having closely similar plating potentials in a given medium. Zirino[477] cites nickel/zinc, silver/copper and tin/lead as examples of such determinand pairs; under acid conditions, hydrogen reduction may interfere with the zinc stripping peak – see, for example, ref. 524. Again the magnitude of the effect will depend upon the nature of the samples and the analytical conditions, and no simple assessment can be offered. Special tests are recommended if the presence of the problem is suspected, and various approaches to reduce the magnitude of any interference are available. These include more complex data reduction,[525] judicious choice of plating conditions,[477] addition of a complexing agent to mask (or shift the reduction potential of) one of the elements[526] and, finally, stripping into a different medium (so called 'medium exchange') thereby separating the peaks.[482,524,527]

Other potential sources of bias which have been observed include the formation of electroinactive Cu^I species on the electrode in the determination of copper in unacidified (but not in acidified) sea-water[528] and the enhancement of the copper response by virtue of oxidation of ferrous iron at the electrode.[529] In the latter case the additional peak current due to iron was independent of 'plating' time, so that a voltammogram run without prior plating permitted a simple correction. Moreover, serious interference on copper determination was only observed at high iron concentrations.

Of course, as with other measurement techniques, many interference effects in ASV may be overcome by the application of a suitable separation technique prior to measurement. This has the disadvantage of increasing the time required for analysis, but may be combined with a preconcentration to give improved detection limits. A recent example is the determination of Cd, Cu, Pb and Zn by ASV after preconcentration and separation of the determinands from the sample using C18-bonded glass beads.[530]

(4) *General*

The range of determinands accessible to ASV is rather limited – Edwards[476] in his review of applications to water analysis, listed the following:

Bi, In, Tl, Co, Ni, Hg, Ag, Au, Mn Cr⎱
Cd, Pb, Zn, Cu.   ⎰   (Direct)

Sn, Fe, As, V.   (Indirect)

(More recently, the determination of molybdenum in sea-water by ASV has also been reported[531]). Of the determinands listed, only Cd, Pb, Cu, Zn and, to a lesser extent, Bi, Tl and Hg are at all frequently determined by the technique. It will be noted, however, that the "direct" list includes some of the most important determinands in water analysis for pollution control – especially Pb, Cd, Cu and Zn (which may often be determined simultaneously) and Hg (for which a gold electrode is used).

It has already been noted that the DPASV technique is, for trace metals, the only one (apart from PSA – see below) capable of direct determination (*i.e.*, without preconcentration) of naturally occurring levels in pristine natural waters. Moreover, the capital and operating costs (consumables) are markedly lower than those of AAS, which does not possess such capability. Additionally, ASV has (limited) simultaneous multi-element capacity – unlike AAS – and, if concentrations are so low that preconcentration (and therefore pretreatment) is required for AAS, the need for pretreatment with ASV no longer constitutes a disadvantage. It is for this reason that the technique has been so widely used in sea-water analysis, a field of application which has been extensively reviewed elsewhere.[471,477] However, the technique may, of course, also be applied to other types of water, including drinking water (*e.g.*, refs. 532, 533) rainwater (*e.g.*, ref. 526) river/lake water (*e.g.*, ref. 512) and snow (*e.g.*, ref. 511).

(iv) Potentiometric stripping analysis

(1) *Limits of Detection*

As with ASV, the range of detection limits attainable using PSA is very wide, depending on the instrumentation and detailed approach adopted (especially plating time and oxidant). Jagner[534] has compared the oxidation rates in samples saturated with dissolved oxygen with those in samples previously deaerated, containing low levels of Hg^{II} as oxidant. He found that, for lead and cadmium, the oxidation rate in the non-deaerated solutions was some 25 times higher, and, for copper, some 12 times higher, than in the deaerated solutions. Consequently, detection limits in deaerated systems will be superior (other things being equal) to those attainable using oxygen itself as oxidant, because of the shorter stripping plateaux which require to be measured in the latter case.

Thus, Jagner[534] reported detection limits, with non-deaerated samples, of 2.5 μg l^{-1}, 2.5 μg l^{-1} and 1.0 μg l^{-1} for Cd, Pb and Cu, respectively, using a 64-min plating time at the mercury-plated GCE and a recorder system with a resolution of 0.1 s. By contrast, Jagner and Aren[490] reported detection limits for Zn, Cd, Pb and Cu of 0.03, 0.03, 0.01 and 0.02 μg l^{-1} respectively, again using a 64-min plating time at the mercury-plated GCE, in deaerated sea-water, using a prototype commercial recorder-based system.

Using the differential PSA technique, even lower detection limits are attainable. Thus, Kryger[488] has reported values for Pb and Cd of less than 0.1 μg l^{-1} and 0.06 μg l^{-1}, respectively, *using only a 60 s plating time*, by means of DPSA. Again, computer processing of the data in PSA can improve detection capability. Jagner, Josefson and Westerlund[491] have reported detection limits of about 1.3 μg l^{-1} (Cu, 4 min plating), 16 ng l^{-1} (Zn, 16 min plating), 12 ng l^{-1} (Pb, 32 min plating) and 5 ng l^{-1} (Cd, 32 min plating) for computerised PSA in non-deaerated sea-water. The signals are stored in the computer, which subtracts the background from the analytical response and then displays the net result on a chart recorder at a slower rate, thereby improving resolution and detection limits.

(2) *Precision*

Jagner[475] has noted that one of the potential advantages of PSA is the fact that time can be measured with better precision than current. The same author has reported relative standard deviations of 2–3 % at high determinand concentrations,[534] and of 3–6% at typical coastal sea-water concentrations.[490] Kryger[488] obtained relative standard deviations of about 5% at concentrations of lead and cadmium of about 1 μg l^{-1} and 0.6 μg l^{-1} respectively (10 times his reported detection limits).

It would appear, therefore, that the precision which can be achieved using PSA is comparable to that attainable using ASV – that is, typically 2–5% relative standard deviation, except near the detection limit.

(3) *Selectivity*

PSA is subject to many of the problems which arise with the other techniques considered in this sub-section. Thus, pretreatment will often be necessary if total concentrations are of interest, in order to convert otherwise electroinactive forms of the determinand.

It has been claimed,[475] however, that PSA is less sensitive to the presence of low concentrations of surface-active agents than ASV, and that (since no current passes through the working electrode during stripping) PSA is "insensitive to interference from electroactive substances"[475] and, as noted previously, better suited to calibration curve use (rather than the standard additions technique) than ASV.[490]

As with anodic stripping, PSA has been shown to be subject to intermetallic compound formation.[535]

(4) *General*

The review of PSA by Jagner[475] shows that PSA has been used for the determination of a variety of elements in a number of matrices, including Zn, Pb, Cd, Cu and Bi^{III} in sea-water, and others (Hg, As^{III}, Mn^{II}, S^{2-} and Se^{2-}) in standard solutions. Moreover, like ASV, it is capable of the direct determination, without preconcentration, of many important trace metals in unpolluted natural waters. The apparatus is relatively simple and cheap[475] and commercial equipment is available. As with ASV, the technique has particular advantages over AAS for the analysis of saline waters.

(d) General Summary of the Three Techniques

The principal disadvantages of polarography, ASV and PSA include: (1) the need in many cases to deaerate samples prior to analysis, (2) the likely need, in many applications, to apply a conversion technique in order to recover all forms of trace metals and (3) the need often to destroy surface-active organic material and/or to employ standard additions calibration.

It would appear, therefore, that those engaged in water analysis have usually elected to use AAS, rather than polarography or ASV, whenever the former technique has permitted direct analysis (and often when it has not). This practice reflects, presumably, the disadvantages of the electroanalytical techniques (noted above) and, probably, a belief that the latter are lacking in robustness in comparison with flame atomic absorption spectrometry. The present authors consider that polarography, ASV and PSA are indeed less robust than flame AAS and solution spectrophotometry, not least because the electroanalytical techniques involve surface-dependent processes not at chemical equilibrium (nor under steady-state conditions).

However, the growing demand for determinations of certain trace metals at even lower concentrations, aided by improvements in the design and ease-of-use of electroanalytical equipment, may lead to an increased application of ASV (and/or PSA) in routine analysis. This is likely to be true of saline water analysis in particular, for which direct determination at very low concentrations (using small sample volumes) is possible by ASV and PSA, but not, in general, by graphite furnace AAS. (Although ASV and PSA are less simple to automate than graphite furnace AAS, the preconcentration/separation step usually necessary when the latter is applied to marine waters negates its advantage in this respect).

An effective assessment of the relative merits of ASV and PSA is not possible at present, given the relatively recent introduction of the latter. PSA would appear to possess some potentially important advantages,[475,490] but further experience of the technique will be required to determine if these can be realised, without other penalties, in practice.

Automated batchwise analysis and discontinuous on-line analysis are possible with these techniques*. Thus, for example, polarographic,[536] ASV[537] and PSA[538] flow-through cells have been described, and on-line monitoring of drinking water accomplished by ASV, using a system in which all analytical operations – including standard additions calibration – were fully automated.[532] Automatic and on-line analysis are dealt with further in Section 12, and the application of electroanalytical techniques to such analysis has been reviewed by Toth *et al.*[539]

Polarographic detection has been used in high performance liquid chromatography, dealt with in Section 11.3.8. The application of polarographic and other electrochemical detectors to HPLC has been extensively reviewed by Brunt.[540]

Future developments in voltammetry may include wider use of electrodes other

---

* Continuous on-line analysis is not, of course, possible because only a part of the analytical cycle produces the response itself.

than mercury. Thus, for example, Izutsu *et al.*[541,542] have described the use of glassy carbon electrodes coated with trioctylphosphine oxide (TOPO) for the determination of uranyl ion. The uranyl ions (but not lead or thallium ions) could be concentrated into the TOPO layer and then reduced by a negative potential scan, the peak current during reduction being used for quantification. Determination of uranyl ions at μg l^{-1} concentrations was reported.[542]

In a related approach (which they call 'Adsorptive Stripping Voltammetry'), Lam *et al.*[543] also determined uranium(VI) in water by adsorption of its pyrocatechol complex on a hanging mercury drop electrode followed by potential scanning.

11.3.8 Chromatographic Techniques

The measurement techniques considered in prevous sub-sections are, for the most part, inapplicable to the identification and quantitative determination of specific organic compounds in the presence of others, particularly when the former are present at only trace concentrations. This topic is of particular importance to water research laboratories, in which the study of the occurrence and behaviour of trace organics in many different types of water is often a major topic, but it is also an increasing requirement in routine water quality assessment.

The identification and determination of individual organic compounds causes particular difficulty. This is mainly a consequence of the many organic substances that may be present in waters and the similarities in the physical and chemical properties of related compounds. (With regard to the numbers of different compounds, Janardan and Schaeffer[544] have used various statistical techniques to predict the total number of distinct organic compounds in waters which might be identified by the techniques – principally GC-MS – which had, up to the time of their study, been used to identify such compounds in waters. This total number of distinct identifiable compounds was predicted to lie in the range 2500–9300. The prediction did not take account of the application of new analytical techniques, such as high-performance liquid chromatography/mass spectrometry (HPLC-MS), which are capable of dealing with the non-volatile component of the total organic content of waters, not amenable to gas chromatography. Since such non-volatile material may account for 80–90% of the total organic carbon content of waters,[545] the total number of identifiable, distinct compounds may greatly exceed the above values. It should also be noted that a large number (300–500) of new formulations reach the stage of commercial production every year,[546] and many of these may ultimately reach the aquatic environment.)

Because of this large number of compounds, and the physico-chemical similarities of many of them, methods of measurement must frequently be preceded by powerful separation techniques so that individual compounds are isolated. Chromatographic techniques (especially GC in the work carried out to date) have proved invaluable in this respect; their use in water analysis is not only well established, but likely also to increase as new variants are applied. Despite the fact that chromatography is strictly a separation, rather than a measurement technique, it is convenient to consider it in the present section.

All chromatographic techniques are based upon the differing distributions of individual compounds (in a mixture) between two immiscible phases as one of the latter (the mobile phase) passes through or over the other (the stationary phase). The mixture is usually added to one end of the stationary phase and the individual compounds are then eluted at the other end by the mobile phase, the compound appearing first being that which has the smallest distribution into the stationary phase.

Chromatographic procedures are classified according to the natures of the phases. GC covers variants in which the mobile phase is a gas, and there are two sub-variants according to the nature of the stationary phase, *i.e.*, gas-liquid and gas-solid chromatography. (The term gas chromatography or GC is, however, used almost exclusively in the modern literature when gas-liquid chromatography is the technique under consideration). Similarly, liquid chromatography covers all those procedures using a liquid mobile phase – including column chromatography (of which HPLC is a variant), paper chromatography and TLC.

Gas chromatography is now used by a number of routine water analysis laboratories and fully merits inclusion as a generally important technique, though it finds even greater application in research and investigational work – particularly in combination with mass spectrometry (see Section 11.4.1). The use of liquid chromatography in water analysis is more limited, though growing rapidly with the advent of HPLC, but its variants will be considered here for convenience. Sections 11.3.8(*a*)–(*c*) will deal specifically with gas chromatography, Sections 11.3.8 (*d*)–(*f*) with HPLC, Section 11.3.8(*g*) with IC Section 11.3.8(*h*) with thin-layer chromatography and, finally, Section 11.3.8 (*i*) will deal with the application of flame ionisation detectors (FID) to the determination of the total organic carbon (TOC) content of waters. (With regard to this last topic, it should be noted that other measurement techniques [*e.g.*, conductivity – see Section 11.3.9(*b*) – and infrared spectrometry – see Section 11.4.3(*e*)] may also be applied to TOC determination. However, it is convenient to deal with this use of FID detectors here, because consideration of their nature and characteristics is also necessary in the treatment of GC).

(a) General Basis of Gas Chromatography

A schematic diagram of a gas chromatograph is shown in Figure 11.8. The sample (usually a gas or a liquid) is injected, directly or indirectly, onto the column, through which it is swept by the carrier gas – commonly, nitrogen or helium. The column is contained within an oven, the temperature being closely controlled at the value* necessary for the required chromatographic separations. The carrier gas elutes the compounds from the column, at times which are related to their distribution coefficients between the stationary phase and the gas; the higher the distribution coefficient, the longer it takes for the compound to be eluted. On emerging from the column, the gas passes through a detector unit sensitive to the eluted compounds, the response of which is fed continuously to an amplifier and thence to a chart recorder (and, increasingly, to a computer data handling system, also). When a compound is eluted, its presence is revealed as a peak – ideally of Gaussian shape – on the recorder chart. A schematic diagram of a typical chromatogram for a sample containing several components is given in Figure 11.9.

This simplified description conceals a number of points of great importance.

---

* Or values, in temperature-programmed mode.

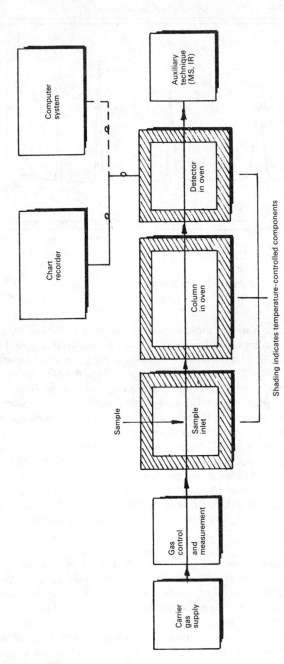

Figure 11.8 *Schematic representation of a gas chromatograph*

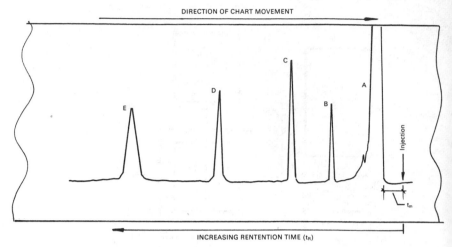

Compounds B, C, D and E are present in the sample, and their distribution into the stationary phase increases from B to E.
The peak A is caused by the solvent in which the sample was dissolved.
t_m = gas hold-up time (see text).

Figure 11.9 *Example of a gas chromatogram*

Though brief mention of the most important points is made below, a detailed examination of this complex and powerful technique cannot be provided here. The reader seeking such detail is referred to one of the many general texts available (see, for example, refs. 547–552), the latest of which[550–552] are devoted to the subject of capillary column GC, which is of great and growing importance. GC as a whole is reviewed biennially (in even years) in *Analytical Chemistry*,[553] and its application to water analysis is considered in the reviews of water analysis which appear, in the same journal, in odd years.[554] The application of high resolution, capillary column GC to water analysis has also been reviewed comprehensively by Wilson and Davis.[555] A short introduction to gas chromatography in water analysis has also appeared in the series of publications produced by the U.K. Standing Committee of Analysts.[556] The *Journal of Chromatographic Science* publishes a useful annual guide to chromatographic equipment available commercially,[557] and occasional reviews of gas-chromatographic instrumentation in particular (see reference 558 for the latest up-date).

(i) The chromatographic column The influence of the column upon the chromatographic process is fundamental, and the performance of columns is assessed mainly on the basis of 'efficiency', that is the ability to achieve a separation in a particular time. Other column properties, of secondary concern, are reactivity (polarity, acidity/basicity) and capacity.

A number of different terms are used in describing the separation efficiency of a column. Some of these derive from the theory of fractional distillation (that process being also concerned with the separation of mixtures of volatile species). A brief summary of the more important terms is given below.

(1) *Number of theoretical plates, n.* This is usually calculated from:

$$n = 16 \, (t_R/w)^2 \qquad (11.5)$$

where t_R is the retention time of the peak (see Figure 11.9) and w is the base width of the peak. Because the width of the peak at half height ($w_{\frac{1}{2}}$) is easier to measure than w, an alternative formula (assuming a Gaussian peak) is:,

$$n = 5.54 \, (t_R/w_{\frac{1}{2}})^2 \qquad (11.6)$$

The same units must be used for the t_R and w (or $w_{\frac{1}{2}}$) measurements. Separation efficiency increases as n increases.

(2) *Number of effective theoretical plates, N.* This includes a correction for the finite time t_m (the 'gas hold-up time') taken for the carrier gas to traverse the column which, although contributing to t_R, does not aid the separation process. N is calculated by substituting the corrected retention time $t'_R = t_R - t_m$ for t_R in equations (11.5) and (11.6), and replacing n by N.

(3) *Height equivalent to a theoretical plate, h.* Longer columns will, other things being equal, have larger numbers of theoretical plates. To take account of column length, the height equivalent to a theoretical plate may be quoted, calculated from:

$$h = L/n \qquad (11.7)$$

where L is the column length. Smaller values of h indicate greater efficiency of separation, for a given column length.

(4) *Height equivalent to an effective theoretical plate, H.* This again allows for the gas hold-up time, and is given by:

$$H = L/N \qquad (11.8)$$

Smaller values of H again indicate greater separation efficiency, for given column length.

(5) *Separation numbers or 'Trennenzahl', TZ.* This is the number of peaks completely separated* between two adjacent members of an homologous series (n-alkanes are normally chosen). It is given by:

$$TZ = \frac{t_{R(Cn+1)} - t_{R(Cn)}}{w_{\frac{1}{2}(Cn+1)} + w_{\frac{1}{2}(Cn)}} - 1 \qquad (11.9)$$

where $t_{R(Cn+1)}$ and $t_{R(Cn)}$ are the retention times, and $w_{\frac{1}{2}(Cn+1)}$ and $w_{\frac{1}{2}(Cn)}$ the peak widths at half peak height, of the two chosen homologues. TZ may be related directly to the chromatographic output, and is a popular means of describing separation efficiency. It is approximately proportional to the square root of column length and – like the previous measures of column efficiency – depends on the chosen conditions and compounds, which should be quoted with the value observed.

The above concepts are all means of *describing* the separation efficiency of columns; the dependence of this efficiency upon a series of factors is described by the van Deemter equation which is given (in simplified form) by:

$$L = A + B/\bar{u} + C\bar{u} \qquad (11.10)$$

* See ref. 559 for further details and a definition of 'complete separation' in this context.

where A is a term relating to the packing geometry of the column, which affects eddy currents in the gas flow, $B/\bar{u}$ is a term relating to longitudinal diffusion (which causes zone spreading) and $C\bar{u}$ relates to the mass transfer between the gas and stationary phases. The term $\bar{u}$ is the average linear velocity of the carrier gas, and is given by:

$$\bar{u} = L/t_m \tag{11.11}$$

Further details of the van Deemter equation, and its value in describing the process of GC, may be found in texts on the technique.

In gas-solid chromatography, the stationary phase is a solid and it is the varying extent of adsorption of different compounds upon the solid which provides the separation. Porous polymers are most widely used as stationary phases in this type of gas chromatography, and may be employed to separate permanent gases or short chain polar organic species such as amines and glycols. Zeolite 'molecular sieves' have also been used in gas-solid chromatography for the separation of permanent gases.

In gas-liquid chromatography, it is the distribution of the compounds of interest to different degrees between the stationary liquid phase and the mobile gas phase which effects the separation. In this variant, adsorption (particularly if irreversible) of the components of samples on the support or other parts of the column can cause serious deterioration of performance (both qualitatively and quantitatively), and should always be minimised.

Two main types of column are in use: packed columns in which the liquid phase is coated on granules of an inert supporting material, and unpacked or 'open tubular columns' (OTCs), in which the liquid phase is coated on the walls of a capillary*. The development of capillary columns for GC has recently been reviewed by Ettre.[560] Fused silica is now widely used for the fabrication of such columns, and the use of this material has been reviewed by Jennings.[561] It has the advantages, over conventional glass materials, of greater inertness and chromatographic reproducibility, together with greater flexibility which greatly aids the practical application of capillary GC.[561]

In earlier stages of capillary column development, the liquid phase was coated upon a porous, supporting layer produced on the inside wall of the column. Columns of this type have been called 'porous-layer open tubular' (PLOT) and 'support-coated open tubular' (SCOT). Their use has been to a large extent superseded by 'wall-coated open tubular' (WCOT) columns†. In these, the liquid phase is either merely coated uniformly upon the inside capillary wall or immobilised thereon (so called 'bonded phases', in which the volatility of the polymeric liquids is reduced by crosslinking[560,562]). The major advantage of open tubular columns over packed columns resides in the far greater separation efficiency of the former. Because of the 'open' nature of capillary columns, much

---

* Various types of capillary have been used, including stainless steel, glass and fused silica or quartz. Fused silica is now widely favoured. The term 'capillary column' is often used as an alternative to 'open tubular column' and the names 'capillary gas chromatography' and 'high resolution gas chromatography' are often applied to GC with open tubular columns.

† Note, however, that the distinction between PLOT and WCOT columns may not be entirely clear-cut, and that increase of surface area without important increase in tube diameter has theoretical justification.[560]

TABLE 11.6

A Comparison of Packed and Wall-coated Open Tubular Columns (adapted from ref. 555)

Column type	Internal diameter mm	Column length m	Separation n (1000's)	Efficiency TZ	Flow rate ml min^{-1}	Maximum sample capacity μg
Packed	2–4	1–4	1–10	2–10	20–80	100–500
WCOT	0.2–0.5	10–50	50–250	25–60	0.5–4	0.1–10

greater column lengths can be used before the pressure drop along the column becomes limiting, with consequent increase in separation efficiency. Other factors are thought to include the lack of eddy currents in the gas flow caused by column packing*, increased phase ratio (volume of gas to volume of liquid) and reduced tendency of the liquid phase to depart from a thin film (which it does at column packing points).[550] Table 11.6, adapted from ref. 555, compares the chromatographic characteristics of packed and WCOT columns. The greater separation efficiency of the latter (higher *n* and *TZ* values) is evident, but it should also be noted that the capacities† of WCOT columns are considerably smaller than those of packed columns.

The separation of two compounds, A and B, depends not only upon the separation efficiency of the column but also upon their separation factor, $X_{A/B}$, which is given by:

$$X_{A/B} = t'_{R,A}/t'_{R,B} \qquad (11.12)$$

where $t'_{R,A}$ and $t'_{R,B}$ are the corrected retention times of A and B (such that $t'_{R,A} > t'_{R,B}$). Column polarity, which is determined chiefly by the nature of the stationary phase, governs the value of $X_{A/B}$.

The selection of stationary phases is discussed in texts on GC; systems of column characterisation have been developed to aid the process, the most widely used being the Rohrschneider[563] approach to polarity, subsequently modified by McReynolds.[564] Basically, the retentions of several representative compounds on a given column are measured and compared to those obtained for the same compounds on a 'standard', non-polar column of squalane (a highly branched saturated hydrocarbon). The difference in retention on the two columns, expressed as the Rohrschneider/McReynolds constant, is a measure of the increased polarity of the test phase over squalane, for compounds similar to the particular test compound. These constants are now often quoted in manufacturers' literature on stationary phases, which normally gives an indication of the likely suitability of the different phases for different purposes. Another source of information on stationary phases used for different determinations is the abstracting service of the U.K.-based 'Chromatography Discussion Group'.[565] For those compounds which are the subject of standard or recommended methods of analysis, information on the performance attained using the different types of column which have proved

---

* That is, the A term in the van Deemter equation [equation (11.10) above] is reduced to zero.

† The capacity of a column refers to the maximum amount of material which can be loaded onto it without serious deterioration of chromatographic performance, indicated by asymmetry of peak shapes.

suitable is often given with the method – typically in the form of example chromatograms or relative retention data (see, for example, refs. 168 and 566–572). For less commonly determined substances, information on columns which have been used in water analysis may be obtained through the reviews in *Analytical Chemistry*[554] and the *Journal of the Water Pollution Control Federation.*[573]

It should be noted, however, that the separation achieved depends on the liquid phase *and* on the length and type of column used. Thus, care should be taken when attempting to select a liquid phase from study of previous work. This is particularly necessary when establishing capillary column GC, in which the separation factor (and hence nature of the liquid phase) plays a less important role than in packed column work.[555] (In the latter, the relatively low separation efficiencies attainable may often necessitate the use of the more retentive, polar liquid phases. In general, capillary columns offer such high separation efficiency that a much smaller range of liquid phases has been used, in comparison with packed columns).

Both packed and capillary columns are commercially available, though the former can be made by the analyst himself quite simply. Production of the (much more expensive) capillary columns is far less straightforward. Fabrication of fused silica columns requires high temperature facilities,[561] and is best regarded as beyond the scope of a typical, routine laboratory. Production of conventional glass capillary columns is simpler, but still requires apparatus costing about £2000 and considerable time and effort;[555] it too is probably best avoided by routine analysis laboratories.

(ii) Sample injection Injection of samples onto the column is relatively straightforward in packed column work, using a calibrated microlitre syringe *via* a septum. (Automatic injectors are also available.) The injection port may be heated to ensure that the liquid sample is vaporised before it meets the column, but on-column injection without heating of the port may be advantageous in some cases, to minimise degradation of thermally sensitive compounds. Injection techniques for packed column work are discussed in general texts.

In capillary column GC, injection is much more complex, because of the low carrier gas flow-rates and limited capacities of the columns. Many different procedures have been devised, and the nomenclature of the subject has become so extensive that 'Even the initiated have become confused.'[574] Pretorius and Bertsh[574] have produced a paper on the terminology and classification of sample introduction in capillary GC, the first in a series intended to deal with the principles of sample introduction, and Schomburg[575] has produced a useful summary of the principal techniques. The following are brief, simplified descriptions of some of the more important approaches for water analysis.

In the early stages of capillary GC, the only available approach was to use a 'split' injection technique; in this, the sample is vaporised into a high flow of carrier gas and a small portion (1/10–1/100) of the resulting mixture is passed to the column, the remainder being vented. This technique is, however, rarely applicable in water analysis, being unsuited for trace components.[555]

In so called 'splitless' injection (more appropriately called 'direct injection with vaporisation and solute focusing using the solvent effect', following the terminology of ref. 574), devised by Grob and Grob,[576,577] the sample is injected and vaporised without splitting, but then condenses at the head of the cooled column. By appropriate selection of conditions (especially of solvent and temperature) the solutes may be 'focused' or concentrated at the head of the column.

Yet another approach involves direct on-column injection of the liquid sample, onto a cool column. (This is the so called 'cold on-column injection' devised by Schomburg *et al.*[578]) A focusing effect is again achieved, and the technique has particular value for the determination of thermally sensitive compounds and compounds having high boiling points. It is now widely applied.

Another sample introduction technique, widely used in water analysis, is 'headspace analysis',[579] which involves the analysis of the gas phase in contact with the sample. Direct headspace analysis has been used,[580] but headspace analysis with trapping of the determinands prior to the GC stage appears to be more commonly employed. Two principal modifications of the headspace/trapping technique are used. In the Grob closed-loop stripping technique, the determinands are stripped from the sample by a recirculating stream of air and are adsorbed on a small carbon filter.[581,582] The trapped compounds are eluted from the carbon with small amounts of a solvent, the resulting eluate being directly injected onto a capillary gas chromatograph.

In the second approach, the stripping step is accomplished using a 'straight through' (rather than recirculating) stream of gas (helium or nitrogen), and the compounds are concentrated on a suitable solid such as Tenax GC (a porous polymer), from which they are subsequently desorbed by heating, directly into the gas chromatograph. This technique has been widely used in water analysis in the U.S.A.,[583–588] whereas the Grob technique has been extensively employed in Europe.[555,581,589,590]

Other preconcentration techniques commonly used with gas chromatography in water analysis are liquid–liquid extraction (*e.g.*, refs. 168,566–569,587, 591–594) and adsorption from solution by macroreticular resins (*e.g.*, refs. 594–599).

The recently published, two volume text on advances in the determination of organic pollutants in water, edited by Keith,[600] contains papers that describe the application of many of the above techniques to water analysis by GC.

(iii) Detection of eluted compounds   On emerging from the column, the carrier gas and any eluted compounds pass through the detector unit so that the presence of these compounds is revealed and their concentrations measured. A wide variety of detectors has been used in GC, and texts are available which deal specifically with the topic (*e.g.*, refs. 601 and 602). The detectors most commonly employed with GC in water analysis are the flame ionisation detector (FID), electron capture detector (ECD), thermionic detectors (TID – nitrogen and phosphorus selective), flame photometric detectors (FPD – phosphorus and sulphur selective) electrolytic conductivity detectors (ElCD – halogen, nitrogen and sulphur selective) and, mainly for permanent gases, the thermal conductivity detector (TCD). The use of the mass spectrometer (MS) as a GC detector in water analysis is considered in detail in Section 11.4.1. The cost of a conventional mass spectrometer is a bar to its widespread use, but a recent development is an inexpensive MS system of limited resolving power designed specifically as a GC detector – the so called ion trap detector (ITD).[603]

The operating principles of the more common GC detectors are described briefly below; fuller details can be obtained elsewhere (refs. 601 and 602).

(1) *Flame ionisation detector (FID)*. The production of ions in the hydrogen/air flame is greatly enhanced when an organic substance enters. When two electrodes are placed near the flame, and maintained at a constant potential difference, the ion current measured permits detection and quantification of the organic material.

(2) *Electron capture detector (ECD)*. Compounds which have an affinity for electrons produce ions by 'capturing' electrons from a β-emitter. This gives rise to an ionisation current which can again be measured using a suitable electrode system.

(3) *Thermionic detector (TID)*. The flame thermionic detector is similar to the FID, but contains a solid alkali metal salt, the presence of which enhances the response to compounds containing nitrogen or phosphorus. The mechanism is complex and imperfectly known.[601,602]

(4) *Flame photometric detector (FPD)*. The emission of characteristic radiations by phosphorus and sulphur species formed by combustion in a hydrogen-rich flame forms the basis of these detectors. The radiation is measured by a photomultiplier, after passage through a suitable optical filter.

(5) *Thermal conductivity detector (TCD)*. The thermal conductivity of the column effluent is affected by the eluted compounds, and provides another basis for detection and measurement. A differential technique is used in which the conductivities of column effluent and pure carrier gas are continuously compared.

(6) *Electrolytic conductivity detector (ElCD)*. In this system of detection, the GC effluent is thermally decomposed and the gaseous reaction products are absorbed in a suitable solvent, the electrical conductivity of which is continuously monitored. Detected compounds must, therefore, decompose to give products which dissolve and ionize in the solvent. By controlling the conditions of thermal decomposition, absorption and conductivity measurement the ElCD may be made selective to nitrogen-, halogen- and sulphur-containing compounds.

Flame ionisation detectors respond to virtually all organic compounds, and are widely used in water analysis. For certain types of compound, most notably halogenated organics, electron capture detection offers much higher sensitivity than FID. The most common applications of ECD in water analysis are to the determination of organochlorine pesticides, polychlorinated biphenyls (PCBs) and halogenated alkanes and alkenes. Electrolytic conductivity detection provides poorer detection power than ECD, but offers excellent selectivity and has been used, for example, in the determination of tri-halogenated methanes.[585] Similarly, thermionic and flame photometric detectors find application when nitrogen-, sulphur- or phosphorus-containing organic compounds are of interest (*e.g.*, sulphur-selective flame photometric detection has been used in GC studies of oil pollution[604,605] and phosphorus-selective flame photometric[606] and thermionic[607] detection have been used in the determination of organophosphorus pesticides).

Thermal conductivity detection is used almost exclusively for the determination of permanent gases.

Atomic absorption and emission spectrometry have been employed as GC detection techniques. Thus, for example, atomic absorption detection has been applied to the determination, by GC, of inorganic and organic tin species,[608,609] organoselenium compounds[610,611] and inorganic and organic lead species,[612-614] and microwave-excited plasma emission detection has been applied to the determination of trialkyl-lead species,[615] alkylmercury species[616] and organohalogens (chlorination products of humic acid).[617]

The application of AAS as a GC detector in trace metal speciation studies has been reviewed by Fernandez,[618] van Loon[619] and Chau and Wong[618] [see also Section 11.3.8(c)]. The potential value of the microwave-excited helium plasma as a GC detector in the determination of organic compounds in water has been discussed by Wilson and Davis.[555]

Fourier-transform infrared spectrometry has also been applied as an on-line GC detector system in water analysis. This is discussed in Section 11.4.3(d), which deals with i.r. spectrometry.

(iv) Identification and quantification It will be apparent from the preceding discussion of GC columns that the retention time of a particular compound on a column is characteristic of that compound for that particular column and other experimental conditions. Thus, measurement of this time and comparison with the retention times of pure compounds provides the basis for qualitative analysis. However, more than one compound may have essentially the same retention time, so that confirmation (or rejection) of a tentative identification, by measuring retention times on several columns having liquid phases of different polarity, is usually desirable.

Because the absolute retention times are applicable over only short periods, under closely controlled experimental conditions, other approaches to recording retention data have been developed. In the simplest of these, the relative retention, $r_{A/B}$, of the substance of interest (A) compared with that of a standard (B) is used:

$$r_{A/B} = t'_{R,A}/t'_{R,B} \qquad (11.13)$$

where $t'_{R,A}$ and $t'_{R,B}$ are the relevant retention times, corrected for the gas hold-up time as described previously. The relative retention is independent of the amount of stationary phase and the gas flow-rate, and less dependent on the temperature than the retention time itself, but is not very accurate, particularly when $t'_{R,A}$ and $t'_{R,B}$ differ markedly, and is too unreliable for exchange of data between different laboratories.[555]

The need for more reliable means of recording retention data has led to the development of retention indices, of which the most widely used is that of Kováts.[621] Retention indices exploit the fact that, for the members of an homologous series under isothermal GC conditions, the logarithm of the corrected retention time (t'_R) is linearly related to molecular weight.

In the Kováts system, the retention behaviour of a compound is related to the retention characteristics of normal paraffins, the Retention Index*, I_b^a being given by equation (11.14):

$$I_b^a = 100P + 100(Q-P) \frac{(\log t'_{R,A} - \log t'_{R,P})}{(\log t'_{R,Q} - \log t'_{R,P})} \tag{11.14}$$

where a is the liquid phase and b is the temperature, and:

$t'_{R,A}$ = corrected retention time of compound of interest;

$t'_{R,P}$ = corrected retention time of normal paraffin of carbon number P, $t'_{R,P}$ being smaller than $t'_{R,A}$;

$t'_{R,Q}$ = corrected retention time of normal paraffin of carbon number Q, $t'_{R,Q}$ being larger than $t'_{R,A}$.

It is usually best to select the n-paraffins such that $Q - P = 1$ (*i.e.*, such that they differ by one in carbon number). The n-paraffins themselves, by definition, have retention indices 100 times their carbon number, on any liquid phase.

Kováts indices may be used for the exchange of information between laboratories, and as a means of identifying compounds in mixtures. Again, the determination of retention indices on two capillary columns of different polarity gives much increased confidence in the identification. The development of capillary GC has, by increasing separation efficiency and reproducibility, greatly enhanced the value of retention data for compound identification. Further discussion of this topic may be found in Chapter 7 of the monograph on capillary GC, by Jennings.[550]

In addition to retention data, the results of additional analytical techniques may be employed to identify the separated compounds. MS (see Section 11.4.1) is the most powerful tool in this respect, and it seems probable that the availability of (relatively) inexpensive mass spectrometric detectors (see, for example, ref. 603) will increase the application of GC-MS in water analysis. However, the use of other detection techniques such as infrared spectrometry [see Section 11.4.3(d)] can also be of value.

For quantitative analysis, the area under the recorded peak is used, being governed by the amount of the particular compound in the sample. Integral or on-line computing systems have superseded mechanical integrators, but it should also be noted that peak height is more simple to measure and is often adequately accurate.

(b) Principal Analytical Characteristics of Gas Chromatography
The many different problems to which GC is applied, together with the wide range of available instrumentation, make it difficult to give an accurate summary of the analytical characteristics of the technique. The approximate nature of the following material must, therefore, be borne in mind.

It should also be noted that it is normal to extract the substances of interest from the water sample prior to application of GC. This is usually undertaken to achieve a preconcentration of the determinands, though it is also true that the performances of many columns and detectors are affected deleteriously by water. Thus, dissolved gases are usually stripped from the sample and injected directly on to the column

---

* Strictly, the Retention Index is defined in terms of corrected retention *volumes* rather than *times*. However, for practical convenience most workers use times and report data in the form shown here.

and dissolved organic materials are usually (and inorganics may be) extracted into a suitable solvent, portions of which are then analysed. Consequently, the performance characteristics of analytical systems will depend upon these extraction procedures as well as upon the chromatographic finish. In the first edition of this book, the discussion of performance was based essentially on analysis of the extracts. Since that time, standard or recommended systems of analysis employing GC have become common, and performance data for complete analytical methods, correspondingly, more readily available. Both the performance of the GC technique and of complete systems based upon it will, therefore, be considered.

(i) *Calibration curve* Provided that the column chosen is appropriate, each of the compounds of interest should be quantitatively eluted. Thus, the relationship between the peak area (or height) on the chromatogram and the amount of substance is governed primarily by the detector. For narrow ranges of concentration this relationship is essentially linear for commonly used detectors. The FID normally maintains a linear response over many orders of magnitude (up to 10^8) and, although other detectors provide more limited ranges, these may often be extended by plotting the data on log–log paper or by suitable signal processing.

To evaluate the amounts of determinands in samples, a calibration curve is used (a procedure sometimes known as 'external standardisation'). This curve is prepared by measurement on standards containing known amounts of the determinand taken through exactly the same procedure as samples. In practice, some GC determinations are so lengthy that standards are routinely taken through only the final GC step, and not through the preliminary stages of extraction, concentration and clean-up which may be applied to samples. It is essential in such cases that at least one standard (in addition to the blank) be taken through the entire procedure with each batch of samples, to check the calibration.

The reader is cautioned, however, that it has become conventional practice in, for example, the determination of organochlorine insecticides and polychlorinated biphenyls not to correct for the variable and low (typically about 50%) recoveries of the determinands through the pre-treatment stages, but to report the results obtained, uncorrected (see, for example, ref. 568 and ref. 571, pages 493–503). The present authors feel strongly that the methods concerned should be modified to allow correction for the low recovery to be made, thus avoiding production of analytical results known to be subject to large biases.

As an alternative to the use of a calibration curve, 'internal' standardisation may be performed, using a substance having similar properties to the determinand (or its derivative). As in the case of 'external' standardisation, the term 'internal standardisation' can refer to an internal standard taken through the entire analytical procedure, or to such a standard taken through only the GC stage. In either case, the use of the internal standard depends crucially upon the similarity in behaviour between it and the determinand. In principle, the use of an internal standard taken through the entire analytical procedure checks more sources of error; deuteriated compounds are commonly used in gas chromatography/mass spectrometry for this purpose (see Section 11.4.1).

It is again recommended that, when either form of internal standardisation is used, at least one standard solution of the determinand (in addition to the blank) be taken through the entire analytical procedure with each batch of samples, as a check. This is recommended, for example, in a recent GC method[168] for the determination of phenols (involving extraction and formation of their trimethylsilyl ethers) in which a higher alcohol (heptanol, nonanol or decanol) is used as an internal standard taken through the derivatisation and GC stages only.

Although calibration of the GC stage itself may seem straightforward, involving merely the injection of suitable standards ('external' or 'internal'), there are a number of potential pitfalls. The subject is discussed further by Shatkay and Flavian[622] and by Shatkay.[623] These authors conclude, *inter alia*, that certain assumptions commonly made when internal standardisation is used may not be always justified; thus, such standardisation may not compensate for fluctuations in GC conditions (temperatures, gas flows and electrical supply) and the concentration of the internal standard – not merely the determinand/standard ratio – may be important.

(ii) Limit of detection For a given column, the limit of detection is governed largely by the detector, assuming ideal column behaviour. However, the use of capillary columns produces much narrower peaks than are obtained with packed columns, and may therefore allow a substantially lower limit of detection, even when using the same detector. Moreover, the practical limit of detection may depend on other factors such as the presence of overlapping peaks and sloping baselines.

These factors should be borne in mind when consulting Table 11.7, which gives approximate limits of detection for the more common GC detectors. It is also emphasised that the exact nature of the compound of interest may markedly affect the limit attained, another reason for taking the values in Table 11.7 as merely a

TABLE 11.7
Approximate Detection Limits for some Gas Chromatography Detectors

Detector	Approximate limit of detection μg	Ref.
Thermal Conductivity (TCD)	1–0.1	601, 624
Thermal Conductivity (Micro TCD)	$10^{-1}-10^{-3}$	601
Flame Ionisation (FID)	$10^{-4}-10^{-5}$	601, 624
Electron Capture (ECD)	$10^{-6}-10^{-7}$	601, 624
Flame Photometric (S,P FPD)	10^{-5}	601
Thermionic (P,N TID)	10^{-6}	601
Electrolytic Conductivity (ElCD)	10^{-5}	625
Mass Spectrometry (Single ion monitoring)	10^{-6}	555
Mass Spectrometry (Complete spectra)	10^{-3}	555

general guide. The values quoted for the flame ionisation, flame photometric, thermionic and electrolytic conductivity detectors show the extremely small limits attainable and illustrate the potential of gas chromatography for the determination of trace organics in water.

Some limits of detection attained using complete methods employing gas chromatography are shown in Table 11.8. Again, the excellent detection capabilities of the technique are evident.

(iii) Precision As with any other analytical technique, the precision attainable depends on the care with which the technique is used and on the complexity of any prior concentration or pre-treatment step.

With regard to the gas chromatographic step itself, the reproducibility of sample injection and vaporisation is of particular importance. Relative standard deviations of 2–5% can be attained routinely, provided that the amounts being determined are not too close to the detection limit.

The precision attained with complete methods depends greatly upon the sample pre-treatment. Relative standard deviations of 2–10% are attainable with determinations involving relatively straightforward and reproducible pretreatments (see, for example, refs. 168, 566, 567), provided that the concentrations are not close to the detection limit. By contrast, if the pretreatment is complicated (involving, perhaps, not only extraction from the water sample but also further concentration and/or 'clean-up' of the extract) the relative standard deviations may be substantially higher – 10–20% or greater – even at concentrations far above the detection limit (see, for example, ref. 568 and ref. 571, pages 493–503).

(iv) Selectivity When the compounds to be separated are known, it is usually possible to select a column and detector such that good selectivity is achieved. However, the determination of particular organic compounds in samples which may contain many unknown organic materials can often present problems that can be overcome by using either a more efficient column or a more specific detector. The advent of capillary columns has assisted greatly in this regard. If, however, these approaches are not fruitful, recourse must be made to preliminary separation procedures to isolate the substance(s) of interest; this is particularly true when polluted waters are to be analysed.

(v) General A wide range of instrumentation for gas chromatography is available commercially (see, for example, refs. 557 and 558). As with many other instrumental techniques, automation and microcomputer control are of increasing importance. A text on the application of a microcomputer to chromatography in general has been recently produced,[644] and Risby *et al.*[553] have reviewed developments over the period 1980–1982.

It should be noted that, although any gas chromatograph can, in principle, be converted to high-resolution, capillary column operation, the results of such conversion may be disappointing; for this reason, it has been recommended that a laboratory starting to use high-resolution GC does so on an instrument specifically designed for operation with capillary columns.[555]

Although the basic operation of a gas chromatograph is not difficult to learn, the

TABLE 11.8

Some Limits of Detection attained using Complete Analytical Methods based on Gas Chromatography

Determinand	Detector	Limit(s) of detection $\mu g\ l^{-1}$	Ref.
Organochlorine pesticides and degradation products. (Various)	ECD	0.003–0.015 0.010–0.025 0.003–0.24	568 571 626
Polychlorinated biphenyls. (Various)	ECD	0.106 0.065	568 626
Chlorinated herbicides. (Various)	ECD	0.002–0.050 0.07–249	571 627
Organophosphorus pesticides. (Various)	FPD FPD/TID	0.04–0.40 0.1–5.0	569 628
Organonitrogen pesticides. (Various)	TID	0.46–3.39	629
Triazine pesticides. (Various)	TID	0.03–0.07	630
Pyrethrins (mixed)	ECD	0.06	566
Permethrin (a synthetic pyrethroid)	ECD	0.24	566
Phenols			
Phenol	ECD	0.1	631
Various	FID	100–900	168
Various	FID	0.14–16.0	632
Chloro- and bromo-trihalogenated methanes (THMs)			
Chloroform	ECD ElCD	0.08–0.9* 0.05	567 633
Bromodichloromethane	ECD ElCD	0.06–0.5* 0.10	567 633
Dibromochloromethane	ECD ElCD	0.03–1.2* 0.09	567 633
Bromoform	ECD ElCD	0.03–1.4* 0.20	567 633
Other volatile halogenated organics (Various)	ElCD	0.02–1.81	633
Purgeable aromatics (Various)	PID†	0.2–0.4	634
Phthalate esters (Various)	ECD	0.29–3.0	635
Nitrosamines (Various)	TID	0.15–0.81	636
Haloethers (Various)	ElCD	0.3–3.9	637
Chlorinated hydrocarbons (Various)	ECD	0.03–1.19	638

TABLE 11.8 (*cont.*)

Determinand	Detector	Limit(s) of detection $\mu g\, l^{-1}$	Ref.
Inorganics and Organometallics			
Total selenium	ECD	0.0008–0.002	639, 640
Inorganic and organic lead species	} AAS	0.1	612
Organomercury species	MPE‡	0.0004–0.4	616
As^{III}	} AMP§	{ 0.2	} 641
Sb^{III}		0.2	
Sn^{II}		0.8 }	
Organoarsenic species	ECD/FID	*ca.* 0.002	642
Hydrazine	TID	0.1	643

* Range of values from different laboratories
† PID – photoionisation detector
‡ MPE – microwave-excited plasma emission detector
§ AMP – amperometric detection

processes involved, from injection to data treatment, are relatively complex. A reasonable knowledge of these processes is essential for efficient trouble-shooting, particularly with high resolution GC using open tubular columns which, with its major advantages, is replacing packed column operation in most applications. Again, both the GC stage itself and the substantial pretreatment of samples which is often required are demanding of operator skill and dexterity.

The sample pretreatment stages can considerably lengthen the analytical process – the GC stage itself, from injection to result, rarely requires more than 30 min, and may be much quicker.

Contamination can present problems because very small amounts of compounds are often involved, and because they are of a volatile nature. Complicated pretreatment steps obviously increase the risk of contamination occurring.

GC is, in general, applicable only to suitably volatile determinands. Although derivatisation can extend to some extent the range of compounds which can be handled, it adds to the complexity of, and time required for, analysis, and is frequently inapplicable. As noted previously, only about 10–20% of the total organic matter in most waters is thought to be amenable to GC, so that the volatility requirement is an important limitation. The technique is also inapplicable to compounds which decompose or adsorb irreversibly on columns.

(c) Applications of Gas Chromatography to Water Analysis
The technique has been widely used in water analysis, for many different determinands. It is not the purpose of this sub-section to provide an exhaustive listing, but rather to illustrate this wide range of application and draw attention to the principal uses. The determination of specific organic compounds is considered first, then the use of whole (or partial) chromatograms is examined. Finally, the determination of permanent gases and inorganic and organometallic species is dealt with.

The use of gas chromatography with mass spectrometry is considered in detail elsewhere (Section 11.4.1).

(i) The determination of specific organic compounds As at the time of publication of the first edition of this monograph, the determination of pesticides in water remains an important application of gas chromatography. GC methods for pesticides are included in many collections of 'standard' or 'recommended' methods for water analysis. Examples include: organochlorine pesticides [refs. 568, 571 (pages 493–503), 626 and 645 (pages 431 442)]; organophosphorus pesticides [refs. 569, 628 and 645 (pages 442–451)]; phenoxyacetic acid and other chlorinated herbicides [refs. 571 (pages 504–507) and 627]; organonitrogen pesticides (ref. 629); triazine pesticides (ref. 630) and pyrethrins (and a synthetic pyrethroid) used in water mains disinfestation (ref. 566).

At the levels of interest, separation of many of the pesticides from one another and from other organic compounds may be difficult, and the use of capillary columns offers advantages in this respect and has been adopted by a number of workers.[646–648] [Polychlorinated biphenyls (PCBs) are determined using similar procedures to those for organochlorine pesticides – see for example, refs. 568, 571 (pages 493–503) and 626 – and, again, capillary columns have found application in their determination[649,650].]

GC has been applied to the determination of many other pesticides, including carbamates[651,652] and the triphenyltin organometallic fungicides and molluscicides, and their derivatives.[653,654] Paraquat and diquat have been determined by gas-chromatographic measurement of their pyrolysis products.[655] The biennial reviews of water analysis published in *Analytical Chemistry*[554] contain sections devoted to pesticide determinations, which should be consulted for further information on this application of the technique.

Phenolic compounds are also determined by gas chromatography, which has the advantage over spectrophotometry of allowing separate determination of individual compounds. Methods have appeared in various authoritative compilations, involving both direct injection of aqueous samples (refs. 571 and 645, which give the same method) and prior extraction (refs. 168 and 632). Flame ionisation detection is usually used, but electron capture detection (with enhanced sensitivity by bromination of the compounds) has been applied to the determination of phenol and chlorophenols.[631,656] ECD has also been applied to the determination of phenols after conversion to their heptafluorobutyryl[657] or pentafluorobenzyl[632] derivatives. The use of capillary columns has been reported in a recent 'recommended' method[168] for phenols.

Since the first edition of this monograph, interest has grown in the formation of chlorinated organics during drinking water chlorination, and in the occurrence of halogenated industrial solvents and other halogenated compounds in potable water supplies and other natural waters. Methods for the determination of halogenated methanes[567,571,633] and other halogenated compounds[571,633,634,637,638] have been published in collections of 'standard' or 'recommended' methods. Both solvent extraction[567,637,638] and purge and trap[571,633,634] pretreatments have been employed. Again, the advent of capillary columns has brought advantages in improved separation capability and sensitivity, and such columns have been used for the determination of halogenated organics in water.[658–662].

Although electron capture or electrolytic conductivity detection is almost always used for the determination of these halogenated compounds, the use of a microwave plasma emission detector (which can selectively monitor the individual halogens) has also been reported[617,663] for the determination of organohalogen compounds.

In addition to the most important classes of organic compound considered above, GC has been applied to the determination of many other specific organic compounds in water. A full listing is beyond the scope of this work, but includes*: acrylonitrile,[664] alcohols,[598,665-667] aldehydes,[598,667] amides,[668-670] amines,[671] amino acids,[672,673] carbohydrates,[674] EDTA,[675] ethers,[665] glycols,[676] hydrocarbons[677-682] (including polynuclear aromatics[683] and their chlorination products[684]), ketones,[598,665-667] nitro-aromatics,[143] nitrosamines,[636] organic acids,[599,665,669,686-690] organic sulphur and phosphorus compounds,[691] phthalates,[598,635] products of ozonolysis,[692] sterols[693] and surfactants.[598,694-696]

(ii)   The use of gas chromatography for 'fingerprinting', screening and monitoring   In addition to its uses for determining specific organic compounds (and some inorganic species) with conventional detectors, and as a separation technique with MS, GC may be used in other ways to characterise and monitor water quality. This aspect of the application of the technique to water analysis has recently been examined (as part of a comprehensive review of high-resolution, capillary column GC) by Wilson and Davis,[555] and the following summary is based upon their treatment.

A number of different approaches can be identified to the utilisation of the information contained in chromatograms of samples containing mixtures of organic compounds. Because these approaches all involve comparing the chromatographic profiles (entire or partial) of different samples, the term 'profiling' is sometimes used to describe them collectively. Specifically, the following individual approaches have been identified:[555]

(1) *Fingerprinting*   comparison of the sample chromatogram with reference chromatograms to seek the best correspondence and thereby determine the origin of the organic material present.
(2) *Screening*   to assess overall similarities and differences in water quality at different locations.
(3) *Monitoring*   chromatograms are compared for samples collected at the same location at different times, to detect changes in water quality.

Because comparison of chromatograms is involved, good reproducibility is essential to the successful application of these approaches. Open tubular columns have overcome problems of varying retention caused by reactivity of solid supports, and standardisation of column preparation and attention to instrumental aspects (including the use of automatic injection) can also improve reproducibility.

The wealth of data in chromatograms obtained by capillary column GC in this manner makes it virtually essential to use automatic data capture and handling to record and archive peak heights or areas and absolute retention times. The latter are valid for comparison only over periods of a few days, even when experimental conditions are closely controlled, so the use of relative retentions or Kováts Indices

---

* Some of these determinations employed mass spectrometry.

[see (*a*) (*iv*) above] is mandatory. The use of the latter, though the most accurate approach available for long-term comparison of retention data, is somewhat hampered by the need, in most applications, to use temperature programming to recover compounds of widely varying volatilities. The error in retention indices is then greater than under isothermal conditions, and the indices themselves may differ from those obtained isothermally.

Various approaches to evaluation of the chromatographic data can be envisaged, depending on the nature and objectives of the profiling operation. These range from simple visual comparison to complicated computer techniques of 'pattern recognition'.

In water analysis, capillary column GC profiling has been used to a substantial extent in the 'fingerprinting' mode only, for the identification and tracing of polluting crude oils and petroleum products, although extension of the approach to other pollutants has been suggested.[697]

A considerable literature has now developed on the use of GC to characterise oils and petroleum products in water, and the review[555] upon which this summary is based should be consulted for details. Relevant papers include refs 604, 605 and 698–704. The fingerprinting technique has been most commonly applied to marine pollution events, in which crude oils figure predominantly,[605,702] but it has also been used in freshwater pollution investigations, in which refined petroleum products (rather than crude oils) are usually involved.[555] Application of the technique to marine oil pollution is complicated by 'weathering', by which changes in, and losses of fractions of, the oil are brought about by processes such as evaporation (leading to loss of 'lighter' fractions), photochemical oxidation, bacterial degradation, adsorption and simple dissolution.[604,702] Inland, oil pollution usually goes undetected for comparatively short periods, so weathering presents little problem except in the case of groundwater contamination.[703]

In the application of profiling techniques to oil and petroleum-product pollution incidents, the basic capillary column/flame ionisation detector (FID) approach may be supplemented by other 'fingerprints' obtained by infrared spectrometry [see Section 11.4.3(*d*)], GC profiles with sulphur-selective flame photometric spectrometric detection, and trace metal determinations.[604] The use of sulphur-selective detection is especially valuable in marine work, because the sulphur-containing organic compounds tend to be degraded more slowly than do hydrocarbons. The GC/FPD 'fingerprints' of weathered samples may therefore correspond more closely to that of the parent material than in the case of the GC/FID approach.

Although the 'fingerprinting' mode of profiling has been most commonly applied in water analysis, as described above, the screening and monitoring approaches also have considerable potential. Thus, for example, it has been suggested[705] that capillary column GC could be used to monitor treatment plants and water abstraction points, so that an early warning could be obtained of changes in water quality and, also, as a means of evaluating the water treatment process. By correlating experience of treatment problems with characteristics of chromatographic profiles, much value could be gained even without identification of detected materials by GC-MS.[555]

The concentration technique used in such work would, of course, exert a strong influence upon the results obtained because different fractions of the total organic content would be recovered by different techniques. Wilson and Davis[555] have concluded that solvent extraction, macroreticular resin adsorption and head-space techniques are likely to be most suitable.

Computer techniques would again be most efficient for handling the data generated, and indeed such techniques have been used to process GC data in profiling work to detect changes in organic pollution in the Rhine.[707]

In summary, it is likely that profiling techniques using capillary column GC will grow in importance, extending beyond their current major application in investigating pollution by oils and petroleum products. The application of modern computer approaches to the collection and processing of data will probably be an important factor in the expansion of these applications of gas chromatography.

(iii) Dissolved gases Gas chromatography is a very valuable technique for the determination of dissolved gases because it allows rapid, precise and selective determination of $\mu g\,l^{-1}$ and higher concentrations. Methods have been evolved for natural and industrial waters, and with appropriate choice of columns a number of gases (*e.g.,* oxygen, nitrogen, methane, carbon monoxide, carbon dioxide) can be sequentially determined in one portion of sample.[708-711] Other gases which have been determined by this technique include hydrogen,[712,713] argon,[714] nitrous oxide[715,716] and hydrocarbons.[717] Although thermal conductivity detection has been widely applied to the GC determination of permanent gases, Hall[718] has described a rapid procedure for dissolved oxygen using electron capture detection.

An account of the application of gas chromatography for the determination of gases in sea-water is given in the review of sea-water analysis by Riley.[719]

(iv) Inorganic and organometallic species Although gas-chromatographic methods have been reported for chloride,[720] bromide,[721] fluoride,[722] phosphate,[723] and cyanide,[724] and although the technique may also be applied to other inorganic anions,[725] the use of GC for such determinations is uncommon. Likewise, although it can be applied to the determination of trace metals (see review by Risby *et al.*[553]), application to water analysis (other than in speciation studies – see below) remains rare. One example is a GC method for the determination of cadmium.[726] Isotope dilution GC/MS has also been applied to the determination of chromium at sub-$\mu g\,l^{-1}$ levels in sea-water.[727]

The technique has, however, been applied to the determination of selenium in water[639,640] and, by combining very low limits of detection with good selectivity, offers advantages over other techniques.

A more common use of GC in connection with trace elements in water is its application to the determination of the solution speciation of such trace elements as arsenic,[642,728-730] lead,[612-615] mercury[616,731] selenium[610,611] and tin.[608,609,652,653] The use of gas chromatography with atomic absorption detection for trace element speciation studies has been reviewed;[618-620] it should be noted, however, that many of the above studies employed other detectors. An indirect GC method for the determination of ferric iron in water has been described by Ditzler and Gutknecht;[732] the method involves GC determination of *o*-hydroxyanisole, whose formation from anisole and hydrogen peroxide is catalysed by Fe[III].

(d) General Basis of High-performance Liquid Chromatography

Classicial liquid chromatography, using columns under atmospheric pressure only, has found very limited application in water analysis – principally as a 'clean-up' procedure in organic determinations. However, developments over the last 10 to 15 years have produced so called 'High-performance Liquid Chromatography' (HPLC)* which has already had some impact upon water analysis, and which seems certain to grow considerably in importance in the future.

Interest in the use of HPLC in water analysis derives, to a great extent, from its applicability to the separation and determination of non-volatile organics not amenable to GC. (It has been noted in the previous sub-section that gas-chromatographic techniques can deal with only approximately 10–20% of the total organic matter in a variety of waters). Indeed, the same factor is largely responsible for the rapid growth of interest in HPLC in analysis in general – it has been estimated that about 70% of all known compounds are 'non-volatile' (*i.e.*, presumably, not amenable to GC) and market studies have identified HPLC as a leading growth area in analytical instrumentation (see ref. 733 and literature cited therein).

Although the use of HPLC in water analysis is relatively new, and does not match in extent that of GC, its great potential for separation of the non-volatile compounds, which make up the major part of the total organic content of many types of water, has secured for HPLC a relatively extensive treatment in this monograph.

The most obvious difference between HPLC and classical column chromatography is probably the use of high-pressure pumps. The fundamental difference, however, lies in the much smaller particle sizes of the column packings used. These have provided greatly increased separation efficiency and, in combination with the use of high pressures and the development of on-line detectors, rapid analysis. An important influence in the development of the technique has also been the evolution of chromatographic theory, fostered initially by the growth of GC, but often applicable to HPLC also.

A simplified, schematic diagram of a high performance liquid chromatograph is shown in Figure 11.10. The chromatographic process is, of course, similar to that of GC but with an eluting solvent (or solvent mixture) replacing the carrier gas of the latter technique. Many separations in liquid chromatography do not require elevated temperatures or very close temperature control, so that ovens or liquid thermostat systems for the injector, column and detector are now optional features, rather than an integral part of the system as ovens are in gas chromatography.

Again, this simplified representation of HPLC conceals many features of great importance and fails to reveal the considerable complexity of much of the equipment, particularly the pumping and solvent-mixing systems. Whilst brief mention of the more important aspects will be made here, a thorough examination of such a complex and powerful technique is beyond the scope of this monograph. The reader requiring further information is referred to one of the numerous texts dealing with HPLC (see, for example, refs. 734–738), to the monograph on HPLC

---

* Various names have been used to describe the technique, including 'High-performance Liquid Chromatography', 'High Speed Liquid Chromatography' and 'Modern Liquid Chromatography'. The name 'High-performance Liquid Chromatography', used here, now appears to be well established.

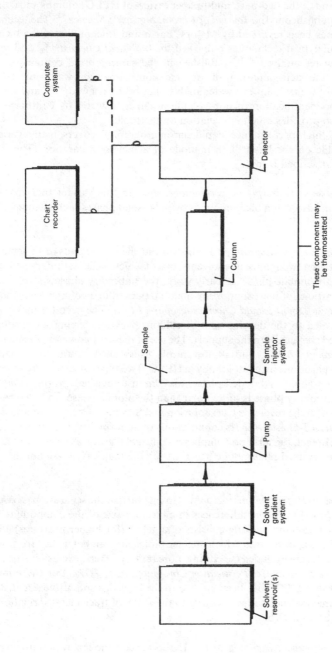

Figure 11.10 *Schematic of basic HPLC system*

instrumentation edited by Huber,[739] to the review of commercial equipment by McNair,[740] and to the two-part, multi-paper review of HPLC columns and column technology published in the *Journal of Chromatographic Science*.[741] The technique as a whole has been reviewed by Majors, Barth and Lochmuller in the biennial reviews of analytical techniques published in *Analytical Chemistry*[733] and recent reviews of water analysis[742,743] published in the same journal contain sections devoted to the determination of organic compounds in water by HPLC. Application of the technique to water analysis has been reviewed[744,745] and a recent review of the identification of non-volatile organics in water by Crathorne and Watts,[746] also provides much information on the topic.

The term 'high-performance liquid chromatography' covers many variants, distinguishable on the basis of their mode of achieving separation. The major variants are described below.

(i) Liquid-solid (adsorption) chromatography   In this variant the separation depends upon adsorption of compounds by polar solid phases such as alumina and silica gel.

(ii) Partition chromatography   In this variant the compounds to be separated partition between two immiscible solvents, one of which is the stationary phase, the other being the mobile phase. In early work, the stationary phase was physically coated on particles of inert supporting material packed in a column – a technique often known as *Liquid-Liquid Chromatography (LLC)*. The need for more stable columns has led to the development of column packings having the stationary phase bonded chemically to the support. The use of these is known as *Bonded Phase Chromatography (BPC)*, and is now firmly established as the most popular approach to partition chromatography in HPLC. Two types of partition chromatography (liquid-liquid or bonded phase) may be distinguished: *Normal Phase*, in which the stationary phase is more polar than the mobile phase, and *Reverse(d)-Phase*, for which the reverse is true. Johnson and Stevenson[734] have remarked that the term reversed-phase, with its connotations of abnormality, is confusing to the uninitiated. In fact, the term now implies no unusual feature, as is indicated by the fact that the reversed-phase mode accounts for the majority of column usage in HPLC.[733]

(iii) Steric exclusion chromatography   In this variant the separation is based (in principle at least) upon the differences in molecular size of the compounds to be separated. The column packing consists of a gel or (for higher pressures) a more rigid solid having pores into which smaller molecules can enter, but from which larger molecules are excluded (so eluting more rapidly). Steric exclusion chromatography [also known as *Gel Permeation Chromatography (GPC)* or *Gel Filtration Chromatograhy (GFC)*] has been applied to the investigation of the size distribution of organic molecules in waters and to the study of trace metal speciation (see below, and Section 11.5.6).

(iv) Ion-exchange chromatography   In this variant the separation involves the exchange of ionic species between the mobile phase and ion-exchange sites on the

column packing. The term *Ion Chromatography (IC)* is commonly used to describe a particular variant of ion-exchange chromatography in which the separated ions are detected conductimetrically, although the term is also used in a wider sense to embrace ion-exchange separation with other modes of detection. For convenience, we have chosen to use a fairly wide definition of ion chromatography and consider that technique separately from HPLC in general [see Section 11.3.8(*g*)] but the arbitrary nature of this distinction should be noted.

Having identified these major variants of high-performance liquid chromatography, we may now turn to consider individual components of HPLC instrumentation.

(v) The chromatographic column As with gas chromatography, the column is of fundamental importance in HPLC. The same basic measures of column efficiency (number of theoretical plates, height equivalent to a theoretical plate) are used in HPLC as in GC, and the concepts of resolution and retention have the same significance. Moreover, the fundamental theoretical relationships and equations are essentially the same for both techniques. The reader is referred to Section 11.3.8(*a*) on GC for details of these aspects.

The major differences between HPLC and GC, apart from the need to pump liquids in the former, are:

(1) The greater variety of solid phases in HPLC (*i.e.*, ion-exchange and exclusion systems as well as liquid-solid and partition modes).

(2) The more limited influence of column temperature in HPLC (though not in the exclusion or ion-exchange modes).

(3) The ability, in HPLC, to vary the eluting strength of the mobile phase. This is known as *gradient elution*, in contrast to *isocratic elution* in which the strength of the mobile phase is held constant*.

With respect to HPLC columns, the variety is such that only a limited, general description is possible here. In addition to texts on HPLC, the special review of columns cited previously[741] and, in particular, the paper by Majors[747] on recent advances (upon which much of the following summary is based) should be consulted for further details.

The efficiency of columns increases as the particle size of the packing decreases. Thus, particle size has fallen considerably during development of the technique as improved separation efficiency has been sought. The so called 'pellicular' packings having diameters of about 40 μm, widely used in the early 1970s, are now used to an important extent in 'guard' columns – columns placed before the analytical column to protect it from irreversible adsorption of undesired components of 'dirty' samples. Typically, modern analytical columns employ 'microparticulate' packings (3–10 μm in diameter), have internal diameters of 2–5 mm and lengths of 5–30 cm. With 5 μm diameter packings, columns having about 80 000 theoretical plates per metre can be prepared.[747] However, such figures are based upon columns of, say, 15 cm length and Majors[747] has warned that it should not be assumed that metre

---

* Strictly, the term isocratic refers to elution under constant eluent composition *and* constant temperature and pressure.

columns could be packed (or prepared by joining shorter columns) to give such plate numbers.

The pressure drop along the column is inversely proportional to the square of the particle diameter; however, for a given number of plates and a given separation time, the pressure needed actually falls with decreasing particle size, down to a limit of about 1 μm.[734] The need for high pressures, it has been stated,[734] has arisen as more complex separations, necessitating higher plate numbers, have been undertaken. Most work is now undertaken using pressures of about 1000–5000 p.s.i.; increasing operating pressure can, of course, permit separations to be achieved in shorter time.

Encouraged, in part at least, by developments in gas chromotography, the use of microbore (0.5–1.0 mm internal diameter) and open-tubular capillary (30–50 μm internal diameter) columns in HPLC has been investigated. Ref. 733 includes a review of such developments. The potential advantage of open-tubular columns in HPLC appears not to be as clear-cut as in the case of GC, however.[733] The smaller flow rates involved with microbore and capillary systems do offer advantage in the interfacing of HPLC with mass spectrometry, and this in itself is a spur to investigation of these techniques (see ref. 733 and Section 11.4.1 of this text).

With respect to the column itself, stainless steel is most frequently employed; glass (or glass-lined) columns have also been used, and may be beneficial or essential in certain applications (*e.g.*, when HPLC is used for the determination of trace metals[748]). PTFE-coating and silanisation of stainless steel columns have also been used.

For liquid-solid (adsorption) chromatography, the most common packings are the microparticulate silicas, though alumina is also used. Adsorption chromatography is the preferred HPLC mode for separating water-sensitive compounds, isomers and multifunctional compounds having similar carbon skeletons, and for performing class separations (*e.g.*, polars/non-polars).[747]

For normal bonded-phase HPLC, siloxane packings (Si–O–Si–R) are standard. Different selectivities can be obtained by varying the type of polar functional group on the organic moiety. Majors[747] gives lists of available packings in three classes – weakly, moderately and highly polar. Normal-phase chromatography is much less widely used than the reversed-phase mode (see below); it is most suited to the separation of moderately polar and non-polar species.[746]

Reversed-phase chromatography is now the most widely used separation mode in HPLC, both in general[733] and in water analysis,[744,746] being capable of separating a very wide range of compounds according (principally) to their polarity.[746] Octadecylsilane (ODS, C-18) packings are predominant, though octylsilane (C-8) and shorter chain types are available and can be advantageous for the separation of more polar compounds. Neutral polystyrene–divinylbenzene (PS-DVB) resins can also be used as non-polar stationary phases in aqueous media, having the advantage, over silica-based reversed phase packings, of being suitable for use at all pH values.[747] Reversed-phase packings in general are discussed by Majors,[747] and the octadecylsilane bonded phases in particular, by Engelhardt *et al.*[749]

Microparticulate packings for ion-exchange chromatography fall into two main categories, namely, resins and bonded phases on silica, both available as anion- and cation-exchangers. The resins are most commonly the PS-DVB type, widely used for the preparation of classical ion-exchange resins, but in small particles. They exhibit better pH stability and (usually) longer lives than their silica-based counterparts, and possess higher ion-exchange capacity – typically by factors of 5–10.[747] However, the chromatographic efficiency of the silica-based packings is higher, and the resin type are usually operated at above-ambient temperatures to improve this aspect of their performance. Most of the commercially available ion-exchange phases (listed by Majors[747]) are strong exchangers, the functional groups being mainly sulphonate (cation) and tetra-alkylammonium (anion). The need for ion-exchange HPLC has been reduced by the introduction of reversed-phase ion-pair chromatography, discussed below, but ion-chromatography remains an important application [see Section 11.3.8 (g)].

Steric (or size) exclusion chromatography is the subject of a book by Yau, Kirkland and Bly,[750] and Majors [747] discusses stationary phases for this technique. Gel Filtration Chromatography (GFC) refers to steric exclusion chromatography in aqueous medium and Gel Permeation Chromatography (GPC) to steric exclusion in organic media. The cross-linked dextrans (soft gels) historically employed by GFC lacked the rigidity necessary for operation at pressures about 150 p.s.i., and the alternative HPLC-compatible silica or glass packings gave rise to adsorption problems. Newer rigid, microparticulate GFC packings have overcome many of the difficulties, being produced by chemical bonding of an aqueous-compatible material onto a silica base. For GPC, both rigid and 'semi-rigid' packings are available for HPLC; the former are controlled pore-size silicas and glasses, whilst the latter are polystyrene gels with varying degrees of cross-linking.

It has been noted that development of ion-exchange chromatography is likely to have been inhibited by the introduction of ion-pair chromatography, a variant of reversed-phase chromatography by means of which the latter can be applied to ionic species. Basically, a counter-ion of opposite charge is used to form (it was thought) an ion-pair with the determinand, which can be chromatographed by HPLC in the reversed-phase mode. Although the explanation of ion-pair formation has given the technique its name, the actual mechanism is now thought to be more complex[751,752] [see (f) below].

(vi) Sample introduction Sample introduction in HPLC may be achieved in a variety of ways. Introduction systems are discussed in HPLC texts and in a chapter[753] of the book on HPLC instrumentation edited by Huber.[739]

Syringe injection *via* a septum port is the simplest approach, but is restricted to low pressures (typically less than 1000 p.s.i.). There are other problems – the septum may be incompatible with the solvent, it may leach plasticiser onto the column and pieces of it may be torn off during injection and cause blockage of the chromatograph. Septumless syringe injection ports can be used to overcome these difficulties, but a more widely used alternative is the loop valve, in which the sample is contained in a cavity in a rotatable valve (or in a separate loop of tubing connected to the valve), being introduced into the main solvent line above the column by rotation of the valve. Automatic injection is possible using an automated loop valve system.

All the systems described above have the advantage of permitting the sample to be introduced without interruption of the solvent flow. The stop-flow injection system, as its name implies, does not; the flow through the column is stopped during injection and the sample is introduced onto the top of the column. This system can be used at high pressures (6000 p.s.i. or more[734]), and is cheap and reliable. However, the flow is disturbed and retention times require correction using an internal standard.[736]

Another approach to sample introduction, of particular relevance to water analysis, is the so called[754] 'trace enrichment' technique. The sample is preconcentrated onto the analytical column itself, or onto another column which is placed in front of the analytical column. Detection limits can be markedly lowered as a result of the preconcentration, but other components of the sample may be simultaneously concentrated and cause interference*. (If the analytical column itself is used, undesired sample components may be irreversibly adsorbed to the detriment of column performance). The trace enrichment technique may be applied 'on-line' (when the enrichment column is attached to the remainder of the chromatograph) or 'off-line' (when it is not). Off-line trace enrichment has the advantage of permitting the sample to be preconcentrated onto the enrichment column at the site of sample collection, which may reduce or eliminate problems of sample instability, and reduce sample transport costs.[755] Further information on the technique can be found in refs. 754–757, and in the reviews by Graham[744] (on the use of HPLC to determine organic compounds in water) and by Cassidy[748] (on the application of HPLC to metal determinations in general).

(vii) Pump systems and gradient elution   As noted previously, exploitation of the high efficiencies of HPLC columns and the accomplishment of complex separations in acceptably short times leads to the requirement for high pressure pumping systems. HPLC pumps are of two main types: constant pressure and constant volume (or displacement). The former have the disadvantage that changes in the resistance to flow in the system (*e.g.*, arising from settlement or swelling of the column packing) cause changes in flow rate which can affect resolution, retention times and peak areas;[734] such pumps are, however, simple and cheap and readily provide pulse-free flow.

Constant displacement pumps provide constant flow rate in the face of increased flow resistance, provided that the pressure increase involved does not exceed their capability. They are more readily incorporated in automatic gradient elution systems (see below), but are generally more expensive than constant pressure pumps. Two main types of constant displacement pumps are in use – single displacement (syringe) pumps and reciprocating pumps. The former are simple and reliable, and offer pulse-free flow. The latter are not restricted with regard to solvent capacity, but require special features to overcome flow pulsing.

Further details of HPLC pumps may be found in textbooks on the technique, and in the review by Martin and Guiochon[758] upon which much of the above summary is based.

It has been noted earlier that, in contrast to GC, the strength of the eluent may be increased during the course of the HPLC separation, in the process known as *gradient elution*. This approach is a useful (if, practically, sometimes inconvenient)

---

* The same may, of course, also occur with other preconcentration procedures (such as liquid/liquid extraction) used with HPLC.

means of optimising the separation of complex mixtures of compounds having widely different retention characteristics, and of reducing analysis times. The eluent strength is changed during the separation by mixing different solvents in different ratios. Very many gradient elution programmes may be used; thus, two or more solvents can be used and the gradient 'profile' (eluent composition *versus* time) may be linear, logarithmic, exponential, parabolic, stepwise or more complex than these.[759] Commonly, however, only two solvents are used. A variety of hydraulic configurations may be employed to produce the gradient; detailed consideration of these is beyond the scope of this monograph and the reader is referred to HPLC texts and to a comprehensive review of the topic (ref. 759) for further information.

Mention has aleady been made of the various solvent mixing 'profiles' available. The *effective gradient*[736] is, of course, determined both by the relative volumes of the eluent at each point in the separation *and* by their relative eluting strengths. A linear effective gradient is optimal with respect to resolution.[736]

The solvents used in gradient elution must be miscible and chemically compatible, and the mixture must also be compatible with the detector used. Refractive index detection cannot, for all practical purposes, be used with gradient elution. Solvent systems transparent to u.v. radiation (and therefore compatible with spectrophotometric detection at such wavelengths) can be devised, though changes in refractive index can cause baseline drift, unless compensated or suppressed. Fluorescence detection may also be compatible with gradient elution.

Solvent purity is of particular importance in gradient elution, and the column must be regenerated – usually by reversing the gradient – after completion of the separation. Although the gradient technique is a powerful means of optimising resolution and speed of analysis, it adds considerable complexity to HPLC and the simpler (and cheaper) gradient systems are often inconvenient to use.

The theoretical basis, and practical application, of gradient elution in reversed-phase HPLC has been reviewed by Snyder *et al.*[760,761]

(viii) Detection of eluted compounds   The role of the detector in an HPLC system is, of course, the same as that of the detector in GC – namely, to reveal the presence of eluted compounds and to allow their concentrations to be measured. Many different types of detector have been used in HPLC, including solution spectrophotometric (mainly at u.v. wavelengths), fluorimetric, refractometric, amperometric, polarographic, flame ionisation, conductimetric, potentiometric, atomic - spectrometric (absorption and emission, including plasma emission), spectrometric and mass spectrometric.

Despite the wide range of possible HPLC detectors, it is considered[733,762] that the detection stage is the 'weakest link' in the HPL chromatograph at present, and much research and development is in progress to remedy this situation. Of the detection systems listed above, three types – solution spectrophotometric (mainly u.v.), fluorimetric and refractometric – accounted for over 90% of detector usage in a survey of HPLC applications (to various fields of analysis) reported by Majors *et al.*[733] (Solution spectrophotometry *alone* accounted for over 70% of detector usage.)

Texts on HPLC contain a chapter or chapters dealing with detectors, three chapters[763–765] of the monograph on HPLC instrumentation edited by Huber[739] are devoted to the topic and a brief review has been produced by McKinley *et al.*[766]

Recent advances in optical detection systems have also been reviewed by Abbot and Tusa,[767] and electrochemical detection is the subject of a review by Brunt.[762] The latest review of HPLC in *Analytical Chemistry*[733] contains a number of references to detector developments, and developments in selective detection, with particular regard to water analysis, have been reviewed by Poppe.[768]

The following brief descriptions of the more important types of detector can serve only as an introduction, and the reader requiring further detail is referred to the above papers and texts, and to other relevant sections of this book [solution spectrophotometry – Section 11.3.3; fluorimetry – Section 11.4.3(*f*); polarography and related techniques – Section 11.3.7; amperometry – Section 11.3.9(*c*); mass spectrometry – Section 11.4.1].

(1) *Solution spectrophotometry.* Many organic molecules have analytically useful absorption spectra in the u.v. (190–400 nm) and/or visible (400–700 nm) regions. A variety of detectors is available to exploit this factor, four principal types having been distinguished:[733] fixed wavelength detectors, commonly employing the 254 nm radiation from a mercury lamp; detectors offering a range of fixed wavelengths, employing a continuum source and suitable optical filters; variable wavelength detectors with a continuum source and a mono-chromator for wavelength detection; and, finally, rapid-scanning variable wavelength detectors usually employing diode arrays or vidicon tubes as light receptors.

Although complete spectra can be obtained with conventional scanning spectrophotometers, the time required – a few hundred seconds[739] – is such that the flow must be stopped to avoid spectral distortion. With rapid-scanning instruments, spectra can be recorded in less than one second;[739] this is short in relation to the width of HPLC peaks, so flow need not be stopped during scanning.

The ability to select the wavelength of operation can greatly increase the selectivity and/or sensitivity of the detection system. The ability to obtain absorption spectra of eluted compounds can be advantageous in dealing with incompletely resolved peaks, and in the identification of eluted compounds, but is not widely employed in routine analysis.

U.v.-visible absorption spectrophotometry is a selective detection tech-nique, but is none the less of wide applicability, particularly with variable wavelength facilities. It has been noted[734] that HPLC is usually applied to the separation of complex organic species, and that as structural complexity increases so does the likelihood that the compounds of interest will contain structural features conferring absorption properties in the u.v. and/or visible region.

Solution spectrophotometry is compatible with many of the solvents commonly used in HPLC (because their spectra exhibit wide 'windows' in the u.v.-visible region) and can be used with gradient elution provided that the solvent system is chosen appropriately.

(2) *Fluorimetry.* Some organic compounds absorb u.v. radiation and re-emit the energy (virtually instantaneously) at a longer wavelength, which may be in the visible region of the spectrum. This property of fluorescence provides the basis of a highly sensitive method of HPLC detection. Excitation radiation may be

obtained using a continuum source and a filter or monochromator, or directly from a mercury lamp. The fluorescent radiation is usually measured at right-angles to the beam of exciting radiation, after passage through a suitable filter (or monochromator) to discriminate against scattered excitation radiation. Detectors employing monochromators are more expensive, and give better sensitivity and selectivity; filter systems can be advantageous, however, in that their greater bandpass can obviate the need to re-tune the detector for each fluorescent compound as it elutes.[734]

Standard, commercial fluorimeters can be used in HPLC when fitted with an appropriate microflow cell,[733,763] though equipment designed specifically for HPLC work is also widely available. The use of lasers for providing excitation radiation, and of fibre-optics technology to conduct the radiation, can produce considerable improvement in the sensitivity of fluorescence detection (see ref. 733 and literature therein).

Though fluorescence is a rare property amongst organic compounds*, such that direct application of fluorescence detection is very restricted,[763] the scope of the technique can be greatly extended by derivatisation. A well known example is the production of the dansyl derivatives of amino acids using 1-dimethylaminonaphthalene-8-sulphonic acid. Such derivatisation may be accomplished pre- or post-column. Post-column derivatisation has the advantage of leaving the chromatographic separation unaffected, but HPLC separation of derivatives prepared pre-column may sometimes be readily achieved, despite the similarity of their molecular properties. Thus, for example, it has been stated[736] that the separation of the dansyl derivatives of amino acids is easily accomplished by HPLC.

Post-column derivatisation is a variant of a general class of detection techniques sometimes known as *post-column reaction detection*. In these techniques the measured entity is either a derivative of the compound of interest (as above) or a product of a reaction involving that compound, other than a derivative. An example of the latter type of post-column detection, involving fluorimetric measurement, is cerate oxidimetry. In this technique, which has not been widely used in water analysis, the compound of interest reduces Ce^{IV} to Ce^{III}, the latter then being measured fluorimetrically. Cerate oxidimetry is considered in Section 11.4.3(f) on molecular luminescence spectrometry, and is not considered further here.

Derivatisation in liquid chromatography is the subject of a book by Lawrence and Frei,[769] and the latter author has provided two reviews[770,771] of reaction detection systems (especially post-column). Post-column reaction detection has also been reviewed by Van der Wal[772] and, in relation to water analysis particularly, by Poppe.[768]

(3) *Refractometry.* The detection of changes in the refractive index of the eluent as compounds are eluted forms the basis of an essentially universal† detection system. Two main types of refractive index detector are used: the 'deflection' type, in which the direction of the refracted beam is measured and the Fresnel type, in which the refracted beam intensity is measured.

---

* Ref. 734 provides a number of basic rules for predicting the fluorescence response of organic molecules.
† It is rare for two liquids to have indistinguishable indices.

Refractometric detection has the major disadvantage of poor detection capability, making it inapplicable to trace analysis. An additional disadvantage is its unsuitability for use with gradient elution.

(4) *Electrochemical.* A number of electrochemical detection systems – including amperometric, polarographic and potentiometric types – have been used in HPLC. The reviews by Brunt[762] and Poppe[764] should be consulted for details. Amperometric detectors are the most common; in these, the working electrode (various types are used[762,764]) is maintained at a constant potential with respect to the reference electrode, and reduction or oxidation of the eluted electroactive species causes a current to flow. The magnitude of the current is a function of the concentration of that species.

Amperometric detection requires that the eluent be electrically conducting. It is, therefore, most commonly used with the reverse-phase and ion-exchange modes of HPLC.

(5) *Other detectors.* The general references on HPLC detectors given previously should be consulted for details which are beyond the scope of this monograph. Note, however, that attempts have been made to apply the FID, noted for its good detection capability and virtually universal response in GC [see Section 11.3.8 (*a*)]. The difficulty is to remove the solvent whilst retaining the eluted compounds. The same basic problem faces attempts to couple HPLC and mass spectrometry. Many approaches to this are possible; a number of papers in refs. 773 and 774 deal with this topic, and it has recently been reviewed by Curry[775] and (with regard to water analysis) by Levsen.[776] See also Section 11.4.1 of this text dealing with mass spectrometry.

(e) Principal Analytical Characteristics of High-performance Liquid Chromatography

The many different variants of HPLC make it difficult to provide an accurate summary of the principal analytical characteristics. Much of the following is of a rather general and approximate nature, a fact which should be kept in mind, especially when quantitative aspects of performance are under consideration.

It should also be borne in mind that an HPLC separation and measurement will often be preceded by an extraction stage for preconcentration and/or for preliminary separation of the compounds of interest, though this is not invariably the case by any means. Much of the following general information refers to analysis of extracts, though details of the performance of entire analytical methods have also been included.

(i) Calibration curve As in GC, the correct choice of column should ensure that the compounds of interest are eluted quantitatively, so that the relationship between peak area (or height) on the chromatogram and the amount of substance is governed primarily by the detector. Refractometric detectors commonly give linear response over about three orders of magnitude, u.v. detectors over about four orders and fluorescence detectors over four to six orders.[734]

To convert the measured response to concentration, a calibration curve may be employed, prepared by measurements on standards containing known amounts of the determinand taken through exactly the same procedure as samples. As in GC, however, the lengthy nature of pretreatment steps may make it more convenient for most standards to be taken only through the final HPLC procedure in some cases. It is then essential that at least two standards (one of them the blank) be taken through the entire procedure with each batch of samples to check that the recovery of determinand is satisfactory.

Standard additions calibration (see Section 6.2.2) and internal standardisation are alternatives to the use of a calibration curve. As with GC, the reader is cautioned that the validity of internal standardisation depends crucially upon the similarity of chromatographic behaviour of the chosen internal standard and the determinand(s) of interest, and appropriate caution should be exercised. Again, it is essential that at least two standard solutions (one of them the blank) be taken through the entire analytical procedure with each batch, as a check.

(ii) Limit of Detection Most HPLC detectors respond to the *concentration* of compounds passing through them. Therefore, the overall detection limit will depend not only upon the detector, but also upon the chromatographic separation. Thus, Engelhardt[736] has noted that, although a u.v. detector may have a microcell volume of 8 μl, the total amount of a separated compound could be distributed over a potentially much larger volume, after elution from the column – perhaps 10–1000 μl or even more. The properties of the compound in question or its derivative (*e.g.*, the extinction coefficient in u.v. absorption) will also exert an important influence.

For all these reasons, any summary of detection limits can serve only as a very rough guide. Moreover, the detection limit attained using a complete analytical

TABLE 11.9

Approximate Detection Limits for Common HPLC Detectors

Detector type	Approx. detection limit in terms of concentration* μg ml^{-1}		Approx. detection limit in terms of amount† μg
U.v. absorption (fixed wavelength)	1×10^{-3}	(734)	10^{-4}
U.v. absorption (variable wavelength)	4×10^{-3}	(734)	4×10^{-4}
U.v. absorption‡	1×10^{-4}	(736)	10^{-5}
Fluorimetry	1×10^{-6}	(734)	10^{-7}
	1×10^{-3}	(736)	10^{-4}
Refractometry	7×10^{-1}	(734)	7×10^{-2}
	1×10^{-1}	(736)	10^{-2}
Wire FID or	3	(734)	3×10^{-1}
'Transport'	1×10^{-1}	(736)	10^{-2}
Polarography	1×10^{-3}	(736)	10^{-4}

* Taken from the reference given in parentheses. Values from ref. 734 are described as minimum detectable concentrations; those from ref. 736 are described as maximum sensitivities in the most favourable case. In neither case is the detailed definition known.
† Assuming stated amount distributed evenly over 100 μl of eluent – see text.
‡ Type unspecified.

method will, of course, be influenced by any preconcentration applied (either by 'trace enrichment' or by solvent extraction, for example) and by the random error introduced by pretreatment processes.

Despite the very approximate nature of the data, a summary of the detection limits of common HPLC detectors is given in Table 11.9. Data in terms of concentrations have been taken from refs. 734 and 736, and also converted to weights of substance by assuming the latter to be distributed evenly through 100 μl of eluent (see above for an estimate of the likely range of such volumes). The table

TABLE 11.10

Some Examples of Detection Limits attained using Complete HPLC Methods

Determinand	Detection limit μg l^{-1}	Ref.
Polynuclear aromatic hydrocarbons (PAH):		
Various	*ca.* 0.001 or less	777–780
Herbicides:		
Eight phenylurea herbicides	*ca.* 10	781
Difenzoquat	2	782
Hexazinone	*ca.* 1	783
2,4-dichlorophenoxyacetic acid and 2-(2,4,5-trichlorophenoxy) propionic acid	*ca.* 0.05	784
Pesticides:		
Polychloro-2-(chloromethyl-sulphonamido) diphenyl ethers (Mothproofers)	0.1–0.2	785
Carbofuran and metabolites	1–5	786
Aldicarb	< 3	787
Phenols:		
Various (including chlorophenols)	0.2–0.7	788
Pentachlorophenol	*ca.* 0.1	789
Various Chlorophenols	0.04–> 3	790
Nitrophenols	0.025–2	791
Other organic compounds:		
Quaternary ammonium compounds	*ca.* 2	792
Alkylbenzenesulphonates	*ca.* 10	793
5-chlorouracil	0.1	794
5-chlorouridine	0.2	794
4-chlororesorcinol	0.5	794
5-chlorosalicylic acid	1	794
Acrylic acid	50	795
Acrylamide	0.2	796
Benzidine and 3,3'-dichlorobenzidine	0.05–0.2	797–799
Phthalate esters (various)	0.05–0.2	757
1,2-diphenylhydrazine	*ca.* 1	797
Benzotriazole and tolyltriazole	1000	800
Inorganic species:		
Cr^{III} and Cr^{VI} species	5–10	801
Cd, Ni, Co, Cu, Bi, Hg	1–12	802
NO$_3^-$, NO$_2^-$	100	803
Se^{IV} and Se^{VI} species	*ca.* 5	804

shows the very low detection limits possible, particularly using u.v. absorption and fluorescence detectors.

Some examples of the detection limits attained using complete HPLC methods (many involving preconcentration steps) are given in Table 11.10, for both organic and inorganic species. Again, these figures illustrate the excellent detection capabilities of HPLC.

(iii) Precision With regard to the HPLC stages alone (*i.e.*, disregarding pretreatment procedures), relative standard deviations of about 1–5% should be attainable routinely, provided that concentrations are not close to the detection limit.

For complete analytical methods, the precision attainable will, as in GC, depend markedly on the pretreatment procedures employed, as well as on the proximity of concentrations to the detection limits. Relative standard deviations of about 1–10% have, however, been reported for a number of determinations (see, for example, refs. 754, 757, 787, 793, 795, 799, 805, 806, 807).

(iv) Selectivity As in GC, problems of interference are most likely to arise when particular organic compounds must be determined in the presence of many other, unknown, compounds. The use of a more efficient column, or of a detection system which responds to the determinand but not to the interfering substance(s), may allow such problems to be overcome; if such measures are impractical or ineffective, recourse must be made to preliminary separation procedures. This approach is most likely to be necessary with polluted surface waters and waste waters.

Identification of eluted compounds in HPLC presents essentially similar features to identification in GC. Retention data may be used, though such data must be obtained with a second separation system (at least) to confirm or reject tentative assignments made on the basis of the correspondence of retention characteristics between standard and unknown on the analytical column. Indeed, it has been pointed out[734] that the availability of widely differing modes of separation in HPLC can afford a valuable tool in identification by retention data, since similarity of retention characteristics (of known and unknown compounds) under very different chromatographic conditions can give considerable confidence in the assignment. Systematisation of HPLC retention data (at least for non-polar phases) in the form of indices similar to the Kováts Indices of GC, has been called for by Engelhardt;[736] Baker and Ma[808] have developed a retention index scale (similar to that of Kováts) for reversed-phase HPLC, based on keto alkanes. It would appear, however, that such indices are not in widespread use at present, at least in the field of water analysis.

Other approaches to compound identification in HPLC include the use of the relative responses of two different detection systems in series and, of course, the use of off-line identification by such techniques as mass spectrometry and i.r. spectrometry, provided that sufficient sample can be collected after separation. [On-line HPLC-MS is at a relatively early stage of development – see above – but its potential for identification and determination is clearly very considerable. Wright[809] has compared its use with that of conventional spectrophotometric

HPLC detection, for 19 carbamate pesticides and Takeuchi *et al.*[810] have demonstrated its application to the determination of polynuclear aromatic hydrocarbons (PAH)].

The accuracy of results obtained using complete HPLC methods will, of course, be determined not only by the chromatographic (and measurement) stages, but also by any pretreatment procedures involved – which may introduce problems such as contamination and interference effects upon extraction of the determinand(s) from the sample. For complete HPLC methods, recoveries of about 90–110% from spiked samples are commonly reported, but the need to run at least one standard (in addition to the blank) through the entire procedure with each batch of analyses is again emphasised.

(v) General   High-performance liquid chromatography is now a well estab-lished analytical technique, and equipment is available from many manufacturers. The cost of an HPLC system depends markedly on the degree of sophistication of the apparatus (*e.g.*, with respect to gradient elution detection and sample introduction capabilities), but is generally similar to that of gas-chromatographic systems.

As with gas chromatography, a reasonably high level of skill and knowledge is required of HPLC operators, particularly to deal with the problems which inevitably occur from time to time and to establish new analytical procedures. A recently published review on the application of HPLC to water analysis[745] contains a useful section on trouble-shooting.

(f) Applications of High-performance Liquid Chromatography to Water Analysis

(i) Organic compounds   To date, the most common uses of HPLC in water analysis have been the determinations of polynuclear aromatic hydrocarbons (PAH),[777–780,811,812] phenols,[788–791,813–816] herbicides[781–784,817] and pesticides.[785–787,818] Other organic determinands to which the technique has been applied include acrylic acid,[795] acrylamide,[796] amino acids,[812] aromatic triazoles (corrosion inhibitors),[800] benzidine and 3,3′-dichlorobenzidine,[797–799] carboxylic acids,[819–821] organochlorine compounds,[794,805] 1,2-diphenylhydrazine,[797] EDTA,[822] *N*-methyl-2-pyrrolidone,[823] phthalate esters,[757] phytoplankton pigments,[824] quaternary ammonium com-pounds[792] and surfactants.[793,807]

It is expected that the range of organic substances to which HPLC is applied will continue to increase rapidly as the technique becomes more widely available to water analysis laboratories. It may also be expected that the technique will, like gas chromatography, continue to be used in the 'profiling' mode [see Section 11.3.8(*c*)]. HPLC has, for example, already been applied to oil-spill identification.[825] A novel use has also been described by Cheh and Carlson,[826] in which HPLC is used to detect mutagens in water. Most direct-acting mutagens are electrophiles, so these workers treated XAD resin extracts from waters with a nucleophile, to form thioethers which could be detected using reversed-phase HPLC.

Steric exclusion chromatography has been applied to the size characterisation of organic matter in water samples (see, for example, refs. 827 and 828). These studies have employed gel filtration on 'soft' column packings under low pressure conditions, and Urano *et al.*[829] have cast doubt upon the mechanism of separation with such gels, reporting that steric (*i.e.*, size) exclusion may often not be the basis of separation. More recently, Miles and Brezonik[830] have emphasised the need to use a high ionic strength eluent in high performance steric exclusion chromatography using a bonded-phase packing if accurate molecular weight distributions are to be obtained.

(ii) Inorganics The application of HPLC in water analysis for the determination of inorganics has involved three main areas. The determination of 'major' ionic species – especially anions such as chloride, sulphate and nitrate – by ion exchange techniques, commonly with conductimetric detection, is dealt with in the next Section [11.3.8(g)] on ion chromatography.

It should be noted, however, that inorganic ions have been determined by HPLC using reversed phases rather than conventional ion-exchange materials (see, for example, refs. 803, 831 and 832). The eluent contains a modifier which is adsorbed on the surface of the reverse phase packing and confers ion exchange capability upon it. This mechanism is now also considered to occur in so called 'ion pair chromatography' as used for organic ions (discussed above), and the basic techniques are therefore the same. Although the mechanism is ion exchange, we have chosen to consider the technique here and cite detection limits for NO_3^- and NO_2^- in Table 11.10, rather than in Section 11.3.8(g) on ion chromatography, because of the nature of the basic stationary phase. The arbitrary character of such a distinction is emphasised, however, and the confused state of terminology in HPLC is illustrated by the fact that this type of separation technique has been known variously as 'ion pair chromatography', 'paired ion chromatography', 'mobile phase ion chromatography' and 'ion interaction chromatography'. The better understanding of separation mechanisms should lead to rationalisation of terminology in due course, however.

The other two main areas of HPLC application to inorganics are the determination of trace elements and the elucidation of trace element speciation. In both cases, a wide variety of HPLC techniques has been employed, but in neither case can the application be regarded as routine. The space afforded to these applications here reflects the potential value of HPLC in these areas of water analysis, rather than the current usage.

The application of HPLC to trace element determination is the subject of a number of reviews.[748,833,834] A particular attraction is the potential of the technique, in some of its variants, for sequential multi-element analysis. The application of HPLC to trace element determination has been undertaken using a variety of separation modes (*e.g.*, normal phase, reversed phase and ion exchange*) and detection has most commonly involved solution spectrophotometry or amperometry. With solution spectrophotometry, the formation of the absorbing species, by reaction of the determinand with a chromogenic reagent, can occur before or after the chromatographic separation (pre-column or post-column derivatisation),

* Strictly, we have adopted a wide definition of ion chromatography [see Section 11.3.8(g)] which could embrace all forms of ion-exchange chromatography. For convenience, the application of ion-exchange separation to trace element determination and speciation will be dealt with here, rather than under ion chromatography [Section 11.3.8(g)].

or *in situ* – that is, on the separation column itself. Sample introduction can be direct, or involve some form of pre-concentration; for example, by solvent extraction of the metal chelates in the case of pre-column derivatisation, or by 'trace enrichment' in the cases of pre- or post-column derivatisation or *in situ* derivatisation, *i.e.*, by concentrating the determinands or their derivatives on the separator column or, more commonly, on a separate concentrator column from which they may be eluted on to the separator column.

The use of low pressure ion-exchange separation of the unreacted trace elements prior to automated spectrophotometric determination of one of them is a fairly common approach to overcoming problems of interference effects. An example is the determination of zinc by automated spectrophotometry of its complex with zincon, after its separation from other trace elements by ion exchange, as described by Matsui.[835] From a chromatographic viewpoint, this type of procedure can be regarded as ion-exchange chromatography with post-column derivatisation (known also as post-column reaction detection). Of greater current interest, however, is the use of HPLC for multi-element determination. Much of the research effort in this field has been devoted to the investigation of suitable chromogenic reagents and chromatographic conditions. A discussion of such research is outside the scope of this book, and the interested reader should consult the comprehensive reviews by Cassidy[748] and O'Laughlin.[834]

The use of HPLC to determine, sequentially, a number of trace elements in water samples is exemplified by the work of Ichinoki *et al.*,[802] who have reported the determination of bismuth(III), cadmium, cobalt, copper, mercury(II) and nickel by reverse phase chromatography of their hexamethylene dithiocarbamato complexes, followed by spectrophotometric detection at 260 nm. As shown in Table 11.10, the detection limits attained by Ichinoki *et al.* lay in the low μg l^{-1} range, in part a consequence of the 50-fold preconcentration achieved by solvent extraction of the metal complexes prior to the chromatographic stage.

Solvent extraction of dithiocarbamate complexes, prior to their separation on a reverse phase column and spectrophotometric detection has also been employed by Bond and Wallace[836] to determine cadmium, cobalt, copper, lead, mercury and nickel in zinc sulphate plant electrolyte. After solvent extraction, zinc complexes and excess ligand were removed on an anion exchange column prior to the main chromatographic stage. Detection limits lay in the range of 0.1–1 mg l^{-1}. Amperometric detection was also investigated; it gave similar detection limits to spectrophotometry except for cobalt, lead and nickel, for which the amperometric limits were 4–10 times higher.

The same workers have described an automated, discontinuous on-line method for the determination of copper and nickel.[837] Solvent extraction was not employed, and the diethyldithiocarbamate complexes were formed *in situ* on the reversed phase column. Detection limits using spectrophotometric detection were (for 10 μl sample injections) 40 μg l^{-1} (Cu) and 20 μg l^{-1} (Ni); using amperometric detection, the above limits were reduced by a factor of two.

In situ formation of dithiocarbamate complexes, with spectrophotometric detection, has also been applied to the reverse phase HPLC determination of cobalt, copper, iron(III) and nickel by Smith and Yankey.[838] Relative standard deviations of about 10% were obtained for copper at concentrations of 50 μg l^{-1} and 200 μg

1^{-1}, though the authors concluded that their procedure had potential for the determination of all four elements in the 1–10 mg 1^{-1} range.

Edward-Inatimi[839] has described the determination of bismuth, cobalt, copper, manganese, mercury, nickel and lead in river water and trade effluents by HPLC of dithizone or dithiocarbamate complexes after extraction in chloroform, at concentrations of 20 μg 1^{-1} and above.

Examples of the application of trace enrichment as a preconcentration technique for metal determinations by HPLC are provided by the work of Cassidy and co-workers.[840,841] By preconcentration from 100 ml sample volumes, Elchuk and Cassidy[840] were able to attain detection limits of about 10 ng 1^{-1} for lanthanide elements, using ion-exchange HPLC with spectrophotometric post-column reaction detection. Cassidy *et al.*[841] later applied the same basic approach to the HPLC determination of cobalt, copper, lead, nickel and zinc at μg 1^{-1} and sub-μg 1^{-1} levels in groundwater samples. Preconcentration of the diethyldithiocarbamate complexes of copper, cobalt and nickel (rather than of the metals themselves) on a reverse phase material has been used by Haring and Ballschmiter,[842] prior to separation of the complexes by reversed phase HPLC with spectrophotometric detection. Using this procedure, the three metals could be determined at concentrations in the range 0.2–10 μg 1^{-1}. Further examples of the trace enrichment approach can be found in the review by Cassidy.[748]

Other examples of the use of HPLC for the determination of trace elements include the determination of selenium[843] and of beryllium,[844] with detection limits of about 0.3 and 1.5 μg 1^{-1}, respectively.

The application of HPLC to trace metal speciation analysis has been discussed briefly in the reviews by Cassidy[748] and O'Laughlin[834] cited previously, and in the review by Van Loon.[845] Additionally, the paper by Chau and Wong[846] on the direct speciation of organometals contains information on the coupling of liquid chromatography with atomic absorption detection – as also does the review by Fernandez[847] (now rather dated). The use of inductively coupled plasma optical emission spectrometry (ICPOES) as an HPLC detector has been reviewed by Krull and Jordan.[848]

For those elements which can exist in water in more than one oxidation state, separate determination of the concentrations present in different states may be possible using HPLC after derivatisation, as described above. Thus, for example, Schwedt[849] has developed a procedure for the determination of Cr^{III} and Cr^{VI} based on reverse phase HPLC of their pyrrolidine dithiocarbamate complexes and Hoffman and Schwedt[850] have described a procedure for the separate determination of Mn^{II} and Mn^{III}, based on reverse phase HPLC of the oxinate complexes.

Most applications of HPLC to trace element speciation have, however, involved chromatographic separation of the species of interest (rather than their derivatives) and the use of detection systems other than solution spectrophotometry, including atomic absorption spectrometry (AAS – flame, hydride and furnace – see Section 11.3.5), plasma optical emission spectrometry [both inductively coupled (ICP) and direct current (DCP) – see Section 11.4.2] and amperometry [see Section 11.3.9(*c*)].

Thus, a combination of HPLC and flame AAS has been used by Jones and Manahan[851] to determine copper chelates, but, as noted by Chau and Wong,[846] flame AAS detection limits are inadequate for environmental speciation studies.

This has prompted the investigation of graphite furnace AAS, with its lower detection limits, for HPLC detection. The HPLC/GFAAS approach has been applied to the determination of organoarsenic,[852,853] organolead,[852,854] organomercury[852] and organotin[852,855] species. It has also been applied to the determination of inorganic Se^{IV} and Se^{VI}.[804]

A difficulty with the interfacing of HPLC with GFAAS is the discontinuous nature of GFAAS measurements (one measurement every few minutes, typically). Thus, if the HPLC system feeds a graphite furnace system directly, the GFAAS response to a narrow chromatographic peak may be very variable, depending on the synchrony or otherwise of the GFAAS cycle and the peak appearance.[846] Storage of the column effluent from the HPLC stage for GFAAS measurement off-line[854] can overcome the difficulty, although the analysis time is substantially increased.

If the element of interest forms hydrides, the hydride generation AAS technique can be interfaced with the HPLC system to provide a continuous measurement of the element in the column effluent, overcoming the problems of direct HPLC/GFAAS coupling described above. This approach has been applied to arsenic and tin speciation.[856,857]

Amperometric detection has been combined with HPLC for the determination of organomercury[858,859] and organotin[859,860] speciation. Chau and Wong[846] have predicted that electrochemical detectors will be more widely used with HPLC for speciation analysis in future.

Optical emission spectrometry using plasma sources has also been used for HPLC detection in speciation studies. Examples include the determination of Cr^{VI} and Cr^{III},[801] Se^{IV} and Se^{VI} [806] and As^{III} and As^{V}.[806,861] The use of a continuous hydride generation system as the interface between the HPLC system and the plasma spectrometer produces much improved detection limits compared with those attained using direct nebulisation of eluent,[861] though at about 50 μg l^{-1} (for arsenic) they are still high in relation to typical environmental concentrations.

Most HPLC/plasma spectrometer studies have employed the plasma spectrometer as a single element detector, for the determination of well defined chemical species. Another approach is to employ the plasma spectrometer as a simultaneous multi-element HPLC detector in the investigation of less well defined species. Thus, for example, Morita *et al.*[862] have used HPLC/ICPOES for the speciation of trace elements in macromolecules (vitamin B_{12} and proteins) and Hausler and Taylor[863,864] have used the technique to investigate trace metal speciation in coal-derived materials.

The use of HPLC in trace element speciation is placed in the context of other speciation techniques in Section 11.5. In that section consideration is also given to the use of ion-exchange resins and gel filtration systems in low pressure – rather than HPLC – configurations for trace element speciation analysis. The distinction is arbitrary, in the sense that these are variants of liquid chromatography, but convenient in that such uses cannot be described as HPLC, which is the primary concern of the present section, and do not require the specialised equipment described above.

(g) Ion Chromatography

Ion chromatography (IC) may be regarded as a form of ion-exchange chromatography [see Section 11.3.8(*d*)] in which a complete instrument is supplied (or made) to integrate the chromatographic and measurement processes. The term ion chromatography was initially applied to instrumental analysis using conductimetric detection, but is now applied when other detection techniques (*e.g.*, amperometry) are employed.

Two main types of ion chromatography may be distinguished: 'suppressed' and 'non-suppressed'. In the former, a 'suppressor' column is placed between the separator column and the detector to suppress the high conductivity of the ionic solution used as eluent, the better to detect the compounds of interest eluted from the separator column (see below). In the latter, the separator column is connected directly to the conductivity (or other) detector, and the eluents chosen are of lower conductivity than those used in suppressed IC.

Ion chromatography is the subject of a recent text by Smith and Chang,[865] and of reviews by Pohl and Johnson[866] and Johnson.[867] The technique and its application to environmental analysis are dealt with in a two-volume collection of papers edited by Sawicki and co-workers,[868,869] and applications of the technique in trace analysis have been reviewed by Small.[870] Recently, a review of the technique in relation to water analysis in particular has been published as part of a U.K. series on recommended methods for water analysis.[745]

The basis of suppressed IC is illustrated in Figure 11.11, using as an example* the separation of anions X^- and Y^- on anion exchange resin $R_1 - OH^-$, with MOH as eluent. The suppressor column resin is the cation-exchange system R_2-H^+. During elution, the effluent from the separator contains M^+, OH^-, X^- and Y^- (the last two, of course, at markedly different concentrations at different times if separation is successful). If this effluent were to be passed directly to the conductivity detector, the response to X^- and Y^- would be swamped by the high background conductivity of the eluent ions, M^+ and OH^-. However, when the separator column effluent passes through the suppressor column, the reaction shown in Figure 11.11 proceeds from left to right. The cationic component of the eluent, M^+, exchanges with hydrogen ion on the resin, and the latter reacts with the hydroxyl ion of the eluent to form water. The effluent from the suppressor column which passes through the conductivity detector contains the determinands X^- and Y^- (separated by virtue of their differing affinities for the separator column resin) in what is, in essence, pure water. (For simplicity, the counter cations to X^- and Y^- have been omitted, but this does not alter the fundamental argument). The application of suppressed IC to the determination of cations proceeds in an analogous fashion, with the separator resin a strong base type and the suppressor resin a strong acid type.

It is not essential that the product of the suppression process be deionised water; the essential feature is that the product be of low conductivity, against which background the compounds of interest can be more readily detected by conductivity measurement. Indeed, the eluent most commonly used in anion chromatography is not a simple alkali, but a mixture of carbonate and bicarbonate, the

* The technique may also be applied to the determination of cations.

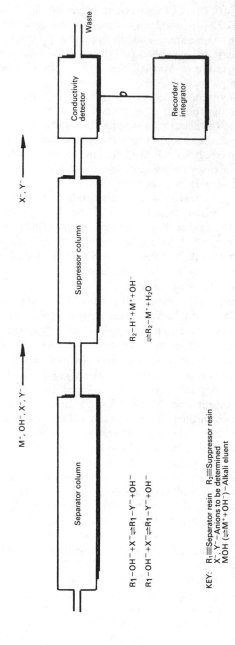

Figure 11.11 *Schematic of ion chromatography*

proportions of which may be varied to suit the particular application. In addition to this facility to vary eluting strength simply by changing the pH (and hence the carbonate/bicarbonate ratio*), the chemicals involved are cheap and non-toxic and the product of the suppression step – carbonic acid – is a weak acid and, therefore, of low conductivity. Conversely, although mineral acids may often be used as eluents in cation chromatography, other materials having greater affinity for the separator resin (and therefore greater eluting strength) are also used – examples include inorganic cations (*e.g.*, Ag^+, Cu^{2+}) and mineral acid salts of organic amines, the protonated amine being responsible for elution (*e.g.*, pyridine, *m*-phenylenediamine).

In non-suppressed IC the eluent must be both effective in the separation process and of sufficiently low conductivity to allow conductimetric detection in the absence of suppression. Solutions of phthalic and other aromatic acids have been commonly used in non-suppressed anion chromatography.[865,871–873]

The basic theory and concepts of chromatography (*e.g.*, height equivalent to a theoretical plate) outlined in Section 11.3.8(*a*) apply to the separation stage of ion chromatography. In suppressed IC, consideration must also be given to the suppressor column and its characteristics. The basic theory of suppressor column operation, relating the lifetime of the column before regeneration is required to various properties of the suppressor and separator columns, may be found in refs. 745 and 874.

To obtain good separations, the separator resins in IC must be of low capacity. (In non-suppressed IC, low separator capacity is especially important in allowing the use of low eluent concentrations, and hence achieving low background conductivities). Typical packings have capacities of 0.005–0.1 meq g^{-1}. For suppressed cation chromatography, pellicular surface-sulphonated polystyrene-divinylbenzene (PS-DVB) resins are employed as separator column packings, and for suppressed anion chromatography, a similar material is converted to an anion-exchanger by coating the surface-sulphonated beads with an aminated latex. Bead sizes in the range 15–25 μm are commonly used, and operating pressures of less than 75 p.s.i. are typical.[865] Separator columns in suppressed ion chromatography are typically between 60 and 500 mm long and have internal diameters in the range 3–6 mm.[865]

In non-suppressed IC, the separator resins are pellicular silica-based materials, in which the ion-exchange groups are introduced by conventional silanisation reactions. They suffer the disadvantage in relation to the PS-DVB type used in suppressed IC, of instability in alkaline media,[745] but patent rights preclude the supply of PS-DVB resins to manufacturers of the high performance liquid chromatography equipment which can be used for non-suppressed IC.[745] In non-suppressed anion chromatography, separator columns are typically 250 mm long and have internal diameters of about 5 mm.[865]

With regard to the suppressor column in suppressed IC, the requirements have been discussed in detail by Pohl and Johnson.[867] Conventional fully porous gel type PS-DVB resins are used having capacities of 2–5 meq g^{-1},[745] and suppressor column dimensions of 6–9 mm (internal diameter) and 60–250 mm (length) are typical.[865] The need periodically to regenerate the suppressor column is an

* The doubly charged carbonate ion is, of course, the stronger eluent.

inconvenience, and so called 'hollow fibre' suppressors for anion chromatography have therefore been introduced, which permit continuous regeneration.[875] These consist of a sulphonated low-density polyethylene tube, through which the effluent from the separator column flows and around which dilute sulphuric acid flows. The eluent cation (*e.g.*, Na^+) is exchanged for H^+ at the membrane, the sulphuric acid allowing continuous regeneration of the latter.

Many ion chromatographic determinations can be carried out without special temperature control of separator column and detector. For highest precision, however, or when ambient temperature fluctuations are high, these components [and the eluent reservoir(s)] may be thermostatted.[745,876] Operating the separator column at above-ambient temperature may be beneficial in improving the resolution of the separation system (see, for example, ref. 876).

The introduction of samples in IC is essentially the same as in other variants of HPLC [see Section 11.3.8(*d*)]. An injection loop system is commonly used, permitting both manual and automatic operation, but manual injection using a syringe is also possible. Small volumes of sample or extract (*e.g.*, 10 μl) may be introduced using a sample loop system, but the use of a concentrator column allows the determinands of interest to be concentrated from much larger volumes of sample, with consequent reduction in detection limits. This is another example of the 'trace enrichment' technique described previously [see Section 11.3.8(*d*)] in relation to HPLC in general. The concentrator column is a short column (relative to the separator column), typically of the same resin as the separator column, located upstream of the separator column in place of the normal sample injection loop. A relatively large volume (typically 10–50 ml) of sample is passed through the concentrator column, which retains the ions of interest. Subsequently, the retained ions are transferred to the separator column by passing the eluent through the concentrator column, and analysis then proceeds in the normal way. Further details of the use of concentrator columns can be found in ref. 865, and examples of their use are included in the sub-section below on limits of detection. It is worth noting that concentration of the sample, using the concentrator column, may be carried out at a location remote from the laboratory, and the column then placed in the IC system for analysis on return to the laboratory. However, it has been reported[865] that the precision of results is then impaired, relative to that attained when the concentration step is carried out in the laboratory.

It has been noted that conductimetric detection is by far the most common detection technique in IC. Other detection systems which have been used include amperometry, u.v./visible spectrophotometry (using post-column reaction with a chromogenic reagent, in the case of species which do not absorb directly), fluorimetry and indirect u.v. absorption.[745] In the last of these, it is the eluent which absorbs the u.v. radiation, the absorption being reduced when the determinands pass the detector.[877]

Recently, Downey and Hieftje[878] have described a technique known as 'replacement ion chromatography with flame photometric detection'. In this, the counter cation associated with the eluting anion is stoicheiometrically replaced by a flame photometrically active cation (*e.g.*, lithium) in a cation exchange column, and detection is then conducted using a flame photometer.

For obvious reasons, detection systems other than conductivity are most commonly encountered with non-suppressed IC.

(i) Principal analytical characteristics of ion chromatography:

(1) *Calibration.* Calibration in IC is achieved by means of a calibration curve prepared from peak area (or height) measurement on standards containing known amounts of the determinands, analysed in exactly the same way as samples. Calibrations in suppressed IC are often linear over a concentration range of four orders of magnitude, but in non-suppressed IC the linear range may cover only one decade change of concentration.[866]

(2) *Limits of detection.* The detection limits attainable by ion chromatography depend markedly upon the chromatographic conditions employed. Moreover, the technique is capable of determining many different ions, sequentially, in the same sample. The separation conditions may, therefore, represent a compromise. Some examples of the detection limits attained using IC, under a variety of conditions, are given in Table 11.11. As the data shows, detection limits can be lowered by using larger sample volumes and a pre-concentrator column, or by performing a detailed optimisation of chromatographic conditions and closely controlling the temperature of the separator column and detector.[876]

For a number of reasons (which are discussed by Pohl and Johnson[866]), including the high noise level of the conductivity detector at high background conductivities, the detection limits attainable by non-suppressed IC with conductivity detection are typically higher than those achievable using the suppressed technique. This is again evident in the data in Table 11.11.

(3) *Precision.* Again, the precision attainable will depend upon the exact conditions employed, as well as upon the care with which the technique is applied. Detailed data on precision, such as are given in refs. 879 and 880, indicate that relative standard deviations between 1 and 5% should be attainable, provided that the concentrations measured are not close to the relevant detection limit.

(4) *Selectivity.* For the ions commonly determined in water samples, it is possible to devise separation conditions such that good selectivity is achieved. Numerous authors (see, for example, refs. 865, 879, 880, 884 and 886) have reported good recoveries (90–110% or better) of added 'spikes' of the major anions determined by IC, and/or good agreement between the results obtained by IC and other, well established techniques such as solution spectrophotometry. However, a large number of inorganic and organic ions may be separated and determined by IC, as discussed below. It follows, therefore, that interference effects from co-eluted species could arise if (for example) methods developed for the determination of the 'common' anions (F^-, Cl^-, Br^-, NO_3^-, PO_4^{3-}) in unpolluted waters were applied to polluted waters or waters containing high concentrations of other anionic species. In such circumstances, recourse would have to be made to more efficient chromatographic separation or to preliminary separation procedures. In either case, longer analysis times would probably be required.

(5) *General.* Commercial instrumentation is available (for both suppressed and

TABLE 11.11 *Examples of Detection Limits attained using Ion Chromatography*

Ion	Detection limit $\mu g\ l^{-1}$	Sample volume ml	Mode and Detector*	Ref.
F^-	5			
CL^-	20			
NO_2^--N	40			
$PO_4^{3-}-P$	100	0.4	S/C	879
Br^-	100			
NO_3^--N	20			
SO_4^{2-}	60			
Cl^-	<100			
NO_3^--N	<150	0.1	S/C	880
SO_4^{2-}	500			
F^-	20			
Cl^-	800			
$PO_4^{3-}-P$	~700	0.1	S/C	881
NO_3^--N	~40			
SO_4^{2-}	300			
Cl^-	600			
NO_3^-	200	0.25	NS/C	882
SO_4^{2-}	200			
Cl^-	200			
NO_3^-	70	0.25	NS/UV	882
SO_4^{2-}	30			
Cl^{-1}	250			
NO_3^--N	~120	0.1	NS/C	873
SO_4^{2-}	1250			

Ion	Detection limit $\mu g\ l^{-1}$	Sample volume ml	Mode and Detector*	Ref.
F^-	1			
Cl^-	2			
Br^-	4	1.5	S/C†	876
$PO_4^{3-}-P$	~13			
NO_3^--N	~3			
SO_4^{2-}	11			
F^-	<10			
Cl^-	<10			
$PO_4^{3-}-P$	<10	10‡	S/C	881
NO_3^--N	<10			
SO_4^{2-}	<10			
Cl^-	2			
$PO_4^{3-}-P$	~3	10‡	S/C	880
NO_3^--N	~1.5			
SO_4^{2-}	5			
Cl^-	~2			
SO_4^{2-}	~2	25‡	NS/C	883
Br^-	10	0.2	NS/A	884
Na^+	~1			
K^+	~1	?	S/C	885
Na^+	5			
K^+	20	0.1	S/C	886
NH_4^+	5			
Na^+	~100			
K^+	~200	0.1	NS/C	887
Mg^{2+}	~250			
Ca^{2+}	~500			

* = suppressed; NS = non-suppressed; C = conductivity detector; UV = ultraviolet absorption detector; A = amperometric detector
† Using detailed optimisation of chromatographic conditions, thermostatic control and re-designed detector electronics
‡ Using preconcentrator column

non-suppressed IC) at a cost broadly comparable with that of GC or HPLC systems and automatic operation is readily achieved. Refs. 865 and 745 contain much useful information on the practical application of the technique, and the latter reference has a check-list for trouble-shooting. The time required for analysis depends markedly upon the number of determinands required, and upon the performance needed. Thus, Slanina *et al.*[876] report that determination of F^-, Cl^-, Br^-, NO_3^-, PO_4^{3-} and SO_4^{2-} at low levels in a rainwater sample required 50 min, but much shorter times may apply if less stringent performance is adequate, if concentrator columns are used or if fewer determinands are measured. Thus, for example, Kapelner *et al.*,[881] using a concentrator column, were able to determine F^-, Cl^-, SO_4^{2-}, PO_4^{3-} and NO_3^-, with detection limits better than 10 μg l^{-1}, in about 15 min when all five ions were present in similar concentration ranges.

Sawicki[888] has tabulated 61 species determined by IC, indicating the potentially wide scope of the technique. Most applications to water analysis reported to date have, however, involved only a limited number of ions – most notably the 'common' anions such as F^-, Cl^-, NO_3^-, SO_4^{2-}, PO_4^{3-}, for which the sequential multi-determinand capability of IC offers an advantage over the more conventional approaches to their determination.

It would seem that ion chromatography is a particularly attractive technique for laboratories which need often to determine most or all of such ions in their samples, but which lack the very high sample throughput needed to justify the greater capital cost of large, automatic analysers based on solution spectro-photometry (see Section 12.2). Ion chromatography also has the advantage of requiring only small sample volumes, which may be of benefit in (for example) the analysis of soil and sediment pore waters.

(ii) Applications of ion chromatography to water analysis Most applications of the technique to water analysis have been concerned with the determination of inorganic species – especially the common anions listed above, but also Br^-, NO_2^-, Na^+, K^+, NH_4^+, Ca^{2+} and Mg^{2+}. Table 11.12 lists some relevant literature dealing with the determination of inorganic ions in natural and treated waters. Ref. 865 also contains a useful part-chapter on the use of IC for water analysis.

The determination of organic species in water samples by IC with conductimetric detection has been much less commonly reported, although the potential exists as noted previously. The use of HPLC with other (more sensitive) detectors, or of GC, may have reduced the application of IC in this area. One interesting application is the determination of the nerve-agent-related isopropyl methylphosphonic acid, ethyl methylphosphonic acid and methylphosphonic acid in water, at concentra-tions of about 0.1–0.2 mg l^{-1} and above, reported by Schiff *et al.*[893]

(h) Thin-layer Chromatography
In thin-layer chromatography (TLC) the stationary phase (*e.g.*, silica gel) is coated on to a glass or plastic plate. A small volume of the sample (typically an extract in water analysis) is 'spotted' on to a small area at one end* of the plate, which is then 'developed' by being placed in contact with a suitable solvent or solvent mixture

* Except in 'circular' and 'anticircular' TLC, see below.

TABLE 11.12

Some Applications of Ion Chromatography to the Determination of Inorganic Ions in Waters

Sample type	Determinands	Ref.
Rainwater	F^-, Cl^-, Br^-, NO_3^-, SO_4^{2-}	876
Rainwater	Na^+, NH_4^+, K^+	886
Simulated rainwater	F^-, Cl^-, NO_3^-. SO_4^{2-}, Na^+, K^+, NH_4^+, Ca^{2+}, Mg^{2+}	889
Rainwater	Cl^-, NO_3^-, SO_4^{2-}	890
Geothermal waters	Cl^-, SO_4^{2-}	891
Pore waters	F^-, Cl^-, Br^-, NO_3^-, SO_4^{2-}	879
Industrial process waters	F^-, Cl^-, NO_3^-, NO_2^-, PO_4^{3-}, SO_3^{2-}, SO_4^{2-}	880
High-purity water	F^-, Cl^-, PO_4^{3-}, NO_3^-, SO_4^{2-}	881
Boiler water	Cl^-, SO_4^{2-}, Na^+, K^+	885
Flue gas desulphurization (FGD) liquors	SO_3^{2-}, SO_4^{2-}, $S_2O_3^{2-}$	892

(the mobile phase) such that the latter moves along the plate. This causes the compounds in the sample spot to move, but at rates inversely related to their retention (by adsorption or partition) by the stationary phase. Thus the sample components are separated into (ideally) a series of spots at different distances from the starting point. The locations of these spots, and the quantification of the compound in them, can be undertaken visually or instrumentally. This may require the formation of light-absorbing or fluorescent derivatives, either by pre- or post-chromatographic derivatisation. The positions of the spots relative to the starting point provide the basis for qualitative analysis and the intensities of their light-absorption or fluorescence permit quantitative analysis. To allow for variations in experimental conditions, it is normal to analyse solutions containing known amounts of the compounds of interest on the same plate as samples.

The separation capabilities of TLC can be enhanced by sequential developments of the plate in two directions at right-angles, using two different developing solvents. This is known as two-dimensional TLC, a technique which has been reviewed by Zakaria *et al.*[894]

As with other variants of chromatography, TLC has undergone considerable improvement in the last decade, leading to the establishment of so called 'High Performance Thin Layer Chromatography' or HPTLC. The developments made have included the introduction of improved solid phases, new sample application techniques and separation procedures and the widespread use of direct photometric evaluation of developed plates. Collectively, these have led not only to improved accuracy of results, but also to more rapid analysis and to greater automation of the technique.

Texts and monographs dealing with the theory and practice of TLC are given in refs. 895–901, and the instrumental evaluation of plates ('densitometry') is also the subject of a specialist monograph.[902] Thin-layer chromatography (and paper chromatography, which it has essentially replaced in water analysis) is reviewed biennially in *Analytical Chemistry*[903] and the use of TLC in water analysis is the subject of reviews by Funk[904] and Hunt.[905] (Paper chromatography in environmental analysis generally has been reviewed by McNally and Sullivan[906]).

The reader is cautioned that developments in TLC have been so rapid in recent years that the older texts cited above are now rather dated in some respects. It is beyond the scope of this monograph to provide a detailed account of these developments, but a brief description of the more important advances is given below. Further information may be obtained from the more recent texts cited above, that is, refs. 897–902 in particular.

With respect to stationary phases, the most significant developments have been the production of particulate coating materials having closely controlled physical characteristics (surface area, pore volume, *etc.*) – which permit increased resolution *and* shorter analysis times – and the introduction of chemically bonded stationary phases for reversed phase* HPTLC (a subject reviewed by Brinkman and de Vries[907]).

Sample introduction systems have also been greatly improved, and now enable very small sample volumes (typically less than 1 μl) to be applied with a high degree of precision. This improves resolution and allows closer placing of sample spots, so that up to about 70 samples may, in favourable cases, be analysed on a single 10 cm × 20 cm plate. Sample introduction has also been aided by the development of plates having 'concentration zones',[908] into which the sample is concentrated (by preliminary 'development') prior to the start of the development proper. The use of this technique improves resolution (by reducing the spreading of the sample in the direction of development) and permits better precision without closer control of the geometry of sample spotting, by ensuring good sample alignment prior to plate development. The application of samples as 'streaks' rather than 'spots' also gives improved resolution and precision.

Plate development systems have also undergone major advances. The classical vertical development procedure has given way to 'flat' development systems requiring only small volumes of mobile phase, which permit closer control of development conditions and consequent improvements in analytical performance. Two new modes of separation have also been introduced: circular and anti-circular development. In the former, the sample spots are deposited in a small circle around the centre of the plate, and developed radially outwards. This gives quicker separation and better resolution than normal 'linear' development, particularly of those sample components that do not travel far, relative to the solvent front, during the development process. Conversely, in anti-circular development the sample spots are placed in a much larger circle away from the centre of the plate, which is then developed radially towards the centre. This gives even faster separation than circular development, but with poorer resolution than in the circular mode, especially for components which do not travel far from the sample location during development. (It should be noted, however, that the instrumental scanning of anti-circular chromatograms – to quantify the separated compounds, see below – may take longer than the scanning of circular chromatograms, so that the advantage of faster separation may be nullified[909]).

The use of instrumental, rather than visual, evaluation of developed plates has

* For a definition of 'reversed phase' chromatography, see Section 11.3.8(*d*).

also led to great improvements in the performance of TLC. Although the name 'densitometry' is often applied to such instrumental evaluation, several different approaches are in use. The absorption of light of a suitable wavelength by the compound(s) of interest can be exploited, either by measuring the transmitted beam or the reflected radiation, or a combination of both. The combined use of reflectance and transmission gives the best detection power (lowest signal to noise ratio) but the use of either technique alone may give adequate performance. Fluorescence of the compounds of interest (or derivatives of them) may also be used for evaluation, giving better sensitivity and linearity of calibration than absorption measurement.

The cumulative effects upon analytical performance of these advances in the various stages of thin-layer chromatography has been considerable. Thus, Funk[904] has noted that modern HPTLC (with instrumental measurement) can provide detection limits – per spot – in the nanogram to picogram range and random and systematic errors of only 1–3%, in contrast to relative errors of up to 30% when visual measurement is employed. (Instrumental evaluation of plates, in addition to providing lower detection limits and better precision, can also reduce interferences arising from incomplete resolution of compounds, by means of selective scanning using different wavelengths of light).

Despite the major developments of recent years, thin-layer chromatography finds relatively little application in water analysis. It seems likely that the technique has been overshadowed both by GC and by HPLC, which have also undergone considerable development in the same period. Both HPTLC and HPLC offer the capability of determing non-volatile or unstable compounds not amenable to GC, but HPLC appears to have found greater application. Given the relatively recent introduction of both HPLC and HPTLC, a final assessment of their relative merits and disadvantages cannot be made. It seems that HPLC offers better sensitivity and detection power, but HPTLC may be advantageous in certain circumstances – it has been claimed,[909] for example, that the ability of HPTLC to deal with many samples simultaneously (or essentially so) may give considerable saving of time, in comparison with HPLC, when large numbers of analyses are required. It remains to be seen if this claimed capability secures for HPTLC a wider usage in water analysis; one area of future application may be the use of conventional TLC (with visual evaluation of plates) described by Sdika *et al.*[910] in a recent paper. These authors employed TLC to produce 'Quality Indices' for the assessment of potential toxicological risk arising from the presence of particular groups of organic compounds in raw and drinking waters. An extract of the sample was separated into three fractions (acid, base and neutral) and various groups of compounds in each fraction were then determined semi-quantitatively by TLC, the results being used to construct indices of the relative qualities of different samples. The authors claimed that this approach is valuable in allowing a simple assessment of water quality (with respect to some important organic contaminants) much more rapidly than would be possible by GC-MS and HPLC. (It should be noted, however, that they specifically stated that the approach is not intended to replace the latter techniques, but rather to complement them.)

In addition to this rather specialised application, TLC has also been applied to water analysis for the determination of polycyclic aromatic hydrocarbons,[911–918] phenols,[919-922] surfactants[923-926] and artificial whiteners,[927] amines,[928] amino acids,[929] chlorinated long-chain paraffins,[930] hydrocarbons,[931,932] statutory chemical markers in duty-free gas oils,[570,933] mercury compounds,[934,935] selenium,[936] herbicides and pesticides,[569,937-946] and plant pigments.[947-949] In common with other chromatographic techniques, TLC has been used for 'fingerprinting' in investigations of oil pollution.[950,951] It may also be used as a preliminary separation technique, for 'clean-up' prior to the application of other chromatographic techniques; an example is the use of TLC for this purpose in the determination of polycyclic aromatic hydrocarbons in water.[952]

Equipment for TLC is available commercially. The cost of establishing the technique varies enormously, depending on the nature of the equipment required. 'Classical' TLC, using vertical development and visual evaluation, can be undertaken very cheaply, but modern HPTLC with instrumental evaluation can involve similar capital expenditure to that required for an HPLC or GC system.

(i) Use of Flame-ionisation Detection to Determine Total Organic Carbon

The mode of operation of the flame-ionisation detector (FID) has been described in 11.3.8(*a*) above. This device has, in addition to its conventional use in gas chromatography, found application in the determination of the total organic carbon (TOC) content of waters.

Unlike non-dispersive i.r. spectrometry, which also finds use in TOC determination [see Section 11.4.3(*e*)], flame-ionisation detection is not suitable for the direct measurement of the carbon dioxide produced by oxidation of the organic matter. The oxidation step is, therefore, followed by a reduction of the evolved carbon dioxide to methane, the latter then being measured by flame-ionisation detection.

A recent publication[48] in an authoritative series of recommended methods for water analysis provides a detailed review of TOC determination, including a summary of commercial instrumentation and its performance. Systems based on FID measurement of methane, as described above, are included.

Small volumes of sample are, typically, injected into a high temperature (*ca.* 900 °C) combustion furnace containing a catalyst such as copper or cobalt oxide, in which oxidation to carbon dioxide occurs. The combustion gases are then mixed with hydrogen and passed through a second, lower temperature furnace (*ca.* 400 °C) containing a nickel catalyst. The carbon dioxide is reduced to methane, which is swept to a FID for measurement. Details of systems of this general type are given in refs. 953–956, and of a similar system employing a boat for sample introduction in ref. 957.

Commercial instruments based on these principles are available, as is also a system employing u.v. photo-oxidation of organic carbon and flame-ionisation detection.[48]

The reported[48] detection limits achieved using these instruments lie in the range 20–100 μg l^{-1}, and the same reference quotes relative standard deviations of *ca.* 1–2%.

Instruments of this type are capable of rapid analysis – a few minutes per sample – but require that inorganic carbon be removed prior to TOC determination, or measured separately and subtracted from the result for total carbon. (The carbon dioxide is simply stripped from an acidified sample, to separate the inorganic and organic carbon. Volatile organic compounds may be lost during the stripping, but can be measured if required – see ref. 48.)

By replacing the high temperature furnace of the above approach with a pyrolysing furnace containing nickel, a system not subject to interference from inorganic carbon has been developed,[958] but it suffers from the disadvantage that the sensitivity of the FID depends to some extent on the organic pyrolysis products being measured. The same disadvantage applies to a system[959] in which the sample is heated in a stream of nitrogen, which is then passed to the FID, though for many organic compounds the variation is not marked. Neither of these approaches appears to be available on a commercial instrument.[48]

11.3.9 Miscellaneous

The importance and widespread application of certain determinations warrants their brief mention here even though the analytical techniques involved are commonly used for rather few determinands. The three techniques chosen for mention below are not related, and are brought together in this section only for convenience.

(a) Nephelometry
This technique is used to determine undissolved particles in suspension and is based on the measurement of the amount of light scattered by the particles (*i.e.*, as in the Tyndall effect). A schematic representation of a nephelometer is shown in Figure 11.12. Most instruments measure the light scattered at right angles to the incident beam but measurement at other angles can be beneficial[960] and is used in some

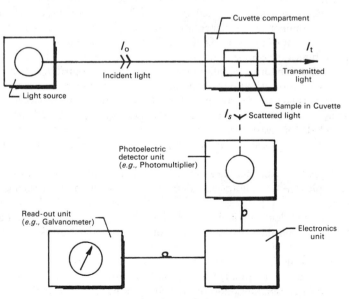

Figure 11.12 *Schematic arrangement of a nephelometer*

instruments. In order that the scatter of light is caused only by particles in the sample, the sample cuvette must be kept scrupulously clean and free from scratches; the presence of bubbles in the sample during measurement must also be avoided. If the sample contains dissolved materials that absorb the light, the intensity of scattered light will be reduced. This error is eliminated in some instruments by measuring the ratio of intensities of scattered and transmitted light, (I_s/I_t). The intensity of the scattered light depends on many factors including the number, concentration, size, shape and refractive index of the particles, the intensity and wavelength of the incident light and the angle of scatter. The theoretical background to the technique is comprehensively discussed by Kerker.[961] In general, for small concentrations of particles of uniform size and shape, I_s is directly proportional to their concentration.

The most important application of nephelometry in water analysis is to the estimation of the relative concentrations of undissolved materials suspended in the water. Given the dependence of the scattered light intensity upon, *inter alia*, the nature of the particles present, the use of nephelometry for this purpose is subject to some important limitations. If the nature of the particles at a particular sampling location remains essentially the same, then the technique may be used to obtain an estimate of the relative concentrations of suspended material from time to time, by means of an empirical relationship between I_s and the concentration of suspended matter determined by gravimetry.

If the nature of the particles varies, the validity of the above approach is impaired but the technique may still be used to give a broad indication of the relative amounts of undissolved substances in different samples, or to provide a measure of the optical clarity of the water if that – rather than the concentration of suspended matter – is the property of interest.

Nephelometry is particularly useful for the measurement of optical clarity or the relative amounts of suspended matter when the latter are low, being much more sensitive than the alternative technique – turbidimetry* – under such conditions.[962] Thus, for example, nephelometry has been recommended[963] as the preferred method for determining the turbidity of water. (Ref. 963 also contains many references to earlier work on turbidimetric methods.) The sensitivity of commercially available nephelometers is such that the light scattered by pure solvents must be taken into account in calibrating the instruments.[964] Calibration is normally made by using suspensions of reproducible concentration and particle size; formazin is recommended as the insoluble material for such suspensions.[963] The question of units of measurement of turbidity (both by nephelometry and turbimetry) is discussed in refs. 963 and 965. The latter also contains much background information on the principles of turbidity measurement.

On-line and *in situ* turbidimeters have been developed and are widely used in water monitoring; see Section 12.3 for further details and references.

In addition to measuring the particles originally present in a sample, nephelo-

---

* The turbidity, τ, of a sample is defined by:

$$\tau = (1/L) \ln (I_0/I_t)$$

and L = length of sample through which the light beam passes. Turbidimetric methods are those which measure transmitted rather than scattered light. There is some confusion in common terminology because the term 'turbidity' is used also in the (not strictly defined) sense of denoting a lack of optical clarity.

metry may also be used to measure the amount of insoluble material formed when appropriate reagents are added to a sample. The most common application in water analysis is the determination of sulphate by adding barium chloride to the sample and measuring the light scattered by the barium sulphate particles formed. Standard methods employing this approach are given in various collections of methods for water analysis (*e.g.*, refs. 963, 966 and 967), and detection limits of 1–10 mg $SO_4 l^{-1}$ are attained. As in all such techniques, great care is needed to ensure that the precipitate is reproducibly formed, otherwise, variations in particle size will lead to poor analytical accuracy. The above methods specify 'conditioning reagents' (consisting of, for example, glycerol, alcohol, acid and sodium chloride) to stabilise the precipitate and minimise interferences. Even lower concentrations of sulphate ($0.01–5$ mg $SO_4 l^{-1}$) can be determined nephelometrically using organic reagents.[968]

(b) Conductivity (Electrical)*

The ability of a solution to conduct an electrical current is governed by the concentrations and natures of the ions present. Most soluble inorganic compounds normally present in waters dissociate in solution to produce ions so that the conductivity of a sample provides an indication of the concentrations of dissolved inorganic materials. The theoretical background to this topic is well covered in any text-book on electrochemistry (see, for example, refs. 969 or 970) and is not discussed here.

In practice, a determination usually involves measurement of the resistance of the solution in a Wheatstone or other bridge, though other methods are used. The measuring instrument is normally calibrated in conductivity units because these are directly related to concentration (rather than inversely, as is resistance). Details of equipment are given in standard methods (see refs. 971, 972 and especially 973). Two main types of cell are used: those employing (usually two) electrodes (of platinum or other conducting material) and those of the electrodeless, inductive type. Polarisation of the electrodes in the former type precludes the use of direct current systems, and is minimised by using an alternating potential across the cell, with frequencies in the range 25–3000 Hz.[973] Electrodeless, inductively coupled cells (which perforce require a.c. excitation) avoid the problem of polarisation and are less subject to fouling problems; they have found widespread application in the determination of the electrical conductivity of brines and brackish waters[973] and of the salinity of sea-water (see below).

For waters having conductivities below 10 μS cm^{-1} (*i.e.*, 1 mS m^{-1} or 10 μmho cm^{-1}), measurements should be made in flow-type conductivity cells[973] to avoid errors arising from sample contamination (*e.g.*, by carbon dioxide).

Many descriptions of procedures refer to the rather high temperature coefficient of conductivity, and values of about 2% per °C are commonly quoted for fresh waters. However, two points should be borne in mind:

(1) The temperature coefficient depends on the nature and concentrations of the ions present (and on the temperature itself); thus, for accurate work it is best to measure the temperature coefficient for each type of water so that an

---

* The older term 'specific conductance' is now rarely used, and should be avoided; conductance (in Siemens) and conductivity (in Siemens per metre) refer to different properties – it is the latter which is used in water analysis. For convenience, most measurements of the electrical conductivity of waters are reported in microSiemens per centimetre (μS cm^{-1}), this unit being equivalent to the previously used μmho cm^{-1}.

appropriate correction can be made,[973] or work at a fixed temperature to eliminate the need for correction[973,974] (but see also below).

(2) For waters of high purity (conductivities < 1 μS cm^{-1}), the temperature coefficient begins to increase and approaches 8% per °C as pure water is approached.

Although the use of a fixed temperature of measurement avoids the need for correction, and its associated uncertainties, it may not always be possible or convenient. Likewise, the establishment of temperature coefficients for all the types of water encountered may often be very time consuming. In these circumstances, recourse may have to be made to the use of some form of 'average' or 'representative' coefficient. If this is to be done, careful consideration must be given to the accuracy required of results, the nature of the waters involved, the temperature range over which correction is to be made and the value of the coefficient to be used. For example, a recent report on conductivity monitoring in U.K. rivers[974] considered all these factors (including data on temperature coefficients for real waters obtained by Smith[975] and Wagner[976]) and concluded that important bias from correction was unlikely to arise (in that application) if a suitable coefficient was chosen, provided that the correction was made over a temperature range not exceeding ± 5 °C from the selected 'standard' temperature (20 °C). Nonetheless, the use of temperature correction in this manner is not ideal, and is better avoided if possible.

Many instruments for conductivity measurement incorporate a facility for automatic temperature correction. The implicit value of the temperature coefficient used may, of course, again be inappropriate to the work in hand and the user should carefully consider the particular requirement of his work in deciding whether or not use of such a facility is appropriate.

Measurement of electrical conductivity is rapid, simple and accurate, and the technique is widely used as a rapid means of detecting changes or differences in the general ionic composition of all types of water. The fact that the measured value depends on all the ions present in the samples usually precludes detailed interpretation of any differences observed (though not in the case of sea-water, discussed below); the need for specific methods of analysis may, however, be rapidly indicated. When waters containing approximately constant ratios of different ions are measured it is often possible to establish a good correlation between conductivity and the 'total dissolved ionisable solids'.[971]

For sea-waters, the constancy (for many practical purposes) of the relative ionic composition has permitted the use of conductivity measurement as a means of determining the salinity and (of great importance) the density of the water. It is beyond the scope of this text to consider this complex and specialised subject in detail. The reader wishing to pursue the topic is referred to two comprehensive reviews by T. R. S. Wilson of the U.K. Institute of Oceanographic Sciences,[977,978] and to literature cited therein.

It should also be noted that conductimetric detection may be employed with liquid chromatography, particularly with ion-chromatographic analysis [see Section 11.3.8(g)]. Conductimetric detection of carbon dioxide produced by the destruction of organic material (using u.v. irradiation) is also employed in a commercially available instrument for the determination of so called 'Total Organic Carbon'.[979] Additionally, conductivity measurement has been used[980] as a basis for the determination of mercury. In this case, it is the conductivity of (solid) gold which is measured, and which changes as the gold amalgamates with mercury. The technique has been employed with the chemical generation of mercury vapour from solution (widely used with atomic absorption spectrometry, see Section 11.3.5) for the analysis of water, and commercial instrumentation is available. The equipment is readily portable, and a detection limit of <0.1 μg l^{-1} has been reported.[980]

(c) Amperometry

Mention has already been made of amperometric titrations (Section 11.3.1), and the present section considers direct amperometry in which no titration is involved. In essence, a pair of electrodes is placed in the sample and a constant voltage applied across them so that the determinand undergoes an electrochemical reaction at one of the electrodes. The voltage is chosen appropriately for the determinand so that the rate of the reaction is limited by the diffusion of the determinand to the electrode [see Section 11.3.7(a)]. Under these conditions the rate of reaction and, therefore, the current is directly proportional to the concentration of the determinand. By suitable choice of the materials of the two electrodes it is sometimes possible to achieve the required reaction without any externally applied voltage.

This technique is very extensively used to determine dissolved oxygen and many instruments and special electrodes are available. It is also applied to the determination of disinfecting agents.

Dealing first with the determination of dissolved oxygen, it may be noted that two broad variants of the amperometric technique have been applied in water analysis.

In the first, the electrodes are immersed directly into the sample. Portable instruments have been described which can be used to determine less than 0.01 mg l^{-1} in high purity boiler feed water[981] or to determine higher concentrations (typically 0.1–20 mg l^{-1}) in natural waters.[982–984] The application to relatively pure waters is generally satisfactory and allows accurate determinations (± 2–5%) to be rapidly and simply made. Troubles may be experienced in other waters because of 'poisoning' of the electrodes and the reaction of other substances at the electrodes,[985] and bare electrodes are not widely used. These problems are overcome to a very large extent by the second type of system, *i.e.*, the widely used membrane-covered electrodes.

Membrane electrodes for dissolved oxygen are the subject of a number of reviews and texts (see, for example, refs. 986–989). In essence, they usually consist of a pair of metal electrodes surrounded by an electrolyte, this assembly being separated from the sample in which it is immersed by a thin plastic membrane permeable to unchanged molecules but not to ionic species. Oxygen diffusing through the membrane into the electrolyte is reduced at the cathode, so that a current is generated, the magnitude of which is proportional to the concentration of dissolved oxygen in the sample.

In systems employing an imposed potential (often referred to as 'polarographic' or 'voltammetric' systems), the cathode is usually gold or platinum and the anode, silver. Systems in which the required potential is generated by the cell reaction itself – known as 'galvanic' – have employed a silver cathode and a lead anode.

Because the membrane is not permeable only to oxygen, but to other uncharged species also, interference from the latter can occur. The permeability of the membranes to sulphur dioxide is much lower than to oxygen,[990] so that the former will seldom be at sufficiently high concentration to be troublesome. Carbon dioxide can interfere by altering the pH of the electrolyte, but its effects have been overcome by improvement in instrument design to allow a faster exchange of the electrolyte film around the cathode with the bulk.[990] Hydrogen sulphide can interfere by reaction with silver, but this effect can be overcome by suitable choice of electrolyte.[990,991] Other gases, capable of reduction at the potential employed, can interfere[75] positively – examples are chlorine and nitrous oxide – but do not normally present serious problems. Overall, therefore, membrane electrode systems offer great practical selectivity for dissolved oxygen and have the additional advantage that they can be used for *in situ* analysis; this is often more convenient and informative and also avoids problems of sample stability.

Many variants of the oxygen membrane electrode system are commercially available (see, for example, refs. 987 and 992 for names of suppliers) and the technique is included in collections of standard or recommended methods (e.g. refs. 75, 992, 993). The limit of detection has been reported to be about 0.2–0.4 mg l^{-1}, and relative standard deviations of less than 1% have been observed at concentrations of 8–10 mg l^{-1}.[75] The response of the electrodes must be calibrated, but straightforward procedures are available.[75,987,993,994] The overall measurement process requires little skill and experience, and can provide a result in a few minutes; on-line measurement of dissolved oxygen is also readily achieved (see Section 12.3). The electrode systems have a rather high temperature coefficient for their response, but automatic correction is commonly incorporated in the instrumentation (see, for example, refs. 987, 995 and 996).

Amperometric oxygen electrodes are now very widely used in water analysis; *e.g.*, to provide oxygen determinations for pollution estimation, for sewage-treatment control and for characterising reservoir stratification. They have allowed measurements to be made which would probably otherwise have remained impossible (*e.g.*, the rapid *in situ* measurement of the dissolved oxygen profiles of large water bodies). They also provide an alternative to the well known Winkler titration (see Section 11.3.1) for the analysis of discrete samples in the laboratory, including the analysis of biochemical oxygen demand (BOD) samples. In the latter application, particularly, problems can arise from fouling of the membrane. Phelan *et al.*[997] have reported a modification of the conventional oxygen membrane electrode to deal with this difficulty; they interspersed a multiplicity of anodes and cathodes so that there was no net consumption of oxygen, and no oxygen flux through the membrane during measurement. By this means, fouling effects (and effects from variations in the sample flow rate around the electrode) were overcome.

Amperometry can also be applied to the determination of chlorine used as a disinfecting agent, being suitable for on-line monitoring. Initially, systems operating without a membrane were developed. Free chlorine (Cl_2, HOCl and OCl^-) was determined after mixing the sample stream with an acetic acid/acetate buffer, and total residual chlorine (free chlorine plus chloramines) after mixing the sample stream with an acetic acid/acetate buffer containing potassium iodide, from which iodine was liberated. Polarisation effects necessitated electrode cleaning in the free chlorine systems, but the cleaning adversely affected the performance.[998] By adding potassium bromide to the buffer, free chlorine could, however, be measured as the bromine analogue.[998] Canelli[999] has reported an examination of the performance of a portable, bare-electrode amperometric system for the determination of free chlorine in potable and swimming pool waters. A detection limit of 0.1 mg l^{-1} was attained, and relative standard deviations were less than 6% in the range 0.25–2.7 mg l^{-1}. However, a number of interferences were reported – including N-chloro compounds such as NH_2Cl, $NHCl_2$ and NCl_3.

Recently, membrane systems have been developed. That produced by Johnson, Edwards and Keeslar[1000] is more sensitive to nitrogen trichloride, nitrogen tribromide, bromine and dibromamine than it is to HOCl and OCl$^-$ (the principal components of free chlorine at normal pH values), but the authors state that nitrogen trichloride is slow to form and therefore not of importance in drinking water, and the bromine species should normally also be unimportant in such water. With regard to the components of free chlorine, they report that the sensitivities decrease in the order $Cl_2 >$ HOCl $>$ OCl$^-$; Cl_2 is usually a minor species at pH values typical of drinking water, and the greater sensitivity of the system to HOCl than to OCl$^-$ is thought to mimic their relative disinfecting powers.[1000]

A recent comparison[1001] of procedures for the determination of chlorine residuals found, however, that amperometric membrane electrodes gave highly variable results, considerably lower than those obtained by amperometric titration.

Dimmock and Midgley[1002,1003] have applied the system of Johnson *et al.*[1000] to the determination of free and total residual chlorine in sea-water. Problems initially encountered[1002] were the interference of chloramines and bromine*, and the effect of varying salinity upon the response. By incorporating bromide (for free chlorine) and iodide (for total residual chlorine) in the cell filling solution, these problems were overcome.[1003] Limits of detection of 0.014–0.018 mg Cl_2 l^{-1} (free chlorine) and 0.018–0.032 mg Cl_2 l^{-1} (total residual chlorine) were reported,[1003] and total standard deviations of 0.01 mg Cl_2 l^{-1} to 0.055 mg Cl_2 l^{-1} (free chlorine, at concentrations in the range 0.1–1 mg Cl_2 l^{-1}) and of 0.01–0.05 mg Cl_2 l^{-1} (total residual chlorine, at concentrations in the range 0.1–0.5 mg Cl_2 l^{-1}).

The production of chlorinated organic compounds during chlorination of drinking water has prompted the study of alternative disinfecting agents. In turn, this has led to the development of amperometric methods of determination for such alternatives as ozone and chlorine dioxide. For example, Dormond-Herrera and Mancy[1004] have developed an amperometric membrane system for the determination of chlorine dioxide, with a detection limit of 15 μg l^{-1}, which employs pulsing of the imposed potential to improve sensitivity. (A similar, pulsed voltage system was previously developed by Schmid and Mancy[1005] for the determination of dissolved oxygen, to improve sensitivity and overcome effects due to the membrane such as high temperature coefficient and drift of sensitivity from accumulation of deposits). Freese and Smart[1006,1007] have also used an amperometric membrane system for the determination of chlorine dioxide. These workers employed a rotating membrane electrode to provide a well defined and more reproducible

* A product of the chlorination of sea-water.

transport of determinand to the membrane, and reported a relative standard deviation of 6.4% at 0.30 mg l^{-1}. Again, Smart *et al.*[1008] and Stanley and Johnson[1009] have developed amperometric membrane instruments for ozone determination at μg l^{-1} concentrations. The latter authors report a detection limit of 62 μg l^{-1}.

Recently, the modification of an amperometric oxygen probe to permit the determination of hydrogen has been reported.[1010,1011] Oxygen electrode systems (as such) also find application as basic components of sensors employing immobilised bacteria or enzymes (see Section 11.4.4). Thus, for example, Okada, Karube and Suzuki[1012] have developed an ammonium ion probe based on an oxygen electrode surrounded by an immobilised microbial membrane. Amperometry has also been used to obtain rapid and sensitive measurements of the total oxygen demand of water samples.[1013]

Another technique of recent development employing amperometric measurement is so called 'pneumoamperometry', in which the determinand (or the product of a reaction involving it) is generated from the sample as a vapour and subsequently measured in an amperometric cell. The technique has been applied to the determination of arsenic,[1014,1015] antimony,[1014] tin,[1014] mercury,[1014,1016] iodide and iodate,[1015] cyanide[1017] and nitrite and nitrate.[1018] The theory of pneumoamperometry has been described by Bruckenstein;[1019] it appears to offer very low detection limits with only small sample volumes – thus Gifford and Bruckenstein[1014] report limits of 0.2 μg l^{-1} (arsenic), 0.2 μg l^{-1} (antimony) and 0.8 μg l^{-1} (tin) using 5 ml sample volumes.

Amperometric detection has been employed with HPLC, though only rarely in water analysis applications [see Section 11.3.8(*d*) on HPLC, and the review of electrochemical detectors by Brunt cited therein]. It has also been employed, instead of conductivity-based detection, in an ion-chromatography [see Section 11.3.8(*g*)] system for the determination of bromide ion.[1020] A detection limit of 0.01 mg l^{-1} was reported,[1020] in contrast to a value of 0.1 mg l^{-1} obtained with conductimetric detection.

Amperometry as a general technique appears to be but poorly served by reviews and texts, perhaps because of the diversity of practical procedures and the limited number of applications. Some information on the technique and its theoretical basis can be obtained from texts on electroanalytical chemistry, *e.g.*, ref. 1021 [see also Section 11.3.7(*a*)]. The review of amperometric procedures in water analysis by Maienthal and Taylor[1013] is now rather dated [though recent reviews on the amperometric determination of oxygen in water are numerous (see above)]. Amperometric titrations are reviewed biennially (even-numbered years) in *Analytical Chemistry*.

11.4 OTHER MEASUREMENT TECHNIQUES

11.4.1 Mass Spectrometry

The instrumentation required for this technique is amongst the most complex, costly and difficult to maintain of all the techniques applied to water analysis. In the first edition of this book it was noted that mass spectrometry (MS) was then used only in research laboratories or central laboratories of large organisations, and that this rather restricted distribution would probably continue for some time. That prediction has proved essentially correct – at least outside the U.S.A. – but application of MS to both research and investigational work has grown markedly over the last decade.

This growth has arisen because the technique, particularly in combination with gas chromatography (GC-MS), has provided a most powerful and sensitive approach to the identification and quantitative determination of organic substances in water, though MS can also be applied to inorganic determinands.

In the U.S.A., particular requirements arising from judicial decisions and environmental legislation have secured an even wider application of MS. Thus, McGuire and Webb[1022] (in 1981) reported that the U.S. Environmental Protection Agency (EPA) possessed 50 GC-MS systems, whereas only a few were available in that organisation about ten years previously.

Because MS appears to be less widely used in other countries – especially in survey and monitoring applications – it has not been treated here as a routine technique for water analysis. However, its scope and sensitivity, growing usage and inherent complexity all require that it be considered in some detail in a monograph of this kind, albeit as a special – rather than routine – technique. Particular emphasis will be placed upon its application to organic substances in sections (a)–(c) below, but the use of MS for inorganic determinands will also be briefly discussed, in section (d).

(a) General Basis of the Technique

Basically, MS consists of three stages: (1) the volatilisation and ionisation of the substances of interest (not necessarily in that order – see below), (2) the separation of the ions on the basis of their mass/charge (m/z) ratios and (3) the detection and recording of the separated ions. Figure 11.13 shows the mass spectrometer in schematic form.

The identification and determination of organic compounds in water is of great and growing importance, especially with regard to the study of pollution by synthetic chemicals. MS has particular value in this context, because it can provide information on both the nature and concentration of organic substances present.

Characterisation of organic compounds by MS is usually performed by exploiting the characteristic pattern of molecular fragmentation* of different compounds during ionisation; the m/z ratios present and their intensities constitute the 'mass spectrum', which may be compared with those of known compounds to identify the substance(s) present in the sample. Less commonly, high resolution MS can provide very accurate estimates of the mass numbers of observed peaks, which may be compared with those calculated for different compounds suspected to be present.

In water analysis, the mass spectrometer is most frequently used to identify and quantify the compounds separated by GC – i.e., as a GC detector [see Sections 11.3.8(a)–(c)]. In addition to providing information on compound identity

---

* When an organic compound is ionised, the ion formed may subsequently break up or 'fragment' into charged species of lower molecular weight.

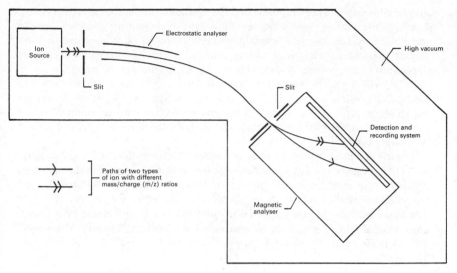

Figure 11.13 *Schematic of a mass spectrometer*

(commonly by the spectrum matching described above), the mass spectrometer provides a most sensitive means of quantifying the substances separated.

Within the simple three-stage outline of MS given above, there exists a great variety of techniques, particularly for the production and separation of the ions, the choice of which has a profound influence on the nature of the results obtained. Given the complexity of MS, only a brief outline of these approaches can be given here. Further details should be sought in general texts (*e.g.*, refs. 1023–1025), in the book on quantitative aspects of MS by Millard[1026] and in the biennial reviews (even-numbered years) in *Analytical Chemistry* (*e.g.*, ref. 1027). Other valuable sources of information include the *Specialist Periodical Reports* of the Royal Society of Chemistry (*e.g.*, ref. 1028) and the quarterly *Mass Spectrometry Reviews*, published since 1982. The important GC-MS technique is the subject of a book by McFadden[1029] and its application to water analysis has been recently reviewed.[1022] Two recent compilations on the determination of trace organic pollutants in water[10,11] also contain useful information on the use of MS, both for qualitative and quantitative analysis, and the subject of contamination effects in MS has been reviewed by Ende and Spiteler.[1030]

(A compendium of mass spectra for 114 organic compounds on the U.S. Environmental Protection Agency's 'Priority Pollutant' list has also been published recently. In addition to giving details of the most abundant and/or analytically useful ions and their intensities, this volume[1031] also provides a selected bibliography).

(i) Sample introduction procedures The introduction of discrete samples into a mass spectrometer is relatively straightforward. Gases are very simply introduced, because the instrument must be maintained at high vacuum, and volatile liquids can

also be allowed simply to evaporate into the instrument under the reduced pressure. Less volatile liquids, and solids of sufficiently high vapour pressure, can be introduced by heating, and involatile materials can be introduced into certain ionisation sources directly, on a suitable probe. Other ionisation techniques require deposition of the sample on a suitable anode, *e.g.*, a fine wire or metal foil.

In GC-MS, the gas flow from the chromatograph may be taken directly to the MS inlet (often possible with capillary GC columns [see Sections 11.3.8(*a*)–(*c*)]) or *via* a suitable separator which transfers the maximum amount of sample, in the minimum of carrier gas. In environmental analysis, the glass jet separator is most commonly used.[1022,1029]

The use of a mass spectrometer as an on-line detector with HPLC is particularly attractive, because the latter, though able (unlike GC) to separate the involatile compounds present in water samples, is hampered by the limitations of the commonly used detectors (u.v. and fluoresence, mainly) – see Section 11.3.8(*d*). The main difficulty in interfacing HPLC and MS is the high eluent flow rate of conventional HPLC; a number of approaches are possible, however. The subject has been reviewed recently by Curry[1032] and by Levsen[1033] [see also Section 11.3.8(*d*)].

(ii) Ionisation Ionisation of the sample molecules may be accomplished in a variety of ways, and this variation adds to the flexibility and scope of the technique.

The most common procedure for producing positive ions is *electron impact ionisation*, in which the vapour is bombarded by an electron beam (the energy of which may be varied). Sufficient energy can be imported to ionise most organic molecules, M:

$$M + e^- \rightarrow M^{+\cdot} + 2e^-$$

Indeed, the energy supplied to the molecules is usually sufficiently in excess of the ionisation energy to bring about fragmentation of the molecular ion, $M^{+\cdot}$. Fragmentation may proceed by different mechanisms, but the products include further ionic species amenable to separation by MS.

When electron impact ionisation is used, the degree of fragmentation is usually large, resulting in complex mass spectra. In some cases, the molecular ion $M^{+\cdot}$ may itself be undetectable. Though such fragmentation may be very valuable in the elucidation of the structures of organic compounds, it greatly complicates the application of MS to qualitative and quantitative analysis of mixtures of compounds. (The traditional application of electron impact MS to structural elucidation has, it has been claimed,[1034] led many analytical chemists to view MS as a means of investigating structure only, rather than as (also) a powerful analytical technique).

Much effort has, therefore, been devoted to the development of the so called 'soft' ionisation[1035] techniques which include *chemical ionisation, field ionisation, field desorption, plasma desorption, fast atom bombardment* and *laser desorption*. All of these impart less energy to the target molecules, thus producing a much smaller degree of fragmentation such that the molecular ion is detectable – and indeed, sometimes the dominant peak (the 'base peak') of the resulting mass spectrum.

Chemical ionisation involves the reaction between the molecules of interest and 'reagent ions', commonly generated from a suitable gas in a high pressure ion source, and present in great excess. Field ionisation employs strong electric fields (attained by using metal anodes of very small radii of curvature – points, blades or wires) to ionise the vaporised molecules.

In field desorption, ionisation precedes volatilisation so that involatile or thermally labile compounds can be dealt with, as well as volatile ones. The sample is deposited on an 'emitter' (commonly a fine wire covered with 'microneedles' of carbon, silicon, nickel or cobalt) which then forms the anode of a field ionisation apparatus. Electrons transfer from the compound(s) of interest to the anode, and the resulting positive molecular ion is strongly repelled from it.

Plasma desorption involves deposition of the sample on a thin metal foil, which is then bombarded with ^{252}Cf fission fragments. When the latter pass through the foil, the local temperature increases to *ca.* 10 000 K and sample molecules are desorbed, so rapidly that thermal disintegration of them is avoided. Both positive and negative ions result.

Fast atom bombardment is a fairly new technique, in which the sample is bombarded with argon atoms of controlled kinetic energy. Although few reports of its utility are yet available, it has a number of advantages (avoidance of the need for sample volatilisation and of thermal effects, ease of sample preparation) and is able to deal with species of high molecular weight.[1036]

Laser desorption involves rapid heating of the sample by laser irradiation, leading to volatilization with reduced thermal fragmentation. Ionisation occurs, primarily, through interaction of the compounds of interest with alkali ions produced by thermal ionisation during the irradiation.

(iii) Separation of ions Many different types of apparatus are available for the separation of the generated ions, and only a very brief description of the main types in common use will be attempted here. Texts on mass spectrometry give much fuller detail; see for example, ref. 1025.

In many instruments, a magnetic field sector is employed to separate the ions, the radius of the ion path through the sector being dependent on the m/z ratio. Resolution is, however, impaired unless steps are taken to deal with the range of kinetic energies of ions leaving the source – this problem is commonly overcome by using, in addition to the magnetic field, an electric field to accomplish 'velocity focusing' of the ions. Figure 11.13 shows a typical configuration of a double-focusing instrument – one in which both directional and velocity focusing occurs, and in which photographic recording is possible. Yet other instruments of the same basic type (electrostatic analyser, followed by magnetic analyser) have only one double-focusing point, and only electrical detection is then possible. The mass spectrum may be scanned by varying the magnetic field strength or the accelerating voltage applied to the generated ions. Other instruments have what is called 'reversed geometry'; in these, the ions first pass through a magnetic analyser and then through an electrostatic analyser to the detector. Such systems may be advantageous in examining the decomposition of particular ions (see below).

In quadrupole mass spectrometers, separation of the ions is accomplished by directing the ion beam along the axis between four parallel rods of circular or hyperbolic cross-sections (usually ceramic), across pairs of which constant d.c.

voltages and r.f. (radiofrequency) potentials are applied. Spectral scanning is achieved by varying the amplitudes of the voltages or by varying the frequency of the r.f. potential. Quadrupole instruments allow fast scanning, but magnetic sector instruments can offer higher resolution; both types are available commercially.

Recently, increased attention has been given to the use of MS in such a way that the need for preliminary chromatographic separation of complex mixtures is avoided. Various approaches are possible, but the basis of them all is mass spectrometric isolation of a particular molecular ion, which is then induced to fragment by collisional activation (*i.e.*, by colliding with suitable gas molecules in a collision chamber).

The simplest approach is to employ a mass spectrometer of reversed-geometry (*i.e.*, the magnetic analyser precedes the electrostatic analyser) and place the collision chamber between the analysers. The molecular ions are isolated, one at a time, by means of the magnetic analyser and their fragmentation products (daughter ions) resulting from the collisional activation are detected by scanning the potential on the electrostatic analyser. This technique is known as 'mass-analysed ion kinetic energy spectroscopy' (MIKES) or 'direct analysis of daughter ions' (DADI). Although it can be undertaken on a single mass spectrometer, the technique offers relatively poor resolution (because the daughter ions entering the electrostatic analyser have very variable translational energies) and is regarded as more suited to the elucidation of structure than to analytical use in the more general sense.[1037]

Of potentially greater application to water analysis is the technique of tandem mass spectrometry (MS/MS). Here the collision chamber is placed between two mass spectrometer systems (mass analyser or quadrupole). The first is used to isolate molecular ions, the daughter ions of which are subjected to conventional MS in the second. (Somewhat confusingly for the non-specialist, the quadrupole system commonly used for MS/MS is referred to as a '*triple* quadrupole' – the third, central quadrupole is used as the collision chamber.)

The value of MS/MS as an alternative to chromatography/MS in water analysis cannot be judged at present, but it has been considered[1037,1038] to be a promising approach, worthy of further investigation. Even if MS/MS does not prove able to avoid the need for prior chromatographic separation, the use of GC-MS/MS may have advantages – for example, in reducing matrix effects.[1039] Response factors* for the determination of a number of compounds by GC-MS and by GC-MS/MS (triple quadrupole) have shown close agreement (within $\pm 15\%$).[1039]

(iv) Detection, recording and data handling   As noted previously, photographic detection or electrical detection (electron multiplier systems) are possible. The latter is now commonly employed and computer techniques are used both for instrument control and data processing. Data processing can involve not merely the acquisition of signals and the construction and storage of mass spectra, but also comparison of 'unknown' spectra with those of known compounds in spectral libraries and the elucidation of molecular structure from mass spectra. The application of computer-aided procedures for interpretation of mass spectra is the subject of a recent general

---

* The response factor is a constant used for quantitative GC-MS analysis with internal standardisation. It is the ratio of the integrated ion current values for the determinand and the internal standard.

review by Martinsen.[1040] This covers data base construction, library searches, pattern recognition (for compound classification/structural features identification) and structure assignment (without library comparison). Library search techniques are also the subject of papers by McLafferty and Venkataraghavan[1041] and Henneberg;[1042] the latter reference deals with various aspects of computer software for GC-MS in water analysis. A brief, generalised description of computer handling of GC-MS data is also given in ref. 1043. The data handling procedures (including library searching) employed by the U.S. Environmental Protection Agency in a large survey of wastewaters have been described by Shackelford *et al.*;[1044] an evaluation of the spectrum matching (library search) procedures employed has also been published recently.[1045] This evaluation showed that the reliability of compound identification by spectrum matching could be improved by using GC retention data, particularly when capillary columns were used [see Section 11.3.8(*a*)]. Reference has already been made to a publication giving mass spectra for 114 organic compounds on the U.S. Environmental Protection Agency's list of priority pollutants.[1031] A recent paper by Cornu[1046] describes a compilation of organic micropollutants mass spectra from European laboratories, and Groll[1047] has reported progress on a computerised collection of mass spectra of organic pollutants, again from laboratories in Europe.

Improved identification of the organic components of mixtures can also be obtained by using GC with *both* mass spectrometric *and* Fourier-transform i.r. spectrometric detection [see Section 11.4.3(*d*)], as demonstrated by Shafer *et al.*[1048]

(v) Selected ion monitoring When GC-MS is used to characterise a large number of unknown organic compounds in a mixture, it is usual for the spectrometer to be set to scan the entire mass range, repetitively throughout the chromatography. A 'reconstructed chromatogram' can then be prepared by plotting the total ion current against time, and the mass spectra obtained by the repeated scans provide information on the compounds present at each point.

However, when GC-MS is used to determine a small number of known compounds, or a single known compound, another procedure may be used, to give improved sensitivity and detection capability. This is selected (or 'specific') ion monitoring, and it involves controlling the spectrometer to 'switch' quickly between a suitable number of prominent ions, choosen appropriately for the compound(s) of interest, rather than to run a complete scan. By so doing, the 'wastage' of a major part of the total ion current involved in monitoring unproductive parts of the spectrum is reduced and improved sensitivity results. Between 1 and 20 ions are commonly used, and the terms 'single ion monitoring' and 'multiple ion monitoring' are also used to describe the approach.

Of course, as the number of ions monitored is reduced, the sensitivity increases but the chance of error caused by other substances (giving ions of the same m/z ratio as does the determinand) increases. The choice of ions to be used must, therefore, maintain an appropriate balance between these two factors, for the particular application.

(b) Principal Analytical Characteristics
MS, in combination with GC, has been most widely used in water analysis as a

means of identifying large numbers of organic compounds in general surveys (*e.g.*, ref. 1043). Quantification if attempted, has usually been crude (". . . at best only semi-quantitative"[1043]). In recent years, greater emphasis has been placed upon quantification of the results obtained in such surveys (see, for example, ref. 10 and literature cited therein). However, a combination of factors makes the undertaking a difficult one; these include (1) the wide range of compound types involved, (2) the large number of individual compounds which may be detected in a single sample and (3) the complexity of the pre-treatments required (*e.g.*, extraction, concentration and separation – see Section 11.2).

The first two of these factors are, of course, absent when a mass spectrometer is used as a sensitive and selective GC detector, for the determination of small numbers of individual compounds whose identity is known in advance. Moreover, as discussed above, selected ion monitoring can often be applied in such circumstances, thereby giving improved sensitivity and detection power. However, as in survey work, the accuracy of analytical results is likely to be greatly influenced by the pretreatment stages. The use of GC-MS for general survey work is far more common in water analysis, not least because our knowledge of the organic compounds in water is so limited and because, therefore, the ability of GC-MS to obtain information on many different compounds is its most attractive feature. Quantitative determination is accomplished by means of a calibration function (usually linear) relating the detector current for a given ion to the concentration of the parent compound in the sample.

(i) Calibration Calibration is most commonly performed using isotopically labelled analogues of the compounds of interest as internal standards added to the samples before extraction and concentration – see, for example, refs. 1043 and 1049. Deuterated materials are commonly used. Isotopically labelled standards have the obvious advantages of very similar behaviour to the corresponding compound of interest under extraction and concentration and of similar GC retention characteristics. The problem in survey work is, of course, the large number of compounds involved. To obtain accurate results, the entire analytical procedure should be the subject of calibration, using the compounds of interest or their deuterated analogues; this has – not surprisingly – proved impossible with general surveys involving maybe 100 or more different compounds, and recourse has been made to the expedient of assuming similarity of behaviour of similar compounds, throughout the extraction/GC-MS procedures. The cost of such expediency is, of course, results of poor accuracy in comparison with those produced by most 'routine' techniques.

(ii) Detection limits These will, of course, vary with the particular compound, with the nature of the pretreatments used and with the conditions employed for both the GC and MS stages. In GC-MS survey work, detection limits (for GC-amenable compounds, of course) are commonly in the sub-μg l^{-1} range, and single-ion monitoring can allow detection limits in the pg l^{-1} range to be attained.

(iii) Precision and bias Relatively little information is available on the analytical performance of extraction/GC-MS procedures for water analysis. Some

data relating to the U.S. Environmental Protection Agency's 'Master Analytical Scheme' have, however, been published,[1050,1051] and these show that both random and systematic errors can be very large when GC-MS is used for survey work – coefficients of variation typically greater than 10% (and frequently in excess of 30%), and recoveries of added spikes frequently less than 70%, even when the results were 'corrected for . . . historical recovery',[1050] and sometimes less than 50%, or greater than 150%.

The reader should not, however, assume that GC-MS cannot produce analytical results more accurate than these. The errors estimated for the U.S. E.P.A. Scheme, described above, reflect the difficulties of quantification in survey work, and the highly compromised calibration procedures which are used in consequence. The use of GC-MS for the determination of small numbers of compounds, under less compromised conditions, should allow much better accuracy to be achieved. Thus, for example, Burleson *et al.*[1052] determined amino acids in wastewaters by GC-MS (packed column, electron impact ionisation) and obtained recoveries of 69–100% (mean 90%).

(iv) General As noted above, GC-MS is a very powerful technique in water analysis, but it is correspondingly complex and expensive. A computerised system could cost up to £100 000 or more, and will require well trained and experienced staff for successful operation. There is now, however, a considerable amount of experience of the application of the technique in water analysis, both in particular organisations in possession of it and amongst instrument manufacturers, so the task of establishing a GC-MS capability is perhaps less daunting now than it was ten years ago. However, modifications and improvements (*e.g.*, ionisation by fast atom bombardment) are constantly being made and this inevitably complicates the selection of systems. Choosing an MS system is the subject of the last chapter of ref. 1025, which gives useful (though inevitably general) advice to the intending buyer.

(c) Applications to Water Analysis
As noted above, the principal application of GC-MS in water analysis has been to the identification of large numbers of organic compounds in water, particularly for the assessment of pollution by synthetic materials and/or of the production of potentially harmful substances during drinking water disinfection.

Reference has already been made to the U.S. E.P.A. 'Priority Pollutants' survey work, which involves some 114 specific organic compounds. Details of the use of GC-MS in this application may be found in various papers in the book edited by Keith,[10] which includes a summary of the results obtained.[1044] Several papers have referred to the cost effectiveness of GC-MS in comparison with GC alone for the determination of these 'Priority Pollutants'.[1053,1054] Recently, Sauter and co-workers have described the advantages of capillary GC-MS using fused silica columns for the determination of these compounds.[1055,1056]

Fielding *et al.*[1043] have reported the results of surveys of U.K. raw and potable waters by GC-MS. In the main survey on treated waters, some 324 compounds were identified in all (from 14 samples). Compounds regularly identified in different samples were often hydrocarbons and halogenated materials, but 41% of the 324 were identified only once.

Coleman *et al.*[1057] examined an extract (by reverse osmosis and ether extraction) from a large volume of drinking water, and identified approximately 460 compounds, including polynuclear aromatic compounds, polychlorinated biphenyls, amines, amides and other halogenated substances.

Other examples of GC-MS surveys of organic compounds in water include studies of the Hayashida, a highly polluted Japanese river,[1058] and of industrial products in the Niagara catchment.[1059] Both of these investigations again revealed the presence of large numbers of organic compounds.

These studies illustrate well the great power of the mass spectrometer, in combination with chromatographic separation, for the investigation of organic substances in water. The large numbers of compounds identified also show how large the problem is, when it is recalled that GC can deal with at best some 20% of the organic compounds likely to be present in water (see Section 11.3.8). It is hoped that the combination of HPLC-MS will allow information to be gained concerning involatile organics, in a manner analogous to that in which GC-MS has been employed in survey work, and research in this direction is already in progress (see, for example, refs. 1060–1062).

As noted above, the application of mass spectrometry in water analysis has mostly involved 'survey' work. Its application to smaller numbers of specified determinands is less common. Reference has already been made to the determination of amino acids in wastewaters reported by Burleson *et al.*;[1052] other studies include the identification of phenyl-urea and carbamate pesticides in HPLC fractions by off-line field desorption MS[1063] and (by GC-MS): the determination of polynuclear aromatic hydrocarbons,[1064] the determination of vinyl chloride,[1065] the determination of various volatile compounds,[1066] the study of faecal steroids in an estuary,[1067] the investigation of substances producing taste and odour problems in drinking water[1068] and the determination of polychlorinated biphenyls in surface water and drinking water.[1069]

(d) Mass Spectrometry for Inorganic Determinands

Although mass spectrometry has found its predominant application to water analysis in the identification and determination of organic compounds, it can also be applied to the determination of inorganic species (for which it has the advantage of simultaneous – or near-simultaneous – multi-element capability). In the past, spark source mass spectrometry (SSMS) has been used for this purpose, but inductively-coupled plasma mass spectrometry (ICP-MS) – a technique of recent development – shows considerable promise as a simpler alternative, offering better overall performance.

Spark source MS is dealt with in the book on trace analysis by MS edited by Ahearn[1070] and has been recently reviewed both in general terms by Ure,[1071] and in relation to water analysis by Buck and Hass.[1072] Typically, an SSMS system comprises a r.f. spark discharge source in a double-focusing spectrometer of conventional geometry, with electronic or photographic detection. Developments in source technology, to produce a narrower range of energies in the ions produced, could permit the use of the cheaper quadrupole spectrometers – an important possibility in view of the high cost (*ca.* £150 000) of present systems.[1071]

The principal problem in applying SSMS to water analysis is the need to incorporate the determinands in (or transfer them to a solid film on) the electrodes forming the source. This may be accomplished by freeze-drying the sample, to which graphite or carbon powder has been added. The dissolved and/or suspended matter is left on the carbon, from which electrodes are formed.[1072] Freeze-drying in the absence of carbon, followed by dry-ashing to remove organic material and mixing with carbon (to fabricate the electrodes) is also practised,[1073] the dry-ashing step being necessary to avoid interferences (from organic fragment ions) with 'dirty' samples. Chelation of the determinands of interest with oxine, and subsequent adsorption of the oxinates on carbon, is also possible;[1074] other techniques include direct application of liquid samples to the electrodes and the formation of electrodes from small volumes of sample by freezing*, but these are not very satisfactory.[1072]

Photographic detection and recording with quantification by microdensitometry is less accurate than is electrical detection; the advantage of a permanent record of the entire spectrum provided by the former has been overtaken by the use of computer data handling in association with electrical detection.

As in MS for organic determinands, calibration has been accomplished in a number of ways. The use of an internal standard in association with 'Relative Sensitivity Factors' (or 'Coefficients') for the determinands of interest, in a representative matrix, is relatively rapid but calibration by isotope dilution provides greater accuracy.[1071] Thus, Ure[1071] reports that standardisation using Relative Sensitivity Coefficients (obtained from analysis of geological 'Certified Reference Materials') allows an 'accuracy' (presumed 'bias' in the usage of this text – see Section 4.4.1) of ± 10–15% to be attained, whereas better than $\pm 5\%$ can be obtained using isotope dilution for standardisation.

The application of SSMS to water analysis is exemplified by the work of Crocker and Merritt[1073] and of Vanderborght and Van Grieken.[1074] The former used freeze-drying to concentrate water samples, followed by low temperature ashing to destroy organic matter prior to mass spectrometric analysis. Using normal scanning, the spectrum from mass number 238 to mass number 7 was obtained in 10 min, with individual peaks measured to a precision of $\pm 30\%$. (The precision could be improved to $\pm 5\%$ by electrostatic peak switching techniques). Results for lake water samples were obtained for over 30 elements, with detection limits of about 0.01 μg l^{-1}. The accuracy of the results was not firmly established, but the authors stated that accuracy within a factor of three was normal. The more recent work by Vanderborght and Van Grieken[1073] employed chelation with oxine and subsequent adsorption of the oxinates on activated carbon, rather than freeze-drying, for preconcentration, but again used low temperature ashing prior to mass spectrometric analysis. Detection limits of about 0.1 μg l^{-1} were reported (for 1 litre samples) with a precision of about 35% and 'errors' (presumably bias) of about 30%. These authors[1073] concluded that the technique, though not suitable for routine analysis, might be considered for multi-element survey work.

Although the relatively poor accuracy of the results obtained in these studies is probably a consequence, in large measure, of the compromised calibration procedures used, the technique suffers from the requirement for preconcentration

---

* The frozen sample is rendered conducting by condensing a conducting layer on the surface, or adding graphite before freezing.

(with its concomitant penalties in terms of sample throughput). For this reason, and on the grounds of capital cost, spark source MS cannot compete effectively with plasma emission spectrometry (see Section 11.4.2) for simultaneous multi-element analysis of waters. A more promising approach to the determination of inorganics in water by mass spectrometry is the use of ICP-MS systems.

The application of a plasma ion source with a mass spectrometer appears to have been first described by Gray.[1075] The use of an inductively coupled plasma in this manner was reported subsequently,[1076] and improved interfacing of the source and spectrometer led to detection limits in the range 0.1–1 μg l^{-1} being obtained for many elements, using conventional (pneumatic) nebulisation, without thermal desolvation [see Section 11.4.2(a)].[1077,1078] Very recently (May 1983), further improvements in the technique have led to even lower detection limits being attained – values of < 0.1 μg l^{-1} being reported for many trace elements.[1079] These developments, and the commercial availability of ICP-MS instrumentation, appear to have brought the technique to a point where its application to water analysis seems particularly attractive.

At the time of writing (August 1983), the authors know of no comprehensive evaluation of ICP-MS with respect to water analysis. The coupling of such an excellent source (the inductively-coupled plasma) with the reliable and fast-scanning quadrupole mass analyser would appear, however, to have produced an instrument capable of rapid multi-element analysis with detection limits superior to those of plasma/optical emission [see Section 11.4.2(b)], and (frequently) rivalling those of graphite furnace atomic absorption [see Section 11.3.5(c)]. Additionally, ICP-MS offers linear calibrations over many orders of magnitude of determinand concentrations, and relative standard deviations of a few percent should be readily attainable except at concentrations near detection limits.

Interference from organic matter in samples, noted as a potentially serious problem in spark source MS, would be expected to give much less difficulty (and perhaps virtually none) in ICP-MS, by virtue of the high temperatures attained in the source, which would be expected to ensure the complete destruction of organic substances. Some 'spectral' interferences are known – for example, the $^{19}OH^+$ ion can interfere seriously in the determination of fluorine and isotopes of two different elements can produce ions of the same m/z ratio. Interferences from fragment ions (such as $^{19}OH^+$) should normally be few in number, however, and those from isotopes of single elements are likely often to be small in magnitude. Nebulisation interferences, similar to those which have been observed in plasma/optical emission spectrometry, are to be expected but these are again likely to be of relatively minor importance with many types of water*. Increased interference could, however, result from the application of hydride generation or electrothermal atomisation techniques to reduce further the detection limits, but those reported using conventional pneumatic nebulisation are already very low in relation to many, if not the majority, of water analysis requirements.

Only a detailed evaluation of this new technique will allow the above predictions to be tested, and its advantages and disadvantages to be properly assessed.

---

* Moreover, the low detection limits reported may allow samples to be diluted in many applications, thus reducing such interference effects as may occur.

However, the potential of the technique should ensure that such an evaluation is not long delayed.

Other examples of the application of MS to inorganic determinands include the work of Schulten,[1080] who determined caesium and thallium by field desorption MS [see Section 11.4.1(*a*)], and the determination of silver, cadmium, lead, zinc and palladium at ng l^{-1} levels in sea-water, by Rosman *et al.*[1081] using thermal ionisation isotope dilution MS. These techniques seem unlikely, however, to offer the potential for water analysis in general of the plasma/MS approach.

(e) Determination of Isotopes
The mass numbers of isotopes of the same element differ and MS can, therefore, be used to determine the isotopic composition of elements. Sometimes the relative abundances of the different isotopes are of interest in themselves, *e.g.*, in providing clues to the origin of substances in water. However, the ability to determine different isotopes also acts as the basis of isotope dilution methods [see Section 11.4.5(*d*)]. The use of deuteriated organic substances as internal standards in GC-MS, referred to in (*b*) above, is in essence an example, though the term isotope dilution does not appear to be used to describe the procedure. The technique is also applied to inorganic analysis by MS, the term 'isotope dilution' being used explicitly in such cases.

Isotope dilution techniques have the potential to form the basis of 'reference methods of analysis'. Such methods, although they usually have little *direct* application to routine analysis, can be important in allowing estimates to be obtained of the errors (especially bias) of analytical results obtained using routine methods. This subject is considered in Section 5.1.2, which deals with the use of reference methods for estimating bias, but it is raised here to draw attention to the potential value of MS, in combination with isotope dilution, as a basis for such methods.

11.4.2 Plasma Source Optical Emission Spectrometry

Plasma spectrometry cannot yet be described as a generally important measurement technique, since its use in water analysis is, at present, restricted to relatively few laboratories in most countries in which it is applied. Nevertheless, the technique has such great potential value to water analysis that it merits more detailed consideration than many others dealt with under the heading 'Other Measurement Techniques'. Moreover, it is likely that plasma spectrometry will be adopted in many more laboratories during the lifetime of this volume – it is a measure of its rapid development and acceptance that, although now regarded as one of the most important future techniques in water analysis, no reference was made to it in the first edition of this work published about ten years ago.

(a) General Basis of the Technique
In common with other variants of atomic emission spectrometry (*e.g.*, flame emission – Section 11.3.4) plasma emission spectrometry involves introduction of the sample into an excitation source, in which the determinand is converted into free atoms. These are then raised, by the energy of the source, into an excited state and

then emit radiation at a characteristic wavelength. The intensity of the emission at that wavelength is proportional to the concentration of the determinand, and quantification involves isolation and measurement of the radiation(s) of interest in a spectrometer. The basic equations of plasma emission spectrometry are, therefore, those of flame emission (see Section 11.3.4).

The excitation source in plasma emission spectrometry is a 'plasma' – that is, a luminescent region in a gas in which a sizeable proportion of the atoms are ionised. Although the design of some plasma sources gives them the appearance of a flame, it should be emphasised that flames and plasmas, despite such occasional superficial similarities, are fundamentally different. Flames are sustained by the energy liberated by the chemical reactions involved in combustion, whereas plasmas (of the type under consideration here) are sustained by an input of electrical power.

The use of plasmas as excitation sources offers improved analytical performance, compared with conventional flame, arc and spark sources, because of a favourable combination of a number of important operating characteristics; these include high temperature, operation in inert atmospheres, excellent stability and good signal-to-background ratios.[1082] Because plasmas produce strong emission from many elements, they are particularly suited to multi-element analysis;[1083] moreover, they are optically 'thin' sources, so that negligible self-absorption occurs and large linear ranges result.

(i) Plasma types   Three main types of plasma have been used in analysis, namely the microwave-excited plasma (MEP), the d.c. plasma jet (DCP) and the radio-frequency inductively-coupled plasma (ILP). The differences between them relate principally to the manner in which electrical energy is supplied to maintain the plasma.

In microwave plasmas (see reviews by Skogerboe and Coleman[1084] and by Zander and Hieftje[1085]), the energy is supplied by a high frequency electric field. Most commonly, energy coupling between the field and the plasma is accomplished by induction, the plasma being maintained within a thin cylindrical tube inside a resonant cavity which is supplied with power from a high frequency generator which provides a standing wave pattern. Capacitive coupling has also been used, in which the microwave from the generator is conducted along an electrode, at the tip of which the plasma is maintained.

In the d.c. plasma jet a d.c. arc of high current density produces the plasma in the argon gas. The simplest arrangement is the two-electrode system shown in Figure 11.14, but a three-electrode configuration offering improved performance is now used on commercial DCP instruments.[1087] Refs. 1087–1089 should be consulted for further details.

In r.f. inductively-coupled systems, the plasma is maintained by an energy input from an oscillating magnetic field produced by a r.f. alternating current in an induction coil. Many types of apparatus have been produced, and the description given (based on ref. 1086) applies to a design commonly used in commercial plasma emission systems. The whole device upon which the plasma is maintained is called a 'plasma torch', and Figure 11.15 shows the typical arrangement. An induction coil, placed around the torch, provides the magnetic field. The torch itself consists of three quartz tubes, arranged concentrically, up which flow three separate gas

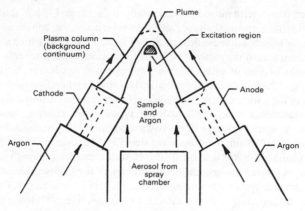

Figure 11.14 *A two-electrode d.c. plasma jet* (after ref. 1086)

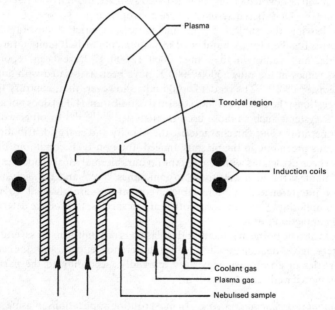

Figure 11.15 *An inductively-coupled plasma torch* (a schematic after ref. 1086)

streams (most commonly, of argon). The coolant gas is used to stabilise the plasma – often being introduced tangentially to aid this function[1090] – and to keep the torch itself cool.

The general shape of the plasma produced is as shown in Figure 11.15, but the detailed form varies with the gas flow rates and frequency of magnetic field oscillation. With regard to the latter, frequencies in the range 5–60 MHz have usually been used*, with output power usually in the range 1–5 kW. The base of the plasma tends to a toroidal shape, an effect which can be enhanced by suitable adjustment of nebulisation gas flow rate (see Figure 11.15) and which aids ingress of nebulised sample into the plasma.[1090] For further information see refs. 1091–1097.

Direct nebulisation of water samples into microwave plasmas leads to plasma instability, and de-solvation of the nebulised sample has therefore been applied with such sources (see, for example, ref. 1098). However, desolvation may give rise to 'matrix' interference effects[1099] and other types of interference have also been reported to be more prevalent in microwave plasmas than in inductively-coupled systems (see ref. 1098 and literature cited therein). Microwave plasma emission has therefore found little application in direct water analysis, though it has found application as an element-specific detection system for GC [see Section 11.3.8(a)].

A spectrometer system employing the d.c. plasma jet source (DCP) is commercially available, and has been evaluated for water analysis.[1100] However, by far the most widely used plasma source – both in water analysis and in analysis generally – is the r.f. inductively-coupled type (ICP). For this reason, the remainder of this section will be devoted mainly to ICP-based systems, though brief reference will also be made to DCP-based systems where appropriate.

Before leaving the subject of plasma sources, a brief discussion of their temperatures is called for. A number of measurements of ICP temperatures have been made, and values in the range 2000–12000 K have been reported.[1101] Typically, values in the range 5000–9000 K have been associated with analytical ICP systems.[1082,1090,1102] The reader should note, however, that, contrary to some early suggestions, the concept of local thermal equilibrium (LTE) does not apply to argon ICP systems such as those used in analysis.[1101,1103,1104] In other words, no single temperature value can characterise the velocity and energy distributions and the reactions prevailing in the plasma. Indeed, the non-LTE conditions of argon ICPs have been associated with some of the favourable analytical properties of such plasmas, namely the high sensitivity to ionic lines[1101,1103] and the lack of major ionisation interferences.[1103] Metastable argon atoms are thought to provide ionisation buffering[1103,1105] and, with other argon species, to produce determinand ions by Penning ionisation.[1103,1106]

It should also be borne in mind that the plasma is not uniform. As several groups of workers have demonstrated,[1107–1111] interference effects and the emission characteristics of an element may vary markedly, according to the part of the plasma being viewed.

(ii) Sample introduction systems   In most routine applications, plasma spectrometry (with ICP or DCP sources) has been employed with pneumatic nebulisation. Both cross-flow (or right-angle) and concentric (or co-axial) nebulisers have been used. The former (see, for example, refs. 1112, 1113) are less prone to clogging problems with samples of relatively high dissolved solids content. Babington-type nebulisers (see, for example, ref. 1114–1116) have also been used for samples of high dissolved solids content, or containing suspended material.

Thermal desolvation of the aerosol has been applied with pneumatic nebulisation and gave improved detection limits for a number of elements in ICP systems – by

* 27–56 MHz in currently available commercial instruments, see ref. 178.

factors of about two (at best) in one case[1117] and five in another.[1118] However, the approach does not appear to have been very widely applied in water analysis by ICP spectrometry – the increased propensity to interferences resulting from the process of desolvation[1118] may have been a factor in this limited usage.

Because of the limited efficiency of pneumatic nebulisers, especially at the low nebulising gas flow rates necessary with ICP sources,[1117,1119] there has been considerable interest in the use of ultrasonic nebulisers. The higher nebulisation efficiencies which such systems offer (and for which they are chosen) demand that thermal desolvation of the aerosol be practised when ultrasonic nebulisation is used, to avoid the entry of excessive amounts of water into the plasma.

Both continuous-flow and batch-type ultrasonic nebulisers have been investigated, but the latter are more inconvenient to use and are prone to memory effects.[1117] Comparisons of pneumatic nebulisation and ultrasonic nebulisation with thermal desolvation have shown[1117,1120] that the use of the latter gives a five- to ten-fold (or greater) improvement in detection limits, for the majority of elements examined. However, ultrasonic nebulisation – or, more specifically, the accompanying desolvation step – has been shown to introduce interference effects from the sample matrix.[1118,1121,1122,1123] (Desolvation interference has also been encountered with a microwave excited plasma[1124].) Probably for this reason, and because ultrasonic nebulisers have tended to be less robust than pneumatic nebulisers in routine use, the direct application of plasma spectrometry (ICP and DCP) to water analysis has usually been undertaken with pneumatic nebulisation without desolvation (*e.g.*, references[1116,1124–1128]). However, ultrasonic nebulisation/thermal desolvation has been used for the ICP determination of trace metals in sea-water after ion-exchange preconcentration,[1129] which provides a simultaneous separation from major matric components.

A number of workers have investigated the use of electrothermal atomisation for sample introduction into an ICP[1130–1133] and others have introduced dried samples on a graphite rod.[1134,1135] Although the use of such techniques can permit low detection limits to be attained using very small volumes of sample (typically 10 μl), interference effects can be serious[1131,1132] and these approaches have not yet found wide application in water analysis by plasma spectrometry.

Detection limits, by direct pneumatic nebulisation, are often adequate for many analysis applications (see below, and refs. 1126 and 1136). However, this is not true for such elements as arsenic, antimony and selenium[1126,1136] and – as in AAS (see Section 11.3.5) – hydride generation techniques have been widely investigated and used as a means of obtaining lower detection limits.[1137–1146] Although very low detection limits can be attained for hydride-forming elements (see below), interferences – most notably from certain metal ions – may be serious[1138,1139,1142,1145] and pretreatment necessary to separate the determinands from the sample matrix.[1138]

(iii) Spectrometers A major advantage of plasma emission spectrometry over AAS is its potential for simultaneous multi-element analysis. It is not surprising, therefore, that commercial exploitation of the technique has resulted in the production of many spectrometers capable of truly simultaneous multi-element analysis (see Figure 11.16), commonly for 50–60 different elements. Systems of this

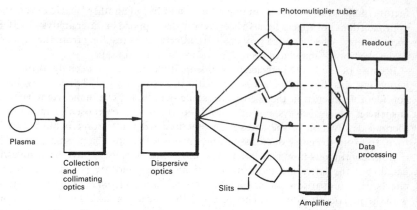

Figure 11.16 *Schematic of a simultaneous (or polychromator) system for plasma emission spectrometry* (after ref. 1083, with modifications)

type are most suitable for applications in which the same, large group of elements is to be determined in large numbers of samples, on a continuing basis (see, for example, ref. 1026). The disadvantage of this configuration is its relative inflexibility – the user must specify the wavelengths of interest, and subsequent changes in requirements (*e.g.*, alternative or additional determinands) will, if they can be accommodated, require modifications to the spectrometer by the manufacturer. For this reason, sequential, scanning spectrometer systems (see Figure 11.17) are also commercially available.

The principle of sequential instruments is that the dispersive element is driven rapidly by a stepper motor, under computer control, from one position to another. At each position, emission at a wavelength of interest impinges on the photomultip-

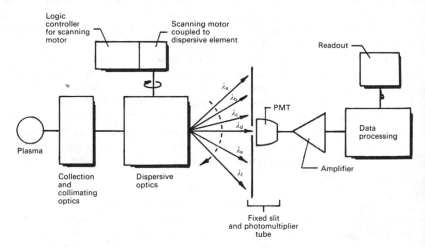

Figure 11.17 *Schematic of a sequential (or scanning monochromator) system for plasma emission spectrometry* (after ref. 1083, with modifications)

lier tube and its intensity is recorded. The motor drive and recording systems are operator-programmable so that different sets of determinands can be selected at will. Because the dispersive element is driven rapidly ('slewed') from one wavelength to another, with a slower movement ('scanning') near the new position, accurately to locate the wavelength of interest, the term 'slewed scan' is sometimes used to describe this type of system. Clearly, time is lost – relative to a simultaneous instrument – in the transfers from wavelength to wavelength, and in the need for separate periods of measurement for each determinand, but the advantage is gained of flexibility with respect to choice of wavelengths of measurement.

To reduce the time taken to change from one wavelength to another, and at the same time achieve very reliable wavelength location, some commercially available scanning monochromator systems employ a fixed grating and photomultiplier tubes which can be rapidly driven from place to place behind an array of closely spaced exit slits.[178] Accurate location of a new wavelength is said to be attained within less than 1–5 s.[178]

Given the advantages, in different applications, of both simultaneous and sequential spectrometer systems, a number of manufacturers offer instruments which incorporate both types of system with a single plasma source. Such instruments offer the advantages of simultaneous measurement in routine, high-throughput applications together with the benefits of sequential measurement for analytical development and research – though inevitably at a cost substantially higher than those of single-spectrometer types.

Most commercially available plasma spectrometers employ a diffraction grating as the dispersive element. The resolving power of the spectrometer – its ability to separate lines of similar wavelength – is of importance in determining which spectral lines can be used without spectral interference. This is of particular relevance because optical emission spectra contain many spectral lines – often closely spaced – when the sample contains many elements capable of excitation in the particular source used; it has already been noted that plasmas are highly efficient excitation sources. Moreover, when diffraction gratings are used for dispersion, each wavelength is brought to a number of foci, one for each spectrum order. It follows that an element giving an emission at a wavelength X in the first-order spectrum may give rise to a spectral interference on elements whose emissions are at (or near) 2X and 3X. The relative intensities of the spectral orders depend on the nature of the diffraction grating, and are influenced by the 'blaze-angle' (the angle of diffraction from the grating at which the radiation tends to be concentrated).

Certain commercial spectrometers, now available with both DCP and ICP sources, do not employ a conventional diffraction grating, but have an echelle grating as the principal dispersive element. Echelle gratings are gratings which produce spectra of high order (100 or more, in contrast to the first or second orders typically exploited with conventional gratings); because dispersion is directly related to spectral order, instruments based on echelle gratings can provide higher dispersion, and therefore resolution, than can instruments of comparable size based on diffraction gratings of conventional type. A prism is used as a secondary dispersive element with the echelle grating to reduce the problem of overlap of spectral orders which would otherwise be very great with an echelle system.

(iv) Data handling The generation of large amounts of data by plasma emission spectrometers and the need to perform data manipulation both for calibration and for interference correction (see below) place considerable demands upon data-handling and transmission systems. Moreover, real-time output of results is required so that problems can be rapidly disclosed and rectified. These requirements can be met in a variety of ways, using local or central computing facilities or a combination of both. Whatever approach is adopted, it is important to recognise the magnitude of the requirements when acquisition of a plasma spectrometer is planned, as has been emphasised by Norton and Orpwood.[1126] Commercial spectrometer systems offer both the hardware and software likely to be required, however, although the need to expend a suitable effort on operator training must be borne in mind.

(b) Principal Analytical Characteristics
As with all complex instrumental techniques of analysis, the analytical performance of plasma emission spectrometry depends upon the detailed design of the instrument and the manner in which it is used, as well as upon the type of source employed. Because it is a complex but relatively new technique, which has undergone rapid development, the effects of instrumental and other factors are not always fully understood. The reviews of plasma emission spectrometry cited in this section should be of value to those considering acquisition of such equipment, but the pace of development and change in commercial systems is such that there can be no substitute for a careful assessment of needs and a detailed examination of available instrumentation (including tests on typical samples) prior to purchase.

Despite the difficulties engendered by the differences in the capabilities of different instruments, an attempt will be made here concisely to summarise the performance of plasma emission spectrometry.

(i) Calibration As has been noted, the basic equations of flame emission spectrometry (see Section 11.3.4) apply. Typically, calibrations are linear over four or five orders of magnitude and this property constitutes a major advantage of the technique.[1082,1086,1093] Detailed calibration procedures vary quite widely, but the stability and linearity of calibrations have permitted the initial establishment of calibration curves, with subsequent daily drift correction using only two standards, with satisfactory results.[1126] Other workers have used a range of procedures, from daily single-point calibrations[1100] to duplicate analyses of three to six standards on each analytical occasion.[1136]

(ii) Limits of detection Table 11.13 shows Criteria of Detection obtained by Norton and Orpwood[1126] (according to the definition given in Section 8.3.1) with a simultaneous ICP system and pneumatic nebulisation. Also included for comparison are 'detection limits' reported in other studies, for an ICP system employing ultrasonic nebulisation/thermal desolvation and for a three-electrode d.c. plasma instrument with pneumatic nebulisation. These latter values represent twice the standard deviation of the background (ICP with ultrasonic nebulisation, ref. 1136) and twice the standard deviation of the intercept of calibration curves (DCP, ref. 1100). Although, therefore, the bases of the values in the table differ slightly, the

TABLE 11.13

Detection Limits* for Plasma Spectrometry with Direct Nebulisation

Detection limits μg l^{-1}

Element	ICP/PN† (Ref. 1126)	ICP/UND† (Ref. 1136)	DCP/PN† (Ref. 1100)
Ag	3	—	—
Al	18	0.4	10
B	1	—	5
Ba	3	—	20
Be	0.1	—	10
Bi	39	—	—
Ca	17	—	200
Cd	2	0.07	10
Co	5	0.1	—
Cr	2	0.08	5
Cu	2	0.04	5
Fe	2	0.5	—
K	35	—	40
Li	2	—	—
Mg	1	—	130
Mn	1	0.01	10
Mo	29	0.3	10
Na	35	—	400
Ni	12	0.3	10
Pb	51	1.0	10
Si	13	—	10
Sr	<1	—	100
Ti	2	—	—
V	2	0.09	—
Zn	3	0.1	—
Zr	4	—	—

* (The figures quoted are effectively 'Criteria of Detection' in the terms of Section 8.3.1 – see text for details).

† ICP/PN   Inductively-coupled plasma with pneumatic nebulisation.

ICP/UND   Inductively-coupled plasma with ultrasonic nebulisation and desolvation.

DCP/PN   3-electrode d.c. plasma with pneumatic nebulisation.

figures are adequate for a broad comparison of these three main variants of plasma emission. It must also be emphasised that all the data presented refer to operation in the simultaneous mode; consequently, the analytical conditions employed were a compromise for a set of determinands, rather than optimal for individual elements. (The issue of 'compromised' or 'optimised' conditions must always be borne in mind when the results of multi-element analysis are being considered, and in particular when performance data are being assessed prior to instrument purchase. A failure to ensure that appropriate measures of performance are evaluated could have undesirable consequences, if a simultaneous spectrometer were to be acquired on the basis of performance under optimal conditions for individual determinands).

It will be seen from Table 11.13 that the detection capability usually follows the order: ICP (Ultrasonic/Desolvation) > ICP (Pneumatic) > DCP (Pneumatic), in

TABLE 11.14

Detection Limits for a Number of Elements by Plasma Emission Spectrometry (ICP) with Pneumatic Nebulisation and with Hydride Generation

		Detection limits ($\mu g\ l^{-1}$)			
	Pneumatic nebulisation	*Hydride generation*			
Element	(Ref. 1138)	(Ref. 1138)	(Ref. 1139)	(Ref. 1142)	(Ref. 1144)
As	440	0.8	—	—	0.02
Bi	46	0.8	—	—	0.03
Ge	—	—	0.3	—	—
Pb	51*	—	—	1.0	—
Sb	150	1.0	—	—	0.02
Se	430	0.8	—	—	0.03
Sn	150*	—	0.2	—	0.05
Te	17	1.0	—	—	—

* Values from ref. 1126.

accord with earlier remarks. The great potential value of plasma emission spectrometry in water analysis will also be apparent from the individual values, particularly when it is borne in mind that the data refer to direct, simultaneous analysis without any preconcentration.

It has been noted that hydride generation has been employed to improve the detection limits for a number of elements by plasma spectrometry. Table 11.14 provides a comparison of the detection limits reported for direct pneumatic nebulisation with those attained using the hydride technique, in all cases with an ICP source. It is clear that the use of hydride generation allows markedly lower detection limits to be attained, and that the detection limits with direct, pneumatic nebulisation are (by contrast) too high for most water analysis applications. As shown by Thompson et al.,[1146] simultaneous generation of the hydrides of a number of elements, to take advantage of the simultaneous multi-element capability of plasma spectrometry, is possible.

(iii) Precision   Following a detailed study of the performance of an ICP system employing pneumatic nebulisation (operating under compromise conditions for the determination of some 26 elements), Norton and Orpwood[1126] reported relative total standard deviations in the range 1–10%. When results for concentrations less than 20 times the observed Criterion of Detection are excluded (together with those for barium, for which a minor system fault was suspected), the range is 1–5%, with the majority of values less than 3%. Broadly similar data have been reported by other workers[1125,1136] using ICP systems (including ultrasonic nebulisation/desolvation conditions[1136]), and the tentative conclusion of Ranson[1086] that the technique offers similar precision to flame atomic absorption is correct.

With regard to the precision attainable using d.c. plasmas, Greenfield et al.[1093] reported that such systems gave poorer precision than ICPs, and the relative standard deviations reported[1100] for a commercial DCP/echelle system also appear to be somewhat inferior to those observed with ICP systems[1125,1126] – though not markedly so.

(iv) Selectivity Interference effects in atomic emission spectrometry may be classified under the following broad headings:

(1) Nebulisation/transport interferences.
(2) Inter-element effects within the excitation source.
(3) Spectral interferences.

Under (1), it is to be expected that factors which lead to changes in the rate of sample uptake to the nebuliser, or to changes in nebulisation efficiency, will lead to interference. Thus, for example, it has been reported[1126] that a decrease in sensitivity of up to 9% accompanied a decrease in the pH of solutions from 2.03 to 0.57, using an ICP with a concentric pneumatic nebuliser. Because solution uptake was found to be essentially independent of pH, the effect was tentatively attributed to viscosity or surface tension differences affecting the aerosol droplet size. The effect can be readily overcome by matching the pH of samples and standards.[1126] Again, deposition of salt particles in the nebuliser[1147] can be a source of interference and various measures can be used to reduce or prevent such build-up – an automatic rinse of the nebuliser tip[1148] and/or humidification of the argon gas supply to the nebuliser. The use of a pump to supply sample to the nebuliser at a constant rate can also ameliorate transport interferences.

Inadequate washout of a previous sample from the nebuliser/spray chamber system can cause memory effects.[1126,1147] Thus, for example, Norton and Orpwood[1126] observed that 50–100 s were required to give 99.7% removal of a multi-element standard of high concentration, though subsequent developments in instrumentation have improved this aspect of performance. The problem appears to be caused mainly by the slow dispersal of the 'fog' in the spray chamber.[1147]

It has already been noted that the use of aerosol desolvation may give rise to matrix interferences. Many studies have shown the existence of such effects,[1099,1118,1121,1122] and the work of Goulden and Anthony[1118] on a pneumatic nebuliser/desolvation system has shown that the magnitude of desolvation interferences under normal conditions of water analysis may be considerable. At concentrations of calcium, magnesium and sodium of 72 μg^{-1}, 16 μg l^{-1} and 20 mg l^{-1}, respectively, present in the solution simultaneously, a 20% depression of plasma sensitivity was observed. Matrix-matching of standards and samples, or standard additions calibration, was necessary in routine analysis.[1118]

With regard to inter-element effects within the excitation source, it is widely agreed that the ICP is remarkably free from interferences of this type (see, for example, refs. 1123, 1149, 1150, 1151). Although interference effects arising from refractory compound formation[1152] and from ionisation[1153] can be observed, they are negligibly small under normal conditions of plasma operation and fresh water matrix composition. In d.c. plasma systems, however, inter-element effects are more serious, as suggested by Boumans.[1154] Urasa and Skogerboe[1155] reported that the presence of a readily ionised element, lithium, caused enhancement of the neutral atom emission of elements with ionisation potentials less than about 7.5 eV. This was attributed to the normal effect of determinand ionisation suppression, but later studies[1100] revealed that both the neutral atom *and* ion emissions were enhanced by the presence of caesium or lithium, and a mechanism involving temperature increase in the presence of these latter elements was considered more likely. To reduce the effects in routine analysis, caesium (at 1000 mg l^{-1}) was added to samples to provide a more uniform matrix and indium was used as an internal standard.[1100] Yet further studies of the interference effects led to explanations in terms of an increase in the metastable argon concentration within the plasma[1156] and improved penetration of determinand atoms into the plasma, coupled with increase of temperature round the edges of the latter.[1157] Regardless of the exact mechanism of the enhanced signals in the presence of readily ionised elements, it is clear that inter-element interferences are greater in d.c. plasmas than in ICPs, and require control in routine analysis.[1100] Spectral interference in plasma emission spectrometry can arise in a number of ways, including overlap of emission lines, stray light in the spectrometer and shifts in the background emission from ion-electron recombination. Direct overlap of emission lines can often be avoided by careful choice of analytical wavelengths, and tables of emission lines in ICP spectrometry have been produced to aid such selection.[1158,1159] The enhanced resolution of echelle grating spectrometers (see above) is also advantageous in respect of such interferences, especially from complex matrices.[1160,1161]

Stray light ("The unwanted radiation that reaches the detector in unintended ways"[1162]) represents an important potential source of bias in plasma spectrometry. The intense emission from the major cations in waters (*e.g.*, calcium and magnesium) may be scattered in the spectrometer and reach the detectors for other determinands (in simultaneous systems), or the sole detector when emission from another element is being measured (in sequential systems), thereby causing positive bias.

The potential importance of spectral interferences (in which is included stray light) is illustrated by the work of Norton and Orpwood,[1126] who observed that, for example, 250 mg Ca l^{-1} produced an effect equivalent to 124 μg Mo l^{-1} and 220 μg Si l^{-1}, and 50 mg Mg l^{-1} produced an effect of 31 μg Pb l^{-1} and 59 μg Cr l^{-1}.

Procedures for reducing the effects of stray light have been reviewed by Larson *et al.*,[1162] and include:

(1) Use of good quality optical components.
(2) Attention to spectrometer design, including making all internal non-optical surfaces with an optically flat black coating, and deploying baffles to trap unwanted radiation.
(3) Judicious choice of detectors.
(4) The mounting of narrow bandpass filters in front of detectors.
(5) Correction for residual effects, by either the 'on-peak' or 'off-peak' background correction techniques.[1163]

Background correction in ICP systems has been discussed by Sobel and Dahlquist.[1163] In the so-called 'off-peak' approach, the background emission is measured adjacent to the emission line of interest, on one or both sides of it. This simply involves a slight displacement of the wavelength of measurement (*e.g.*, by rotating the grating on a sequential instrument or by moving the primary slit of a simultaneous instrument). In the 'on-peak' system, effects of one element on the determination of another are stored in the system's computer in the form of mathematical relationships relating the effect and the concentration of the interfering element. When, for samples, the latter is measured, the value obtained is used, with the stored relationship, to calculate a correction for the interference. 'On-peak' correction has the advantages[1163] of being able to correct for spectral overlap and of rapidity, since only one measurement is required at each analytical wavelength (once the relationships have been established), but can only correct for background shifts (such as in the ion–electron recombination continuum) when the latter are caused by a determinable element. 'Off-peak' correction can correct for such shifts even when the interferent cannot be determined, but is slower, cannot correct for spectral overlap and is itself subject to a greater risk of error from unanticipated lines at off-peak wavelengths.[1163] Both types of correction may be applied together, and both may equally well be applied to DCP systems as to ICP systems.

Provided that the determinand is in solution, the high temperatures of plasma sources should ensure that different forms of it are recovered equally. For drinking water, acidification of samples to pH 1 (for preservation) has been a sufficient pretreatment.[1126]

Analysis of reference water samples has been used to check the accuracy achieved with both ICP[1136] and DCP[1100] systems, with good results in both cases. Recovery tests have also been conducted[1126] with a simultaneous ICP system, using soft upland water (SUW), a hard mixed surface and ground water (MSGW) and a hard ground water (HGW). Data were obtained for 24 elements for SUW and MSGW, and for 21 elements for HGW, and recoveries lay in the following ranges: SUW, 90.6–112.2%; MSGW, 94.0–106.8%; HGW, 90.0–99.8%. Thus, though there appeared to be evidence of matrix interferences giving low recoveries from the hard ground water, the effects were not large.

With respect to hydride generation/plasma emission spectrometry, it has already been noted that serious interference effects may occur, particularly from trace metals. Again, special precautions are usually necessary to ensure that different forms of the elements concerned (*e.g.*, different oxidation states) give equal responses. These problems are discussed in detail in Section 11.3.5(*d*) on hydride generation/AAS, and will not be re-examined here; it will merely be noted that their solution is likely to be more difficult if, as is desirable in plasma spectrometry, simultaneous generation of a number of different hydrides is attempted.

Pre-concentration techniques can, of course, be applied to the determination of other elements by plasma spectrometry. Thus, for example, evaporation,[1164] ion exchange[1129] and solvent extraction[1165–1167] have all been used. Such techniques can, of course, introduce interference problems and difficulty in recovering all forms of the determinands.

(v) General Plasma emission spectrometry has now been quite widely used in water analysis, though its high capital cost (in excess of £60 000 for a typical ICP system) has usually restricted its use to central and research laboratories. The technique obviously possesses great potential for water analysis, combining good detection power for many elements of interest with relative freedom from serious interference effects and (simultaneous) multi-element capability. ICPs appear to offer significant advantages over DCPs, including better detection limits, precision and selectivity.

Greenfield[1095] has compared the detection limits attainable by ICP emission spectrometry (with pneumatic nebulisation) and by flame AAS [see Section 11.3.5(b)], for 59 elements. For the vast majority (53 of the 59), lower limits can be attained by the plasma technique, typically by one to two orders of magnitude. Coupled with the other advantages of plasma emission noted above, this makes the technique a strong competitor with flame atomic absorption for certain applications – *e.g.*, drinking water analysis – at least in larger laboratories in which the high sample throughput justifies the much greater cost (*ca.* four-fold) of a plasma instrument. It should be noted, however, that the detection limits attainable by plasma spectrometry with pneumatic nebulisation may be inadequate, for a number of elements, in certain environmental monitoring applications. Graphite furnace atomic absorption [see Section 11.3.5(c)] offers detection limits much lower than those attainable by ICP spectrometry with pneumatic nebulisation (typically by two to three orders of magnitude[1095]), and may therefore be preferable in such applications. The development of electrothermal atomisation/plasma spectrometry and/or probe introduction systems for ICPs (see above) may change this position, however.

Many different plasma emission spectrometers are available commercially, offering a range of capabilities in respect of speed and convenience of analysis, in particular. The reliability of a commercial ICP system has been rated similar to that of an atomic absorption spectrometer[1126], and it seems likely that the use of the technique in water analysis will grow, unless curtailed by developments in ICP-MS [see Section 11.4.1(d)]. The latter technique appears to offer many of the advantages of ICP emission (essentially simultaneous multi-element capability, relative freedom from interferences and large linear range), together with detection limits comparable, in many cases, to those attainable by graphite furnace AAS, albeit at much higher cost. The commercial availability of plasma fluorescence spectrometry [see Section 11.4.3(b)] may also provide an alternative to plasma emission in water analysis.

Apart from alternative sample introduction systems to pneumatic nebulisation, future developments in ICP emission spectrometry seem likely to include the commercial production of instruments using smaller torches to reduce the consumption of argon, which can cost several thousand pounds *per annum* with current instruments. Work in this area is already well advanced.[1168]

Useful sources of information concerning plasma emission spectrometry are the 'Annual Reports on Analytical Atomic Spectroscopy' published by the Royal Society of Chemistry (*e.g.* ref. 178), the *ICP Information Newsletter*[1169] and the recent text by Thompson and Walsh.[1091]

11.4.3 Other Spectrometric Techniques

(a) Emission Spectrometry using Arc or Spark Sources

This technique has never been widely used in water analysis, requiring as it does specialised equipment and expertise. Since the first edition of this book was published (in 1974), its role in water analysis has declined markedly – virtually to the point of extinction – because arc and spark discharges in emission spectrometry have been superseded (for solution analysis) by plasma sources. Plasma source optical emission spectrometry is dealt with in the previous section, and only a very brief account is given here of emission spectrometry using arcs and sparks, in keeping with its essentially obsolete status in water analysis.

A schematic diagram of the instrumentation required is shown in Figure 11.18. The sample is introduced into the source (an arc or high-voltage spark electrical discharge), in which its constituents are vapourised and atomised and in which the resulting atoms are then excited to emit radiation at wavelengths characteristic of each element.

The radiation from the source is focused by a lens so that it passes through the narrow entrance slit of a spectrometer, in which it is dispersed into its constituent wavelengths by the diffracting effect of a diffraction grating or prism, the intensities of the dispersed radiations being measured by photomultiplier tubes. The signals from the latter are subsequently amplified, processed and read out. [As an alternative to photoelectric detection and recording, the dispersed radiation can be focused onto a photographic plate so that, after development, a spectrogram is produced. This consists of thin black lines – images of the entrance slit – at positions along the plate corresponding to the emitted wavelengths, and therefore characteristic of the excited elements. The blackness of each of the lines is a measure of the amount of the corresponding element, and can be assessed with a photoelectric microphotometer (densitometer) to provide quantitative analysis.]

The technique is capable of determining 60–70 elements with varying degrees of sensitivity. Although methods for the direct analysis of solutions are available, the limits of detection attainable ($0.1–10$ mg l^{-1}) are inadequate for most purposes and pre-concentration procedures have, therefore, normally been applied when analysing natural waters; with such procedures, detection limits between 0.1 and 100 μg l^{-1} can be attained. As in AAS (see Section 11.3.5), hydride generation and cold vapour procedures have been used to improve limits of detection for certain elements.[1170–1172]

Although the technique has been applied to the determination of one or a few elements at a time, the multi-element capability of the technique (especially before the introduction of plasma source spectrometers) represented its major value in water analysis. When used for multi-element quantitative analysis, optimum precision cannot usually be achieved for all determinands and relative standard deviations of 5–20% have been observed. Additionally, the use of pre-concentration procedures may contribute to an important degree to the random error of analytical results.

Systematic error may arise if the general compositions of samples and calibrating standards differ appreciably and, as in the case of random error, bias can also arise from the pre-concentration procedures used – from interference effects or from a failure to recover all forms of a determinand.

The speed of analysis can be very rapid (excluding the pre-concentration step) when photoelectric detection is used, but photographic detection and densitometry is inevitably much slower – although it has the advantage of providing a permanent record of the spectrum, which can be re-examined subsequently if required.

The technique may be regarded as one of the less accurate in use in water analysis, but its multi-element capability found for it a limited application in water analysis prior to the advent of plasma-source optical emission spectrometry (see previous Section), which has essentially displaced it entirely. Very few reports of the use of

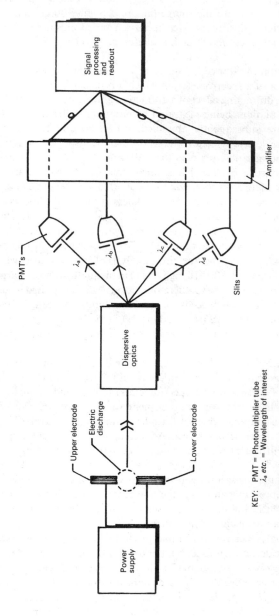

KEY: PMT = Photomultiplier tube
λ_a etc. = Wavelength of interest

Figure 11.18 *Schematic arrangement of emission spectrometer with arc or spark source*

emission spectrometry using arc and spark sources have appeared in the 'Annual Reports on Analytical Atomic Spectroscopy,[1173] and virtually none from Europe or North America in the last 5–10 years.

Texts on the technique are given in refs. 1174–1176. A recent review[1177] of emission spectrometric multi-element analysis of water and related materials includes discussion of its principles and practice. The papers by Kopp and Kroner[1178] and by Steiner et al.[1179] exemplify its use for simultaneous multi-element analysis.

(b) Atomic Fluorescence Spectrometry

This technique is closely related to atomic emission spectrometry and AAS. In essence, the sample is vapourised and atomised, and the cloud of atoms produced is irradiated with a beam of light from a lamp emitting the spectrum of the element to be determined. Part of this radiation is absorbed by the atoms of the determinand, and a fraction of the absorbed energy is then re-emitted as fluorescent radiation at wavelengths characteristic of the determinand. The intensity of this fluorescent radiation is measured, and is directly proportional to the concentration of the determinand in the sample – provided that the concentration is not too great. Flames, non-flame atomisers [see Section 11.3.5(c)] and, more recently, ICP (see Section 11.4.2) have all been used for sample vaporisation and atomisation. Radiation sources used include lasers, xenon arc lamps, and, most commonly, HCLs and EDLs. When line sources are used, a relatively low resolution monochromator or filter – or even a non-dispersive configuration – can be employed. Conversely, with a continuum source high resolution is required.

A schematic diagram of an atomic fluorescence spectrometer is shown in Figure 11.19. The fluorescent radiation is separated from the primary light beam by measuring the former at right angles to the latter. Measurement of direct emission from the flame or other atom source can be avoided by modulating the source of the primary, exciting radiation and using a frequency-selective amplifier. Thus, the instrumentation required for single-element atomic fluorescence spectrometry (AFS) is very similar to that required for atomic absorption, and most commercial atomic absorption spectrometers could be adapted for fluorescence work. Although some commercial atomic absorption instruments have been fitted with fluorescence facilities, no such instrument is specifically identified in a recent listing of commercially available AAS equipment.[1173] Until recently, no commercially available multi-element AFS systems were available (though a 6-element system had been marketed in the early 1970s); however, a 12-channel plasma-based instrument has recently been developed[1180,1181] (see below) and a non-dispersive atomic fluorescence spectrometer for determining hydride-forming elements and mercury has recently been produced commercially in China.[1182]

AFS has been the subject of reviews by Omenetto and Winefordner,[1184] by Ullman[1185] and by Van Loon.[1186] The technique is also dealt with in texts by Sychra et al.,[1187] Kirkbright and Sargent,[1188] Thompson and Reynolds,[1189] Grove[1190] and Dean and Rains.[1191]

Despite a number of potential advantages over AAS (including greater sensitivity, increased linear range of calibration and simultaneous multi-element capability), the prediction in the first edition of this book that AFS was likely to find

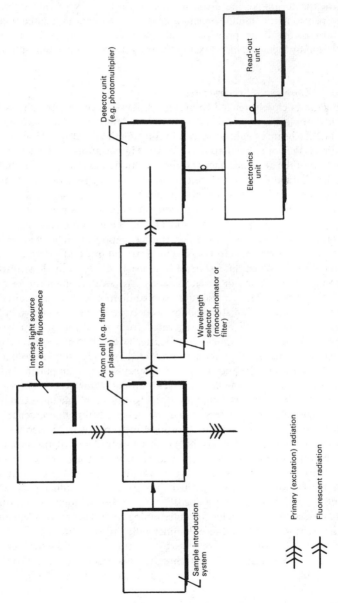

Figure 11.19 *Schematic arrangement of an atomic fluorescence spectrometer*

increased application in water analysis has barely proved correct. Application of the technique has been restricted largely to the determination of mercury,[1192-1197] for which it offers lower detection limits (by a factor of *ca.* 10[1197]) than does AAS. (Other uses have included the determination of zinc[1198] and magnesium[1199] in boiler feedwaters, and of antimony in waste water[1200]). Reasons for this restricted usage probably include the limited development of AFS instrumentation, the widespread acceptance of AAS for routine analysis (see Section 11.3.5) and the introduction of plasma emission spectrometry for simultaneous multi-element analysis (see Section 11.4.2).

The position may change with the introduction of a commercially available, simultaneous multi-element atomic fluorescence spectrometer using, as atom cell, an ICP (with HCLs as sources of excitation radiation), based on the work of Demers and Allemand.[1180] For most of the elements examined by these workers, detection limits by ICP-AFS were similar to those attainable with ICP-OES [see Section 11.4.2(*b*)]. However, detection limits for a number of refractory elements were markedly higher by AFS. The major advantages claimed for plasma AFS over plasma emission spectrometry are the greater freedom of the former technique from spectral and background interferences.[1180] To the present authors' knowledge, however, no detailed assessment of the relative merits of the two techniques with respect to water analysis has yet been reported, although the application of multi-element AFS to water analysis has been demonstrated by analysis of a reference water sample.[1181] Both plasma fluorescence and plasma emission can allow the simultaneous, routine determination of a number of elements, though it might prove simpler to change the measured elements in plasma AFS than with some direct-reading plasma emission systems.

(c) X-Ray Fluorescence Spectrometry

In some respects, the general characteristics of this technique are similar to those of emission spectrography [see Section 11.4.3(*a*)] and plasma emission spectrometry [see Section 11.4.2]. Thus, a single sample can be used for the simultaneous (or rapid sequential) determination of a large number of elements, but the technique requires specialised instrumentation which is not available in the majority of routine water analysis laboratories.

The principle of the technique, and the main instrumental components, are shown in Figure 11.20. The sample is irradiated with a beam of *X*-rays* which are absorbed by its constituent atoms. The atoms so excited then emit fluorescent *X*-rays at wavelengths (energies) characteristic of the elements present, these rays then being separated according to wavelength ('wavelength dispersion') or energy ('energy dispersion'), and their intensities measured. Scanning the spectrum of fluorescent radiation thus indicates which elements are present and calibration of the system with known amounts of the elements permits quantitative analysis using the measured intensities at the characteristic wavelengths (energies).

A number of texts dealing specifically with *X*-ray fluorescence spectrometry (XRF) are available (*e.g.*, refs. 1201–1205), and the technique is reviewed biennially in *Analytical Chemistry*.[1206] Application of the technique to the determination of trace elements in general has been reviewed by Gonsior and Roth;[1207] Dzubay has

---

* In some closely related techniques, irradiation with protons or α-particles is used.

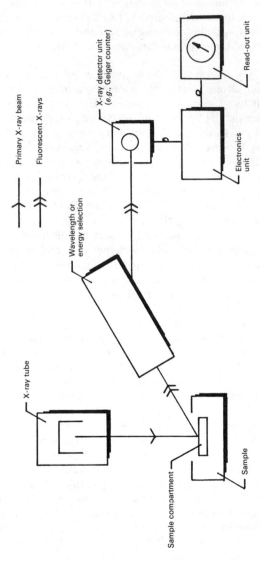

Figure 11.20 *Schematic arrangement of an X-ray fluorescence spectrometer*

edited a collection of papers dealing with the use of XRF in environmental analysis[1208] and application to water analysis is the subject of a comprehensive review by Leyden.[1209] A review of multi-element techniques for drinking water analysis by Ranson[1201] also includes a section on X-ray fluorescence.

Commercially available instrumentation provides options on the natures of the three main functions – X-ray generation, signal selection and signal measurement. As noted above, the characteristic fluorescent X-rays may be separated according to wavelength or energy; in wavelength-dispersive systems the fluorescent radiation from the sample impinges upon an analysing crystal which diffracts the radiation into its constituent wavelengths, the intensities of which are then measured by the detectors – gas flow proportional counters for X-rays of lower energy and scintillation counters for X-rays of higher energy. In energy-dispersive instruments, the separation of the characteristic fluorescent radiations is not achieved by a distinct dispersive element, but by the detector system. The detector is a lithium-drifted silicon semiconductor detector, which yields a signal proportional to the energy of the incident fluorescent X-ray photon. In combination with a suitable electronic data-handling system, this detector provides a spectrum of fluorescent X-ray intensity as a function of photon energy, the latter being characteristic of the element fluorescing.

With regard to the source of the excitation X-rays, in wavelength-dispersive instruments this is normally an X-ray tube in which electrons produced at a heated tungsten filament cathode are accelerated under a high potential (typically up to 100 kV) to bombard a metal anode. Excitation of the metal atoms of the anode produces the X-radiation, which is then focused on the sample. In energy-dispersive systems, radioactive sources have been used as an alternative to X-ray tubes – their photon yield is much lower, however, and this restricts their application to systems of the energy-dispersive type (wavelength-dispersive spectrometers have inadequate sensitivity).

The above description is necessarily extremely simplified, and the reader desiring further information on the details of XRF spectrometer design and construction is referred to the above texts. It is important, however, to note that the type and detailed design of an XRF instrument may have a considerable impact upon the analytical performance. The following brief summary of performance can, therefore, serve only as a general guide. Again, the texts and reviews cited above should be consulted for fuller information.

Using commercially available instrumentation, the technique may be applied to the determination of elements having atomic numbers greater than about 11. For elements lighter than this, the sensitivities and detection limits are so poor that, even with the use of pre-concentration procedures, it is doubtful that they could be determined at their typical concentrations in waters by XRF.[1210] For many of the elements to which XRF is applicable, detection limits (absolute) of 0.01–1 μg are attainable. The detection limits achieved depend, *inter alia*, upon the thickness of the sample and values as low as 1 ng may be attainable for some elements if very thin samples are employed (*e.g.*, precipitates on filters).[1209]

Calibration normally involves the analysis of appropriate standards with each batch of analyses. When (as is usual in water analysis – see below) a pre-concentration technique is employed, it is recommended that at least one standard solution be taken through the entire procedure in each batch, to check the recovery through the pre-treatment. Calibrations are typically linear over several orders of magnitude, though deviations from linearity can arise from the presence of high concentrations of certain elements and from the presentation of thick samples to the spectrometer.[1210]

Although direct analysis of solutions by XRF is possible, it is inconvenient and, most importantly, detection limits are, at best, about 1 mg l^{-1}, too high to be widely useful in water analysis. [Although the use of ion-induced X-ray spectrometry would permit lower detection limits to be attained for many elements, the improvements (commonly about ten-fold[1211]) would be insufficient to allow direct analysis in all applications. Moreover, the required instrumentation is costly and not normally available outside certain research laboratories.] Because of the inadequacy of the detection limits attainable by direct XRF analysis of solutions, pre-concentration procedures are now almost invariably employed when the technique is applied to water analysis. It is beyond the scope of this text to examine these pre-concentration procedures in detail; the reader is referred to Section 11.2.2 for a general discussion of such procedures, and to the review by Leyden[1209] for information on their use with XRF in water analysis. It will, however, be noted here that a variety of different techniques has been used – *e.g.*, solvent extraction,[1212] carbon adsorption,[1213] ion exchange[1214-1220] and precipitation/co-precipitation.[1221-1226] The possibility of using electrodeposition has been noted by Leyden,[1209] and evaporation has also been investigated[1227,1228] but may be subject to problems of determinand volatility, adsorption losses during evaporation, non-homogeneous residues (adversely affecting the XRF analysis) and spectral interferences in the XRF stage from co-evaporated major components.[1209,1210] Thus, Leyden[1209] has concluded that evaporation has little promise as a pre-concentration technique for water analysis by XRF.

Although some of these pre-concentration techniques are quite complex, others – for example, passage of the water through a filter loaded with ion exchanger, the latter then being analysed by XRF[1214,1219] – are fairly simple. By combining XRF with pre-concentration, detection limits within the range 0.01–1 μg l^{-1} have been attained for many elements (see, for example, refs. 1212, 1213, 1217, 1218, 1221, 1224, 1226).

Amongst the reported disadvantages of energy-dispersive XRF is the lower detection power, compared with wavelength-dispersive systems.[1210] However, the differences in the detection limits between the two types of system can be relatively small if the energy-dispersive systems are equipped with secondary sources of exciting radiation, and other features to improve counting efficiency.[1209,1211] In any case, as Ranson[1210] has remarked, the need to employ pre-concentration procedures with both types of instrument reduces the practical importance of any differences in detection power. The inferior resolution attainable using energy-dispersive systems is likely to be of greater concern – see below.

X-Ray fluorescence spectrometry is itself capable of producing results of excellent precision, relative standard deviations of 1% often being attainable with amounts of determinand substantially in excess of the detection limit.[1202] Inevitably, however, precision may be markedly impaired by the application of a pre-concentration step; thus, relative standard deviations of about 1–10% have been observed when XRF is coupled with pre-concentration techniques for water analysis (see, for example, refs. 1212, 1213, 1220, 1222, 1224).

Following Ranson,[1210] we may distinguish four potentially important sources of systematic error in XRF: absorption and enhancement interferences, particle size and surface effects, effects due to the chemical form of the determinand and spectral interference. Because the use of XRF in water analysis requires that pre-concentration be applied, the sample placed in the spectrometer is usually in the form of a thin film or layer of constant thickness, so the first of these sources of systematic error is not normally a problem. For the same reason, particle size and surface effects do not usually cause serious difficulty, though they may do so when pre-concentration by evaporation is used. Different oxidation states of an element may give rise to small differences in the wavelengths of the fluorescent X-rays produced. Such differences are not regarded[1210] as an important problem, though the ability of wavelength-dispersive instruments to resolve them has been noted as a potential advantage of such systems over those employing energy-dispersion.[1229]

Spectral interferences present more serious problems, particularly with energy-dispersive instruments, the resolution of which is markedly inferior to that of wavelength-dispersive systems.[1229] Although the pre-treatment procedures which are necessary to achieve adequately small detection limits may eliminate many interferences, some may still remain.[1221] Moreover, although spectral overlap in energy-dispersive instruments can be corrected, these processes may themselves introduce considerable error – and, in some cases, may not be feasible.[1229]

Before leaving the subject of systematic error, it should be noted that the pre-concentration procedure may itself be a source of bias; for example, because it fails to recover all forms of the determinand or because it is subject to interference effects.

With respect to time for analysis, this will obviously depend markedly on the pre-concentration technique employed. Additionally, the time taken for the XRF stage will depend upon the instrumentation; with simultaneous wavelength-dispersive or energy dispersive systems, a sample can, in many cases, be analysed in about 5 min or less[1209] though longer times have also been used.[1210] Multi-element analysis using wavelength-scanning instruments is, of course, a much slower process, requiring counting times of about 2–15 min *per element* in many cases.[1210]

Refs. 1212–1228 give some examples of XRF procedures for the analysis of water; many other reports of this application of the technique may be found in the aforementioned reviews,[1207–1210] and in the general reviews of water analysis published annually in the *Journal of the Water Pollution Control Federation* (*e.g.*, ref. 1230) and, in odd-numbered years, in *Analytical Chemistry* (*e.g.*, ref. 1231). Although virtually all reported applications of the technique involve the analysis of discrete samples, Ho and Lin[1219] have recently described an XRF system for the automatic analysis of wastewater streams. This employs ion-exchange paper for collection of the determinands of interest, analysis of which is subsequently performed with an energy-dispersive instrument.

Examination of the literature suggests that X-ray fluorescence is much more widely used in water analysis than is emission spectrography [see Section 11.4.3(*a*)]. However, of the well established multi-element techniques, ICP-OES (see Section 11.4.2) is better suited to the analysis of solutions than is XRF – and likely, therefore, to be preferred in most potential survey applications in which it can be used without prior separation and/or concentration of the determinands. (It should be noted, however, that XRF may be advantageous in the analysis of water-related solids, such as sewage sludge or sediment.)

(d) Infrared Spectrometry – General use for the Identification and Determination of Materials

This technique is yet another which has found relatively limited use in routine water analysis laboratories. Although not uncommonly used for the determination of oil in water, it has otherwise mainly been applied in research and central laboratories, for the identification of organic substances.

The principle of the technique is exactly the same as that of solution spectrophotometry. Radiation from an i.r. source is passed through the sample, and absorption occurs at certain wavelengths which are characteristic of particular chemical groups (*e.g.*, $-CO_2H$, $-NH_2$, $-CH_3$, $-OH$). In conventional i.r. spectrometry, the wavelength of the radiation passing through the sample is varied by scanning with a prism or grating monochromator, and the degree of absorption is plotted continuously with a chart recorder so that the absorption spectrum of the sample is produced for the range 2–15 μm*. In recent years, however, instruments employing interferometry and mathematical transformation techniques to process the signals have become quite widely used in general analysis. These instruments – known as Fourier Transform infrared (FT-i.r.) spectrometers – offer much improved signal-to-noise ratios and resolution, though at a cost substantially higher than conventional systems. Although it is usually preferable for samples to be dissolved in a suitable solvent for i.r. analysis, solids and gases can be handled if necessary. Indeed, FT-i.r. spectrometers are now being employed as detectors in GC (see below).

The presence of absorption peaks at particular wavelengths in the i.r. spectrum indicates the chemical groupings present, and the degree of absorption depends on the concentrations of those groups. (The technique is of value in identifying molecular groupings – absorptions characteristic of elements occur at much shorter wavelengths.) In practice, 'fingerprinting' techniques are often used, *i.e.*, the spectrum of the 'unknown' is compared with spectra of known compounds, commonly obtained from collections of reference spectra. Although aqueous solutions can be examined by i.r. spectrometry, they show strong absorption (by water itself) at certain wavelengths, and if samples in liquid form are to be used it is preferable that organic solvents are employed which do not absorb radiation appreciably at the wavelengths of interest.

General texts on i.r. spectrometry are given in refs. 1232–1234; ref. 1235 deals with FT-i.r. spectrometry, and ref. 1236 describes the application of FT-i.r. spectrometry as a detection system in capillary gas chromatography. Infrared spectrometry is reviewed (in even-numbered years) in *Analytical Chemistry*,[1237] and its application to water analysis, though limited, can be followed in the reviews of that subject published (in odd-numbered years) in the same journal.

As noted above, i.r. spectrometry has found its primary application in water

---

* Instruments giving smaller wavelength ranges are sometimes used and may be adequate for certain problems.

analysis in the identification and determination of hydrocarbons in water. In addition to being used for the quantitative determination of oil in water (following its extraction by a suitable solvent),[1239-1242] i.r. spectrometry has also been used for the characterisation and identification of oils in pollution incidents – see, for example, refs. 1243–1249. In this latter application, pattern recognition and related mathematical techniques have been employed to make the most effective use possible of the i.r. spectra in the characterisation and identification of the polluting oil, see refs. 1246, 1248 and 1249. I.r. spectrometry has also been applied to the measurement of surfactants[1250-1253] and pesticides.[1254-1256]

As noted above, FT-i.r. spectrometry is now being investigated as an on-line detector for GC – and, indeed, for HPLC. In addition to the individual references given here, the reader is referred to the special issue of the *Journal of Chromatographic Science* given in ref. 1257, in which a number of papers appeared dealing with GC/FT-i.r. and HPLC/FT-i.r.

The use of GC/FT-i.r. for water analysis has been investigated recently by Shafer *et al.*,[1258] who concluded that the technique might usefully complement GC-MS (see Section 11.4.1) because it showed greatest selectivity to polar compounds, whereas GC-MS showed greatest selectivity to non-polar compounds. A later study by Gurka, Laska and Titus[1259] determined the minimum identifiable quantities of fifty-five toxic substances by GC/FT-i.r., and concluded that the technique showed greatest sensitivity for aliphatic and aromatic compounds containing carbonyl or other oxygen-containing functional groups, and poorest sensitivity for alkyl halides and aromatic hydrocarbons. The minimum identifiable quantities of the substances examined lay in the range 0.5–10 μg, approximately, though further reductions by a factor of at least 5 were thought feasible.

The two studies considered above were undertaken or funded by the United States Environmental Protection Agency; details of a GC/FT-i.r. system established by that Agency have been published by Azarraga and Potter.[1260] It remains to be seen exactly how valuable the use of GC/FT-i.r., as an adjunct to GC-MS, will be in water analysis, but the decreasing cost of Fourier Transform spectrometers and the ability to connect both an FT-i.r. spectrometer and a mass spectrometer on-line to the same chromatographic column[1261] are both favourable to increased use of the technique. The necessary computer procedures for chromatogram reconstruction and library searching have also been developed.[1262-1264]

Some applications – albeit rather limited – of GC/FT-i.r. in water analysis have already been reported. Keith *et al.*[1265] employed the technique to confirm the identity of 'Alachlor' (a chloroacetanilide derivative) in a study of organic compounds in drinking water, and Worley *et al.*[1266] have also employed the technique for the separation and identification of a number of α-chloroacetanilides and related compounds at low μg l^{-1} levels in water.

(e) Infrared Spectrometry – Use for the Determination of Carbon Dioxide, Total Organic Carbon and Total Oxygen Demand

Simple, non-dispersive filter instruments can be used for the sensitive determination of carbon dioxide in the gas phase, because of its strong absorption at a wavelength of 4.4 μm. Carbon monoxide can also be determined by this means (absorption at 4.6 and 4.7 μm). For carbon dioxide, detection limits of the order of 10 ng can be simply and rapidly achieved, and commercial instrumentation is available. This approach has been used to determine total carbon dioxide in water after acidification and extraction of the gas from the sample, and to determine the partial pressure of free carbon dioxide in the water by i.r. measurement of the carbon dioxide content of inert carrier gas (*e.g.*, nitrogen or air) in equilibrium with the liquid phase. Both these types of determination have been reviewed by Skirrow.[1267]

An extremely useful application of the non-dispersive i.r. measurement of carbon dioxide is the determination of the total organic carbon content of waters. The basic approach is to convert the organic carbon into carbon dioxide, which is then measured by the i.r. technique. (Other measurement techniques are also used – most notably flame ionisation detection of methane from the reduction of carbon dioxide [see Section 11.3.8(*i*)] and conductimetric measurement of released carbon dioxide, following dissolution of the latter in water [see Section 11.3.9(*b*)].

The use of i.r. spectrometry for the determination of total organic carbon (TOC) is discussed in refs. 1268, 48 and 1269. Ref. 48, a publication in an authoritative collection of recommended methods for water analysis, includes details of commercial equipment and its performance, and ref. 1269 considers the determination of TOC in sea-water (more difficult than in fresh waters, because of the lower ratio of total organic carbon to total inorganic carbon in typical sea-water samples, see below).

The oxidation of the organic carbon to carbon dioxide can be undertaken in several different ways: persulphate oxidation,[1270,1271] combustion in the presence of a catalyst[1268,1272–1274] or u.v. photo-oxidation.[1271,1275,1276] In order to determine the total organic carbon content, the total inorganic carbon must first be removed by acidification/inert gas purging, or it must be separately determined. The latter is accomplished in some combustion systems[1272] by performing a low temperature combustion simultaneously with a high temperature combustion (for total carbon) and determining TOC as the difference. The use of a purging step to remove total inorganic carbon will cause loss of volatile organic carbon, but by suitable elaboration of the procedures, this fraction can also be determined if desired.[48]

The limits of detection achieved using the various types of system are tabulated in ref. 48. They may be summarised as follows: persulphate oxidation, $10–20~\mu g\,l^{-1}$; combustion, $50–1000~\mu g\,l^{-1}$; u.v. photo-oxidation, $1–10~\mu g\,l^{-1}$. It should be noted, however, that bias from the TOC content of the blank water (see Sections 5.4.4 and 10.7) may be an important limiting factor in the determination.[48] With regard to the precision of analytical results, data in refs. 48 and 1274 suggest standard deviations of a few percent or better at concentrations distant from the detection limit. For unfiltered samples, the precision may be markedly worse (5–10%) because of the difficulties of representatively sampling particulate matter.[1274]

Non-dispersive i.r. measurement is also employed in an instrumental procedure for determining the total oxygen demand of water samples.[1277] In this approach, carbon dioxide is used in place of oxygen in the combustion step, and the oxidation of carbon, nitrogen and other compounds leads to the formation of carbon monoxide, which is determined by i.r. absorption. The instrumentation is available commercially, and a limit of detection of 8 mg $O_2\,l^{-1}$ has been quoted.[48]

(f) Molecular Luminescence Spectrometry

It has already been seen that in atomic fluorescence spectrometry [Section 11.4.3(b)] atoms are excited to higher energy states by absorption of radiation, and the excited atoms then emit fluorescent radiation at characteristic wavelengths. Molecules may demonstrate the same effect when irradiated and this forms the basis of some of the techniques considered in this section. Molecular luminescence may be caused by processes other than absorption of radiation, e.g., chemiluminescence can be induced by the absorption of energy released by a chemical reaction. A number of analytical methods involving chemiluminescence have been applied to water analysis, and these are considered later in this section.

Two types of photoluminescence* can occur, and are termed fluorescence and phosphorescence. There are important differences between them (see below), but, for simplicity, they are treated as one in this section. The technique of molecular-photoluminescence spectrometry (hereafter referred to more simply as 'fluorimetry') stands in the same relationship to spectrophotometry (Section 11.3.3) as atomic fluorescence does to atomic absorption. As in the case of atomic fluorescence, fluorimetry – despite possessing some potential advantages over spectrophotometry – has, until recently, found limited application in water analysis. The first edition of this book, in 1974, noted signs that interest in fluorimetry was increasing; indeed, the technique is now finding much greater usage, particularly as a basis for detection in HPLC (see below).

A schematic representation of the technique is shown in Figure 11.21. A high

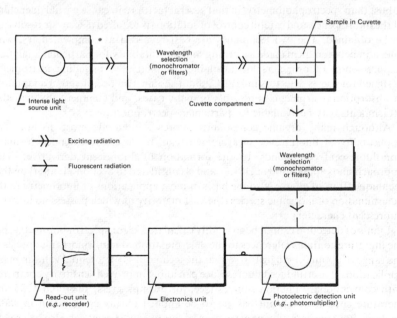

Figure 11.21 *Schematic arrangement of a fluorimeter*

* Luminescence induced by irradiation of the sample.

intensity continuum source (*e.g.*, a xenon-arc lamp) is often used for the primary exciting beam of radiation, but the use of lasers is increasing. The intensity of the fluorescence is directly proportional to the intensity of the exciting radiation, but it is also affected by the wavelength of the latter, which has therefore to be selected from the continuum either by optical filters or by a monochromator. The primary radiation passes through the sample solution, held in a suitable cuvette, and the fluorescent radiation is usually observed at right angles to the primary beam. The intensity of the fluorescent radiation is a maximum at a characteristic wavelength which is selected by means of a second filter or monochromator. Finally, the intensity is measured photoelectrically and displayed on a suitable read-out device. At suitably small concentrations of the determinand, the intensity of the fluorescent radiation is directly proportional to concentration.

The reader seeking further details of fluorimetry is referred to one of the many texts (or part-texts) available (see for example, refs. 1278–1283), to the annual reviews published in *Analytical Chemistry* (see ref. 1284) and, in relation to water analysis, to the recent review by Seitz.[1285] The use of fluorescence detection in HPLC (see below) has been described in some detail in Section 11.3.8(*d*).

Calibration curves in fluorimetry are commonly linear over a wider range of concentration than those in spectrophotometry. However, the main advantages claimed for fluorimetry over spectrophotometry are the greater sensitivity of the former (typically, it has been claimed,[1285] by factors of 10–1000) and the possibility of greater selectivity conferred by the ability to select both the wavelengths of the exciting and fluorescent radiation employed. However, fluorimetry is probably less robust than spectrophotometry, in that several factors can cause partial quenching of the fluorescence, and careful control of solutions is required if accurate results are to be obtained. The two techniques are best regarded as complementary, with selection between them made according to their suitability for particular analytical requirements. Fluorimeters are available commercially, including instruments designed for use as detectors in HPLC, but it should also be noted that adaptation of absorption instruments is possible in many cases, and commercially available attachments may be available for particular spectrophotometers.

Although many organic compounds fluoresce intensely, most do not. The applicability of fluorescence measurement can, however, be extended to many non-fluorescent compounds, by the formation of fluorescent derivatives. This approach finds application in HPLC, and is considered in Section 11.3.8(*d*) on that technique. It is, of course, also the basis of most applications of fluorimetry to the determination of inorganic species (the vast majority of which possess no inherent fluorescent character).

Limits of detection cannot be so readily calculated as for spectrophotometry, but the literature contains references to the determination of many inorganic species in the range of about $0.1–10 \mu g \, l^{-1}$, so that the technique is of potential value in many applications. Laser-induced fluorescence can allow improved sensitivity compared with conventional radiation sources (see, for example, refs. 1284 and 1285 and literature cited therein), but has yet to be applied to any great extent in water analysis. Pre-concentration or extraction of fluorescent species can also be used to lower detection limits, in analogous fashion to spectrophotometry.

With respect to organic species, the primary application of fluorimetry is in combination with HPLC [see Section 11.3.8(*d*)]. The detection limits attained depend markedly upon the nature of the chromatographic separation and upon any pre-concentration applied, but amounts of determinand as low as $10^{-4}-10^{-7}$ μg may be detectable when distributed evenly over 100 μl of column eluent (a fairly typical volume). Corresponding values for u.v. spectrophotometric detection are $10^{-4}-10^{-5}$ μg, demonstrating the greater sensitivity and detection power of the fluorescence technique. The reader is referred to Section 11.3.8(*d*) for further details. Determinands detected by HPLC/fluorimetry include amino acids[812] and (a common application) polynuclear aromatic hydrocarbons.[777,780] (It should also be noted that fluorescence detection is used with TLC [see Section 11.3.8(*h*)] for the detection and measurement of separated species. Determination of polynuclear aromatic hydrocarbons is again a common application.)[1286,1287]

Another fluorescence-based approach to detection in HPLC is cerate oxidimetry. This technique has not been very widely applied in water analysis, and is not considered in detail in Sections 11.3.8(*d*) and 11.3.8(*e*). A short description only is given here – for further detail, the review by Seitz[1285] should be consulted.

Cerate oxidimetry involves measurement of the fluorescence from Ce^{III}, which is a product of the reduction of Ce^{IV} by reducing substances (eluting from the chromatography column, for example). The technique may be applied to the wide range of organic substances amenable to oxidation by Ce^{IV}. Sensitivity can rival that of u.v. spectrophotometric detection.[1285] Cerate oxidimetry has been applied to the chromatographic determination of phenols[1288] and sugars[1289] in waters.

Although most applications of fluorimetry to the determination of organic compounds in water involve the prior use of HPLC, this is not always the case. Thus, the fluorescence of natural waters is well correlated with the TOC content[1290] and humic substances may be determined directly by fluorimetry[1291] (and indeed, may interfere with other fluorimetric determinations – see for example, ref. 1292). Fluorescence has also been used to characterise and determine oil (from spillages) in water – see, for example, refs. 1293–1295. Chlorophyll and related plant pigments[1296] have also been determined by fluorimetry (after extraction from phytoplankton), and *in vivo* fluorescence has been determined as a potential measure of plant biomass.[1297] Similarly, the determination of DNA (deoxyribonucleic acid) in micro-organisms filtered from water has been undertaken by a fluorimetric method.[1298]

Recent papers describing the use of fluorimetry for the determination of inorganic species include the following: aluminium,[1299,1300] selenium,[1301,1302] mercury,[1303] ammonia,[1304] nitrate[1305] and uranyl ion.[1306] Laser-induced fluorescence has been applied to the determination of uranium[1307,1308] and aluminium.[1309] An indirect method for the determination of hypochlorous acid and hypochlorite, based on measurement of their reduction of fluorescein fluorescence, has been reported by Williams and Robertson.[1310]

Chemiluminescence, as an analytical technique, has been reviewed by Seitz and Neary;[1311] application of the technique to water analysis has also been included in the review of luminescence techniques by Seitz.[1285] A number of methods for the determination of inorganic species have been developed. These include: chromium(III),[1312,1313] chromium(VI),[1314] cobalt(II),[1315,1316] iron(II)[1317] and hypochlorite.[1318,1319] Detection limits in the range 5 ng l^{-1}–5 μg l^{-1} were reported.

Interference from other components of water samples may be a problem with a number of these methods, however. Thus, for example, calcium in particular is a potentially important interference in the determination of cobalt[1315,1316] and hypochlorite (when determined by its reaction with hydrogen peroxide[1318]). To overcome the problem in the case of cobalt, separation by chelating ion-exchange has been employed;[1315] in the case of hypochlorite, the use of the more sensitive luminol reaction[1319] permits the use of sample dilution, which is reported to overcome most interference problems.

When a chemiluminescence reaction derives from a living system, it is referred to as an example of *bioluminescence*. Adenosine triphosphate (ATP), the principal 'energy store' in living organisms, has been determined in micro-organisms filtered from water by exploiting its essential role in the reactions responsible for firefly bioluminescence.[1320] Instrumentation adapted for this specific determination is available commercially.

11.4.4 Catalytic/Kinetic Procedures

In the first edition of this book, it was noted that neither of these two types of procedure was then widely used in water analysis, but that interest in them was increasing. Although that increased interest has led to the development of a number of methods in the intervening decade, the use of catalytic and kinetic procedures in water analysis remains limited.

With the exception of flow injection analysis* (characterised by Stewart[1321] in a recent review as "... a kinetic measurement system in which the system is normally not completely mixed and the radial and axial sample concentration profiles are time-dependent functions."), kinetic techniques have – as predicted in the first edition – found much less application than catalytic techniques in water analysis. The latter will therefore be emphasised here.

Kinetics in analysis (including catalysis) is the subject of texts by Mark and Rechnitz[1322] and by Yatsimirskii;[1323] the subject is also reviewed biennially in *Analytical Chemistry* – for the most recent review, by Mottola and Mark, see reference.[1324] Other recent reviews are those on catalytic procedures by Mueller[1325] and on catalytic and differential rate procedures† by Mottola.[1326] (Reviews on flow injection analysis are given in Section 12.2).

Certain analytical procedures employ enzymes, which may be regarded as a special class of catalysts. We have chosen to include enzyme procedures in this treatment of catalytic procedures, but to consider non-enzymatic and enzymatic procedures separately.

(i) Non-enzyme procedures   In most catalytic procedures applied to water the determinand acts as a catalyst or inhibitor of a chemical reaction, experimental conditions being so arranged that the rate of the reaction is governed by the concentration of the determinand. Usually, the reaction rate is measured by determining the concentration of one of the reactants or products after a fixed time,

---

* Flow injection analysis is considered in Section 12.2 on automated procedures.

† Differential rate procedures are those in which different determinands are separately determined, in the same analysis, by exploiting reaction rate differences.[1326]

but other approaches are possible and may be preferred in particular cases.[1322,1323,1325,1326] Thus, procedures usually consist of adding reagents to the sample and making a measurement of some concentration at a later time. In principle, any analytical measuring technique can be used, but spectrophotometric procedures are the most common. Catalytic procedures employing spectrophotometric measurement could, of course, be regarded as examples of indirect spectrophotometry; however, it is more usual to regard them as part of a separate technique – catalysis.

The most useful feature of catalytic procedures from the standpoint of water analysis is their rather high sensitivity and low detection limits. Detection limits in the range 0.1–1 μg l^{-1} are common, and even lower values have been attained in some cases [Section 11.4.3(f) describes the low detection limits which have been attained using catalytic methods involving luminescence measurement]. Table 11.15 lists the detection limits reported for a number of determinands in water using catalytic methods.

Poor selectivity is the most common problem encountered with catalytic methods of analysis – many of the reactions employed are catalysed by a number of substances. However, by suitable choice of experimental conditions (reactions, pH, the use of masking reagents or separation techniques, *etc.*), it is often possible to achieve adequate selectivity for particular applications. It should also be noted that

TABLE 11.15
Some Examples of the use of Catalytic Methods in Water Analysis

Determinand	Detection limit* μg l^{-1}	Ref.
Arsenic	0.02	1327
Bromide	0.03	1328
	10	1329
Chloride	0.25	1330
Chromium(III)	0.025	1331
	0.2	1332
Chromium(VI)	0.015–0.3	1333
Cobalt	0.1	1334
	<0.5	1335
Copper	0.5	1336
	<10	1337
Iodide	0.3	1338
Iron(II)	0.005	1339
Iron(III)	0.25	1340
Manganese	0.2	1341
	<0.5	1342
Molybdenum	0.0001	1343
	0.25	1344
Rhenium	0.5	1345
Vanadium	2.5	1346
	<1.0	1347
	1.0	1348
	0.03	1349

* Not necessarily defined as in Section 8.3.1

many of the procedures for trace metals are specific for particular oxidation states, so that appropriate pretreatment procedures would have to be applied if the total metal concentration was of interest.

A further disadvantage of catalytic procedures is that very close control of experimental conditions is necessary, accurate timing and temperature control (± 0.1–$0.5\,^{\circ}$C) usually being essential. Although catalytic methods for organic compounds have been described, those which have found application in water analysis are almost exclusively for inorganic determinands – the exceptions being certain enzymatic procedures, discussed below.

It appears, therefore, that normal spectrophotometric or atomic absorption procedures would usually be preferable, except where catalytic procedures are able to give greater sensitivity without recourse to more complex and costly equipment and/or time-consuming preconcentration procedures. This is borne out by the fact that few catalytic methods appear in the collections of standard or recommended methods produced by national or international bodies. Thus only bromide (*e.g.*, ref. 1328), iodide (*e.g.*, ref. 1338) and vanadium (*e.g.*, ref. 1346) are the subjects of catalytic methods given in the collections of methods produced by the American Public Health Association and the American Society for Testing and Materials.

(ii) Enzyme procedures A special variant of catalytic procedures (considered by some authors as a separate technique) involves the use of enzymes which themselves act as catalysts, often very selectively. Two broad types of enzymatic procedure can be distinguished. In the first, the rate of an enzyme-catalysed reaction is measured, the conditions being chosen so that the enzyme is either activated or inhibited by the determinand. Examples of applications to water analysis include the determination of arsenic,[1350] the determination of certain toxic nerve gases[1351] and the determination of adenosine triphosphate (ATP)[1320] – see also Section 11.4.3(*f*). In the second type of method, an enzyme is used to catalyse selectively a reaction of the determinand to form a substance that can be determined easily and/or selectively. Thus, for example, urease has been used to convert urea to ammonia in samples of sea-water; by measuring ammonia in samples before and after the addition of urease, the concentration of urea can be determined.[1352] Similarly, Stevens[1353] has determined orthophosphate in fresh waters using its reaction with glyceraldehyde 3-phosphate in the presence of the enzyme, glyceraldehyde 3-phosphate dehydrogenase and NAD (nicotinamide adenine dinucleotide) to form 1,3-diphosphoglycerate and reduced NAD. The latter is measured colorimetrically, and its concentration may be simply related to that of the orthophosphate present originally. The same principle has been used to produce numerous 'enzyme electrodes' or 'biosensors' in which an ion-selective electrode is surrounded by an enzyme. The electrode responds to a product of the enzyme reaction, so that its signal is a measure of the concentration of the substrate of interest. An example is the urea electrode of Guilbault and Nagy,[1354] which exploits an ammonium ion-selective electrode and the reaction described previously. Enzyme electrodes have been recently reviewed by Guilbault.[1355]

Enzyme methodology does not appear to have had a major impact on water analysis, possibly because of a lack of robustness of enzyme systems and electrodes and (in the case of methods of the first type described above) inadequate selectivity.

The propensity of certain enzyme reactions to inhibition by a relatively large number of different 'toxic' substances provides, however, a basis for their potential use as broad-spectrum pollution monitors for the protection of drinking water intakes from chemical spillages.[1356] This subject is, however, outside the scope of the present text.

11.4.5 Radiochemical Techniques

The term 'radiochemical' is used here to denote all those techniques in which measurement of natural or induced radioactivity is involved. Many techniques are available and four of those of greatest general interest in water analysis are considered in (b)–(e) below. For simplicity, the common basis of all these techniques, *i.e.*, the measurement of radioactivity, is first briefly considered in (a).

(a) Measurement of Radioactivity

The following description is intended only to indicate the principles involved and is over-simplified. Full details of theory and practice can be found in texts on nuclear chemistry and radiochemistry (see, for example, refs. 1357–1362). The last-mentioned of these, by Faires and Boswell,[1362] provides a succinct, yet comprehensive, general introduction to radiochemical techniques for analysts lacking previous experience of them. A useful glossary of terms used in nuclear analytical chemistry has been produced by de Bruin,[1363] and a book dealing with the application of radioanalytical techniques to environmental problems has also been published recently.[1364]

The atomic nuclei of certain naturally occurring isotopes of some elements are unstable, *e.g.*, ^{238}U, ^{40}K; unstable isotopes can also be produced artificially, *e.g.*, ^{90}Sr and ^{137}Cs from the fission of uranium atoms in nuclear reactors. Unstable isotopes (radionuclides) spontaneously disintegrate and, in doing so, emit radiation of one or more types, *e.g.*, alpha-particles (particles of mass 4 and charge $+2$), neutrons, beta-particles (electrons) and gamma rays. The last two of these find the greatest use in water analysis and attention will be devoted mainly to them in this account. As an example of radioactive decay, consider the disintegration of ^{90}Sr by emission of electrons (β-particles):

$$^{90}Sr \longrightarrow {}^{90}Y + e$$

The rate of decay of an isotope is given by the equation:

$$-(dN/dt)_t = 0.693 \, N_t/T$$

where $-(dN/dt)_t$ is the rate of decay of the number, N_t, of atoms of the isotope present at time t, and T is the half-life of the isotope, *i.e.*, the time taken for half the atoms originally present to decay. The value of T is a characteristic constant for a particular isotope; for most of the isotopes of interest in water analysis, T ranges from hours to millions of years. As an indication of the sensitivity of radiochemical techniques, a simple calculation shows that 1 ng of ^{90}Sr ($T=28.5$ years) initially disintegrates at a rate of approximately 3×10^5 atoms min^{-1}. Thus, if the emitted electrons are detected with an efficiency of 10%, and are counted for 1 min, a count of 3×10^4 will be obtained. Such a count could be made with a relative standard deviation of approximately 0.6%. (Radioactive decay is a random process; provided that the dominant source of error is the random variation in the number of disintegrations, the standard deviation of a count is equal to the square root of the number of counts. Thus, if C counts are recorded, the relative standard deviation is given by $100\sqrt{C}/C = 100/\sqrt{C}$).

The radiations emitted during the decay process have energies that are characteristic of a particular isotope. Electrons (β-particles) show a spread of energies up to a characteristic maximum value, but γ-rays are essentially mono-energetic. This property of γ-radiation has allowed the development of the very powerful technique of γ-ray spectrometry. In this technique, the counts are recorded as a function of energy so that a spectrum is displayed as a series of peaks. The energies of the peaks serve to identify the isotopes present and the counts for each peak provide a measure of the amount of the isotope.

A number of devices are available for counting the emitted radiation. These include Geiger-Muller counters, proportional counters, scintillation counters and semiconductor detectors. The general radiochemical texts cited above give further information, but specialist texts and part-texts are also available: refs. 1365–1368 deal with detection in general, refs. 1369 and 1370 with liquid scintillation counting and refs. 1371 and 1372 with γ-ray spectrometry. A text by Nicholson[1373] has as its subject the electronics of radiation detection.

The simpler detection systems, such as proportional counters, which may be used for measuring total α- or β-radiation (see below) are convenient to use and are relatively cheap – no more expensive than, say, a spectrophotometer. For more specialised applications, *e.g.*, low-level counting and γ-ray spectrometry, more refined procedures and/or equipment are required.

The basis of radiochemical analysis can thus be seen from the above description. The sample is placed in the counting equipment and the number of counts obtained in a known time is a measure of the amount of a radionuclide. In practice, this amount is usually calculated by comparing the counts obtained from the sample and from a known amount of the radionuclide in question. Confirmation of the identity of the radionuclide can be obtained in several ways. For example, the half-life can be estimated from a series of counts at different times, or the energy of the emitted radiation can be measured, *e.g.* by using a pulse-height discriminator in a scintillation detector or, better still, by using a γ-ray spectrometer.

When samples contain a number of radionuclides, whose separate determination is required, some means of resolving the total emitted radiation is necessary. Adequate resolution can sometimes be obtained by a purely instrumental approach, *e.g.*, by γ-ray spectrometry or by recording the count at different times. However, preliminary chemical separation may sometimes be required, the appropriate fractions then being counted separately. Special procedures are often needed in such separations, and refs. 1364, 1374 and 1375 should be consulted for further details. The use of γ-ray spectrometry in neutron activation analysis, with high resolution semiconductor detection, has reduced the need for radiochemical (or 'post-irradiation') separations in that technique, though they may still be necessary in some applications – to deal with samples containing high concentrations of NaCl, for example. Sub-section (*e*) below should be consulted for further details.

For those who may wish to adopt radiochemical techniques, laboratory design and safety considerations are of great importance. The book by Faires and Boswell[1362] gives valuable guidance on these topics, and includes a discussion of the regulations applying to the use of radioactive substances and the names and addresses of regulatory agencies in a number of countries, as well as advice on laboratory construction and hazard control in general. The Radiochemical Manual,[1376] now out of print, remains a useful source of information on the properties, availabilities and uses of radioactive materials, and comprehensive information on the properties of radionuclides may be obtained from the 'Table of Isotopes' edited by Lederer and Shirley,[1377] now in its seventh edition.

The four selected radiochemical techniques are described in the following sub-sections.

(b) Determination of Radionuclides in Water

The determination of radionuclides in natural and industrial waters involves direct application of the general principles outlined in (a) above, and is not discussed in detail here. The biennial reviews of water analysis published in *Analytical Chemistry*[743] contain a section which deals with the determination of radioisotopes, and collections of standard methods for water analysis include methods for the most common determinands. The latter include gross α- and gross β-activity and selected radionuclides, including strontium, radium, tritium, caesium, iodine and barium – see, for example, refs. 1378–1381. These texts give a short background to the methods, and ref. 1378 cites the various guidelines on maximum levels of radionuclides in United States potable waters. The book by Holden[1382] also gives a short account of the significance and measurement of radioactivity in natural and potable waters. Burton[1383] has reviewed the occurrence and measurement of radionuclides in sea-water, and has also described how they may be used for tracing and dating purposes.

The use of measurements of radionuclides in water for tracing, dating and related purposes is widespread, but a detailed description of such applications lies beyond the scope of the present text. The reader wishing to obtain further information is referred to reviews by White[1384,1385] and Evans,[1386] to the above-mentioned review by Burton,[1383] to the proceedings of conferences on various aspects of the topic, organised by the International Atomic Energy Agency (see, for example, refs. 1387 and 1388) and to general texts on hydrology and hydrometry (*e.g.*, refs. 1389 and 1390). The radionuclides exploited in this manner may be naturally occurring, man-made, but not introduced for tracing purposes (*e.g.*, tritium from nuclear weapons tests) or man-made, and introduced deliberately for the tracing or other application (*e.g.*, ^{82}Br for time-of-travel studies in rivers).

Before moving on to radiometric methods of analysis, it should also be noted that an important application of ^{14}C measurements in aquatic science is the measurement of primary production by phytoplankton.[1391] Carbonate, labelled with ^{14}C, is introduced into water samples contained in both clear and darkened bottles. These are incubated (normally *in situ*, at the point of collection of the samples) and, after a defined period, the phytoplankton are collected by filtration, fumed with hydrochloric acid to remove inorganic radiocarbon, and the residual radioactivity on the filters (*i.e.*, ^{14}C taken up by the phytoplankton) determined. The total carbon fixed by photosynthesis is proportional to the fraction of added ^{14}C taken up by the organisms. The dark bottle count acts as a 'blank', and is subtracted from the light bottle count in calculating the productivity.

(c) Radiometric Methods
These methods consist essentially of the addition to the sample of a radionuclide which reacts with the determinand to form a product that can be separated from the sample, *e.g.*, a precipitate or solvent-extractable compound. After separating this product its activity is counted, which provides a measure of the amount of determinand originally present. For example, in a procedure for the determination of sulphate in water,[1392] the determinand is precipitated with ^{133}Ba within a dialysis bag. Excess ^{133}Ba is removed by soaking the bag in deionised water for 12 h, and the activity of the precipitate is then measured by placing the bag directly in the well of a scintillation counter. Other procedures of this general nature have been used for chloride[1393,1394] and anionic surfactants.[1395]

   A variant of this technique is to allow the sample to react with an insoluble form of a radioactive nuclide so that the amount of this nuclide released into the sample is governed by the amount of determinand. Separation of the sample from the insoluble radioactive material and counting the sample then provides a measure of the determinand. This approach has been used to determine dissolved oxygen by passing the sample through a column of thallium metal chippings (containing ^{204}Tl) and counting the ^{204}Tl released into solution by reaction with the oxygen.[1393,1396] Similarly, cyanide has been determined by passing samples through a column of radiolabelled silver iodide.[1397]

   A further variant of the technique involves the displacement of a metallic radionuclide from a complex by another metal, the determinand. This approach has been used by DuBois and Sharma[1398] to determine copper in sea-water. Copper displaces ^{59}Fe bound by a substoicheiometric amount of bathophenanthroline, the residual ^{59}Fe complex being separated from the solution by vortexing the samples in tubes. The complex coats the sides of the tubes, and the decanted solution contains the displaced ^{59}Fe which is counted. Interference from ferric ions in the sample is overcome by complexing them with desferrioxamine.

   Radiometric methods are not commonly used in water analysis. Their main advantage is that they often offer great sensitivity, as a result of the general sensitivity of radiochemical procedures. Thus, for example, the procedure for the determination of copper in sea-water noted above[1398] was reported to allow a detection limit of 2 μg l^{-1} to be attained using only 5 ml of sample. It should be noted, however, that the sensitivity of the radiometric measurement is advantageous only if the chemical stages of the procedure do not introduce such variability as to limit the accuracy and limit of detection.

   As was noted in the first edition of this text, it would seem that radiometric procedures have little part to play in normal water analysis, but their use may be helpful for special problems, especially in laboratories which already possess the necessary equipment.

(d) Isotope Dilution Methods
In these procedures,[1357–1362,1399,1400] a radioactive isotope of the element to be

determined is added to the sample which is chemically treated so that the natural and radioactive isotopes are equilibrated. The determinand is then separated as a suitable compound, *e.g.*, by precipitation, whose weight and activity are determined. The weight and activity obtained by a similar analysis of the pure radioactive solution are also determined and a simple calculation allows the weight of the determinand in the sample to be calculated. Procedures for compounds as well as elements are possible.

The advantages of these procedures are that quantitative separation of all the determinand in a sample is not necessary, and that they are capable of high sensitivity. The disadvantages are that the determinand must be separated in a pure form, and that vigorous chemical procedures are necessary to ensure equilibration between the radioactive and natural isotopes when the latter are present in samples in kinetically inert forms.

When determinands are present in trace concentrations, as is often the case in water analysis, problems arise in weighing the small amounts involved. The technique of 'substoicheiometry' described by Ruzicka and Stary[1401] avoids this difficulty. Basically, the procedure is arranged so that, when the material is taken from the samples and standards for counting, equal weights are always removed, and the actual weight need then not be known.

Thus, for example, Das *et al.*[1402] have determined fluoride in tap water by ^{18}F isotope dilution using a substoicheiometric extraction with trimethylchlorosilane, into benzene. Other applications to water analysis include the determination of iron in sea-water using substoicheiometric amounts of bathophenanthroline[1403] and the determination of zinc in water[1404] using a resin loaded with a substoicheiometric quantity of 8-quinolinol-5-sulphonic acid.

(e) Activation Analysis
The general principle of this technique is to irradiate the sample with energetic particles (*e.g.*, neutrons, protons) or radiation (*e.g.*, γ-rays) which, by interaction with the nuclei of atoms in the sample, lead to the formation of radionuclides. The natures and amounts of these radionuclides may be determined, and indicate the elements present in the original sample, and their concentrations. In water analysis, neutron activation analysis (NAA) has been used almost exclusively, and attention will here be confined to that mode of activation. Similarly, γ-ray spectrometry is almost always used for the identification and quantitative measurement of the radionuclides produced because they can often be resolved and detected by that means without the need for chemical separations. The name 'Instrumental Neutron Activation Analysis' (INAA) is sometimes used to describe the use of γ-ray spectrometry in this manner.

Activation analysis is the subject of a number of texts[1374,1405-1407] and part-texts,[1408,1409] some of which (*e.g.*, refs. 1374, 1409) are devoted to neutron activation specifically. The book by Das *et al.*[1364] contains much information on the application of activation analysis to environmental samples (including waters). Application of neutron activation analysis to water and related materials has also been reviewed recently by Guinn,[1410] and activation analysis in general is treated in the biennial reviews published (in even-numbered years) in *Analytical Chemistry*, most recently by Lyon and Ross[1411] under the general topic of 'Nucleonics'.

When irradiated with thermal neutrons*, almost all elements undergo a neutron-capture reaction, in which a neutron (n) is absorbed and 'prompt' γ-rays emitted. This is known as an (n, γ) reaction. The products of such reactions may be one or more radioactive isotopes of the same element, or a mixture of stable and radioactive isotopes of the same element, depending on the natural isotopic distribution of the element in question. As a simple example, consider the behaviour of cobalt, which occurs in nature monoisotopically as stable ^{59}Co:

$$^{59}\text{Co} + \text{n} \longrightarrow {}^{60}\text{Co} + \gamma$$

The sole product, ^{60}Co, is radioactive, decaying with a half-life of 5.2 years according to the reaction:

$$^{60}\text{Co} \longrightarrow {}^{60}\text{Ni} + \text{e} + \gamma \ (1.17 \text{ and } 1.33 \text{ MeV})$$

In principle, therefore, it is only necessary to measure the induced radioactivity (*e.g.*, by counting the γ-radiation emitted by the ^{60}Co) in order to estimate the amount of ^{59}Co originally present in the sample. Quantitative analysis is normally achieved by comparing the counts obtained from the sample with those obtained from a known amount of the determinand irradiated simultaneously. In practice, irradiation of water samples leads to the formation of many different radionuclides so that either chemical separation (before or after the irradiation) and/or instrumental resolution of the emitted radiation is necessary.

Some elements produce only a radionuclide decaying by emission of a β-particle, but most produce radionuclides which decay with emission of γ-rays and/or X-rays, which may be detected by γ-ray spectrometry. The latter technique is now the favoured means of identifying and quantifying the radionuclides produced by neutron activation, because it provides an ability to resolve the emitted radiations from complex matrices (such as water samples) without, in many cases, the need for chemical separations, and so confers an essentially simultaneous multi-element capability upon neutron activation analysis. Gamma-spectrometers are composed of a detector unit giving electrical pulses (the amplitude of which is governed by the energy of the radiation) and an electronic system able to count the pulses of different amplitudes, and produce some form of output of the resulting spectrum (number of pulses as a function of energy). Two main types of detector are available: thallium-activated sodium iodide scintillation detectors [NaI(Tl) detectors] and lithium-drifted germanium semiconductor detectors [Ge(Li) detectors]. In the former, the emitted γ-photon produces a sintillation which is in turn converted to an electrical pulse by means of a photomultiplier. In the latter, the photon produces an electrical pulse directly. The Ge(Li) detector offers markedly better resolution than the NaI(Tl) type, which makes it particularly appropriate for multi-element neutron activation analysis, but at some cost in reduced detection efficiency.

It is quite simple to calculate reasonable estimates of the limits of detection attainable for different elements – details are given in texts on the technique, and the results have been summarised elsewhere – see, for example, ref. 1410. The detection

---

* *i.e.*, Neutrons with a kinetic energy distribution in equilibrium with the surrounding temperature

limits are inversely proportional to the neutron flux used for irradiation of the sample, and for trace analysis large fluxes (10^{12}–10^{14} neutrons cm^{-2} s^{-1}) are employed. For a flux of 10^{13} neutrons cm^{-2} s^{-1}, an irradiation period not exceeding 5 h, zero decay time (see below) and counting in a 40 cm^3 Ge(Li) detector for a period not exceeding 100 min, Guinn[1410] quotes absolute limits of detection for 68 elements ranging from 10^{-7} to 10 μg, with a median of about 2×10^{-4} μg*. For 1 g samples, these values (in μg) are numerically identical to the parts per million (mg l^{-1}) detection limits in concentration units. Thus, under the conditions described, many elements can – in principle – be determined with detection limits better than 1 μg l^{-1}, using only 1 g of sample.

Neutron fluxes of 10^{12}–10^{14} neutrons cm^{-2} s^{-1} are, however, attainable only in nuclear reactors. Although smaller isotopic or accelerator sources can be installed in laboratories, these give much lower fluxes and are unsuitable for general trace analysis. This need for reactor facilities is a major disadvantage of neutron activation, since few laboratories have direct access to a suitable installation. However, organisations exist which provide such facilities for irradiation on a commercial basis – e.g., the Atomic Energy Research Establishment at Harwell in the U.K. Even so, the need then to arrange transport and irradiation of samples is an inconvenience detracting from the general usefulness of the technique, particularly in view of the availability of several other techniques offering comparable or better detection power for many elements. A further disincentive to general adoption of the technique is the need for special shielding and residue disposal arrangements to cope with the high activities which may be induced in samples.

The outline of neutron activation given above may suggest that application of the technique to water analysis presents no difficulties, provided that suitable facilities and equipment are available. Indeed, it may be noted that the technique, being based upon nuclear processes, is (in its purely 'instrumental' form) not subject to bias arising from variable response to different forms of the elements determined, and is in this regard superior to many other techniques used in trace analysis. In practice, however, the complexity of water samples and the resulting formation of many different radionuclides can cause interference problems, the control of which may greatly increase the complexity and cost of the approach and, in some cases, destroy the independence of response and chemical form of the determinand.

Thus, for example, interference effects can arise when the γ-rays emitted by different radionuclides have either the same energics, or cannot be resolved adequately by the γ-spectrometer. It is often possible to control such effects to a sufficient extent by suitable selection of irradiation, 'cooling' and counting periods. That is, elements producing radionuclides having short half-lives can be determined by using short irradiation and counting periods, with minimal interference from elements producing radionuclides having much longer half-lives. Conversely, elements producing radionuclides having long half-lives can be determined, with minimal interference from elements producing radionuclides having much shorter half-lives, by using longer irradiation periods with a suitable 'cooling' period between irradiation and counting to allow decay of the undesired (short-lived)

* Incorrectly recorded as 2×10^{-4} g in ref. 1410

radionuclides to a tolerable level. Such considerations can lead to fairly elaborate analytical strategies, in which perhaps three or more separate irradiation/'cooling' measurement stages are employed for the determination of different groups of elements in water samples (see, for example, refs. 1412, 1413).

The application of such approaches may not always be effective in reducing interference effects, and recourse must then be made to separation procedures to separate the interfering radionuclide(s) from the radionuclide(s) to be measured ('post-irradiation separation') or to separate their precursors ('pre-irradiation separation'). Such procedures are used, for example, in sea-water analysis by neutron activation to control interference effects from the high levels of salts present in that medium (see below). Post-irradiation separation may consist of group separations to isolate the trace elements of interest (*e.g.*, refs. 1414, 1415) or specific procedures to remove particular interferences – for example, the removal of ^{24}Na and ^{42}K (in sea-water analysis) using a column of hydrated antimony pentoxide and of $^{80/82}Br$ by carbon tetrachloride extraction (following pre-irradiation addition of iodide to prevent the formation of organo-bromine compounds) used by Ammann and Knoechel.[1416] Antimony pentoxide has also been used for the *pre*-irradiation removal of sodium in sea-water analysis by neutron activation,[1417] and chlorine and bromine were removed with nitric acid and hydrogen peroxide in the same procedure.

Pre-irradiation separation is often achieved in combination with a pre-concentration step undertaken to improve detection limits in trace analysis. Numerous pre-treatment techniques of this kind have been employed with neutron activation for water analysis, including chelating or ion-exchange resins,[1415,1418–1423] foams[1424] and filters,[1421] chelation/coprecipitation,[1425] chelation/solvent extraction,[1426] chelation/adsorption[1427] and freeze-drying.[1413,1429]

Reference has already been made to the detection limits attainable by purely 'instrumental' neutron activation analysis. The precision of analytical results depends, *inter alia*, upon the counting statistics of the γ spectrometry. Guinn[1410] cites relative precisions of 5–10% as common in multi-element neutron activation analysis, and 'accuracies' (presumed 'biases') of similar magnitude, although he notes that better performance (precisions and biases of 1–3% or better) can be attained with the technique. However, the use of elaborate chemical concentration and separation techniques can result in poorer precision and increased systematic error. Thus, for example, Jørstad and Salbu[1430] report relative standard deviations of 5–10% for sea-water analysis by neutron activation after freeze drying, with electrolytic post-irradiation separation, Hamilton and Chatt[1413] quote relative standard deviations of about 2–12% for analysis of freeze-dried rainwater samples and Lenvik *et al.*[1431] report relative standard deviations of 5–20% for the determination of a number of trace elements in fresh waters by neutron activation based on anion-exchange separation followed by precipitation.

The reader seeking further information on applications of neutron activation to water analysis is referred particularly to the review by Guinn,[1410] to the book by Das *et al.*[1364] and to the 'Multiconstituents' sections of the biennial reviews of water analysis published in *Analytical Chemistry* (*e.g.*, ref. 743).

11.5 SPECIATION ANALYSIS

11.5.1 Introduction

Many elements and substances may exist in water in a variety of physico-chemical forms. These different forms will often exhibit different physical, chemical and biochemical properties, and it has been increasingly recognised – in recent years particularly – that a proper understanding of the transport, behaviour, effects and fate of many elements and substances in water systems will usually require more than a knowledge of their total concentrations alone. The term 'speciation' has become widely accepted as appropriate to describe this distribution of an element between different physico-chemical forms or 'species', and it will be used here.

Obvious and familiar examples of the existence of different species of an element in water are provided by the classical micronutrient elements nitrogen, phosphorus and silicon; however, particular attention is being paid to the aqueous speciation of trace metals and metalloids. This reflects the importance of these elements as common pollutants, the potent effects which they may exert upon aquatic and other ecosystems* and the complexity of their speciation – and therefore behaviour and properties – in water. Numerous authors have drawn attention to the need for an improved knowledge of trace metal speciation, and of the manner in which it depends on general aspects of water quality (see, for example, refs. 1432–1437), although the need to bear in mind the importance of non-metals (such as the micronutrients) has also been emphasised.[1438]

Reflecting the general concern about trace metal speciation in particular, and the difficulties of trace metal speciation analysis, the remainder of this section will be devoted largely to analytical approaches to the speciation of such elements. Although by no means a routine activity in water analysis laboratories, the growing interest in, and problems of, the topic are such that it merits reasonably detailed consideration here.

It should be noted that three different basic approaches have been used to investigate the speciation of trace elements:[1439] (1) study of the behaviour and reactions of an element in a model water system, (2) prediction of the species distribution using thermodynamic modelling, and (3) the determination of species or groups of species in real samples. The last of these approaches will be emphasised here.

The speciation of trace elements in water has been reviewed by Florence and Batley,[1437] and is the subject of a volume of conference proceedings edited by Leppard.[1440] Analytical approaches to the characterisation of trace element speciation have been reviewed by Florence,[1441] by Buffle[1442] and by Braman;[1443] particular aspects have also been reviewed by Steinnes[1444] and by de Mora and Harrison[1445] (physical separation procedures), by Neubecker and Allen[1446] (complexation capacity and conditional stability constant measurements) and by Chau and Wong[1447] (determination of well defined organometallic compounds). The reader interested in the use of thermodynamic data to predict trace element speciation is referred to the conference proceedings edited by Jenne,[1448] in which

---

* Both as toxic substances and as materials essential for the healthy growth of organisms.

many papers on the topic are included, and to texts on aquatic chemistry by Stumm and Morgan[1449] and Morel.[1450]

Although the use of thermodynamic data to predict trace element speciation will not be discussed in detail here (because this text is concerned with the chemical *analysis* of water) it represents a most important facet of trace element speciation studies. Despite a number of problems – most notably, the inaccuracies of, and gaps in, stability constant data – thermodynamic calculation of equilibrium species distributions provides a necessary theoretical basis for speciation analysis, and can give information on simple inorganic species (in particular), which are not usually separable by analytical means.

11.5.2 The Forms of Trace Elements in Waters

Given the multiplicity of forms in which these elements can exist in natural and treated waters, some form of classification scheme is essential. Various schemes are possible, but that shown in Figure 11.22 (based upon a similar scheme in ref. 1432) is felt to represent a good compromise between detail and excessive complexity.

Because analytical schemes are not available which can separate the total element concentration unambiguously into the classes shown, these classes are concepts developed from our overall knowledge of trace elements in water – that is, from theoretical calculations, analytical evidence, general chemical knowledge, inferences from laboratory studies of model systems, *etc.*

11.5.3 Potential Applications of Speciation Data

Though it is outside the scope of this book to describe in any detail the potential applications of speciation data, the following brief summaries are included as an introduction for the reader unfamiliar with the subject, to set the subsequent discussion of speciation analysis in context.

(a) Pollution Control
The transport and fate of trace elements in water bodies is of importance to the control of pollution, and may depend markedly upon the speciation of the element – particularly upon its distribution between the water itself and suspended particulate matter within it[1434,1443] and on the volatility or otherwise of the species present.[1443]

(b) Bioavailability and Toxicity to Organisms
Numerous authors have reported that the availability and/or toxicity of trace elements to aquatic organisms (from bacteria to fish) is dependent upon the form in which the species is present. The literature has been reviewed by Smies,[1451] by Babich and Stotzky[1452] (with respect to toxicity to the microbiota) and by Borgmann[1453] (with particular reference to toxicity of free metal ions relative to complexes).

For divalent metals such as Cu, Cd and Zn, the toxicity of the free metal ion is thought often to be greater than that of complexes.[1453] Borgmann[1453] has noted, however, that this (tentative) conclusion has been drawn mainly from studies of

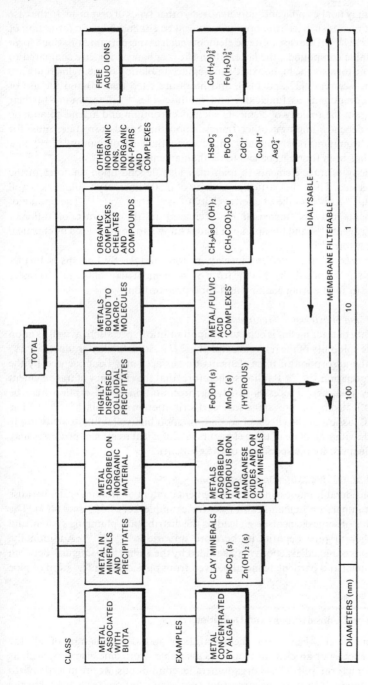

Figure 11.22 *Classification scheme for trace metal (and metalloid) species in aqueous systems*

algae, and may not be applicable universally, to other types of organism. In the case of trivalent metals such as iron, uptake may well be enhanced by the formation of complexes.[1441,1451] In the case of those elements such as mercury and tin which form organometallic compounds, the latter often show enhanced toxicity compared to the inorganic forms – the high toxicity and bioaccumulation of methylmercury are well known (see ref. 1437 and 1443, and literature cited therein) and the use of organotin compounds as fungicides and molluscicides[1454] is testimony to their toxicity. Again the toxicity of elements such as chromium and arsenic to aquatic organisms depends, amongst other things, upon their oxidation state; thus, for example, chromium(VI) is generally more toxic than chromium(III)[1441,1451] and arsenic(V) has been found to be less toxic than arsenic(III).[1441,1455]

The toxicity of trace elements to man may also depend upon the form of the element ingested. Thus, the particular toxicity of methylmercury to humans is well established,[1456] as is also the greater toxicity of As^{III} than As^V.[1457] The uptake of trace elements may be increased or decreased in the presence of different complexing agents,[1456] and the uptake of lead is known to be influenced by chemical and physical form.[1458]

The influence of trace element speciation upon uptake by humans is further complicated by the possibility that the form of the element may influence its transfer from water to food during cooking and food preparation.

(c) Water and Wastewater Treatment
The solubility of trace metals is dependent upon solution speciation, as well as upon the intrinsic solubility of the relevant solid phase (*i.e.*, its solubility product).[1459,1460] Likewise, the adsorption of trace elements on surfaces may be influenced by the dissolved speciation.[1461,1462] It follows, therefore, that the efficiency of trace element removal by treatment processes involving precipitation or adsorption may be markedly affected by the prevailing species distribution in solution.[1434,1459,1463] (The previous discussion of the effects of speciation upon bioavailability/toxicity imply that the behaviour and fate of trace elements in biological treatment processes may also be influenced by the speciation of the element).

(d) Trace Metal Acquisition in Distribution
There is considerable interest in the trace element content of drinking water because of possible impact on human health (see, for example, refs. 1463 and 1464). The major source of some elements (*e.g.*, lead) is the distribution/plumbing system, and the solubility of pipe deposits is therefore important.[1460,1465,1466] Additionally, investigations of remedial actions may be aided by the ability to distinguish between dissolved metal and particulate metal derived from physical disintegration of pipe deposits.

11.5.4 General Considerations and Definitions   .

Ideally, analytical schemes for the separation and determination of all the important species of an element would be available. For some elements, such as arsenic, which form well defined organometallic compounds, we are indeed able to determine many, at least, of the important species in solution (see below), but the ideal is otherwise unattainable at present, and likely to remain so given the complexity of natural waters. Thus, for many trace elements, the best that can be expected from analysis, in the foreseeable future, is a fractionation of the total concentration into practically useful groups of species. (Of course, the application of analytical techniques to model systems, or to the determination of thermodynamic data such as stability constants, may yield predictive information about individual species of elements in real waters but, as noted above, speciation analysis of real samples will be emphasised here.)

In this review, the fractions or groups of species identified will be regarded as the 'determinands' and the overall analytical procedure for their measurement will be the 'speciation (or analytical) scheme'. Although much has been written about analytical approaches to the elucidation of trace element speciation, many publications do not give such detailed descriptions of the procedures used as warrant the term 'analytical method', in the accepted sense of a complete written description of the steps to be followed by the analyst. The term 'speciation scheme' used here should not, therefore, be taken to imply an integrated group of well defined methods, but rather an outline of the techniques and procedures used. It may seem that this implied criticism of speciation work is harsh, given the difficulties faced in this field, but it must be recognised that, with some exceptions, the determinands of speciation schemes are non-specific – that is, the analytical method itself defines the determinand. Consequently, an analyst cannot be certain that he is dealing with the same determinands as another worker whose scheme he is following, unless all details of procedure likely to affect the result have been specified. Progress in trace element speciation will obviously be hindered if such transfer of procedures is prevented.

11.5.5 The Construction of Speciation Schemes

A large number of analytical schemes for trace element speciation have been described. These are of widely differing levels of complexity, and any overview of the subject must have some framework within which the diversity of approaches can be accommodated. The most logical approach is to consider the basic *techniques* from which the schemes are constructed. This, in addition to permitting a review of a large topic in a relatively small space, has the advantage of placing particular emphasis upon the analytical features and upon their inherent limitations – which sometimes appear to have received inadequate attention.

Following the classification used throughout this section of the text, three basic types of technique can be recognised:

(a) Measurement Techniques (see Table 11.16)
These measure the concentration of one or more forms of the element of interest. They may, of course, have no *useful* inherent ability to respond only to particular forms, in which case they will be described, for convenience, as 'total' techniques*.

* This should not, however, be taken to imply that they are capable of responding equally to all forms of an element in a water sample without the application of pre-treatment techniques. Rather, it reflects the fact that any differential response to particular forms is not sufficiently great and/or well defined to be of value in speciation analysis.

TABLE 11.16

Principal Measurement Techniques used for Trace Element Speciation

Technique	Inherent speciation capability
Atomic absorption spectrometry	
Plasma emission spectrometry	'Total' element techniques, having no practically useful ability to measure only a limited fraction of the total concentration of a trace element
Neutron activation analysis	
Solution spectrophotometry	
Anodic stripping voltammetry and related electroanalytical techniques	Techniques with some ability to measure only a fraction of the total element concentration, but differentiation is usually difficult to define and of an operational character
Ion-selective electrodes	The only measurement technique responsive to a single, well defined metal species

TABLE 11.17

Principal Separation Techniques used for Trace Element Speciation

Technique	Application
Filtration (cellulose membrane and etched-track polycarbonate membranes, mainly)	Separation of species of differing size (ideally; in practice, other factors, such as charge and shape can influence the separation achieved).
Ultrafiltration Dialysis	Filtration widely used to separate the so-called 'particulate' and 'dissolved' fractions, but colloidal material may contribute to 'dissolved' (*i.e.*, filterable) fraction.
Gel filtration (Size exclusion) chromatography	Ultrafiltration, dialysis and gel filtration can also separate high M.W. forms from smaller dissolved species.
Chromatography (various types, including simple selective volatilisation from cold traps)	Nature of separation varies widely according to exact technique and nature of species. Well defined separation of organic compounds of some elements (*e.g.*, arsenic, antimony, lead, tin) possible.
Chelating resins, chelation/ solvent extraction	Separation of forms by exploitation of differential uptake by the chelation system. Usually operational, but may allow well defined separation of different oxidation states.
Adsorption/coprecipitation	Can be used to separate different oxidation states of certain trace elements (*e.g.*, chromium).
Selective generation of volatile species	Can be used to separate different oxidation states of certain trace elements (*e.g.*, arsenic, selenium, antimony).

TABLE 11.18

Principal Conversion Techniques used in Trace Element Speciation

Technique	Application
Acidification	To release metal, bound either organically or inorganically (including adsorbed forms). When used alone, may only release an ill defined fraction of all the bound metal in some types of sample. Can be used with heating for more vigorous attack.
Wet oxidation	Destruction/degradation of organic material to release complexed or chelated metal.
UV photolysis	Oxidation may also be used to alter
Ozonolysis	oxidation state of element of interest (see below).
Oxidation/reduction	Used with separation techniques capable of separating different oxidation states, to convert non-reactive species to a reactive oxidation state.

(b) Separation Techniques (see Table 11.17)
These separate species or groups of species. In conventional analysis to determine total concentrations, separation techniques are employed to separate the determinand from other elements and their compounds which would otherwise interfere in the determination. Although the same basic objective (separation of species which would otherwise interfere with the determination of the form(s) of interest) applies in speciation analysis, the separation – involving different forms of the same element – may have to exploit weaker differences in chemical behaviour and properties, and be correspondingly more difficult to perform*. Separation procedures in speciation analysis include those which separate on the basis of size, which are not subject to the limitation noted above.

(c) Conversion Techniques (see Table 11.18)
These convert groups of metal species into other forms. The techniques available are often those employed in 'total' concentration work to convert all forms to (most commonly) one, measurable form.

11.5.6 An Assessment of the Techniques Available

(a) Measurement Techniques
The range of applicable measurement techniques is limited by the low concentrations of trace elements which are commonly encountered, even in treated or polluted waters. For many types of sample (*e.g.* open-ocean sea water, uncontaminated snow or ice and certain unpolluted freshwaters) the very low total trace element concentrations ($< 1 \ \mu g \, l^{-1}$) further restrict the choice of technique.

(i) Atomic absorption spectrometry, plasma emission spectrometry, and neutron activation analysis These techniques have no practically useful inherent ability to

* Separation from other elements may, of course, also be necessary in speciation work.

respond differentially to forms of trace elements. They must, therefore, be used in conjunction with suitable separation (and perhaps conversion) techniques in speciation work.

(ii) Solution spectrophotometry   This has some potential as a species-selective measurement technique, because chelated or complexed metal (in which category may be included adsorbed metal) may not be available to react with the chromogenic reagent*. Thus, application of the measurement technique alone would yield a 'labile' or 'free' (unbound and 'weakly' bound) fraction of the element, whereas its use with a suitable conversion technique would yield the total metal concentration. Such an approach was widely used in early studies of trace metal speciation in the late 1960s and early 1970s. For example, the speciation of copper in sea-water was investigated by Alexander and Corcoran[1468] using neocuproine as chromogenic reagent, with perchloric acid oxidation as the conversion technique, and by Williams[1469] and Foster and Morris[1470] using diethyldithiocarbamate and u.v. photolysis. (To obtain the necessary detection limits, the complexes were extracted into an organic solvent prior to spectrophotometric measurement. It could, therefore, be argued that the speciation ability is conferred by a chelation/solvent extraction separation, rather than by the measurement technique. Whilst recognising the validity of the argument, we have chosen to consider this approach to speciation under solution spectrophotometry, because it is the chromogenic reagent which confers the differential response, regardless of whether or not the coloured complex is extracted from the sample prior to measurement. As a corollary, the same limitations which apply to the use of solution spectrophotometry as a means of differentiating between different species of an element apply to chelation/solvent extraction as a separation technique – see below).

Solution spectrophotometry has also been used to characterise the speciation of aluminium in fresh waters. For example, Smith[1471] has employed a ferron–orthophenanthroline procedure to determine three aluminium fractions – 'monomeric', 'polynuclear' and 'solid'. In this case, conversion techniques were not employed, rather, the extent of colour formation was monitored as a function of time and the resulting absorbance/time curves used to estimate the concentrations of the three aluminium fractions.

Because chromogenic reagents specific for particular oxidation states of an element are often available, solution spectrophotometry has also found application in schemes for the elucidation of the speciation of such metals as chromium and iron. For example, Koenings[1472] employed bathophenanthroline (BPN) – specific for ferrous iron – combined with various conversion techniques to yield four iron fractions in acid bog lake samples. These four fractions were: 'reactive ferrous iron' (BPN, no pre-treatment), 'total ferrous iron' (BPN, 10 min pre-treatment with 4M

---

* Solution spectrophotometry could, in principle, provide a species-selective measurement without the use of a chromogenic reagent, if the absorption spectrum of a single species itself was sufficiently displaced from that of other species of the element of interest. This would have the advantage that no disturbance of existing equilibria by a chromogenic reagent would occur (see main text below). The approach has been used by Arah and McDuffie[1467] to determine the $HgCl_4^{2-}$ ion in the presence of $CH_3HgCl_2^-$ ions but, as Braman[1443] has remarked, the approach lacks the necessary sensitivity for trace analysis, though it might be useful in laboratory studies. Lack of adequate sensitivity seems likely to limit the value of this otherwise attractive approach of spectrophotometry without chromogenic reagents.

HCl), 'total reactive iron' (BPN, 5 min pretreatment with hydroxylamine hydrochloride) and 'total iron' (BPN, 15 min pretreatment with 4M HCl and hydroxylamine hydrochloride at 100 °C).

The nature of the separation achieved by solution spectrophotometry merits further consideration. Kamp-Nielsen[1473] calculated that, for the diethyldithiocarbamate technique[1474] as used by Foster and Morris[1470]: "...if 1 μmol of a (natural) chelator is present in the actual extraction procedure, 99% of the copper is determined if the stability constant for the Cu-chelate does not exceed 10^{30}." Consequently, Kamp-Nielsen[1473] concluded, this[1470] and similar speciation schemes were all likely to underestimate the organically associated copper fraction, and were certainly recovering all inorganic complexes in the 'free' fraction. The calculations made by Kamp-Nielsen[1473] illustrate that (1) such speciation analysis involves disturbance of existing equilibria and (2) the chromogenic reagent will usually be a much stronger complexer* than the natural materials present in the sample. Although, therefore, the separation achieved by solution spectrophotometry could reflect the competition of chromogenic reagent and natural ligands, it will, in practice, usually reflect kinetic rather than thermodynamic factors. If a naturally occurring metal complex *does not* dissociate rapidly, in relation to the time-scale of the analysis, it will be recovered in the 'bound' or 'non-labile' fraction (*i.e.*, it will be recovered only after the pre-treatment to determine the total metal); this will occur regardless of the fact that, in most cases, the equilibrium position overwhelmingly favours the formation of the chromogenic complex or chelate over the natural one. Conversely, if the naturally occurring metal complex *does* dissociate rapidly during the time-scale of measurement, it will be recovered in the 'free' or 'labile' fraction, provided that, as will usually be the case, the equilibrium attained overwhelmingly favours the formation of the complex or chelate of the metal with the chromogenic reagent. Of course, if the rate of dissociation of the natural complex (in relation to the time-scale of the analysis) lies between these two extremes, then the natural complex will be recovered in important amounts in both the 'free' and 'bound' fractions. (The same result would also obtain if the dissociation kinetics were rapid in relation to the time-scale of analysis, but the competitive equilibrium established between the natural ligand and the chromogenic reagent did not overwhelmingly favour formation of the chromogenic complex. This outcome, as noted above, will usually be unlikely because of the higher concentration and/or stability constant of the chromogenic reagent.)

Before leaving the subject of solution spectrophotometry, it is worth noting that it is, of course, widely used for the determination of different nitrogen species in water (nitrate, nitrite and ammonia), and, combined with various pretreatment procedures, for the determination of various forms of silicon and phosphorus. In the case of the nitrogen species, the different forms are chemically well defined but in the cases of silicon and phosphorus, the usual spectrophotometric procedures employed allow determination of so called 'reactive' fractions, defined by the chosen analytical approach. Such procedures may be combined with a variety of pretreatment steps, such as filtration and hydrolysis, to allow determination of a range of different fractions of the elements – see, for example, refs. 152 and 153.

---

* In terms both of the stability constant of its complex with the element of interest, and of its concentration.

(iii) Ion-selective Electrodes (see Section 11.3.6). As noted in Table 11.16, ion-selective electrodes represent (essentially) the only measurement technique responding to a single, well defined species – namely the free aquo ion. The cupric ion-selective electrode has been most widely used in trace element speciation, but zinc, lead, mercury and cadmium electrodes are also available.

Such electrodes can respond to free ion activity, in Nernstian manner (see Section 11.3.6), to very low levels provided that the total metal concentration is sufficiently high – that is, provided that the solution acts to buffer the free ion activity. Thus, for example, Stiff[1475] stated that "In the presence of copper-complexing ligands, provided that the total copper concentration is not less than 50 μg l^{-1} Nernstian response to cupric ion is maintained to very low concentrations of Cu^{2+}", and he reported free Cu^{2+} concentrations as low as 0.8 ng l^{-1} (*ca.* 10^{-11} M). Similarly, using a cadmium electrode, Gardiner[1476] found that Nernstian response was maintained to 500 μg l^{-1}, provided that the *total* concentration of cadmium remained above 1000 μg l^{-1}. However, he also noted that the electrode specifications stated that it was " . . . sensitive to [Cd^{2+}] down to 10 ng l^{-1} provided that the total cadmium concentration . . . is greater than 10 μg l^{-1}". Again, Kivalo *et al.*,[1477] using two different types of lead electrode, observed Nernstian response to about 10^{-10}M Pb^{2+} in metal ion buffers prepared after the manner of Hansen and Ružička.[1478]

In the absence of buffering, Nernstian response ceases at much higher concentrations. For example, with standard lead solutions containing only 0.1M KNO$_3$, Kivalo *et al.*[1477] found Nernstian response only to about 10^{-4}M Pb^{2+}. Similarly, Stiff[1475] encountered non-Nernstian response to copper below about 50 μg l^{-1} (*ca.* 10^{-6}M) in 0.005M KNO$_3$ solutions, and below about 1000 μg l^{-1} (*ca.* 2 × 10^{-5}M) in 0.005M HNO$_3$. This inability to obtain Nernstian response even at relatively high concentrations in unbuffered solutions is attributed[1477] to the difficulty of controlling free ion concentrations of this magnitude in such media, by virtue of adsorption and/or contamination processes. (As discussed in Section 11.3.6, the latter may arise from dissolution of the electrode membrane itself).

Given the low metal concentrations prevailing in most natural and treated waters, the use of ion-selective electrodes to determine free ion concentrations will usually require that the free ion level is appropriately buffered by complexing equilibria in the sample.[1475] Additionally, the calibration step presents difficulties. Stiff,[1475] for example, calibrated his electrode system at high copper concentrations and then had to extrapolate the calibration over many orders of magnitude because of the low Cu^{2+} ion concentrations in real samples. The development of metal ion buffers (see above) should allow this difficulty to be overcome.

Other speciation studies which have employed ion-selective electrodes include those of Gardiner,[1476] Ramamoorthy and Kushner,[1479] Sunda and Hanson[1480] and Swallow *et al.*[1481] In the last three of these, samples were titrated with the metal ion of interest and, at each stage of the titration, the free metal ion concentration (determined by electrode) was compared with the total concentration. In some cases,[1479,1480] the technique was combined with various separation (filtration, ultra-filtration) and conversion (u.v., irradiation, pH adjustment) techniques in an attempt to gain more detailed speciation data. The work of Sunda and Hanson[1480] involved the use of certain check procedures – comparison with calculations using thermodynamic data, application of the scheme to test solutions – and is therefore particularly interesting. Total copper concentrations employed lay in the range 10^{-4}–10^{-8}M, and reported free Cu^{2+} concentrations lay in the range 10^{-8}–10^{-12}M. However, the check procedures revealed divergences between theoretical predictions and experimental observations at low copper concentrations, and the authors remarked that their experience suggested that 10^{-7}M was often ". . . the lower limit for accurate cupric ion measurement . . .". The lower limit depended upon the complexing characteristics of the solution, and in a test solution containing histidine (as an organic complexing agent) there was good agreement between the predicted and observed Cu^{2+} concentrations as low as 10^{-10}M. This is again an illustration of the importance of metal ion buffering in the application of ion-selective electrodes to speciation. A somewhat similar approach to that of Sunda and Hanson[1480] has been used by McCrady and Chapman,[1482] again to examine copper complexing. The free Cu^{2+} ion activity measured after adding spikes of copper to samples was compared with values calculated using thermodynamic data and differences between the two values interpreted as the result of complexing capacity exerted by organic ligands.

In addition to the problems of working in poorly buffered media, ion-selective electrodes are also subject to interferences (see Section 11.3.6). Thus, for example, Swallow *et al.*[1481] were unable to use the cupric ion-selective electrode in sea-water because of chloride interference,[1483] and Gardiner[1476] had to remove copper from one of his river water samples prior to spiking with cadmium because of the interference of the former with the cadmium electrode.

It is apparent that the application of ion-selective electrodes to trace metal speciation faces many difficulties. Their inability to measure accurately at the low levels frequently encountered in natural waters has been noted by Florence[1441] as a major limitation. However, the fact that they represent essentially the only measurement technique responsive to a single metal species should assure their continued role in speciation studies, albeit perhaps predominantly at relatively high metal levels and in well buffered media such as can be produced in model systems. Thus, such applications as the determination of equilibrium constants (*e.g.*, refs. 1476, 1484–1486) and the investigation of the effects of metal speciation on toxicity in laboratory experiments (*e.g.*, refs. 1487–1489) seem most likely to be fruitful.

Finally, it should be noted that ion-selective electrodes may be capable of indirect use to elucidate metal speciation. Driscoll[1490] has recently reported the use of free and total fluoride determinations (by fluoride ISE) with thermodynamic data to calculate the distribution of aluminium species in water. He notes, however, that this procedure, which exploits the strong complexing tendency of aluminium and fluoride ions, requires further evaluation. A similar approach has also been described by LaZerte.[1491]

(iv) Anodic stripping voltammetry ASV has received considerable attention in speciation studies, because it is a measurement technique which has some inherent differential response to different metal species, and because it is capable of determining a number of metals of environmental interest (*e.g.*, Cu, Cd, Pb, Zn) in water at low concentrations (μg l^{-1}–ng l^{-1}), without prior concentration. The basic principles of ASV (and related techniques) have been described in Section 11.3.7.

The basis for the differential response of ASV to different species of a trace metal has been outlined by a number of workers (see, for example, Buffle *et al.*[1492] and Davison[1493]). Consider a metal (M) and a ligand (L) forming a series of complexes:

$$M \underset{\rightleftharpoons}{\overset{L}{}} ML \underset{\rightleftharpoons}{\overset{L}{}} ML_2 \underset{\rightleftharpoons}{\overset{L}{}} \dots ML_n$$

(for clarity, charges have been omitted).

The electroactive species is not necessarily only the free ion, and Davison[1493] refers to cases where reduction proceeds *via* a more readily reduced species. However, for systems which are electrochemically reversible, the species reduced cannot be distinguished and it is customarily considered to be the free ion. The system will be reversible if the rates of exchange of the train of equilibria are rapid in relation to the timescale of the measurement, in which case the complexes are said to be labile (it follows, of course, that the lability or otherwise of a complex will depend on the conditions of measurement). When this condition is met, the stripping current will be virtually identical to that for a simple aqueous sample of the free metal ion (assuming that the plating step is performed at a sufficiently negative potential) – any slight difference will be the result of different diffusion coefficients. However, there will be a shift in the potential at which reduction occurs; this is a thermodynamic effect which can be exploited to evaluate the stability constant(s) of the complex(es). In polarography, the shift in half-wave potential is employed in this manner, using the Lingane[1494] equation. In ASV, the shift in stripping potential has been used directly,[1495] or 'pseudo-polarograms' have been constructed by performing ASV cycles with different plating potentials, and the half wave potential shifts so obtained used in the Lingane equation.[1496]

Although the use of voltammetric and related techniques (such as differential pulse polarography) to measure stability constants in this way cannot be regarded as speciation analysis, it is important to speciation studies in that the stability constants so obtained may be used to predict the equilibrium distribution of metal species under various conditions of water quality. Recent work on lead solubility in relation to lead in drinking water[1460,1465,1466] and on the speciation of heavy metals in sea and lake waters[1497] exemplifies the approach.

If the kinetics of ligand exchange are slow in relation to the timescale of measurement, the complexes are said to be non-labile (under the particular conditions) and the observed current will be – in the extreme – negligible in comparison with that produced by labile complexes. The stability constant of such an inert species cannot be measured by the approach described above, for the reduction of the complex usually takes place at a sufficiently negative potential, relative to that at which the free ion and any labile complexes present are reduced, for the two species to have separate reduction waves.[1492,1498] In principle, it is possible to measure the stability constant of such inert (non-labile) complexes by monitoring *either* the decrease in the free metal concentration *or* the increase in the complex concentration, during titrations of metal and ligand.[1492,1498] Valenta[1498] has noted that monitoring the free metal concentration is usually to be preferred, because reduction of the inert complex proceeds irreversibly.

Turning to the use of ASV in speciation analysis (rather than stability constant measurement), it may be noted that the approach adopted exploits the fact that non-labile complexes will not be determined by the technique unless a suitable pre-treatment is employed. Thus, for example, Allen, Mattson and Mancy[1499] interpreted an increase in the ASV current after the sample had been acidified in terms of release of metal bound in non-labile complexes at the natural pH. Many workers have used this basic approach, though the detailed procedures have varied. Labile metal has been determined in acetate buffer[1500] for example, and total metal has been determined after acid oxidation[1501] or u.v. photochemical oxidation in acid medium,[1439,1502] but the basic principle remains. The approach may also be combined with separation techniques, such as chelating resin separation (see below) to provide even more detailed fractionation.[1500]

The use of ASV in speciation studies, either to measure stability constants or in speciation analysis of real samples, is not as straightforward as the above summary might seem to imply, for a number of reasons. First, organic materials present in natural waters can give rise to interference effects which can change the peak currents and potentials, and even give rise to entirely spurious peaks (see ref. 1441 and literature cited therein, and also refs. 1503 and 1504). Because such effects may vary with, for example, pH[1503] and because they will not occur if the organic matter is destroyed in a determination of total metal, they can bias the measurements of both the labile and non-labile fractions.[1503]

A second problem in the use of ASV for speciation studies resides in the nature of the separation achieved. As has already been remarked, the concepts of lability and non-lability must be defined in terms of the timescale of measurement (*i.e.*, the diffusion layer thickness in ASV[1493]). Additionally, as Buffle *et al.*[1492] have emphasised, the concepts are limiting cases. Behaviour intermediate between that of labile complexes undergoing reversible reduction and non-labile complexes undergoing irreversible reduction may be encountered, and such quasi-reversibility of the electrode reactions of certain complexes can prejudice the interpretation of ASV signals. This topic has been discussed recently by Laxen and Harrison,[1439] who have drawn attention to the potential difficulties of calibrating the ASV response to the 'labile' metal fraction. This may in principle be accomplished by the method of standard additions, in which the sample is titrated with the metal ion until, after the (non-labile) complexation capacity of the sample has been exhausted, a region of higher current/concentration slope is attained, which slope is then used to quantify the original response to 'labile' metal. This, Laxen and Harrison[1439] point out, may be invalid if the 'labile' species have changed during the titration process because complexes undergoing quasi-reversible reduction (which will contribute to the 'labile' response) each require a separate calibration. They also suggest that, if quasi-reversible reduction of certain complexes is occurring, the inflexion in titration curves may represent a change in 'labile' speciation rather than a saturation of (non-labile) complexation capacity. The potential at which the stripping peak occurs may be monitored during titration as a diagnostic – if it changes, alteration in the 'labile' speciation may be suspected.[1439]

Similar considerations to those outlined above apply also to the use of ASV (and related techniques such as differential pulse polarography) for measurement of complex stability constants, as Buffle *et al.*[1492] have warned. They advise that the electrochemical process must be known before the ASV results are interpreted.

It will be apparent from the above discussions that the use of ASV in speciation *analysis*, particularly*, is subject to considerable problems and limitations. As in the case of spectrophotometry dealt with previously and separation techniques such as solvent extraction and chelating resins discussed below, the nature of the separation achieved may be far from ideal. Again, we see that quasi-reversible complexes may contribute to both the 'labile' and 'non-labile' fractions distinguished by ASV.

Despite the operational nature of the separation achieved by ASV, it has been reported[1505] that the 'labile' fraction determined by the technique has correlated with the toxicity of copper to shrimp larvae. However, this may reflect a further correlation between ASV-labile metal and free metal ion, and therefore not *necessarily* be inconsistent with the suggestion that the latter is often the most toxic form.[1453]

(v) Fluorescence   The fluorescence exhibited by humic substances is quenched by heavy metals. This effect, it is believed, arises mainly as a result of direct interaction (complexation) of the metals with the salicylate groups of the humic material, which act as fluorophors.[1506] Measurement of this quenching effect has been used to characterise the interaction of copper and lead with humic material.[1507]

(b) Separation Techniques
The separations so far considered have been accomplished by the measurement technique itself (except the use of chelation/solvent extraction in combination with spectrophotometry). Here we consider separation techniques which have no measurement capabilities.

(i) Filtration   As noted in Section 11.2.1, membrane filtration using a filter of 0.45 μm or similar pore-size has been widely used for the separation of so called 'dissolved' and 'particulate' metal fractions, better referred to as 'filterable' and 'non-filterable' respectively. Such filtration has often been used as the only separation in a speciation scheme, but it has also been widely employed as the first step in more complex analytical schemes.

The problems associated with filtration – most notably retention characteristics, contamination and adsorption – have been dealt with in Section 11.2.1 and will not be explored again here. A comprehensive review of filtration in relation to trace metal speciation analysis has been provided by de Mora and Harrison.[1445]

It was noted in Section 11.2.1 that filters of a wide range of pore-sizes are available, and this has been exploited by Laxen and Harrison[1439] who have devised a speciation scheme in which filtration through etched-track polycarbonate filters (see Section 11.2.1) of various pore-sizes (12, 1.0, 0.4, 0.08 and 0.015 μm) plays a major part. Adsorption losses were minimised by pretreatment of the filters and filtration cells using calcium solutions, and filter clogging – a potentially serious problem with filters of this type – was reduced by using stirred cells. It has been pointed out by de Mora and Harrison[1445] that parallel, rather than sequential,

* In measuring stability constants, the nature of the medium is within experimental control.

filtration should be used whenever possible to minimise trace metal adsorption errors in size fractionation, though this may not be feasible for the entire set of filtrations in waters of high particulate content.

For size fractionation work of this kind, polycarbonate etched-track filters are the only suitable type at present, since alternatives (such as cellulose acetate membrane filters) lack the necessary correspondence of nominal and effective pore-size – see ref. 1445 and literature cited therein. The relative expense and limited capacity of such filters are drawbacks, but they are less prone to adsorb trace metals than are cellulose-based membranes (see Section 11.2.1).

Given the ability of polycarbonate filters to provide a size differentiation of suspended particulate material without unacceptably large adsorptive losses of trace metal, their use in multiple-filtration speciation schemes of the kind described[1439] seems very attractive – especially since much of the trace metal content of natural waters appears to be associated with suspended matter.[1437,1439,1441] However, the need for speciation of the dissolved trace metal content remains of great importance because, as Batley has remarked,[1508] a primary reason for undertaking speciation studies is the importance of speciation to toxicity, and toxicity – to algae at least[1453] – appears to reside particularly in the dissolved phase.

(ii) Ultrafiltration 'Ultrafiltration' has been used[1445] to describe filtration through membranes having a pore-size of less than 15 nm, but this definition appears not to be universal, Steinnes referring[1444] to a pore-size of less than 100 nm. In practice, the use of polycarbonate filters (see above) at a pore-size of 15 nm has been found preferable to the use of ultrafiltration at a similar pore-size,[1439] so the former cut-off may possess some practical significance, and will be used here.

Ultrafilters consist of a filtration membrane on a porous support to provide strength. Because of their small pore-sizes (down to *ca.* 1 nm), the use of ultrafilters requires high pressures. Ref. 1445 provides a summary of the characteristics of a number of ultrafilters from one manufacturer, from which the illustrative data in the following table have been taken:

Type	Nominal pore-size nm	Nominal molecular weight cut-off (Daltons)
XM-300	14	300 000
XM-100A	5.1	100 000
YM-50	3.1	50 000
PM-30	2.2	30 000
PM-10	1.6	10 000
UM-2	1.4	1 000
UM-05	1.1	500

The molecular weight cut-off is defined in terms of 90% retention of a globular solute,[1445] but the molecular size and shape of the moiety being filtered will determine the fractionation achieved.[1445,1509] The potential value of ultrafiltration can, however, be seen in relation to the estimated diameters of species in natural waters (taken from ref. 1510): water molecule, 0–2 nm; solvated cupric ion, *ca.* 0.8 nm; humic substances*, 2–6 nm; colloidal species, 1–1000 nm.

The problems of applying ultrafiltration to trace element speciation are those associated with 'conventional' filtration – retention characteristics, adsorption and contamination – compounded by the greater practical difficulties in handling and usage of ultrafilters.

The size discrimination achieved by ultrafilters applied to natural organic material appears to be poor, possibly by virtue of such factors as adsorption, aggregation and shape-influenced retention (see ref. 1445 and literature cited therein). Contamination, both by trace metals[1512,1513] and by organic matter[1514] leached from the membranes, has been reported, but the latter can be eliminated and the former is thought not to be a major difficulty.[1445] By contrast, adsorption of metal species which should otherwise pass the membrane is a serious potential problem.[1444] Guy and Chakrabarti[1509,1510] pointed out that the porous backing of ultrafilters, with its channel structure, exposes a considerable area to the filtrate and they observed considerable adsorptive losses of copper, lead and manganese (at 20 μg l^{-1} in synthetic media) when using UM-2 – but not PM-10 – ultrafilters. Similarly, LaZerte[1491] has recently reported loss of aluminium by adsorption on UM-2 ultrafilters, at pH values as low as 3 and despite the addition of background electrolyte. The UM series of ultrafilters possess ionic sites, and may be susceptible to adsorption for this reason,[1445] but losses by sorption onto PM-10 membranes have been reported for several metals in river water by Beneš and Steinnes,[1512] who compared results obtained by ultrafiltration and *in situ* dialysis – see below.

Ultrafiltration has, despite these problems, found application in several speciation studies – see, for example, refs. 1509–1515. Laxen and Harrison[1439] however, found inconsistencies between the results obtained in the different size fractions separated by ultrafiltration applied to river waters, and reported serious metal adsorption losses to the smallest pore-size ultrafilters used (MW cut-off, 500). It would appear that the technique requires further investigation before its value in speciation analysis can be finally judged, but the problems in its use appear to be considerable.

In addition to its application in speciation analysis of the natural metal content of samples, ultrafiltration has been combined with complexation capacity measurements (see below) to assess in which molecular weight ranges the main metal binding capacity of waters lies. This approach has been used, for example, by Ramamoorthy and Kushner,[1479] Smith[1514] and Lee.[1516]

(iii) Dialysis In dialysis, which is similar in some respects to ultrafiltration, colloidal particles are separated from smaller species by virtue of the fact that they are unable to pass through the dialysis membrane. The nominal pore sizes of dialysis membranes lie in the range 1–5 nm, corresponding to molecular weight cut-offs of 1000–10 000 Daltons.[1444,1445] No pressure is applied to the sample, the permeable species diffusing across the membrane from the sample to the dialysate liquid on the other side. At equilibrium, the concentration of the metal of interest in the dialysate should, in principle, be the same as that of the dialysable forms of that metal in the original sample. However, in practice a number of complicating factors need to be considered.

---

* Fulvic acids are probably excluded, because they are expected to pass the PM-10 ultrafilter,[1511] having molecular weights of *ca.* 1000, which would be impossible if they had a diameter of greater than 2 nm.

(1) When a sample is dialysed in the laboratory, the dialysate volume must be small compared to that of the sample to avoid gross disturbance of equilibria in the latter, and an unrepresentative concentration of dialysable species in the dialysate, relative to that in the original sample.[1509] The availability of measurement techniques requiring only small volumes of sample (*i.e.*, dialysate) means that this requirement is not a serious limitation. In the case of *in situ* dialysis, in which a dialysis membrane bag containing deionised water is placed in the water body of interest and allowed to equilibrate, the problem disappears for all practical purposes.

(2) Adsorptive loss to the dialysis membrane (usually negatively charged[1444,1445]) may occur,[1444] though serious problems do not appear to have been encountered.[1445,1491] Again, in *in situ* dialysis the problem should not arise because the adsorption capacity can be saturated.

(3) Because dialysis membranes usually bear negative charge, anionic species will tend to be repelled and thus require extensive periods (up to several days[1512]) to equilibrate.

(4) In principle, Donnan effects* could cause the concentration of 'dialysable' species in the dialysate to be lower than that in the sample, if non-dialysable anions are present.[1491,1509,1510] Correction for such effects may be possible.[1491]

Dialysis has been employed in a small number of speciation studies, both laboratory[1491,1509–1511,1513] and *in situ*,[1512] and has also been used with titration to determine complexing capacities.[1518] The technique, particularly *in situ* dialysis, would seem to merit more detailed consideration than it has so far received.

(iv) Chromatographic techniques Gel filtration chromatography [GFC – see Section 11.3.8(*d*)] provides yet another means of separating organometallic species on the basis (in principle, at least) of molecular size. Its fundamental principles, and practical limitations in relation to the fractionation of organic matter in general, have been discussed in the above-mentioned section and will not be described again here.

As de Mora and Harrison[1445] have noted, the main potential advantage of GFC over ultrafiltration in speciation analysis is its ability to provide a continuous, rather than discrete, size separation. However, its application involves high dilutions of the species of interest during the elution step and (since preconcentration of samples cannot be performed†) this means that the technique has found few applications in speciation work. Bender, Matson and Jordan[1519] used GFC to separate organic species of copper and lead after spiking sewage effluents with the metals, in order to identify the molecular weight ranges of the major complexing substances in the effluents. Similarly, Sugai and Healy[1520] used the technique to study the major complexing factions of various marine samples, again with reference to copper and lead.

The use of GFC to characterise metal complexes in unspiked samples of bacterial

---

* For a discussion of the Donnan effect, see reference 1517 and the other references cited.
† It would almost certainly change the speciation if say, the sample was evaporatively concentrated.

growth media, and activated sludge process effluent, has been described by Sterritt and Lester.[1521,1522]

Recently, de Mora and Harrison[1523] have published the results of a critical evaluation of GFC for lead speciation analysis. This showed that lead adsorption was a severe problem, leading to poor recovery of the metal, asymmetric peak shapes and high partition coefficients not consistent with a size exclusion mechanism. The use of a lanthanum salt as eluent reduced these problems (by competing for the adsorption sites) but could not be recommended for routine use because of its likely competition for lead complexing sites within the sample, as well as for adsorption sites on the stationary phase. These authors concluded that only qualitative information could be gained from application of GFC to lead speciation in tap waters.

Given the difficulties of using GFC to separate organic substances on the basis of their molecular weight [see Section 11.3.8(*f*)], compounded by the adsorption and dilution difficulties involved in application to trace element speciation, it is hardly surprising that reviews of physical separation techniques in speciation[1444,1445] appear to agree that GFC offers little promise in this field. It seems possible, however, that size exclusion chromatography [of which GFC is a variant – see Section 11.3.8(*d*)] may yet have a larger role in speciation as new, more rigid stationary phases are developed for use with HPLC [see Section 11.3.8(*d*)]. The possibilities of more rapid separation using smaller columns than in conventional GFC *may* allow the adsorption and dilution problems to be reduced.

HPLC with other variants of chromatographic separation (*e.g.*, reverse phase and ion exchange) has also found application to trace element speciation analysis, particularly for well defined species such as organic compounds and/or different oxidation states of arsenic, chromium, lead, selenium and tin. In such work, HPLC separation has commonly been followed by graphite furnace atomic absorption detection, though other detection systems (*e.g.*, plasma emission spectrometry, solution spectrophotometry and amperometry) have also been employed.

Such applications of HPLC to trace element speciation analysis have been considered in Section 11.3.8(*f*), which gives details of reviews and individual references.

GC has also found application in trace element speciation, though it is restricted to suitably volatile species and volatile derivatives. Thus, for example, GC has been applied to the separation of organic compounds of arsenic, antimony, lead, selenium and tin. For details of individual papers and review articles, Section 11.3.8(*c*) should be consulted; a recent paper by Andreae[1524] also gives an excellent account of the application of such techniques to the speciation of the hydride-forming elements arsenic, antimony and tin (see below).

In such applications of GC to speciation, both atomic absorption detection and electron capture detection have been applied [see Section 11.3.8(*c*)]. In the simplest cases, the chromatographic separation consists solely of the selective volatilisation of the volatile species (substituted hydrides of arsenic[1525–1527] or antimony,[1528] for example) from a cold trap. In other cases, separation of the volatile species is accomplished by means of a more conventional GC step – for example, the determination of organo-arsenic,[1527,1529] organo-selenium,[1530,1531] organo-lead[612–614] and organo-tin.[608,609]

(v) Chelation and ion-exchange As in the case of solution spectrophotometry, the selective reaction of added chelating reagents with particular forms of trace elements can be exploited as a separation technique in speciation analysis. Although both chelating resins (in column or batch mode) and chelation/solvent extraction systems can, in principle, be employed in this manner, the former appear to have been most widely used.

Florence and Batley[1532,1533] drew attention to the fact that chelating ion-exchange resins (commonly used for trace element preconcentration) will generally not recover all forms of trace metals from natural water samples without appropriate pretreatment (such as u.v. irradiation[1533]). They and other workers have exploited this behaviour by using resins to separate weakly bound ('labile' or resin-reactive) and strongly bound ('non-labile' or resin-unreactive) metal fractions in water samples – see, for example, refs. 1439, 1500, 1513 and 1534. Chelation/solvent extraction[1491,1535,1536] and conventional ion-exchange resins[1490,1537,1538] have been used in a similar fashion.

Conventional ion-exchange resins have also been used to separate cationic and anionic forms of trace metals (most notably chromium) in water – see, for example, refs. 1539–1541. It has also been noted, in *(iv)* above, that ion-exchange techniques may be used in HPLC procedures for trace element speciation.

It must again be emphasised that, as in the cases of solution spectrophotometry and anodic stripping voltammetry, the fractionation achieved by chelating resins (and chelation/solvent extraction systems and conventional ion-exchange resins) may be far from ideal. As discussed previously in relation to solution spectrophotometry, both kinetic and thermodynamic factors may be involved, although the former will usually determine the separation achieved, and the same natural complex or chelate could contribute to both the 'labile' (or 'weakly bound') and 'non-labile' (or 'strongly bound') fractions.

(vi) Selective adsorption/coprecipitation Adsorption and coprecipitation processes, once widely used for concentrating and separating determinands prior to measurement, may exhibit a useful selectivity with respect to different forms of trace elements. This approach has, for example, been exploited in the determination of Cr^{III} and Cr^{VI}. Thus, Cranston and Murray[1542] determined Cr^{III} by coprecipitation with ferric hydroxide at pH B, and Cr^{III} *and* Cr^{VI} by coprecipitation with ferrous hydroxide; oxidation of the latter to ferric hydroxide reduced Cr^{VI} to Cr^{III}. Nakayama *et al.*[1543] determined Cr^{III} in the same manner, but used hydrated bismuth oxide at pH 8 to collect both Cr^{III} *and* Cr^{VI}.

(vii) Selective generation of volatile species The determination of such elements as arsenic, antimony, selenium and tin at low concentrations is often undertaken by generating the volatile hydrides of the elements, which are subsequently measured by various techniques including plasma optical emission spectrometry (Section 11.4.2) and – most commonly – AAS [Section 11.3.5(*d*)]. As noted in Section 11.3.5(*d*), the generation of the hydride from different oxidation states of the elements may vary (in rate and extent) according to the conditions of reduction, such that selective determination of the different oxidation states may be possible. This approach has been applied to the determination of As^{III} and As^{V} (*e.g.*

refs. 1525–1527), Sb[III] and Sb[V] (*e.g.*, ref. 1528), and Se[IV] and Se[VI] (*e.g.*, ref. 1531). In combination with the ability to determine individual organic compounds of these same elements by chromatographic separation of the corresponding substituted hydrides [see (*iv*) above] this differential response to different oxidation states permits a detailed speciation scheme to be developed. The recently published paper by Andreae,[1524] cited above, should again be consulted for further details.

A similar approach can be adopted in the case of mercury, using cold vapour techniques [most commonly with AAS measurement – see Section 11.3.5(*e*)]. For example, Goulden and Anthony[1544] have described a cold vapour AAS procedure in which the conditions of reduction are varied to obtain measurements of inorganic mercury, inorganic mercury *and* aryl mercury compounds or total mercury (including methylmercury). A somewhat similar procedure, but for the determination of inorganic and organic mercury fractions only, has been described by Ingle and Oda.[1545]

(c) Conversion Techniques

The conversion techniques used in speciation analysis to convert one species (or group of species) into another are those used in 'conventional' analysis to ensure that all forms of an element are recovered by the procedures for total element determination. These are acidification, oxidation and reduction. Reduction is used in dealing with elements which can exist in different oxidation states, and in hydride generation and cold-vapour procedures. Acidification and/or oxidation have been widely used in speciation analysis to release 'bound' (*i.e.*, complexed or adsorbed) element for measurement, as in conventional analysis; oxidation may be performed using a chemical oxidising agent (*e.g.*, persulphate) or by u.v. photolysis or ozonolysis (possibly in combination with a chemical oxidant).

Acidification and oxidation have been considered in Section 11.2.3 on pretreatment processes, and will not be considered in detail again here. It will be noted, however, that u.v. photolysis, widely used in speciation analysis to release organically bound metal, has been reported to induce precipitation of hydrous ferric oxide (which may adsorb the trace elements of interest) in some samples – see refs. 1439 and 1441 for a discussion of the problem and for additional references. The alternative ozonolysis approach is not without potential problems either – organic matter may be incompletely destroyed[1502] and Pb[II] may be oxidised and precipitated as Pb[IV].[1439]

11.5.7 Speciation Schemes

The available measurement, separation and conversion techniques (the most important of which have been discussed above) may be combined in many different ways to yield a multiplicity of speciation schemes. These range from a simple scheme for the determination of only two fractions (*e.g.*, the separation of 'filterable' and 'non-filterable' fractions, or of 'labile' and 'non-labile' fractions with respect to spectrophotometry, ASV or resin exchange) to complicated schemes intended to distinguish many more fractions – up to seven in some cases. Amongst these more complex schemes are those of Laxen and Harrison,[1439] Stiff,[1475] Batley and Florence,[1500] Hart and Davis[1511] and Figura and McDuffie.[1546] In some instances (*e.g.*, Stiff,[1475] LaZerte[1491]), speciation analysis may be combined with thermodynamic calculation (using available stability constants) to yield the final results, but this approach has not been widely practised.

A detailed examination of the numerous speciation schemes which have been devised by different workers is outside the scope of this book, and would add little to the evaluation of their component parts made above. Two particular points are worth making, however:

(1) Given that only one widely used measurement technique (the ion-selective electrode) is capable of responding to a single, well defined fraction of the total metal concentration, and that most of the separation techniques used in speciation analysis do not produce a clear-cut separation of well defined species*, the majority of the fractions distinguished by speciation schemes are non-specific in nature. That is, they are only defined by the experimental conditions. This is well illustrated by a listing of the seven fractions of the scheme described by Batley and Florence:[1500] (*a*) ASV-labile species removed by chelating resin, (*b*) ASV-labile organic species not removed by chelating resin, (*c*) ASV-labile inorganic species not removed by chelating resin, (*d*) ASV-non-labile organic complexes removed by chelating resin, (*e*) ASV-non-labile inorganic complexes removed by chelating resin, (*f*) ASV-non-labile species not removed by chelating resin and (*g*) ASV-non-labile inorganic species not removed by chelating resin. (Both lability under ASV, and the removal or otherwise by chelating resin, are determined by the precise conditions of application of the technique concerned.)

This non-specific nature of the fractions in turn means that: (a) all factors likely to affect the results obtained must be specified in methods if comparable results are to be obtained by other workers, and (b) to be of value, the fractions measured must be identifiable with particular species (or groups thereof), or with particular 'effects of interest' (such as toxicity).

(2) As the number of fractions distinguished increases, so does the complexity of the analytical procedures. In turn, the increased sample manipulation enhances the risks of loss of determinands (by adsorption) and of contamination. Moreover, the time required for analysis also increases (to about 8 h per sample in the Batley and Florence scheme,[1508] for example). The long analysis times then militate against the careful experimental work necessary to establish the correlation between the speciation measurements and the particular effects of interest.

Perhaps in part because of this last-mentioned factor, there appears to have been a tendency in recent years for simpler speciation schemes to be employed. Moreover, as Batley[1508] has remarked, in relation to the Batley and Florence scheme: "The classifications afforded by this scheme are possibly more complex than is required to define biological availability, although to date this has not been demonstrated by toxicity testing".

---

* The use of selective generation of volatile species, and of chromatography to separate well defined organic compounds of trace elements, constitute the exceptions to this statement.

11.5.8 Complexation Capacity Measurements

Given that the toxicity of many trace metals to aquatic organisms is diminished by complexation, the ability of a particular water to complex added free or inorganic metal species is clearly a property of considerable interest. A detailed examination of complexation capacity measurements is beyond the scope of this book, but the topic is dealt with in a detailed review by Neubecker and Allen,[1446] and other reviews of speciation in general (*e.g.*, ref. 1441) often devote attention to the subject. A recent paper by Sterritt and Lester,[1547] which reviews the speciation of heavy metals in waste waters, pays particular attention to the determination of complexing capacities and complex stability constants.

The basic approach to measuring complexation capacity involves titration of the sample with the free metal ion, with some measurement of the resulting free metal ion concentration (or other 'labile' metal fraction) being made after each addition. By plotting the measured free ion or labile metal concentration against the total added metal concentration, a titration curve is obtained. In the simplest case, the curve will have the form of an initial region of relatively low slope, in which the incremement of total metal produces a smaller increment of free ion or labile metal, by virtue of complexation, followed by (at higher concentrations of added metal) a line of higher slope. The latter reflects the absence of complexing above the point of inflection (the projection of which on the total metal concentration axis represents the complexation capacity), and the slope is then unity – each increment of added metal producing an equal increment of free ion or labile metal.

The case described above refers to the case when a single complexing substance is being titrated. In the case of mixtures of ligands, sequential titration of each will be apparent provided that the products of their stability constants (for complex formation with the added metal) and concentrations are sufficiently different.[1446]

From the titration curve data, it is possible in principle to obtain not only the complexing capacity but also the conditional stability constant(s) of the complex(es), though the difficulty of estimating the latter increases as the number of ligands increases.

A very wide variety of techniques has been used in measurements of complex capacity (and of conditional stability constants). Ion-selective electrode measurement has the advantage, as in all speciation work, of allowing the free metal ion concentration to be followed during the titration, but cannot be used at low metal concentrations, such as are necessary to titrate the low complexing capacity of most natural waters.[1446] Other techniques [*e.g.*, ion-exchange resin competition with the natural ligand(s), solubilisation of metal salts by the natural ligand(s), ASV measurement of the 'labile' metal] have been used to attain lower concentrations. However, their use normally imparts an operational character to the complexation capacity (just as, for example, the use of ASV does in general speciation analysis), so that different techniques measure different 'complexation capacities'.[1446]

In addition to the above problem of the operational character of most complexation capacity measurements, the latter may also vary according to the metal used as titrant. Thus, for example, copper normally has a greater propensity for complex formation than does cadmium, so that – other things being equal – complexation capacities measured by copper titration will be higher than those measured by cadmium titration.

Despite the evident problems of measuring and interpreting complexation capacities, they have found considerable application as a means – albeit imperfect – of describing the likely fate (and, by implication, effects) of trace metals added to natural waters. As in other areas of speciation analysis, however, further work is required to make the measurements of greater practical value, *e.g.*, as a means of predicting the likely biological impact of an increase in the total trace metal content of a natural water.

11.5.9 Trace Element Speciation Analysis of Particulate Matter

It is beyond the scope of this text to deal with the speciation analysis of sediment in great detail, but some consideration must be given to this topic because the techniques used may be applied to suspended particulate matter concentrated from water samples by filtration or centrifugation.

The speciation analysis of sediment is commonly undertaken by selective chemical digestion procedures. Förstner and Wittman,[1548] in their comprehensive text on metal pollution in aquatic systems, have summarised many of the schemes devised by different workers. Although there is wide variation in the details of the procedures, certain basic extraction techniques are common to many of the speciation schemes.

Thus, displacement by magnesium or barium salts, or extraction with ammonium acetate, has been used to determine 'adsorbed' or 'exchangeable' metal, hydroxylamine/acid digestion has been used to extract metals associated with iron and manganese oxides, acetic acid/acetate buffer has been used to extract metals present in carbonate phases and hydrogen peroxide has been used to extract metals present in sulphide and organic phases. Mineral acid digestion, as a final step, extracts 'residual' metal strongly bound within the sediment, not released by the less vigorous treatments. Schemes based on these basic steps have been devised by, for example, Tessier, Campbell and Bisson[1549] and Salomons and Förstner.[1550]

None of the extraction steps used for trace metal speciation in sediments is, however, specific for a particular phase. Careful assessment of the degree of specificity attained using different schemes, using 'model' phases, has revealed that considerable overlap may occur between the metal fractions obtained by the various extraction steps. Thus, for example, Rapin and Förstner[1551] have found that the use of ammonium acetate to extract 'adsorbed' or 'exchangeable' metal also caused release from the carbonate phase and that iron is extracted, to some extent, from ferrous sulphide by a sodium acetate/acetic acid mixture intended to attack carbonates. Likewise, Meguellati *et al.*[1552] found copper and lead associated with humic acid to be substantially extracted by sodium acetate. Both these groups of workers also found the hydroxylamine extraction step ineffective for the dissolution of certain oxides – well crystallised iron oxide was resistant in the study by Rapin and Förstner,[1551] and manganese dioxide in the study by Meguellati *et al.*[1552]

These few examples reveal very clearly that the speciation analysis of sediments, like that of waters, is fraught with difficulties, and that the fractions 'separated' are again operationally defined. The importance, in this context, of unambiguous definition of procedures to the attainment of comparable analytical results is again emphasised. Because of the lack of specificity of extraction steps, the order in which they are performed can have an important impact on the results obtained – thus, for example, Meguellati et al.[1552] recommend that peroxide/acid digestion precede attack on carbonates (by acetic acid/acetate) and on hydroxides (by hydroxylamine/acetic acid) because the use of peroxide/acid to extract metal in organic and sulphide phases has little effect on the carbonate and hydroxide phases, whereas the carbonate and hydroxide extraction steps both affect the organic/sulphide phases.

11.5.10 The Storage of Water Samples for Trace Element Speciation Analysis

It is well recognised that trace elements, including heavy metals such as lead, copper and zinc, are liable to be lost from solution during storage of samples, as a result of adsorption to the container walls and/or of precipitation, unless suitable preservation is employed. This topic has been considered in Section 3.6.3, in which details of recommend preservation procedures are given.

The use of preservatives, such as mineral acids to prevent heavy metal adsorption and hydrolysis, cannot be undertaken if the speciation of the metal, rather than its total concentration, is to be determined. Acidification will radically alter the distribution of the metal between its different forms, and thereby invalidate subsequent speciation analysis. The same is true of other preserving agents which may be employed in 'total' metal determinations (e.g., oxidising agents in the case of mercury).

The inability to use preserving agents in trace element speciation work carries potentially serious implications, and the subject has been discussed by Florence[1441] in his recent general review of trace element speciation analysis. Freezing of water samples, he points out, may also alter the speciation by concentrating the element in the unfrozen liquid during the freezing process, which may in turn lead to irreversible reactions (such as polymerisation of iron hydrolysis products at the elevated concentrations). He concludes that storage at 4 °C '. . . appears to be the safest for both fresh water and sea-water . . .' though freezing of the latter could be acceptable because the high chloride concentration may prevent hydrolysis reactions occurring.[1441]

Storage of samples at low temperatures will reduce the rate of bacterial and enzymatic processes, but loss of metal by adsorption onto container walls may be little affected by holding at 4 °C. The problem of such loss in unacidified samples has also been discussed by Florence.[1441] He concludes that, with one exception, workers who found serious adsorption losses had used synthetic samples, whereas those who had conducted tests using real samples had found losses of most metals to be negligible. This result was attributed[1441] to the fact that natural waters are in equilibrium with particulate materials of high adsorbing capacity, in relation to the polyethylene walls of a sample bottle, and their containment in the latter therefore induces little adsorption and alteration in speciation. Though this may be widely true of natural waters having low concentrations of trace elements in relation to the adsorption capacity of the particulate material present, it may well not apply to waters containing fairly high concentrations of trace elements, but only low concentrations of suspended matter (e.g., drinking water samples containing lead or copper from dissolution of pipe deposits). In any event, insufficient knowledge exists of the chemical behaviour of trace elements (in relation to adsorption on surfaces) to allow assumptions to be made about the stability or otherwise of their speciation in storage, and experimental tests (see Section 3.6.4) are advisable whenever speciation work is undertaken.

It has been noted that acid treatment of storage vessels, to reduce trace metal contamination, may in fact promote adsorption of the metals from samples by converting adsorption sites to the hydrogen form, thereby activating them.[1441,1553] One way to resolve this conflict of priorities may be to check the contamination from bottles which have not been acid washed, using a solution of acid of lower pH than the samples. If negligible contamination occurs, it may be possible to use the bottles without an acid wash and thereby reduce the ability of the surface to adsorb. However, care would have to be taken to ensure that a sufficiently large sub-sample of the bottle stock was tested. The best approach to the problem is probably to acid wash the bottles, but rinse them thoroughly many times with the water being sampled before the final filling, in an attempt to equilibrate the bottle surface with the sample being collected.[1441]

It should be clear from the above that the problems of storing unpreserved samples for speciation analysis may be considerable, and the subject remains poorly understood.

In some cases, it may be possible to eliminate, or at least reduce, potential problems of the kind described by appropriate choice of technique and/or procedure. Thus, for example, *in situ* dialysis [see Section 11.5.6(*b*)] effectively overcomes adsorption and contamination by allowing the dialysis membrane to equilibrate with the sample. Again, it may be possible to conduct a separation step (*e.g.* using ion-exchange or chelating resins) immediately upon sample collection, thereby avoiding the storage of potentially unstable samples. Elution of the resin, and measurement of the exchanged trace element, can then be accomplished on return to the laboratory. The use of automatic, on-line speciation systems may also circumvent the problems of sample storage. However, care would then need to be taken to check that the speciation was not altered by adsorption of the element of interest on the sampling pipework.

11.5.11 Speciation Analysis – Current Position and Prospects

The success of conventional analysis, to determine the *total* concentrations of trace elements, usually depends upon the conversion of the various chemical species of the element of interest into one particular form (either by pretreatment or as part of the measurement stage itself) before the appropriate property is measured. Such conversion is normally unnecessary only when the measurement technique exploits a property of the element which is not affected by chemical form (*e.g.*, nuclear properties, as in NAA). In most conventional analysis, therefore, disturbance of the (often delicately poised) network of equilibria involving the determinand in the original sample is a necessary part of the procedure.

By contrast, the ideal speciation scheme would permit measurement of all the important species of the element of interest, without perturbation of the existing equilibria in such a manner as to change the concentrations of the different species. In relatively few cases, however, can this ideal even be approached at present.

Speciation analysis has undoubtedly been most straightforward and successful in cases where the species are well defined and relatively stable in relation to the timescale of separation. Thus, for example, it has proved very successful in allowing selective determination of the different oxidation states and organic compounds of the 'hydride-forming' elements such as arsenic, selenium and antimony, and of the organic compounds of lead.

With respect to the commonly available measurement techniques for dealing with less straightforward systems, only ion-selective electrodes respond to a single, well defined metal species. Others, such as ASV and solution spectrophotometry, are not selective for a single species and exhibit a complex response to mixtures of metal species, depending upon the dissociation kinetics of the complexes in relation to the timescale of measurement, and upon the relevant stability constants and ligand concentrations.

Among the separation techniques, filtration is widely used and can provide a useful, operational separation; it may, however, be subject to problems of adsorption which can bias the determination of both the filterable and non-filterable fractions. Ultrafiltration, though capable of separating much smaller entities, suffers from even more severe adsorption problems. Adsorption of trace elements on gel filtration chromatography columns has also proved to be a severe limitation to the application of that technique, which is capable – in principle – of providing a continuous (rather than discrete) size separation. Factors affecting the separation achieved, other than molecular weight, also complicate use of the gel filtration technique.

Of the size-based separation techniques, dialysis appears (in its *in situ* modification, especially) to offer some important advantages over ultrafiltration, most notably the freedom from adsorption effects conferred by lengthy equilibration with large volumes of sample. The technique would seem to merit greater attention than it has hitherto received in trace element speciation work.

It has been noted that, to be useful, the trace element fractions distinguished by speciation analysis must be: (*a*) identifiable with chemically or physically defined species or groups of species or (*b*) capable of association with effects of interest, such as toxicity. In most speciation schemes, the operational nature of the separations achieved (*i.e.*, the non-specific nature of the determinands) precludes outcome (*a*), and the usefulness or otherwise of the separation achieved must be judged on the extent to which it allows effects of interest to be predicted.

It is perhaps hardly to be expected that empirical trace metal fractions distinguished by speciation analysis will provide perfect predictions of effects of interest, such as toxicity to an aquatic organism. For this to occur with, for example, a separation based on a chelating ion exchange resin, the interaction of the latter with the trace metal species in a water sample would have to mimic exactly the interaction of the organism of interest with the same metal species in the same water. Such a close parallel between a chemical and a biological system seems most unlikely. However, it cannot be concluded from this that operational or non-specific measures of trace element speciation are always valueless. Provided that the operational measurement gives a better prediction of the toxicity effect than does the total metal concentration, over the range of water quality conditions of interest, the replacement of determinations of total metal concentration by determinations of the relevant 'fraction' using speciation analysis may be practically useful.

It is therefore to be expected that, in future, studies of analytical speciation techniques will increasingly be carried out in conjunction with investigations of effects of interest. The trend is already apparent, in that studies of trace metal speciation and of toxicity are now often undertaken in concert.

Further progress in understanding trace element speciation, and its influence upon the behaviour and effects of trace elements in aquatic systems, will come about through the application of many different approaches – thermodynamic calculation, speciation analysis, laboratory investigations of model systems and concerted studies of effects such as toxicity. The authors do not expect dramatic progress in speciation analysis in the near future, in the sense of new techniques becoming available which can extend our ability to determine individual, well defined chemical species. Rather, we expect a greater emphasis on seeking associations between operationally defined trace element fractions and effects of interest, for the reasons already stated. It also seems likely that the speciation schemes used will be relatively modest, in terms of numbers of fractions, to facilitate the study of such associations. Moreover, we can perhaps expect an increasing use of a smaller number of selected measurement and separation techniques, particularly those which can be used in the field to circumvent sample storage problems.

Expectations of speciation analysis have been tempered by a growth in understanding of the inherent limitations of the techniques available, but the potential benefits of a knowledge of trace element speciation in aquatic systems remain considerable.

11.6 REFERENCES

[1] G. W. Ewing, 'Instrumental Methods of Chemical Analysis', 4th Edition, McGraw-Hill, New York, 1975.

[2] R. L. Pecsok, L. D. Shields, T. Cairns and I. G. McWilliam, 'Modern Methods of Chemical Analysis', Wiley, New York, 1976.

[3] H. H. Bauer, G. D. Christian, and J. E. O'Reilly, 'Instrumental Analysis', Allyn and Bacon, Boston, 1978.

[4] G. D. Christian, 'Analytical Chemistry', 3rd Edition, Wiley, New York, 1981.

[5] M. Pinta, 'Modern Methods for Trace Element Analysis', Ann Arbor Science Publishers, Ann Arbor, 1978.

[6] K. H. Mancy, ed., 'Instrumental Analysis for Water Pollution Control', Ann Arbor Science Publishers, Ann Arbor, 1971.

[7] L. L. Ciaccio, ed., 'Water and Water Pollution Handbook', Dekker, New York, Vol. 1, 1971; Vol. 2, 1971; Vol. 3, 1972; Vol. 4, 1973.

[8] H. E. Allen, N. T. Mitchell, J. A. P. Steel, and K. K. Kristensen, in M. J. Suess, ed., 'Examination of Water for Pollution Control', Vol. 1, Pergamon Press, Oxford, 1982, pp. 125–210.

[9] H. B. Mark and J. S. Mattson, ed., 'Water Quality Measurement. The Modern Analytical Techniques', Dekker, New York, 1981.

[10] L. H. Keith, ed., 'Advances in the Identification and Analysis of Organic Pollutants in Water', Ann Arbor Science Publishers, Ann Arbor, Vol. 1 and Vol. 2, 1981.

[11] A. Bjørseth and G. Angeletti, ed., 'Analysis of Organic Micropollutants in Water', D. Reidel, Dordrecht, 1982.

[12] G. Angeletti and A. Bjørseth, ed., 'Analysis of Organic Micropollutants in Water', D. Reidel, Dordrecht, 1984.

[12a] L. G. Hutton, 'Field Testing of Water in Developing Countries', Water Research Centre, Medmenham, Bucks., 1983.

[13] K. Grasshoff, M. Ehrhardt, and K. Kremling, ed., 'Methods of Seawater Analysis', 2nd Ed., Verlag Chemie, Weinheim, 1983.

[14] J. P. Riley, in J. P. Riley and G. Skirrow, ed., 'Chemical Oceanography', 2nd Ed., Academic Press, London, 1975, Vol. 3, pp. 193–514.

[15] D. T. E. Hunt, 'Filtration of Water Samples for Trace Metal Determinations', Technical Report TR 104, Water Research Centre, Medmenham, Bucks., 1979.

[16] S. J. de Mora and R. M. Harrison, *Water Res.*, 1983, **17**, 723.

[17] Environmental Protection Agency, 'Methods for Chemical Analysis of Water and Wastes', EPA-600/4-79-020, The Agency, Cincinnati, 1979.

[18] American Public Health Association, *et al.*, 'Standard Methods for the Examination of Water and Wastewater', 15th Ed., APHA, Washington, 1980.

[19] H. L. Golterman, R. S. Clymo, and M. A. M. Ohnstad, 'Methods for Physical and Chemical Analysis of Fresh Waters', 2nd Ed., Blackwell Scientific Publications, Oxford, 1978.

[20] R. W. Sheldon, *Limnol. Oceanogr.*, 1972, **17**, 494.

[21] P. E. Biscaye and E. Eittreim, in R. J. Gibbs, ed., 'Suspended Solids in Water', Plenum Press, New York, 1974, pp. 227–260.

[22] J. D. H. Strickland and T. R. Parsons, 'A Practical Handbook of Seawater Analysis', Fisheries Research Board of Canada, Ottawa, 1969.

[23] M. W. Skougstad and G. F. Scarbro, *Environ. Sci. Technol.*, 1968, **2**, 298.

[24] G. E. Batley and D. Gardner, *Water Res.*, 1977, **11**, 745.

[25] A. Zirino and M. L. Healey, *Limnol. Oceanogr.*, 1971, **16**, 773.

[26] K. T. Marvin, R. R. Proctor, and R. A. Neal, *Limnol. Oceanogr.*, 1970, **15**, 320.

[27] D. W. Spencer and F. T. Manheim, 'Ash Content and Composition of Millipore HA Filters', *U.S. Geol. Surv. Prof. Pap. No. 650-D*, 1969, pp. 288–290.

[28] D. W. Spencer and P. G. Brewer, *CRC Crit. Rev. Solid State Sci.*, 1970, **1**, 409.

[29] H. R. Watling and R. J. Watling, *Water S.A.*, 1975, **1**, 28.

[30] M. J. Gardner and D. T. E. Hunt, *Analyst*, 1981, **106**, 471.

[31] M. J. Gardner, 'Adsorption of Trace Metals from Solution during Filtration of Water Samples', Technical Report TR 172, Water Research Centre, Medmenham, Bucks., 1982.

[32] J. Gardiner, *Water Res.*, 1974, **8**, 157.

[33] H. S. Posselt, 'Environmental Chemistry of Cadmium in Aqueous Systems', Ph.D. Thesis, University of Michigan, 1971, pp. 59–60.

[34] G. A. Parks, in J. P. Riley and G. Skirrow, ed., 'Chemical Oceanography', 2nd Ed., Academic Press, London, 1975, Vol. 1, pp. 241–308.

[35] H. W. Nürnberg, P. Valenta, L. Mart, B. Raspor, and L. Sipos, *Fresenius' Z. Anal. Chem.*, 1976, **282**, 357.

[36] R. E. Truitt and J. H. Weber, *Anal. Chem.*, 1979, **51**, 2057.

[37] T. M. Florence, *Talanta*, 1982, **29**, 345.

[38] L. Mart, *Fresenius' Z. Anal. Chem.*, 1979, **296**, 350.

[39] J. B. Andelman and S. C. Caruso, in L. L. Ciaccio, ed., ref. 7, pp. 483–591.

[40] B. Orpwood, 'Concentration Techniques for Trace Elements–A Review', Technical Report, TR 102, Water Research Centre, Medmenham, Bucks., 1979.

[41] J. S. Fritz, in E. Reid, ed., 'Trace-Organic Sample Handling', Methodological Surveys – Sub-series (A): Analysis, Vol. 10, Ellis Horwood Ltd., Chichester, England, 1981, pp. 68–75.

[42] I. Wilson, in E. Reid, ed., ref. 41, pp. 76–85.

[43] C. D. Watts, B. Crathorne, R. I. Crane, and M. Fielding, in L. H. Keith, ed., ref. 10, Vol. 1, pp. 383–397.

[44] American Public Health Association *et al.*, ref. 18, p. 141.

[45] Methods for the Examination of Waters and Associated Materials, 'Manganese in Raw and Potable Waters by Spectrophotometry. 1977', Her Majesty's Stationery Office, London, 1978.

[46] Methods for the Examination of Water and Associated Materials, 'Acid Soluble Aluminium in Raw and Potable Waters by Spectrophotometry. 1979', Her Majesty's Stationery Office, London, 1980.

[47] R. J. Oake, 'A Review of Photo-Oxidation for Determination of Total Organic Carbon in Water', Technical Report TR 160, Water Research Centre, Medmenham, Bucks., 1981

[48] Methods for the Examination of Waters and Associated Materials, 'The Instrumental Determination of Total Organic Carbon, Total Oxygen Demand and Related Determinands. 1979', Her Majesty's Stationery Office, London, 1980.

[49] R. J. Oake and R. L. Norton, 'The Photochem Organic Carbon Analyser: An Evaluation', Technical Report TR 186, Water Research Centre, Medmenham, Bucks., 1982.

[50] Methods for the Examination of Waters and Associated Materials, 'Arsenic in Potable and Sea Waters by Spectrophotometry. 1978', Her Majesty's Stationery Office, London, 1979.

[51] R. G. Clem and A. T. Hodgson, *Anal. Chem.*, 1978, **50**, 102.

[52] J. Yates, 'Determination of Trace Metals (Cadmium, Copper, Lead, Nickel and Zinc) in Filtered Saline Water Samples', Technical Report TR 181, Water Research Centre, Medmenham, Bucks., 1982.

53 Methods for the Examination of Waters and Associated Materials, 'Iron in Raw and Potable Waters by Spectrophotometry (using 2,4,6-tripyridyl-1,3,5-triazine), 1977', Her Majesty's Stationery Office, London, 1978.

54 W. Wagner and C. J. Hall, 'Inorganic Titrimetric Analysis: Contemporary Methods', Dekker, New York, 1971.

55 A. I. Vogel, 'A Text-book of Quantitative Inorganic Analysis', 3rd Ed., Longmans, London, 1964.

56 C. L. Wilson and D. W. Wilson, ed., 'Comprehensive Analytical Chemistry', Volume 1B, Elsevier, Amsterdam, 1960.

57 C. L. Wilson and D. W. Wilson, ed. ref. 56, Volume 1C, Elsevier, Amsterdam, 1962.

58 I. M. Kolthoff and V. A. Stenger, 'Volumetric Analysis', Volume I, 2nd Ed., Interscience, New York, 1942.

59 I. M. Kolthoff and V. A. Stenger, 'Volumetric Analysis', Volume II, Interscience, New York, 1964.

60 I. M. Kolthoff and R. Belcher, 'Volumetric Analysis', Volume III, Interscience, New York, 1957.

61 E. Bishop, ed., 'Indicators', International Series of Monographs in Analytical Chemistry, Volume 51, Pergamon, Oxford, 1972.

62 G. Schwarzenbach and H. Flaschka, 'Complexometric Titrations', Translated by H.M.N.H. Irving, Second English Ed., Methuen, London, 1969.

63 J. B. Headridge, 'Photometric Titrations', International Series of Monographs in Analytical Chemistry, Volume 4, Pergamon, New York, 1961.

64 J. T. Stock, 'Amperometric Titrations', Wiley-Interscience, New York, 1965.

65 H. J. W. Tyrell and A. E. Beezer, 'Thermometric Titrimetry', Chapman and Hall, London, 1968.

66 American Public Health Association, *et al.*, ref. 18, pp. 483–489.

67 Methods for the Examination of Waters and Associated Materials, 'The Determination of Alkalinity and Acidity in Water. 1981', Her Majesty's Stationery Office, London, 1982.

68 Methods for the Examination of Waters and Associated Materials, 'Biochemical Oxygen Demand. 1981', Her Majesty's Stationery Office, London, 1983.

69 Methods for the Examination of Waters and Associated Materials, 'Boron in Waters, Effluents, Sewage and some Solids, 1980', Her Majesty's Stationery Office, London, 1981.

70 Methods for the Examination of Waters and Associated Materials, 'Bromide in Waters. High Level Titrimetric Method. 1981', Her Majesty's Stationery Office, London, 1981.

71 Methods for the Examination of Waters and Associated Materials, 'Total Hardness, Calcium Hardness and Magnesium Hardness in Raw and Potable Waters by EDTA Titrimetry. 1981', Her Majesty's Stationery Office, London, 1982.

72 Methods for the Examination of Waters and Associated Materials, 'Chemical Oxygen Demand (Dichromate Value) of Polluted and Waste Waters, 1977', Her Majesty's Stationery Office, London, 1978.

73 Methods for the Examination of Waters and Associated Materials, 'Chloride in Waters, Sewage and Effluents. 1981', Her Majesty's Stationery Office, London, 1982.

74 Methods for the Examination of Waters and Associated Materials, 'Chemical Disinfecting Agents in Water and Effluents, and Chlorine Demand. 1980', Her Majesty's Stationery Office, London, 1980.

75 Methods for the Analysis of Waters and Associated Materials, 'Dissolved Oxygen in Natural and Waste Waters. 1979', Her Majesty's Stationery Office, London, 1980.

76 Methods for the Analysis of Waters and Associated Materials, 'Ammonia in Waters. 1981', Her Majesty's Stationery Office, London, 1982.

77 Methods for the Analysis of Waters and Associated Materials, 'Oxidised Nitrogen in Waters. 1981', Her Majesty's Stationery Office, London, 1982.

78 Methods for the Analysis of Waters and Associated Materials, 'Sulphate in Waters, Effluents and Solids. 1979', Her Majesty's Stationery Office, London, 1980.

79 Methods for the Analysis of Waters and Associated Materials, 'Sulphide in Waters and Effluents. 1983', Her Majesty's Stationery Office, London, 1983.

80 G. L. Baughman, B. T. Butler, and W. M. Sanders, *Water Sewage Works*, 1969, **116**, 359.

81 A. Montiel and J. J. Dupont, *Trib. Cebedeau*, 1974, **27**, 27.

82 E. Barbolani, G. Piccardi, and F. Pantani, *Anal. Chim. Acta*, 1981, **132**, 223.

83 F. Ehrenberger, *Fresenius' Z. Anal. Chem.*, 1981, **306**, 24.

84 H. Clysters, F. Adams, and F. Verbeek, *Anal. Chim. Acta*, 1976, **83**, 27.

85 S. Virojanavat and C. O. Huber, *J. Water Pollut. Contr. Fed.*, 1979, **51**, 2941.

86 E. P. Scheide and R. A. Durst, *Anal. Lett.*, 1977, **10**, 55.

87 M. Mascini, *Analyst*, 1973, **98**, 325.

88 P. Bruno, M. Caselli, A. Di Fano, and A. Traini, *Anal. Chim. Acta*, 1979, **104**, 379.

89 D. Chakraborti and F. Adams, *Anal. Chim. Acta*, 1979, **109**, 307.

90 K. Ohzeki, E. Schumacher, and F. Umland, *Fresenius' Z. Anal. Chem.*, 1978, **293**, 18.

91 E. Jacobsen and G. Tandberg, *Anal. Chim. Acta*, 1973, **64**, 280.

92 E. C. Potter and J. F. White, *J. Appl. Chem.*, 1957, **7**, 459.

93 F. J. Millero, S. R. Schrager, and L. Hansen, *Limnol. Oceanogr.*, 1974, **19**, 711.

588   *Chemical Analysis of Water*

94 A. Graneli and T. Anfalt, *Anal. Chim. Acta*, 1977, **91**, 175.
95 T. Anfalt and A. Graneli, *Anal. Chim. Acta*, 1976, **86**, 13.
96 D. Jagner, *Anal. Chim. Acta*, 1973, **68**, 83.
97 H. Sato and K. Mamoki, *Anal. Chem.*, 1972, **44**, 1778.
98 W. J. MacKellar, R. S. Wiederanders, and D. E. Tallman, *Anal. Chem.*, 1978, **50**, 160.
99 C. Woodward and H. N. Redman, 'High-Precision Titrimetry', Society for Analytical Chemistry, London, 1973.
100 'I. R. S. Wilson, in M. Whitfield and D. Jagner, ed., 'Marine Electrochemistry. A Practical Introduction', Wiley, Chichester, 1981, pp. 145–185.
101 G. W. C. Milner and G. Phillips, 'Coulometry in Analytical Chemistry', Pergamon, Oxford, 1967.
102 R. T. Moore and J. A. McNulty, *Environ. Sci. Technol.*, 1969, **3**, 741.
103 D. K. Albert, R. L. Stoffer, I. J. Oita, and R. H. Wise, *Anal. Chem.*, 1969, **41**, 1500.
104 Y. Kumagai and T. Nakata, *Kogyo Yosui*, 1971, (158), 48.
105 J. Hissel and J. Price, *Bull. Cebedeau*, 1959, (44), 76.
106 T. J. Rohm, H. C. Nipper, and W. C. Purdy, *Anal. Chem.*, 1972, **44**, 869.
107 H. L. Golterman, R. S. Clymo, and M. A. M. Ohnstad, 'Methods for Physical and Chemical Analysis of Fresh Waters', IBP Handbook No. 8, 2nd Ed., Blackwell Scientific Publications, Oxford, 1978, p. 146.
108 F. Strafelda, *Sb. Vys. Sk. Chem. Technol., Praze, Anal. Chem.*, 1970, (6), 31.
109 E. Schumacher and B. Hackmann, *Fresenius' Z. Anal. Chem.*, 1981, **305**, 21.
110 M. C. Van Grondelle, F. Van de Craats, and J. D. Van der Laarse, *Anal. Chim. Acta*, 1977, **92**, 267.
111 G. De Groot, P. A. Greve, and R. A. A. Maes, *Anal. Chim. Acta*, 1975, **79**, 279.
112 E. E. Brull and G. S. Golden, *Anal. Chim. Acta*, 1979, **110**, 167.
113 P. A. Greve and P. J. A. Haring, *SchrReihe Ver. Wass.-Boden-u. Lufthyg.*, 1972, **37**, 59.
114 J. Solomon and J. F. Uthe, *Anal. Chim. Acta*, 1974, **73**, 149.
115 W. H. Glaze, G. R. Peyton, and R. Rawley, *Environ. Sci. Technol.*, 1977, **11**, 685.
116 R. C. C. Wegman and P. A. Greve, *Sci. Total Environ.*, 1977, **7**, 235.
117 W. Jäger and H. Hagenmaier, *Z. Wasser Abwasser Forsch.*, 1980, **13**, 66.
118 U. Fritschi, G. Fritschi, and H. Kussmaul, *Z. Wasser Abwasser Forsch.*, 1978, **11**, 165.
119 M. R. Jekel and P. V. Roberts, *Environ. Sci. Technol.*, 1980, **14**, 970.
120 Y. Takahashi, R. T. Moore, and R. J. Joyce, in W. J. Cooper, ed., 'Chemistry in Water Reuse', Ann Arbor Science Publishers, Ann Arbor, 1981, Vol. 2, Chapter 7, pp. 127–146.
121 R. A. van Steenderen, *Lab. Pract.*, 1980, **29**, 380.
122 R. C. C. Wegman, in A. Bjørseth and G. Angeletti, ed. ref. 11, pp. 249–263.
123 I. Sekerka and J. F. Lechner, *Int. J. Environ. Anal. Chem.*, 1982, **11**, 43.
124 W. Kühn and H. Sontheimer, *Vom Wasser*, 1973, **41**, 65.
125 W. Kühn, H. Sontheimer, L. Steiglitz, D. Maier, and R. Kurz, *J. Am. Water Works Assoc.*, 1978, **70**, 326.
126 F. Zuercher, in A. Bjørseth and G. Angeletti, ed. ref. 11, pp. 272–276.
127 R. C. Dressman and E. F. McFarren, *J. Chromatogr. Sci.*, 1977, **15**, 69.
128 A. I. Vogel, 'A Text-Book of Quantitative Inorganic Analysis', 3rd Ed., Longmans, London, 1964.
129 C. L. Wilson and D. W. Wilson, ed., 'Comprehensive Analytical Chemistry', Volume 1C, Elsevier, Amsterdam, 1962.
130 C. L. Wilson and D. W. Wilson, ed. ref. 129, Volume 1A, Elsevier, Amsterdam, 1959.
131 L. Erdey, 'Gravimetric Analysis', Volumes I–III, Pergamon, Oxford, 1965.
132 I. M. Kolthoff, E. B. Sandell, E. J. Mechan, and S. Bruckenstein, Quantitative Chemical Analysis', 4th Ed., Collier-Macmillan, New York, 1970.
133 American Public Health Association, *et al.*, ref. 18, pp. 74, 76, 90, 211, 427, 436, 460, 475.
134 J. P. Riley, in J. P. Riley and G. Skirrow, *ed.*, 'Chemical Oceanography', 2nd Ed., Volume 2, Academic Press, London, 1975, Chapter 19, pp. 193–514.
135 American Public Health Association, *et al.*, ref. 18, pp. 90–99.
136 American Public Health Association, *et al.*, ref. 18, pp. 460–466 and 475–482.
137 Methods for the Examination of Water and Associated Materials, 'The Determination of Material Extractable by Carbon Tetrachloride and of Certain Hydrocarbon Oil and Grease Components in Sewage Sludge. 1978', Her Majesty's Stationery Office, London, 1980.
138 American Public Health Association *et al.*, ref. 18, pp. 211–213.
139 American Public Health Association *et al.*, ref. 18, pp. 427–429.
140 American Public Health Association *et al.*, ref. 18, pp. 436–439.
141 American Public Health Association *et al.*, ref. 18, pp. 74–78.
142 E. B. Sandell and H. Onishi, 'Photometric Determination of Traces of Metals', 4th Ed., Wiley, London and New York, 1978.
143 Methods for the Examination of Waters and Associated Materials, 'Ultraviolet and Visible Solution Spectrophotometry and Colorimetry 1980. An Essay Review', Her Majesty's Stationery Office, London, 1981.

[144] Methods for the Examination of Waters and Associated Materials, 'Oxidised Nitrogen in Waters, 1981', Her Majesty's Stationery Office, London, 1982.

[145] P. J. Rennie, A. M. Sumner, and F. B. Basketter, *Analyst*, 1979, **104**, 837.

[146] A. L. Wilson, *J. Appl. Chem.*, 1959, **9**, 501.

[147] Department of the Environment, 'Analysis of Raw, Potable and Waste Waters', Her Majesty's Stationery Office, London, 1972, pp. 26–28.

[148] R. A. Dobbs, R. H. Wise, and R. B. Dean, *Water Res.*, 1972, **6**, 1173.

[149] R. Briggs and K. V. Melbourne, *J. Soc. Water Treatment Examin.*, 1968, **17**, 107.

[150] A. T. Palin, *J. Am. Water Works Assoc.*, 1967, **59**, 255.

[151] American Public Health Association, *et al.*, ref. 18, pp. 332–334 and 337–340.

[152] Methods for the Examination of Waters and Associated Materials, 'Silicon in Waters and Effluents, 1980', Her Majesty's Stationery Office, London, 1981.

[153] Methods for the Examination of Waters and Associated Materials, 'Phosphorus in Waters Effluents and Sewages, 1980', Her Majesty's Stationery Office, London, 1981.

[154] Methods for the Examination of Waters and Associated Materials, 'Chromium in Raw and Potable Waters and Sewage Effluents, 1980', Her Majesty's Stationery Office, London, 1981.

[155] D. F. Boltz and J. A. Howell, 'Colorimetric Determination of Non-Metals', 2nd Ed., Wiley, London & New York, 1978.

[156] G. F. Lothian, 'Absorption Spectrophotometry', 3rd Ed., Adam Hilger, London, 1969.

[157] J. R. Edisbury, 'Practical Hints on Absorption Spectrometry (Ultraviolet and Visible)', Hilger and Watts, London, 1966.

[158] A. B. Calder, 'Photometric Methods of Analysis,' Adam Hilger, London, 1969.

[159] C. D. Snell and C. T. Snell, 'Colorimetric Methods of Analysis', Vol. I, 1948; Vol. II, 1950; Vol. III, 1953; Vol. IV, 1954; Vol. IIA, 1959; Vol. IIIA, 1961; Vol. IVA, Van Nostrand, New York, 1966.

[160] International Union of Pure and Applied Chemistry, 'Spectrophotometric Data for Colorimetric Analysis', Butterworth, 1963.

[161] J. A. Howell and L. G. Hargis, *Anal. Chem.*, 1982, **54**, 171R.

[162] F. Feigl and V. Anger, 'Spot Tests in Inorganic Analysis', 6th Ed., Elsevier, Amsterdam, 1972.

[163] L. C. Thomas and G. J. Chamberlain, 'Colorimetric Chemical Analytical Methods', 9th Ed., revised by G. Shute, Wiley, Chichester 1980 (in conjunction with the Tintometer Co. Ltd., Salisbury).

[164] A. L. Wilson, *Analyst*, 1962, **87**, 884.

[165] Methods for the Examination of Waters and Associated Materials, 'Hydrazine in Waters. Spectrophotometric Method, 1981', Her Majesty's Stationery Office, London, 1981.

[166] J. A. Tetlow and A. L. Wilson, *Analyst*, 1964, **89**, 442.

[167] Methods for the Examination of Waters and Associated Materials, 'Nickel in Potable Waters. 1981', Her Majesty's Stationery Office, London, 1982.

[168] Methods for the Examination of Waters and Associated Materials, 'Phenols in Waters and Effluents by Gas-Liquid Chromatography, 4-aminoantipyrine or 3-methyl-2-benzothiazolinone hydrazone. 1981'. Her Majesty's Stationery Office, London, 1982

[169] Methods for the Examination of Waters and Associated Materials, 'Fluoride in Waters, Effluents, Sludges, Plants and Soils. 1982', Her Majesty's Stationery Office, London, 1983.

[170] Methods for the Examination of Waters and Associated Materials, 'Analysis of Surfactants in Waters, Wastewaters and Sludges, 1981', Her Majesty's Stationery Office, London, 1982.

[171] G. Svehla, *Talanta*, 1966, **13**, 641.

[172] Methods for the Examination of Waters and Associated Materials, 'Air Segmented Continuous Flow Automatic Analysis in the Laboratory. 1979. An Essay Review', Her Majesty's Stationery Office, London, 1980.

[173] Mavrodineau, R., ed., 'Analytical Flame Spectroscopy', Macmillan, London, 1970.

[174] C. T. J. Alkemade and R. Herrmann, 'Flame Spectroscopy', Adam Hilger, Bristol, 1978.

[175] J. A. Dean and T. C. Rains, ed., 'Flame Emission and Atomic Absorption Spectrometry', Marcel Dekker, New York, Vol. 1, 1969; Vol. 2, 1971; Vol. 3, 1975.

[176] K. C. Thompson and R. J. Reynolds, 'Atomic Absorption, Fluorescence and Flame Emission Spectrometry. A Practical Approach', 2nd Ed., Charles Griffin, London, 1978.

[177] C. Th. Alkemade, Tj. Hollander, W. Snelleman, and P. J. Th. Zeegers, 'Metal Vapours in Flames', Pergamon Press, Oxford, 1982.

[178] M. S. Cresser and L. Ebdon, ed., 'Annual Reports on Analytical Atomic Spectroscopy. Volume 12, reviewing 1982', The Royal Society of Chemistry, London, 1983.

[179] W. J. Boyko, P. N. Keliher and J. M. Patterson, *Anal. Chem.*, 1982, **54**, 188R.

[180] G. Horlick, *Anal. Chem.*, 1982, **54**, 276R.

[181] L. I. Pleskach and G. D. Chirkova, *Zavod. Lab.*, 1971, **37**, 168.

[182] G. H. Wagner and K. F. Steele, *Am. Lab.*, 1982, **14**, 12.

[183] Methods for the Examination of Waters and Associated Materials, 'Dissolved Sodium in Raw and Potable Waters. Tentative Methods 1980', Her Majesty's Stationery Office, London, 1981.

[184] Methods for the Examination of Waters and Associated Materials, 'Dissolved Potassium in Raw and Potable Waters. Tentative Methods, 1980', Her Majesty's Stationery Office, London, 1981.

[185] W. A. Crandall and W. Nacovsky, *Proc. Am. Power Conf.*, 1958, **20**, 726.

[186] H. M. Webber and A. L. Wilson, *Analyst*, 1969, **94**, 569.

[187] W. D. Basson and J. F. van Staden, *Fresenius' Z. Anal. Chem.*, 1980, **302**, 370.

[188] L. Ebdon, R. C. Hutton, and J. Ottaway, *Anal. Chim. Acta*, 1978, **96**, 63.

[189] M. Suzuki, K. Ohta, and T. Yamakita, *Anal. Chem.*, 1981, **53**, 1796.

[190] J. Marshall, D. Littlejohn, J. M. Ottaway, J. M. Harnly, N. J. Miller-Ihli, and T. C. O'Haver, *Analyst*, 1983, **108**, 178.

[191] L. Bezur, J. Marshall, J. M. Ottaway, and R. Fakhrul-Aldeen, *Analyst*, 1983, **108**, 553.

[192] J. M. Ottaway, 'Carbon Furnace Atomic Emission Spectrometry – Past, Present and Future', Presented at the 23rd Colloquium Spectroscopicum Internationale, Amsterdam, June/July 1983. (Abstract in *Spectrochim. Acta, Part B*, 1983, **38** (Supplement), 253).

[193] H. Falk, E. Hoffmann, and Ch. Luedke, *Spectrochim. Acta, Part B*, 1981, **36**, 767.

[194] H. Falk, E. Hoffmann, and Ch. Luedke, 'Multielement Analysis of Microliter Samples in the PPB-Range using Furnace Non-Thermal Excitation Spectrometry', Presented at the 23rd Colloquium Spectroscopicum Internationale, Amsterdam, June/July 1983. (Abstract in *Spectrochim. Acta, Part B*, 1983, **38** (Supplement), 94.

[195] J. V. Sullivan, *Prog. Anal. Atom. Spectrosc.*, 1981, **4**, 311.

[196] K. C. Thompson and R. J. Reynolds, 'Atomic Absorption, Fluorescence and Flame Emission Spectroscopy. A Practical Approach', 2nd Ed., Charles Griffin, London, 1978.

[197] C. W. Fuller, 'Electrothermal Atomization for Atomic Absorption Spectrometry', Analytical Sciences Monograph No. 4, The Chemical Society, London, 1977.

[198] R. A. Newstead, W. J. Price, and P. J. Whiteside, *Prog. Anal. Atom. Spectrosc.*, 1978, **1**, 267.

[199] E. Grassam and J. B. Dawson, *Euro. Spectrosc. News*, 1978, **21**, 27.

[200] K. Yasuda, H. Koizumi, K. Ohishi, and T. Noda, *Prog. Anal. Atom. Spectrosc.*, 1980, **3**, 299.

[201] L. De Galan, *TRAC, Trends Anal. Chem.*, 1982, **1**, 203.

[202] M. T. C. De Loos-Vollebregt and L. De Galan, *Spectrochim. Acta, Part B*, 1978, **33**, 495.

[203] J. J. Sotera and H. L. Kahn, *Am. Lab.*, 1982, **14**(11), 100.

[204] G. R. Carnrick, W. Slavin, and D. C. Manning, *Anal. Chem.*, 1981, **53**, 1866.

[205] E. Pruszkowska, G. R. Carnrick, and W. Slavin, *Anal. Chem.*, 1983, **55**, 182.

[206] W. Slavin, D. C. Manning, G. Carnrick, and E. Pruszkowska, *Spectrochim. Acta, Part B*, 1983, **38**, 1157.

[207] C. Th. J. Alkemade and R. Herrmann, 'Fundamentals of Analytical Flame Spectroscopy', Adam Hilger, Bristol, 1979.

[208] W. J. Price, 'Spectrochemical Analysis by Atomic Absorption', Heyden, London, 1979.

[209] M. Slavin, 'Atomic Absorption Spectroscopy', 2nd Ed., Wiley, New York, 1979.

[210] J. A. Dean and T. C. Rains, ed., 'Flame Emission and Atomic Absorption Spectrometry', Volume 1, Dekker, New York, 1969.

[211] J. A. Dean and T. C. Rains, ed. ref. 210, Volume 2, Dekker, New York, 1971.

[212] J. A. Dean and T. C. Rains, ed. ref. 210, Volume 3, Dekker, New York, 1975.

[213] B. V. L'vov, 'Atomic Absorption Spectrochemical Analysis', 2nd Ed., Adam Hilger, London, 1970.

[214] M. S. Cresser and L. Ebdon, ed., 'Annual Reports on Analytical Atomic Spectroscopy. Volume 12. Reviewing 1982', The Royal Society of Chemistry, London, 1983.

[215] G. Horlick, *Anal. Chem.*, 1982, **54**, 276R.

[216] Methods for the Examination of Waters and Associated Materials, 'Atomic Absorption Spectrophotometry 1979 Version, An Essay Review', Her Majesty's Stationery Office, London, 1980.

[216a] S. A. Katz and S. W. Jenniss, 'Regulatory Compliance Monitoring by Atomic Absorption Spectroscopy', Verlag Chemie International, Deerfield Beach, Florida, 1983.

[217] G. F. Kirkbright and H. N. Johnson, *Talanta*, 1973, **20**, 433.

[218] S. S. M. Hassan, 'Organic Analysis using Atomic Absorption Spectrometry', Ellis Horwood, Chichester, 1984.

[219] K. C. Thompson and O. Hydes, *Anal. Newslett.*, (Water Research Centre, Medmenham, Bucks, U.K.), 1982, **2**, 2.

[220] S. R. Koirtyohann, *Anal. Chem.*, 1980, **52**, 736A.

[221] M. S. Cresser and R. F. Browner, *Appl. Spectrosc.*, 1980, **34**, 364.

[222] R. J. Watling, *Water S.A.*, 1977, **3** (4), 218.

[223] R. J. Watling, *Anal. Chim. Acta*, 1977, **94**, 181.

[224] B. Milner, *Am. Lab.*, 1983, **15** (11), 74.

[225] H. T. Delves, *Analyst*, 1970, **95**, 431.

[226] J. D. Kerber and F. J. Fernandez, *At. Absorpt. Newslett.*, 1971, **10**, 78.

[227] H. L. Kahn, G. E. Peterson, and J. E. Shallis, *At. Absorpt. Newslett.*, 1968, **7**, 35.

[228] I. S. Maines, K. M. Aldous, and D. G. Mitchell, *Environ. Sci. Technol.*, 1975, **9**, 549.

[229] H. Berndt and J. Messerschmidt, *Spectrochim. Acta, Part B*, 1979, **34**, 241.

230 B. Orpwood, 'Concentration Techniques for Trace Elements: A Review', Technical Report, TR 102, Water Research Centre, Medmenham, Bucks, 1979.
231 C. G. Farnsworth, *Water Sewage Works*, 1972, **119**, 52.
232 Methods for the Examination of Waters and Associated Materials, 'Copper in Potable Waters by Atomic Absorption Spectrophotometry. 1980', Her Majesty's Stationery Office, London, 1981.
233 Methods for the Examination of Waters and Associated Materials, 'Zinc in Potable Waters by Atomic Absorption Spectrophotometry. 1980', Her Majesty's Stationery Office, London, 1981.
234 Methods for the Examination of Water and Associated Materials, 'Cobalt in Potable Waters. 1981', Her Majesty's Stationery Office, London, 1982.
235 J. P. Riley and D. Taylor *Anal. Chim. Acta*, 1968, **40**, 479.
236 T. M. Florence and G. E. Batley, *Talanta*, 1976, **23**, 179.
237 B. Orpwood, 'Use of Chelating Ion-exchange Resins for the Determination of Trace Metals in Drinking Waters', Technical Report TR 153, Water Research Centre, Medmenham, Bucks., 1980.
238 M. S. Cresser, 'Solvent Extraction in Flame Spectroscopic Analysis', Butterworths, London, 1978.
239 J. Korkisch and A. Sorio, *Anal. Chim. Acta*, 1975, **79**, 207.
240 T. N. Tweeten and J. W. Knoeck, *Anal. Chem.*, 1976, **48**, 64.
241 H. E. Allen and R. A. Minear, in M. J. Suess, ed., 'Examination of Water for Pollution Control. A Reference Handbook', Published on behalf of the World Health Organisation Regional Office for Europe by Pergamon Press, Oxford, 1982, Vol. 2, pp. 43–168.
242 D. Y. Hsu and W. O. Pipes, *Environ. Sci. Technol.*, 1972, **6**, 645.
243 Y. K. Chau and K. Lum-Shue-Chan, *Anal. Chim. Acta*, 1969, **48**, 205.
244 Methods for the Examination of Waters and Associated Materials, 'Lead in Potable Waters by Atomic Absorption Spectrophotometry. 1976', Her Majesty's Stationery Office, London, 1977.
245 Methods for the Examination of Waters and Associated Materials, 'Cadmium in Potable Waters by Atomic Absorption Spectrophotometry. 1976', Her Majesty's Stationery Office, London, 1977.
246 R. R. Brooks, R. J. Presley, and I. R. Kaplan, *Talanta*, 1967, **14**, 809.
247 P. G. Brewer, D. W. Spencer, and C. L. Smith, *Am. Soc. Test Mater., Spec. Tech. Publ. No. 443*, 1969, pp. 70–77.
248 S. L. Sachdev and P. W. West, *Environ. Sci. Technol.*, 1970, **4**, 749.
249 J. D. Kinrade and J. C. Van Loon, *Anal. Chem.*, 1974, **46**, 1894.
250 D. Goulden, P. Brooksbank, and J. F. Ryan, *Am. Lab.*, 1973, **5**, 10.
251 I. Rubeška and J. Musil, *Prog. Analyt. Atom. Spectrosc.*, 1979, **2**, 309.
252 K. C. Thompson, *Analyst*, 1978, **103**, 1258.
253 K. C. Thompson and K. Wagstaff, *Analyst*, 1979, **104**, 224.
254 K. C. Thompson and K. Wagstaff, *Analyst*, 1980, **105**, 1252.
255 G. Horlick, *Anal. Chem.*, 1980, **52**, 290R.
256 L. Ranson, 'Non-Flame Devices in Atomic Absorption Spectrometry', Technical Report TR 1, Water Research Centre, Medmenham, Bucks., 1975.
257 F. Dolinšek and J. Stupar, *Analyst*, 1973, **98**, 841
258 J. G. T. Regan and J. Warren, *Analyst*, 1976, **101**, 220.
259 K. C. Thompson, K. Wagstaff, and K. C. Wheatstone, *Analyst*, 1977, **102**, 310.
260 L. Ranson and B. Orpwood, 'An Evaluation of an Electrothermal Device for the Determination of Lead and Cadmium in Potable Water', Technical Report TR 49, Water Research Centre, Medmenham, Bucks., 1977.
261 Water Research Centre, unpublished data.
262 J. P. Matousek, *Prog. Anal. Atom. Spectrosc.*, 1981, **4**, 247.
263 M. P. Bertenshaw, D. Gelsthorpe, and K. C. Wheatstone, *Analyst*, 1981, **106**, 23.
264 R. D. Ediger, *At. Absorpt. Newslett.*, 1975, **14**, 127.
265 D. J. Hydes, *Anal. Chem.*, 1980, **52**, 959.
266 K. Matsusaki, T. Yoshino, and Y. Yamamoto, *Anal. Chim. Acta*, 1980, **113**, 247.
267 R. Guevremont, *Anal. Chem.*, 1980, **52**, 1574.
268 D. J. Hodges, *Analyst*, 1977, **102**, 66.
269 K. Ohta and M. Suzuki, *Fresenius' Z. Anal. Chem.*, 1979, **298**, 140.
270 W. Frech and A. Cedergren, *Anal. Chim. Acta*, 1977, **88**, 57.
271 C. L. Chakrabarti, C. C. Wan, and W. C. Li, *Spectrochim. Acta, Part B*, 1980, **35**, 93.
272 D. T. E. Hunt, R. L. Norton, B. Orpwood, and D. A. Winnard, 'The Determination of Lead in Drinking Water by Graphite Furnace Atomic Absorption Spectrometry', Enquiry Report ER 710, Water Research Centre, Medmenham, Bucks, 1979.
273 A. Anderson, *At. Absorpt. Newslett.*, 1976, **15**, 71.
274 M. P. Bertenshaw, D. Gelsthorpe, and K. C. Wheatstone, *Analyst*, 1982, **107**, 163.
275 I. J. Fletcher, *Anal. Chim. Acta*, 1983, **154**, 235.
276 M. Bengtsson, L.-G. Danielsson, and B. Magnusson, *Anal. Lett.*, 1979, **12**, 1367.
277 K. Johansson, W. Frech, and A. Cedergren, *Anal. Chim. Acta.*, 1977, **94**, 245.
278 B. V. L'vov, *Spectrochim. Acta, Part B*, 1978, **33**, 153.

[279] D. C. Gregoire and C. L. Chakrobarti, *Anal. Chem.*, 1977, **49**, 2018.
[280] L. Hageman, A. Mubarak, and R. Woodriff, *Appl. Spectrosc.*, 1979, **33**, 226.
[281] L. R. Hageman, J. A. Nichols, P. Viswanadham, and R. Woodriff, *Anal. Chem.*, 1979, **51**, 1406.
[282] D. A. Winnard and D. T. E. Hunt, 'The Woodriff Furnace – An Illustration of the Benefits of Constant Temperature Atomisation', Proceedings of the International Conference on Heavy Metals in the Environment, Heidelberg, September 1983, CEP Consultants Ltd., Edinburgh, 1983, pp. 214–217.
[283] G. Lundgren, L. Lundmark, and G. Johansson, *Anal. Chem.*, 1974, **46**, 1028.
[284] G. Lundgren and G. Johansson, *Talanta*, 1974, **21**, 257.
[285] W. Frech, *Talanta*, 1974, **21**, 565.
[286] W. Frech, *Anal. Chim. Acta*, 1975, **77**, 43.
[287] F. J. Fernandez and J. Iannarone, *At. Absorpt. Newlett.*, 1978, **17**, 117.
[288] C. L. Chakrabarti, H. A. Hamed, C. C. Wan, W. C. Li, P. C. Bertels, D. C. Gregoire, and S. Lee, *Anal. Chem.*, 1980, **52**, 167.
[289] C. L. Chakrabarti, C. C. Wan, H. A. Hamed, and P. C. Bertels, *Nature*, 1980, **288**, 246.
[290] C. L. Chakrabarti, C. C. Wan, H. A. Hamed, and P. C. Bertels, *Anal. Chem.*, 1981, **53**, 444.
[291] B. V. L'vov, L. A. Pelieva, and A. I. Sharnopolskii, *Zh. Prikl. Spectrosk.*, 1977, **27**, 395.
[292] W. Slavin and D. C. Manning, *Anal. Chem.*, 1979, **51**, 261.
[293] W. Slavin, D. C. Manning, and G. R. Carnrick, *At. Spectrosc.*, 1981, **2**, 137.
[294] M. W. Routh, *Appl. Spectrosc.*, 1982, **36**, 585.
[295] D. C. Manning, W. Slavin, and G. R. Carnrick, *Spectrochim. Acta, Part B*, 1982, **37**, 331.
[296] W. K. Oliver, S. Reeve, K. Hammond, and F. B. Basketter, *J. Inst. Water Eng. Sci.*, 1983, **37**, 460.
[297] S. R. Koirtyohann, R. C. Giddings, and H. E. Taylor, *Spectrochim. Acta, Part B*, 1984, **39**, 407.
[298] C. L. Chakrobarti, S. Wu, and P. C. Bertels, *Spectrochim. Acta, Part B*, 1983, **38**, 1041.
[299] C. L. Chakrabarti, S. Wu, R. Karwowska, J. T. Rogers, L. Haley, P. C. Bertels, and R. Dick, *Spectrochim. Acta, Part B*, 1984, **39**, 415.
[300] T. M. Rettberg and J. A. Holcombe, *Spectrochim. Acta, Part B*, 1984, **39**, 243.
[301] D. C. Manning, W. Slavin, and S. Myers, *Anal. Chem.*, 1979, **51**, 2375.
[302] D. C. Manning and W. Slavin, *Anal. Chim. Acta*, 1980, **118**, 301.
[303] S. K. Giri, D. Littlejohn, and J. M. Ottaway, *Analyst*, 1982, **107**, 1095.
[304] S. K. Giri, C. J. Shields, D. Littlejohn, and J. M. Ottaway, *Analyst*, 1983, **108**, 244.
[305] D. Littlejohn, J. Marshall, J. Carroll, W. Cormack, and J. M. Ottaway, *Analyst*, 1983, **108**, 893.
[306] D. Littlejohn, S. Cook, D. Durie, and J. M. Ottaway, *Spectrochim. Acta, Part B*, 1984, **39**, 295.
[307] H. L. Kahn, M. K. Conley, and J. J. Sotera, *Am. Lab.*, 1980, **12** (8), 72.
[308] J. J. Sotera, L. C. Cristiano, M. K. Conley, and H. L. Kahn, *Anal. Chem.*, 1983, **55**, 204.
[309] D. J. Churella and T. R. Copeland, *Anal. Chem.*, 1978, **50**, 309.
[310] J. A. Krasowksi and T. R. Copeland, *Anal. Chem.*, 1979, **51**, 1843.
[311] R. E. Sturgeon and C. L. Chakrabarti, *Prog. Anal. Atom. Spectrosc.*, 1978, **1**, 5.
[312] *Spectrochim. Acta, Part B*, 1984, **39**, Parts 2 and 3.
[313] B. V. L'vov, *Spectrochim. Acta, Part B*, 1984, **39**, 149.
[314] L. De Galan, M. T. C. De Loos-Vollebregt, and R. A. M. Oosterling, *Analyst*, 1983, **108**, 138.
[315] R. P. Mitcham, *Analyst*, 1980, **105**, 43.
[316] G. F. Kirkbright, S. Hsiao-Chuan, and R. D. Snook, *At. Absorpt. Newslett.*, 1980, **1**, 85.
[317] V. B. Stein, E. Canelli, and A. H. Richards, *At. Absorpt. Newslett.*, 1980, **1**, 61.
[318] V. B. Stein, E. Canelli, and A. H. Richards, *At. Absorpt. Newslett.*, 1980, **1**, 133.
[319] A. M. Gunn, 'The Determination of Arsenic and Selenium in Raw and Potable Waters by Hydride Generation/Atomic Absorption Spectrometry – A Review', Technical Report TR 169, Water Research Centre, Medmenham, Bucks, 1981.
[320] P. N. Vijan and G. R. Wood, *Analyst*, 1976, **101**, 966.
[321] R. B. Robbins and J. A. Caruso, *Anal. Chem.*, 1979, **51**, 889A.
[322] K. C. Thompson and D. R. Thomerson, *Analyst*, 1974, **99**, 595.
[323] H. W. Sinemus, M. Melcher, and B. Welz, *At. Absorpt. Newslett.*, 1981, **2**, 81.
[324] W. A. Maher, *Analyst*, 1983, **108**, 305.
[325] M. O. Andreae, J. F. Asmode, P. Foster, and L. Van't dack, *Anal. Chem.*, 1981, **53**, 1766.
[326] A. M. Gunn, 'An Automated Hydride Generation Atomic Absorption Spectrometric Method for the Determination of Total Arsenic in Rain and Potable Waters', Technical Report TR 191. Water Research Centre, Medmenham, Bucks., 1983.
[327] C. Cremisini, M. Dall'Aglio, and E. Ghiara, Proceedings of 'International Conference on Management and Control of Heavy Metals in the Environment', London, September 1979, p. 341.
[328] B. Welz and M. Melcher, *Analyst*, 1983, **108**, 213.
[329] M. O. Andreae and P. N. Froelich, *Anal. Chem.*, 1981, **53**, 287.
[330] F. D. Pierce, T. C. Lamoureaux, H. R. Brown, and R. S. Fraser, *Appl. Spectrosc.*, 1976, **30**, 38.
[331] P. D. Goulden and P. Brooksbank, *Anal. Chem.*, 1974, **46**, 1431.
[332] Methods for the Examination of Water and Associated Materials', Arsenic in Potable Waters by

Atomic Absorption Spectrophotometry (semi-automatic method) 1982' Her Majesty's Stationery Office, London, 1983.

333 M. J. Suess, ed. ref. 241, pp. 175–179.
334 V. F. Hodge, S. L. Siedel, and E. D. Goldberg, *Anal. Chem.*, 1979, **51**, 1256.
335 P. N. Vijan and R. S. Sadana, *Talanta*, 1980, **27**, 321.
336 R. Smith, *At. Absorpt. Newslett.*, 1981, **2**, 155.
337 A. E. Smith, *Analyst*, 1975, **100**, 300.
338 A. Meyer, C. F. Hofer, G. Tolg, S. Raptis, and G. Knapp, *Fresenius' Z. Anal. Chem.*, 1979, **296**, 337.
339 G. F. Kirkbright and M. Taddia, *At. Absorpt. Newslett.*, 1979, **18**, 68.
340 G. F. Kirkbright and M. Taddia, *Anal. Chim. Acta*, 1978, **100**, 145.
341 J. Azad, G. F. Kirkbright, and R. D. Snook, *Analyst*, 1979, **104**, 232.
342 Y. Yamamoto and T. Kumamaru, *Fresenius' Z. Anal. Chem.*, 1976, **282**, 139.
343 M. O. Andreae, *Anal. Chem.*, 1977, **49**, 820.
344 G. A. Cutter, *Anal. Chim. Acta*, 1978, **98**, 59.
345 American Public Health Association, *et al.*, ref. 18, p. 160.
346 P.-K. Hon, O.-W. Lau, and M.-C. Wong, *Analyst*, 1983, **108**, 64.
347 W. R. Hatch and W. L. Ott, *Anal. Chem.*, 1968, **40**, 2085.
348 A. M. Ure, *Anal. Chim. Acta*, 1975, **76**, 1.
349 R. J. Watling and H. R. Watling, *Water S. A.*, 1975, **1**, 113.
350 R. C. Rooney, *Analyst*, 1976, **101**, 678.
351 R. Braun and A. P. Husbands, *Spectrovision*, 1971, **26**, 2.
352 Methods for the Examination of Water and Associated Materials, 'Mercury in Waters, Effluents, and Sludges by Flameless Atomic Absorption Spectrophotometry. 1978', Her Majesty's Stationery Office, London, 1978.
353 E. Yamaguchi, S. Tateishi, and A. Hashimoto, *Analyst*, 1980, **105**, 491.
354 W. F. Fitzgerald, W. B. Lyons, and C. D. Hunt, *Anal. Chem.*, 1974, **46**, 1882.
355 Zs. Wittmann, *Talanta*, 1981, **28**, 271.
356 G. Topping and J. M. Pirie, *Anal. Chim. Acta*, 1972, **62**, 200.
357 L. A. Nelson, *Anal. Chem.*, 1979, **51**, 2289.
358 H. Agemian and A. S. Y. Chau, *Anal. Chem.*, 1978, **50**, 13.
359 B. J. Farey, L. A. Nelson, and M. G. Rolfe, *Analyst*, 1978, **103**, 656.
360 M. J. Suess, ed. ref. 241, p. 125.
361 B. Welz, M. Melcher, H. W. Sinemus, and D. Maier, *At. Spectrosc.*, 1984, **5**, 37.
362 G. Mattock, 'pH-Measurement and Titration', Heywood, London, 1961.
363 G. Eisenman, 'Glass Electrodes for Hydrogen and Other Cations', Edward Arnold, London, 1967.
364 R. G. Bates, 'Determination of pH – Theory and Practice', 2nd Ed., Wiley-Interscience, New York, 1973.
365 A. K. Covington, P. D. Whalley, and W. Davison, *Analyst*, 1983, **108**, 1528.
366 D. Midgley and K. Torrance, *Analyst*, 1979, **101**, 833.
367 D. Midgley and K. Torrance, *Analyst*, 1979, **104**, 63.
368 R. A. Durst, ed., 'Ion-Selective Electrodes', National Bureau of Standards Special Publication 314, U.S. Department of Commerce, Washington, 1969.
369 G. J. Moody and J. D. R. Thomas, 'Selective Ion-Sensitive Electrodes', Merrow, Watford, 1971.
370 D. Midgley and K. Torrance, 'Potentiometric Water Analysis', Wiley-Interscience, Chichester, 1978.
371 P. L. Bailey, 'Analysis with Ion-Selective Electrodes', 2nd Ed., Heyden, London, 1980.
372 J. Koryta and K. Štulik 'Ion-Selective Electrodes', 2nd Ed., Cambridge University Press, Cambridge, 1983.
373 A. Hulanicki and M. Trojanowicz, *Ion-Selective Electrode Rev.*, 1980, **1**, 207.
374 H. B. Herman, in H. B. Mark and S. Mattson, ed., ref. 8, pp. 151–185.
375 D. J. G. Ives and G. J. Janz, 'Reference Electrodes. Theory and Practice', Academic Press, New York, 1961.
376 A. K. Covington and M. J. F. Rebelo, *Ion-Selective Electrode Rev.*, 1983, **5**, 93.
377 B. Jaselskis and M. K. Bandemer, *Anal. Chem.*, 1969, **41**, 855.
378 A. Homola and R. O. James, *Anal. Chem.*, 1976, **48**, 776.
379 J. W. Ross and M. S. Frant, *Anal. Chem.*, 1969, **41**, 967.
380 E. P. Scheide and R. A. Durst, *Anal. Lett.*, 1977, **10**, 55.
381 A. Hulanicki, R. Lewandowski, and A. Lewenstam, *Analyst*, 1976, **101**, 939.
382 G. Ouzounian and G. Michard, *Anal. Chim. Acta*, 1978, **96**, 405.
383 W. Selig, *Mikrochim. Acta*, 1970, 564.
384 M. Mascini, *Ion-Selective Electrode Rev.*, 1980, **2**, 17.
385 D. Midgley and K. Torrance, *Analyst*, 1972, **97**, 626.
386 M. J. Beckett and A. L. Wilson, *Water Res.*, 1974, **8**, 333.
387 W. H. Evans and B. F. Partridge, *Analyst*, 1974, **99**, 367.
388 T. R. Gilbert and A. M. Clay, *Anal. Chem.*, 1973, **45**, 1757.

389 I. Sekerka and J. F. Lechner, *Analyt. Lett.*, 1974, **7**, 463.
390 R. F. Thomas and R. L. Booth, *Environ. Sci. Technol.*, 1973, **7**, 523.
391 J. Mertens, P. Van den Winkel, and D. L. Massart, *Anal. Chem.*, 1975, **47**, 522.
392 R. J. Stevens, *Water Res.*, 1976, **10**, 171.
393 L. R. McKenzie and P. V. W. Young, *Analyst*, 1975, **100**, 620.
394 J. H. Lowry and K. H. Mancy, *Water Res.*, 1978, **12**, 471.
395 C. Garside, G. Hull, and S. Murray, *Limnol. Oceanogr.*, 1978, **23**, 1073.
396 L. Braunstein, K. Hochmueller, and K. Spengler, *Vom Wasser*, 1980, **54**, 307.
397 R. C. Harriss and H. H. Williams, *J. Appl. Meteorol.*, 1969, **8**, 299.
398 M. E. Thompson and J. W. Ross, *Science*, 1966, **154**, 1643.
399 G. J. Moody, R. B. Oke, and J. D. R. Thomas, *Analyst*, 1970, **95**, 910.
400 A. Hulanicki and M. Trojanowicz, *Anal. Chim. Acta*, 1974, **68**, 155.
401 E. H. Hansen, J. Růžička, and A. K. Ghose, *Anal. Chim. Acta*, 1978, **100**, 151.
402 W. Back, *J. Am. Water Works Assoc.*, 1960, **52**, 923.
403 J. C. Van Loon, *Anal. Chim. Acta*, 1971, **54**, 23.
404 I. Sekerka, J. F. Lechner, and R. Wales, *Water Res.*, 1975, **9**, 633.
405 K. Torrance, *Analyst*, 1974, **99**, 203.
406 K. Tomlinson and K. Torrance, *Analyst*, 1977, **102**, 1.
407 M. Vandeputte, L. Dryon, and D. L. Massart, *Anal. Chim. Acta*, 1977, **91**, 113.
408 G. B. Marshall and D. Midgley, *Analyst*, 1978, **103**, 438.
409 G. B. Marshall and D. Midgley, *Analyst*, 1979, **104**, 55.
410 J. H. Riseman, *Am. Lab.*, 1972, **4**, 63.
411 I. Sekerka and J. F. Lechner, *Water Res.*, 1976, **10**, 479.
412 R. A. Durst, *Anal. Lett.*, 1977, **10**, 961.
413 N. T. Crosby, A. L. Dennis, and J. G. Stevens, *Analyst*, 1968, **93**, 643.
414 S. J. Patterson, N. G. Bunton, and N. T. Crosby, *Water Treatment Examin.*, 1969, **18**, 182.
415 J. E. Harwood, *Water Res.*, 1969, **3**, 273.
416 T. B. Warner and D. J. Bressan, *Anal. Chim. Acta*, 1973, **63**, 165.
417 I. Sekerka and J. F. Lechner, *Talanta*, 1973, **20**, 1167.
418 D. E. Erdmann, *Environ. Sci. Technol.*, 1975, **9**, 252.
419 S. F. Deane, M. A. Leonard, V. McKee, and G. Svehla, *Analyst*, 1978, **103**, 1134.
420 R. C. Reynolds, *Water Resourc. Res.*, 1971, **7**, 1333.
421 I. Sekerka and J. F. Lechner, *Anal. Lett.*, 1974, **7**, 399.
422 I. Sekerka and J. F. Lechner, *Talanta*, 1975, **22**, 459.
423 E. Pungor, *Anal. Chem.*, 1967, **39**, 29A.
424 L. P. Rigdon, G. J. Moody, and J. W. Frazer, *Anal. Chem.*, 1978, **50**, 465.
425 N. G. Bunton and N. T. Crosby, *Water Treatment Examin.*, 1969, **18**, 338.
426 D. Capon, *Water Treatment Examin.*, 1975, **24**, 333.
427 H. C. Brinkhoff, *Environ. Sci. Technol.*, 1978, **12**, 1392.
428 E. W. Baumann, *Anal. Chem.*, 1974, **46**, 1345.
429 E. Mor, V. Scotto, G. Marcenaro, and G. Alabiso, *Anal. Chim. Acta*, 1975, **75**, 159.
430 H. Guterman, S. Ben-Yaakov, and A. Abeliovich, *Anal. Chem.*, 1983, **55**, 1731.
431 H. M. Webber and A. L. Wilson, *Analyst*, 1969, **94**, 209.
432 E. L. Eckfeldt and W. E. Proctor, *Anal. Chem.*, 1971, **43**, 332.
433 A. A. Diggens, K. Parker, and H. M. Webber, *Analyst*, 1972, **97**, 198.
434 E. L. Eckfeldt and W. E. Proctor, *Anal. Chem.*, 1975, **47**, 2307.
435 G. I. Goodfellow, D. Midgley, and H. M. Webber, *Analyst*, 1976, **101**, 848.
436 American Public Health Association *et al.*, ref. 18, pp. 335–337.
437 American Public Health Association *et al.*, ref. 18, pp. 369–370.
438 American Public Health Association *et al.*, ref. 18, pp. 362–363.
439 G. P. Evans and D. Johnson 'Detection of Pollution at Drinking Water Intakes' Water Research
     Centre Conference on Environmental Protection: Standards, Compliance and Costs, Keele
     University, 1983, Paper 16. The Water Research Centre, Medmenham, Bucks., 1983.
440 G. G. Guilbault, *Ion-Selective Electrode Rev.*, 1982, **4**, 187.
441 J. Janata and R. J. Huber, *Ion-Selective Electrode Rev.*, 1979, **1**, 31.
442 L. P. Rigdon, G. J. Moody, and J. W. Frazer, *Anal. Chem.*, 1978, **50**, 465.
443 N. A. Dimmock and D. Midgley, *Talanta*, 1982, **29**, 557.
444 D. Midgley, *Talanta*, 1983, **30**, 547.
445 G. B. Marshall and D. Midgley, *Analyst*, 1983, **108**, 701.
446 D. Midgley, *Anal. Chem.*, 1977, **49**, 1211.
447 D. Midgley, *Analyst*, 1980, **105**, 417.
448 R. Jain and J. R. Schultz, *Anal. Chem.*, 1984, **56**, 141.
449 International Union of Pure and Applied Chemistry, *Pure Appl. Chem.*, 1976, **48**, 127.
450 D. Midgley, *Analyst*, 1979, **104**, 248.

451 W. J. van Oort and E. J. J. M. van Eerd, *Anal. Chim. Acta*, 1983, **155**, 21.
452 R. Stella and M. T. Ganzerli Valentini, *Anal. Chim. Acta*, 1983, **152**, 191.
453 K. Nicholson and E. J. Duff, *Anal. Lett.*, 1981, **14**, 493.
454 M. Trojanowicz and R. Lewandowski, *Fresenius' Z. Anal. Chem.*, 1981, **308**, 7.
455 International Union of Pure and Applied Chemistry, Analytical Division, Commission on Electroanalytical Chemistry, *Pure Appl. Chem.*, 1975, pp. 81–97.
456 D. R. Turner and M. Whitfield, in M. Whitfield and D. Jagner, ed. 'Marine Electrochemistry', Wiley, Chichester, 1981, pp. 67–97.
457 H. Schmidt and M. Von Stackelberg, 'Modern Polarographic Methods', Academic Press, New York, 1963.
458 L. Meites, 'Polarographic Techniques', 2nd Ed., Interscience, New York, 1965.
459 J. Heyrovsky and J. Kuta, 'Principles of Polarography', Academic Press, New York, 1966.
460 D. T. Sawyer and J. L. Roberts, 'Experimental Electrochemistry for Chemists', Wiley-Interscience, New York, 1974.
461 Z. Galus, 'Fundamentals of Electrochemical Analysis', Wiley, New York, 1976.
462 A. M. Bond, 'Modern Polarographic Techniques in Analytical Chemistry', Marcel Dekker, New York, 1980.
463 F. Vydra, K. Štulik, and E. Juláková, 'Electrochemical Stripping Analysis', Ellis Horwood, Chichester, 1976.
464 E. Barendrecht, 'Stripping Voltammetry', in A. J. Bard, ed., 'Electroanalytical Chemistry', Volume 2, Edward Arnold, London, 1967, pp. 53–109.
465 I. Shain, in I. M. Kolthoff and P. J. Elving, ed., 'Treatise on Analytical Chemistry', Volume 4, Part 1, Interscience, New York, 1967, p. 2533–2568.
466 J. B. Flato, *Anal. Chem.*, 1972, **44**, 75A.
467 D. K. Roe and P. Eggiman, *Anal. Chem.*, 1976, **48**, 9R.
468 D. K. Roe, *Anal. Chem.*, 1978, **50**, 9R.
469 D. C. Johnson, *Anal. Chem.*, 1982, **54**, 9R.
470 C. J. Lind, in P. N. Cheremisinoff and H. J. Perlis, ed., 'Analytical Measurements and Instrumentation for Process and Pollution Control', Ann Arbor Science Publishers, Ann Arbor, 1981, pp. 175–222.
471 M. Whitfield, in J. P. Riley and G. Skirrow, ed., 'Chemical Oceanography', Volume 4, Academic Press, London, 1975, pp. 1–154.
472 W. F. Smyth, ed. 'Electroanalysis in Hygiene, Environmental, Clinical and Pharmaceutical Chemistry', Elsevier, Amsterdam, 1980.
473 W. R. Heineman and P. T. Kissinger, *Anal. Chem.*, 1978, **50**, 166R.
474 J. Wang, *Environ. Sci. Technol.*, 1982, **16**, 104A.
475 D. Jagner, *Analyst*, 1982, **107**, 593.
476 C. A. Edwards, 'Anodic Stripping Voltammetry – A Review', Technical Report TR 22, Water Research Centre, Medmenham, Bucks, 1976.
477 A. Zirino, in M. Whitfield and D. Jagner, ed., ref. 456, pp. 421–503.
478 J. Osteryoung, in H. B. Mark and J. S. Mattson, ed., ref. 8, pp. 85–123.
479 W. R. Heineman, in H. B. Mark and J. S. Mattson, ed. ref. 8, pp. 125–150.
480 M. D. Ryan and G. S. Wilson, *Anal. Chem.*, 1982, **54**, 20R.
481 J. P. Riley and H. Gu, *Anal. Chim. Acta*, 1981, **130**, 199.
482 L. Sipos, J. Golimowski, P. Valenta, and H. W. Nürnberg, *Fresenius' Z. Anal. Chem.*, 1979, **298**, 1.
483 P. H. Davis, G. R. Dulude, R. M. Griffin, W. R. Matson, and E. W. Zink, *Anal. Chem.*, 1978, **50**, 137.
484 W. M. Peterson, *Am. Lab.*, 1979, **11**, 69.
485 A. M. Bond and R. D. Jones, *Anal. Chim. Acta*, 1980, **121**, 1.
486 L. Kryger and D. Jagner, *Anal. Chim. Acta*, 1975, **78**, 251.
487 S. D. Brown and B. R. Kowalski, *Anal. Chim. Acta*, 1979, **107**, 13.
488 L. Kryger, *Anal. Chim. Acta*, 1980, **120**, 19.
489 H. W. Nürnberg, *Electrochim. Acta*, 1977, **22**, 935.
490 D. Jagner and K. Årén, *Anal. Chim. Acta*, 1979, **107**, 29.
491 D. Jagner, M. Josefson, and S. Westerlund, *Anal. Chim. Acta*, 1981, **129**, 153.
492 J. G. Osteryoung and R. A. Osteryoung, *Int. Lab.*, 1972, (Sept./Oct.), 10.
493 A. M. Bond, *Talanta*, 1973, **20**, 1139.
494 J. T. Kinard, 'Determination of Trace Metals in Effluents by Differential Pulse Anodic Stripping Voltammetry', U.S. Environmental Protection Agency Report EPA-600/4-77-034, The Agency, Cincinnati, July, 1977.
495 C. J. Flora and E. Nieboer, *Anal. Chem.*, 1980, **52**, 1013.
496 D. J. Myers and J. Osteryoung, *Anal. Chem.*, 1973, **45**, 267.
497 M. I. Abdullah and L. G. Royle, *Anal. Chim. Acta*, 1972, **58**, 283.
498 H. N. Heckner, *Fresenius' Z. Anal. Chem.*, 1972, **261**, 29.
499 W. Davison and C. D. Gabbut, *J. Electroanal. Chem.*, 1979, **99**, 311.
500 W. J. Masschelein and M. Denis, *Water Res.*, 1981, **15**, 857.

501 W. J. Masschelein, M. Denis, and R. Ledent, *Anal. Chim. Acta*, 1979, **107**, 383.
502 H. Hu, *Anal. Chim. Acta*, 1979, **107**, 387.
503 M. P. Colombini and R. Fuoco, *Talanta*, 1983, **30**, 901.
504 J. D. McLean, V. A. Stenger, R. E. Reim, M. W. Long, and T. A. Hiller, *Anal. Chem.*, 1978, **50**, 1309.
505 J. R. Herring and P. S. Liss, *Deep-Sea Res.*, 1974, **21**, 777.
506 K. Grasshoff in M. Whitfield, and D. Jagner, ed., ref. 456, pp. 327–420.
507 G. W. Luther, A. L. Meyerson, and A. D'Addio, *Mar. Chem.*, 1978, **6**, 117.
508 B. Ćosović and V. Vojvodić, *Limnol. Oceanogr.*, 1982, **27**, 361.
509 H. W. Nürnberg, in P. Ahlberg and L.-O. Sundelöf, ed., 'Structure and Dynamics in Chemistry', Proceedings from symposium held at Uppsala, Sweden, 22–27 September, 1977, Acta Universitatis Upsaliensis, Symposia Universitatis Upsaliensis, Annum Quingentesimum Celebrantis 12, Uppsala, 1978. (Distributed by Almqvist and Wiksell Int., Stockholm).
510 G. Gillain, G. Duyckaerts, and A. Disteche, *Anal. Chim. Acta*, 1979, **106**, 23.
511 M. P. Landy, *Anal. Chim. Acta*, 1980, **121**, 39.
512 W. Davison, *Freshwater Biol.*, 1980, **10**, 223.
513 R. J. O'Halloran, *Anal. Chim. Acta*, 1982, **140**, 51.
514 C. Brihaye and G. Duyckaerts, *Anal. Chim. Acta*, 1982, **143**, 111.
515 T. M. Florence, *J. Electroanal. Chem.*, 1974, **49**, 255.
516 L. Mart, *Fresenius' Z. Anal. Chem.*, 1979, **296**, 350.
517 L. Mart, *Fresenius' Z. Anal. Chem.*, 1979, **299**, 97.
518 L. Mart, H. W. Nürnberg, and P. Valenta, *Fresenius Z. Anal. Chem.*, 1980, **300**, 350.
519 P. L. Brezonik, P. A. Brauner, and W. Stumm, *Water Res.*, 1976, **10**, 605.
520 T. R. Copeland, R. A. Osteryoung, and R. K. Skogerboe, *Anal. Chem.*, 1974, **46**, 2093.
521 R. W. Gerlach and B. R. Kowalski, *Anal. Chim. Acta*, 1982, **134**, 119.
522 J. A. Wise, D. A. Roston, and W. R. Heineman, *Anal. Chim. Acta*, 1983, **154**, 95.
523 E. Ya. Neiman, L. G. Petrova, V. I. Ignatov, and G. M. Dolgopolova, *Anal. Chim. Acta*, 1980, **113**, 277.
524 J. Wang and B. Greene, *Water Res.*, 1983, **17**, 1635.
525 J. H. Liu, *Analyst*, 1980, **105**, 939.
526 R. G. Dhaneshwar and L. R. Zarapkar, *Analyst*, 1980, **105**, 386.
527 M. Ariel, U. Eisner, and S. Gottesfeld, *J. Electroanal. Chem.*, 1964, **7**, 307.
528 R. J. Siebert and D. N. Hume, *Anal. Chim. Acta*, 1981, **123**, 335.
529 J. E. Bonelli, R. K. Skogerboe, and H. E. Taylor, *Anal. Chim. Acta*, 1978, **101**, 437.
530 S. Taguchi, T. Yai, Y. Shimada, and K. Goto, *Talanta*, 1983, **30**, 169.
531 H. Monien, R. Bovenkerk, K. P. Kringe, and D. Rath, *Fresenius' Z. Anal. Chem.*, 1980, **300**, 363.
532 P. Valenta, H. Ruetzel, P. Krumpen, K. H. Salgert, and P. Klahre, *Fresenius Z. Anal. Chem.*, 1978, **292**, 120.
533 P. Klahre, P. Valenta, and H. W. Nürnberg, *Vom Wasser*, 1978, **51**, 199.
534 D. Jagner, *Anal. Chem.*, 1979, **51**, 3.
535 B. W. Woodget and K. R. Franklin, *Analyst*, 1981, **106**, 1017.
536 M. D. Booth, B. Fleet, S. Win, and T. S. West, *Anal. Chim. Acta*, 1969, **48**, 329.
537 J. Wang and M. Ariel, *Anal. Chem. Acta*, 1978, **99**, 89.
538 D. Jagner, M. Josefson and K. Åren, *Anal. Chim. Acta*, 1982, **141**, 147.
539 K. Tóth, G. Nagy, Z. Fehér, G. Horvai, and E. Pungor, *Anal. Chim. Acta*, 1980, **114**, 45.
540 K. Brunt, in J. F. Lawrence, ed., 'Trace Analysis', Vol. 1, Academic Press, New York, 1981, p. 47–120.
541 K. Izutsu, T. Nakamura, R. Takizawa, and H. Hanawa, *Anal. Chim. Acta*, 1983, **149**, 147.
542 K. Izutsu, T. Nakamura, and T. Ando, *Anal. Chim. Acta*, 1983, **152**, 285.
543 N. K. Lam, R. Kalvoda, and M. Kopanica, *Anal. Chim. Acta*, 1983, **154**, 79.
544 K. G. Janardan and D. J. Schaeffer, *Water Resourc. Res.*, 1981, **17**, 243.
545 M. Fielding, T. H. Gibson, H. A. James, K. McLoughlin, and C. P. Steel, 'Organic Micropollutants in Drinking Water', Technical Report TR 159, Water Research Centre, Medmenham, Bucks., 1981.
546 W. Giger and P. V. Roberts, in R. Mitchell, ed., 'Water Pollution Microbiology', Volume 2, Wiley, New York, 1978, pp. 135–175.
547 H. Purnell, 'Gas Chromatography', Wiley, New York, 1967.
548 A. B. Littlewood, 'Gas Chromatography. Principles, Techniques and Applications', 2nd Ed., Academic Press, New York, 1970.
549 R. L. Grob, ed., 'Modern Practice of Gas Chromatography', Wiley, New York, 1977.
550 W. G. Jennings, 'Gas Chromatography with Glass Capillary Columns', 2nd Ed., Academic Press, New York, 1980.
551 W. Bertsch, W. G. Jennings, and R. E. Kaiser, ed., 'Recent Advances in Capillary Gas Chromatography', Huethig, New York, Vol. 1, 1981, Vol. 2, 1982, and Vol. 3, 1982.
552 P. Sandra, 'Basic Capillary Gas Chromatography', Huethig, New York, 1983.
553 T. H. Risby, L. R. Field, F. J. Yang, and S. P. Cram, *Anal. Chem.*, 1982, **54**, 410R.
554 M. J. Fishman, D. E. Erdmann, and J. R. Garbarino, *Anal. Chem.*, 1983, **55**, 102R.

555 I. Wilson and I. Davis, 'The Application of High-Resolution Gas Chromatography to Water Analysis', Technical Report TR 168, Water Research Centre, Medmenham, Bucks., 1981.

556 Methods for the Examination of Waters and Associated Materials, 'Gas Chromatography – an Essay Review, 1982', Her Majesty's Stationery Office, London, 1983.

557 *J. Chromatogr. Sci.*, 1984, **22** (2); 1983, **21** (2).

558 H. M. McNair, *J. Chromatogr. Sci.*, 1983, **21**, 529.

559 L. S. Ettre, *Chromatographia*, 1975, **8**, 291.

560 L. S. Ettre, *Chromatographia*, 1982, **16**, 18.

561 W. Jennings, *J. High. Resolut. Chromatogr. Chromatogr. Commun.*, 1980, **3**, 601.

562 G. Schomburg, *J. Chromatogr. Sci.*, 1983, **21**, 97.

563 L. Rohrschneider, *J. Chromatogr.*, 1966, **22**, 6.

564 W. O. McReynolds, *J. Chromatogr. Sci.*, 1970, **8**, 685.

565 Chromatography Discussion Group, Trent Polytechnic, Burton Street, Nottingham, NG1 4BU, U.K.

566 Methods for the Examination of Waters and Associated Materials, 'Pyrethrins and Permethrin in Potable Waters by Electron-Capture Gas Chromatography. 1981', Her Majesty's Stationery Office, London, 1982.

567 Methods for the Examination of Waters and Associated Materials, 'Chloro- and Bromo-Trihalogenated Methanes in Water, 1980', Her Majesty's Stationery Office, London, 1981.

568 Methods for the Examination of Waters and Associated Materials, 'Organochlorine Insecticides and Polychlorinated Biphenyls in Waters, 1978', Her Majesty's Stationery Office, London, 1979.

569 Methods for the Examination of Waters and Associated Materials, 'Organo-Phosphorus Pesticides in River and Drinking Water, 1980', Her Majesty's Stationery Office, London, 1983.

570 Methods for the Examination of Waters and Associated Materials, 'Gas Chromatographic and Associated Methods for the Characterization of Oils, Fats, Waxes and Tars, 1982', Her Majesty's Stationery Office, London, 1983.

571 American Public Health Association *et al.*, ref. 18, pp. 493–503 (Organochlorine Pesticides), pp. 504–507 (Chlorinated Phenoxy Acid Herbicides), pp. 514–522 (Phenols) and pp. 538–549 (Halogenated Methanes and Ethanes).

572 J. E. Longbottom and J. J. Lichtenberg, ed., 'Methods for Organic Chemical Analysis of Municipal and Industrial Wastewater', Report EPA-600/4-82-057, U.S. Environmental Protection Agency, Cincinnati, July, 1982.

573 F. B. DeWalle, C. Lo, J. Sung, D. Kalman, E. S. K. Chian, M. Giabbai, and M. Ghosal, *J. Water Pollut. Contr. Fed.*, 1982, **54**, 555.

574 V. Pretorius and W. Bertsch, *J. High Resolut. Chromatogr. Chromatogr. Commun.*, 1983, **6**, 64.

575 G. Schomburg, *J. Chromatogr. Sci.*, 1983, **21**, 97.

576 K. Grob and G. Grob, *J. Chromatogr. Sci.*, 1969, **7**, 584.

577 K. Grob and G. Grob, *J. Chromatogr. Sci.*, 1969, **7**, 587.

578 G. Schomburg, H. Behlau, R. Dielmann, F. Weeke, and H. Husmann, *J. Chromatogr.*, 1977, **142**, 87.

579 H. Hachenberg and A. P. Schmidt, 'Gas Chromatographic Headspace Analysis', Heyden, London, 1977.

580 E. A. Dietz and M. F. Singley, *Anal. Chem.*, 1979, **51**, 1809.

581 K. Grob and G. Grob, *J. Chromatogr.*, 1974, **90**, 303.

582 K. Grob and F. Zürcher, *J. Chromatogr.*, 1976, **117**, 285.

583 W. Bertsch, E. Anderson, and G. Holzer, *J. Chromatogr.*, 1975, **112**, 701.

584 B. Dowty and J. L. Laseter, *Anal. Lett.*, 1975, **8**, 25.

585 T. A. Bellar and J. J. Lichtenberg, *J. Am. Water Works Assoc.*, 1974, **66**, 739.

586 J. M. Symons, T. A. Bellar, J. K. Carswell, J. DeMarco, K. L. Kropp, G. G. Robeck, D. R. Seegar, C. J. Slocum, B. L. Smith, and A. A. Stephens, *J. Am. Water Works Assoc.*, 1975, **67**, 634.

587 H. J. Brass, *Am. Lab.*, 1980, **12**(7), 23.

588 W. C. Schnute and D. E. Smith, *Am. Lab.*, 1980, **12**(7), 87.

589 L. Stieglitz, W. Roth, W. Kühn, and W. Leger, *Vom Wasser*, 1976, **47**, 347.

590 W. Giger, M. Reinhard, C. Schaffner, and F. Zürcher, in L. H. Keith, ed. 'Identification and Analysis of Organic Pollutants in Water', Ann Arbor Science Publishers, Ann Arbor, 1976, pp. 433–452.

591 J. F. J. Van Rensburg, J. J. Van Huyssteen, and A. J. Hassett, *Water Res.*, 1978, **12**, 127.

592 M. M. Varma, M. R. Siddique, K. T. Doty, and A. Machis, *J. Am. Water Works Assoc.*, 1979, **71**, 389.

593 R. C. Dressman, A. A. Stevens, J. Fair, and B. Smith, *J. Am. Water Works Assoc.*, 1979, **71**, 392.

594 I. Wilson, in E. Reid, ed., 'Trace-Organic Sample Handling', Ellis Horwood, Chichester, 1981, pp. 76–85.

595 E. E. McNeil, R. Otson, W. F. Miles, and F. J. M. Rajabalee, *J. Chromatogr.*, 1977, **132**, 277.

596 P. Van Rossum and R. G. Webb, *J. Chromatogr.*, 1978, **150**, 381.

597 V. N. Mallet, J. M. Francoeur, and G. Volpe, *J. Chromatogr.*, 1979, **172**, 388.

598 R. Shinohara, A. Kido, S. Eto, T. Hori, M. Koga, and T. Akiyama, *Water Res.*, 1981, **15**, 535.

599 W. A. Prater, M. S. Simmons, and K. H. Mancy, *Anal. Lett.*, 1980, **13**, 205.

[600] L. M. Keith, ed., 'Advances in the Identification and Analysis of Organic Pollutants in Water', Volumes 1 and 2, Ann Arbor Science Publishers, Ann Arbor, 1981.

[601] D. J. David, 'Gas Chromatographic Detectors', Wiley-Interscience, New York, 1974.

[602] J. Sevcik, 'Detectors in Gas Chromatography', Journal of Chromatography Library – Volume 4, Elsevier, Amsterdam, 1976.

[603] G. C. Stafford, P. B. Kelley, and D. C. Bradford, *Am. Lab.*, 1983, **15**(6), 51.

[604] A. P. Bentz, *Anal. Chem.*, 1976, **48**, 454A.

[605] E. R. Adlard, L. F. Creaser, and P. II. D. Matthews, *Ibid.*, 1972, **44**, 64.

[606] W. Krijgsman and C. G. van de Kamp, *J. Chromatogr.*, 1976, **117**, 201.

[607] J. Hild, E. Schulte and H.-P. Thier, *Chromatographia*, 1978, **11**, 397.

[608] Y. K. Chan, P. T. S. Wong, and G. A. Bengert, *Anal. Chem.*, 1982, **54**, 246.

[609] R. J. Maguire and R. J. Tkacz, *J. Chromatogr.*, 1983, **268**, 99.

[610] G. A. Cutter, *Anal. Chim. Acta*, 1978, **98**, 59.

[611] Y. K. Chau, P. T. S. Wong, and P. D. Goulden, *Anal. Chem.*, 1975, **47**, 2279.

[612] Y. K. Chau, P. T. S. Wong, and O. Kramar, *Anal. Chim. Acta*, 1983, **146**, 211.

[613] Y. K. Chau, P. T. S. Wong, and P. D. Goulden, *Anal. Chim. Acta*, 1976, **85**, 421.

[614] B. Radziuk, Y. Thomassen, L. R. P. Butler, Y. K. Chau, and J. C. van Loon, *Anal. Chim. Acta*, 1979, **108**, 31.

[615] S. A. Estes, P. C. Uden, and R. M. Barnes, *Anal. Chem.*, 1981, **53**, 1336.

[616] K. Chiba, K. Yoshida, K. Tanabe, H. Haraguchi, and K. Fuwa, *Anal. Chem.*, 1983, **55**, 450.

[617] B. D. Quimby, M. F. Delaney, P. C. Uden, and R. M. Barnes, *Anal. Chem.*, 1980, **52**, 259.

[618] F. J. Fernandez, *At. Absorpt. Newslett.*, 1977, **16**, 33.

[619] J. C. van Loon, *Am. Lab.*, 1981, **13**(5), 47.

[620] Y. K. Chau and P. T. S. Wong, in G. G. Leppard, ed. 'Trace Element Speciation in Surface Waters and its Ecological Implications', Plenum Press, New York, 1983, pp. 87–103.

[621] E. Kováts, *Adv. Chromatogr.*, 1965, **1**, 229.

[622] A. Shatkay and S. Flavian, *Anal. Chem.*, 1977, **49**, 2222.

[623] A. Shatkay, *Anal. Chem.*, 1978, **50**, 1423.

[624] T. A. Gough and E. A. Walker, *Analyst*, 1970, **95**, 1.

[625] R. C. Hall, *J. Chromatogr. Sci.*, 1974, **12**, 152.

[626] U. S. Environmental Protection Agency, Method 608, 'Organochlorine Pesticides and PCB's', in ref. 572, pp. 608-1–608-11.

[627] T. A. Pressley and J. E. Longbottom, 'The Determination of Chlorinated Herbicides in Industrial and Municipal Wastewater. Method 615', Report EPA-600/4-82-005, U.S. Environmental Protection Agency, Cincinnati, January, 1982.

[628] T. A. Pressley and J. E. Longbottom, 'The Determination of Organophosphorus Pesticides in Industrial and Municipal Wastewater. Method 622', Report EPA-600/4-82-008, U.S. Environmental Protection Agency, Cincinnati, January, 1982.

[629] T. A. Pressley and J. E. Longbottom, 'The Determination of Organonitrogen Pesticides in Industrial and Municipal Wastewater, Method 633', Report EPA-600/4-82-013, U.S. Environmental Protection Agency, Cincinnati, January, 1982.

[630] T. A. Pressley and J. E. Longbottom, 'The Determination of Triazine Pesticides in Industrial and Municipal Wastewater, Method 619', Report EPA-600/4-82-007, U.S. Environmental Protection Agency, Cincinnati, January 1982.

[631] P. J. Rennie, *Analyst*, 1982, **107**, 327.

[632] U.S. Environmental Protection Agency, Method 604, 'Phenols', in ref. 572, pp. 604-1–604-9.

[633] U.S. Environmental Protection Agency, Method 601, 'Purgeable Halocarbons', in ref. 572, pp. 601-1–601-10.

[634] U.S. Environmental Protection Agency, Method 602, 'Purgeable Aromatics', in ref. 572, pp. 602-1–602-10.

[635] U.S. Environmental Protection Agency, Method 606, 'Phthalate Esters', in ref. 572, pp. 606-1–606-8.

[636] U.S. Environmental Protection Agency, Method 607, 'Nitrosamines', in ref. 572, pp. 607-1–607-9.

[637] U.S. Environmental Protection Agency, Method 611, 'Haloethers', in ref. 572, pp. 611-1–611-8.

[638] U.S. Environmental Protection Agency, Method 612 'Chlorinated Hydrocarbons', in ref. 572, pp. 612-1–612-8.

[639] C. I. Measures and J. D. Burton, *Anal. Chim. Acta*, 1980, **120**, 177.

[640] Y. Shimoishi and K. Toei, *Anal. Chim. Acta*, 1978, **100**, 65.

[641] P. R. Gifford and S. Bruckenstein, *Anal. Chem.*, 1980, **52**, 1028.

[642] M. O. Andreae, *Anal. Chem.*, 1977, **49**, 820.

[643] S. Selim and C. R. Warner, *J. Chromatogr.*, 1978, **166**, 507.

[644] R. E. Kaiser and A. Rackstraw, 'Computer Chromatography', Volume 1, Hüthig, Heidelberg, 1983.

[645] M. J. Suess, ed., 'Examination of Water for Pollution Control. A Reference Handbook', Volume 2, Published on behalf of the World Health Organisation, Pergamon Press, Oxford, 1982.

[646] N. V. Brodtman and W. E. Koffskey, *J. Chromatogr. Sci.*, 1979, **17**, 97.

647 M. Suzuki, Y. Yamato, and T. Watenabe, *Environ. Sci. Technol.*, 1977, **11**, 1109.
648 I. R. DeLeon, P. C. Remele, S. K. Miles, and J. L. Laseter, *Anal. Lett.*, 1980, **13**, 503.
649 P. E. Mattson and S. Nygren, *J. Chromatogr.*, 1976, **124**, 265.
650 G. L. LeBel and D. T. Williams, *Bull. Environ. Contam. Toxicol.*, 1980, **24**, 397.
651 K. M. S. Sundaram, S. Y. Szeto, and R. Hindle, *J. Chromatogr.*, 1979, **177**, 29.
652 K. Nagasawa, H. Uchiyama, A. Ogamo, and T. Shinozuka, *J. Chromatogr.*, 1977, **144**, 77.
653 C. J. Soderquist and D. G. Crosby, *Anal. Chem.*, 1978, **50**, 1435.
654 B. W. Wright, M. L. Lee, and G. M. Booth, *J. High Resol. Chromatogr. Chromatogr. Commun.*, 1979, **2**, 189.
655 A. J. Cannard and W. J. Criddle, *Analyst*, 1975, **100**, 848.
656 O. Soerensen, *Vom Wasser*, 1978, **51**, 259.
657 L. L. Lamparski and T. J. Nestrick, *J. Chromatogr.*, 1978, **156**, 143.
658 G. T. Piet, P. Slingerland, F. E. de Grunt, M. P. M. van der Heuvel, and B. C. J. Zoeteman, *Anal. Lett.*, 1978, **11**, 437.
659 W. Giger and E. Molnar-Kubicka, *Bull. Environ. Contam. Toxicol.*, 1978, **19**, 475.
660 A. Bjørseth, G. E. Carlberg, and M. Møller, *Sci. Total Environ.*, 1975, **11**, 197.
661 B. G. Oliver and K. D. Bothen, *Anal. Chem.*, 1980, **52**, 2066.
662 A. R. Trussell, M. D. Umphres, L. Y. C. Leong, and R. R. Trussell, *J. Am. Water Works Assoc.*, 1979, **71**, 385.
663 B. D. Quimby, M. F. Delaney, P. C. Uden, and R. M. Barnes, *Anal. Chem.*, 1979, **51**, 875.
664 U.S. Environmental Protection Agency, Method 603, 'Acrolein and Acrylonitrile', in ref. 572, pp. 603-1–603-8.
665 K. Urano, K. Ogura, and H. Wada, *Water Res.*, 1981, **15**, 235.
666 E. S. K. Chian, P. P. K. Kuo, W. J. Copper, W. F. Cowen, and R. C. Fuentes, *Environ. Sci. Technol.*, 1977, **11**, 282.
667 P. P. K. Kuo, E. S. K. Chian, and F. B. DeWalle, *Water Res.*, 1977, **11**, 1005.
668 B. T. Croll and G. M. Simkins, *Analyst*, 1972, **97**, 281.
669 J. J. Richard and J. S. Fritz, *J. Chromatogr. Sci.*, 1980, **18**, 35.
670 A. Hashimoto, *Analyst*, 1976, **101**, 932.
671 F. I. Onuska, *Water Res.*, 1973, **7**, 835.
672 C. R. Lytle and E. M. Perdue, *Environ. Sci. Technol.*, 1981, **15**, 224.
673 J. L. Burleson, G. R. Peyton, and W. H. Glaze, *Environ. Sci. Technol.*, 1980, **14**, 1354.
674 M. Ochiai, *J. Chromatogr.*, 1980, **194**, 224.
675 J. Gardiner, *Analyst*, 1977, **102**, 120.
676 A. Di Corcia and R. Samperi, *Anal. Chem.*, 1979, **51**, 776.
677 B. V. Stolyarov, *Chromatographia*, 1981, **14**, 699.
678 G. Matsumoto and T. Hanya, *Water Res.*, 1981, **15**, 217.
679 J. Teply and M. Dressler, *J. Chromatogr.*, 1980, **191**, 221.
680 J. Drozd, J. Novak and J. A. Rijks, *J. Chromatogr.*, 1978, **158**, 471
681 W. E. May, S. N. Chesler, S. P. Cram, B. H. Gump, H. S. Hertz, D. P. Enagonio, and S. M. Dyszel, *J. Chromatogr. Sci.*, 1975, **13**, 535.
682 Z. Voznakova, M. Popl, and M. Berka, *J. Chromatogr. Sci.*, 1978, **16**, 123.
683 W. Giger and C. S. Schaffner, *Anal. Chem.*, 1978, **50**, 243.
684 P. P. K. Kuo, E. S. K. Chian, F. B. DeWalle, and J. H. Kim, *Anal. Chem.*, 1977, **49**, 1023.
685 A. Hashimoto, H. Sakino, E. Yamagami, and S. Tateishi, *Analyst*, 1980, **105**, 787.
686 D. Klockow, W. Bayer, and W. Faigle, *Fresenius' Z. Anal. Chem.*, 1978, **292**, 385.
687 K. Urano, H. Maeda, K. Ogura, and H. Wada, *Water Res.*, 1982, **16**, 323.
688 R. A. Larson and A. L. Rockwell, *J. Chromatogr.*, 1977, **139**, 186.
689 G. Matsumoto, R. Ishiwatari, and T. Hanya, *Water Res.*, 1977, **11**, 693.
690 W. R. White and J. A. Leenheer, *J. Chromatogr. Sci.*, 1975, **13**, 386.
691 L. N. Puijker, G. Voenendaal, H. M. J. Janssen, and H. Griepink, *Fresenius' Z. Anal. Chem.*, 1981, **306**, 1.
692 R. E. Sievers, R. M. Barkley, G. A. Eiceman, R. H. Shapiro, H. F. Walton, K. J. Kolonko, and L. R. Field, *J. Chromatogr.*, 1977, **142**, 745.
693 R. M. Goodfellow, J. Cardoso, G. Eglinton, J. P. Dawson, and G. A. Best, *Mar. Poll. Bull.*, 1977, **8**, 272.
694 H. Hon-nami and T. Hanya, *J. Chromatogr.*, 1978, **161**, 205.
695 H. Hon-nami and T. Hanya, *Water Res.*, 1980, **14**, 1251.
696 V. T. Wee, in L. H. Keith, ed., ref. 10, Vol. 1, pp. 467–479.
697 R. Jeltes, *J. Chromatogr. Sci.*, 1974, **12**, 599.
698 R. D. Adlard, L. F. Creaser, and P. H. D. Matthews, *Anal. Chem.*, 1972, **44**, 64.
699 F. K. Kawahara, *J. Chromatogr. Sci.*, 1972, **10**, 629.
700 H. A. Clark and P. C. Jurs, *Anal. Chem.*, 1975, **47**, 374.
701 D. V. Rasmussen, *Anal. Chem.*, 1976, **48**, 1562.

702 O. C. Zafiriou, *Est. Coast, Mar. Sci.*, 1973, **1**, 81.
703 F. Zürcher and M. Thüer, *Environ. Sci. Technol.*, 1978, **12**, 838.
704 R. Alexander, M. Cumbers, R. Kagi, M. Offer, and R. Taylor, in J. Albaiges, R. W. Frei and E.
      Merian, ed., 'Chemistry and Analysis of Hydrocarbons in the Environment', Gordon and Breach,
      New York, 1983, pp. 273–297.
705 T. H. Suffet and E. R. Glaser, *J. Chromatogr. Sci.*, 1978, **16**, 12.
706 J. R. Popalisky and F. W. Pogge, *J. Am. Water Works Assoc.*, 1972, **64**, 505.
707 K. Haberer and F. Schredelseker, *Vom Wasser*, 1975, **45**, 253.
708 P. Cukor and H. Madlin, in L. L. Ciaccio, *ed.* ref. 7, Vol. 4, pp. 1557–1615.
709 J. A. J. Walker and E. D. France, *Analyst*, 1969, **94**, 364.
710 H. L. Golterman, R. S. Clymo, and M. A. M. Ohnstad, 'Methods for Physical and Chemical Analysis
      of Fresh Waters', 2nd Ed., Blackwell, Oxford, 1978, p. 181.
711 H. P. Kollig, J. M. Falco, and F. E. Stancil, *Environ. Sci. Technol.*, 1975, **9**, 957.
712 J. Hissel, *Trib. Cebedean*, 1963, **16**, 60.
713 R. T. Williams and A. B. Bainbridge, *J. Geophys. Res.*, 1973, **78**, 2691.
714 L. P. Atkinson, *Anal. Chem.*, 1972, **44**, 885.
715 J. Hahn, *Anal. Chem.*, 1972, **44**, 1889.
716 J. W. Elkins, *Anal. Chem.*, 1980, **52**, 263.
717 J. W. Swinnerton and V. J. Linnenbom, *J. Gas Chromatogr.*, 1967, **5**, 570.
718 K. C. Hall, *J. Chromatogr. Sci.*, 1978, **16**, 311.
719 J. P. Riley, in J. P. Riley and G. Skirrow, ed., 'Chemical Oceanography', 2nd Ed., Volume 3, Academic
      Press, London, 1975, pp. 193–514.
720 R. Belcher, J. R. Majer, J. A. Rodriguez-Vasquez, W. I. Stephen, and P. C. Uden, *Anal. Chim. Acta*,
      1971, **57**, 73.
721 G. Nota, G. Vernassi, A. Accampora, and N. Sannolo, *J. Chromatogr.*, 1979, **174**, 228.
722 R. Bock and H. J. Semmler, *Fresenius' Z. Anal. Chem.*, 1967, **230**, 161.
723 D. R. Matthews, W. D. Shutts, and M. R. Guerin, *Anal. Chem.*, 1971, **43**, 1582.
724 G. Nota, C. Improta, U. Papale, and C. Terretti, *J. Chromatogr.*, 1983, **262**, 360.
725 T. P. Mawhinney, *Anal. Lett.*, 1983, **16**, 15.
726 B. Flegler, V. Gemmer-Colos, A. Dencks, and R. Neeb, *Talanta*, 1979, **26**, 761.
727 K. W. M. Siu, M. E. Bednas, and S. S. Berman, *Anal. Chem.*, 1983, **55**, 473.
728 S. Fukui, T. Hirayama, M. Nohara, and Y. Sakagami, *Talanta*, 1981, **28**, 402.
729 S. Fukui, T. Hirayama, M. Nohara, and Y. Sakagami, *Talanta*, 1983, **30**, 89.
730 Y. Odanaka, N. Tsuchiya, O. Matano, and S. Goto, *Anal. Chem.*, 1983, **55**, 929.
731 J. A. Rodriguez-Vasquez, *Talanta*, 1978, **25**, 299.
732 M. A. Ditzler and W. F. Gutknecht, *Anal. Chem.*, 1980, **52**, 614.
733 R. E. Majors, H. G. Barth, and C. H. Lochmüller, *Anal. Chem.*, 1982, **54**, 323R.
734 E. L. Johnson and R. Stevenson, 'Basic Liquid Chromatography', Varian, Palo Alto, 1978.
735 P. A. Bristow, 'Liquid Chromatography in Practice', hetp, Wilmslow, 1976.
736 H. Engelhardt, 'High Performance Liquid Chromatography. Chemical Laboratory Practice',
      Springer-Verlag, Berlin, 1979.
737 L. R. Snyder and J. J. Kirkland, 'Introduction to Modern Liquid Chromatography', 2nd Ed.,
      Wiley-Interscience, New York, 1979.
738 Horváth, Cs., ed., 'High Performance Liquid Chromatography. Advances and Perspectives', Vols. 1
      and 2, Academic Press, New York, 1980.
739 J. F. K. Huber, ed., 'Instrumentation for High Performance Liquid Chromatography', Journal of
      Chromatography Library, Vol. 13, Elsevier, Amsterdam, 1978.
740 H. L. McNair, *J. Chromatogr. Sci.*, 1982, **20**, 537.
741 *J. Chromatogr. Sci.*, 1980, **18**, 393–486 and 487–582.
742 M. J. Fishman, D. E. Erdmann, and T. R. Steinheimer, *Anal. Chem.*, 1981, **53**, 182R.
743 M. J. Fishman, D. E. Erdmann, and J. R. Garbarino, *Anal. Chem.*, 1983, **55**, 102R.
744 T. R. Graham, in J. F. Lawrence, ed., 'Trace Analysis', Vol. 1 Academic Press, New York, 1981, pp.
      1–46.
745 Methods for the Examination of Waters and Associated Materials, 'High Performance Liquid
      Chromatography, Ion Chromatography, Thin Layer and Column Chromatography of Water
      Samples. 1983', Her Majesty's Stationery Office, London, 1984.
746 B. Crathorne and C. D. Watts, in A. Bjørseth and G. Angeletti, ed. ref. 11, pp. 159–173.
747 R. E. Majors, *J. Chromatogr. Sci.*, 1980, **18**, 488.
748 R. M. Cassidy, in J. F. Lawrence, ed. 'Trace Analysis,' Vol. 1 Academic Press, New York, 1981, pp.
      121–192.
749 H. Engelhardt, B. Dreyer, and H. Schmidt, *Chromatographia*, 1982, **16**, 11.
750 W. W. Yau, J. J. Kirkland, and D. D. Bly, 'Modern Size-Exclusion Liquid Chromatography. Practice
      of Gel Permeation and Gel Filtration Chromatography', Wiley-Interscience, New York, 1979.
751 B. A. Bidlingmeyer, *J. Chromatogr. Sci.*, 1980, **18**, 525.

752 B. A. Bidlingmeyer, S. N. Deming, W. P. Price, B. Sachok, and M. Petrusek, *J. Chromatogr.*, 1979, **186**, 419.
753 J. C. Kraak, in J. F. K. Huber, ed., ref. 739, pp. 63–74.
754 J. N. Little and G. J. Fallick, *J. Chromatogr.*, 1975, **112**, 389.
755 A. W. Wolkoff and C. Creed, *J. Liq. Chromatogr.*, 1982, **4**, 1459.
756 J. A. Graham and A. W. Garrison, in L. H. Keith, ed., ref. 10, Vol. 1, pp. 399–432.
757 H. P. M. Van Vliet, T. C. Bootsman, R. W. Frei, and U. A. Th. Brinkman, *J. Chromatogr.*, 1979, **185**, 483.
758 M. Martin and G. Guiochon, in J. F. K. Huber, ed., ref. 739 pp. 11–40.
759 M. Martin and G. Guiochon, 'Solvent Gradient Systems', in J. F. K. Huber, ed., ref. 739, pp. 41–62.
760 L. R. Snyder, J. W. Dolan, and J. R. Gant, *J. Chromatogr.*, 1979, **165**, 3.
761 J. W. Dolan, J. R. Gant, and L. R. Snyder, *J. Chromatogr.*, 1979, **165**, 31.
762 K. Brunt, in J. F. Lawrence, ed., 'Trace Analysis', Vol. 1, Academic Press, New York, 1981, pp. 47–120.
763 H. Poppe, in J. F. K. Huber, ed., ref. 739, pp. 113–129.
764 Poppe, H. in J. F. K. Huber, ed., ref. 739, pp. 131–149.
765 P. Markl, 'Radiometric Detectors', in J. F. K. Huber, ed., ref. 739, pp. 151–161.
766 W. A. McKinley, D. J. Popovich, and T. Layne, *Am. Lab.*, 1980, **12** (8), 37.
767 S. R. Abbot and J. Tusa, *J. Liq. Chromatogr.*, 1983, **6** (Supplement 1), 77.
768 H. Poppe, 'Developments in Selective Detectors for HPLC' in ref 11, pp. 141–148.
769 J. F. Lawrence and R. W. Frei, 'Chemical Derivatisation in Liquid Chromatography', Elsevier, Amsterdam, 1976.
770 R. W. Frei, *J. Chromatogr*, 1979, **165**, 75.
771 R. W. Frei, *Chromatographia*, 1982, **15**, 161.
772 Sj. Van der Wal, *J. Liq. Chromatogr.*, 1983, **6** (Supplement 1), 37.
773 First Workshop on Liquid Chromatography–Mass Spectrometry, Montreux, October 1981, *J. Chromatogr.*, 1982, **251** (2).
774 Second Workshop on Liquid Chromatography–Mass Spectrometry and Mass Spectrometry–Mass Spectrometry, Montreux, October 1982, *J. Chromatogr.*, 1983, **271** (1).
775 Z. F. Curry, *J. Liq. Chromatogr.*, 1982, **5** (Supplement 2), 257.
776 K. Levsen, in A. Bjørseth, and G. Angeletti, ed. ref. 11, pp. 149–158.
777 K. Ogan, E. Katz, and W. Slavin, *J. Chromatogr. Sci.*, 1978, **16**, 517.
778 N. T. Crosby, D. C. Hunt, L. A. Philp, and I. Patel, *Analyst*, 1981 **106** 135.
779 R. K. Sorrell and R. Reding, *J. Chromatogr.*, 1979, **185**, 655.
780 K. Ogan, E. Katz, and W. Slavin, *Anal. Chem.*, 1979, **51**, 1315.
781 D. S. Farrington, R. G. Hopkins, and J. H. A. Růžička, *Analyst*, 1977, **102**, 377.
782 I. Ahmad, *Anal. Lett.*, 1982, **15**, 27.
783 D. C. Bouchard and T. C. Lavy, *J. Chromatogr.*, 1983, **270**, 396.
784 C. W. Vaughan, *Anal. Chim. Acta*, 1981, **131**, 307.
785 D. E. Wells and S. J. Johnstone, *J. Chromatogr. Sci.*, 1981, **19**, 137.
786 P. H. Cramer, A. D. Drinkwine, J. E. Going, and A. E. Carey, *J. Chromatogr.*, 1982, **235**, 489.
787 J. Spalik, G. E. Janauer, M. Lau, and A. T. Lemley, *J. Chromatogr.*, 1982, **253**, 289.
788 D. N. Armentrout, J. D. McLean, and M. W. Long, *Anal. Chem.*, 1979, **51**, 1039.
789 R. E. Shoup and G. S. Mayer, *Anal. Chem.*, 1982, **54**, 1164.
790 C. E. Werkhoven-Goewie, U. A. Th. Brinkman, and R. W. Frei, *Anal. Chem.*, 1981, **53**, 2072.
791 B. Schultz, *J. Chromatogr.*, 1983, **269**, 208.
792 V. T. Wee and J. M. Kennedy, *Anal. Chem.*, 1982, **54**, 1631.
793 A. Nakae, K. Tsuji, and M. Yamanaka, *Anal. Chem.*, 1980, **52**, 2275.
794 B. Crathorne, C. D. Watts, and M. Fielding, *J. Chromatogr.*, 1979, **185**, 671.
795 L. Brown, *Analyst*, 1979, **104**, 1165.
796 L. Brown and M. Rhead, *Analyst*, 1979, **104**, 391.
797 R. M. Riggin and C. C. Howard, *Anal. Chem.*, 1979, **51**, 210.
798 D. N. Armentrout and S. S. Cutié, *J. Chromatogr. Sci.*, 1980, **18**, 370.
799 J. R. Rice and P. T. Kissinger, *Environ. Sci. Technol.*, 1982, **16**, 263.
800 G. C. Hawn, P. A. Diehl, and C. P. Talley, *J. Chromatogr. Sci.*, 1981, **19**, 567.
801 I. S. Krull, K. W. Panaro, and L. L. Gershman, *J. Chromatogr. Sci.*, 1983, **21**, 460.
802 S. Ichinoki, T. Morita, and M. Yamakazi, *J. Liq. Chromatogr.*, 1983, **6**, 2079.
803 S. H. Kok, K. A. Buckle, and M. Wootton, *J. Chromatogr.*, 1983, **260**, 189.
804 D. Chakraborti, D. C. J. Hillman, K. J. Irgolic, and R. A. Zingaro, *J. Chromatogr.* 1982, **249**, 81.
805 R. Kummert, E. Molnar-Kubica, and W. Giger, *Anal. Chem.*, 1978, **50**, 1637.
806 J. P. McCarthy, J. A. Caruso, and F. L. Fricke, *J. Chromatogr. Sci.* 1983, **21**, 289.
807 F. Smedes, J. C. Kraak, C. F. Werkhoven-Goewie, U. A. Th. Brinkman, and R. W. Frei, *J. Chromatogr.*, 1982, **247**, 123.
808 J. K. Baker and C.-Y. Ma, *J. Chromatogr.*, 1979, **169**, 107.

809 L. H. Wright, *J. Chromatogr. Sci.*, 1982, **20**, 1.
810 T. Takeuchi, D. Ishii, A. Saito, and T. Ohki, *J. High Resolut. Chromatogr. Chromatogr. Commun.*, 1982, **5**, 91.
811 R. Fischer, *Fresenius' Z. Anal. Chem.*, 1982, **311**, 109.
812 D. Kasiske, K. D. Klinkmüller, and M. Sonneborn, *J. Chromatogr.*, 1978, **149**, 703.
813 J. C. Hoffsomer, D. J. Glover, and C. Y. Hazzard, *J. Chromatogr.*, 1980, **195**, 535.
814 G. Blo, F. Dondi, A. Betti, and C. Bighi, *J. Chromatogr.*, 1983, **257**, 69.
815 P. A. Realini, *J. Chromatogr. Sci.*, 1981, **19**, 124.
816 E. Tesařova and V. Pacáková, *Chromatographia*, 1983, **17**, 269.
817 B. Crathorne and C. D. Watts, *J. Chromatogr.*, 1979, **169**, 436.
818 E. Grou, V. Rădulescu, and A. Csuma, *J. Chromatogr.* 1983, **260**, 502.
819 M. J. Barcelona, H. M. Liljestrand, and J. J. Morgan, *Anal. Chem.*, 1980, **52**, 321.
820 D. A. Hullett and S. J. Eisenreich, *Anal. Chem.*, 1979, **51**, 1953.
821 J. Pempkowiak, *J. Chromatogr.*, 1983, **258**, 93.
822 J. Harmsen and A. van den Toorn, *J. Chromatogr.*, 1982, **249**, 379.
823 D. A. Frick, *J. Liq. Chromatogr.*, 1983, **6**, 445.
824 J. K. Abaychi and J. P. Riley, *Anal. Chim. Acta*, 1979, **107**, 1.
825 J. Grimalt and J. Albaiges, *J. High Resolut. Chromatogr. Chromatogr. Commun.*, 1982, **5**, 255.
826 A. M. Cheh and R. E. Carlson, *Anal. Chem.*, 1981, **53**, 1001.
827 R. Ishiwatari, H. Hamana, and T. Machihara, *Water Res.*, 1980, **14**, 1257.
828 E. S. K. Chian and F. B. DeWalle, *Environ. Sci. Technol.*, 1977, **11**, 158.
829 K. Urano, K. Katagiri, and K. Kawamoto, *Water Res.*, 1980, **14**, 741.
830 C. J. Miles and P. L. Brezonik, *J. Chromatogr.*, 1983, **259**, 499.
831 W. E. Barber and P. W. Carr, *J. Chromatogr.*, 1983, **260**, 89.
832 R. M. Cassidy and S. Elchuk, *J. Chromatogr. Sci.*, 1983, **21**, 454.
833 G. Schwedt, *Chromatographia*, 1979, **12**, 613.
834 J. W. O.Laughlin, *J. Liq. Chromatogr.*, 1984, **7** (Supplement 1), 127.
835 H. Matsui, *Anal. Chim. Acta*, 1973, **66**, 143.
836 A. M. Bond and G. G. Wallace, *J. Liq. Chromatogr.*, 1983, **6**, 1799.
837 A. M. Bond and G. G. Wallace, *Anal. Chem.*, 1983, **55**, 718.
838 R. M. Smith and L. E. Yankey, *Analyst*, 1982, **107**, 744.
839 E. B. Edward-Inatimi, *J. Chromatogr.*, 1983, **256**, 253.
840 S. Elchuk and R. M. Cassidy, *Anal. Chem.*, 1979, **51**, 1434.
841 R. M. Cassidy, S. Elchuk, and J. O. McHugh, *Anal. Chem.*, 1982, **54**, 727.
842 N. Haring and K. Ballschmiter, *Talanta*, 1980, **27**, 873.
843 G. Schwedt and A. Schwarz, *J. Chromatogr.*, 1978, **160**, 309.
844 G. Schwedt, *Fresenius' Z. Anal. Chem.*, 1981, **309**, 359.
845 J. C. Van Loon, *Am. Lab.*, 1981, **13**(5), 47.
846 Y. K. Chau and P. T. S. Wong, in G. G. Leppard, ed. ref. 620, pp. 87–103.
847 F. J. Fernandez, *At. Absorpt. Newslett.*, 1977, **16**, 33.
848 I. S. Krull and S. Jordan, *Am. Lab.*, 1980, **12**(10), 21.
849 G. Schwedt, *Fresenius' Z. Anal. Chem.*, 1979, **295**, 382.
850 B. W. Hoffman and G. Schwedt, *J. High Resolut. Chromatogr.*, *Chromatogr. Commun.*, 1982, **5**, 439.
851 D. R. Jones and S. E. Manahan, *Anal. Chem.*, 1976, **48**, 502.
852 F. E. Brinckman, W. R. Blair, K. L. Jewett, and W. P. Iverson, *J. Chromatogr. Sci.*, 1977, **15**, 493.
853 R. H. Fish, F. E. Brinckman, and K. L. Jewett, *Environ, Sci. Technol.*, 1982, **16**, 174.
854 T. M. Vickrey, H. E. Howell, and M. T. Paradise, *Anal. Chem.* 1979, **51**, 1880.
855 K. L. Jewett and F. E. Brinckman, *J. Chromatogr. Sci.*, 1981, **19**, 583.
856 G. R. Ricci, L. S. Shepard, G. Colovos, and N. E. Hester, *Anal. Chem.*, 1981, **53**, 610.
857 D. T. Burns, F. Glockling, and M. Harriott, *Analyst*, 1981, **106**, 921.
858 W. A. MacCrehan and R. A. Durst, *Anal. Chem.*, 1978, **50**, 2108.
859 W. A. MacCrehan, *Anal. Chem.*, 1981, **53**, 74.
860 W. A. MacCrehan and R. A. Durst, *Anal. Chem.*, 1981, **53**, 1700.
861 D. S. Bushee, I. S. Krull, P. R. Demko, and S. B. Smith, *J. Liq. Chromatogr.*, 1984, **7**, 861.
862 M. Morita, T. Uehiro, and K. Fuwa, *Anal. Chem.* 1980, **52**, 349.
863 D. W. Hausler and L. T. Taylor, *Anal. Chem.*, 1981, **53**, 1223.
864 D. W. Hausler and L. T. Taylor, *Anal. Chem.*, 1981, **53**, 1227.
865 F. C. Smith and R. C. Chang, 'The Practice of Ion-Chromatography', Wiley, Chichester, 1982.
866 C. A. Pohl and E. L. Johnson, *J. Chromatogr. Sci.*, 1980, **18**, 442.
867 E. L. Johnson, *Am. Lab.*, 1982, **14**(2), 98.
868 E. Sawicki, J. D. Mulik, and E. Wittgenstein, ed. 'Ion Chromatographic Analysis of Environmental Pollutants', Vol. 1., Ann Arbor Science Publishers, Ann Arbor, 1978.
869 J. D. Mulik and E. Sawicki, ed., 'Ion Chromatographic Analysis of Environmental Pollutants', Vol. 2., Ann Arbor Science Publishers, Ann Arbor, 1979.

870 H. Small, in J. F. Lawrence, ed., 'Trace Analysis', Volume 1, Academic Press, New York, 1981, pp. 267–322.
871 D. T. Gjerde, J. S. Fritz, and G. Schmuckler, *J. Chromatogr.*, 1979, **186**, 509.
872 D. T. Gjerde, J. S. Fritz, and G. Schmuckler, *J. Chromatogr.*, 1980, **187**, 35.
873 J. A. Glatz and J. E. Girard, *J. Chromatogr. Sci.*, 1982, **20**, 266.
874 H. Small, in E. Sawicki, J. D. Mulik, and E. Wittgenstein, ed., ref. 868, pp. 2–21.
875 T. S. Stevens, J. C. Davis, and H. Small, *Anal. Chem.*, 1981, **53**, 1488.
876 J. Slanina, W. A. Lingerak, J. E. Ordelman, P. Borst, and F. P. Bakker, in J. D. Mulik and E. Sawicki, ed., ref. 869, pp. 305–317.
877 H. Small and T. E. Miller, *Anal. Chem.*, 1982, **54**, 462.
878 S. W. Downey and G. M. Hieftje, *Anal. Chim. Acta*, 1983, **152**, 1.
879 G. S. Pyen and M. J. Fishman, in J. D. Mulik and E. Sawicki, ed., ref. 869, pp. 235–244.
880 J. A. Rawa, in J. D. Mulik and E. Sawicki, ed., ref. 869, pp. 245–269.
881 S. M. Kapelner, J. C. Trocciola, and M. S. Freed, in J. D. Mulik and E. Sawicki, ed., ref. 869, pp. 345–360.
882 J. A. Hern, G. K. Rutherford, and G. W. van Loon, *Talanta*, 1983, **30**, 677.
883 K. M. Roberts, D. T. Gjerde, and J. S. Fritz, *Anal. Chem.*, 1981, **53**, 1691.
884 G. S. Pyen and D. E. Erdmann, *Anal. Chim. Acta*, 1983, **149**, 355.
885 M. A. Fulmer, J. Penkrot, and R. J. Nadalin, in J. D. Mulik and E. Sawicki, ed., ref. 869, pp. 381–400.
886 D. C. Bogen and S. J. Nagourney, 'Ion Chromatographic Analysis of Cations at Baseline Precipitation Stations', in, J. D. Mulik and E. Sawicki, ed., ref. 869, pp. 319–328.
887 J. S. Fritz, D. T. Gjerde, and R. M. Becker, *Anal. Chem.* 1980, **52**, 1519.
888 E. Sawicki, in J. D. Mulik and E. Sawicki, ed., ref. 869, pp. 1–15.
889 S. Y. Tyree, J. M. Stouffer, and M. Bollinger, in J. D. Mulik and E. Sawicki, ed., ref. 869, pp. 295–304.
890 J. Crowther and J. McBride, *Analyst*, 1981, **106**, 702.
891 R. P. Lash and C. J. Hill, *Anal. Chim. Acta*, 1979, **108**, 405.
892 L. J. Holcombe, B. F. Jones, E. E. Ellsworth, and F. B. Meserole, in J. D. Mulik and E. Sawicki, ed., ref. 869, pp. 401–412.
893 L. J. Schiff, S. G. Pleva, and E. W. Sarver, in J. D. Mulik and E. Sawicki, ed., ref. 869, pp. 329–344.
894 M. Zakaria, M.-F. Gonnord, and G. Guiochon, *J. Chromatogr.*, 1983, **271**, 127.
895 E. Stahl, ed., 'Thin-Layer Chromatography: A Laboratory Handbook', 2nd Ed., Springer, New York, 1969.
896 J. C. Touchstone, ed., 'Quantitative Thin Layer Chromatography', Wiley-Interscience, New York, 1973.
897 J. G. Kirchner, ed., 'Thin-Layer Chromatography', 2nd Ed., Wiley, New York, 1978.
898 A. Zlatkis and R. E. Kaiser, ed., 'High Performance Thin-Layer Chromatography,' Elsevier, Amsterdam, 1977.
899 J. C. Touchstone and D. Rogers, 'Thin Layer Chromatography: Quantitative Environmental and Clinical Applications', Wiley-Interscience, New York, 1980.
900 W. Bertsch, S. Hara, R. E. Kaiser, and A. Zlatkis, ed., 'Instrumental HPTLC', Hüthig, Heidelberg, 1980.
901 B. Fried and J. Sherma, 'Thin-Layer Chromatography: Techniques and Applications', Dekker, New York, 1982.
902 J. C. Touchstone and J. Sherma, ed., 'Densitometry in Thin Layer Chromatography. Practice and Applications', Wiley-Interscience, New York, 1979.
903 J. Sherma and B. Fried, *Anal Chem.*, 1982, **54**, 45R.
904 W. Funk, *Vom Wasser*, 1980, **54**, 207.
905 G. T. Hunt, in R. L. Grob, ed., 'Chromatographic Analysis of the Environment', 2nd Ed. Dekker, New York, 1983, pp. 297–344.
906 M. E. P. McNally and J. F. Sullivan, in R. L. Grob, ed. ref. 905, pp. 651–690.
907 U. A. Th. Brinkman and G. de Vries, *J. High Resolut. Chromatogr., Chromatogr. Commun.*, 1982, **5**, 476.
908 H. Halpaap, and K.-F. Krebs, *J. Chromatogr. Sci.*, 1977, **142**, 823.
909 D. E. Jänchen, in W. Bertsch, S. Hara, R. E. Kaiser, and A. Zlatkis, ed., ref. 900, pp. 133–164.
910 A. Sdika, R. Cabridenc, and C. Hennequin, in A. Bjørseth and G. Angeletti, ed., ref. 11, pp. 24–37.
911 J. Borneff and H. Kunte, *Arch. Hyg.*, 1969, **153**, 220.
912 H. Woidich, W. Pfannhauser, G. Blaicher, and K. Tiefenbacher, *Chromatographia*, 1977, **10**, 140.
913 D. Basu and J. Saxena, *Environ. Sci. Technol.*, 1978, **12**, 791.
914 H. Hellman, *Fresenius' Z. Anal. Chem.*, 1979, **295**, 24.
915 R. I. Crane, B. Crathorne, and M. Fielding, in B. K. Afghan and D. Mackay, ed., 'Hydrocarbons and Halogenated Hydrocarbons in the Aquatic Environment', Plenum, New York, 1980, pp. 161–172.
916 D. J. Futoma, S. Ruven Smith, T. E. Smith, and J. Tanaka, 'Polycyclic Aromatic Hydrocarbons in Water Systems', CRC Press, Boca Raton, 1981, pp. 79–84.

917 M. L. Lee, M. V. Novotny, and K. D. Bartle, 'Analytical Chemistry of Polycyclic Aromatic Compounds', Academic Press, New York, 1981, pp. 135–143.

918 M. J. Suess, ed., ref. 645, pp. 464–471.

919 M. G. Zigler and W. F. Phillips, *Environ. Sci. Technol.*, 1967, **1**, 56.

920 O. M. Aly, *Water Res.*, 1968, **2**, 587.

921 P. Chambon, R. Chambon-Mougenot, and J. Bringuier, *Rev. Inst. Pasteur Lyon*, 1970, **3**, 395.

922 H. Thieleman, *Z. Wass. Abwass. Forsch.*, 1970, **3**, 122.

923 H. Hellmann, *Z. Wasser Abwasser Forsch.*, 1978, **293**, 359.

924 H. Hellmann, *Fresenius' Z. Anal. Chem.*, 1980, **304**, 129.

925 H. Hellmann, *Fresenius' Z. Anal. Chem.*, 1982, **310**, 224.

926 M. J. Suess, ed., ref. 645, pp. 494–497.

927 A. Akemi and Y. Hiroshi, *Water Res.*, 1979, **13**, 1111.

928 E. Hohaus, *Fresenius' Z. Anal. Chem.*, 1982, **310**, 70.

929 C. D. Litchfield and J. M. Prescott, *Limnol. Oceanogr.*, 1970, **15**, 250.

930 J. I. Hollies, D. F. Pinnington, and A. J. Handley, *Anal. Chim. Acta*, 1979, **111**, 201.

931 H. Hellmann, *Vom Wasser*, 1977, **48**, 129.

932 L. Hunter, *Environ. Sci. Technol.*, 1975, **9**, 241.

933 P. J. Whittle, *Analyst*, 1977, **102**, 976.

934 K. Osawa, K. Fuzikawa, and K. Imaeda, *Bunseki Kagaku*, 1980, **29**, 431.

935 K. Osawa, K. Fuzikawa, and K. Imaeda, *Bunseki Kagaku*, 1981, **30**, 305.

936 W. Funk, R. Kerler, J. Th. Schiller, V. Damman, and F. Arndt, *J. High Resolut. Chromatogr., Chromatogr. Comm.*, 1982, **5**, 534.

937 R. W. Frei, J. F. Lawrence, and D. S. LeGay, *Analyst*, 1973, **98**, 9.

938 F. I. Onuska, *Anal. Lett.*, 1974, **7**, 327.

939 J. Sherma and J. Koropchack, *Anal. Chim. Acta*, 1977, **91**, 259.

940 A. H. M. T. Scholten, C. Van Buren, J. F. Lawrence, U. A. Th. Brinkman, and R. W. Frei, *J. Liq. Chromatogr.*, 1979, **2**, 607.

941 M. F. Dobbins and J. C. Touchstone, *in* J. C. Touchstone, ed., ref. 896, pp. 293–304.

942 R. D. Stephens and J. J. Chan, in J. C. Touchstone, and D. Rogers, ed., ref. 899, pp. 363–369.

943 Y. Volpe and V. N. Mallet, *Anal. Chim. Acta*, 1976, **81**, 111.

944 V. N. Mallet and Y. Volpe *Anal. Chim. Acta*, 1978, **97**, 415.

945 K. Berkane, G. E. Caissie, and V. N. Mallet, *J. Chromatogr.*, 1977, **139**, 386.

946 J. Sherma, K. E. Klopping, and M. E. Cetz, *Am. Lab.* 1977, **9** (12), 66.

947 C. Garside and J. P. Riley, *Anal. Chim. Acta*, 1969, **46**, 179.

948 S. W. Jeffrey, *Limnol. Oceanogr.*, 1981, **26**, 191.

949 H. L. Golterman, R. S. Clymo and A. M. M. Ohnstad, 'Methods for Physical and Chemical Analysis of Fresh Waters', IBP Handbook No. 8, 2nd Ed., Blackwell, Oxford, 1978, pp. 167–168.

950 P. Matthews, *J. Appl. Chem.*, 1970, **20**, 87.

951 W. Saner and G. E. Fitzgerald, *Environ. Sci. Technol.*, 1976, **10**, 893.

952 D. J. Futoma, S. R. Smith, T. E. Smith, and J. Tanaka, in ref. 916, pp. 73–75.

953 R. A. Dobbs, R. H. Wise, and R. B. Dean, *Anal. Chem.*, 1967, **39**, 1255.

954 F. R. Cropper, D. M. Heinekey, and A. Westwell, *Analyst*, 1967, **92**, 436 and 443.

955 B. T. Croll, *Chem. Ind.*, 1972, 386.

956 B. T. Croll, 'Determination of Total Organic Carbon in Water', Technical Memorandum TM 124, The Water Research Centre, Medmenham, Bucks., 1976.

957 Y. Takahashi, R. T. Moore, and R. J. Joyce, *Am. Lab.*, 1972, **4**, 31.

958 I. Lysyj, K. H. Nelson, and P. R. Newton, *Water Res.*, 1968, **1**, 233.

959 F. T. Eggertsen and F. H. Stross, *Anal. Chem.*, 1972, **44**, 709.

960 A. P. Black and S. Hannah, *J. Am. Water Works Assoc.*, 1965, **57**, 901.

961 M. Kerker, 'The Scattering of Light and other Electromagnetic Radiation', Academic Press, New York, 1969.

962 D. G. Talley, J. A. Johnson, and J. E. Pilzer, *J. Am. Water Works Assoc.*, 1972, **64**, 184.

963 American Public Health Association *et al.*, ref. 18, pp. 131–134.

964 D. W. Eichner and C. C. Hach, *Water Sewage Works*, 1971, **120**, 299.

965 W. R. McCluney, *J. Water Pollut. Contr. Fed.*, 1975, **47**, 252.

966 American Society for Testing and Materials, '1981 Annual Book of ASTM Standards. Part 31. Water', ASTM, Philadelphia, 1981, pp. 516–518.

967 U.S. Environmental Protection Agency, 'Methods for the Chemical Analysis of Water and Wastes', 1979, EPA-600/4-79-020, The Agency, Cincinnati, pp. 375.4-1–375.4-3.

968 W. I. Stephen, *Anal. Chim. Acta*, 1970, **50**, 413.

969 H. S. Harned and B. B. Owen, 'The Physical Chemistry of Electrolytic Solutions', 3rd. Ed., Reinhold, New York, 1958.

970 R. A. Robinson and R. H. Stokes, 'Electrolyte Solutions', 3rd. Ed., Butterworths, London, 1970.

971 Methods for the Examination of Waters and Associated Materials, 'The Measurement of Electrical

Conductivity and the Laboratory Determination of the pH value of Natural, Treated and Waste Waters. 1978', Her Majesty's Stationery Office, London, 1979.

972 American Public Health Association *et al.*, ref. 18, pp. 70–73.

973 American Society for Testing and Materials, 'Annual Book of ASTM Standards, Part 31, Water', The Society, Philadelphia, 1981, pp. 138–146.

974 The Committee for Analytical Quality Control (Harmonised Monitoring), 'Accuracy of Determination of the Electrical Conductivity of River Waters', Technical Report TR 190, The Water Research Centre, Medmenham, Bucks., 1983.

975 S. Smith, *Limnol. Oceanogr.*, 1962, **7**, 330.

976 R. Wagner, Z. *Wasser Abwasser Forsch.*, 1980, **13**, 62.

977 T. R. S. Wilson, in J. P. Riley and G. Skirrow, ed., 'Chemical Oceanography', 2nd Ed., Vol. 1, Academic Press, London, 1975, pp. 365–413.

978 T. R. S. Wilson, in M. Whitfield and D. Jagner, ed., 'Marine Electrochemistry. A Practical Introduction', Wiley, Chichester, 1981, pp. 145–185.

979 R. J. Oake and R. L. Norton, 'The Photochem Organic Carbon Analyser; An Evaluation', Technical Report, TR 186, Water Research Centre, Medmenham, Bucks., 1982.

980 J. Murphy, *Anal. Chem.*, 1979, **51**, 1959.

981 C. F. McCourt and R. G. H. Watson, Brit. Pat., 894 370, Nov. 24, 1959.

982 K. H. Mancy, D. A. Okun, and C. N. Reilly, *J. Electroanal. Chem.*, 1962, **4**, 65.

983 R. Briggs, G. V. Dyke, and G. Knowles, *Analyst*, 1958, **83**, 304.

984 E. Føyn, in H. L. Golterman and R. S. Clymo, ed., 'Chemical Environment in the Aquatic Habitat', Royal Netherlands Academy of Sciences, Amsterdam, 1967, pp. 127–132.

985 K. H. Mancy and D. A. Okun, *Anal. Chem.*, 1960, **32**, 108.

986 E. Gnaiger and H. Forstner, ed., 'Polarographic Oxygen Sensors', Springer-Verlag, Berlin, 1983.

987 K. Grasshoff, in M. Whitfield, and D. Jagner, ed., ref. 978, pp. 327–420.

988 M. Whitfield in J. P. Riley and G. Skirrow, ed., 'Marine Chemistry', 2nd Ed., Vol. 4, Academic Press, London, 1975, pp. 1–154.

989 M. L. Hitchman, 'Measurement of Dissolved Oxygen', Wiley, New York, 1978.

990 R. Bucher, in E. Gnaiger and H. Forstner, ed., ref. 986, pp. 66–72.

991 J. M. Hale, in E. Gnaiger and H. Forstner, ed., ref. 986, pp. 73–75

992 H. L. Golterman, R. S. Clymo, and M. A. M. Ohnstad, 'Methods for Physical and Chemical Analysis of Fresh Waters', IBP Handbook No. 8, 2nd Ed., Blackwell Scientific Publications, Oxford, 1978, pp. 178–180.

993 American Public Health Association *et al.*, *op. cit.* ref. 18, pp. 395–399.

994 M. L. Hitchman, in E. Gnaiger and H. Forstner, ed., ref. 986, pp. 18–30.

995 H. Forstner, in E. Gnaiger and H. Forstner, ed., ref. 986, pp. 90–101.

996 I. Bals and J. M. Hale, in E. Gnaiger and H. Forstner, ed., ref. 986, pp. 102–107.

997 D. M. Phelan, R. M. Taylor, and S. Fricke, *Am. Lab.*, 1982, **14**(7), 65.

998 J. J. Morrow and R. N. Roop, *J. Am. Water Works Assoc.*, 1975, **67**, 184.

999 E. Canelli, *Water Res.*, 1980, **14**, 1533.

1000 J. D. Johnson, J. W. Edwards, and F. Keeslar, *J. Am. Water Works Assoc.*, 1978, **70**, 341.

1001 W. J. Cooper, M. F. Mehran, R. A. Slifker, D. A. Smith, J. T. Villate, and P. H. Gibbs, *J. Am. Water Works Assoc.*, 1982, **74**, 546.

1002 N. A. Dimmock and D. Midgley, *Water Res.*, 1979, **13**, 1101.

1003 D. Midgley and N. A. Dimmock, *Water Res.*, 1979, **13**, 1317.

1004 R. Dormond-Herrera and K. H. Mancy, *Anal. Lett.*, 1980, **13**, 561.

1005 M. Schmid and K. H. Mancy, *Schweiz. Z. Hydrol.*, 1970, **32**, 328.

1006 J. W. Freese and R. B. Smart, *Anal. Chem.*, 1982, **54**, 836.

1007 R. B. Smart and J. W. Freese, *J. Am. Water Works Assoc.*, 1982, **74**, 530.

1008 R. B. Smart, R. Dormond-Herrera, and K. H. Mancy, *Anal. Chem.*, 1979, **51**, 2315.

1009 J. H. Stanley and J. D. Johnson, *Anal. Chem.*, 1979, **51**, 2144.

1010 V. S. Srinivasan and G. P. Tarcy, *Anal. Chem.*, 1981, **53**, 928.

1011 L. W. Niedrach and W. H. Stoddard, *Anal. Chem.*, 1982, **54**, 1651.

1012 T. Okada, I. Karube, and S. Suzuki, *Anal. Chim. Acta*, 1982, **135**, 159.

1013 E. J. Maienthal and J. K. Taylor, in L. L., Ciaccio, ed, ref. 7, Vol. 4, pp. 1751–1800.

1014 P. R. Gifford and S. Bruckenstein, *Anal. Chem.*, 1980, **52**, 1028.

1015 P. R. Gifford and S. Bruckenstein, *Anal. Chem.*, 1980, **52**, 1024.

1016 D. Nygaard, *Anal. Chem.*, 1980, **52**, 358.

1017 P. Beran and S. Bruckenstein, *Anal. Chem.*, 1980, **52**, 1183.

1018 D. Nygaard, *Anal. Chim. Acta*, 1981, **130**, 391.

1019 S. Bruckenstein, *Anal. Chem.*, 1980, **52**, 2207.

1020 G. S. Pyen and D. E. Erdmann, *Anal. Chim. Acta*, 1983, **149**, 355.

1021 J. J. Lingane, 'Electroanalytical Chemistry', 2nd Ed., Interscience, New York, 1958.

1022 J. M. McGuire and R. G. Webb, in H. B. Mark and J. S. Mattson, ed., ref. 8, pp. 1–31.

606 *Chemical Analysis of Water*

1023 K. Levsen, 'Fundamental Aspects of Organic Mass Spectrometry', Progress in Mass Spectrometry, Vol. 4, Verlag Chemie, Weinheim, 1978.
1024 U. P. Schlunegger, 'Advanced Mass Spectrometry. Applications in Organic and Analytical Chemistry', Pergamon Press, Oxford, 1980.
1025 I. Howe, D. H. Williams, and R. D. Bowen, 'Mass Spectrometry. Principles and Applications', 2nd Ed., McGraw-Hill, New York, 1981.
1026 B. J. Millard, 'Quantitative Mass Spectrometry', Heyden, London, 1978.
1027 A. L. Burlingame, A. Dell, and D. H. Russell, *Anal. Chem.*, 1982, **54**, 363R.
1028 R. A. W. Johnstone, ed., 'Mass Spectrometry', (Specialist Periodical Reports), Vol. 6, The Chemical Society, London, 1981.
1029 W. H. McFadden, 'Techniques of Combined Gas Chromatography/Mass Spectrometry. Applications in Organic Analysis', Wiley, New York, 1981.
1030 M. Ende and G. Spiteler, *Mass Spectrom. Reviews*, 1982, **1**, 29.
1031 B. S. Middleditch, S. R. Missler, and H. B. Hines, 'Mass Spectrometry of Priority Pollutants', Plenum, New York, 1981.
1032 Z. F. Curry, *J. Liq. Chromatogr.*, 1982, **5** (Supplement 2), 257.
1033 K. Levsen, in A. Bjørseth and G. Angeletti, ed., ref. 11, pp. 149-158.
1034 J. G. Dorsey, in H. B. Mark and J. S. Mattson, ed., ref. 8, pp. 63-83.
1035 H. R. Morris, 'Soft Ionisation Biological Mass Spectrometry', Heyden, London, 1981.
1036 M. Barber, R. S. Bordoli, and R. G. Sedgwick, in H. R. Morris, ed., ref. 1035, pp. 137-152.
1037 A. Cornu, in A. Bjørseth and G. Angeletti, ed., ref. 11, pp. 201-216.
1038 R. E. Finnigan, M. S. Story, and D. F. Hunt, in L. H. Keith, ed., ref. 10, Vol. 2, pp. 555-570.
1039 A. D. Sauter, L. D. Betowski, and J. M. Ballard, *Anal. Chem.*, 1983, **55**, 116.
1040 D. P. Martinsen, *Appl. Spectrosc.*, 1981, **35**, 255.
1041 F. W. McLafferty and R. Venkataraghavan, *J. Chromatogr. Sci.*, 1979, **17**, 24.
1042 D. Henneberg, in A. Bjørseth and G. Angeletti, ed., ref. 11, pp. 219-230.
1043 M. Fielding, T. M. Gibson, H. A. James, K. McLoughlin, and C. P. Steel, 'Organic Micropollutants in Drinking Water', Technical Report TR159, Water Research Centre, Medmenham, Bucks. 1981.
1044 W. M. Shackelford, D. M. Cline, L. Burchfield, L. Faas, G. Kurth, and A. D. Sauter, in Keith, L.H., ed., ref. 10, Vol. 2, pp. 527-554.
1045 W. M. Shackelford, D. M. Cline, L. Faas, and G. Kurth, *Anal. Chim. Acta*, 1983, **146**, 15.
1046 A. Cornu, in A. Bjørseth and G. Angeletti, ed., ref. 11, pp. 242-245.
1047 P. Groll in A. Bjørseth and G. Angeletti, ed., ref. 11, pp. 238-241.
1048 K. H. Shafer, M. Cooke, F. DeRoos, R. J. Jakobsen, O. Rosario, and J. D. Mulik, *Appl. Spectrosc.*, 1981, **35**, 469.
1049 A. W. Garrison, A. L. Alford, J. S. Craig, J. J. Ellington, A. F. Haeberer, J. M. McGuire, J. D. Pope, W. M. Shackelford, E. D. Pellizzari, and J. E. Gebhart, in L. H. Keith, ed., ref. 10, Vol. 1, pp. 17-30.
1050 K. B. Tomer, J. T. Bursey, L. C. Michael, L. S. Sheldon, E. D. Pellizzari, A. L. Alford, J. D. Pope, and A. W. Garrison, in L. H. Keith, ed., ref. 10, Vol. 1, pp. 67-85.
1051 L. C. Michael, R. Wiseman, L. S. Sheldon, J. T. Bursey, K. B. Tomer, E. D. Pellizzari, T. A. Scott, R. Coney, A. W. Garrison, J. E. Gebhart, and J. F. Ryan, in L. H. Keith, ed., ref. 10, Vol. 1, pp. 87-114.
1052 J. L. Burleson, G. R. Peyton, and W. H. Glaze, *Environ. Sci. Technol.*, 1980, **14**, 1354.
1053 W. C. Schnute and D. E. Smith, *Am. Lab.*, 1980, **12**(7), 87.
1054 R. E. Finnigan, D. W. Hoyt, and D. E. Smith, *Environ. Sci. Technol.*, 1979, **13**, 534.
1055 A. D. Sauter, L. D. Betowski, T. R. Smith, V. A. Strickler, R. G. Beimer, B. N. Colby, and J. E. Wilkinson, *J. High Resolut. Chromatogr. Chromatogr. Commun.*, 1981, **4**, 366.
1056 A. D. Sauter, P. E. Mills, and W. L. Fitch, *J. High Resolut. Chromatogr. Chromatogr. Commun.*, 1982, **5**, 27.
1057 W. E. Coleman, R. G. Melton, F. C. Kopfler, K. A. Barone, T. A. Aurand, and M. G. Jellison, *Environ. Sci. Technol.*, 1980, **14**, 576.
1058 A. Yasuhara, H. Shiraishi, M. Tsuji, and T. Okuno, *Environ. Sci. Technol.*, 1981, **15**, 570.
1059 B. L. Proctor, V. A. Elder, and R. A. Hites, in L. H. Keith, ed., ref. 10, Vol. 2, pp. 1017-1036.
1060 B. Crathorne, C. D. Watts, and M. Fielding, *J. Chromatogr.*, 1979, **185**, 671.
1061 C. D. Watts, B. Crathorne, R. I. Crane, and M. Fielding, in L. H. Keith, ed., ref. 10, Vol. 1, pp. 383-397.
1062 B. Crathorne and C. D. Watts, in A. Bjørseth and G. Angeletti, ed., ref. 11, pp. 159-173.
1063 I. Stöber and H.-R. Schulten, *Sci. Total Environ.*, 1980, **16**, 249.
1064 W. Giger and C. S. Schaffner, *Anal. Chem.*, 1978, **50**, 243.
1065 T. Fujii, *Anal. Chem.*, 1977, **49**, 1985.
1066 R. D. Lingg, R. G. Melton, F. C. Kopfler, W. E. Coleman, and D. E. Mitchell, *J. Am. Water Works Assoc.*, 1977, **69**, 605.
1067 R. M. Goodfellow, J. Cardoso, G. Eglinton, J. P. Dawson, and G. A. Best, *Mar. Pollut. Bull.*, 1977, **8**, 272.
1068 S. W. Krasner, C. J. Hwang, and M. J. McGuire, in L. H. Keith, ed., ref. 10, Vol. 2, pp. 689-710.

1069 M. Brinkman, K. Fogelman, J. Hoeflein, T. Lindh, M. Pastel, W. C. Trench, and D. A. Aikens, in L. H. Keith, ed., ref. 10, Vol. 2, pp. 1001–1015.

1070 A. J. Ahearn, ed., 'Trace Analysis by Mass Spectrometry', Academic Press, New York, 1972.

1071 A. M. Ure, *Trends Anal. Chem.*, 1982, **1**, 314.

1072 R. P. Buck and J. R. Hass, in H. B. Mark and J. S. Mattson, ed., ref. 8, pp. 33–61.

1073 I. H. Crocker and W. F. Merritt, *Water Res.*, 1972, **6**, 285.

1074 B. M. Vanderborght and R. E. Van Grieken, *Talanta*, 1980, **27**, 417.

1075 A. L. Gray, *Proc. Soc. Anal. Chem.*, 1974, **11**, 182.

1076 R. S. Houk, V. A. Fassel, G. D. Flesch, H. J. Svec, A. L. Gray, and C. E. Taylor, *Anal. Chem.*, 1980, **52**, 2283.

1077 A. L. Gray and A. R. Date, *Int. J. Mass Spec. Ion Phys.*, 1983, **46**, 7.

1078 A. R. Date and A. L. Gray, *Analyst*, 1983, **108**, 159.

1079 A. L. Gray and A. R. Date, 'Continuum Flow Ion Extraction from an ICP', Presented at the 31st Annual Conference of The American Society for Mass Spectrometry, Boston, May 9–13, 1983.

1080 H. R. Schulten, in H. R. Morris, ed., ref. 1035, pp. 6–38.

1081 K. J. R. Rosman, J. R. De Laeter, and A. Chegwidden, *Talanta*, 1982, **29**, 279.

1082 R. M. Ajhar, P. D. Dalager, and A. L. Davison, *Am. Lab.*, 1976, **8**(3) 72.

1083 J. D. Winefordner, J. J. Fitzgerald, and N. Omenetto, *Appl. Spectrosc.*, 1975, **29**, 369.

1084 R. K. Skogerboe and G. N. Coleman, *Anal. Chem.*, 1976, **48**, 611A.

1085 A. T. Zander and G. M. Hieftje, *Appl. Spectrosc.*, 1981, **35**, 357.

1086 L. Ranson, 'Multi-element Analysis of Drinking Waters: A Review of Relevant Techniques and Applications', Technical Report TR 81, Water Research Centre, Medmenham, Bucks., 1978.

1087 J. Reednick, *Am. Lab.*, 1979, **11**(3), 53.

1088 A. T. Zander, *Ind. Res. Dev.*, 1982, **24**, 146.

1089 S. Greenfield, H. McD. McGeachin, and P. B. Smith, *Talanta*, 1975, **22**, 1.

1090 R. N. Kniseley, V. A. Fassel, and G. C. Butler, *Clin. Chem.*, 1973, **19**, 807.

1091 M. Thompson and J. N. Walsh, 'A Handbook of Inductively Coupled Plasma Spectrometry,' Blackie, Glasgow and London, 1983.

1092 S. Greenfield, H. McD. McGeachin, and P. B. Smith, *Talanta*, 1975, **22**, 553.

1093 S. Greenfield, H. McD. McGeachin, and P. B. Smith, *Talanta*, 1976, **23**, 1.

1094 P. W. J. M. Boumans, *Fresenius' Z. Anal. Chem.*, 1979, **299**, 337.

1095 S. Greenfield, *Analyst*, 1980, **105**, 1032.

1096 P. F. E. Montfort and J. Agterdenbos, *Talanta*, 1981, **28**, 637.

1097 J. P. Robin, *Prog. Anal. At. Spectrosc.*, 1982, **5**, 79.

1098 G. F. Larson and V. A. Fassel, *Anal. Chem.*, 1976, **48**, 1161.

1099 P. W. J. M. Boumans and F. J. de Boer, *Spectrochim. Acta, Part B*, 1976, **31**, 355.

1100 G. W. Johnson, H. E. Taylor, and R. K. Skogerboe, *Spectrochim. Acta, Part B*, 1979, **34**, 197.

1101 G. R. Kornblum and L. de Galan, *Spectrochim. Acta, Part B*, 1977, **32**, 71.

1102 D. J. Kalnicky, R. N. Kniseley, and V. A. Fassel, *Spectrochim. Acta, Part B*, 1975, **30**, 511.

1103 P. W. J. M. Boumans and F. J. de Boer, *Spectrochim. Acta, Part B*, 1977, **32**, 365.

1104 J. Jarosz, J. M. Mermet, and J. P. Robin, *Spectrochim. Acta, Part B*, 1978, **33**, 55.

1105 H. Uchida, K. Tanabe, Y. Nojiri, H. Haraguchi, and K. Fuwa, *Spectrochim. Acta, Part B*, 1981, **36**, 711.

1106 P. W. J. M. Boumans, *Spectrochim. Acta, Part B*, 1982, **37B**, 75.

1107 S. R. Koirtyohann, J. S. Jones, C. P. Jester, and D. A. Yates, *Spectrochim. Acta, Part B*, 1981, **36**, 49.

1108 M. W. Blades and G. Horlick, *Spectrochim. Acta, Part B*, 1981, **36**, 861.

1109 M. W. Blades and G. Horlick, *Spectrochim. Acta, Part B*, 1981, **36**, 881.

1110 J. P. Rybarczyk, C. P. Jester, D. A. Yates, and S. R. Koirtyohann, *Anal. Chem.*, 1982, **54**, 2162.

1111 H. Kawaguchi, T. Ito, and A. Mizuike, *Spectrochim. Acta, Part B*, 1981, **36**, 615.

1112 Anonymous, *ICP Inf. Newslett.*, 1981, **6**, 488 (see also ref. 1169).

1113 Anonymous, *ICP Inf. Newslett.*, 1981, **6**, 509 (see also ref. 1169).

1114 C. W. Fuller, R. C. Hutton, and B. Preston, *Analyst*, 1981, **106**, 913.

1115 J. F. Wolcott and C. B. Sobel, *Appl. Spectrosc.*, 1982, **36**, 685.

1116 J. R. Garbarino and H. E. Taylor, *Appl. Spectrosc.*, 1980, **34**, 584.

1117 K. W. Olson, W. J. Haas, and V. A. Fassel, *Anal. Chem.*, 1977, **49**, 432.

1118 P. D. Goulden and D. H. J. Anthony, *Anal. Chem.*, 1982, **54**, 1678.

1119 R. N. Kniseley, H. Amenson, C. C. Butler, and V. A. Fassel, *Appl. Spectrosc.*, 1974, **28**, 285.

1120 C. E. Taylor and T. L. Floyd, *Appl. Spectrosc.*, 1981, **35**, 408.

1121 M. H. Abdallah, J. M. Mermet, and C. Trassy, *Anal. Chim. Acta*, 1976, **87**, 329.

1122 P. W. J. M. Boumans and F. J. de Boer, *Spectrochim. Acta, Part B*, 1976, **31**, 355.

1123 P. W. J. M. Boumans, F. J. de Boer, J. Dahmen, H. Hoelzel, and A. Meier, *Spectrochim. Acta, Part B*, 1975, **30**, 449.

1124 R. K. Skogerboe and G. N. Coleman, *Appl. Spectrosc.*, 1976, **30**, 504.

1125 J. R. Garbarino and H. E. Taylor, *Appl. Spectrosc.*, 1979, **33**, 220.

1126 R. L. Norton and J. R. Orpwood, 'Multi-element Analysis using an Inductively-Coupled Plasma Spectrometer', Part 1, Commissioning and Preliminary Evaluation of Performance', Technical Report TR 141, Water Research Centre, Medmenham, Bucks., 1980.
1127 E. Janssens, P. Schutyser, and R. Doms, *Environ. Tech. Lett.*, 1982, **3**, 35.
1128 M. Thompson, M. J. Ramsey, and B. Pahlavanpour, *Analyst*, 1982, **107**, 1330.
1129 S. S. Berman, J. W. McLaren, and S. N. Willie, *Anal. Chem.*, 1980, **52**, 488.
1130 A. M. Gunn, D. L. Millard, and G. F. Kirkbright, *Analyst*, 1978, **103**, 1066.
1131 D. L. Millard, H. C. Shan, and G. F. Kirkbright, *Analyst*, 1980, **105**, 502.
1132 K. C. Ng and J. A. Caruso, *Anal. Chem.*, 1983, **55**, 1513.
1133 H. M. Swaidan and G. D. Christian, *Anal. Chem.*, 1984, **56**, 120.
1134 G. F. Kirkbright and S. J. Walton, *Analyst*, 1982, **107**, 276.
1135 Z. Li-Xing, G. F. Kirkbright, M. J. Cope, and J. M. Watson, *Appl. Spectrosc.*, 1983, **37**, 250.
1136 R. K. Winge, V. A. Fassel, R. N. Kniseley, E. DeKalb, and W. J. Haas, *Spectrochim. Acta, Part B*, 1977, **32**, 327.
1137 M. Thompson, B. Pahlavanpour, S. J. Walton, and G. F. Kirkbright, *Analyst*, 1978, **103**, 568.
1138 M. Thompson, B. Pahlavanpour, S. J. Walton, and G. F. Kirkbright, *Analyst*, 1978, **103**, 705.
1139 M. Thompson and B. Pahlavanpour, *Anal. Chim. Acta*, 1979, **109**, 251.
1140 R. C. Fry, M. B. Denton, D. L. Windsor, and S. J. Northway, *Appl. Spectrosc.*, 1979, **33**, 399.
1141 J. A. C. Broekaert and F. Leis, *Fresenius' Z. Anal. Chem.*, 1980, **300**, 22.
1142 M. Ikeda, J. Nishibe, S. Hanada, and R. Tujino, *Anal. Chim. Acta*, 1981, **125**, 109.
1143 T. Nakahara, *Anal. Chim. Acta*, 1981, **131**, 73.
1144 P. D. Goulden, D. H. J. Anthony, and K. D. Austen, *Anal. Chem.*, 1981, **53**, 2027.
1145 D. D. Nygaard and J. H. Lowry, *Anal. Chem.*, 1982, **54**, 803.
1146 M. Thompson, B. Pahlavanpour, and L. T. Thorne, *Water Res.*, 1981, **15**, 407.
1147 D. E. Dobb and D. R. Jenke, *Appl. Spectros.*, 1983, **37**, 379.
1148 N. R. McQuaker, P. D. Kluckner, and G. N. Chang, *Anal. Chem.*, 1979, **51**, 888.
1149 R. L. Dahlquist and J. W. Knoll, *Appl. Spectrosc.*, 1978, **32**, 1.
1150 P. W. J. M. Boumans and F. J. de Boer, *Spectrochim. Acta, Part B*, 1975, **30**, 309.
1151 G. F. Larson, V. A. Fassel, R. H. Scott, and R. N. Kniseley, *Anal. Chem.*, 1975, **47**, 238.
1152 G. R. Kornblum and L. de Galan, *Spectrochim. Acta, Part B*, 1977, **32**, 455.
1153 J. P. Rybarczyk, C. P. Jester, D. A. Yates, and S. R. Koirtyohann, *Anal. Chem.*, 1982, **54**, 2162.
1154 P. W. J. M. Boumans, *Fresenius' Z. Anal. Chem.*, 1976, **279**, 1.
1155 I. T. Urasa and R. K. Skogerboe, *Appl. Spectrosc.*, 1978, **32**, 527.
1156 G. W. Johnson, H. E. Taylor, and R. K. Skogerboe, *Appl. Spectrosc.*, 1980, **34**, 19.
1157 D. D. Nygaard and T. R. Gilbert, *Appl. Spectrosc.*, 1981, **35**, 52.
1158 P. W. J. M. Boumans, 'Line Coincidence Tables for Inductively-Coupled Plasma Atomic Emission Spectrometry', Vols. I and II, Pergamon Press, Oxford, 1980.
1159 N. L. Parsons, A. Forster, and D. Anderson, 'An Atlas of Spectral Interferences in ICP Spectroscopy', Plenum, New York, 1980.
1160 S. S. Berman and J. W. McLaren, *Appl. Spectrosc.*, 1978, **32**, 372.
1161 R. I. Botto, *Spectrochim. Acta, Part B*, 1983, **38**, 129.
1162 G. F. Larson, V. A. Fassel, R. K. Winge, and R. N. Kniseley, *Appl. Spectrosc.*, 1976, **30**, 384.
1163 H. R. Sobel and R. C. Dahlquist, *Am. Lab.*, 1981, **13**(11), 152.
1164 M. Thompson, M. H. Ramsey, and B. Pahlavanpour, *Analyst*, 1982, **107**, 1330.
1165 A. Miyazaki, A. Kimura, K. Bansho, and Y. Umezaki, *Anal. Chim. Acta*, 1982, **144**, 213.
1166 C. W. McLeod, A. Otsuki, K. Okamoto, H. Haraguchi, and K. Fuwa, *Analyst*, 1981, **106**, 419.
1167 A. Sugimae, *Analyt. Chim. Acta*, 1980, **121**, 331.
1168 G. M. Hieftje, *Spectrochim. Acta, Part B*, 1983, **38**, 1465.
1169 *ICP Information Newsletter*, available from Dr. R. M. Barnes, Department of Chemistry, University of Massachusetts, Amherst, Massachusetts 01003, U.S.A.
1170 R. S. Braman, L. L. Justen, and C. G. Foreback, *Anal. Chem.*, 1972, **44**, 2195.
1171 F. E. Lichte and R. K. Skogerboe, *Anal. Chem.*, 1972, **44**, 1480.
1172 F. E. Lichte and R. K. Skogerboe, *Anal. Chem.*, 1972, **44**, 1321.
1173 M. S. Cresser and L. Ebdon, ed., 'Annual Reports on Analytical Atomic Spectroscopy', Vol. 12, reviewing 1982, The Royal Society of Chemistry, London, 1983.
1174 E. L. Grove, ed., 'Analytical Emission Spectroscopy', Vol. 1, Part 1, Dekker, 1971.
1175 E. L. Grove, ed., 'Analytical Emission Spectroscopy', Vol. 1, Part II, Dekker, 1972.
1176 M. Slavin, 'Emission Spectrochemical Analysis', Wiley-Interscience, New York, 1971.
1177 Methods for the Examination of Waters and Associated Materials', 'Emission Spectrophotometric Multielement Methods of Analysis for Waters, Sediments and other Materials of Interest to the Water Industry. 1980', Her Majesty's Stationery Office, London, 1982.
1178 J. F. Kopp and R. C. Kroner, *J. Appl. Spectrosc.*, 1965, **19**, 155.
1179 R. L. Steiner, J. B. Austin, and D. W. Lander, *Environ. Sci. Technol.*, 1969, **3**, 1192.
1180 D. R. Demers and C. D. Allemand, *Anal. Chem.*, 1981, **53**, 1915.

[1181] D. R. Demers, D. A. Busch, and C. D. Allemand, *Am. Lab.*, 1982, **14**(3), 167.

[1182] W.-H. Du, L. Lieu, and Z.-S. Wung, 'Study and application of a new commercial non-dispersive AF spectrometer', presented at the 9th International Conference on Atomic Spectroscopy and XXII Colloquium Spectroscopicum Internationale, September, 1981, Tokyo, Japan. (Further details are given in ref. 1183).

[1183] M. S. Cresser and B. L. Sharp, ed., 'Annual Reports on Analytical Atomic Spectroscopy', Vol. 11, reviewing 1981, the Royal Society of Chemistry, London, 1982.

[1184] N. Omenetto and J. D. Winefordner, *Prog. Anal. At. Spectrosc.*, 1979, **2**, 1.

[1185] A. H. Ullman, *Prog. Anal. At. Spectrosc.*, 1980, **3**, 87.

[1186] J. C. Van Loon, *Anal. Chem.*, 1981, **53**, 332A.

[1187] V. Sychra, V. Svoboda, and I. Rubeska, 'Atomic Fluorescence Spectroscopy', Van Nostrand Reinhold, London, 1975.

[1188] G. F. Kirkbright and M. Sargent, 'Atomic Absorption and Fluorescence Spectroscopy', Academic Press, London, 1974.

[1189] K. C. Thompson and R. J. Reynolds, 'Atomic Absorption, Fluorescence and Flame Emission Spectrometry. A Practical Approach, 2nd Ed., Charles Griffin, London, 1978.

[1190] E. L. Grove, ed., 'Analytical Emission spectroscopy', Vol. 1 of 'Analytical Spectroscopy Series', Parts I and II, Dekker, New York, 1971 and 1972. (See especially Chapter 7, pp. 255–394, 'Flame Spectrometry' by T. J. Vickers and J. D. Winefordner).

[1191] J. A. Dean and T. C. Rains, ed., 'Flame Emission and Atomic Absorption Spectrometry', Dekker, New York, Vol. 1, 1969, Vol. 2, 1971, Vol. 3, 1975. (See especially: Chapter 8, Vol. 2, 'Atomic Fluorescence Spectrometry by A. Syty).

[1192] K. C. Thompson and R. G. Godden, *Analyst*, 1975, **100**, 544.

[1193] B. J. Farey, L. A. Nelson, and M. G. Rolph, *Analyst*, 1978, **103**, 656.

[1194] R. C. Hutton and B. Preston, *Analyst*, 1980, **105**, 981.

[1195] R. Ferrara, A. Serritti, C. Barghigiani, and A. Petrosino, *Anal. Chim. Acta*, 1980, **117**, 391.

[1196] A. Seritti, A. Petrosino, R. Ferrara, and C. Barghigiani, *Env. Technol. Lett.*, 1980, **1**, 50.

[1197] M. P. Bertenshaw and K. Wagstaff, *Analyst*, 1982, **107**, 664.

[1198] G. B. Marshall and A. C. Smith, *Analyst*, 1972, **97**, 447.

[1199] H. D. Fleming, *Anal. Chim. Acta*, 1980, **117**, 241.

[1200] K. Tsujii, *Anal. Lett., Part A*, 1981, **14**, 181.

[1201] L. S. Birks, 'X-Ray Spectrochemical Analysis', 2nd Ed., Wiley, New York, 1969.

[1202] R. Jenkins and J. L. de Vries, 'Practical X-Ray Spectrometry', London, Macmillan, 1972.

[1203] R. O. Muller, (transl. K. Keil), 'Spectrochemical Analysis by X-Ray Fluorescence', Adam Hilger, London, 1972.

[1204] E. P. Bertin, 'Principles and Practice of X-Ray Spectrometric Analysis', 2nd Ed., Plenum, New York, 1975.

[1205] R. Jenkins, 'An Introduction to X-Ray Spectrometry', Heyden, London, 1976.

[1206] G. L. Macdonald, *Anal. Chem.*, 1982, **54**, 150R.

[1207] B. Gonsior and M. Roth, *Talanta*, 1983, **30**, 371.

[1208] T. G. Dzubay, ed., 'X-Ray Fluorescence Analysis of Environmental Samples', Ann Arbor Science Publishers, Ann Arbor, 1977.

[1209] D. E. Leyden, in H. B. Mark and J. S. Mattson, ed., ref. 8, pp. 271–306.

[1210] L. Ranson, 'Multi-element Analysis of Drinking Waters: A Review of Relevant Techniques and Applications', Technical Report TR 81, Water Research Centre, Medmenham, Bucks., 1978.

[1211] J. M. Jaklevic and R. L. Walter, in T. G. Dzubay, ed., ref. 1208, pp. 63–75.

[1212] A. W. Morris, *Anal. Chim. Acta*, 1968, **42**, 397.

[1213] H. Robbrecht, R. van Grieken, M. van Sprundel, D. Vandenberghe, and H. Deelstra, *Sci. Total Environ.*, 1983, **26**, 163.

[1214] W. J. Campbell, E. F. Spano, and J. E. Green, *Anal. Chem.*, 1966, **38**, 987.

[1215] D. E. Leyden and G. H. Luttrell, *Anal. Chem.*, 1975, **47**, 1612.

[1216] D. E. Leyden, G. H. Luttrell, A. E. Sloan, and N. J. de Angelis, *Anal. Chim. Acta*, 1976, **84**, 97.

[1217] P. Burba, W. Dyck, and K. H. Lieser, *Vom Wasser*, 1980, **54**, 227.

[1218] H. Kingston and P. A. Pella, *Anal. Chem.*, 1981, **53**, 223.

[1219] J. S. Y. Ho and P. C. L. Lin, *Am. Lab.*, 1982, **14**, 58.

[1220] D. E. Leyden, J. A. Patterson, and J. J. Alberts, *Anal. Chem.* 1975, **45**, 733.

[1221] H. Watanabe, S. Berman, and D. S. Russell, *Talanta*, 1972, **19**, 1363.

[1222] E. Bruninx and E. Van Meyl, *Anal. Chim. Acta*, 1975, **80**, 85.

[1223] M. G. Vanderstappen and R. E. Van Grieken, *Talanta*, 1978, **25**, 653.

[1224] A. J. Pik, A. J. Cameron, J. M. Eckert, E. Sholkovitz, E. Williams, and K. L. Williams, *Anal. Chim. Acta*, 1979, **110**, 61.

[1225] R. V. Moore, *Anal. Chem.*, 1982, **54**, 895.

[1226] P. C. Cole, J. M. Eckert, and K. L. Williams, *Anal. Chim. Acta*, 1983, **153**, 61.

[1227] R. G. Stone, *Analyst*, 1963, **88**, 56.

[1228] F. A. Rickey, K. Mueller, P. C. Simms, and B. D. Michael, in T. G. Dzubay, ed., ref. 1208, pp. 135–143.
[1229] L. S. Birks, in T. G. Dzubay, *ed.*, ref. 1208, pp. 57–62.
[1230] D. S. Polcyn, *J. Water Pollut. Control Fed.*, 1983, **55**, 555.
[1231] M. J. Fishman, D. E. Erdmann, and J. R. Garbarino, *Anal. Chem.*, 1983, **55**, 102R.
[1232] N. L. Alpert, W. E. Keiser, and H. A. Szymanski, 'IR-Theory and Practice of Infrared Spectroscopy', 2nd Ed., Plenum, New York, 1970.
[1233] J. E. Stewart, 'Infrared Spectroscopy. Experimental Methods and Techniques' Dekker, New York, 1970.
[1234] F. Scheinhmann, ed., 'An Introduction to Spectroscopic Methods for the Identification of Organic Compounds', Vol. 1, Pergamon, Oxford, 1970.
[1235] P. R. Griffiths, 'Chemical Infrared Fourier Transform Spectroscopy, Wiley-Interscience, New York, 1975.
[1236] P. R. Griffiths, J. A. de Haseth, and L. V. Azarraga, *Anal. Chem.*, 1983, **55**, 1361A.
[1237] R. S. McDonald, *Anal. Chem.*, 1982, **54**, 1250.
[1238] M. J. Fishman, D. E. Erdmann, and J. R. Garbarino, *Anal. Chem.*, 1983, **55**, 102R.
[1239] M. Uchiyama, *Water Res.*, 1978, **12**, 299.
[1240] J. Mallevialle, *Water Res.*, 1974, **8**, 1071.
[1241] American Public Health Association *et al.*, ref. 18, pp. 462–463.
[1242] M. Gruenfeld, *Environ. Sci. Technol.*, 1973, **7**, 636.
[1243] P. F. Lynch and C. W. Brown, *Environ. Sci. Technol.*, 1973, **7**, 1123.
[1244] P. F. Lynch, S.-Y. Tang, and C. W. Brown, *Anal. Chem.*, 1975, **47**, 1696.
[1245] C. W. Brown, P. F. Lynch, M. Ahmadjian, and C. D. Baer, *Am. Lab.*, 1975, **7**, 59.
[1246] F. K. Kawahara and Y. Y. Yang, *Anal. Chem.*, 1976, **48**, 651.
[1247] M. Ahmadjian, C. D. Baer, P. F. Lynch, and C. W. Brown, *Environ. Sci. Technol.*, 1976, **10**, 777.
[1248] J. S. Mattson, C. S. Mattson, M. J. Spencer, and S. A. Starks, *Anal. Chem.*, 1977, **49**, 297.
[1249] J. S. Mattson, C. S. Mattson, M. J. Spencer, and F. W. Spencer, *Anal. Chem.*, 1977, **49**, 500.
[1250] H. Hellmann, *Fresenius' Z. Anal. Chem.*, 1978, **293**, 359.
[1251] H. Hellmann, *Fresenius' Z. Anal. Chem.*, 1979, **297**, 102.
[1252] K. Oba, K. Miura, H. Sekiguchi, R. Yagi, and A. Mori, *Water Res.* 1976, **10**, 149.
[1253] American Public Health Association *et al.*, ref. 18, pp. 533–536.
[1254] J. T. Chen and W. R. Benson, *J. Assoc. Off. Anal. Chem.*, 1966, **49**, 412.
[1255] W. F. Ulrich, *Ind. Res.*, 1971, **13**, 52.
[1256] V. D. Chmil', *Zh. Anal. Khim.*, 1979, **34**, 2067.
[1257] *J. Chromatogr. Sci.*, 1979, **17** (8), 70 pp.
[1258] K. H. Shafer, M. Cooke, F. DeRoos, R. J. Jakobsen, O. Rosario, and J. D. Mulik, *Appl. Spectrosc.*, 1981, **35**, 469.
[1259] D. F. Gurka, P. R. Laska, and R. Titus, *J. Chromatogr. Sci.*, 1982, **20**, 145.
[1260] L. V. Azarraga and C. A. Potter, *J. High Resolut. Chromatogr.*, *Chromatogr. Commun.*, 1981, **4**, 60.
[1261] C. L. Wilkins, G. N. Giss, G. M. Brissey, and S. Steiner, *Anal. Chem.*, 1981, **53**, 113.
[1262] J. A. de Haseth and T. L. Isenhour, *Anal. Chem.*, 1977, **49**, 1977.
[1263] A. Hanna, J. C. Marshall, and T. L. Isenhour, *J. Chromatogr. Sci.*, 1979, **17**, 434.
[1264] B. A. Hohne, G. Hangac, G. W. Small, and T. L. Isenhour, *J. Chromatogr. Sci.*, 1981, **19**, 283.
[1265] L. H. Keith, A. W. Garrison, F. R. Allen, M. H. Carter, T. L. Floyd, J. D. Pope, and A. D. Thruston in L. H. Keith, ed., 'Identification and Analysis of Organic Pollutants in Water', Ann Arbor Science Publishers, Ann Arbor, 1976, Chapter 22, pp. 329–373.
[1266] J. W. Worley, M. L. Rueppel, and F. L. Rupel, *Anal. Chem.*, 1980, **52**, 1845.
[1267] G. Skirrow, in J. P. Riley and G. Skirrow, ed., 'Chemical Oceanography', 2nd Ed., Vol. 2, Academic Press, London, 1975, pp. 1–192.
[1268] T. J. Kehoe, *Environ. Sci. Technol.*, 1977, **11**, 137.
[1269] P. J. Wangersky and R. G. Zika, 'The Analysis of Organic Compounds in Sea Water', NRCC No. 16566, National Research Council Canada, Halifax N.S., 1978, pp. 41–60.
[1270] D. W. Menzel and R. F. Vaccaro, *Limnol. Oceanogr.*, 1964, **9**, 138.
[1271] P. D. Goulden and P. Brooksbank, *Anal. Chem.*, 1975, **47**, 1943.
[1272] C. E. Van Hall and V. A. Stenger, *Anal. Chem.*, 1967, **39**, 503.
[1273] R. Briggs, J. W. Schofield, and P. A. Gorton, *J. Inst. Water Pollut. Contr.*, 1976, **75**, 47.
[1274] American Public Health Association *et al.*, ref. 18, pp. 471–475.
[1275] C. D. Baker, P. D. Bartlett, I. S. Farr, and G. I. Williams, *Freshwater Biol.*, 1974, **4**, 467.
[1276] R. J. Oake, 'A Review of Photo-Oxidation for the Determination of Total Organic Carbon in Water', Technical Report TR 160, Water Research Centre, Medmenham, Bucks., 1981.
[1277] V. A. Stenger and C. E. Van Hall, *Anal. Chem.*, 1967, **39**, 206.
[1278] G. G. Guilbault, 'Practical Fluorescence, Theory, Methods and Techniques', Dekker, New York, 1972.
[1279] C. A. Parker, 'Photoluminescence of Solutions', Elsevier, Amsterdam, 1968.

1280 J. D. Winefordner, S. G. Schulman, and T. O'Haver, 'Luminescence Spectrometry in Analytical Chemistry', Wiley-Interscience, New York, 1972.
1281 C. White, 'Fluorescence Analysis: A Practical Approach', Dekker, New York, 1970.
1282 W. R. Seitz, in 'Treatise on Analytical Chemistry', 2nd Ed., P. J. Elving, E. J. Meehan, and I. M. Kolthoff, ed., Wiley, New York, 1981, Part 1, Vol. 7, pp. 159–248.
1283 E. L. Wehry, 'Modern Fluorescence Spectroscopy', Plenum, New York, 1981, Vols. 3 and 4.
1284 E. L. Wehry, *Anal. Chem.*, 1982, **54**, 131R.
1285 W. R. Seitz, in H. B. Mark and J. S. Mattson, ed., ref. 8, pp. 307–334.
1286 J. Borneff and H. Kunte, *Arch. Hyg. Bakteriol.*, 1969, **220**, 153.
1287 R. I. Crane, B. Crathorne, and M. Fielding, in B. K. Afghan and D. Mackay, ed., 'Hydrocarbons and Halogenated Hydrocarbons in the Aquatic Environment', Plenum, New York, 1980, pp. 161–172.
1288 A. W. Wolkoff and R. H. Larose, *J. Chromatogr.*, 1974, **99**, 731.
1289 W. W. Pitt, R. L. Jolley, and S. Katz, in L. H. Keith ed. 'Identification and Analysis of Organic Pollutants in Water', Ann Arbor Science Publishers, Ann Arbor, 1976, pp. 215–231.
1290 P. L. Smart, B. L. Finlayson, W. D. Rylands, and C. M. Ball, *Water Res.*, 1976, **10**, 805.
1291 G. L. Brun and D. L. D. Milburn, *Anal. Lett.*, 1977, **10**, 1209.
1292 R. E. Carlson and J. Shapiro, *Limnol. Oceanogr.*, 1981, **26**, 785.
1293 E. M. Levy, *Water Res.*, 1971, **5**, 723.
1294 P. John and I. Soutar, *Anal. Chem.*, 1976, **48**, 520.
1295 P. John, E. R. McQuat, and I. Soutar, *Analyst*, 1982, **107**, 221.
1296 M. E. Loftus and J. H. Carpenter, *J. Marine Res.*, 1971, **29**, 319.
1297 M. G. Tunzi, M. Y. Chu, and R. C. Gain, *Water Res.*, 1974, **8**, 623.
1298 J. H. Paul and B. Myers, *Appl. Environ. Microbiol.*, 1982, **43**, 1393.
1299 D. J. Hydes and P. S. Liss, *Analyst*, 1976, **101**, 922.
1300 J. M. Cano-Pavon, M. L. Trujillo, and A. Garcia de Torres, *Anal. Chim. Acta*, 1980, **117**, 319.
1301 J. M. Rankin, *Environ. Sci. Technol.*, 1973, **7**, 823.
1302 I. I. Nazarenko and I. V. Kislova, *Zh. Anal. Khim.*, 1978, **33**, 157.
1303 J. Holzbecher and D. E. Ryan, *Anal. Chim. Acta*, 1973, **64**, 333.
1304 W. S. Gardner, *Limnol. Oceanogr.*, 1978, **23**, 1069.
1305 B. K. Afghan and J. F. Ryan, *Anal. Chem.*, 1975, **47**, 2347.
1306 R. Kaminski, F. J. Purcell, and E. Russavage, *Anal. Chem.*, 1981, **53**, 1093.
1307 E. R. Hinton and L. E. White, *Anal. Lett.*, 1981, **14**, 947.
1308 D. L. Percy, S. M. Klainer, H. R. Bowman, F. P. Milanovich, T. Hirschfeld, and S. Miller, *Anal. Chem.*, 1981, **53**, 1048.
1309 G. R. Haugen, L. L. Steinmetz, T. B. Hirschfeld, and S. M. Klainer, *Appl. Spectrosc.*, 1981, **35**, 568.
1310 P. M. Williams and K. J. Robertson, *J. Water Pollut. Control Fed.*, 1980, **52**, 2167.
1311 W. R. Seitz and M. P. Neary, *Anal. Chem.*, 1974, **46**, 188A.
1312 W. R. Seitz, W. W. Suydam, and D. M. Hercules, *Anal. Chem.*, 1972, **44**, 957.
1313 C. A. Chang, H. H. Patterson, L. M. Mayer, and D. E. Bause, *Anal. Chem.*, 1980, **52**, 1264.
1314 D. F. Marino and J. D. Ingle, *Anal. Chem.*, 1981, **53**, 294.
1315 D. F. Marino and J. D. Ingle, *Anal. Chem.*, 1981, **53**, 292.
1316 R. J. Miller and J. D. Ingle, *Talanta*, 1982, **29**, 303.
1317 W. R. Seitz and D. M. Hercules, *Anal. Chem.*, 1972, **44**, 2143.
1318 D. F. Marino and J. D. Ingle, *Anal. Chim. Acta*, 1981, **123**, 247.
1319 D. F. Marino and J. D. Ingle, *Anal. Chem.*, 1981, **53**, 455.
1320 O. Holm-Hansen and C. R. Booth, *Limnol. Oceanogr.*, 1966, **11**, 510.
1321 K. K. Stewart, *Anal. Chem.*, 1983, **55**, 931A.
1322 H. B. Mark and G. A. Rechnitz, 'Kinetics in Analytical Chemistry', Interscience, New York, 1968.
1323 K. B. Yatsimirskii, 'Kinetic Methods of Analysis', Pergamon, Oxford, 1966.
1324 H. A. Mottola and H. B. Mark, *Anal. Chem.*, 1982, **54**, 62R.
1325 H. Mueller, *CRC Crit. Rev. Anal. Chem.*, 1982, **13**, 313.
1326 H. A. Mottola, *CRC Crit. Rev. Anal. Chem.*, 1975, **4**, 229.
1327 L. V. Markova and T. S. Maksimenko, *Zh. Anal. Khim.*, 1970, **25**, 1620.
1328 American Society for Testing and Materials, '1981, Book of ASTM Standards, Part 31, Water', ASTM, Philadelphia, 1981, pp. 438–440.
1329 G. S. Pyen, M. J. Fishman, and A. G. Hedley, *Analyst*, 1980, **105**, 657.
1330 B. K. Afghan, R. Leung, A. V. Kulkani, and J. F. Ryan, *Anal. Chem.*, 1975, **47**, 556.
1331 W. R. Seitz, W. W. Suydam, and D. M. Hercules, *Anal. Chem.*, 1972, **44**, 957.
1332 C. A. Chang, H. H. Patterson, L. M. Mayer, and D. E. Bause, *Anal. Chem.*, 1980, **52**, 1264.
1333 D. F. Marino and J. D. Ingle, *Anal. Chem.*, 1981, **53**, 294.
1334 G. E. Batley, *Talanta*, 1971, **18**, 1225.
1335 R. J. Miller and J. D. Ingle, *Talanta*, 1982, **29**, 303.
1336 V. I. Rychkova, I. F. Dolmanova, and V. M. Peshkova, *Teploenergetika*, 1972, (2), 66.

[1337] J. L. Ferrer-Herranz and D. Pérez-Bendito, *Anal. Chim. Acta*, 1981, **132**, 157.
[1338] American Public Health Association *et al.*, ref. 18, pp. 344–346.
[1339] W. R. Seitz and D. M. Hercules, *Anal. Chem.*, 1972, **44**, 2143.
[1340] M. A. Ditzler and W. F. Gutknecht, *Anal. Chem.*, 1980, **52**, 614.
[1341] I. F. Dolmanova, N. T. Yatsimirskaya, V. P. Poddubienko, and V. M. Peshkova, *Zh. Anal. Khim.*, 1971, **26**, 1540.
[1342] D. P. Nikolelis and T. P. Hadjiiannou, *Anal. Chim. Acta*, 1978, **97**, 111.
[1343] Yu. A. Klyachko and N. Petukhova, *Zavod. Lab.*, 1972, **38**, 921.
[1344] M. Otto and H. Müller, *Talanta*, 1977, **24**, 15.
[1345] N. Iordanov, M. Pavlova, and S. Stefanov, *Talanta*, 1978, **25**, 389.
[1346] American Public Health Association *et al.*, ref. 18, pp. 237–239.
[1347] B. G. Zhelyazkova, A. L. Tsvetanova, and K. B. Yatsimirskii, *Zh. Anal. Khim.*, 1972, **27**, 795.
[1348] A. T. Basson and P. L. Kempster, *Water S.A.*, 1980, **6**, 88.
[1349] T. Fukasawa and T. Yamane, *Anal. Chim. Acta*, 1977, **88**, 147.
[1350] S. R. Goode and R. J. Matthews, *Anal. Chem.*, 1978, **50**, 1608.
[1351] H. O. Michel, E. C. Gordon, and J. Epstein, *Environ. Sci. Technol.*, 1973, **7**, 1045.
[1352] J. J. McCarthy, *Environ. Sci. Technol.*, 1970, **15**, 309.
[1353] R. J. Stevens, *Water Res.*, 1979, **13**, 763.
[1354] G. G. Guilbault and G. Nagy, *Anal. Chem.*, 1973, **45**, 417.
[1355] G. G. Guilbault, *Appl. Biochem. Biotechnol.*, 1982, **7**, 85.
[1356] G. P. Evans and D. Johnson, 'Detection of Pollution at Drinking Water Intakes', Paper 16, Water Research Centre Conference, 'Environmental Protection: Standards Compliance and Costs', Keele University, September, 1983. 19 pp.
[1357] G. Friedlander, J. W. Kennedy, and J. M. Miller, 'Nuclear and Radiochemistry', 2nd Ed., Wiley, New York, 1964.
[1358] G. R. Choppin and J. Rydberg, 'Nuclear Chemistry, Theory and Applications', Pergamon Press, Oxford, 1980.
[1359] H. A. C. McKay, 'Principles of Radiochemistry', Butterworths, London, 1971.
[1360] D. J. Malcolme-Lawes, 'Introduction to Radiochemistry', Macmillan, London, 1979.
[1361] I. M. Kolthoff and P. J. Elving, ed., 'Treatise on Analytical Chemistry. Part I. Theory and Practice', Wiley-Interscience, New York, 1971, Volume 9, Section D6, pp. 5385–5641.
[1362] R. A. Faires and G. G. J. Boswell, 'Radioisotope Laboratory Techniques', 4th Ed., Butterworths, London, 1981.
[1363] M. de Bruin, *Pure Appl. Chem.*, 1982, **54**, 1533.
[1364] H. A. Das, A. Faanhof, and H. A. van der Sloot, 'Environmental Radioanalysis', Studies in Environment Science No. 22, Elsevier, Amsterdam, 1983.
[1365] G. F. Knoll, 'Radiation Detection and Measurement', Wiley, New York, 1979.
[1366] W. H. Tait, 'Radiation Detection', Butterworths, London, 1982.
[1367] G. G. Eichholz and J. W. Poston, 'Principles of Nuclear Radiation Detection', Ann Arbor Science Publishers, Ann Arbor, 1979.
[1368] J. D. Hemingway, in A. G. Maddock, ed., 'International Review of Science. Inorganic Chemistry.' Series Two. Volume 8. 'Radiochemistry', Butterworths, London, 1975, pp. 105–196.
[1369] A. Dyer, 'Liquid Scintillation Counting Practice', Heyden, London, 1980.
[1370] K. D. Neame and C. A. Homewood, 'Liquid Scintillation Counting', Butterworths, London, 1974.
[1371] C. E. Crouthamel, 'Applied Gamma Ray Spectrometry', 2nd Ed., (revised and enlarged by F. Adams and R. Dams), Pergamon Press, Oxford, 1970.
[1372] P. Quittner, 'Gamma-ray Spectroscopy with Particular Reference to Detector and Computer Evaluation Techniques', Hilger, London, 1972.
[1373] P. W. Nicholson, 'Nuclear Electronics', Wiley, London, 1974.
[1374] D. De Soete, R. Gijbels, and J. Hoste, 'Neutron Activation Analysis', Wiley-Interscience, London, 1972.
[1375] H. M. N. H. Irving, *J. Radioanal. Chem.*, 1976, **33**, 287.
[1376] B. J. Wilson, ed., 'The Radiochemical Manual', 2nd Ed., The Radiochemical Centre, Amersham, 1966.
[1377] C. M. Lederer and V. S. Shirley, ed., 'Table of Isotopes', 7th Ed., Wiley, New York, 1978.
[1378] American Public Health Association *et al.*, ref. 18, pp. 557–612.
[1379] N. T. Mitchell, 'Radiological Examination', in M. J. Suess, ed., ref. 645, Volume 2, Chapter 5, pp. 500–538.
[1380] J. Rodier, 'L'Analyse de L'eau', 16th Ed., Dunod, Paris, 1978, pp. 79–106.
[1381] American Society for Testing and Materials, '1984 Annual Book of ASTM Standards', The Society, Philadelphia, 1984, Section 11, 'Water and Environmental Technology', Volume 11.02, 'Water. (II)', pp. 325–495.
[1382] W. S. Holden, ed., 'Water Treatment and Examination', Churchill, London, 1970.
[1383] J. D. Burton, in J. P. Riley and G. Skirrow, ed., ref. 14, Volume 3, pp. 91–191.

[1384] K. E. White, *Chem. Br.*, 1976, **12**, 375.
[1385] K. E. White, *Water Pollut. Contr.*, 1981, **80**, 498.
[1386] G. V. Evans, *Int. J. Appl. Radiat. Isotopes*, 1983, **34**, 451.
[1387] International Atomic Energy Agency, 'Isotope Hydrology', *Proc. Int. Symp. Isotope Hydrology, Neuherberg, June 1978*, Volumes 1 and 2, The Agency, Vienna, 1979.
[1388] International Atomic Energy Agency, 'Isotopes in Lake Studies', *Proc. Advisory Grp. Meeting, Vienna, August/September 1977*, The Agency, Vienna, 1979.
[1389] J. C. Rodda, ed., 'Facets of Hydrology', Wiley, London, 1976.
[1390] R. W. Herschy, ed., 'Hydrometry', Wiley, Chichester, 1978.
[1391] R. A. Vollenweider, 'A Manual on Methods for Measuring Primary Production in Aquatic Environments', International Biological Programme Handbook No. 12, Blackwell, Oxford, 1969.
[1392] R. J. Rosenbauer and J. L. Bischoff, *Limnol. Oceanogr.*, 1979, **24**, 393.
[1393] C. Beaudet and J. Rygaert, *J. Radioanal. Chem.*, 1968, **1**, 153.
[1394] J. K. Johannesson, *J. Radioanal. Chem.*, 1970, **6**, 27.
[1395] C. G. Taylor and J. Waters, *Analyst*, 1972, **97**, 533.
[1396] H. G. Richter and A. S. Gillespie, *Anal. Chem.*, 1962, **34**, 1116.
[1397] H. J. M. Bowen, *Isotope Radiat. Technol.*, 1971, **9**, 44.
[1398] H. R. DuBois and G. M. Sharma, *Anal. Chem.*, 1979, **51**, 1702.
[1399] T. T. Gorsuch, 'Radioactive Isotope Dilution Analysis', Radiochemical Centre Review No. 2., The Centre, Amersham, 1968.
[1400] J. Tölgyessy, T. Braun, and M. Kyrs, 'Isotope Dilution Analysis', Pergamon Press, Oxford, 1972.
[1401] J. Růzicka and J. Stary, 'Substoichiometry in Radiochemical Analysis', Pergamon Press, Oxford, 1968.
[1402] H. A. Das, W. H. Köhnemann, G. D. Wals, and J. Zonderhuis, *J. Radioanal. Chem.*, 1975, **25**, 261.
[1403] G. M. Sharma and H. R. DuBois, *Anal. Chem.*, 1978, **50**, 516.
[1404] H. Akaiwa, H. Kawamoto, and K. Ogura, *Talanta*, 1977, **24**, 394.
[1405] W. S. Lyon, ed., 'Guide to Activation Analysis', Van Nostrand, Princeton, 1964.
[1406] H. J. M. Bowen and D. Gibbons, 'Radioactivation Analysis, Clarendon Press, Oxford, 1963.
[1407] P. Kruger, 'Principles of Activation Analysis', Wiley-Interscience, New York, 1971.
[1408] V. P. Guinn, 'Activation Analysis', in I. M. Kolthoff and P. J. Elving, ed., 'Treatise on Analytical Chemistry', Part 1, Volume 9, Wiley, New York, 1971, pp. 5583–5641.
[1409] V. P. Guinn, 'Neutron Activation Analysis', in A. Weissberger and B. W. Rossiter, ed., 'Physical Methods of Chemistry', Part IIID, Vol. 1, Wiley, New York, 1972, pp. 447–500.
[1410] V. P. Guinn, 'Neutron Activation Analysis', in H. B. Mark and J. S. Mattson, ed., ref. 9, pp. 187–220.
[1411] W. S. Lyon and H. H. Ross, *Anal. Chem.*, 1982, **54**, 227R.
[1412] L. Salmon, 'Instrumental Neutron Activation Analysis in Environmental Studies of Trace Elements', Atomic Energy Research Establishment, Harwell, June 1975.
[1413] E. P. Hamilton and A. Chatt, *J. Radioanal. Chem.*, 1982, **71**, 29.
[1414] S. Bigot and M. Treuil, *J. Radioanal. Chem.*, 1980, **59**, 341.
[1415] T. M. Tanner, L. A. Rancitelli, and W. A. Haller, *Water, Air, Soil Pollut.*, 1972, **1**, 132.
[1416] K. Ammann and A. Knoechel, *Fresenius' Z. Anal. Chem.*, 1981, **306**, 161.
[1417] S. K. Nyarku and A. Chatt, *J. Radioanal. Chem.*, 1982, **71**, 129.
[1418] A. Hirose, K. Kobori, and D. Ishii, *Anal. Chim. Acta*, 1978, **97**, 303.
[1419] R. R. Greenberg and H. M. Kingston, *J. Radioanal. Chem.*, 1982, **71**, 147.
[1420] C. Lee, N. B. Kim, I. C. Lee, and K. S. Chung, *Talanta*, 1977, **24**, 241.
[1421] C. Bergerioux, J. P. Blanc, and W. Haerdi, *J. Radioanal. Chem.*, 1977, **37**, 823.
[1422] J. Akaiwa, H. Kawamoto, K. Ogura, and S. Kogure, *Radioisotopes*, 1979, **28**, 681.
[1423] S. Sava, L. Zikovsky, and J. Boisvert, *J. Radioanal. Chem.*, 1980, **57**, 23.
[1424] T. Braun, M. N. Abbas, A. Elek, and L. Bakas, *J. Radioanal. Chem.*, 1981, **67**, 359.
[1425] T. Fujinaga, Y. Kusaka, M. Koyama, H. Tsuji, T. Mitsuji, S. Imai, J. Okuda, T. Takamatsu, and T. Ozaki, *J. Radioanal. Chem.*, 1973, **13**, 301.
[1426] Y. Kusaka, H. Tsuji, S. Imai, and S. Ohmori, *Radioisotopes*, 1979, **28**, 139.
[1427] B. M. Vanderborght and R. E. Van Grieken, *Anal. Chem.*, 1977, **49**, 311.
[1428] H. Akaiwa, H. Kawamoto, K. Ogura, and T. Koizumi, *Radioisotopes*, 1981, **30**, 535.
[1429] J. Schneider and R. Geisler, *Fresenius Z. Anal. Chem.*, 1973, **267**, 270.
[1430] K. Jørstad and B. Salbu, *Anal. Chem.*, 1980, **52**, 672.
[1431] K. Lenvik, E. Steinnes, and A. C. Pappas, *Anal. Chim. Acta*, 1978, **97**, 295.
[1432] W. Stumm and H. Bilinski, in S. H. Jenkins, ed., 'Advances in Water Pollution Research', Proc. 6th Int. Conf. (Jerusalem, 18–23 June 1972), Pergamon, Oxford, 1973, pp. 39–52.
[1433] J. Gardiner, *Water Res.*, 1974, **8**, 23.
[1434] A. L. Wilson, 'Concentrations of Trace Metals in Rivers Waters: A Review', Technical Report TR 16, Water Research Centre, Medmenham, Bucks., 1976.
[1435] T. M. Florence, *Water Res.*, 1977, **11**, 681.
[1436] H. W. Nürnberg, *Electrochim. Acta*, 1977, **22**, 935.

[1437] T. M. Florence and G. E. Batley, *CRC Crit. Rev. Anal. Chem.*, 1980, **9**, 219.
[1438] G. G. Leppard, in G. G. Leppard, ed., 'Trace Element Speciation in Surface Waters and its Ecological Implications', Plenum Press, New York, 1983, pp. 1–15.
[1439] D. P. H. Laxen and R. M. Harrison, *Sci. Total Environ.*, 1981, **19**, 59.
[1440] G. G. Leppard, ed., 'Trace Element Speciation in Surface Waters and its Ecological Implications', Plenum Press, New York, 1983.
[1441] T. M. Florence, *Talanta*, 1982, **29**, 345.
[1442] J. Buffle, *Trends Anal. Chem.*, 1981, **1**, 90.
[1443] R. S. Braman, in D. F. S. Natusch and P. K. Hopke, ed., 'Analytical Aspects of Environmental Chemistry', Wiley-Interscience, New York, 1983, pp. 1–59.
[1444] E. Steinnes, in G. G. Leppard, ed., ref. 1440 pp. 37–47.
[1445] S. J. de Mora and R. M. Harrison, *Water Res.*, 1983, **17**, 723.
[1446] T. A. Neubecker and H. E. Allen, *Water Res.*, 1983, **17**, 1.
[1447] Y. K. Chau and P. T. Wong, in G. G. Leppard, ed., ref. 1440 pp. 87–103.
[1448] E. A. Jenne, ed., 'Chemical Modeling in Aqueous Systems', ACS Symposium Series No. 93, American Chemical Society, Washington, 1979.
[1449] W. Stumm and J. J. Morgan, 'Aquatic Chemistry; An Introduction Emphasising Chemical Equilibria in Natural Waters', 2nd Ed., Wiley, New York, 1981.
[1450] F. M. Morel, 'Principles of Aquatic Chemistry', Wiley, New York, 1983.
[1451] M. Smies, in G. G. Leppard, ed., ref. 1440 pp. 177–193.
[1452] H. Babich and G. Stotzky, in J. O. Nriagu, ed., 'Aquatic Toxicology', Wiley-Interscience, New York, 1983, pp. 1–46.
[1453] U. Borgmann, in J. O. Nriagu, ed., ref. 1452, pp. 47–72.
[1454] C. W. Heckman, in J. O. Nriagu, ed., ref. 1452 pp. 355–400.
[1455] J. B. Best and M. Morita, in J. O. Nriagu, ed., ref. 1452 pp. 137–154.
[1456] E. J. Underwood, *Philos. Trans. R. Soc. London, Ser. B*, 1979, **288**, 5.
[1457] E. J. Moynahan, *Philos. Trans. R. Soc. London, Ser. B*, 1979, **288**, 65.
[1458] D. Barltrop, *Philos. Trans. R. Soc. London, Ser. B*, 1979, **288**, 205.
[1459] J. W. Patterson, H. E. Allen, and J. J. Scala, *J. Water Pollut. Control Fed.*, 1977, **49**, 2397.
[1460] D. T. E. Hunt and J. D. Creasey, 'The Calculation of Equilibrium Trace Metal Speciation and Solubility in Aqueous Systems by a Computer Method, with Particular Reference to Lead', Technical Report TR 151, Water Research Centre, Medmenham, Bucks., 1980.
[1461] I. Sanchez and G. F. Lee, *Water Res.*, 1973, **7**, 587.
[1462] H. A. Elliott, Y. T. Lin and C. P. Huang, in 'Proceedings, International Conference on Management and Control of Heavy Metals in the Environment, London, September 1979. CEP Consultants, Edinburgh, 1979, pp. 476–479.
[1463] A. L. Wilson, *Philos. Trans. R. Soc. London, Ser. B*, 1979, **288**, 25.
[1464] A. Britton and W. N. Richards, *J. Inst. Water. Eng. Sci.*, 1981, **35**, 349.
[1465] I. Sheiham and P. J. Jackson, *J. Inst. Water Eng. Sci.*, 1981, **35**, 491.
[1466] M. R. Shock, *J. Am. Water Works Assoc.*, 1980, **72**, 695.
[1467] R. O. Arah and B. McDuffie, *Anal. Chem.*, 1976, **48**, 195.
[1468] J. E. Alexander and E. F. Corcoran, *Limnol. Oceanogr.*, 1967, **12**, 236.
[1469] P. M. Williams, *Limnol. Oceanogr.*, 1969, **14**, 156.
[1470] P. Foster and A. W. Morris, *Deep Sea Res.*, 1971, **18**, 231.
[1471] R. W. Smith, in 'Non-equilibrium Systems in Natural Water Chemistry', Advances in Chemistry Series, 106, American Chemical Society, Washington, 1971, pp. 250–279.
[1472] J. P. Koenings, *Limnol. Oceanogr.*, 1976, **21**, 674.
[1473] L. Kamp-Nielsen, *Deep-Sea Res.*, 1972, **19**, 899.
[1474] J. D. Strickland and T. R. Parsons, *Bull. Fish. Res. Bd. Canada*, 1967, **167**, 115.
[1475] M. J. Stiff, *Water Res.*, 1971, **5**, 585.
[1476] J. Gardiner, *Water Res.*, 1974, **8**, 23.
[1477] P. Kivalo, R. Virtanen, K. Wickström, M. Wilson, E. Pungor, G. Horvai, and K. Tóth, *Anal. Chim. Acta*, 1976, **87**, 401.
[1478] E. Hansen and J. Růžička, *Anal. Chim. Acta*, 1974, **72**, 365.
[1479] S. Ramamoorthy and D. J. Kushner, *J. Fish. Res. Bd. Canada*, 1975, **32**, 1755.
[1480] W. G. Sunda and P. J. Hanson, in E. A. Jenne, ed., ref. 1448, p. 147.
[1481] K. C. Swallow, J. C. Westall, D. M. McKnight, N. M. L. Morel, and F. M. M. Morel, *Limnol. Oceanogr.*, 1978, **23**, 538.
[1482] J. K. McCrady and G. A. Chapman, *Water Res.*, 1979, **13**, 143.
[1483] J. C. Westall, F. M. Morel, and D. N. Hume, *Anal. Chem.*, 1979, **51**, 1792.
[1484] M. J. Stiff, *Water Res.*, 1971, **5**, 171.
[1485] J. Buffle, F. L. Greter, and W. Haerdi, *Anal. Chem.*, 1977, **49**, 216.
[1486] W. T. Bresnahan, C. L. Grant, and J. H. Weber, *Anal. Chem.*, 1978, **50**, 1675.
[1487] P. Zitko, W. V. Carson, and W. G. Carson, *Bull. Environ. Contam. Toxicol.*, 1973, **10**, 265.

[1488] J. P. Giesy, G. J. Leersee, and D. R. Williams, *Water Res.*, 1977, **11**, 1013.
[1489] W. G. Sunda, D. W. Engel, and R. M. Thuotte, *Environ. Sci. Technol.*, 1978, **12**, 409.
[1490] C. T. Driscoll, *Int. J. Environ. Anal. Chem.*, 1984, **16**, 267.
[1491] B. D. LaZerte, *Can. J. Fish. Aquat. Sci.*, 1984, **41**, 766.
[1492] J. Buffle, F. L. Greter, G. Nembrini, J. Paul, and W. Haerdi, *Fresenius' Z. Anal. Chem.*, 1976, **282**, 339.
[1493] W. Davison, *J. Electroanal. Chem.*, 1978, **87**, 395.
[1494] J. J. Lingane, *Chem. Rev.*, 1941, **29**, 1.
[1495] R. Ernst, H. E. Allen, and K. H. Mancy, *Water Res.*, 1975, **9**, 969.
[1496] H. W. Nürnberg, *Electrochim. Acta*, 1977, **22**, 935.
[1497] H. W. Nürnberg, in G. G. Leppard, ed., ref. 1440, pp. 211–230.
[1498] P. Valenta, in G. G. Leppard ed., ref. 1440, pp. 49–69.
[1499] H. E. Allen, W. R. Matson, and K. H. Mancy, *J. Water Pollut. Contr. Fed.*, 1970, **42**, 573.
[1500] G. E. Batley and T. M. Florence, *Anal. Lett.*, 1976, **9**, 379.
[1501] Y. K. Chau and K. Lum-Shue-Chan, *Water Res.*, 1974, **8**, 383.
[1502] G. E. Batley and Y. J. Farrar, *Anal. Chim. Acta*, 1978, **99**, 283.
[1503] P. L. Brezonik, P. A. Brauner, and W. Stumm, *Water Res.*, 1976, **10**, 605.
[1504] K. S. Subramanian, *Curr. Sci.*, 1979, **48**, 576.
[1505] J. S. Young, J. M. Gurtisen, C. W. Apts, and E. A. Crecelius, *Marine Environ. Res.*, 1979, **2**, 265.
[1506] W. R. Seitz, *Trends Anal. Chem.*, 1981, **1**, 90.
[1507] R. A. Saar and J. H. Weber, *Anal. Chem.*, 1980, **52**, 2095.
[1508] G. E. Batley, in G. G. Leppard, ed., ref. 1440, pp. 17–36.
[1509] R. D. Guy and C. L. Chakrabarti, *Chem. Can.*, 1976, **28**, 26.
[1510] R. D. Guy and C. L. Chakrabarti, *Proc. Intl. Conf. on Heavy Metals in the Environment*, Toronto, 1975, pp. 275–294.
[1511] B. T. Hart and S. H. Davies, *Interact. Seds. Fresh Water, Proc. Intl. Symp. (1976)*, Junk, The Hague, 1977, pp. 398–402.
[1512] P. Beneš and E. Steinnes, *Water Res.*, 1974, **8**, 947.
[1513] B. T. Hart and S. H. Davies, *Aust. J. Marine Freshwater Res.*, 1977, **28**, 105.
[1514] R. G. Smith, *Anal. Chem.*, 1976, **48**, 74.
[1515] M. R. Hoffmann, E. C. Yost, S. J. Eisenreich, and W. J. Maier, *Environ. Sci. Technol.*, 1981, **15**, 655.
[1516] J. Lee, *Water Res.*, 1983, **17**, 501.
[1517] W. J. Moore, 'Physical Chemistry', 4th Ed., Longmans, London, pp. 760–761.
[1518] R. E. Truitt and J. H. Weber, *Environ. Sci. Technol.*, 1981, **15**, 1204.
[1519] M. E. Bender, W. R. Matson, and R. A. Jordan, *Environ. Sci. Technol.*, 1970, **4**, 520.
[1520] S. F. Sugai and M. L. Healy, *Marine Chem.*, 1978, **6**, 291.
[1521] R. M. Sterritt and J. N. Lester, *Bull. Environ. Contam. Toxicol*, 1980, **24**, 196.
[1522] R. M. Sterritt and J. N. Lester, *Environ. Pollut. Ser. A*, 1982, **27**, 37.
[1523] S. J. de Mora and R. M. Harrison, in 'Environmental Contamination', Proceedings of an International Conference, London, July 1984, CEP Consultants Ltd., Edinburgh, 1984, pp. 124–130.
[1524] M. O. Andreae, in C. S. Wong, E. Boyle, K. W. Bruland, J. D. Burton, and E. D. Goldberg, ed., 'Trace Metals in Sea Water', Plenum Press, New York, 1983, pp. 1–19.
[1525] R. S. Braman, D. L. Johnson, C. C. Foreback, J. M. Ammons, and J. L. Bricker, *Anal. Chem.*, 1977, **49**, 621.
[1526] A. U. Shaikh and D. E. Tallman, *Anal. Chim. Acta*, 1978, **98**, 251.
[1527] M. O. Andreae, *Anal. Chem.*, 1977, **49**, 820.
[1528] M. O. Andreae, J. F. Asmodé, P. Foster, and L. Van't dack, *Anal. Chem.*, 1981, **53**, 1766.
[1529] Y. Talmi and D. T. Bostick, *Anal. Chem.*, 1975, **47**, 2145.
[1530] Y. K. Chau, P. T. S. Wong, and P. D. Goulden, *Anal. Chem.*, 1975, **47**, 2279.
[1531] G. A. Cutter, *Anal. Chim. Acta*, 1978, **98**, 59.
[1532] T. M. Florence and G. E. Batley, *Talanta*, 1975, **22**, 201.
[1533] T. M. Florence and G. E. Batley, *Talanta*, 1976, **23**, 179.
[1534] P. G. C. Campbell, M. Bisson, R. Bougie, A. Tessier, and J. P. Villeneuve, *Anal. Chem.*, 1983, **55**, 2246.
[1535] R. B. Barnes, *Chem. Geol.*, 1975, **15**, 177.
[1536] H. M. May, P. A. Helmke, and M. L. Jackson, *Chem. Geol.*, 1979, **24**, 259.
[1537] A. Henriksen, O. K. Skogheim, and B. O. Rosseland, *Vatten*, 1984, **40**, 246.
[1538] K. Matsunaga, M. Negishi, S. Fukase, and K. Hasebe, *Geochem. Cosmochim. Acta*, 1980, **44**, 1615.
[1539] J. F. Pankow and G. E. Janauer, *Anal. Chim. Acta*, 1974, **69**, 97.
[1540] J. F. Pankow, D. P. Leta, J. W. Lin, S. E. Ohl, W. P. Shum, and G. E. Janauer, *Sci. Total Environ.*, 1977, **7**, 17.
[1541] M. S. Shuman and J. H. Dempsey, *J. Wat. Pollut. Control. Fed.*, 1977, **49**, 2000.
[1542] R. E. Cranston and J. W. Murray, *Anal. Chim. Acta*, 1978, **99**, 275.
[1543] E. Nakayama, T. Kuwamoto, H. Tokoro, and T. Fujinaga, *Anal. Chim. Acta*, 1981, **131**, 247.
[1544] P. D. Goulden and D. H. J. Anthony, *Anal. Chim. Acta*, 1980, **120**, 129.
[1545] J. D. Ingle and C. E. Oda, *Anal. Chem.*, 1981, **53**, 2305.

[1546] P. Figura and B. McDuffie, *Anal. Chem.*, 1980, **52**, 1433.

[1547] R. M. Sterritt and J. N. Lester, *Sci. Total Environ.*, 1984, **34**, 117.

[1548] U. Förstner and G. T. W. Wittmann, 'Metal Pollution in the Aquatic Environment', Springer-Verlag, Berlin, 1981, pp. 238–244.

[1549] A. Tessier, P. G. C. Campbell, and M. Bisson, *Anal. Chem.*, 1979, **51**, 844.

[1550] W. Salomons and U. Förstner, *Environ. Technol. Lett.*, 1980, **1**, 506.

[1551] F. Rapin and U. Förstner, in 'Proc. Int. Conf. on Heavy Metals in the Environment', Heidelberg, September 1983. CEP Consultants Ltd., Edinburgh, 1983, pp. 1074–1077.

[1552] N. Meguellati, D. Robbe, P. Marchandise, and M. Astruc, in ref. 1551, pp. 1090–1093.

[1553] H. W. Nürnberg, P. Valenta, L. Mart, B. Raspor, and L. Sipos, *Anal. Chem.*, 1976, **282**, 357.

12 Automatic and On-line Analysis

12.1 INTRODUCTION

12.1.1 Definition of Terms and Scope of the Section

Several definitions have been proposed for 'automatic'[1-3] and 'on-line'[3] but both terms tend to be used with a variety of senses by different authors; see for example, the discussions of the term 'automation' in refs. 4 and 5. To avoid confusion, the meanings used in this book are as follows.

Automatic analysis denotes those procedures in which the analyst presents a number of discrete 'samples for analysis' [see Section 3.5.1(*a*)(*ii*)] and, when necessary, appropriate calibration standards directly to a piece or set of equipment which then carries out without further human involvement all steps of the analysis up to and including production of the analytical results in a written format. The only human treatment of the solutions for analysis is that of transferring portions of them to appropriate containers. Automatic analysis in this sense clearly has no direct effect on a number of activities involved in the measurement of water quality, *e.g.*, the collection and transport of samples, the receipt of, and 'book-keeping' arrangements for, samples in a laboratory, the preparation of calibration standards, high-purity water, and any reagents that may be necessary, the preparation of special samples and standards for analytical quality control purposes and the evaluation of the results from such tests. Such operations should not be overlooked when considering the introduction – and its effects – of automatic analysis into a laboratory.

We use the term *semi-automatic analysis* to denote those systems in which at least one step of the analytical procedure – after appropriate portions have been taken of the solutions for analysis – requires involvement of the analyst. Of course, *manual analysis* corresponds to the special case where all steps of the analytical procedure involve the analyst. Given these definitions, it should be noted that 'automatic' is commonly used in the literature to describe systems that we speak of as 'semi-automatic'.

The equipment used to automate analytical procedures has usually not simulated the precise operations of the analyst in carrying out these procedures. Recently, however, growing attention seems to be being paid to electromechanical, robotic devices that do simulate human operations in procedures such as dispensing samples from sample-containers, diluting samples, and other more complex operations. Refs. 6 to 8b provide examples of such applications, and consider the advantages of robots; see also the comments of Stockwell[9] on these devices and his description of a Low-cost Robot Devices Club, whose aims include facilitating the development and appropriate application of robotic systems. These devices may allow mechanisation of some operations in automatic analytical systems that, at present, involve the analyst, *e.g.*, sample dispensing, reagent preparation, though their general value for such purposes in water analysis seems to us rather doubtful. In any event, the use of robotic systems appears to be only just beginning to be explored, and we have not further considered this approach; interested readers may consult the cited references.

We use the term *on-line analysis* to denote those analytical systems in which either: (1) the sensor of an automatic analytical system is placed directly in the water of interest at the desired sampling point, or (2) an automatic analytical system is automatically presented with a sample stream (continuous or discontinuous) from the sampling point in the water of interest. With both (1) and (2), a series (continuous or discontinuous) of analytical results is obtained automatically. Given this definition, note that on-line analytical systems generally require calibration, and that the calibration standards for this may be presented either manually or automatically [see Section 12.3.4(f)(i)]. Of course, on-line analysis avoids human involvement in sample collection and transport.

The terms *in-line* or *in situ* analysis are commonly used[3,10] to denote analytical systems in which the analytical sensor is placed directly in the water of interest. This type of analysis has the very valuable feature of eliminating the various problems that may arise through the use of sample-collection devices (see Section 3.5). *In situ* systems, however, also have two important restrictions. First, they may be more subject than the other type of on-line analysis to problems arising from the 'fouling' of the analytical sensor by materials present in the water [see Section 12.3.4(f)(vi)]. Second, the accurate measurement of most determinands of common interest requires at present some form of sample treatment before measurement, *e.g.*, the addition of one or more reagents. This need usually precludes the use of *in situ* analysis for all but a few determinands, *e.g.*, dissolved oxygen, pH, electrical conductivity, optical absorption. We have, therefore, not further specifically considered this type of analysis, but its potential value is always worth bearing in mind for those determinands that may be measured without any sample treatment.

One other type of analysis should be briefly mentioned here though it falls outside the scope of this book. This is the use of '*remote-sensing*' techniques which may be used to investigate certain aspects of the qualities of large, natural bodies of water. Spectral techniques with the source and/or detector mounted in an airborne system are generally used for remote sensing, but they are suitable for rather few determinands and are not commonly available to most laboratories. Kish[11] has recently provided a useful, introductory review of the capabilities of remote-sensing systems; and see also ref. 10.

Since the first edition of this book, the application of microprocessors and computers to analytical systems has grown enormously for two broad purposes: (1) the monitoring and control of experimental parameters in analytical systems and (2) the automatic processing of analytical responses and the calculation and presentation of analytical results. Such applications have allowed many analytical systems to be converted to semi-automatic or automatic forms with resultant improvement in important aspects of their performance. Computing systems, however, have other applications of value in analytical laboratories and in the measurement and control of water quality. We thought it more useful, therefore, to deal with all the relevant uses of such systems in Section 13, and they are not dealt with specifically here. This division between sections should not conceal the essential role that these techniques play in automatic analytical systems.

On the above basis, the four main types of analysis that we distinguish are manual, semi-automatic, automatic, and on-line. The relative roles of these approaches are considered in the following section.

12.1.2 Relative Roles of the Different Types of Analytical System

In the past, the manual collection of samples and their manual analysis after transport to a laboratory have been very widely used in the measurement of water quality. This particular combination of sampling and analytical systems, however, has increasingly – and especially during the last ten years or so – been found to have a number of limitations that may prevent the required information on water quality from being obtained. These limitations may be summarised as follows.

(1) The delay between sample collection and the reporting of the analytical result may be too long, particularly when either control or knowledge of water quality is required on a continuous or very frequent basis, *e.g.*, in the automatic control of treatment plant.

(2) Difficulties may arise when, as will often be necessary, samples have to be collected and/or analysed outside normal working hours.

(3) The man-power and laboratory facilities (*e.g.*, space) available to a laboratory may be insufficient to allow collection and/or analysis of all the required samples.

(4) The accuracy of analytical results can be markedly dependent on the skill of the analyst and the conscientiousness and accuracy with which specified analytical procedures are followed in routine, manual analysis.

(5) It may be difficult to ensure adequately small instability of samples for certain determinands (see Section 3.6), particularly when samples must be collected at locations remote from the laboratory.

In addition to these limitations, the need to reduce the costs of sampling and analysis is being felt increasingly by many laboratories. Finally, it is worth mentioning, as Mitchell[12] has noted, that the replacement of manual analytical systems by the latest and most sophisticated automatic may confer some prestige to a laboratory, and this may be of value for a number of purposes. For all these reasons, the use of automatic and semi-automatic sampling and analytical systems has grown very markedly in the last decade or so. The question of which types of sampling and analytical system represent the optimal combination for a given determinand and measurement programme involves a number of factors that must be considered in detail by each laboratory. Some broad generalisations, however, can be suggested.

For example, limitation (5) may be reduced or overcome by making the analyses manually in the field at or near the sampling points. This approach is commonly recommended and applied for relatively unstable determinands for which robust and portable analytical equipment is available, *e.g.*, dissolved oxygen, pH, electrical conductivity, residual chlorine. For many stable or stabilisable determinands, however, limitations (1) to (4) may still apply to manual analysis in the field. The use of mobile laboratories can reduce or overcome these four limitations since they may be equipped with virtually any manual, semi-automatic or automatic analytical equipment. This latter approach can be of considerable value when there is special interest in just one sampling point,[13,14] but it clearly becomes generally impracticable for measurement programmes involving a number of sampling points so widely separated that a number of mobile laboratories would be required to operate simultaneously. For relatively stable determinands not subject to limitations (1) to (4), it is normal practice, of course, to transport all samples to a central laboratory.

Similarly, when limitations (1) and (2) are important, on-line analysis is generally the only suitable approach. Such analysis may even be justified on the purely economic base of the reduction in effort required for manual sample collection and transport to a laboratory, though Holland[15] has made the important point that the manual collection and transport of samples needs less skilled man-power than that required for maintaining on-line analytical instrumentation. On-line analysis generally eliminates or markedly reduces the importance of limitations (5). The impact of this type of analysis on limitations (3) and (4), however, is less clear-cut. For example, though sampling and analytical man-power are generally much reduced, if not eliminated, appropriate skills must be provided for the installation and maintenance of on-line systems. Similarly, though there is generally economy in laboratory facilities, the potential need for many on-line systems when a number of sampling points are of interest may make this approach economically impracticable. Equally, though the human factor is largely eliminated from limitation (4), on-line systems may be subject to other effects that are essentially absent in the laboratory analysis of discrete samples. These and other points concerning the use of on-line analytical systems are considered in more detail in Section 12.3.4.

Finally, when limitations (3) and/or (4) rather than (1) and (2) are of prime concern, automatic or semi-automatic analysis in a laboratory will often be the preferred approach. Such analysis is today usually capable of much greater rates of analysis of samples than the corresponding manual analytical procedures so that one analyst with an automatic analyser can often handle the number of samples that would otherwise have required several analysts, greater amounts of laboratory space, and possibly more than one of the appropriate measuring instruments. Further, the automation of an analytical procedure under laboratory conditions commonly ensures greater consistency in the execution of the procedure and correspondingly more precise analytical results. This is, however, not necesarily true also of the bias of analytical results, and this aspect of accuracy should always be carefully considered. Given the faster rates of automatic and semi-automatic analysis, their use may also allow the delay between receipt of samples and reporting of results to be reduced. Of course, such types of analysis in the laboratory are subject to problems of sample instability. It can be seen that an ideal application of laboratory-based, automatic analysis is the analysis of samples collected from many different sampling locations at frequencies smaller than would justify the much greater cost of installing on-line analysers at each location. These and other aspects of the use of automatic and semi-automatic analysis in the laboratory are considered in more detail in Section 12.2.4.

The above points suffice to indicate the nature of the considerations involved in deciding, in principle, the type of sampling and analytical systems preferred for a given purpose. Appropriate decisions can clearly be reached only if the requirements for information on water quality have been carefully and precisely defined. The importance of such definitions has already been stressed in Section 2, and many authors have made the same point in discussing both automatic and also on-line analytical systems; see for example, refs. 4, 5, 16 and 17 for the former, and refs. 18 to 22a for the latter. The ability effectively to apply automatic and/or on-line analysis depends on the availability of suitable equipment as well as a number of other factors; these aspects are considered in Sections 12.2 and 12.3, respectively.

12.2 AUTOMATIC ANALYSIS OF DISCRETE SAMPLES

12.2.1 Introduction

There is a vast array of commercially available, automatic analysers and semi-automatic analytical equipment. This array changes relatively rapidly as equipment of new design is introduced, and as various features of existing instrumentation are modified by manufacturers. In addition, the literature contains many papers describing automatic and semi-automatic systems assembled by their authors. In this position, we have attempted neither to list all such systems nor to describe in detail the designs and methods of operation of particular instruments and components. Comprehensive lists of equipment manufacturers are available – see, for example, the Laboratory Buyer's Guide Edition of the journal *International Laboratory* – and the book by Foreman and Stockwell,[24] though now somewhat out-of-date, still provides a valuable and detailed discussion of many relevant instruments. The manufacturers of commercially available equipment are, of course, generally the best source of up-to-date information on the nature and operation of their equipment. With these points in mind, our main aim here is to describe what seems to us generally important considerations concerning the availability and use of automatic analytical systems for the analysis of discrete samples.

Whether or not the costs involved in the use of automatic analytical systems are justified depends on a number of factors [see Section 12.2.4(*c*)]. When the costs of full automation cannot be justified, partial automation can be very valuable, and much equipment for this purpose is commercially available. This ranges from relatively simple devices such as automatic pipettes and sample-diluters to more complex items such as fully automatic measuring instruments, *e.g.*, spectrophotometers, atomic emission and absorption spectrometers. Virtually any stage of any analytical procedure can be performed automatically, and each such use of automatic components can give substantial savings in time and effort, and can also often help to improve the precision of analytical results. Since semi-automatic analytical systems are subject to essentially the same general considerations as fully automatic systems, we restrict attention hereafter to the latter.

There has been a very marked growth during the last 20 years or so in the use of laboratory-based, automatic systems for the analysis of discrete samples, and many instruments for this purpose are commercially available. Much of the considerable activity in this field and many of the instruments are primarily devoted to the field of clinical analysis, and at present rather few of the many types of instrument appear to find widespread, routine use in the analysis of waters. This position, however, seems to be changing, and – as many authors have noted – most if not all of the automatic systems used in clinical analysis are potentially applicable to water analysis. Accordingly, it seems to us useful to attempt to describe briefly the various types of automatic analytical system that are commercially available whether or not they are currently employed for the routine analysis of waters. This description is restricted to commercially available equipment because we are primarily concerned here with routine analysis [see Section 12.2.4(*d*)].

There are so many types of automatic analytical systems that some scheme for classifying them is desirable. Various terms and classifications have been used by many authors for this purpose; see, for example, refs. 12 and 24 to 29. For present purposes, we use a somewhat different scheme of classification which, of course, makes use of these other schemes. Our scheme is based on the main components of equipment in automatic analytical systems; these components and their functions are shown in Table 12.1. We should stress that the division into the separate components of Table 12.1 is to some extent, arbitrary, and that in some systems one piece of equipment may execute more than one function. Note also that our scheme does not include the function of sample-identification. Notwithstanding such points, the table allows a useful, initial classification of automatic systems into two broad classes:

(1) those not involving sample treatment
(2) those involving sample treatment, and of these a number of different types can usefully be distinguished.

The different types of automatic systems are considered on this basis in Section 12.2.2, their applications to water analysis are briefly reviewed in Section 12.2.3, and finally in Section 12.2.4 we discuss several generally important aspects of the use of automatic analytical systems.

12.2.2 Types of Automatic Analytical System

(a) Systems not Involving Sample Treatment
This type of system is appropriate whenever the analytical procedure simply requires direct presentation of solutions for analysis to an analytical sensor unit. Automation of such procedures then requires only units 1, 2, 4 and 5 of Table 12.1, and the sample-transport unit (2) commonly consists of a simple tube connecting units 1 and 4. Units for automatic sample presentaion (1) and calculation and print-out of results (5) are available as optional components for most modern measuring instruments (unit 4) so that automation is readily achieved when no sample treatment is required.

The most common method of automatic sample-presentation involves a unit which sequentially and at pre-set intervals presents containers holding the solutions to be analysed to a sampling probe. The sampling probe then dips into a solution, withdraws a portion of the solution, and passes this portion into the sample-transport system. A peristaltic pump is commonly used for this withdrawal of solution, but other means are also employed. The containers of the solutions for analysis are

TABLE 12.1

Main Components of Automatic Analysers

Component	Function	Examples
1. **Sample presentation unit**	To select suitable volumes of the solutions for analysis from each of a set of pre-filled containers, and to present these sub-samples to unit 2 at chosen times	Automatically controlled turntable on which the sample-containers are mounted. Automatic syringe for controlling the volume of sub-samples.
2. **Sample transport unit**	To move the sub-samples from 1 through units 3 and 4	Sub-sample containers mounted on an automatically controlled moving belt
3. **Sample treatment unit**	To effect any required chemical and physical treatments of the sub-samples	Automatic addition of reagents, heating, concentration and separation procedures
4. **Analytical-sensor unit**	To provide an analytical response (usually an electrical signal) to each treated sub-sample whose magnitude is governed by the concentration of the determinand	Automatic spectrophotometer
5. **Calculation and print-out unit**	To calculate the calibration function from the responses of standards, and to calculate and print-out the analytical results for the original samples	On-line computing system and printer

mounted on an automatically controlled moving belt or turntable, calibration standards being placed in appropriately positioned containers to allow automatic calibration. Thus, having made any necessary settings of the measuring instrument, sample-presentation, and calculation units, the analyst has only to fill the containers with samples and standards and to press a 'start' button, and is not further involved until all the analytical results have been printed-out. To reduce cross-contamination of successive solutions caused by residual solution remaining in and on the sampling probe, it is usual to clean the probe automatically between successive samples, *e.g.*, by placing it in an appropriate wash-liquid in the period between completion of withdrawal of one sample and the beginning of withdrawal of the next and then drying it. It is worth noting that this approach to automatic sampling presentation results, in many automatic analysers without sample treatment, in brief periods where no solution is flowing into the analytical-sensor unit, *i.e.*, those periods when the sampling probe is not immersed in a solution. This discontinuity can be avoided by using a low-dispersion, flow-injection system [see Section 12.2.2(*b*)(*ii*)] in place of the direct injection of the solutions to be analysed into the analytical-sensor unit. Several authors have recently noted that avoidance of discontinuity of flow into this unit may lead to more stable operation of the sensor and more precise analytical results; see, for example, refs. 30 and 31 (inductively-coupled plasma spectrometry) and 32 (atomic absorption spectrometry). In passing, it is also worth noting that flow-injection systems have also been reported to be of value in facilitating the calibration of such systems; see, for example, refs. 31 to 34.

In principle, the analytical-sensor unit may consist of any analytical measuring instrument provided it is equipped either with a flow-through cell through which the solutions to be measured pass or with a means of disposing of each solution after its analytical response has been measured. Thus, automatic systems without sample-treatment are potentially capable of a very wide range of analyses including titrimetic procedures for which a number of automatic titrators are commercially available. Such systems are sometimes suitable for the analysis of waters, and their applications are noted in Section 12.2.3(*b*). Commonly, however, some form of sample treatment is required, and in (*b*) below we consider automatic systems with this capacity.

(b) Systems Involving Sample Treatment

(i) Classification of systems and its scope As noted above, a high proportion of the determinands of common importance in water analysis requires some form of sample treatment. Several different approaches are used in the many commercially available instruments to effect sample treatment, and it is useful, therefore, to base a classification of automatic systems on the means by which this treatment is effected. This classification is shown below, and each of the types of system and their general characteristics are briefly considered in (*ii*) and (*iii*) below.

Sequential analysers

(1) Continuous-flow systems

    (1.1) Gas-segmented streams

    (1.2) Unsegmented streams

(2) Discrete-sample systems

    (2.1) 'Wet-reagent' systems

    (2.2) 'Dry-reagent' systems

(3) Stopped-flow systems

Parallel analysers

(1) Centrifugal systems

Before considering these different systems, the following points should be noted.

(1)   The sample treatment and transport systems are always very closely inter-related. For simplicity, therefore, we use the term 'sample treatment' hereafter to denote both systems except where specifically noted otherwise.

(2)   In principle, virtually any type of analytical sensor may be used in conjunction with some – though not all – types of sample treatment system. Commercially available automatic analysers differ, however, in the range of analytical sensors that can be used.

(3)   The mode of operation of a given type of analytical sensor may differ among instruments employing the same type of sample treatment system. Such differences may affect some aspects of the performance of the instruments, *e.g.*, the degree of cross-contamination between successive samples. With a few exceptions, however, we have not discussed this effect.

(4)   Some commercially available automatic analysers are so designed that they can be applied to many different determinands requiring different types of sample-treatment. Other instruments provide a fixed sample-treatment system intended for one or more determinands for which the treatment is appropriate. In general, we do not distinguish these two types of instrument.

(5)   The simultaneous measurement of a number of determinands requiring different sample treatments may, in general, be achieved by simple multiplication of all the necessary components. Many approaches, however, are used in a number of commercially available automatic analysers to reduce the degree of multiplication required, *e.g.*, by making use of common components. Thus, single-, dual-, and multi-channel (up to twenty determinands) instruments can be purchased. Apart from the time saving afforded by simultaneous analysis (for a given type of instrument), multi-channel systems may be of value when the volumes of samples available are restricted.

(6)   Information concerning the points noted in (2) to (5) above is best obtained from the manufacturers of relevant instruments.

(ii) Sequential analysers   With these analysers, the solutions for analysis are treated and measured one after the other. The total number of solutions that can be analysed with one loading of the sample-presentation unit depends primarily on the capacity of that unit. Various capacities are available, but units holding between 40 and 200 solutions are perhaps most commonly used. The three main types of sequential analytical systems are described in (1) to (3) below.

(1) *Continuous-flow systems* – The distinguishing feature of these systems is that they consist, in essence, of a single flow-through conduit from the sample presentation unit to the analytical sensor and thence to waste. Appropriate volumes of the solutions for analysis are passed in sequence and at suitable intervals along this conduit, a variety of means being used to select appropriate volumes and transport them along the conduit, *e.g.*, pumps, gas pressure, hydrostatic pressure. The required treatment of the solutions is effected as they flow along the conduit.

It should be noted that, on the above basis, automatic, column-chromato-graphic analysis may be considered as a form of continuous-flow analysis. The automation of such procedures is well established and appropriate equipment is commercially available for both GC and HPLC, but fully automatic procedures find – with a few exceptions noted in Section 12.2.3 – little use in the routine analysis of waters at present. Further, the general nature of such automatic systems will be readily apparent from the description of the relevant, chromatographic techniques in Section 11.3.8. Accordingly, we do not further consider automatic chromatographic systems here.

Many types of sample treatment are possible in continuous-flow analysis. For example, a reagent may be added to the main stream by pumping a reagent solution through a tube connected by a suitably shaped and positioned T-piece to the main conduit. The proportions of reagent and main stream can be controlled by their respective rates of pumping. These rates are varied when peristaltic pumps are used by using appropriate diameter tubing in the pump, and one such pump may then be used to pump many different liquid streams at the desired rates. Thus, a series of reagents may be added, and appropriate times allowed for chemical reactions by varying the length and/or diameter of the main conduit between the points of one reagent addition and the next. Dilution of solutions for analysis is readily achieved by using a diluent in place of a reagent solution. To speed the rates of reactions, the stream in the main conduit can be heated by placing an appropriate length (usually as a coil) of the conduit in a heating bath. Suitably designed components can also be added to the main conduit to allow other physical and chemical operations, *e.g.*, filtration, dialysis, solvent-extraction, ion-exchange, distillation, acid-diges-tion, u.v. irradiation. The sample treatment system required to automate very many manual, analytical procedures can, with relatively little experience, be easily assembled from a series of standard components such as time-delay coils, T-pieces, connecting tubing, peristaltic-pump tubing, *etc.* Spectrophoto-metric measurement is most commonly used in the analytical sensor unit, but any type of sensor suitable for a continuously flowing stream can, in principle, be used; for example, fluorometry, emission and absorption spectrometry, direct potentiometry using ion-selective electrodes, voltammetry, conducto-metry and other techniques have all been used.

Continuous-flow systems generally represent the most versatile approach for automating relatively complex sample-treatment procedures and they are widely used for such purposes. An important, potential limitation is that, since all solutions for analysis flow through the same conduit, cross-contamination of successive solutions occurs. In practice, the magnitude of this effect is restricted to tolerable values by various means (see below) which limit the rate at which solutions can be analysed. Cross-contamination also tends to increase as the complexity of the sample treatment increases, and it is difficult to make valid and useful generalisations on achievable rates of analysis (see below), particularly since they will also depend on the accuracy required of results. Note, however, that discrete-sample treatment systems are not generally subject to this effect. A further, potential disadvantage – though of no importance in many applications – when peristaltic pumps are used is the incompatibility of the materials used for pump-tubes with certain reagents and solvents. Means of overcoming such problems are available, but tend to be rather cumbersome. The gradual wear of peristaltic-pump tubes also requires their replacement from time to time to maintain the required pumping-rates; such replacement is simple and rapid. On the other hand, continuous-flow systems are simpler mechanically than some other types of automatic, sample treatment systems.

Developments in continuous-flow analysis in the last five to ten years make it important to distinguish two types of system characterised by the presence or absence of segmentation of the main liquid stream by bubbles of gas.

In gas-segmented systems (air is usually but not always employed), the gas is pumped at a controlled rate into the sample-stream as it enters the main flow-through conduit so that this stream becomes an alternating sequence of liquid segments and gas bubbles. The purposes of the bubbles are to reduce the amount of cross-contamination between successive samples and to promote mixing of the samples with subsequently added reagents. This type of system was originally devised by Skeggs[35] in the late 1950s, and since then has seen very many developments and applications. A number of manufacturers produce automatic analysers incorporating this form of sample treatment. Industrial applications, including water analysis, began in the early to middle 1960s, and such systems have been the type most commonly applied to water analysis since then (see Section 12.2.3). Two books[36,37] discuss automatic analysis using this type of sample treatment system; the former is particularly comprehensive and includes nearly a thousand references. The topic has also been reviewed by many authors; see, for example, refs. 10 and 38 which deal with the analysis of waters.

The second type of continuous-flow analysis – unsegmented-flow systems – has seen a very rapid growth of interest and application in the last 5 to 10 years. In fact, a number of variants of this approach have been developed for various purposes, and their natures and historical development are reviewed by Stewart[39] and Růžička and Hansen.[25] All these variants have the common feature that portions of samples are injected sequentially and at pre-set intervals into an unsegmented 'carrier-stream' that flows through the main conduit into the analytical-sensor unit and thence to waste. This type of analysis is commonly referred to as *flow-injection analysis*, the term used by Růžička and Hansen[40] in their first description of their particular variant. At about the same time, Stewart, Beecher, and Hare[41] described a very similar approach. It is the variant of unsegmented-flow analysis developed by these two groups of workers that has so far received the most attention, found most application, and been shown to be capable of greatest versatility in the automation of sample treatment. We shall, therefore, restrict attention to this variant in subsequent paragraphs, and shall speak of it as flow-injection analysis. Before leaving the other variants of unsegmented-flow analysis, we should mention that their developers have argued that they are advantageous for certain purposes; see, for example, Pungor et al.[42] In addition to the references cited by Stewart and by Růžička and Hansen for these variants, see, for example, refs. 42 to 44 (variants due to Pungor and co-workers), ref. 45 (variant of Mottola's group), and ref. 46 for a very recent variant called 'rapid-flow analysis' by Alexander and co-workers.

Many studies have shown that portions of samples injected into a flowing liquid-stream tend to disperse within that stream as it flows through the main conduit of the sample treatment system. Such dispersion will, of course, tend to cause cross-contamination between successive samples with resultant worsening of the accuracy of analytical results. The gas bubbles introduced into gas-segmented systems are intended to reduce the magnitude of this effect. The essential achievement (in the present context) of the developers of flow-injection analysis was to show that such cross-contamination could be controlled to adequately small levels by suitable choice of factors such as the internal diameter of the conduit, volume of each sample injected, and flow-rate of the carrier-stream. Segmenting bubbles were not necessary, and their elimination offers some advantages; at the same time, rates of sample analysis equal to or greater than those commonly achieved with gas-segmented systems could be achieved. The developers and subsequent users of the technique have shown that flow-injection analysis is a versatile means of automating sample treatment, and there has been a very rapid development in (1) the number of different chemical and physical operations that can be effected by the sample treatment system, and (2) the number of different types of analytical sensor that may be used. Several automatic analysers incorporating flow-injection analysis systems have become commercially available in the last few years, and interest in, and applications of, this approach seem certain to continue to grow. Růžička in particular has stressed a number of exciting possibilities for new approaches to analysis and general chemical investigations that are offered by flow-injection systems. Such considerations are beyond the scope of this section, though we should mention the capability of what amounts to automatic titrations – though see also Pardue and Fields'[47,48] comments on the use of the term 'titration' as well as Růžička and Hansen's[49] views on these comments. Readers interested in reading in detail about this relatively new approach to automatic analysis should consult the book by Růžička and Hansen,[26] a number of recent reviews[50-54] (none of which is directly concerned with water analysis), and the proceedings of two conferences on flow-analysis,[55,56] both of which contain many papers on the theory and applications of flow-injection analysis. A number of automated, sample treatment procedures for the analysis of waters have been reported, and – to judge from the literature – laboratories in a number of different countries have adopted such procedures for routine water analysis. We know, however, of few routine applications in this country.

Since the development of flow-injection analytical systems there has been considerable discussion and a little polemic in the literature concerning their relative advantages and disadvantages in comparison with gas-segmented systems. It seems to us too soon to attempt any firm assessment of such points because: (1) flow-injection analysis is still developing rapidly and manufacturers and users of gas-segmented systems are responding to the challenge by various improvements to these latter systems and (2) there appear to be no detailed evaluations of the routine, analytical performances of equally optimised automatic systems involving the two approaches. A detailed and largely theoretical comparison has led one author[57] to conclude that, though both types of system can be applied for many sample treatment procedures, flow-injection analysis is better suited to procedures involving simple treatments and requiring short residence-times (5 s or less) within the flow-through system; otherwise, gas-segmented systems are to be preferred. Other authors have expressed broadly similar views though the dividing lines tend to vary substantially from author to author. For example the residence-time criterion has been suggested as 2 min[51] and 5 min.[53] It is also worth noting that Rocks and Riley[53] have predicted that flow-injection systems will increasingly replace gas-segmented systems in clinical laboratories.

Finally, mention should be made of a variant of unsegmented-flow analysis in which a reagent is injected into a continuously flowing stream of sample. This approach could be described as 'reverse flow-injection analysis', but appears at present to have received little attention.[58,59]

(2) Discrete-sample systems – The distinguishing feature of these systems is that each solution for analysis is held in or on its own special container throughout the sample-treatment system. Cross-contamination between successive samples during their treatment is thereby essentially eliminated. The sample transport system generally consists of an automatically controlled moving-belt or turntable on which the sample containers are mounted. These containers are loaded with appropriate volumes of the solutions for analysis either manually or automatically using an automatic sample presentation unit. Two main types of discrete-sample system can be distinguished depending on the means by which solutions for analysis are mixed with reagent solutions.

In the first type – 'wet-reagent' systems – chosen volumes of reagents are added automatically to the solutions for analysis which are held in small containers (disposable or washable), and each solution is so treated as its container is moved by the sample transport system. One or more reagents (the number possible varies from instrument to instrument) may be added in sequence and with appropriate delays after each addition. The reagents are usually added by some type of mechanically operated syringe, but see below for a very different approach. This type of system commonly has facilities for heating the solutions in order to speed the desired reactions, but few if any other physical and chemical operations are commonly available. At least three different approaches are used to measure the analytical responses of the treated solutions: (1) measurement is made on the individual sample-containers, (2) portions of the treated solutions are transferred to individual containers on which measurements are then made, and (3) portions of the treated solutions are transferred sequentially to the same container (flow-cell) in the analytical-sensor unit. Which of these approaches is used, and the precise method of operation of a given approach can affect the level of cross-contamination between successive solutions. Provided this effect is adequately small during the measurement stage, its absence during sample treatment means that such systems are potentially capable of very high rates of sample analysis. Most determinands for which these systems are used are measured spectrophotometrically, but other types of sensor are sometimes available and used. 'Wet-reagent' systems are mechanically more complex (but see next paragraph) than continuous-flow systems, and are much less versatile with respect to the treatment procedures that can be automated. They find application in many commercially available instruments which are commonly used in clinical laboratories, and the application of such automatic analysers to water analysis – though apparently rather rare in the past – appears to be increasing (see Section 12.2.3). More detailed descriptions of such instruments are given in refs. 24 and 25.

Two particular variants of 'wet-reagent' systems are worth noting. In the first, the solutions for analysis are placed in special containers which are supplied by the instrument manufacturer already containing sealed portions of the reagent solutions required for the determinand of interest. These reagents are released automatically and sequentially at pre-set intervals during sample treatment. This approach avoids much of the mechanical complexity usually present in 'wet-reagent' systems, and the need for reagent preparation by laboratories is eliminated. Spectrophotometric measurement is most often used, though a few other types of analytical sensor may be used. A limitation of this approach is that the determinands that can be measured, and the analytical procedures used for them, are controlled by the manufacturer of the equipment. More detailed descriptions of a commercially available instrument incorporating this approach are given in refs. 4 and 60. This instrument finds routine use in clinical laboratories, but we know of no applications to water analysis.

The second variant is a very recent development, and its novel feature is that the analytical-sensor unit consists, in essence, of a rapidly rotating, spectrophotometric measurement system so arranged as to scan repeatedly a circularly arranged set of sample containers to which one or two reagents may be added. The characteristics of this sytem are similar in several respects to centrifugal analysers [see (*iii*) below], but it combines with these several desirable features of sequential analysers, *e.g.*, much larger numbers of samples can be analysed without human intervention. This approach is so new that its success in the clinical field cannot yet be judged, and we know of no applications to water analysis. More detailed descriptions of an instrument that is or will be commercially available are given in refs. 4 and 61 to 63.

The second type of discrete-sample system – 'dry-reagent' systems – is also a very recent development. In essence, this approach consists of the placing of small portions of the solutions for analysis onto strips of film or other material consisting of layers in which are impregnated the reagents required for an analytically useful reaction with the determinand. The solution percolates through the film – at a rate that can be controlled by its composition – reacting as it does so with reagents in the appropriate sequence to produce a coloured or fluorescent compound that is subsequently measured directly and automatically on the film by a suitable, optical-measurement system. Direct potentiometry using specially prepared types of ion-selective electrode is also available for a few determinands. Systems of this type have been described by several terms of which 'dry-reagent' is probably the most common. They are mechanically very simple, capable of high rates of sample analysis, do not require preparation of reagents, and, for these reasons, require minimal analytical skill in operation and essentially no specialised laboratory facilities. The main disadvantages are the rather small range of sample-treatment procedures that can be effected, and the absolute dependence on the manufacturer for the choice and preparation of appropriate reagents and films for the determinands of interest. Several automatic analysers are commercially available; they have been developed for clinical applications and a number of relevant determinands can now be measured. Such systems appear to be at the stage of evaluation by clinical laboratories, and we know of no applications to water analysis. Recent reviews of this approach are provided by refs. 4 and 64 to 66.

(3) Stopped-flow systems – The distinguishing characteristic of this type of system is that it is so designed that very rapid mixing of a solution for analysis and a reagent is achieved – usually in substantially less than a second. Such rates of mixing allow the rates of many chemical reactions to be measured accurately even when these rates are relatively rapid. By suitable choice of the composition of the reagent, the ratio solution for analysis/reagent, and the times of measurement of a reaction-product, the rate of a reaction can be made proportional to the concentration of the determinand. In essence, therefore, stopped-flow systems consist of: (1) an automatically controlled, sample presentation unit which holds the solutions for analysis in their individual containers, (2) a means of withdrawing a portion of a solution and rapidly mixing it with a reagent in a flow-cell in which the analytical response is then measured, and (3) a dedicated microcomputer for calculating analytical results and controlling the operation of the analyser. Spectrophotometric measurement is most commonly used with such systems though other types of sensor can be used. Although such systems can be applied to the rapid measurement of equilibrium analytical responses resulting from very fast reactions, their most useful application appears to be the measurement of reaction rates since several advantages are claimed for this approach.[67,68] In the context of water analysis, one of these advantages is particularly worth noting. Since, in measuring reaction rates, one is effectively concerned with the difference in analytical responses at two different times, potentially interfering substances whose effects are independent of time become of little or no importance.

There is, of course, the corresponding potential disadvantage that substances which alter the rate of reaction but not the equilibrium analytical response may cause greater interference in this type of system.

With this type of system, the analysis of one solution may be completed in periods as short as 10 s. Since all solutions for analysis pass through the one mixing unit and flow-cell in the analytical-sensor unit, cross-contamination must be controlled by careful design of these units and by their flushing with each solution before measurements are made. Rates of analysis up to 360 solutions per hour have been claimed for such a system, but the approach has been applied only to determinands where solutions need to be mixed with only one reagent. Detailed investigation and application of this type of automatic analyser seems to be restricted mainly to the research group led by Malmstadt; the construction of this group's instrument has been described in detail,[69] and procedures for nitrate[70] and nitrite[71] in waters have been reported. As far as we know, however, this instrument is not commercially available, and we are not aware of any applications of this type of system to the routine analysis of waters. Several stopped-flow spectrophotometers are commercially available, and may be converted to automatic analytical systems; see, for example, ref. 72.

A fully automated, commercially available instrument which has similar features to those described above – as well as other capabilities – is described in refs. 73 and 74. The design philosophy behind this instrument seems rather different from that of Malmstadt's group, and the authors mention appreciably smaller rates of analysis in the range 20 to 60 samples per hour.

(iii) Parallel analysers   The distinguishing feature of this type of system is that all solutions for analysis are analysed simultaneously in parallel. Centrifugal systems represent the only fully automatic application of parallel analysis, though see the rapid-scanning system mentioned in *(ii)*(2) above.

Centrifugal systems were first described by Anderson.[75] They consist, in essence, of a circular plate within or on which are mounted a number of radial channels. One places into each channel a small volume of a solution for analysis (near the centre of the plate) and a small volume of an appropriate reagent (nearer the circumference of the plate); mixing of solution and reagent is prevented at this stage by a small lip built into each channel. The plate is mounted in a centrifuge and spun at a relatively slow rate until it reaches a desired temperature. It is then briefly spun rapidly so that centrifugal force causes the solutions for analysis to flow over the lips in the channels and into the reagent. Rapid deceleration of the plate rapidly mixes the solutions and reagent, and it is then spun again for a time sufficient to achieve the desired extent of reaction. During this latter period, the analytical response for each treated solution is measured each time it rotates past the analytical sensor units so that the rate of change of the analytical response or the equilibrium response for each solution can be calculated by an on-line computing system. Spectrophotometric measurement is most commonly used, but some systems also allow fluorometric measurement, and an investigation of the possibility of amperometric measurement has recently been reported.[76]

Centrifugal systems are rather complex mechanically and in relation to the measurement of analytical responses and calculation of results they are also limited to very simple forms of sample treatment. The number of channels on a plate varies between about 10 and 40, but – since the analyses from one plate can be completed in favourable cases within a few minutres, the repeated analysis of plates allows high rates of sample analysis. Some cross-contamination of adjacent solutions has been reported[77] though the size of this effect is usually unimportant. The rapid mixing of solutions and reagents as well as the ability to make rapidly many measurements of analytical response for each solution facilitates the use of procedures involving the measurement of reaction rate [see (*ii*)(3) above]. A number of centrifugal analysers are commercially available, and are used routinely in clinical laboratories; we know of no applications to routine water analysis. Two books[78,79] discuss centrifugal analysis in detail, and refs. 68 and 80 give two recent reviews of its application in clinical chemistry.

12.2.3 Applications of Automatic Analysers to Routine Analysis of Waters

(a) Purpose of this Section

From the discussion above, it is evident that, if a laboratory wishes, virtually any manual analytical procedure may be automated and applied to water analysis. There seems to us, therefore, little value in a comprehensive review of the literature in order to provide what would be a very long list of such applications, and we have not attempted such a compilation. Rather, we have tried to describe the current role of automatic analysis for determinands of general interest in laboratories concerned with the routine analysis of natural and treated waters. This approach faces the problems that, to some extent, our judgements are subjective and their accuracy is governed by our fields of experience. None the less, we have attempted this assessment in (*b*) and (*c*) below, and in reading these sections several points should be noted.

(1) The size, staffing, and sample throughput of a laboratory may all have an important effect on the extent to which automatic analytical techniques are justifiable [see Section 12.2.4(*c*)]. We have largely ignored such factors since our views are intended to represent general tendencies in water analysis rather than developments affecting all laboratories.

(2) Although many publications describe applications of automatic (or potentially automatic) analysers assembled by their authors, we have considered below only commercially available analysers since the latter seem to us generally of crucial importance for routine laboratories [see Section 12.2.4(*d*)].

(3) Given that a fully automatic version of a particular analyser is available, we have not distinguished below the use of semi-automatic and fully automatic systems provided that the former lacked only an automatic sample presentation unit and/or a calculation and print-out unit. This emphasis on full automation should not be construed as implying the view that it is always preferable to semi-automatic analysis; in some circumstances, it may well be that semi-automation is the better approach, and this decision must generally be made by each laboratory for each analytical requirement; see also Section 12.2.4.

A broad indication of the role played by automatic [see (3) above] analytical systems in modern water analysis would be provided by the proportion of determinations that are made in this way. We do not have precise figures for this proportion, which will in any event depend on the number and nature of samples and determinands that a laboratory must analyse. We would estimate, however, that of all determinations approximately 70 to 90% are made semi- or fully automatically in large, well equipped, routine laboratories dealing with samples of natural and treated waters. Of this proportion, there is a steadily increasing fraction of measurements that are completely automatic.

(b) Automatic Analysis without Sample Treatment
As noted in Section 12.2.2(*a*), rather few determinands of common interest in water analysis can be reliably measured without any form of sample treatment. Further, of these determinands, several are relatively unstable, and it is usually recommended that they be measured immediately after sampling, *e.g.*, dissolved oxygen, residual chlorine, pH (see Section 3.6). This requirement will usually preclude the use of automatic analytical systems. Thus, automatic analysis without sample treatment is not in common use for a wide range of determinands, but it is a very valuable approach that has found common application for those determinands for which it is suitable.

The approach is probably most commonly used for the determination of metals by flame atomic emission and absorption spectrometry since, for many purposes, direct injection of solutions for analysis is all that is required. Electrothermal atomisation may also be effected automatically, but in practice this technique commonly requires some form of sample treatment and/or the use of standard addition methods of calibration which may not always be possible in a fully automatic mode with commercially available equipment. Direct sample-injection is frequently appropriate also for plasma spectrometers (see Section 11.4.2), and laboratories employing this technique usually use automatic analytical systems; see, for example, ref. 81. The potential advantages of flow-injection systems [see Section 12.2.2(*b*)(*ii*)(1)] for transport of samples to flame and plasma spectrometric analysers should be noted; see refs. 30 to 34.

Although automatic spectrophotometers are commercially available, the only generally feasible use of direct measurement of the absorbances of solutions would be in measuring the colours of waters or their absorption in the ultraviolet. The latter is sometimes used as a means of on-line estimation of organic carbon [see Section 12.3.3(*a*)], but we know of no general use of this approach for the analysis of discrete samples in laboratories; automatic analysers for the direct measurement of organic carbon are readily available for laboratory-based analysis [see (*c*) below]. Automatic, electroanalytical measurements may also be made, *e.g.*, electrical conductivity and direct potentiometry using ion-selective electrodes. Although the former is sometimes effected automatically, we are not aware of any widespread use of this approach, and the direct measurement with ion-selective electrodes is not generally applicable since, for most determinands of common interest, sample-treatment is usually required before measurement. Of course, pH is one determinand where sample treatment is not generally required, but we have already noted the common need to measure pH directly after sampling. Automatic titrimetry without sample treatment may be and has been used for the measurement of alkalinity, and many automatic titrators are commercially available. Our impression, however, is that automatic measurement of this determinand is more commonly made spectrophotometrically after sample treatment as noted in (c) below. Chloride may also be determinable by automatic titrimetry, and a number of special purpose automatic analysers based on this approach are commercially available. Again, however, it appears that automatic spectrophotometric procedures are more commonly used; see (c) below.

To summarise, therefore, it seems that the most widespread and useful applications of automatic analysis without sample treatment concern the use of the various atomic emission and absorption techniques.

(c) Automatic Analysis with Sample Treatment

This type of analysis has found steadily increasing application over the last twenty years or so. In the first half of this period, the move towards routine automatic analysis was primarily concerned with inorganic determinands, but the rapidly growing interest in organic compounds since then has led to increasing attention being devoted to their automatic measurement. It is our impression that, up to the present, a very large proportion of routine applications have involved the use of automatic analysers capable of many different types of sample-treatment; such equipment has been applied mainly, but not exclusively, to inorganic determinands. In recent years, however, increasing numbers of instruments intended essentially to automate just one type of sample-treatment have become available, and their use is growing, particularly for organic determinands. We describe these two types of instrument as 'general-purpose' and 'special-purpose', respectively, and note their applications in (i) and (ii) below.

(i) General-purpose analysers Most applications have involved the use of gas-segmented, continuous-flow systems, and this is probably because of the early availability of suitable and reliable equipment coupled with the versatility of this type of system for automating a wide range of sample treatment procedures. Single-, dual-, and multi-channel analysers have found widespread use for many determinands. Although various types of analytical-sensor unit can be and are used, e.g., flame-photometric for sodium and potassium, potentiometric using a fluoride ion-selective electrode for fluoride, spectrophotometric measurement has been by far the most commonly used approach. Commonly measured determinands include alkalinity, ammonia, calcium, chloride, cyanide, fluoride, magnesium, nitrate, nitrite, phenolic compounds, phosphate, silicate, sulphate, surfactants (anionic), and total hardness, but many other substances can be determined. Spectrophotometric procedures for many trace metals are also available and sometimes used, but they have tended to be replaced by atomic absorption procedures. In general, since virtually any type of sample treatment may be automated by gas-segmented, continuous-flow systems, most determinands of interest can be determined automatically if desired, particularly given the availability of automatic forms of virtually all the relevant measuring instruments. This type of automatic analysis has become so well established that such procedures for a number of determinands are included in many collections of 'Standard/Recommended' methods; see, for example, refs. 10, 38 and 83 to 91. There has been general agreement that the appropriate use of such automatic systems [see also Section 12.2.4(c)] provides, in comparison with manual analysis, both substantially

greater sample throughputs for a given amount of man-power and also reduced costs per analysis. A number of manufacturers produce automatic analysers of this type, and can provide details of the procedures for particular determinands.

Within the last five years or so, the increasing development, experience, and commercial availability of other types of general-purpose automatic analyser coupled with growing emphasis on the cost-effectiveness of analytical laboratories had led to increasing interest in the potential advantages of these other types. For example, the versatility of gas-segmented, continuous-flow analysers is of no direct benefit for determinands measurable spectrophotometrically after the simple addition of one or a few reagents. Less versatile analysers may be equally applicable to such procedures, and may also allow more rapid and/or cheaper analyses. Although many of the types of automatic system noted in Section 12.2.2 are potentially applicable to water analysis, we are aware of only two of those types which are finding routine application at present: (1) discrete-sample, 'wet-reagent' systems, and (2) flow-injection analysis systems. For example, a large Water Authority laboratory in the United Kingdom has recently reported[92] its adoption for routine analysis of an analyser of type (1) in place of an air-segmented, continuous-flow system. The analyser measures ammonia, total oxidised nitrogen, chloride, phosphate, sulphate, calcium, magnesium, total hardness, silicate, and nitrite, and the authors quote the results from a detailed evaluation of its performance. They concluded that this analyser showed a number of important advantages for the quoted determinands, but also noted its inability to measure determinands requiring more complex, sample treatment procedures, *e.g.*, surfactants require solvent-extraction, cyanide and phenol require distillation. Other comparisons of type (1) analysers with gas-segmented, continuous-flow systems are reported in refs. 93 to 95. Similarly, the literature contains reports of the adoption of flow-injection analysis for the routine measurement of a number of determinands; see, for example, the reviews in refs. 26 and 96. Given that appropriate equipment and procedures for this technique have become available only within the last few years, it seems likely that it will find increasing application.

In view of these developments, we feel that the use of general-purpose automatic analysers in water analysis is now at a transitional phase in which the previously almost complete reliance on gas-segmented, continuous-flow analysis is being abandoned. Ideally, a critical, comparative assessment of all the many potentially applicable types of automatic analyser is required in order to identify the optimal systems for particular purposes. Such an assessment, however, faces several important difficulties.

(1) Technological developments and competition among the manufacturers of instruments lead to quite frequent and important changes in design and performance of automatic analysers.

(2) There are great practical difficulties – because of the number of relevant analytical systems and the amount of time and effort involved – in properly assessing the detailed performance of different analysers with respect to factors such as accuracy, speed, cost, reliability, and amount of maintenance required.

(3) The relative advantages and disadvantages of different analysers depend on factors such as the number and nature of samples to be analysed, the

determinands to be measured, and the accuracy required of analytical results. Such factors tend to vary from laboratory to laboratory.

We do not know of any such comprehensive assessment, and, indeed, it would be a formidable task to attempt one. It seems likely, therefore, that individual laboratories will have to make the best appraisal possible of available equipment. The experiences gained from the use of a variety of systems – and the detailed publication of any evaluations of particular analysers as in ref. 92 – should, however, help to provide a body of knowledge that will allow firmer guidelines on the choice of optimal systems in the future; see also Section 12.2.4.

(ii) Special-purpose analysers Many analysers of this type are commercially available; some are intended for the measurement of just one determinand, while others may be used for a number of determinands all measurable by the one sample treatment procedure. The latter type is commonly though not always based on the use of chromatographic procedures, and the distinction between this type and relatively unversatile, general-purpose analysers is somewhat arbitrary and indistinct. Of the many available analysers, our impression is that rather few play a generally important role in water analysis at present. We shall, therefore, be very selective in the following remarks.

One of the most important developments has been the commercial availability of a number of analysers for measuring the organic carbon content of waters. Various sample treatment procedures and types of analytical sensor are used in these instruments, and refs. 97 and 98 provide useful reviews of such systems. All involve the treatment of samples to convert organic carbon to carbon dioxide, and our impression is that the use of u.v. irradiation of solutions to effect this conversion is finding increasing application, probably because it is a simpler process to automate. A recent review of this oxidation procedure is given in ref. 99, and recent evaluations of such instruments for water analysis have been reported.[100-103] A number of precautions must be observed in applying these analysers – for these, see references cited – but they have found very widespread use in water analysis. It is also worth mentioning here that a few manufacturers supply automatic analysers for a related determinand – the total oxygen demand of a water. These instruments seem not to have found such common use, but they may be of considerable value for some laboratories. Reviews are provided again in refs. 97 and 98, and see also refs. 104 and 105 for recent developments and applications. Under favourable circumstances, the correlation between determinands such as organic carbon or total oxygen demand and biochemical or chemical oxygen demands may be sufficiently good that direct measurement of the latter two may be replaced – with gain of analytical time and effort – by automatic measurement of one or other of the former two. Of course, careful consideration is necessary before adopting such an approach, and an extended period of measurement will generally be required to establish the correlation; the approach is, however, well worth consideration.

Over the last decade or so there has been a great growth of interest in the concentrations of many organic compounds in waters, *e.g.*, pesticides, herbicides, polynuclear aromatic hydrocarbons, phenolic compounds, haloforms and other chlorinated hydrocarbons. The identification and quantitative measurement of such substances is nearly always based on the use of chromatographic procedures in which appropriate analytical sensors (detectors) are used, and on preliminary concentration and separation procedures before the chromatographic stage (see Section 11.3.8). Although fully automatic gas-chromatographs and high-performance liquid chromatographs are available from many manufacturers, the relative complexity of the preliminary concentration and separation procedures commonly required makes reliable automation rather cumbersome and difficult. Few fully automatic systems for particular determinands are commercially available, and fully automatic analysis seems not to be in common use at present. In this position, we simply mention below some possibilities that seem worth noting. Of course, automatic analysis is readily achieved in principle whenever direct injection of a portion of a sample onto the chromatographic column is feasible. Such procedures are used in gas chromatography, *e.g.*, for phenolic compounds[82] and haloforms,[106] but may be subject to various problems including insufficiently low detection limits caused by the need to restrict the volume of sample applied to the column. Head-space analysis (see Section 11.3.8) overcomes such problems for suitably volatile compounds, and automatic systems – for volatile halogenated hydrocarbons – based on this technique are supplied by a number of manufacturers of gas-chromatographic equipment. Such systems allow up to 40 samples to be analysed automatically for a number of determinands. HPLC is also potentially applicable for many organic compounds, and an instrument based on this technique has recently become available for the determination of phenolic compounds in waters;[107] see also the review of column-switching techniques in ref. 108.

Another column-chromatographic procedure that should be mentioned is ion-chromatography [see Section 11.3.8(*g*)]. Although many of the reported applications of this technique have not been fully automatic, such equipment is available. In the context of routine analysis of waters, the most useful potential application is the determination of anions such as nitrite, nitrate, chloride, bromide, fluoride, sulphate and phosphate;[109] under favourable circumstances, many or all of these determinands can be measured in one small portion of sample – often to concentrations in the μg l^{-1} range. Again, our impression is that the routine, automatic measurement of such anions more commonly involves general-purpose analysers and spectrophotometric procedures, but the potential value of ion chromatography for appropriate applications is worth bearing in mind.

The determination of trace metals is a very important task in many laboratories. As mentioned above, automatic analytical systems for this purpose are usually based on atomic emission or absorption techniques. In general, atomic-absorption systems measure only one determinand at a time, and though plasma emission spectrometers can measure many metals simultaneously, the equipment required is relatively very expensive. A number of electroanalytical techniques (*e.g.*, polarography and its variants, stripping voltammetry and potentiometry and their variants – see Section 11.3.7) are capable of measuring several metals of common interest (*e.g.*, lead, cadmium, copper and zinc) essentially simultaneously on one portion of a sample. The literature has for a number of years contained a succession of papers reporting automatic and semi-automatic electroanalytical procedures for such purposes, and a variety of appropriate equipment is commercially available and is substantially cheaper than plasma spectrometers. It is our impression, however, that the various electroanalytical techniques are rarely used for the routine, automatic analysis of waters. For those wishing to read more of these techniques. refs. 43 and 44 provide comprehensive reviews of automatic electroanalytical systems, and see also refs. 110 to 113 for recent developments.

Finally, mention should be made of the general approach whereby an automatic measuring instrument is converted – permanently or only as required – to a special-purpose, automatic analyser by the addition of appropriate, commercially available sample presentation, treatment, and transport units. This approach has been commonly used for a number of trace elements for which special sample treatment procedures are required before measurement. The most common applications are probably for mercury [by cold-vapour atomic absorption or fluorescence – see Sections 11.3.5(e) and 1.4.3(b)], and for arsenic and selenium[114] [by hydride formation and atomic absorption or emission – see Sections 11.3.5(d) and 11.4.2].

12.2.4 General Aspects of Automatic Analysis

(a) Purpose of this Section
Whatever the precise analytical requirements of a laboratory, several considerations are generally important in: (1) assessing whether or not automatic analysis represents the optimal approach, (2) selecting the optimal automatic analytical system, and (3) evaluating the performance of, and routinely applying, automatic systems. Many of the relevant points have already been dealt with in previous sections of this book where we discussed analytical systems in general, but we think it useful to note here a number of points that particularly concern automatic systems. The various considerations have, for simplicity, been grouped under a number of headings in (b) to (h) below, but we would stress the great inter-relatedness of these aspects.

(b) Definition of Analytical Requirements
As already emphasised in Sections 2 and 12.1.12, an essential preliminary to considering the suitability of any analytical system(s) is a precise specification of all important requirements for the analyses of concern. Thus, one should firmly establish factors such as the number and nature of samples, the determinands, the maximum tolerable bias and standard deviation of analytical results, and the maximum tolerable time between receipt of samples and reporting of analytical results. This definition of requirements is discussed in Section 2, and is not further considered here.

(c) Justification of Automatic Analysis
For brevity, we consider only the comparison of manual and automatic analysis, but this does not imply that semi-automatic analysis is without value; the general considerations that follow may also be applied to this type of analysis.

In considering the replacement of a manual by an automatic analytical system, the cost of the equipment for the latter must be balanced against any benefits stemming from automatic analysis. Many authors have discussed the various benefits that may be achieved; see, for example, refs. 4, 12, 16, 24, 25, 115 and 116.

Four main benefits are most commonly quoted as potential justifications for automatic analysis:

(1) the number of samples that can be analysed by a given number of analysts is increased – see (*i*) below,
(2) the period between receipt of samples and reporting of analytical results is decreased – see (*i*) below,
(3) the cost of analytical results is decreased – see (*ii*) below,
(4) the precision of analytical results is improved – see (*iii*) below.

Other types of benefits may be important; for example, a laboratory making extensive use of automatic analysis may achieve prestige[12] or save laboratory space.[117] In the following, however, we consider only the benefits (1) to (4).

The use of automatic analytical systems may, of course, introduce problems as well as benefits. The natures of potential problems are discussed in Sections 10.3.4, 10.4.2, 10.4.3, and in (*iii*) below.

(i) Time taken for analysis Two aspects are generally important:

(1) the total time, T_t, elapsing between the beginning of the analysis of n samples and the production of the corresponding analytical results,
(2) the time, T_1, during that period for which an analyst's attention is required.

Both T_t and T_1 may, of course, be estimated for any analytical system by detailed consideration of all the activities involved in the analysis of n samples. Such estimates may then be used to compare manual and automatic analytical systems. A simple, mathematical model may also be helpful for more general consideration of such comparisons. Thus, both T_t and T_1 may, to a first approximation, be divided into two components, one being independent of the number of samples analysed on any one occasion when an analytical system is applied and the other being directly proportional to the number of samples analysed on that occasion:

$$T_t = \alpha_t + \beta_t n \tag{12.1}$$

and

$$T_1 = \alpha_1 + \beta_1 n \tag{12.2}$$

where α_t, α_1, β_t and β_1 are constants for a given analytical system*.

As an example of the use of this model, consider the use of a manual (denoted by subscript m) and an automatic (subscript a) system for measuring the same determinand. When $T_{tm} = T_{ta}$, it follows from equation (12.1) that the number of samples, n_c, is given by

$$n_c = (\alpha_{ta} - \alpha_{tm})/(\beta_{tm} - \beta_{ta}) \tag{12.3}$$

Thus, when the number of samples to be analysed, n, is smaller than n_c, manual analysis will provide the analytical results more quickly while the converse applies when $n > n_c$. A similar conclusion can be derived from equation (12.2) for the analyst time. Thus, there is, in general, some minimal number of samples, n_c, that

---

* In practice, α may be a stepwise function of n as a result, for example, of the need to re-calibrate an analytical system after a given number of samples have been analysed. For present purposes, this complication may be ignored.

must be analysed on one occasion if automatic analysis is to involve less total or analyst time than manual analysis. For a given manual system, the values of α_{ta}, α_{1a}, β_{ta} and β_{1a} for a given automatic analyser are, therefore, important in determining the value of n_c. Estimation of the values of these parameters for different automatic systems will allow their relative advantages and disadvantages with respect to total and analyst times to be assessed. In this connection, one should note the potential importance of the time taken for an automatic analyser to reach a sufficiently stable response that the analysis of standards and samples may be started. This time has been reported to vary substantially from one type of analyser to another, and may also be much greater than for equipment used in manual analytical systems. This stabilisation period may, however, have little if any effect on α_1; it is α_t that is generally affected. Equally if special procedures are required to shut-down an automatic analyser after the samples have been analysed, the α values will, in general, be affected.

Given the possibility that at least n_c samples must be analysed on each analytical occasion if automatic analysis is to be less time demanding, the analysis of samples less frequently but in larger sets may be helpful in this respect. The feasibility of this approach depends, of course, on the stability of the determinand and on the maximum tolerable period between sample receipt and reporting of results.

The above considerations were concerned with only one determinand; they are easily extended to a number of determinands using the appropriate forms of equations (12.1) and (12.2) for each determinand. For example, as would be expected, the value of n_c tends to become smaller as the number of determinands measured simultaneously by an automatic system increases. If, on the other hand, the same, general-purpose, automatic analyser is to be used sequentially for the determinands measured one at a time, the change-over time between one determinand and another may become an additional, important factor affecting the values of α_a for the different determinands. This change-over time – as it increases in relation to the time during which samples are analysed – may also have important effects on the number of determinands that can be analysed in a given period such as a working day.

The above discussion indicates the nature of the considerations involved in considering the analytical times of different analytical systems. Clearly, the dependences of T_t and T_1 on n are important characteristics of the relevant systems. A related characteristic is the number of samples that can be analysed for given determinands by a given amount of man-power. This parameter reflects to some extent the efficiency of an analytical system or systems, and is equal to n/T_1. When equation (12.2) is valid, $n/T_1 = 1/[\alpha_1/n) + \beta_1]$, so that the efficiency in this sense depends on α_1, β_1 and n. It should be noted, however, that n/T_1 is affected by two broad factors: (1) the inherent efficiency of an analytical system, *i.e.*, the smallest achievable values of α_1 and β_1 for a system, and (2) the efficiency with which that system is operated. Thus, an efficient analytical system may be operated inefficiently, and *vice versa*. Care may be necessary, therefore, in comparing the reported efficiencies of analytical systems or laboratories.

(ii) Cost of analysis Worth[118] has stated 'As capital and running costs of instruments increase it becomes more important to know the relative cost of

carrying out analyses by alternative instrumentation before purchasing new equipment'. One indication of the increasing attention being paid to the costing of analysis is the Symposium on this topic at a recent, international congress on automation in clinical chemistry. The papers from that Symposium[119] provide a useful introduction to the variety of topics involved, and we note below only a few points that seem to us especially worth mention.

The calculation of the total costs of the production of analytical results is rather involved because of the number of contributing items. Some items, *e.g.*, reagents, can be relatively easily costed, but others require prior decisions on factors such as amortisation period of equipment, costing of man-power, the extent to which general laboratory overheads are included. Such decisions may well vary from author to author, and comparison of any reported values for analytical costs should be made with care. Very detailed examples of costing in a clinical laboratory have been described by Worth[118] who also compared the costs of a number of automatic analysers. Some data on the relative costs of manual and gas-segmented, continuous-flow, automatic analysis in water analysis laboratories are provided in refs. 120 to 122.

Costing may be simplified when a laboratory wishes only to consider the relative costs of two or more analytical systems since, under these conditions, certain costs such as general laboratory overheads may reasonably be ignored in that they will be approximately the same for any analytical system. Foreman and Stockwell[24] describe a very simple approach to such relative costing that is valid when the dominant component of cost is analyst time.

Several authors – see, for example, refs. 115 and 116 – have described a general model in which the cost, C, of the analysis of n samples on one occasion is divided into components: (1) fixed-costs independent of n, and (2) costs directly proportional to n, *i.e.*,

$$C = \gamma + \delta n, \tag{12.4}$$

where γ and δ are constants for a given analytical system. This model is of exactly the same nature as that mentioned in (*i*) above, and similar conclusions may be derived from it. For example, a minimum number of samples must be analysed on any one occasion if C_a is to be less than C_m. Note, however, that γ and δ now depend, in general, on the number of analytical occasions during the life of an automatic analyser since its depreciation must be allowed for in the values of γ and δ. Thus, in addition to the minimum number of samples on each analytical occasion, there is a general need for a minimum number of analytical occasions if automatic analysis is to be cheaper than manual. In other words, some minimal sample-throughput must be maintained to justify the cost of automatic analytical equipment.[116] These minimum numbers depend on γ and δ, and vary, in general, from one type of automatic analyser to another. Again, therefore, estimation of the values of γ and δ allows conclusions to be reached on the relative suitabilities of different types of automatic analyser for particular purposes.

(*iii*) *Accuracy of analytical results*   Many authors have reported that the inherent reproducibility with which analytical processes are effected by automatic analysers leads to more precise analytical results than those achieved routinely by manual analytical systems. This better precision may well also provide lower limits of detection (see Section 8.3.1). Although we would generally agree with such views, there are at least three reasons why a particular automatic system may give less precise results than the corresponding manual analyses.

(1) The concentration range of an automatic system may be greater than that of the manual system, and this may substantially worsen the discrimination and precision of the former (see Section 4.3.1).
(2) Automatic analytical systems commonly employ much smaller volumes of samples than manual systems. Whenever the determinand is not distributed homogeneously throughout each sample, automatic systems may, therefore, lead to less precise results [see Section 3.5.3(*c*)(3)].
(3) The calibration parameters of an automatic system may be subject to greater drift [see Section 5.4.3(*d*)(*ii*)] than those of a manual system. Provided that the drift is reasonably uniform, this difficulty can generally be overcome by adjustment of the frequency with which the parameters are estimated, but this, of course, increases the times and cost of automatic analysis.

For these reasons (and see also below), we would suggest caution in regarding the precision of analytical results as necessarily improved by the adoption of automatic analysis.

Even when the precision of automatic analysis is better than that of manual, it might be argued that – so long as both approaches achieve the required precision – this does not justify the use of automatic analysis. We have seen in Sections 5 and 6.4, however, that the estimation of bias and routine analytical quality control are both facilitated as precision improves. Thus, improved precision is, in general, a further justification for automatic analysis.

Of course, the accuracy of analytical results is governed not only by precision but also by bias. Care is, therefore, essential to ensure that any bias that may be introduced into analytical results by an otherwise suitable analytical system is tolerably small. General sources of bias and their control and estimation have been considered in Section 5, and the discussion there applies equally to automatic systems. Three general points are, however, worth emphasis for automatic analysis.

(1) The solutions for analysis in some automatic analysers may be exposed to possible sources of contamination (*e.g.*, the laboratory atmosphere, the sample containers) for longer periods than in manual analysis. Appropriate precautions against such effects should be taken when necessary (see Section 10.8).
(2) When analysis involves treatment of the samples to convert all forms of the determinand to a measurable form (see Section 5.4.5), the difficulty of automating some treatments of this type may lead to a potential source of bias. It may be impracticable to avoid this error if automatic analysis is used, and this is a very important consideration in deciding the suitability of automatic systems.
(3) We would also note here the general comments on the use of 'black-box' analytical systems made in Section 10.4.3.

(d) Choice of Appropriate Automatic Analytical Systems
In principle, this choice is straightforward. The analytical requirements [see (*b*)

above] and any other important needs (*e.g.*, capital cost, running cost, analyst time) are compared with the performance achievable by all relevant automatic analysers, and the most suitable system meeting the analytical requirements is selected. Of course, the optimal instrument for a given purpose may vary from laboratory to laboratory depending on factors such as the rate of sample throughput and the relative weights attached to capital cost, running cost, and analyst time. In practice, however, this simple approach commonly meets the problem that all the relevant performance characteristics of all the relevant automatic analysers may not be known. This difficulty has been recognised for some time in clinical chemistry, and there has been considerable activity in that field in developing and applying standardised schemes for evaluating the performance of the relevant instruments [see (*e*) below]; we are aware of no such organised approach in the field of water analysis.

In this situation, we can only suggest that, for the analysis of waters, the primary aim should be to attempt to identify analysers capable of meeting the requirements of the user of analytical results (see Section 9), and thereafter for each laboratory to make the best choice possible given any other requirements on costs, analyst time, *etc.* As the performances of different automatic analysers become better characterised for water analysis, the selection of the optimal system for a particular purpose should be facilitated. A number of general suggestions concerning the selection of instruments have been made in Section 10.4.3(*d*), and a collection of papers[17] on 'Decision criteria for the selection of analytical instruments used in clinical chemistry' also provides helpful ideas and suggestions.

The above comments refer only to commercially available automatic analysers. There is, of course, the possibility of a laboratory designing and constructing its own automatic analyser for a particular purpose, and the attractive features of this approach are noted in refs. 4 and 16 and discussed in detail in ref. 123. These references also stress the considerable resources necessary if this approach is adopted; the costs involved will generally favour the use of a commercially available instrument whenever it is capable of meeting the analytical requirements. We have, therefore, concentrated our attention on commercial equipment. The intermediate approach – purchase and assembly of a series of individual units to form an automatic analyser – may often be useful, and is subject to essentially the same considerations as for complete automatic analysers.

(e) Definition and Evaluation of Performance Characteristics of Automatic Analytical Systems

In Section 9.4 we described what we mean by the term 'performance characteristics of an analytical system'. The points made in Section 9 concerning the selection of analytical systems are equally applicable to automatic and manual systems. This selection is facilitated when quantitative estimates of performance characteristics concerning the precision and bias of analytical results achievable by the relevant analytical systems are available. Such information must also be obtained by any laboratory when – having selected an analytical system – it estimates the precision and bias achieved in routine analysis (see Sections 4.7, 5.1, 6.1 and 6.3). For both purposes, therefore, the estimation of such performance characteristics is required, and the procedures described in Sections 5 and 6 are generally applicable to both manual and also automatic analytical systems. The particular nature of automatic systems, however, also means that the consideration of their relative suitabilities is greatly aided by quantitative data on a number of other characteristics of their operation. The questions of: (1) the performance characteristics of automatic analytical systems that should generally be estimated, and (2) procedures to be used for such estimations have been discussed by many authors in the field of clinical chemistry; see, for example, refs. 124 to 130. The United States National Committee for Clinical Laboratory Standards* has also produced a number of draft proposals (PSEP-2, -3 and -4) for evaluation procedures that have been widely used and discussed; see also ref. 131 on work in this field by other clinical organisations. Given these publications and our own discussions in previous sections, we have simply noted below a number of points that seem to us especially relevant to the publication of performance characteristics of automatic systems.

(i) Definition of the analytical system   If published data are to meet their intended purposes, they should clearly refer unambiguously to a particular analytical system. For some automatic systems, it may suffice to quote the manufacturer and name of an instrument. Often, however, more detailed information will be required, *e.g.*, details of the analytical procedures that have been automated, rate of analysis of samples, frequency and nature of calibration, *etc.*

(ii) Precision of analytical results   As noted in Section 6.2, repeated analysis of one or more solutions is necessary to estimate the precision of analytical results. We also stressed in the same section the need to define criteria to be used in deciding how the observed variations in results are to be assigned to random and systematic errors, precision being characterised only by the former. Our suggestions for such criteria are also given in Section 6.2, but it must be noted that other practices and terminologies are used (see refs. 124 to 130). Thus, as noted in Section 9.2.3(*b*)(*i*), in publishing data on precision and in interpreting such data, it is essential to make very clear the sources of random error included in the quoted data for precision.

We discussed in Section 6.3 the various factors to be considered in planning tests to estimate precision, and these remain valid for automatic analytical systems. In view, however, of the relative ease of making large numbers of analyses with such systems, we would generally suggest the approach of Section 6.3.5. Note also that the tests often adopted in clinical laboratories for estimating precision involve batches of analyses on each of twenty days.

(iii) Bias of analytical results   The considerations in Section 5 are generally valid. In the clinical field, it appears to be usual to include in evaluation tests a detailed comparison of results from the automatic analytical system of interest and another analytical procedure. The problems of this approach are considered in Section 5.3 (see also Section 5.2), and have also been noted by clinical workers. We generally prefer the approach described in Section 5.4 for the estimation of bias.

(iv) Cross-contamination between successive solutions   We saw in Section 12.2.2

* 771 E. Lancaster Avenue, Villanova, PA 19085, U.S.A.

that this effect – also commonly referred to as 'carry-over' – may affect the accuracy of analytical results and limit the rate at which solutions can be analysed. Since the liability to this effect varies from one type of automatic analyser to another, the magnitude of the effect is an important aspect of such analysers. As many authors have recommended, therefore, estimation of the magnitude of cross-contamination should generally be included in evaluations of automatic analytical systems. An approach very often adopted for this purpose is to analyse in succession three portions of a solution with a high concentration of the determinand followed by three portions of a low-concentration solution.[125] Note, however that Dixon[132] has suggested the need to modify the usual equation employed to calculate cross-contamination from the results obtained. Further, as noted in Section 12.2.2, a given analyser may include more than one source of cross-contamination, and Broughton *et al.*[125] have noted the possible need to characterise the magnitude of cross-contamination from each such source separately.

(v) Analytical times   In view of the general importance of the total time, T_t, and analyst time, T_1 required to analyse a number of samples [see (c)(i) above], we suggest that both these times should be estimated and quoted as performance characteristics. Whenever possible, both T_t and T_1 should be expressed as a function of n using equations (12.1) and (12.2) or other more appropriate functions, and both T_t and T_1 should be on the basis that the instrument has to be switched on and, if necessary, achieve adequate stability before analysis of solutions begins; any times involved in the shut-down of the instrument should also be included. If the initial stabilisation and shut-down periods are appreciable, their equivalent values of T_t and T_1 should be quoted separately. For analysers intended to deal sequentially with a number of determinands, the values of T_t and T_1 for any change-over operations should also be quoted.

Published evaluations seldom give any quantitative data on these aspects, but the need for such data seems to us to be becoming greater and greater.

(vi) Analytical costs   Again, the cost aspects of the use of automatic analytical systems are generally important [see (c)(ii) above]. Ten years ago Broughton *et al.*[125] noted the importance of costs of automatic analysis, and recommended that evaluations should include the estimation of the direct costs attributable to the use of a particular analyser; indirect costs were excluded because of problems arising from the different methods used by various organisations for assigning such costs. Their paper gives some examples of costing procedures, and also notes the possible dependence of costs on the work-load of an analyser. In addition, however, to the problems noted by those authors it seems to us that the financial costs face other difficulties if they are to be used as a performance characteristic. For example, the relative costs of analyst time, consumables (*e.g.*, reagents) and services (*e.g.*, power) may vary from laboratory to laboratory and country to country. For this reason, the simplest and best approach would seem to be to estimate and quote the amounts of all resources required to operate an analyser in the analysis of n samples. Given this information, any laboratory may convert such resources to equivalent costs using their own cost-factors.

(vii) Availability and maintenance of automatic analytical systems Clearly, lack of availability of an expensive instrument through its malfunction or any other reason is highly undesirable. Similarly, the amount and nature of the maintenance required in attempting to prevent such unavailability is of interest. As Broughton *et al.*[125] recommended, therefore, it is useful to give as quantitative an account as possible of both these aspects when attempting to characterise the performance of any automatic analytical system.

(f) Application of Automatic Systems to Routine Analysis
Having selected and installed an automatic analyser, preliminary tests are generally required to ensure that the precision and bias of analytical results meet the necessary values. Principles and procedures for these tests are considered in detail in Sections 5, 6.3, and (*e*) above. If the precision and bias are found to be satisfactory, the routine analysis of samples may be started, and for such analysis it is essential to maintain a regular programme of special tests to ensure the rapid detection of any deterioration in precision and/or bias. Principles and procedures for this type of error estimation are discussed in detail in Section 6.4, and see also Section 10.4.3(*f*).

12.3 ON-LINE ANALYSIS

12.3.1 Introduction

The last twenty years or so have seen a large increase in the use of on-line analysis in the measurement and control of the quality of many types of natural and treated waters. In this period, many instruments and associated techniques have been developed and applied with varying degrees of success, a voluminous but rather scattered literature has evolved, and many problems have been encountered. The severity of some of these problems depends on the particular type of water, but in our view the general approach to, applicability of, and types of problem likely to be met in on-line analysis are almost entirely independent of the nature of the water of interest. The following discussion, therefore, is broadly intended to apply to all types of waters; points specially concerning particular types are made as appropriate.

It should be noted that several other terms, *e.g.*, 'automatic water-quality monitoring', 'continuous water-quality monitoring', 'process-stream analysis' are often used by other authors with the same sense as 'on-line analysis' as employed here (see Section 12.1.1). Further, in the literature with which we are familiar, greatest attention is paid to the on-line analysis of natural, surface waters. A high proportion of the references cited below concern such waters, but we believe that they are generally relevant to other types of waters.

The middle to late 1960s saw a rather uncontrolled enthusiasm to apply on-line analysis, and the many problems encountered then often led to some disenchantment. By the early to middle 1970s the true natures of most of these problems had been appreciated, and a number had been overcome or reduced to tolerable proportions. Since then, although new instruments have appeared, the period seems to us to have been largely one of consolidation so far as routine application of on-line analysis is concerned, emphasis being placed on achieving greater reliability in the operation of on-line equipment. Refs. 133 to 142 provide useful reviews on the progress of on-line analysis and of the main problems encountered. It is interesting to note that in a review[133] of the topic made at the end of the 1970s for the water industry of the United Kingdom, it was concluded that the greatest need was for engineering-type development work to improve the reliability of existing instrumentation. This conclusion was agreed[134] by a Working Party established by the Department of the Environment and the National Water Council*. With these points in mind, our main aim here is to attempt a critical summary of generally important points concerning the applicability of on-line analysis (Section 12.3.4). First, however, we briefly consider the main types of on-line analytical instrumentation in Section 12.3.2, and common applications in water analysis are noted in Section 12.3.3.

12.3.2 Types of On-line Analysers

The equipment involved in on-line analytical systems may be regarded as composed of a number of units, each with a particular function as shown in Table 12.2. Of course, the physical realisation of these functions is achieved in a variety of ways, one physical unit may serve more than one function, and some equipment may not include all functions. The scheme of Table 12.2 is, however, useful for present purposes. For example, comparison of Tables 12.1 and 12.2 shows the close similarity in many functional respects between automatic and on-line analysers. In principle, therefore, most automatic analysers intended for the analysis of discrete samples in a laboratory may be applied to on-line analysis by provision of suitable sampling lines and sample-conditioning and -presentation units. In practice, however, such adaptation is generally much simpler for continuous-flow than for discrete-sample or parallel analysers.

Many on-line analytical systems have been described and applied, and the measurement of many determinands has been accomplished. We know of few texts that deal comprehensively with all these many systems. Refs. 29 and 143 deal with many on-line analysers for various applications including water analysis, but they are both somewhat out-dated. Ref. 144 is the first of a multi-volume work on the topic; it deals only with general aspects of on-line analysis, but the subsequent volumes will describe and discuss the many instruments. Various approaches to the classification of on-line analysers can be followed; see, for example, refs. 29 and 144. In (*a*) to (*e*) below, we simply note five features of such instruments that are useful in classifying them into various types, but we do not attempt any rigid scheme of classification.

(*a*) Presence or Absence of Sample Treatment

At present, although many determinands can be measured in waters when sample treatment is employed, rather few can be measured on-line without such treatment. Experience indicates that analysers not involving sample treatment tend to be the

---

* This Working Party has been disbanded, but a new group has been established – the Water Industry Steering Group on Instrumentation, Control and Automation. On-line analysis of waters forms only one component of this group's activities, and further information on the work of the Steering Group can be obtained from: The Secretary, ICA Steering Group, WRc, Frankland Road, Blagrove, Swindon, Wilts.

TABLE 12.2

Equipment Components of an On-line Sampling and Analytical System

Component	Function	Examples
1. Sampling line	To convey a portion of the water of interest from the sampling point to the on-line analyser	See Section 3.5
2. Sample conditioning unit	To ensure that the properties of the sample entering unit 3 do not cause deterioration in the performance of units 3 to 6	Heat-exchangers to control temperature, filters to remove undissolved materials
3. Sample presentation unit	To select appropriate volumes of sample or to ensure a suitable flow-rate of sample to be presented to the following units	Automatic pipettes, peristaltic pumps
4. Sample transport unit	To convey the sample from unit 3 through units 5 and 6 and thence to waste	Peristaltic pump and connecting conduits
5. Sample treatment unit	To effect any chemical and physical processes required before measurement in unit 6	Reagent addition, heating the sample, gas stripping
6. Analytical-sensor unit	To provide a response (usually an electrical signal) whose magnitude is governed by the concentration of the determinand in the sample	Automatic spectrophotometer, automatic voltmeter with ion-selective electrode
7. Local calculation and read-out unit	To provide a print-out of analytical results and/or some other indication of the concentration of the determinand: installed at the same position as units 3 to 6	Chart-recorder, galvanometer, automatic printer
8. Remote calculation and read-out unit	To provide an indication of the concentration of the determinand and/or a print-out of analytical results: installed at a position remote from units 3 to 7	Computer/printer, chart-recorder

most reliable, and most suitable, therefore, for unattended operation. In general, as the complexity of sample treatment procedures increases so the reliability tends to decrease. Thus, adequate reliability is commonly also achieved when treatment involves, for example, the simple addition of a reagent solution to the sample. Procedures for adding certain reagents without the need for preparation, storage, and dispensing of reagent solutions are also being developed[145] to improve still further the reliability of analysers. Of course, the reliability of an analytical system may be affected by components other than sample treatment, *e.g.*, the sample conditioning and analytical-sensor units may be crucial in some applications. Further, what is considered acceptable reliability is governed by the period of unattended operation that is required of an on-line analyser. Each application clearly requires individual consideration, but it seems reasonable to conclude that – other things being equal – analysers involving the minimal amount of sample treatment are generally to be preferred.[141,146-149]

(b) Types of Analytical-sensor Unit

Many types of analytical sensor have been used for the on-line analysis of waters, and applications of virtually all the measurement techniques described in Section 11 are to be found in the literature and commercially available instrumentation. From the viewpoints of simplicity and reliability, two broad types of sensor seem generally preferable:

(1) Probe-type sensors in which the sensing element is placed directly in the sample, and which produces an electrical signal whose size is governed by the determinand concentration, *e.g.*, ion-selective electrodes.
(2) Sensors which measure some property of the sample when it is held within a cell suitably positioned with respect to the sensor, *e.g.*, spectrophotometric measurement.

Given also the points made in (*a*), sensors requiring no, or the minimal amount of, sample treatment are of special value, and act as the basis of many of the most commonly applied and reliable analysers. Thus, there is widespread use of electrical conductivity cells, amperometric membrane-electrodes [see Section 11.3.9(*c*)], and ion-selective electrodes (see Section 11.3.6), and new sensors of this nature continue to be studied and developed; see, for example, refs. 146, 147, 150 to 152. Nephelometric or spectrophotometric sensors are also in frequent use, and very recent investigations[153-155] are attempting to develop probe-type sensors to improve the simplicity and reliability of the latter type in on-line analysis.

Analytical-sensor units may also be divided into two types depending on whether they allow the measurement of only one or more than one determinand. The most commonly applied systems in water analysis are of the former type, but the latter can be used though they generally involve greater instrumental complexity. Examples of multi-determinand systems are provided by on-line chromatographic and emission spectrometric analysis, both of which may measure many determinands; anodic stripping voltammetry (see Section 11.3.7) is sometimes used to measure metals such as lead, cadmium and copper.

In recent years, the use of biological organisms or species as the analytical sensor has been increasingly applied [see Section 12.3.3(*a*)]. Such sensors are beyond both the scope of this book and our competence to discuss in detail; see, however, refs. 156 and 157 for reviews of the various types of biological sensing systems.

The reliability and accuracy of on-line analysers may often be markedly worsened by deterioration or change in the characteristics of the analytical sensor. In the last few years, attempts to reduce such problems by employing two or more identical sensors for each determinand within the one analyser have shown some success. The bases of this approach are described in refs. 147 and 158, and the application of an on-line system using dual sensors for river-water analysis is described in ref. 159.

(c) Single- or Multi-determinand Analysers*

Some analysers are designed and applied solely for the measurement of one determinand while others are intended for the simultaneous or sequential measurement of a number of determinands. The latter type employs either a multi-determinand analytical-sensor unit or, in effect, a number of single-determinand sensor units are appropriately assembled within the one instrument. Many such assemblies are commercially available, and are commonly used in water analysis.

(d) Frequency of Analysis

Some analysers are designed to provide continuous analysis of a sample stream, and these are based on one of the variants of continuous-flow, automatic analysis [see Section 12.2.2(*b*)(*ii*)]. Other instruments provide analyses of discrete samples at regular time intervals commonly varying from a few minutes to an hour or so. In this latter type, the smallest possible interval between successive measurements is equivalent to the time required for the treatment, transport, and measurement of each portion of sample. Note, however, that there may be advantage in operating a continuously measuring analyser in a discontinuous manner, *e.g.*, to decrease the rate of 'fouling' of the sample treatment and transport and the analytical-sensor units [see Section 12.3.4(*f*)(*vi*)]. Clearly, the purposes for which measurements are required govern the frequency with which analyses should be made. The apparently very high frequency of continuous analysers may often not be essential, and it should also be borne in mind that such analysers are not necessarily capable of accurate measurement of rapid variations in water quality [see Section 12.3.4(*d*)(*i*)(4)].

(e) Mobility

Some analysers or assemblies of analysers are large and heavy or require special services or housing, *e.g.*, to provide a controlled-temperature environment; analysers of this type can be used satisfactorily only in a permanent installation, and are called here 'fixed-site' analysers. Where on-line analysis is required for process control, the use of 'fixed-site' analysers is generally acceptable. For other applications, however, many authors have stressed the value of transportability or mobility of analysers since they are often required for only a relatively short period at any one sampling location; see, for example, refs. 14, 134, 136, 142, 160 and 161. Some analysers can be mounted and used in caravans, boats, *etc.*, and a few others

* Many authors use 'parameter' in place of 'determinand'.

can be carried by hand; both types are termed here 'mobile'. In addition, self-contained analysers have been developed that can be completely immersed in the water body of interest, and moved within it if required.[162-165] These analysers are usually battery operated, and contain a unit for printing the analytical results at regular intervals in either numerical or computer compatible form. Such instruments are commonly called 'submersible analysers', and can be of considerable value for short-term purposes though at present they are limited to rather few, simply measured determinands such as dissolved oxygen, electrical conductivity, pH, turbidity.

12.3.3 Applications of On-line Analysis for Waters

We note in (*a*) to (*c*) below our (admittedly subjective) impressions of the more common, current applications of on-line analysis to natural, drinking, and industrial waters, respectively. We have restricted the considerations mainly to commercially available analysers for the reasons noted in Section 12.2.4(*d*).

(a) Natural Waters

By far the most common application of on-line analysis concerns the measurement of the quality of fresh, surface waters, rivers being of particular interest. A number of recent reviews[134,141,146,147,166-168] deal with the equipment that is currently available and with the determinands that are commonly or can be measured. The proceedings of several conferences[169-175] also contain many relevant papers, and recent experiences with particular fixed-site analysers are described in refs. 159, 160 and 176 to 179. Commonly measured determinands are: dissolved oxygen (membrane electrode), chloride, nitrate, and pH (ion-selective electrodes), ammonia (ion-selective membrane electrode or less often spectrophotometric measurement of indophenol blue), electrical conductivity, and turbidity (nephelometry). With appropriate precautions, (see Section 12.3.4) such analysers seem capable of approximately 1 week's unattended operation. Redox potential (platinum electrode), hardness (ion-selective electrode or automatic titrator), and dissolved carbon dioxide (ion-selective membrane electrode) are capable of similar reliability though their measurement is less frequently of interest. The main interest in turbidity lies in its use to estimate the suspended solids (non-filtrable material) concentration using a previously established, experimental correlation between the two determinands. The accuracy of this approach is governed by the stability of the relationship between turbidity and suspended solids, and caution is necessary because many factors may affect this relationship; see, for example, refs. 147 and 180 to 185. A nephelometric instrument designed to overcome such problems has been described.[186] Although a number of on-line instruments for measuring organic carbon are available [see Section 12.2.3(*c*)(*ii*) and refs. 187 to 190] and their use seems to be increasing, they involve relatively complex sample treatment procedures. Accordingly, there has been considerable interest in, and application of, measurement of light absorption by the untreated sample (usually at 250 nm) since many authors have shown that this absorption may correlate well with the total organic carbon concentration; see, for example, refs. 141, 147, 178, 181 and 191 to 194. Again, as for turbidity/suspended solids, caution is necessary to ensure that variations in the relationship between absorption and organic carbon do not lead to unacceptable accuracy for the latter.

A topic to which much attention has been and continues to be paid is the use of so called 'water-quality monitoring networks' for river systems. The geographical sizes of such systems make the measurement and control of their quality very difficult if done purely on the basis of manual collection of samples. On the other hand, installation of the very many on-line analysers that would be needed to replace manual sampling completely – for the determinands for which on-line analysis is suitable – is not economically feasible. Accordingly, many countries have adopted the approach whereby certain key (crucial) sampling locations are identified in a river system, and fixed-site on-line analysers are installed there. These measurements are then supplemented by manual sampling and the short-term use of mobile, on-line analysers at other locations. Accounts of such monitoring networks are given in many papers; see, for example, refs 160, 161, 169, 170, 177 and Sections 3.2 to 3.4.

Another important subject concerns the protection of water-treatment plant producing drinking water from the effects of abstracting surface water of undesirable quality as a result of accidental pollution. The growing use of rivers as sources of drinking water and the danger of their pollution have led to increasing emphasis on the use of on-line analysis as a continuous means of detecting such problems. By siting the sampling location upstream of the water intake, it is possible, in principle, to provide a sufficiently early warning of pollution so that the abstraction of water can be stopped until the pollution has passed. This topic of 'intake-protection' is reviewed in ref. 149. One of the main difficulties is that the potential pollutants of greatest interest (*e.g.*, trace toxic metals and organic substances) with respect to the quality of drinking water cannot all generally be measured with adequate reliability and precision. Over the last ten years or so, therefore, increasing attention has been devoted to the development of equipment in which the response(s) of biological species to the water are measured continuously with the aim of determining the overall 'toxicity' of the water. This approach is reviewed in refs. 156 and 157. Various physiological responses of fish such as trout have been commonly used,[156,157,196] but other organisms also find application, *e.g.*, nitrifying bacteria[197,198] and mussels.[199]

As noted in Section 3.2.4(*a*)(*vi*), mobile on-line analysers installed on a boat and with the inlet of the sampling line moved through a water body may be of considerable value for short-term investigations. A much wider range of determinands is feasible for such purposes since long periods of unattended operation are not generally involved.

Finally, we would note that though the sampling line and sample conditioning system are always important for any type of water, they tend to be particularly crucial when natural waters are analysed; see Section 12.3.4(*e*).

(b) Drinking Waters

For this type of water and for those at intermediate stages in water-treatment plant, the needs for, and current capabilities of, on-line analysis are reviewed in ref. 134. Determinands that are commonly measured on-line include chlorine and other disinfecting materials such as chlorine dioxide and ozone (usually by amperometric membrane electrodes but see also refs. 200 and 201 for the application of ion-selective electrodes), pH and, when fluoridation is practised, fluoride (ion-selective electrodes), and turbidity (nephelometry); alkalinity (automatic titrator) and hardness (ion-selective electrode or automatic titrator) may also be measured on-line when there is interest in these determinands.

Generally speaking, it will usually be practicable to use on-line analysers with rather more complex sample treatment and analytical-sensor units than those used for natural waters since the long periods of unattended operation required for such waters will often not be essential for analysers installed in a water-treatment plant. Thus, spectrophotometric measurement of determinands such as iron and aluminium has been applied to their on-line determination, and certain toxic metals (lead, cadmium, copper) have been measured by differential pulse anodic-stripping voltammetry[147,202,203] (see Section 11.3.7) though ref. 203 involves manual calculation of analytical results. It is also worth noting that gas-chromatographic head-space analysis (see Section 11.3.8) has been used for the on-line determination of volatile, trace organic compounds.[204] As far as we are aware, none of these more complex systems finds common application at present, but their use may grow in the future.

Again, it is worth noting that mobile on-line analysers may be of value in short-term investigations of quality affecting processes in drinking-water distribution systems.

(c) Industrial Waters

The remarks in (a) and (b) provide a broad indication of on-line analytical capabilities, but each industry tends to have its own particular needs and applications so that generalisation is difficult. For example, the modern use of high-pressure boilers in the generation of electricity requires the concentrations of several determinands in the water in the steam–water circuit to be controlled at very small values, *e.g.* sodium and silicon. On-line analysers for these substances have, therefore, been developed, and find widespread use in power-stations; see, for example, refs. 23 and 205 to 207.

12.3.4 General Aspects of On-line Analysis

Whatever the application, a number of considerations are generally involved in assessing the need for, selecting the optimal equipment for, and applying, on-line analysis. Many of these aspects have been dealt with in previous sections where we discussed such questions for analytical systems in general. In the following sections, we have noted points that are particularly relevant to on-line analysis. We would stress that though these points have, for simplicity, been treated under different headings, they are very closely inter-related.

(a) Objectives of Analysis

As stressed in Section 2 and 12.1.2, the user of analytical results must define precisely what information he needs, and – in collaboration with the analyst – agree the corresponding essential analytical requirements. This information will, in turn, then form the basis of a detailed specification to be satisfied by any on-line analysers. Ten years ago, the first edition of this book included the statement 'recent literature [shows] that the application of fixed-site, on-line analysers has often been far from optimal', and Ballinger's[208] views that ' . . . many analysers are operating inefficiently because of poor design or installation at the wrong places. They are gathering too much of the wrong kinds of data . . . ' were quoted. Ballinger was speaking of river waters, but our experience is that similar difficulties may arise for any type of water. Such problems seem to have been somewhat reduced during the last decade, but authors continue to stress the need for critical consideration of the purposes of, and requirements for, on-line analysis; see, for example, refs. 23, 133, 134, 136, 138 and 149. The vital role of precisely and realistically defined specifications for on-line analysers required by the water industry is rightly stressed in particular in ref. 133.

In defining objectives and requirements for instrumentation, it is useful to distinguish the broad purposes of quality control and quality characterisation [see Section 3.3.2(*g*)]. For the former, it is essential from the outset to regard the on-line analyser(s) as an integral part of any control system, and on-line analysis should, therefore, be applied only when this system has been decided and will be implemented; see also (*d*) below. For quality characterisation purposes, it is similarly suggested that on-line analysis be used only when a system has been decided and will be established for handling the vast number of analytical results that will usually be produced. Such suggestions may seem restrictive, but they will certainly help to avoid situations where expensive on-line analytical and sampling equipment is installed and then largely ignored or allowed to fall into disuse, or whose output of *information* becomes lost amongst myriad analytical results.

Quality control applications will usually require fixed-site analysers. For quality characterisation, however, it is always worth careful consideration whether a fixed-site or a mobile analyser is appropriate. The use of the latter for short-term investigations may show a number of advantages.

As noted in Section 12.3.3(*a*), the installation of large numbers of analysers at many locations of a water body will generally meet important financial limitations. One approach to this problem is well known; one on-line analyser is installed at a central location, and sampling lines from a number of locations are brought to it. A manifold/valve system at the inlet to the analysers then allows sequential analysis of each of the sample streams. Note, however, that this approach may meet severe difficulties caused by long sampling lines, *e.g.*, changes in the concentrations of determinands (see Sections 3.5.2 and 3.5.3), and increases in the response-time of the sampling–analytical system [see (*d*) below].

(b) Justificaton of On-line Analysis

The general types of situation for which on-line analysis tends to be the preferred approach have already been noted in Section 12.1.2, but each potential application clearly requires detailed consideration to assess the justifiable extent and/or cost of on-line analysis. For quality control purposes, the penalties incurred if a process goes out of control may sometimes be so great that approximate considerations will suffice to justify purchase of the necessary equipment and its running and maintenance costs. In other situations, and for quality characterisation, much more careful assessment of the relative costs of the different approaches to sampling and analysis may be necessary. Rather little detailed information on the costs of on-line analysis appears to have been published, and the costs included and the bases on which they are calculated vary from author to author. Some recent data are given in refs. 160, 176 and 177, but – for the above reasons – we suggest caution in their use.

We would also repeat the suggestion made in the first edition that it is generally desirable to approach the use of on-line analysis from the standpoint that it minimises the increase in man-power needed to gain information required on water quality rather than that it allows reduction in man-power; Toms[136] has also stressed this point, and we believe that it has a number of important consequences. See also (*f*) below.

(c) Selection of On-line Analysers

Selection of an analyser for a particular requirement is, in principle, straightforward. The specification of the desired performance characteristics and any other necessary features is compared with the characteristics of all relevant, commercially available analysers, and the most suitable instrument selected. (Of course, re-consideration of the objectives of the measurements and/or of the specification for the on-line analyser is necessary if no reasonably suitable equipment is available). In practice, this simple approach may well meet serious problems that arise primarily through incompleteness, inaccuracy and ambiguity in the initial specification and through similar failings or simple absence in the data supplied by instrument manufacturers and users of relevant equipment. These aspects are considered in (*d*) below, and we would stress only three points here.

(1) Data supplied by manufacturers on performance characteristics should be viewed critically to decide their validity and appropriateness for the particular purpose under consideration.

(2) Great care is generally necessary to ensure that, given a suitable instrument, the sampling line and any sample-conditioning not included in the instrument will not cause serious deterioration in the performance of the total sampling–analytical system [see (*e*) below].

(3) Reliability will usually be a very important, desired feature of any on-line analyser, but it is very difficult to estimate experimentally for a given type of analyser [see (*d*) below]. The reliability can generally be improved by suitable design of a number of features of an instrument, and these are noted at various points in (*d*), (*e*) and (*f*) below. Selection of instruments should include evaluation of such aspects of design, and Rowse[138] has considered this and related points in detail.

Some general advice on the selection of on-line analysers is given in ref. 23, and a number of points concerning instrumentation in general are made in Section 10.4.

(d) Definition and Evaluation of Performance Characteristics of On-line Analysers

Many features of on-line analysers are or may be important, *e.g.*, cost, size, weight, power and other service requirements, ease of maintenance. Such features are relatively easily assessed for any particular purpose, and we consider here only those features directly indicative of the performance of instruments with respect to the accuracy of analytical results, the speed with which results are obtained, and the reliability of analysers in providing this accuracy and speed; these features are referred to as performance characteristics. Such characteristics can be and are defined and evaluated in a number of ways, different terms are used to denote the same characteristic, and sometimes, indeed, no clear definition attaches to quoted performance characteristics. The difficulties that such non-uniformity may cause – in specifying the performance required of an analyser, in comparing specifications with manufacturer's claims, and in pooling experiences with analysers – are obvious, and have been noted by many authors; see, for example, refs. 23, 133, 134, 138, 140, 144, 148 and 209. In (*i*) and (*ii*) below, therefore, we briefly consider the definition and experimental estimation of the relevant performance characteristics, respectively. The potential effects of sampling lines and sample conditioning units on these characteristics are noted in (*e*) below.

(i) Definition of performance characteristics We know of no authoritative and internationally accepted set of definitions for the various characteristics of interest that is applicable to all on-line analysers. At least one international standard[210,211] and a number of national standards (*e.g.*, refs. 212 and 213) as well as recommendations by authoritative organisations (*e.g.*, ref. 23) are available. The terminology and precise definitions of equivalent characteristics, however, show differences, and none of these references discusses in any detail the reasons why their definitions are as they are. In this situation, we suggest that interested readers consult these and/or other standards, and derive from them the information that will be of help. It is worth noting that the Central Electricity Generating Board has developed a standard[213a] concerning instruments for the on-line analysis of waters from the steam–water circuit of power stations. This standard defines: (1) the performance characteristics of interest, (2) the required numerical values of those characteristics for a number of determinands, and (3) the procedures to be used in estimating the performance characteristics achieved when particular instruments are used. It should be noted that refs. 23 and 210 213 deal primarily with continuously analysing equipment; for instruments providing a series of discontinuous results, rather different views may need to be taken of the definition of certain performance characteristics, *e.g.*, response time [see (4) below]. Some years ago, one of us made a number of suggestions on important performance characteristics and definitions of them, and the report[209] includes some discussion of the reasons behind these suggestions. Clearly, we mention this not as a competitor to existing standards, but its discussion may be of some help to those exploring the topic in more detail. Given this background, several points seem to us worth emphasis, and these are noted in (1) to (6) below.

(1) Given the lack of universally accepted definitions of terms such as accuracy, precision, bias, stability, noise, response time, *etc.*, users' specifications should always make clear exactly what their terms mean. Equally, in considering manufacturers' and other users' claims for performance, the precise meaning of their statements should always be ascertained. The response of on-line analysers for a fixed concentration of the determinand tends always to vary with time. This variation can appear as random changes, cyclic fluctuations, and systematic trends or any combination of these. The nature and magnitude

of temporal variations of response may also be affected by variations in several factors that commonly affect the responses of instruments, *e.g.*, variations in ambient and sample temperatures and mains voltage. Care is necessary, therefore, when specifying a required performance, to define the ranges of such factors to which an instrument will be subject. Similarly, estimated performance characteristics should be accompanied by clear statements on the values of such factors during the tests that produced the estimates; see, for example, refs. 209 to 213.

(2) Virtually all the remarks in Section 5.4 on sources of bias in analytical results apply to on-line analysers. The effects of interfering substances and inability to determine all forms of the determinand need particularly careful consideration. The various points made in Sections 5.4.3 and 5.4.4 on calibration identify general precautions that are equally relevant to on-line analysis. In particular, we would suggest that calibration procedures not involving the presentation of solutions of known determinand concentration to an instrument should be regarded with suspicion until the potential user is satisfied of their accuracy for his purpose.

(3) Performance-characteristics concerning accuracy can be markedly dependent on the nature of the samples/solutions analysed. It is important, therefore, in considering quoted values of these characteristics, to ascertain the type(s) of sample/solution used in any tests from which these values were derived.

(4) No instrument responds instantaneously to a change in the concentration of a determinand in the sample at the inlet to the instrument*, and this lag is sometimes appreciable. The rate of response to a change in concentration is usually governed by two main factors: (1) the time period, T_o, after a stepwise change in concentration until the first detectable change in response, and (2) the time period, T_e after the first detectable response until the equilibrium response to the new concentration is achieved. T_o is approximately equal to the time taken by an element of the sample to pass from the inlet of the instrument to the analytical sensor and there to be measured. The effect of T_o is to produce a simple lag between the true concentration and the instrument response. This lag may be important in quality control applications, but is normally unimportant for quality characterisation. The value of T_e is approximately equal to the time required for sample at the new concentration to replace essentially all that of the old concentration in the equipment from the inlet to the instrument up to and including the analytical sensor unit. When T_e is appreciable compared to the period T_p, over which changes in concentration occur, the changes in the response of an instrument will be distorted versions of the true concentration changes, the degree of distortion increasing as T_e/T_p increases. Thus, not only the rate but also the magnitude of concentration changes may be inaccurately indicated by the instrument response. Consequently, the maximum tolerable value for T_e is governed by the natures of the changes that occur in the water of interest, and by the extent to which these changes must be controlled or accurately measured. For many purposes, a

---

* The lag between such a change and its indication by an instrument is normally negligibly small when the analytical sensor is immersed directly in the water of interest (*in situ* analysis).

value of T_e/T_p not exceeding 0.1 will be suitable,[214] but detailed, quantitative discussion of the various points involved will be found in ref. 215.

When standard solutions are used to calibrate an on-line analyser, each such solution should be presented to the analyser in place of the sample for a period at least somewhat greater than T_e; otherwise, the estimates of calibration parameters (see Section 5.4.2) will tend to be biased. The value of T_e is, therefore, also important in governing the time required for calibration. Clearly, very large values of T_e are undesirable because they cause substantial reduction of the periods during which the sample is analysed.

It is also important to note that the value of T_e may depend on the values of the determinand concentrations and on the direction of concentration changes, *i.e.*, large to small or small to large concentrations. In practice, there are often practical difficulties in estimating the value of T_e (or $T_o + T_e$), and all the standards mentioned above[23,210–213] specify measurement of the time to achieve 90 rather than 100% of the equilibrium response. In view of the above considerations, however, it seems to us generally useful to specify and attempt to estimate both T_o and T_e in addition to any other intermediate response times that may be of interest.

This discussion shows that hydraulic aspects of the design of analysers, *e.g.*, flow-rate of sample, volume and configuration of the sample treatment and transport and analytical sensor units, are generally important factors affecting T_o and T_e. It should also be stressed that similar factors arise in the sampling line and sample conditioning unit, and can markedly increase the values of T_o and T_e above their values for the instrument alone. Factors of this type and precautions against them have been discussed in Section 3.5.

(5) The estimation of performance characteristics of on-line analysers is generally a lengthy and expensive procedure [see (*ii*) below]. Commonly therefore, only one such evaluation is made for a given type of analyser, and the question of the representativeness of the estimated characteristics for all such analysers then arises. Of course, one would hope that in the long-term this problem will decrease as the experience of users is properly quantified and pooled.[176] Until then, however, it is clearly prudent to regard quoted values of performance characteristics rather as expressions of the performance achieved by a particular instrument of a given type rather than as the performance necesarily achieved by all instruments of the same type. This situation is not confined solely to on-line analysis, and a number of relevant considerations are noted in Section 9. Detailed appraisal of design features of an on-line analyser as suggested by Rowse[138] will also be useful in assessing the extent to which the performances of instruments of the same type might be expected to vary.

(6) Finally, we wish to touch on the difficult question of 'reliability'. The need for this aspect of performance is stressed by all authors in the many papers already cited on on-line analysis, though none includes a precise definition of what is meant by 'reliability'. Various connotations of the term are possible, and each leads to a corresponding definition with its implication on how a performance characteristic could be quantified and estimated. For example, a definition might be based on the proportion of time for which an analyser achieved a specified performance or on the mean times between failures and to repair. We

shall not consider the various possibilities and their difficulties because, at present, there seems to have been very little quantitative investigation of such aspects in the field of on-line analysis of waters. Further, reasonably representative quantification of 'reliability' seems to us extremely difficult for on-line analysers of the type considered here for two reasons.[138,209,216] First, the desirable duration of tests will usually be impracticably long. Second, large variations of 'reliability' among instruments of the same type may occur so that tests on a number of instruments of a given type would usually be required but will commonly be impracticable. It seems to us, therefore, better at present not to regard 'reliability' as a performance characteristic to be estimated and quantitatively expressed for all on-line analysers. Note also that none of the standards mentioned above[23,210–213] includes reliability or related concepts. Of course, as already mentioned above, it is not only performance characteristics that are important in assessing the suitability of instruments for given purposes. The users of analysers should, therefore, describe as quantitatively and unambiguously as possible their experiences concerning any aspects related to 'reliability'. The pooling of such information should help to place this topic on a firmer basis.

For those wishing to pursue 'reliability', there is a comprehensive British Standard,[217] and refs. 138 and 158 provide some simple examples of calculations of the reliability of instruments for water analysis.

(ii) Estimation of performance characteristics The experimental estimation of performance characteristics may be required for several purposes: (1) to decide the performance to be claimed for a given type of analyser, (2) to check the validity of a claim for the performance of an analyser, (3) to check that the performance of a particular analyser is satisfactory before putting it into routine operation, and (4) to check that satisfactory performance is maintained in routine operation. Tests for purposes (1) and (2) are generally included in the process commonly referred to as 'evaluation' of on-line analysers, but this also includes assessment of various other design and operational features of instruments; see, for example, refs. 135, 138 and 216. Clearly, the general nature of tests to estimate performance characteristics depends on the characteristics that are considered of interest, the definitions adopted for them, and the purpose of the tests. Appropriate experimental designs may be derived by considering the various points discussed in Sections 4, 5 and 6, and we shall not, therefore, discuss this topic in detail. Generally speaking, tests for purposes (1) and (2) need to be the most extensive while tests for purpose (3) may be selected as appropriate from those for (1) and (2) [see $(f)(viii)$ below]; tests for purpose (4) are briefly considered in $(f)(x)$ below. Detailed suggestions on tests for purposes (1) and (2) are given in refs. 135, 210 to 213 and 218, the last being based on ref. 209 and including rather more discussion of the reasons underlying the suggestions than the other references. Four general points concerning these various suggestions seem to us worth special note.

(1) Refs. 135 and 210 to 213 seem to us not to emphasise sufficiently the general importance of the various sources of bias considered in Section 5.4. Of course, such bias may be inappreciable for some analysers so that little or no testing of

this is required. In general, however, very careful consideration of all potential sources of bias is essential as are also special tests to estimate their magnitude.

(2) Performance characteristics are numbers describing the quantitative performance of an analyser. They must be determined experimentally since they cannot, in general, be predicted *a priori*. The experimental measurements, from which the numerical values of performance characteristics are derived, are subject to random errors so that the numerical values obtained are subject to uncertainty (see Section 4). We speak, therefore, of the estimation of performance characteristics, and it is clearly desirable to conduct the experiments so that their uncertainties are held to tolerable values. Thus, sufficient replication of each type of test is generally necessary to allow the confidence limits of an estimate to be calculated and made adequately small. The uncertainties of estimates of performance characteristics must also be borne in mind when comparing such estimates either with required values or with other estimates.

(3) One type of test that is commonly recommended and applied in assessing the accuracy of results from an on-line analyser is to compare its results with those obtained for samples analysed by a different ('reference') analytical system. This approach is useful if the bias and precision of results provided by the reference system are known to be satisfactory for the purpose as discussed in Section 5.3. There seems to us, however, little point in such comparisons if the bias of results from the reference system is not known. Further, in water analysis, it may well be the case that the precison of results from the reference system is similar to, or, possibly worse than, that for the on-line analyser. Under these circumstances, statistical evaluation of the significance of any apparent differences is essential if misleading or false conclusions are to be avoided (see Section 5.3). A second, vital precaution, of course, is to ensure that the determinand concentrations are essentially identical in the samples whose results are to be compared.

(4) The tests required to estimate all performance characteristics are usually long and expensive, and generally call for special test-laboratory facilities. The amount of time, effort and resources should not be underestimated, and organisations and laboratories wishing to test the performances of analysers for a number of determinands will usually have to establish a special unit for such work; see, for example, ref. 133. Small organisations and individual laboratories may find this impracticable, but several organisations operate collaborative schemes of instrument evaluation to ease the problem; see, for example, ref. 216.

(e) Sampling Line and Sample Conditioning
The many problems that may arise through inappropriate design, dimensions, materials, installation and operation of sampling lines have been sufficiently discussed in Section 3.5. Here, we simply repeat that virtually every critical review of on-line analysis stresses the vital importance of the points noted in that section; see also (*d*)(*i*)(4) above.

'Sample conditioning', in the sense used here, refers to any operations intended to change the properties of the sample in order to minimise deterioration in the performance of an on-line analyser. Two properties are of special and general importance, namely, the temperature of the sample at the inlet to an instrument, and the concentration of materials (biotic and abiotic) that cause deposits within the sample treatment and transport and analytical-sensor units [see $(f)(vi)$ below]. On-line analysers sometimes include components for controlling both these aspects, but they are also commonly installed at appropriate points within the sampling line. For example, cooling coils are generaly required when high-temperature waters are of interest. Devices to reduce deposition of materials within the sampling line and analyser often involve some form of filter placed at the inlet to the sampling line; see $(f)(vi)$ below. Reviews of sample conditioning units are provided in refs. 219 and 220, and see also Section 3.5

(f) Various Design, Installation and Operational Aspects
We have noted below a number of points that seem to us of special relevance to the performance and reliability of on-line analysers for waters.

(i) Calibration   The values of calibration parameters tend to vary with time so that periodic re-calibration is generally necessary. The appropriate period depends on the determinand, the instrument, and the accuracy required of analytical results, and frequencies ranging from several times a day to once every month or so are quoted in the literature. Procedures for deciding a suitable frequency are noted in Section 5.4.3$(d)(ii)$, and see also ref. 221 for a detailed example concerned with water analysis. The commissioning tests mentioned in $(viii)$ below will provide data from which the desired frequency can be calculated, and whenever there is doubt it is clearly better to err on the side of a greater rather than smaller frequency.

Automatic calibration is now available for many on-line analysers, and – as many authors have stressed – is a highly desirable feature; indeed, as the required period of unattended operation increases, automatic calibration usually becomes essential. Many automatic calibration units also arrange that the instrument response to a given concentration of the determinand is automatically adjusted to the same value during each calibration. This also is a valuable feature, but it is useful if the initial response of the instrument is available before such adjustment; this provides a regular check on the stability of the calibration parameters. Of course, automatic calibration may meet difficulties if the calibrating solutions are not stable and must, therefore, be frequently prepared and supplied to an instrument.

(ii) Installation   On-line analysers must usually be installed in positions where they are adequately protected from environmental effects. In addition to factors such as ambient temperature and humidity, vibration, dust and constancy of voltage of power supplies also need consideration. Special housing of instruments is commonly required when they are installed in water-treatment and industrial plant, and is essential when they must be sited in the field, *e.g.*, on a river bank. In the latter case particularly, substantially built and reasonably sized housing with appropriate facilities is needed to provide good protection against climatic effects and vandalism, and to allow maintenance, repair and investigational work.

(iii) Contamination of samples   It is clearly important that the materials of

construction of an analyser with which samples and calibrating solutions come into contact are chosen so that they do not cause appreciable and variable contamination by substances affecting the analytical response. Consideration of even apparently minor components of equipment is necessary from this point of view (see Section 10.8).

(iv) Separation of hydraulic and electrical circuits In most analysers some component of the hydraulic circuit is likely to fail at some time and lead to leakage of sample and/or reagents. The analyser should be so assembled that such leakage cannot cause permanent damage to the electrical circuits and components. Fortunately, a high proportion of analysers are now designed to avoid such problems, but the point is worth bearing in mind.

(v) Use of more than one analytical sensor for each determinand We noted in Section 12.3.2(*b*) that the use of more than one sensor for each determinand can lead to substantial improvements in the reliability of on-line analysers. Clearly, therefore, as the required period of unattended operation increases so the use of multi-sensor systems tend to become more attractive.

(vi) 'Fouling' We use the term 'fouling' to denote the formation of deposits and films of undissolved and/or biological materials on the internal surfaces of the hydraulic circuits of an analyser. Such deposits may cause flow-obstructions and their consequent effects, *e.g.*, leakage, affect the concentrations of determinands, and impede or completely prevent operation of the analytical sensor. It is generally agreed that fouling is one of the most important – if not the most important – sources of unreliability and inaccuracy in on-line analysis. Abiotic materials alone may cause fouling, but it is often most problematic when biota are also present since these may multiply and grow within an analyser; fouling is, therefore, commonly a special problem for surface waters. Three main approaches to reducing the extent of fouling are used [see (1) to (3) below]. We have no information on their quantitative advantages and disadvantages, and can only suggest that their consideration is worthwhile when assessing the relative suitabilities of different types of analyser.

(1) Various devices are included in the sample conditioning unit to reduce the amounts of foulants entering the analyser. Filters of various types are usually employed (see, for example, references 139, 148, 219 and 220), but other devices have been used, *e.g.*, ultraviolet irradiation for killing biota.
(2) Regular cleaning of the internal surfaces of the hydraulic circuits and of the analytical sensor is also often used. Various approaches have been developed, and include periodic flushing with solutions of biocides, mechanical cleaning of sensor surfaces, use of high-velocity water jets, *etc*. Automatic cleaning procedures are now included in many analysers, and are a useful feature. Usually, such procedures occupy only a small proportion of the total operation time of an analyser. When, however, the frequency of sample analysis is less important than the reliability, the proportion of time devoted to cleaning may be increased with advantage.[137]

(3) The hydraulic circuits of an analyser can be designed to minimise the likelihood of formation and development of fouling, *e.g.*, by maintaining high flow-velocities, by minimising 'dead-volumes' and obstructions to smooth flow. An interesting approach for protecting the surfaces of optical sensors is described in ref. 222. In this, the sample stream is prevented from contacting these surfaces by 'jacketing' it with a concentric stream of clean water.

(vii) Means of detecting mal-operation   In any application of on-line analysers, it is highly desirable to detect any malfunction as rapidly as possible so that the least amount of operation is lost, and so that the chance of one minor problem inducing much more serious difficulties is minimised. To avoid the need for frequent visits to an analyser, it is clearly of value to transmit its response to a central point where staff are always present. This approach has, of course, been common for many years for process-control instruments, but it was seldom used in initial applications of fixed-site analysers installed in the field – often at remote and rather inaccessible sites – for the analysis of natural waters. Experience with these systems quickly showed the importance of telemetry for transmission of analyser responses to a central point. It is now usual for individual analysers or networks of analysers to employ telemetry for this purpose, and the importance of this approach in improving the availability of analysers has been stressed by many authors. See also (*g*) below.

(viii) Commissioning tests   Given that the performance required of an on-line analyser has been decided, and that a suitable type of instrument has been chosen and installed, there remains the need to check that the performance of the particular instrument obtained meets the requirements before it is put into routine operation. There will always be an understandable desire to reduce the time and effort involved in such checks to the absolute minimum, but we would urge caution in any such reduction. Failure to detect inadequate performance at this stage may lead to various problems arising from the use of inadequately accurate results, and rectification of such performance may well prove more difficult subsequently. The exact nature of commissioning tests depends on many factors: the determinand, the required performance, the performance claimed for the instrument by the manufacturer, the extent to which this latter performance has been achieved by other users of the type of instrument, the design and principles of operation of the instrument, and the nature(s) of the sample(s) to be analysed. The environmental conditions to which the instrument will be subject may also be very important, but these can probably be ignored if the instrument has been installed and provided with services as recommended by the manufacturer.

Given all these relevant factors, we know of no simple scheme of tests that can readily be summarised here and that will be generally valid whatever the definitions of performance characteristics adopted. The detailed nature of commissioning tests appears to have received little consideration in the literature though ref. 218 makes a number of suggestions. Clearly, each application should, ideally, receive individual consideration, and our broad suggestions may be summarised as follows.

(1) Decide those performance characteristics crucial in judging if the performance of the analyser is satisfactory, and which cannot reasonably be assumed to

have essentially the same numerical values as those claimed by the manufacturer [see (*d*)(*i*) above].

(2) Make experimental estimates of the performance characteristics selected in (1) [see (*d*)(*ii*) above].

(ix) Maintenance All authors stress two aspects concerning the maintenance of on-line analysers. First, a planned and regular programme of preventative maintenance and inspection should be established and rigidly implemented. Second, high-quality staff experienced in the various skills and disciplines involved in on-line analysers are required to ensure that installation, maintenance, and fault repair are properly and efficiently executed. Many authors note, however, a shortage of suitable specialists, and this problem needs to be considered when planning the establishment of on-line analytical systems. It seems to us prudent to err on the side of more rather than less frequent maintenance in the initial stages of routine operation; the frequency may always be reduced if experience shows this to be acceptable. We would also stress the need for at least one member of staff to be thoroughly conversant with the principles and mechanisms of all processes involved in the production of analytical results by an on-line analyser (see Section 10.4.3).

(x) Routine tests of performance A number of factors may lead to deterioration in the performance of an on-line analyser once it has begun routine operation. Just as for any other analytical system, therefore, regular tests of performance should be made (see Section 6.4). It will, however, be generally unacceptable to take an analyser out of operation for the periods of time required for tests such as those in (*d*)(*ii*) and (*f*)(*viii*) above. Simple tests involving the minimal disturbance of the normal operating regime of an analyser are required, but we know of no comprehensive discussion of the many possible checks that could be devised. For many purposes, it seems to us that the responses observed for calibrating solutions can provide useful data provided that: (1) each such solution is analysed for a period equal to approximately 1.5 T_e, and (2) any automatic readjustment of the equilibrium response is not made until the end of this analysis period. The responses observed for such solutions then provide information on the stability of calibration parameters, on short-term variability of the response when the concentration of determinand is constant, and on response times. This approach could be extended for some determinands by arranging that one or more of the calibrating solutions is analysed at twice the frequency chosen for calibration. The results from these additional measurements could then be used in exactly the same way as described in Section 6.4.3. Tests of this type provide little or no information on some sources of bias, but this lack may be remedied by regular comparison of results from the analyser and a reference analytical system [see (*d*)(*ii*)(3) above] when a suitable reference system is available.

(g) Calculation and Presentation of Analytical Results
Our definition of on-line analysis (see Section 12.1.1) requires that the desired*

---

* Of course, all the results that an on-line analyser is capable of producing may well not be required. For example, even when a continuous analyser is installed, one may well require only hourly results to be printed. The purposes of the analysis and the information required will define the required analytical results.

analytical results are calculated and printed automatically. Given also the value of transmitting the analyser's response to a central control point [see (f)(vii) above], it is logical that the required calculations and print-outs are made at that point, and this approach is being increasingly adopted for all types of on-line analysis. Central computing facilities are conveniently and economically installed at such control points, and provide an efficient approach to servicing a number of analysers that may be installed at widely separated locations. The computer may then be used for many other tasks such as editing the analytical results, archiving them, deriving further information from them (see Section 13), and controlling the transmission of results and information to yet other locations. Approaches of this type represent the 'remote read-out' unit of Table 12.2. A 'local read-out' unit, however, may also be of value for purposes such as commissioning tests and fault finding and repair, and should generally be provided. Chart-recorders are commonly and usefully employed for such purposes.

12.4 REFERENCES

[1] IUPAC Commission on Automation, *IUPAC Inf. Bull.*, 1978, No. 3, 233.
[2] Working Party on Control Systems for the Water Industry, DOE/NWC Standing Technical Committee Reports No. 11, National Water Council, London, 1978.
[3] British Standards Institution, BS 6068: Part 1: Section 1.2: 1982 (= ISO 6107/2-1981), B.S.I., Milton Keynes, Buckinghamshire, 1982. (See 'automatic sampling'.)
[4] P. B. Stockwell, *Talanta*, 1980, **27**, 835.
[5] J. B. de Haan, in J. K. Foreman and P. B. Stockwell, ed. "Topics in Automatic Chemical Analysis I", Ellis Horwood, Chichester, 1979, pp. 208–236.
[6] G. L. Hawk, J. N. Little, and F. H. Zenie, *Am. Lab.*, 1982, **14(6)**, 96.
[7] G. D. Owens and R. J. Eckstein, *Anal. Chem.*, 1982, **54**, 2347.
[8] G. D. Owens, *et al.*, *Anal. Chem.*, 1983, **55**, 1232A.
[8a] R. E. Dessy, ed., *Anal. Chem.*, 1983, **55**, 1100A; *ibid.*, 1983, **55**, 1232A.
[8b] J. N. Little, *Trends Anal. Chem.*, 1983, **2**, 103.
[9] P. B. Stockwell, *J. Autom. Chem.*, 1983, **5**, 169.
[10] K. H. Mancy and H. E. Allen, in M. J. Suess, ed., "Examination of Water for Pollution Control, Volume 1", Pergamon, Oxford, 1982, pp. 78–124.
[11] T. Kish, *J. Water Pollut. Control Fed.*, 1981, **53**, 420.
[12] F. L. Mitchell, *J. Autom. Chem.*, 1980, **2**, 23.
[13] P. J. McNelis, in L. L. Ciaccio, ed. "Water and Water Pollution Handbook, Volume 4", Dekker, New York, 1973, pp. 1389–1430.
[14] K. H. Mancy and H. E. Allen, in M. J. Suess, ed, ref. 10, pp. 1–22.
[15] G. J. Holland, in "Practical Aspects of Water Quality Monitoring Systems. Proceedings of a Water Research Centre Seminar, 7 December, 1977", Water Research Centre, Medmenham, Buckinghamshire, 1977, Paper 4.
[16] J. K. Foreman, *J. Autom. Chem.*, 1980, **2**, 11.
[17] *J. Autom. Chem.*, 1980, **2**, 22–33. (A collection of papers).
[18] G. C. Gunnerson, in "Water Quality Parameters", (ASTM Special Technical Publication 573), American Society for Testing and Materials, Philadelphia, 1975, pp. 456–486.
[19] C. E. Hamilton, ed., "Manual on Water", Fourth Edition, (ASTM Special Technical Publication 442A), American Society for Testing and Materials, Philadelphia, 1978.
[20] Environmental Resources Ltd. and Watson Hawksley, "Water Industry Control Systems: A report on a U.K. strategy for meeting the requirements of the water industry for instrumentation, control and automation equipment systems from U.K. sources", (Prepared for the Department of Industry), Environmental Resources Ltd., London, 1979.
[21] W. D. Meredith and C. A. Rushton, *Intern. Environ. Safety*, 1981, Dec., 10.
[22] Working Party on Control Systems for the Water Industry, DOE/NWC Standing Technical Committee Reports No. 27, National Water Council, London 1981.
[23] American Society for Testing and Materials, "1983 Annual Book of ASTM Standards. Volume 11.01: Water (I)", A.S.T.M., Philadelphia 1983.
[24] J. K. Foreman and P. B. Stockwell, "Automatic Chemical Analysis", Ellis Horwood, Chichester, 1975.

25 J. K. Foreman and P. B. Stockwell, ed., "Topics in Automatic Chemical Analysis I", Ellis Horwood, Chichester, 1981.
26 J. Růžička and E. H. Hansen, "Flow Injection Analysis", Wiley, Chichester, 1981.
27 P. Bonini, *Pure Appl. Chem.*, 1982, **54**, 2017.
28 J. T. van Gemert, *Talanta*, 1973, **20**, 1045.
29 J. Váňa, "Gas and Liquid Analysers", (Volume XVII of Wilson and Wilson's Comprehensive Analytical Chemistry, ed. G. Svehla, Elsevier, Oxford, 1982). (Note that this book is a translation of an earlier Czechoslovakian book.)
30 A. O. Jacintho. *et al.*, *Anal. Chim. Acta*, 1981, **130**, 243.
31 S. Greenfield, *Spectrochim. Acta, Part B*, 1983, **38**, 93.
32 J. F. Tyson, J. M. H. Appleton, and A. B. Idris, *Anal. Chim. Acta*, 1983, **145**, 139.
33 J. F. Tyson, J. M. H. Appleton, and A. B. Idris, *Analyst*, 1983, **108**, 153.
34 J. F. Tyson and J. M. H. Appleton, *Talanta*, 1984, **31**, 9.
35 L. T. Skeggs, *Am. J. Clin. Pathol.*, 1957, **28**, 311.
36 W. B. Furman, "Continuous Flow Analysis", Dekker, New York, 1976.
37 W. A. Coakley, "Handbook of Automated Analysis. Continuous Flow Techniques", Dekker, New York, 1981.
38 K. W. Petts, "Air Segmented Continuous Flow Automatic Analysis in the Laboratory. 1979". H.M.S.O., London, 1980.
39 K. K. Stewart, *Talanta*, 1981, **28**, 789.
40 J. Růžička and E. H. Hansen, *Anal. Chim. Acta*, 1975, **78**, 145.
41 K. K. Stewart, G. R. Beecher, and P. E. Hare, *Anal. Biochem.*, 1976, **70**, 167.
42 E. Pungor, *et al.*, *Anal. Chim. Acta*, 1979, **109**, 1.
43 E. Pungor, Z. Feher, G. Nagy, and K. Toth, *CRC Crit. Rev. Anal. Chem.*, 1983, **14**, 53.
44 E. Pungor, Z. Feher, G. Nagy, and K. Toth, *CRC Crit. Rev. Anal. Chem.*, 1983, **14**, 175.
45 H. A. Mottola and A. Hanna, *Anal. Chim. Acta*, 1978, **100**, 167.
46 P. W. Alexander and A. Thalib, *Anal. Chem.*, 1983, **55**, 497.
47 H. L. Pardue and B. Fields, *Anal. Chim. Acta*, 1981, **124**, 39.
48 H. L. Pardue and B. Fields, *Anal. Chim. Acta*, 1981, **124**, 65.
49 J. Růžička and E. H. Hansen, *Anal. Chim. Acta*, 1983, **145**, 1.
50 C. B. Ranger, *Anal. Chem.*, 1981, **53**, 20A.
51 K. K. Stewart, *Anal. Chem.*, 1983, **55**, 931A.
52 J. Růžička, *Anal. Chem.*, 1983, **55**, 1040A.
53 B. Rocks and C. Riley, *Clin. Chem. (Winston-Salem, NC)*, 1982, **28** 409.
54 R. A. Leach, J. Růžička, and J. M. Harris, *Anal. Chem.*, 1983, **55**, 1669.
55 *Anal. Chim. Acta*, 1980, **114**.
56 *Anal. Chim. Acta*, 1983, **145**.
57 L. R. Snyder, *Anal. Chim. Acta*, 1980, **114**, 3.
58 K. S. Johnson and R. L. Petty, *Anal. Chem.*, 1982, **54**, 1185.
59 A. G. Fogg and N. K. Bsebsu, *Analyst*, 1984, **109**, 19.
60 T. M. Craig, *J. Autom. Chem.*, 1981, **3**, 63.
61 M. Snook, *et al.*, *J. Autom. Chem*, 1979, **1**, 72.
62 F. L. Mitchell, *Anal. Proc.*, 1981, **18**, 257.
63 D. Tarlow, *J. Phys. E*, 1982, **15**, 611.
64 J. Greyson, *J. Autom. Chem.*, 1981, **3**, 66.
65 A. Zipp, *J. Autom. Chem.*, 1981, **3**, 71.
66 B. Walter, *Anal. Chem.*, 1983, **55**, 498A.
67 H. V. Malmstadt, D. L. Krottinger, and M. S. McCracken, in J. K. Foreman and P. B. Stockwell, ed, ref. 25, pp. 95–137.
68 B. W. Renoe, G. Savory, M. R. Wills, and J. Savory, *J. Clin. Lab. Autom.*, 1981, **1**, 47.
69 M. A. Koupparis, K. M. Wolczak, and H. V. Malmstadt, *J. Autom. Chem.*, 1980, **2**, 66.
70 M. A. Koupparis, K. M. Wolczak, and H. V. Malmstadt, *Anal. Chim. Acta*, 1982, **142**, 119.
71 M. A. Koupparis, K. M. Wolczak, and H. V. Malmstadt, *Analyst*, 1982, **107**, 1309.
72 J. A. Alcover, *et al.*, *J. Autom. Chem.*, 1983, **5**, 83.
73 P. E. Walser and H. A. Bartels, *Am. Lab.*, 1982, **14**(2), 113.
74 H. Bartels and P. Walser, *Fresenius' Z. Anal. Chem.*, 1983, **315**, 6.
75 N. G. Anderson, *Science*, 1969, **166**, 317.
76 H. K. Cho, *et al.*, *Clin. Chem. (Winston-Salem, NC)*, 1982, **28**, 1956.
77 P. Henry, *J. Autom. Chem.*, 1979, **1**, 195.
78 J. Savory and R. E. Cross, ed., "Methods for the Centrifugal Analyzer", American Association for Clinical Chemistry, Washington, 1978.
79 C. P. Price and K. Spencer, ed., "Centrifugal Analysis in Clinical Chemistry", Praeger, New York, 1980.
80 C. P. Price, *Anal. Proc.*, 1981, **18**, 259.
81 J. R. Garbarino and H. E. Taylor, *Spectrochim. Acta, Part B*, 1983, **38**, 323.

668 *Chemical Analysis of Water*

[82] American Public Health Association et al., "Standard Methods for the Examination of Water and Wastewater", Fifteenth Edition, American Public Health Association, Washington, 1981.
[83] American Society for Testing and Materials," 1983 Annual Book of ASTM Standards. Volumes 11.01 and 11.02. Water (I and II)", A.S.T.M., Philadelphia, 1983.
[84] Standing Committee of Analysts, "Silicon in Waters and Effluents, 1980", H.M.S.O., London 1981.
[85] Standing Committee of Analysts, "Boron in Waters, Effluents, Sewages and Some Solids, 1980", H.M.S.O., London, 1981.
[86] Standing Committee of Analysts, "Oxidised Nitrogen in Waters, 1981", H M S O., London, 1982.
[87] Standing Committee of Analysts, "The Determination of Alkalinity and Acidity in Waters, 1981", H.M.S.O., London, 1982.
[88] Standing Committee of Analysts, "Ammonia in Waters, 1981", H.M.S.O., London, 1982.
[89] Standing Committee of Analysts, "Analysis of Surfactants in Waters, Wastewaters and Sludges, 1981", H.MS.O., London, 1982.
[90] Standing Committee of Analysts, "Chloride in Waters, Sewage and Effluents, 1981", H.M.S.O., London, 1982.
[91] Standing Committee of Analysts, "Arsenic in Potable Waters by Atomic Absorption Spectrophotometry (Semi-automated method), 1982", H.M.S.O., London, 1983.
[92] S. Wilson and A. Green, *J. Autom. Chem.*, 1984, **6**, 33.
[93] V. C. Marti and D. R. Hale, *Environ. Sci. Technol.*, 1981, **15**, 711.
[94] R. B. Willis and G. L. Mullins, *Anal. Chem.*, 1983, **55**, 1173.
[95] J. Hilton and E. Rigg, *Analyst*, 1983, **108**, 1026.
[96] H. F. R. Rijnders, P. H. A. M. Melis, and B. Griepink, *Fresenius' Z. Anal. Chem.*, 1983, **314**, 627.
[97] Standing Committee of Analysts, "The Instrumental Determination of Total Organic Carbon, Total Oxygen Demand and Related Determinands. 1979", H.M.S.O., London, 1980.
[98] J. Bortlisz, *Gewässerschutz-Wasser-Abwasser*, 1977, **26**, 87.
[99] R. J. Oake, Technical Report TR 160, Water Research Centre, Medmenham, Buckinghamshire, 1981.
[100] R. J. Oake, Technical Report TR 186, Water Research Centre, Medmenham, Buckinghamshire, 1982.
[101] H. Mueller and W. M. Bandaranayake, *Marine Chem.*, 1983, **12**, 59.
[102] A. Wilhelms, *Z. Wasser Abwasser Forsch.*, 1982, **15**, 53.
[103] E. Keck and U. Grunwald, *Vom Wasser*, 1979, **53**, 86.
[104] D. E. Lueck, R. A. Dishman, and R. B. Thayer, *ISA Trans.*, 1981, **20**, 67.
[105] C. Legrand B. Capdeville, H. Roques, and M. Angles, *Int. J. Environ. Anal. Chem.*, 1982, **11**, 283.
[106] A. A. Nicholson, O. Meresz, and B. Lemyk, *Anal. Chem.*, 1977, **49**, 814.
[107] Anon., *J. Autom. Chem.*, 1982, **4**, 201.
[108] R. E. Majors, H. G. Barth, and C. H. Lochmüller, *Anal. Chem.*, 1982, **54**, 323R.
[109] A. G. Hedley and M. J. Fishman, U.S. Geol. Surv., Water Resources Inventory, 1982, 81–78.
[110] P. Valenta, et al., *Fresenius Z. Anal. Chem.*, 1982, **312**, 101.
[111] A. Hu, R. E. Dessy, and A. Granéli, *Anal. Chem.*, 1983, **55**, 320.
[112] D. Jagner, *Trends Anal. Chem.*, 1983, **2**, 53.
[113] J. Wang and H. D. Dewald, *Anal. Chem.*, 1983, **55**, 933.
[114] R. W. Ward and P. B. Stockwell, *J. Autom. Chem.*, 1983, **5**, 193.
[115] R. Haeckel, *J. Autom. Chem.*, 1983, **5**, 68.
[116] T. M. Craig, *J. Autom. Chem.*, 1980, **2**, 31.
[117] T. M. Craig, *J. Autom. Chem.*, 1983, **5**, 210.
[118] H. G. J. Worth, *J. Autom. Chem.*, 1980, **2**, 125.
[119] *J. Autom. Chem.*, 1983, **5**, 68–82, 210–211.
[120] R. L. McAvoy, in "Advances in Automated Analysis, Volume 8", Mediad Incorporated, Tarrytown, New York, 1973, pp. 37–42.
[121] M. J. McGuire, in ref. 120, pp. 51–54.
[122] E. L. Henn and D. O. Bassett, in ref. 120., pp. 55–57.
[123] D. G. Porter and P. B. Stockwell, in J. K. Foreman and P. B. Stockwell, ed ref. 25, pp. 44–72.
[124] A. Mather, in M. Stefanini, ed. "Progress in Clinical Pathology, Volume 6", Grune and Stratton, New York, 1975, p.1.
[125] P. M. G. Broughton, A. H. Gowenlock, J. J. McCormack, and D. W. Neill, *Ann. Clin. Biochem.*, 1974, **11**, 207.
[126] J. O. Westgard, *CRC Crit. Rev. Clin. Lab. Sci.*, 1981, **13**, 283.
[127] H. M. Barbour, K. Virapen, T. F. Woods, and D. Burnett, *J. Autom. Chem.*, 1981, **3**, 173.
[128] D. Burnett, H. M. Barbour, and T. F. Woods, *J. Autom. Chem.*, 1981, **3**, 178.
[129] D. Burnett, et al., *J. Clin. Chem. Clin. Biochem.*, 1982, **20**, 207.
[130] A. S. McLelland, A. Fleck, and R. F. Burns, *Ann. Clin. Biochem.*, 1978, **15**, 12.
[131] F. L. Mitchell, *J. Autom. Chem.*, 1981, **3**, 51.
[132] K. Dixon, *Ann. Clin. Biochem.*, 1982, **19**, 224.
[133] Environmental Resources Ltd. and Watson Hawksley, "Water Industry Control Systems", Environmental Resources Ltd., London, 1979.

[134] Working Party on Control Systems for the Water Industry, "Final Report", (DOE/NWC Standing Technical Committee reports No. 27), National Water Council, London, 1981.

[135] A. A. Rowse, in "Instruments and Control Systems: Water Research Centre Conference, 15–17 September, 1975", Water Research Centre, Medmenham, Bucks, 1975, Paper No. 8.

[136] R. G. Toms, in "Water Research Seminar on Practical Aspects of Water Quality Monitoring Systems, 7 December, 1977", Water Research Centre, Medmenham, Bucks, 1977, Paper No. 1.

[137] H. R. S. Page, in ref. 136., Paper No. 7.

[138] A. A. Rowse, *Sci. Total Environ.*, 1980, **16**, 193.

[139] D. E. Caddy and P. G. Whitehead, *Effl. Water Treat. J.*, 1981, **21**, 407.

[140] R. Briggs, *Water (N. W. C.)*, 1981, No. 40, 19.

[141] R. Briggs, in A. Porteous, ed., "Developments in Environmental Control and Public Health – 2", Applied Science Publishers, London, 1981, pp. 153–216.

[142] G. A. Best, *Public Health Eng.*, 1983, **11**, 26.

[143] K. J. Clevett, "Handbook of Process Stream Analysis", Ellis Horwood, Chichester, 1973.

[144] D. J. Huskins, "General Handbook of On-line Process Analysers, Volume I", Ellis Horwood, Chichester, 1982.

[145] A. A. Diggens and S. Lichtenstein, in "Proc. 39th Intern. Water Conf., Engrs. Soc. Western Pennsylvania", Engineers Society of Western Pennsylvania, 1978, pp. 83–88.

[146] R. Briggs, *Anal. Proc.*, 1982, **19** 71.

[147] R. Briggs, *ISA Trans.*, 1982, **21**, 59.

[148] W. D. Meredith and C. A. Rushton, *Int. Environ. Safety*, 1981, Dec., 10.

[149] Standing Technical Advisory Committee on Water Quality, "Second Biennial Report, February 1977 January 1979", (DOE/NWC Standing Technical Committee reports No. 22), National Water Council, London, 1979.

[150] J. Janata, *Anal. Proc.*, 1982, **19**, 65.

[151] F. Leroy, P. Gareil, and R. Rosset, *Analusis*, 1982, **10**, 351.

[152] H. Wohltjen, *Anal. Chem.*, 1984, **56**, 87A.

[153] W. R. Seitz, *Anal. Chem.*, 1984, **56**, 16A.

[154] G. F. Kirkbright, R. Narayanaswamy, and N. A. Welti, *Analyst*, 1984, **109**, 15.

[155] F. K. Kawahara, *et al.*, *Anal Chim. Acta*, 1983, **151**, 315.

[156] J. Cairns, "Biological Monitoring in Water Pollution", Pergamon, Oxford, 1982.

[157] G. P. Evans and D. Johnson, in "Water Research Centre Conference on Environmental Protection, September, 1983", Water Research Centre, Medmenham, Bucks, 1983, Paper No. 16.

[158] J. W. Schofield, in M. J. Stiff, "River Pollution Control", Ellis Horwood, Chichester, 1980, pp. 188–193.

[159] J. F. Wallwork, in Stiff, M. J., ed., ref. 158, pp. 175–187.

[160] J. S. Fenlon and D. D. Young, *Water Pollut. Contr. (London)*, 1982, **81**, 343.

[161] P. Van Der Borght, *Tech. Eau Assainissement*, 1982, No. 423, 19.

[162] R. Briggs and K. V. Melbourne, *Water Treat. Exam.*, 1968, **17**, 107.

[163] M. D. Palmer and J. B. Izatt, in J. E. Kerrigan, ed., "Proc. Nat. Symp. Data and Information for Water Quality Management", University of Wisconsin, Madison, 1970, pp. 406–413.

[164] K. N. Birch, *et al.*, in "Instruments and Control Systems: Water Research Centre Conference, 15–17 September, 1975", Water Research Centre, Medmenham, Buckinghamshire, 1975, Paper No. 9.

[165] D. G. Benham and D. G. George, *Freshwater Biol.*, 1981, **11**, 459.

[166] J. Chèze and M. Dutang, *Aqua*, 1980, No. 9, 25.

[167] R. Louboutin, *Tech. Eau Assainissement*, 1983, Nos. 433/434, 15.

[168] IHD-WHO Working Group on Quality of Water, "Water Quality Surveys", UNESCO/WHO, Paris/Geneva, 1978.

[169] J. E. Kerrigan, ed., "Proc. Nat. Symp. Data and Information for Water Quality Management", University of Wisconsin, Madison, 1970.

[170] P. A. Krenkel, ed., "Proc. Specialty Conf. Automatic Water Quality Monitoring in Europe", Vandebilt University, 1971.

[171] "Instruments and Control Systems: Water Research Centre Conference, 15–17 September, 1975," Water Research Centre, Medmenham, Buckinghamshire, 1975.

[172] "Water Research Centre Seminar on Practical Aspects of Water Quality Monitoring Systems, 7 December 1977", Water Research Centre, Medmenham, Buckinghamshire, 1977.

[173] "Water Quality Parameters", (ASTM Spec. Tech. Pub. 573), American Society for Testing and Materials, Philadelphia, 1975, pp. 367–486.

[174] *Water Sci. Technol.*, 1981, **13**:4, 631–719.

[175] *Prog. Water Technol.*, 1977, **9**:5/6, 1–227.

[176] D. C. Hinge, *J. Inst. Water Eng. Sci.*, 1980, **34**, 546.

[177] Y. Deleu and C. de Gand, *Trib. Cebedeau*, 1982, **35**, 427.

[178] R. T. Heslop, *Public Health Eng*, 1982, **10**, 81.

[179] L. Freeman, *Water Bull.*, 1983, 1 April, 12.

[180] W. R. McChuney, *J. Water Pollut. Control Fed.*, 1975, **47**, 252.
[181] G. V. Winters and D. E. Buckley, *Est. Coastal Marine Sci.*, 1980, **10**, 455.
[182] M. Cathelain and M. Robbe, *Trib. Cebedeau*, 1980, **33**, 251.
[183] R. D. Vanous and R. E. Heston, *J. Ind. Water Works Assoc.*, 1980, **12**, 95.
[184] K.-R. Nippes, *Dtsch. Gewässerkundliche Mitt.*, 1981, **25**, 150.
[185] L. Reinemann, H. Schemmer, and M. Tippner, *Dtsch. Gewässerkundliche Mitt.*, 1982, **26**, 167.
[186] P. J. Clack and F. L. Williams, *Prog. Water Technol.*, 1977, **9**:5/6, 93.
[187] S. A. Michalek, R. A. Dishman, and R. B. Thayer, *Prog. Water Technol.* 1977, **9**:5/6, 87.
[188] D. E. Lueck, R. A. Dishman, and R. B. Thayer, *ISA Trans.*, 1981, **20**, 67.
[189] R. Rosset, M. Caude, P. Sassiat, and M. Dutang, *Int. J. Environ. Anal. Chem.*, 1982, **13**, 19.
[190] H. Mueller and W. M. Bandaranayake, *Marine Chem.*, 1983, **12**, 59.
[191] R. Briggs, J. W. Schofield, and P. A. Gorton, *Water Pollut. Contr. (London)*, 1976, **75**, 47.
[192] J. M. Reid, M. S. Cresser, and D. A. MacLeod, *Water Res.*, 1980, **14**, 525.
[193] H. de Haan, T. de Boer, H. A. Kramer, and J. Voerman, *Water Res.*, 1982, **16**, 1047.
[194] M. Mrkva, *Water Res.*, 1983, **17**, 231.
[195] World Health Organization, "The Optimization of Water Quality Monitoring Networks", W.H.O. (Regional Office for Europe), Copenhagen, 1977.
[196] W. S. G. Morgan, P. C. Kuhn, B. Allais, and G. Wallis, *Water Sci. Technol.*, 1982, **14**:4/5, 151.
[197] G. J. Holland and A. Green, *Water Treat. Exam.*, 1975, **24**, 81.
[198] K. C. G. Stroud and D. B. Jones, *Water Treat Exam.*, 1975, **24**, 100.
[199] W. Slooff, D. de Zwart, and J. M. Marquenie, *Bull Environ. Contam. Toxicol.*, 1983, **30**, 400.
[200] N. A. Dimmock and D. Midgley, *Talanta*, 1982, **29**, 557.
[201] D. Midgley, *Talanta*, 1983, **30**, 547.
[202] P. H. A. Hoogweg, *et al.*, *H2O*, 1982, **15**, 102.
[203] P. Valenta, *et al.*, *Fresenius' Z. Anal. Chem.*, 1982, **312**, 101.
[204] G. Horner, W. Kühn, and D. Leonhardt, *Vom Wasser*, 1982, **59**, 115.
[205] J. K. Rice and T. O. Passell, in "Proc. 39th Intern. Water Conf. Engrs Soc. Western Pennsylvania", Engineers' Society of Western Pennsylvania, 1978, pp. 11–19.
[206] D. R. Colman and R. Siev, in ref. 205, pp. 447–454.
[207] "Power Plant Instrumentation", (ASTM Spec. Tech. Pub. 742), American Society for Testing and Materials, Philadelphia, 1981.
[208] D. G. Ballinger, in B. Westley, ed., "Proc. Intern. Symp. Identification and Measurement of Environmental Pollutants", National Research Council of Canada, Ottawa, 1971, pp. 158–162.
[209] A. L. Wilson, "Technical Memorandum TM81", Water Research Centre, Medmenham, Buckinghamshire, 1973.
[210] International Electrotechnical Commission, "Expression of Performance of Electrochemical Analysers. Part 1: General", [Publication 746-1 (1982)], I.E.C., Geneva, 1982.
[211] Ref. 210, "Part 2: pH Value", [Publication 746-2 (1982)], I.E.C., Geneva, 1982.
[212] British Standards Institution, "Method of Evaluating Analogue Chart Recorders and Indicators for Use in Process Control Systems", (BS 4671:1971), B.S.I., Milton Keynes, Buckinghamshire, 1971.
[213] British Standards Institution, "Method of Expression of Performance of Air Quality Infra-red Analysers", [BS 5849:1980 = I.E.C. 528 (1975)], B.S.I., Milton Keynes, Bucks, 1980.
[213a] Central Electricity Generating Board, CEGB Standard 500123, Water Analysis Equipment, Generation Development and Construction Division, CEGB, 2 August, 1983.
[214] D. L. Massart, A. Dijkstra, and L. Kaufman, "Evaluation and Optimization of Laboratory Methods and Analytical Procedures", Elsevier, Amsterdam, 1978.
[215] G. Kateman and F. W. Pijpers, "Quality Control in Analytical Chemistry", Wiley, Chichester, 1981.
[216] D. C. Cornish, *Prog. Water Technol.*, 1977, **9**:5/6, 131.
[217] British Standards Institution, "Reliability of Systems, Equipments and Components", (BS 5760: Part 1: 1979, : Part 2 : 1981, : Part 3: 1982), B.S.I., Milton Keynes, Buckinghamshire, 1979, 1981, 1982.
[218] A. L. Wilson, "Technical Memorandum TM 82", Water Research Centre, Medmenham, Buckinghamshire, 1973.
[219] C. E. Hamilton, ed., "Manual on Water", (ASTM Spec. Tech. Pub. 442A), American Society for Testing and Materials, Philadelphia, 1978.
[220] J. W. Sugar and J. H. Brubaker, in J. W. Scales, ed., "Water Quality Instrumentation, Volume 2", Instrument Society of America, Pittsburgh, 1974, pp. 38–66.
[221] D. Midgley and K. Torrance, *Analyst*, 1973, **98**, 217.
[222] P. John, E. R. McQuat, and I. Soutar, *Analyst*, 1982, **107**, 221.

13 Computers in Water Analysis

13.1 INTRODUCTION

The first edition of this book, written over a decade ago, ended with a section on data-handling – a large part of which was devoted to the use of computers for that purpose. In the intervening period, the role of computers in analysis has extended rapidly. Therefore, the current edition has, instead, this section on the use of computers in analysis, of which only a small part will be devoted to data-handling of the kind described in the first edition (*i.e.*, archiving, processing and retrieval of 'finished' analytical results).

It is emphasised that the purpose of the section is not to provide a detailed coverage of the topic, for two reasons: first, the use of computer techniques in analysis is now so widespread that a detailed description, even in relation to water analysis specifically, would require more space than is available in a book of this kind; second, the rate of change of computer technology is so great that detailed descriptions of equipment and systems would very soon become outdated. The intention has, therefore, been to provide an overview of the uses of computer techniques, with emphasis on broad principles, trends and possible future developments.

The use of computers in analytical chemistry is the subject of a comprehensive book by Barker[1] and of a part-text by Eckschlager *et al.*[2] Papers describing the applications of computers to analysis are now a common feature of many analytical journals, but specific mention may be made of *Analytica Chimica Acta, Analytical Chemistry, American Laboratory, Computer Techniques in the Laboratory*, the *Journal of Automatic Chemistry, Talanta* and *Trends in Analytical Chemistry*. Since 1982, *Analytical Chemistry* has published a series of 'tutorials' on various aspects of computing in analysis, under the general title 'A/C Interface'. The series has covered the following topics: local area networks,[3] laboratory information management systems,[4] computer languages for laboratory applications,[5] operating systems for laboratory computing,[6] robots in analysis,[7] voice recognition and computer graphics[8] and management of the electronic laboratory.[9] A general description of the subject area is followed by examples provided by particular analytical laboratories. Computer systems for analysis have also been reviewed by Ziegler.[10]

13.2 THE SCOPE FOR COMPUTING IN WATER ANALYSIS AND RELATED ACTIVITIES

The application of analysis to characterise or control a system or process (*e.g.* a river system or a water-treatment process) is illustrated, in general terms, in Figure 13.1. The end-product of the analysis may be a written report or data summary (which may affect the system or process indirectly, by causing actions to be initiated

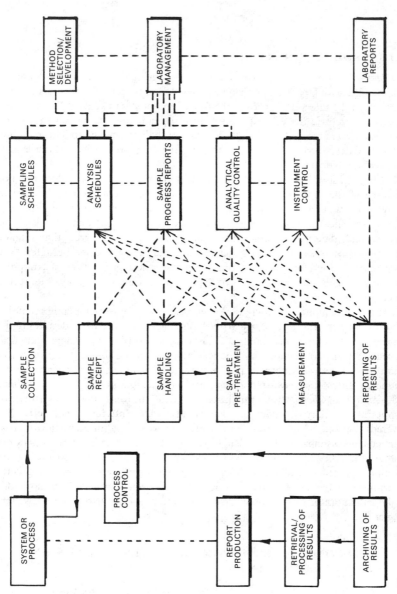

Figure 13.1 *The scope for computer systems in water analysis and related activities*

subsequently), or it may be direct control of the process. The analytical work itself must be managed – this will involve, amongst other things, sample receipt, work scheduling, Analytical Quality Control activities and data reporting (for many analytical systems and samples) in a large laboratory or, more simply, the control, calibration, checking and maintenance of an on-line system for process control.

Figure 13.1 is not intended to be exhaustive, and somewhat different diagrams could be produced by other individuals or organisations, depending on their own roles and responsibilities in water analysis. It serves, however, to identify the major areas of application – current and potential – of computers in analysis, and thus provides a framework for the following sub-sections.

13.3 THE USE OF COMPUTERS IN WATER ANALYSIS

13.3.1 Handling Analytical Results

Historically, computers found first application as means to store, retrieve, process and summarise relatively large quantities of analytical results, such as might be obtained by routine monitoring of a river system. Such applications are, of course, well suited to the large 'main-frame' computers available centrally in many organisations since the 1960s.

The use of computers in this manner for regional, national and international monitoring schemes in well established. Advantages of computer techniques over manual data-handling include: rapid retrieval of data, particularly when only selected values (*e.g.* results exceeding a water quality standard) are required, rapid and accurate calculations (*e.g.* to provide summary statistics or to check for trends) and the capability to produce graphical summaries of large quantities of data, quickly and cheaply.

The application of computers to water quality data storage, retrieval and processing has been reviewed by Deininger and Mancy.[11] These authors give details of national water quality data banks operated by several countries: the STORET (STOrage and RETrieval) system of the U.S. Environmental Protection Agency, the NAQUADAT (NAtional water QUALity DATa bank) of the Canadian Department of the Environment and two systems used in Sweden. A variant of NAQUADAT, GLOWDAT (GLObal Water DATa) has been used for the World Health Organisation's Global Environmental Monitoring System (GEMS).[12]

Although the above-mentioned applications all involve large monitoring programmes and data bases, computer storage is also effective for much smaller local or regional data bases, particularly as the costs of storage hardware have declined dramatically in recent years through advances in technology.

In the United Kingdom, the 'Water Archive System' has been developed to handle a wide range of data relating to water – chemical, biological, hydrological, meteorological, *etc.* – for both regional and national purposes. The system actually consists of a number of separate archives, each controlled by a Water Authority or Central Government Department, using its own computer, but structured on a common basis to allow ready interchange of data. This structured assemblage of archives, and the freedom given to users to determine the nature of the data they wish to store, is claimed to give much greater flexibility than earlier archive systems for water data. For further details of the approach, see ref. 13 and literature cited therein.

A rather different example of the use of computer data storage and handling is seen in the storage of raw gas chromatography/mass spectrometry (GC/MS) data from a survey of wastewaters conducted by the U.S. Enviromental Protection Agency.[14] As well as allowing confirmation and collation of the results obtained in the various laboratories involved, the storage of the raw data on magnetic tape allowed later computer examination to check for the presence of contaminants less important than the 'priority pollutants' with which the survey was primarily concerned.

Before leaving the subject of computerised water quality data bases, two points are worth emphasis, in relation to desirable capabilities for the storage and retrieval of quantitative analytical results. First, the data base should be so designed that actual results can always be recorded – even negative ones, which may be obtained if the determinand concentration is at or near zero, by virtue of random analytical error. This is preferable to entering such results as 'less than X' where X is, say, the limit of detection, since it introduces no bias in subsequent calculation of summary statistics (*e.g.* mean values), such as may occur if all 'less than X' results are taken to be zero or some fixed fraction of X; see Sections 8.1 and 8.3.2. Second, the data-base should have a 'flagging' facility to indicate whether or not the relevant analyses have been subject to Analytical Quality Control to meet known accuracy requirements. Failure to incorporate such a facility may prejudice subsequent use of the data.

13.3.2 Instrument Control and Signal Processing

Re-examination of Figure 13.1 shows that the use of computer techniques to facilitate storage, retrieval and processing of analytical results represented only a limited incursion of computer techniques into water analysis and related activities, albeit an important one. The advent of microprocessors, and the widespread introduction of these devices into analytical instruments (during the 1970s, in particular[15]) brought computer techniques to bear upon the measurement stage of the analysis itself, both for control of the instrument and for processing the analytical signals obtained from it.

According to Barker,[1] a microprocessor is a single integrated circuit (IC, or 'chip' in common parlance) whereas a microcomputer consists of a microprocessor with appropriate external circuitry, memory and peripheral devices (*e.g.*, keyboard). In practice, the incorporation of a microprocessor into an analytical instrument usually involves these additional components, so that distinction between micro-processor-based and microcomputer-based instruments is hardly possible. The former term appears to be often used when the microprocessor system is built into the instrument during manufacture, and the latter when a separate microcomputer (such as a 'personal' computer) is interfaced with the instrument. The term 'microprocessor-based' will here be used to cover both cases*.

---

* With regard to terminology, it should also be noted that the terms 'microcomputer', 'minicomputer' and 'mainframe computer' are used, without clear definition, to refer to small, medium and large systems respectively. Barker[1] uses 'microcomputer' to mean a single-user system whose word length does not exceed 16 bits (typically costing less than £2000), 'minicomputer' to mean a more sophisticated (and expensive)

The introduction of microprocessors to analytical instrumentation has been reviewed by Betteridge and Goad,[16] and is also discussed in some detail by Barker.[1] By now, microprocessors have been incorporated in almost all types of analytical instrument – including semi-automatic micropipettes, automatic titrators, pH meters, balances, spectrometers, polarographs and related electroanalytical devices and chromatographs – and a detailed examination of the facilities afforded by the microprocessor in each case is beyond the scope of this text. Barker[1] has identified the roles of the microprocessor system as providing the instrument with the capability to: (i) store data (both control settings and responses), (ii) process numerical data and (iii) communicate readily with the operator and/or other equipment. As a result of these capabilities, the advantages conferred by the microprocessor typically include the following: storage of instrument operating conditions, automatic set-up of selected conditions, control of conditions during measurement, automatic calculation of calibration function, automatic correction for baseline drift, storage of calibration function and conversion of responses to analytical results, diagnosis of certain operator errors and equipment malfunctions. In combination with an automatic sampling device, itself under the control of the microprocessor system, such features permit, in many instances, truly automatic analysis (in the sense used in Section 12). Computer control is particularly attractive for certain analytical techniques – voltammetry being a good example, as Jagner[17] has noted. He points out that the computer can be used to produce a wide variety of potential ramps and to measure the transient current responses which provide quantitative measurement, and can thus allow new analytical procedures to be developed which were not available previously. In most cases, however, it would seem that the microprocessor has been used to make an instrument easier to use (rather than to alter radically the underlying measurement technique) and, in particular, to facilitate automatic analysis. See Section 12 for a general discussion of the advantages and disadvantages of various forms of automatic analysis, and of the precautions generally required for such analysis.

13.3.3 Laboratory Computer Systems

In the previous sub-section, attention was directed towards the use of 'dedicated' microprocessors and microcomputers used with individual analytical instruments – see Figure 13.2. Although, as Ziegler[10] has noted, the use of a dedicated computer system is normally successful*, it may not always be optimal. Disadvantages of single-instrument computer systems may include: the expense of providing hardware – especially peripherals such as printers – for each instrument, the limited computing power afforded by a simple microprocessor or microcomputer system, limited data storage capacity and the difficulty of implementing computer-based laboratory management.[10] For these reasons, other configurations of laboratory computing have also been used.

One possible approach to an integrated laboratory computing system involves

system offering a wider range of software and capable of supporting multiple users simultaneously and 'mainframe' to mean large, multi-user systems having large data storage facilities and word lengths exceeding 32 bits.

* Indeed, the use of a dedicated system may be essential with some instruments – for example, mass spectrometers, which require very high data acquisition rates.[10]

(a) Dedicated system

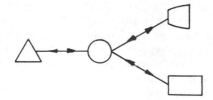

(b) Centralised system

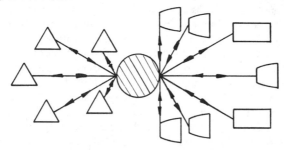

(c) Hierarchical system

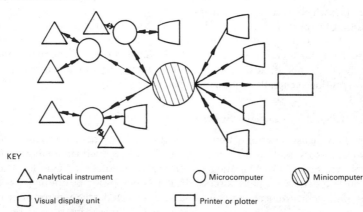

KEY

△ Analytical instrument   ◯ Microcomputer   ⊘ Minicomputer

▢ Visual display unit   ▢ Printer or plotter

Figure 13.2 *Laboratory computer systems*

the connection of a number of different analytical instruments to a central computer, using suitable interfaces (*e.g.*, analogue-to-digital converters[15]) – see Figure 13.2. For obvious reasons, this is referred to as a 'centralised system'[10] or as 'centralised processing'.[15] As Ziegler[10] notes, centralised systems offered the possibility, some 10–15 years ago, of sharing the (then) expensive computer hardware between a group of instruments, but the approach suffered from a number of problems. The operating systems of the computers were not well suited to the time-sharing required to service a number of separate instruments and there were difficulties in performing instrument control actions at specific, critical times and in accomodating high rates of data acquisition; problems were also encountered in connecting instruments distant from the computer, and the lack of commercially available systems exposed the user to financial risks in development.[10]

The advent of microcomputers, and the sharply decreasing cost of computer hardware in general, now permits a different approach – the 'hierarchical system'[10] or 'distributed processing'.[15] In this approach, illustrated in Figure 13.2, the real-time tasks (instrument control, signal collection) and initial processing are undertaken by microprocessor/microcomputer systems suited to the particular instruments to which they are connected. More complicated data processing, data archiving and the production of reports of analytical results are carried out by a central minicomputer to which the individual instruments' microcomputers are linked.

Distributed processing permits the benefits of the dedicated microcomputer to be combined with those of a large, central computer.[10] The assignment of real-time tasks to the dedicated microcomputers avoids problems of time-sharing and instrument control encountered with centralised systems.[10] The individual analyst has the advantage of the flexibility of the dedicated computer, and of the more extensive software for analytical signal processing available for microcomputers.[15] Last, but by no means least, the incorporation of 'buffer' data storage on disks within the dedicated microcomputers, together with the 'redundancy' of processors, allows analysis to continue without interruption in the event of a failure of the main computer, at least for some time.[15]

Reber and Coates[15] note, however, that the centralised system retains some advantages – it is often cheaper because each analogue-to-digital converter is cheaper than a microcomputer, for example. However, this advantage seems likely to diminish as the tendency grows for all instruments to be produced with an in-built capability to output signals (or analytical results) in digital form, and major instrument makers involved with laboratory computer systems appear to be actively engaged in development of 'distributed processing' or 'hierarchical' systems.

It is beyond the scope of this section to consider in detail the interfacing of instruments to computer systems and the construction of 'networks' to link such 'intelligent' instruments to the main laboratory computer ('local networks') or to link the local network to more distant computer systems such as the organisation's main-frame or commercial data bases ('global networks'). Interfacing instruments to computers has been discussed by Barker[1] and is the subject of an article by Liscouski;[18] local networks are discussed by Dessy.[3] A lack of standardisation amongst microprocessor-based and microcomputer based analytical instruments has represented a serious obstacle to the more rapid development of integrated laboratory computing systems, but instrument manufacturers are conscious of the problem and moves towards standardisation are in progress.[19]

13.3.4 Laboratory Information Management

The establishment of an integrated laboratory computing system, on either the centralised or the distributed processing pattern, paves the way for the next area to be discussed, the implementation of a Laboratory Information Management System (LIMS). The management of a large analytical laboratory, involving conduct of the activities shown on the right hand side of Figure 13.1 for all the determinations taking place, is a complex task. The function of LIMS is to allow many of these activities to be carried out automatically or semi-automatically by providing the laboratory manager and laboratory staff with the necessary information to optimise application of the laboratory's resources. Laboratory information management by computers has been reviewed by Dessy[4] and by Reber and Coates.[15] The activities to which LIMS can be applied include:[4,15]

(1) Sample log-in and labelling (including the production and use of bar-code labels, most familiar to us on packaged foods).
(2) Production of work-lists advising individual analysts of the required analyses.
(3) Analytical Quality Control.
(4) Production of status reports (*e.g.*, on backlogs or on the progress of individual samples through the laboratory).
(5) Filing of results and production of reports of results for customers.
(6) Production of laboratory performance information.
(7) Word processing for report writing, letters, memoranda, *etc.*
(8) General management filing.
(9) Filing and handling of financial data, including automatic billing of customers.
(10) Workload planning.

The particular value of LIMS, in addition to its capacity to assist day-to-day laboratory operations, is its potential for helping laboratory managements plan how to meet changing analytical requirements and how best to take advantage of new developments in analytical techniques and instrumentation.[4,15]

13.3.5 Robotics

Figure 13.1 shows that the applications of computers in analysis discussed so far leave one main area untouched – sample handling and pre-treatment. Although some pre-treatments (*e.g.*, acidification, u.v. irradiation) can be conducted automatically – in flowing sample streams, for example – others are not so readily incorporated in conventional automatic systems (*e.g.*, acid digestion involving evaporation to low volumes, co-precipitation). Additionally, there will always remain a need to dispense aliquots of sample from the storage or collection vessel for analysis (except, of course, with on-line instruments).

It is in these areas that computer-controlled robots offer the possibility of automation. Such devices have been considered in Section 12, which should be consulted for further details. We have there expressed the view that the likely value of robots for sample handling and pre-treatment in water analysis is limited. Robots may, however, ultimately find some application in large, central laboratories having a high sample throughput and a limited number of rapid, automatic analysers.

13.3.6 Computer Data Bases for Analytical Information

Reference has already been made to the use of computers for handling data bases of analytical results, for example for a programme of river water monitoring. Computer data bases are, of course, also used to store analytical information of a more general nature, such as mass spectrometric and infrared spectrometric data for particular organic compounds. Data bases of this kind may be held in an individual laboratory (libraries of mass spectra are commonly available with commercial computer 'packages' offered for use with mass spectrometer systems), or, in the case of the most comprehensive types, they may be held by a central organisation for access by individual subscribers *via* public telecommunication systems.[20] An example of the large, central data base is the 'Chemical Information Service' (CIS) developed since 1972 by government organisations in the United States (National Bureau of Standards, National Institutes of Health and Enviromental Protection Agency). It is, in fact, a collection of computer data bases relating to mass spectrometry, infrared spectrometry, nuclear magnetic resonance spectrometry and X-ray spectrometry, available world-wide *via* telecommunication networks.[20]

The availability of computerised literature systems is now well known, and their use is growing as the volume of scientific literature increases and as the cost of printed publications escalates.[20] Two examples of relevance to water analysis are the 'Chemical Abstracts Service' (CAS), available on a number of computer services such as Dialog, CAS Online and the IRS–Esrin service of the European Space Agency, and the 'Aqualine' system for water-related literature developed by the Water Research Centre in the United Kingdom,[21,22] available on Pergamon InfoLine, Dialog and IRS–Esrin. Increasingly, computers are also being used to handle personal or laboratory literature records, in preference to index cards.[1]

Another computer-based system for communicating analytical information, developed by Settle and co-workers at the Virginia Military Institute,[23] is the 'Scientific Instrumentation Information Network and Curricula Project' (SIINC). This consists of a 'series of instructional/information modules' on selected analytical techniques, containing an overview, a general description of the instrumentation, definition of relevant terms, a description of applications, practical information on suppliers, sources of information and data bases, and information on the most recent developments in hardware and methodology.[23] Intended to fill the gap between major literature data-bases, such as the Chemical Abstracts system, and Computer-Aided Instruction (CAI), the modules are produced by a task group led by an expert on the technique concerned, and are available on disk for use through personal microcomputers. Techniques dealt with by 1982 were gas chromatography and gas chromatography/mass spectrometry.[23]

13.3.7 Implementation of Laboratory Computer Systems

Many analysts and laboratory managers with experience of laboratory automation and laboratory computer systems may consider that the descriptions given in previous sub-sections present a falsely rosy picture of the capabilities of laboratory computing. This thought is suggested by the frequently expressed concern in the analytical literature that laboratory automation and computing projects often lead to disappointment amongst their intended beneficiaries (see, for example, refs. 9 and 24). Thus, for example, Dessy[9] suggests that perhaps half of the attempts to automate laboratories fail, in the sense that they do not allow reasonably full exploitation of the resources made available. In part, this may be the result of technical problems – the disadvantages of the early, 'centralised' laboratory computer systems have been described in Section 13.3.3 above, for example – but Dessy[9] considers otherwise. He feels that implementation of laboratory computing fails so often because of human weaknesses, not technological ones. He summarises such failings as bad management, and cites the following as major factors:

(1) Inadequate analysis and misunderstanding of the problems.
(2) Political and personal ('ego') tensions.
(3) Insensitivity to the needs and motivations of the users of the computing system.
(4) Poor diplomacy.
(5) Inadequate provision of educational programmes.

Stockwell[24] also identifies accurate and full specification of requirements and assessment of the needs of the organisation and staff as the principal problem areas. The 'A/C Interface' articles in *Analytical Chemistry* (especially ref. 9), the review by Ziegler[10] and the editorial by Stockwell[24] all provide useful advice on approaches to minimise or avoid these and other difficulties, and should be read by anyone concerned with implementing a laboratory computing programme. Amongst the steps and precautions they recommend are:

(1) Education of laboratory staff in the way computers work, so that they can more readily express their needs in terms accessible to computer experts.[9]
(2) Careful teamwork to identify laboratory needs, including use of questionnaires and personal interviews, and ranking of facilities desired.[9]
(3) Wariness of 'experts' on laboratory computing – they may be expert in computing itself but not expert in analysis, and their experience of the subject may be very limited.[24]
(4) Co-ordination of computer personnel, laboratory management and system users 'in a representative democracy'.[9]
(5) Responsibility for installation to be in the same hands as final system specification, cost justification and purchase.[24]
(6) Appreciation of the possibilities and limitations of laboratory computing, of the difficulties of doing things by computer which may be done easily by the analyst, and of the high cost of specialised software.[24]
(7) Preservation of the inherent performance of instruments by ensuring adequate time resolution and dynamic range of data collection systems.[10]
(8) Provision of sufficient data storage capacity, which is often underestimated.[10]
(9) Provision for subsequent modification and expansion.[10,24]

A question which will often arise in considering laboratory computing and automation is whether to undertake the task in-house or employ the services of a specialist supplier of the necessary hardware and software. As noted earlier, the lack of commercial suppliers of laboratory computer packages once forced users to follow the in-house route, but this is no longer true, and many instrument makers and computer manufacturers are active in the field. The potential advantage of in-house development is the ability to tailor the system exactly to the organisation's requirements, but this advantage will be offset – and may be completely nullified – by the potential pitfalls. The latter relate mainly to item (6) above, and concern the high cost of software development and the possibility that the difficulties of converting to computer operation will be underestimated. A commercial supplier with experience of particular types of analytical instrument will have already encountered the problems of converting to a computer system, and will be able to spread his software development costs over a number of installed systems because many requirements will be common to them all. Even if the 'packages' he offers do not meet all the requirements of the user, they will probably meet the great majority of his needs, will be immediately available and will, if the supplier is correctly chosen, be supported by appropriate testing and good documentation. Moreover, the supplier may be able to tailor his packages to particular requirements, and may be better placed to do this, by virtue of his experience, than in-house computer and laboratory staff. If in-house implementation is adopted, and if the problems of conversion, software development, system testing and documentation are underestimated, costs may soon escalate beyond budget. Then, both the organisation's management and its analysts may become disenchanted with laboratory computing in general, and enjoyment of the benefits it can bring may be delayed not only by the immediate failure, but also by the loss of enthusiasm and impetus in its wake.

For all these reasons, we incline to the view of Dessy[9] that in-house development is not usually advisable, unless no suitable system can be obtained commercially. That is not to imply that in-house development cannot be successful, but rather that it requires a special blend of skills and knowledge which may not be readily available to all water analysis laboratories. On the other hand, we do not wish either to imply that going to a specialist supplier absolves the laboratory's analytical and computer staff from the responsibility of specifying the requirements precisely and completely, and examining the commercial solutions offered with great care. Although no reputable supplier would wish to sell an unsuitable system knowingly, only the close involvement of the purchaser's staff can ensure a successful outcome. In particular, the purchaser's staff need to ensure that all signal-processing operations used to estimate the calibration function and to calculate analytical results are those considered appropriate – see also Section 10.4.3, especially sub-section (*c*).

If, for any reason, it is decided to adopt in-house development despite its potential problems, attention should still be given to the possibility of using bought-in hardware and software for specific parts of the complete system. Such an approach may now be easier than in the past, because of distributed processing, and could offer cost savings whilst retaining the flexibility for which, presumably, the in-house development would be selected.

13.3.8 Specific Applications in Water Analysis

With the exception of sub-section 13.3.1, dealing with handling of analytical results, the references cited in this section so far have been to general texts, reviews and tutorial articles on laboratory applications of computing. Such literature is most suited to providing a wide, up-to-date picture of the capabilities of laboratory computing, and to identifying the paths to be followed, and hazards to be avoided, in its implementation – which vary little with the type of analysis to be undertaken, unless it is highly specialised. In this sub-section, however, we briefly note a number of publications dealing with the application of computing to water analysis specifically.

With regard to computing and automation in the analytical laboratory, papers by Crowther *et al.*,[25] Woods *et al.*,[26] Walker,[27] Taylor *et al.*,[28] Morley *et al.*,[29] Thomson[30] and Jones[31] describe the application of computer systems in routine water analysis laboratories. Particular points to be drawn from these publications include: the value of automatic data capture even with a labour-intensive determination such as BOD,[25] the importance of including Analytical Quality Control procedures in laboratory computing,[26,28,29] the need to maintain the analyst's ability to check peak shapes whilst relieving him of the tedium of measuring peak heights,[27,30] the undesirability of microprocessors introducing correction factors unknown to the analyst,[30] the savings in time and effort, and improvements in quality of results, which computer systems can provide in water analysis laboratories[25-31] and the possibility of providing sampling officers with portable computers to record details of sample collection in the field.[31]

Papers dealing with computers in on-line monitoring and process control include those by Caddy and Whitehead,[32] Grombach,[33] Alla and Guirkinger[34] and Wallwork and Ellison,[35] together with a number published in the proceedings of an international workshop on control and automation in wastewater treatment and water resources management.[36] In this area, the problems appear to lie less with the computer and control systems themselves than with the available sensors, which cover a limited range of determinands and which often suffer from inadequate reliability.

13.4 SUMMARY

It is clear from the previous sub-sections that the introduction of computer technology into water analysis laboratories offers many potential benefits, across the entire range of analytical activities. It should also be clear that there are numerous pitfalls to be avoided, especially in the area of laboratory automation and integrated laboratory computing. Guidance on how to avoid such pitfalls is, however, available, so that the organisation contemplating the introduction of computer techniques can benefit from the experiences, good and bad, of those who have already done so. Given the high cost of integrated laboratory computing, and the numerous disappointments which appear to have occurred, it is of great importance that such guidance be fully exploited.

There is no doubt that the use of computers (including microprocessors) in chemical analysis has brought about considerable changes in the roles and attitudes of analysts, and will continue to do so. Whilst analysts and their organisations must be ready to grasp the opportunities which computer techniques may offer for increased productivity, improved precision of results and avoidance of tedious, repetitive tasks, they must also remember that 'chemistry should control the way computers are used, rather than the reverse'[9]; see also Section 10.4.3. An appropriate cautionary note on which to end this section is provided by that most eminent analyst, I. M. Kolthoff. Reporting an interview with him, Shepherd[37] noted that Kolthoff, though appreciative of the potential benefits of computer techniques, was '... quick to point out that there are potential drawbacks in an approach to analytical problems in which too much emphasis is placed on apparatus and data processing and not enough on the chemical principles behind the measurement being made'.

13.5 REFERENCES

[1] P. Barker, 'Computers in Analytical Chemistry', Pergamon Press, Oxford, 1983.
[2] K. Eckschlager, *et al.*, 'Applications of Computers in Analytical Chemistry', in G. Svehla, ed., 'Comprehensive Analytical Chemistry', Volume XVIII, Elsevier, Amsterdam, 1983, pp. 251–444.
[3] R. E. Dessy, ed., *Anal. Chem.*, 1982, **54**, 1167A (Part I); *Ibid.*, 1982, **54**, 1295A (Part II).
[4] R. E. Dessy, ed., *Anal. Chem.*, 1983, **55**, 70A (Part I); *Ibid.*, 1983, **55**, 277A (Part II).
[5] R. E. Dessy, ed., *Anal. Chem.*, 1983, **55**, 650A (Part I); *Ibid.*, 1983, **55**, 765A (Part II).
[6] R. E. Dessy, ed., *Anal. Chem.*, 1983, **55**, 883A.
[7] R. E. Dessy, ed., *Anal. Chem.*, 1983, **55**, 1100A (Part I); *Ibid.*, 1983, **55**, 1232A (Part II).
[8] R. E. Dessy, ed., *Anal. Chem.*, 1984, **56**, 68A (Part I); *Ibid.*, 1984, **56**, 303A (Part II).
[9] R. E. Dessy, ed., *Anal. Chem.*, 1984, **56**, 725A (Part I); *Ibid.*, 1984, **56**, 855A (Part II).
[10] E. Ziegler, *Trends Anal. Chem.*, 1983, **2**, 148.
[11] R. A. Deininger and K. H. Mancy, in M. J. Suess, ed., 'Examination of Water for Pollution Control. A Reference Handbook', Published on behalf of the World Health Organisation Regional Office for Europe by Pergamon Press, Oxford, 1982, Volume 1, pp. 211–264.
[12] S. Barabas, in L. H. Keith, ed., 'Advances in the Identification and Analysis of Organic Pollutants in Water', Ann Arbor Science Publishers, Ann Arbor, 1981, Volume 2, pp. 481–494.
[13] D. H. Newsome, in M. J. Stiff, ed., 'River Pollution Control', published for the Water Research Centre by Ellis Horwood, Chichester, 1980, pp. 209–222.
[14] W. M. Shackelford, D. M. Cline, L. Burchfield, L. Faas, G. Kurth, and A. D. Sauter, in L. H. Keith, ed., ref. 12, Volume 2, pp. 527–554.
[15] S. A. Reber and J. P. Coates, *J. Auto. Chem.*, 1984, **6**, 67.
[16] D. Betteridge and T. B. Goad, *Analyst*, 1981, **106**, 257.
[17] D. Jagner, in M. Whitfield and D. Jagner, ed., 'Marine Electrochemistry. A Practical Introduction', Wiley, Chichester, 1981, pp. 123–141.
[18] J. G. Liscouski, *Anal. Chem.*, 1982, **54**, 849A.
[19] S. A. Borman, *Anal. Chem.*, 1984, **56**, 408A.
[20] S. R. Heller, R. Potenzone, G. W. A. Milne, and C. Fisk, *Trends Anal. Chem.*, 1981, 62.
[21] P. Russell and B. Wilkinson, *Water*, 1981, **37**, 26.
[22] J. G. Smith, comp., and P. J. Russell, ed., 'Aqualine Thesauraus', Published for the Water Research Centre by Ellis Horwood, Chichester, 1980.
[23] P. A. Settle and M. Pleva, *Trends Anal. Chem.*, 1982, **1**, 242.
[24] P. B. Stockwell, *J. Auto. Chem.*, 1983, **5**, 1.
[25] J. M. Crowther, J. F. Dalrymple, T. Woodhead, P. Coackley, and I. Hamilton, *Water Pollut. Control.*, 1978, **77**, 525.
[26] D. R. Woods, M. D. Frayn, and R. S. A. Walker, *Water Pollut. Control*, 1980, **79**, 42.
[27] D. E. Walker, *Water Pollut. Control*, 1983, **82**, 535.
[28] G. S. Taylor, R. G. H. Tyers, and F. B. Basketter, *J. Inst. Water Eng. Sci.*, 1981, **35**, 412.
[29] P. J. Morley, P. R. Musty, and J. Cope, *J. Inst. Water Eng. Sci.*, 1980, **34**, 256.
[30] D. G. Thomson, *Water Pollut. Control.*, 1982, 81, 193.
[31] J. G. Jones, *Aqua*, 1983, (6), 321.
[32] D. E. Caddy and P. G. Whitehead, *Effluent Water Treat. J.*, 1982, **22**, 179.
[33] P. Grombach, *Aqua*, 1983, (6), 285.
[34] P. Alla and B. Guirkinger, *Aqua*, 1981, (1), 46.
[35] J. F. Wallwork and G. Ellison, *Aqua*, 1983, (6), 313.
[36] Proceedings of an International Workshop of the International Association on Water Pollution Research on 'Practical Experiences of Control and Automation in Wastewater Treatment and Water Resources Management', Munich and Rome, June 1981, *Water Sci. Technol.*, 1981, **13** (8)–(12).
[37] P. T. Shepherd, *Trends Anal. Chem.*, 1981, **1**, 1.